Nuclear-Reactor Analysis

Nuclear-Reactor Analysis

Allan F. Henry

The MIT Press
Cambridge, Massachusetts, and London, England

This book was printed on Finch Title 93
and bound in Columbia Milbank Linen MBL-4310
by The Colonial Press Inc.
in the United States of America

Library of Congress Cataloging in Publication Data

Henry, Allan F
Nuclear-reactor analysis.

Includes bibliographies and index.
1. Nuclear reactors. 2. Neutron transport theory.
I. Title.
QC786.5.H46 539.7 74–19477
ISBN 0–262–08081–8

In Memory of Sidney Krasik

Contents

Preface

A more informative title for this book would be *The Theoretical Foundations of Reactor-Design Methods*. But that would be too many words. Also it sounds forbidding and a bit pretentious. It would, however, be precise. It is my thesis that a firm theoretical foundation actually exists and that it can be made intelligible in a rigorous and systematic manner, starting with very simple physical concepts and avoiding, on the one hand, the more esoteric aspects of transport theory and, on the other, the many prescriptions and loosely defined terms that served so well in the early days of reactor design.

The attempt to prove this thesis has resulted in a book that omits some parts of what is traditionally called "reactor physics." Instead the emphasis is primarily on the underpinnings of the calculational methods applicable to three-dimensional heterogeneous reactors. Of necessity the notation becomes complex, and, although the only prior mathematical training assumed is in the areas of differential equations and elementary matrix theory, some of the mathematical manipulations and compact matrix notation introduced in the last few chapters demand very careful attention.

The foregoing remarks may lead those already familiar with reactor physics to conclude that I am about to begin with the Smoluchowski equation or, at best, the Boltzmann transport equation. Examination of the table of contents will show, however, that the presentation is much more conventional.

Nevertheless the book does depart from the standard format. It is assumed, for example, that the reader is already generally familiar with the conventional model of an atom, the meaning of atomic and molecular weight, and the radioactive-decay law. Only a very qualitative picture of nuclear physics is given, the prime intent being to provide a precise definition of nuclear cross sections and reaction rates.

In addition the order in which material is presented is somewhat unconventional. The chief departure from tradition is the treatment of the energy part of the problem before the spatial part. Thus there is no mention (except a derogatory one) of "one-speed" diffusion theory. Fick's Law is first presented as a plausible assumption and not derived until Chapter 8. The conventional—and, to my mind, misleading—one-speed derivation is avoided.

Procedures for solving the equations associated with particular physical models are also presented in an untraditional order. Computing-machine-oriented numerical methods are described first. Then, for simple cases, in order to bring out the physical characteristics of the solutions, the analytical solutions are discussed.

Since the aim of the book is to provide a basic understanding of theoretical methods currently applied to the design and analysis of power reactors, certain topics not germane to this objective are omitted. There is, for example, no mention of the Milne problem since it is here irrelevant, and there is no discussion of analytical procedures for determining thermal utilization since they are nowadays little used. Other topics are omitted because of arbitrary decisions to limit the scope of the book. For example no mention is made of noise analysis or frequency-response analysis in the chapter on reactor kinetics. Similarly, although throughout the book the viewpoint is adopted that equations describing reactor behavior will in practice be solved on digital computing machines, there is no discussion of how such numerical procedures are carried out. In particular there is no discussion of the convergence and acceleration of iterative procedures for solving numerical problems.

It is in keeping with the aim of providing understanding of the theoretical foundations of reactor methods to omit many details that are important from a practical viewpoint but are not essential to a basic understanding. In Chapter 5, for example, the collision-theory methods for predicting resonance capture in a fuel rod are developed without accounting for the cladding of the fuel rod. My feeling is that this complexity would add confusion to what is already a very involved development. Accordingly the treatment of cladding is omitted from the text (but covered in a reference at the end of the chapter). The work is thus by no means a reactor-physics design manual.

At present the book is the basis of a sequence of three one-term graduate courses in reactor theory at MIT. The first course covers the material through the derivation of the two-group equations, the second ends with the chapter on reactor kinetics, and the third begins with Chapter 8 on transport theory. The first course appears to be able to stand alone (although it may be advisable to include part of Chapter 7 on kinetics if this is done). The first two courses make a rather complete package for students whose interest in reactor physics is secondary to such specialties as reactor engineering or fuel management. The numbers of students taking the three courses have been in the approximate ratio 4:3:2. No prior knowledge of either nuclear physics or reactor physics is assumed, and the mathematical level is about that provided by standard undergraduate engineering training. Juniors and seniors frequently take the first two courses in the sequence and occasionally the third, and they appear to have no great difficulty with the level of the material. Mathematical procedures that may not be a part of the standard curriculum—for example, the weighted-residual method and various variational procedures—are explained, to the extent necessary, as they are encountered.

One major mathematical problem, already mentioned, is that of notation. If one wishes to represent, for some particular isotope, the P_1 component of the macroscopic cross section for elastic scattering from one energy group to another at some mesh point in a three-dimensional region of space, a symbol with many superscripts and subscripts is bound to result. I have tried to be consistent and precise in this matter, but, occasionally, maintaining a convention that was convenient fifty pages back would have resulted in such a clumsy notation that I have cheated. I have, however, made a major effort never to break one rule: All terms introduced are defined precisely.

Much of what I am trying to convey about reactor theory has been learned from persons with whom I have worked. I cannot begin to make a complete list, but I do want to acknowledge the great debt I owe to Bob Hellens, Ely Gelbard, Stan Kaplan, and Kent Hansen. It has been a privilege to work, argue, and learn with them.

It is my hope that this book will also be of use to nuclear designers and analysts working in the field. They in particular must have a clear idea of the virtues and limitations inherent in the theoretical methods they apply. Familiarity with a computer program through having used it for hundreds of cases seems somehow to make us feel that it must be correct, and if it predicts something that is actually observed experimentally, our faith tends to become boundless. But good luck is not a lasting substitute for good science. A computer program should be evaluated on the basis of the nuclear data and reactor theory that underlie it. It is ironic that many of those who claim that our state of knowledge is such that efforts to develop more accurate methods of analysis can be reduced and that clean, unambiguous experiments are no longer necessary really don't believe the predictions of the theoretical models in present use. If it can possibly be managed financially, they always seem to want a full mock-up of any proposed design before making a decision to accept it. My conviction is that it is both uneconomical and unnecessary to allow this situation to persist.

1 The Starting Point and the Goal

1.1 Introduction

The central problem of reactor physics can be stated quite simply. It is to compute, for any time t, the characteristics of the free-neutron population throughout an extended region of space containing an arbitrary, but known, mixture of materials. Specifically we wish to know the number of neutrons in any infinitesimal volume dV that have kinetic energies between E and $E + dE$ and are traveling in directions within an infinitesimal angle of a fixed direction specified by the unit vector $\mathbf{\Omega}$.

If this number is known, we can use the basic data obtained experimentally and theoretically from low-energy neutron physics to predict the rates at which all possible nuclear reactions, including fission, will take place throughout the region. Thus we can predict how much nuclear power will be generated at any given time at any location in the region.

The problem of determining the characteristics of the neutron population is greatly simplified by the fact that, in most cases of interest, that population is so large that the neutrons can be treated as a fluid. Hence, just as with an ordinary gas, we can speak of the neutron "density in phase space" and do not have to face the problem of following the life history of each individual neutron. The quantity of prime interest then becomes the density in phase space $N(\mathbf{r}, \mathbf{\Omega}, E, t)$, which is defined by the statement that $N(\mathbf{r}, \mathbf{\Omega}, E, t)\, dV\, dE\, d\Omega$ is the number of neutrons which, at time t, are located in an infinitesimal volume dV containing the point $\mathbf{r}$, have kinetic energy in an infinitesimal energy range dE about E, and are traveling in a direction contained in the infinitesimal cone of directions $d\Omega$ about $\mathbf{\Omega}$.

When the neutron "fluid" is present inside a material medium, the neutrons will interact with that material, and it is by describing such interactions mathematically that we derive an equation which the function $N(\mathbf{r}, \mathbf{\Omega}, E, t)$ must obey. Specifically, by writing down in mathematical terms a statement of the physical fact that the rate of increase in $N(\mathbf{r}, \mathbf{\Omega}, E, t)$ at time t is the difference between the rate at which neutrons are added to the phase volume $dV\, dE\, d\Omega$ (by being born in a fission process, by being scattered from other energies and directions, or by crossing the boundary from outside to inside dV) and the rate at which they are removed from the phase volume $dV\, dE\, d\Omega$ (by being absorbed in material inside dV, by scattering to other energies and directions, or by crossing the boundary from inside to outside dV), we can derive an integrodifferential equation for $N(\mathbf{r}, \mathbf{\Omega}, E, t)$. This equation is the Boltzmann transport equation. If we were able to solve it, we could then design and predict the behavior of reactors with an accuracy limited only by uncertainties of the basic nuclear data that prescribe the probabilities of what will happen to a neutron of a given energy when it interacts with an atom of a given material.

In view of this state of affairs one might expect to start the study of reactor physics by first deriving the Boltzmann equation and then proceeding to examine ways of solving it. However this approach would introduce so many complicated physical processes at the very beginning that it seems best to avoid it, and we shall, therefore, introduce the various complications associated with describing neutron behavior in a reactor one at a time.

Our first concern will be with neutron populations in homogeneous media of infinite extent containing a spatially uniform, isotropic source emitting neutrons at a constant rate. If, at first, we neglect the neutrons produced by fission, the density of neutrons in such media is independent of position and time and, at all energies and locations, the number traveling in any given direction $\mathbf{\Omega}$ is the same as the number traveling in any other direction. Thus the neutron density in phase space becomes a function of energy alone. Study of this idealized situation will permit us to examine in its most simple form the physics of the chain-reacting process in which a neutron, born in a fission event, slows down and is eventually captured, thereby possibly (i.e., if the capturing material is fissionable) precipitating another fission to continue the chain reaction.

Next we shall attack the much more difficult problem of how neutrons behave in a medium of finite size made up of many regions having different material compositions. The first approach to this problem will be made in terms of the diffusion-theory approximation to the transport equation. This approximation has been found to be quite adequate for most reactor-design analysis, and we shall use it to develop many of the standard methods employed in reactor design (including the analysis of time-dependent phenomena). Only when this is done will we return to the beginning of the subject, derive the Boltzmann transport equation, discuss some of the methods for applying it to reactor-design problems, and show how the diffusion-theory approximation can be derived from it.

Finally, with this complete overview of the field available, we shall discuss approximation strategies in a systematic way, showing how weighted-residual, variational, and finite-element procedures can be used to provide detailed and accurate descriptions of reactor characteristics in a very efficient numerical way.

Before we begin this design-oriented development of reactor physics, it may be helpful to review some of the concepts and terms of low-energy nuclear physics and to present a qualitative discussion of nuclear reactors. Such is the purpose of this first chapter.

1.2 A Qualitative Review of Nuclear Physics

It is a general convention that reactor physics starts with the theory and data of nuclear physics as given. This is not to say that the present knowledge of the nuclear properties of materials is in a satisfactory state for reactor-design purposes. For some types of reactors it is barely adequate. Nevertheless the improvement of this situation, important as it is for reactor design, is not a problem in the area of reactor physics. Accordingly our discussion of nuclear physics will be extremely elementary and qualitative. We shall use classical (i.e., non-quantum-mechanical) concepts of shapes and sizes and avoid any mathematics. The prime intent will be to introduce certain terms and, in particular, to clarify qualitatively the notion of nuclear cross sections so that those who have had no prior exposure to nuclear physics will be able to understand the meaning of the expressions for reaction rates that are the foundation of reactor physics.

A Qualitative Picture of the Nucleus

For most investigations in the area of reactor physics it is suitable to picture the neutron as a very small sphere and to ignore its wave characteristics. In keeping with this approximation we can also picture protons and electrons as small spheres (the former massive and positively charged, the latter light and negatively charged). Models of atoms and their nuclei can then be constructed in terms of these pictures. We can thus describe the nucleus of an atom as a collection of neutrons and protons that are in violent motion but are held together by a very strong, short-range, attractive force, which is called the *nuclear force*. Electrons float around the outside of this nucleus in certain fixed orbits, the number of these orbital electrons exactly equaling the number of protons in the nucleus so that the atom as a whole is electrically neutral.

The number of protons in the nucleus determines the chemical identity of the atom. Except for ordinary hydrogen, the nucleus of which is the proton itself, and a very rare form of helium containing two protons and one neutron in its nucleus, all naturally occurring atoms have nuclei in which the number of neutrons present is equal to or greater than the number of protons present. Atoms having nuclei containing the same number of protons but different numbers of neutrons are called *isotopes*. For example there are three isotopes of silicon found in nature; all have 14 protons in their nuclei, but they may have 14, 15, or 16 neutrons. The number of protons in the nucleus is called the *atomic number* (symbol, Z) of the isotope. The total number of protons plus neutrons is called its *mass number* (symbol, A). Thus (with the two exceptions noted) naturally occurring elements will have $A \geq 2Z$. Moreover A is never very *much* greater than $2Z$; specifically we find (again with the exception noted) that only isotopes such that $2.6Z > A \geq 2Z$—and not all of these—are stable. The explanation for this behavior (as well as the reason that bismuth—$Z = 83$, $A = 209$—is the heaviest completely stable nucleus) involves consideration of the nuclear and electromagnetic forces present within a nucleus. However even a qualitative description of the interaction concerned is beyond the level of the present discussion. We simply note that many isotopes with atomic numbers in the range $2Z$–$2.6Z$, and all isotopes (except hydrogen and He^3) outside that range, are unstable. Since they relieve this instability by emitting a nuclear particle, these isotopes are said to be *radioactive*.

Some radioactive nuclei are only slightly unstable. For example it takes 4.5×10^9 years for half of a given amount of U^{238} to decay. Others, particularly those created artificially, have much shorter *half-lives* (time for half of the material present to decay). For example the silicon isotope Si^{27}, containing 14 protons and 13 neutrons, decays with a 4.2-second half-life by emitting a positively charged electron (positron) and leaving behind a nucleus of aluminum containing 13 protons and 14 neutrons.

The fact that a nucleus containing only neutrons and protons can emit a positron is one conceptual deficiency of the simple nuclear model being described. We must thus add to the model the assumption that one way an unstable nucleus can become stable is for one of its constituent protons to emit a positron, leaving behind a neutron. For nuclei containing too many neutrons for stability, the converse reaction (a neutron emitting a negatively charged electron and leaving behind a proton) is very common. Thus Si^{31}, containing 14 protons and 17 neutrons, becomes stable by emitting an electron and leaving behind P^{31} (15 protons, 16 neutrons). This process of emitting either a positively or negatively charged electron to attain stability is called *β decay*. Sometimes a series of β decays is necessary before a nucleus becomes stable. For example U^{239} emits two electrons, becoming first Np^{239} and then Pu^{239}. Even this latter atom is not stable. With a 24,360-year half-life, it releases an *α particle* (helium nucleus) to become U^{235} (still slightly unstable).

One further complexity must be added to the picture of β decay. It has been found that a particle called a *neutrino*, having no charge and essentially no mass, generally accompanies a β decay. We shall have no concern with neutrinos other than to lament the fact that they carry away about 6 percent of the energy released when a nucleus fissions.

To sharpen the scale of sizes associated with the elementary picture of the atom being discussed, we consider Table 1.1.

Table 1.1 Classical Characteristics of Constituents of Atoms

Object	Mass (g)	Charge (coul)	"Radius" (cm)
Electron	9.11×10^{-28}	-1.6×10^{-19}	1.88×10^{-13}
Proton	1.67×10^{-24}	$+1.6 \times 10^{-19}$	10^{-16}
Neutron	1.67×10^{-24}	0	10^{-16}
Nucleus	$A \times 1.67 \times 10^{-24}$	$Z \times 1.6 \times 10^{-19}$	10^{-13}
Atom	$A \times 1.67 \times 10^{-24}$	0	10^{-8}

The numbers in the table are very approximate. For example the decrease in mass of the nucleus associated with the energy released when the nucleus is formed from neutrons and protons has not been included; nor has the mass of the electrons been added into the mass of the atom. The radii of the electron, proton, and neutron are particularly approximate since picturing these elementary-particle states as spheres is a decided oversimplification in the first place. Nevertheless the numbers in the table permit us to form a rough picture of the environment in which a free neutron traveling in a material medium finds itself.

The first thing to note is that, from the viewpoint of the neutron as we are picturing it, most of the material region through which it is traveling is "unoccupied." Thus, if the nucleus were 30 cm in radius, the outer electrons of the atom would be ~ 30 km away, and the next nearest nucleus would be ~ 60 km away. Hence, even in a material such as solid uranium, the neutron is moving in a virtual vacuum. Only about one part in 10^{15} of "solid matter" is "occupied." It is only because the neutrons move so fast (8000 to 80,000,000 km/hr) and because there are so many nuclei in a given volume ($\sim 5 \times 10^{22}$ nuclei/cc) that reactions between neutrons and nuclei take place so quickly (10^{-6} to 10^{-3} sec).

If we interpret the figures in the table literally, the nucleus is also far from "solid." Its "size" is about 1000 times that of its constituent particles. We picture these particles as being confined to a very small region of space by a very strong, very short-range, attractive "nuclear force." This force is thought to be the same between a neutron and a neutron, between a proton and a proton, or between a neutron and a proton. The force must be very strong indeed since it overcomes the repulsive Coulomb force between two protons in a nucleus. (Two "touching" protons of radius 10^{-16} cm will, if Coulomb's Law is applicable at such distances, repel each other with a force of two billion grams.) Yet the nuclear force must have a very short range. (Otherwise all the protons and neutrons in the universe would collapse into a single lump.)

Actually the range of the nuclear force is $\sim 10^{-13}$ cm, roughly the size of the nucleus. A neutron traveling through a medium doesn't "feel" the presence of a nucleus until it gets within this range. Thus it will continue traveling in a straight line until it gets within 10^{-13} cm of some nucleus. On the other hand a nucleus, if for some reason it starts to move through the medium, will be repelled by the positive charges of neighboring nuclei as soon as it begins to penetrate the electron clouds surrounding those nuclei. This is because electrical forces (Coulomb forces) are long-range. (It is, after all, the positive charge of the nucleus which, by operation of Coulomb's Law (plus some quantum principles), keeps the electrons in their stable "orbits" outside the nucleus.) Thus, with distances again scaled up by a factor of $\sim 3 \times 10^{14}$, we conclude that if electrical forces reach out 30 km from a nucleus, nuclear forces reach out only 30 cm.

Nuclear Reactions

When a neutron does get within range of the nuclear force of a nucleus, it is pulled into that nucleus very abruptly. To put it another way, the potential energy it had by virtue of being outside the range of the nuclear force is suddenly transformed into kinetic energy when it enters the nucleus. Thus the neutron suddenly picks up great speed and then, by "collisions" with the other neutrons and protons making up the nucleus, increases the average kinetic energy of all the particles in the nucleus. The nucleus thus gets energized to an unstable *excited state*. Because of the presence of the extra neutron it is said to be a *compound nucleus*.

Excited states of a nucleus can persist anywhere from 10^{-14} seconds to years depending on the nucleus and the state in question. Yet even the short time (10^{-14} sec) is around a hundred million times longer than the time required for a neutron to make one trip across the nucleus. Thus it is proper to think of the excited nucleus as being in a well-defined state.

When the excited nucleus does give up its excess energy, it can do so in a great variety of ways, of which the following are examples.

Elastic Scattering. A neutron is emitted and the nucleus returns to its initial ground state. The emitted neutron need not be the same one that originally struck the nucleus; but, nevertheless, the elastic-scattering process is in an important sense analogous to the collision of one billiard ball with another. The lifetime of the excited compound nucleus is so short ($\sim 10^{-12}$ sec) that, for the purposes of reactor calculations, it may be neglected and the process of elastic scattering may be analyzed as a billiard-ball collision.

This type of elastic scattering, called *resonance scattering* and involving the formation of a compound nucleus, is actually very rare. A far more common form is *potential scattering*, in which the impinging neutron *does* behave exactly like one billiard ball striking another. It interacts with the nucleus as a whole without entering and forming a compound nucleus. Reaction times can thus be on the order of 10^{-22} seconds.

Thus, whether the elastic scattering is anomalous resonance scattering or the more common potential scattering, we can picture the original neutron as striking the nucleus, imparting some of its momentum and kinetic energy to the nucleus, and then moving off in a direction different from its original flight path. Elastic scattering is an extremely important process for nuclear reactors since it is one of the chief mechanisms by which the high-energy neutrons born in fission lose their kinetic energy. We shall analyze the process in detail in Chapter 2.

Inelastic Scattering. A neutron is emitted from the compound nucleus but the nucleus still remains in an excited state. Again the process is effectively instantaneous (10^{-12}

sec) with respect to the time-scale of processes important to reactor physics. Thus we can still picture the initially incident neutron as being scattered by the nucleus and losing an unusually large amount of energy in the process.

Charged-Particle Emission. The excited compound nucleus becomes de-excited by emitting a charged particle (proton, deuteron, α particle—occasionally an electron). From the viewpoint of the free-neutron population in the system, the process results in the loss of a neutron and hence is equivalent to a neutron capture.

Neutron Capture. The compound nucleus de-excites itself by emitting a γ *ray*, a high-energy photon or quantum of electromagnetic energy. The energy E of a γ ray, (as is the case for all photons) is related to its frequency ν by $E = h\nu$, where h (Planck's constant) $= 6.6 \times 10^{-27}$ erg-sec. Typical frequencies for γ rays are around 2×10^{20} Hz, whereas those for the photons of visible light are around 6×10^{4} Hz.

The effect on the free-neutron population of a neutron capture is the same as in charged-particle emission, namely, a neutron disappears. Often the nucleus is still not stable after the γ-ray emission, and a β decay follows.

(n, 2n) **Reaction.** The excited nucleus relaxes by emitting two neutrons. This is more likely to happen if the kinetic energy of the initial neutron is quite high and if the target nucleus is made up of a large number of particles. Even then, however, it is not a highly probable event.

From the reactor-physics viewpoint, an (n, 2n) reaction increases the free-neutron population and is thus similar to a fission event (except that a varying number of neutrons can be emitted in fission and the probability of the occurrence of an (n, 2n) reaction in a reactor is usually about two orders of magnitude less than the probability of occurrence of a fission).

1.3 Nuclear Fission

The de-excitation process of greatest importance for reactor physics is nuclear fission, the splitting of the nucleus into two large fragments along with the essentially instantaneous emission of, generally, from one to three free neutrons and numerous γ rays. The process occurs with greatest probability in very heavy elements. In fact, for odd-numbered isotopes such as U^{233}, U^{235}, Pu^{239}, or Pu^{241}, fission can occur with high probability if even a very slowly moving neutron strikes the nucleus. For even-numbered isotopes such as Th^{232}, U^{238}, or Pu^{240}, however, the incident neutron must have some kinetic energy—and usually a sizeable amount—before fission can occur. The fission reaction for isotopes of this latter kind is thus said to have a *threshold energy*.

A Qualitative Classical Picture of the Fission Process

The behavior of a nucleus about to fission is analogous to the behavior of a drop of mercury about to subdivide into two smaller drops. The time involved is very much shorter for the fission process, but the phenomenon of surface tension is of central importance in both cases. This can be seen by the following classical (i.e., non-quantum-mechanical and hence approximate) argument.

In a large nucleus containing many (> 200) nucleons (neutrons or protons), a given nucleon is to some extent always being acted upon by all of its near neighbors, since they are all within the range of the nuclear force. The nucleon in question, as well as all its near neighbors, are in a state of violent motion. At any given instant there will thus be a net force on the nucleon made up of the vector sum of the attractions of each of its neighbors; but, because all the nucleons involved are constantly moving, the average force over a few hundred-trillionths of a second, for a nucleon in the interior of the nucleus, tends to zero.

For a nucleon near the surface of the nucleus, however, this is not the case, since, by definition of the "surface" of the nucleus, there are no nucleons "outside" balancing the attractive forces of those "inside." Thus a surface nucleon is, on the average, subject to a net force pulling it toward the center of the nucleus. As a consequence, on the average the nucleus has a spherical shape (just as any liquid that is subject to free fall in a vacuum will tend to form a sphere).

Now, even in the classical model, the nucleus has its spherical shape only in the sense of an average over a long time period. To understand this statement, picture the nuclear particles as being sources of light. If we could photograph the nucleus with an exposure time of 10^{-30} seconds, the resultant picture would show (according to our classical picture) a collection of individual nucleons each $\sim 10^{-16}$ cm in radius clustered in a localized region of space $\sim 10^{-12}$ cm in radius. If we did a "time exposure" of the film for, say, 10^{-8} seconds, the resultant photograph would show a spherical blur about 10^{-12} cm in radius. This fact is the basis for the statement that the nucleus is spherical in a *long* time-average sense.

If we were now to expose the film for an intermediate time, say 10^{-18} seconds, we would again get a blur (it takes only $\sim 10^{-22}$ seconds for a nucleon to travel one nuclear diameter), but its shape would be ellipsoidal rather than spherical. Moreover, if we then took another 10^{-18}-second picture, we would get a differently shaped ellipse. Thus, on an *intermediate* time-average basis (10^{-18} sec), the average positions of the nucleons (and hence the region of space where the nuclear force is significant) forms a shape more like an ellipse than a sphere. As time goes on the shape of this ellipse oscillates much like the surface of a large soap bubble. If the amplitude of this oscillation gets too large, the nucleus, just like the soap bubble, can split in two.

Certain heavy isotopes (U^{238} and Pu^{240}, for example) fission spontaneously even when unexcited by some outside influence. However this is a very rare event likely to happen to a given nucleus only after trillions of years. The probability of fissioning for a given nucleus increases as energy is added to the nucleus from some outside source. This increase is very slight until a certain "critical energy" (different for each nucleus) is added, at which point the probability of fission increases in almost a step fashion. Thus the fission process has, to a good approximation, a threshold energy. A nucleus excited below this threshold is quite unlikely to de-excite itself by fissioning; one excited above the threshold is quite likely (the odds are about 4 to 1) to de-excite by fissioning.

The critical energies for nuclei starting with thorium and going on up in atomic weight run from about 4 to 6 million electron volts (MeV). For isotopes with atomic weights below thorium they increase significantly (E_{crit} for Pb^{208} is $\sim$20 MeV).

Any phenomenon that leads to excitation of a nucleus to an energy above the critical energy is likely to precipitate a fission event. High-energy γ rays, for example, can cause fission (a phenomenon called *photo fission*). However it is rare that the energy of γ rays created in a reactor exceeds 4 MeV; moreover a high-energy γ ray is far more likely to lose its energy in ways that do not excite the nucleus. Hence photo fission is not an important process in reactor physics.

Neutrons are a different matter. As was mentioned earlier, when a neutron gets within the range of the nuclear-force field of a nucleus, it is swept into the nucleus, and the potential energy it had when it was free is quickly converted to kinetic energy. For certain isotopes (U^{233}, U^{235}, Pu^{239}, Pu^{241}, and many of the very rare, heavier isotopes) the excitation energy supplied by the incoming neutron is sufficient to cause fission even if the neutron strikes the nucleus while traveling at very slow speed (thus bringing with it no extra kinetic energy). These isotopes are called *fissile*. They are rare and expensive. Of the four, only U^{235} occurs naturally. It constitutes 0.71 percent of the atoms of natural uranium and costs about $14 per gram when separated in purified form (93 percent U^{235}). Uranium^{233} comes from neutron capture in Th^{232}, which first emits a γ ray to form Th^{233} and then further de-excites itself by emitting an electron (β particle). Plutonium^{239} comes from neutron capture in U^{238} followed by two β decays. Plutonium^{241} comes from neutron capture in Pu^{240}, which in turn comes from neutron capture in Pu^{239}. (A neutron striking a Pu^{239} nucleus will be captured—rather than leading to fission—about 35 percent of the time.) Isotopes such as Th^{232}, U^{238}, and Pu^{240}, which yield fissile nuclei after they capture a neutron, are called *fertile*.

Many other heavy elements will fission if a neutron of high kinetic energy strikes them, thus contributing its former kinetic energy as well as its potential energy to the excitation. For example Th^{232} or U^{238} will fission if struck by a neutron having a kinetic energy in excess of $\sim$1.5 MeV. Such isotopes are called *fissionable*. The term, however, is customarily used only for isotopes that will fission in an operating reactor, that is, when struck by a neutron having an energy in the range characteristic of neutrons emitted during fission itself.

1.4 The Consequences of Fission: Particles Emitted

Fission Fragments

With very rare exceptions a nucleus that fissions splits into two fragments.

Because of certain nuclear stability effects having to do with the numbers of particles in the nucleus, the two fragments are unlikely to be of equal mass. In fact, curves of the probable distribution of masses of fission fragments (see Figure 1.1) have a decided dip in them when the masses of the fragments are equal.

As can be seen, only for the case of fission by high-energy neutrons is there a significant probability of an equal split.

The fission fragments, as they appear initially, are unstable with respect to the number of neutrons in their nuclei. Most commonly they become stable again through a series

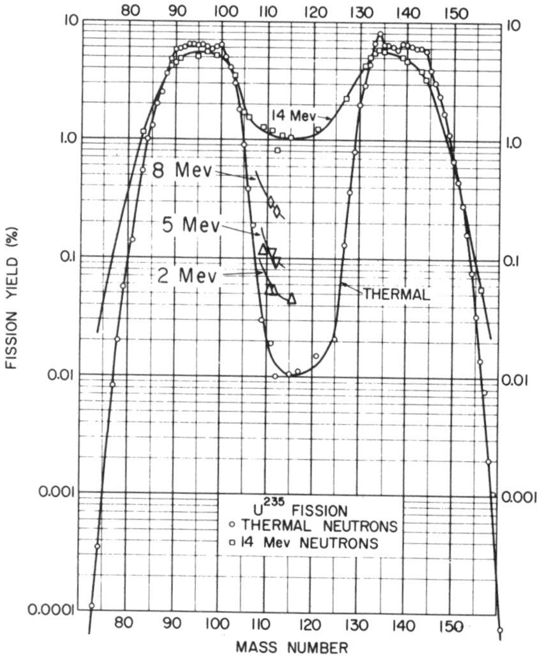

Figure 1.1 Mass-yield curve for fission of U^{235} by thermal, ~2-MeV (fission-spectrum), 5-MeV, 8-MeV, and 14-MeV neutrons. (Thermal and 14-MeV data are from the compilation of Katcoff; intermediate-energy data in the mass valley are from Ford and Gilmore.) From Keepin (1965).

of β decays (emissions of electrons) which, in effect, transform the excess neutrons in the nucleus into protons. Thus, in the course of their radioactive decay to a stable end product, the fission fragments assume a sequence of chemical identities, each successive one having a nucleus containing an additional positive charge. Some of the members of these decay chains are nuclei that have a high probability for capturing neutrons. Such nuclei represent a "poison" in the system in that they compete for neutrons with the fissile material. Thus the details of the splitting curves shown in Figure 1.1 are of some importance. It is unfortunate in this regard that Xe^{135}, which belongs to the mass-135 decay chain, is the isotope that has the highest probability for capturing low-energy neutrons. Examination of Figure 1.1 shows that the mass-135 chain is one of the most likely products of a fission event.

Neutron Emission

Neutrons are also emitted in fission—on the average ~ 2.5 of them, the particular number for any given fission event depending on the particular split that occurs. The average total number of neutrons emitted per fission is symbolized by ν. This quantity differs for the different fissionable materials and is slightly dependent on energy. Thus a more precise notation for the neutrons emitted per fission in, say, U^{235} is $\nu^{235}(E')$, where E' is the kinetic energy of the neutron causing the fission.

The fraction of the ν neutrons emitted in fission that are emitted between the energies E and $E + dE$ is symbolized by $\chi^j(E)\,dE$, the quantity $\chi^j(E)$ being called the *fission spectrum* for isotope j. Strictly speaking $\chi^j(E)$ also depends on the energy of the neutron causing the fission. However this dependence is slight and can, if necessary, be accounted for by letting the $\chi^j(E)$ depend on position within the reactor. In fact even the superscript j is a bit superfluous; the $\chi^j(E)$ for different isotopes are almost indistinguishable. An empirical expression for the fission spectrum $\chi^j(E)$ that can be used in most applications for all isotopes j, regardless of the energy of the neutron causing the fission, is

$$\chi(E) = 0.453\, e^{-1.036E} \sinh \sqrt{(2.29E)}, \tag{1.4.1}$$

where E is in MeV. Since $\chi(E)$ is a probability density, $\int_0^\infty \chi(E)\,dE = 1$. The curve (1.4.1) is plotted as Figure 1.2. Note that the most probable energy of an emitted neutron is slightly below 1 MeV and that very few neutrons are emitted with an energy greater than 10 MeV.

The neutrons emitted in fission may be assumed to emerge with equal probability in all directions. In other words fission neutrons are produced *isotropically*. They preserve no "memory" of the direction of travel $\mathbf{\Omega}$ of the neutron that caused the fission.

Delayed Neutrons

Not all the ν neutrons emitted in an average fission event appear at once. A small

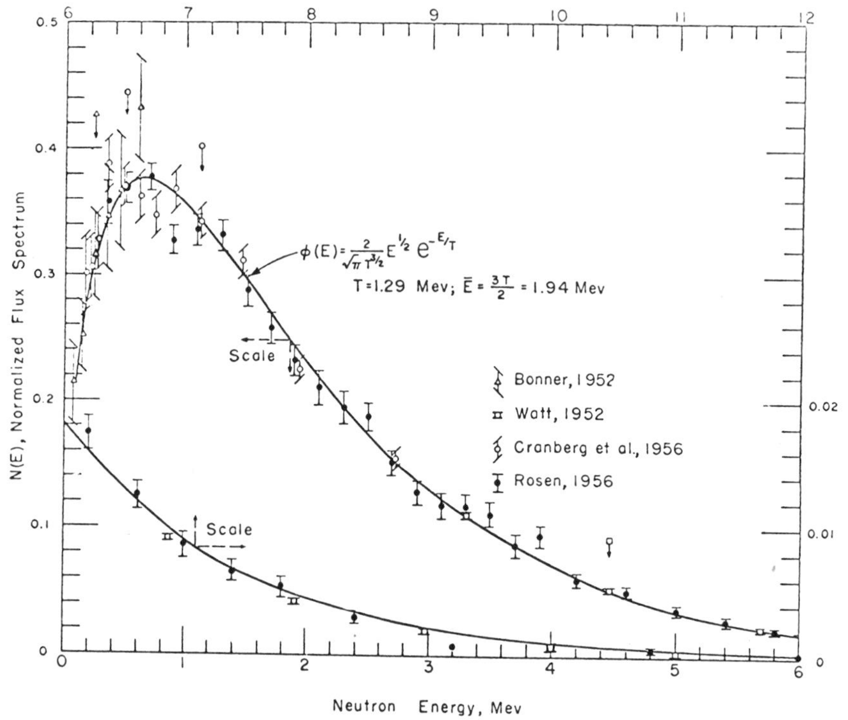

Figure 1.2 Experimental neutron energy spectrum from thermal-neutron-induced fission of U^{235}. Methods of measurement were: Bonner, cloud chamber; Watt, proton recoil; Cranberg, time-of-flight; Rosen, nuclear emulsion. The solid line shows the best-fitted Maxwellian spectral function. Arrows indicate the normalization points for each of the sets of data. From Keepin (1965).

fraction (0.65 percent for U^{235}) are emitted from certain of the fission fragments after a β decay makes these fragments unstable with respect to their neutron content. The β decay that precedes the emission of these *delayed neutrons* is a relatively slow process, the half-lives of the unstable fission fragments involved being between ~0.2 sec and ~54 sec. There are believed to be about twenty fission fragments that emit delayed neutrons. Such fission fragments are called *delayed-neutron precursors.* Each precursor has a unique half-life for β decay (after which the delayed neutron is emitted essentially instantaneously). The amount of precursor formed on the average as the result of a fission is, however, not unique; it depends both on the isotope undergoing fission and on the energy of the neutron causing that fission. Thus the number $C(t)$ of delayed-neutron precursors present at a time t after a single, average fission of fissionable isotope j is given by

$$C(t) = \sum_{i=1}^{20} \nu^j(E')\beta_i^j(E')e^{-\lambda_i t}, \tag{1.4.2}$$

where $\nu^j(E')$ is the total number of neutrons emitted when a neutron of energy E' causes fission in isotope j, $\beta_i^j(E')$ is the fraction of $\nu^j(E')$ that will be emitted from precursor i, and λ_i is the decay constant ($\lambda_i = 0.693$/half-life) of the fission fragment that constitutes the ith precursor.

In practice it is found that an adequate representation of $C(t)$ can be made if only six "effective" precursor "groups" are used (rather than the twenty physically real fragments). When this approximation is made, however, the six "equivalent" λ_i's are no longer physical quantities and are, therefore, slightly dependent on j and E'.

Delayed neutrons (those coming from the precursors, as distinct from *prompt* neutrons which are emitted at essentially the instant the fission takes place) are extremely important in determining the time-dependent behavior of a reactor under non-steady-state conditions (when the number of neutrons created per second through fission differs from the number used up per second by absorption and leakage).

γ Rays, β Particles, and Neutrinos

In addition to producing fission fragments and neutrons, a fission event also results in the release of γ rays, β particles (often still referred to as "β rays"), and neutrinos.

The γ rays come from three sources: those released at the time of the fission (called *prompt* γ*'s*), those emitted during the radioactive decay of the fission fragments (*delayed* γ*'s*), and those released as the result of subsequent capture of some of the neutrons emitted in fission (*capture* γ*'s*).

It has already been pointed out that the β rays released as a result of the fission process come from the radioactive decay of the fission fragments and that neutrino release also accompanies this radioactive decay. In fact the neutrino, the existence of which was originally postulated to account theoretically for the observed characteristics of β decay, was first observed experimentally with a nuclear reactor acting as the neutrino source.

1.5 The Consequences of Fission: Energy Released

The energy emitted as the result of a fission is transmitted to the local environment by the particles and photons released as a consequence of the fission. Thus, to see how this energy is converted into heat (the form desired for present power-generation methods), we shall follow the history of these emanations.

The Energy from Fission Fragments

The two fragments into which a fissioning nucleus splits quickly re-form into roughly spherical shapes, after which, since they are separated by a distance greater than the range of the nuclear force and are highly charged, they are pushed apart by a very great Coulomb repulsive force. Hence they pick up speed and go careening through the medium, repelling (by Coulomb interaction) other charged nuclei and thereby transmitting kinetic energy to them. These charged nuclei, in turn, push their neighbors. The net result is that the initial kinetic energy of the fission fragments ($\sim$168 MeV) is transmitted via Coulomb interaction in a cascade fashion to the nuclei in the vicinity (10^{-3} cm) of the site where the original fission took place. Thus the average kinetic energy of these neighboring nuclei (i.e., their temperature) increases in a time which, for purposes of reactor calculations, is instantaneous. About 82 percent of the energy released in fission is converted to a local increase in temperature in this manner.

Disposition of the Neutron Kinetic Energy

The total kinetic energy of neutrons emitted in a fission is $\sim$5 MeV. This energy, corresponding to a speed of $\sim 80 \times 10^6$ km/hr, is converted into heat as the neutrons collide with the nuclei in the reactor, thereby increasing the average kinetic energy of these nuclei. If they don't leak out of the reactor and are not absorbed, the neutrons thus slow down until their average energy approaches that of the atoms of the material making up the medium, $\sim$0.025 eV, which corresponds to a neutron speed of $\sim$8000 km/hr. Eventually (in 10^{-3} to 10^{-7} seconds, depending on the medium) the neutrons are absorbed.

β- and γ-Ray Energy-Decay Heat

The prompt and delayed γ rays emitted at the time of fission and later on from the fission products each contribute about 7 MeV of energy. The amount of γ energy released as the result of neutron capture depends on the material in which the capture takes place but is usually in the range 3–12 MeV. Strictly speaking this capture-γ energy should not be counted as part of the energy released by fission since it really requires active contributions from other nuclear species. Nevertheless, since its release always accompanies the fission release and since it contributes to the overall reactor power level, we shall include it.

Since γ rays are not charged particles, they are not subject to Coulomb interaction. Instead they lose their energy by other kinds of electromagnetic interactions with charged particles. Specifically three mechanisms are involved: the photoelectric effect (most important for γ rays of low energy, < 0.3 MeV), the Compton effect (most important at intermediate to high energies, 0.3–10 MeV), and pair production (most important at energies > 10 MeV). All these mechanisms result in an energy transfer to a charged particle, which then, because of Coulomb interactions, moves only a short distance before its energy is converted to heat by the cascade effect mentioned earlier. Thus, once a γ ray interacts with matter, the energy it transfers is deposited locally. However the γ ray itself may move a number of centimeters, or, occasionally, meters, before losing all its energy. In this respect it is like the neutron, which also moves a number of centimeters, on the average, before being absorbed. Except in very small reactors neither γ-ray nor neutron leakage out of the reactor amounts to more than a few percent. However, even though most of the energy of these particles is absorbed within the reactor, the small leakage can be a severe biological hazard, and it is therefore necessary to provide a radiation shield for reactors.

The β rays released from radioactive decay of the fission fragments contribute an energy of about 8 MeV per fission, and, β rays being charged, this energy is immediately converted to heat locally.

Because radioactive decay can be a slow process, the energy of β and γ rays coming from fission fragments and from capture γ rays is transformed into heat relatively slowly. On the average the fragments from a single fission decay in time approximately as $t^{-1.2}$ ($1 \text{ sec} \leq t \leq 10^6$ sec). To be specific the average decay power (energy released per second) following a *single* fission is

$$P_d(t) = 2.66\, t^{-1.2} \text{ MeV/sec}, \quad t > 1 \text{ sec}, \tag{1.5.1}$$

where t is the time (in seconds) since the fission took place. If a reactor has been at a constant power corresponding to F fissions per second for a very long time and then at some time t_0 (prior to t) has been turned off, we must add the average decay power from *all* past fission to get the *overall* average decay power. To do this we note that the contribution to the total decay power at time t due to fissions which occurred in time dt' equals the number of fissions $F\,dt'$ that occurred between t' and $t' + dt'$ times $P_d(t - t')$. As a result the total decay power at t is

$$\int_{-\infty}^{t_0} (F\,dt')\, 2.66(t - t')^{-1.2} = \frac{2.66}{0.2} F(t - t_0)^{-0.2}, \quad (t - t_0) > 1 \text{ sec}. \tag{1.5.2}$$

Thus the *total* decay power following sustained, constant power operation of a reactor falls off only as the 1/5th power of the time after shutdown. Only a few percent of the total reactor power is involved in this decay. However, for a reactor which operates at, say, 2400 megawatts, this is a substantial, long-lived decay power. Thus a power reactor must be both cooled and shielded after shutdown.

Neutrino Energy

About 12 MeV of the energy released in a fission is tied up in neutrinos. However, since these particles are not charged and have zero rest mass, the probability that they will interact with any of the atoms of material constituting the reactor is negligibly small. Hence the 12 MeV of neutrino energy out of the ~200 MeV released by a fission is lost to the reactor. To some extent, however, this energy loss is made up by the energy released by the capture γ's. Thus 200 MeV is a reasonably accurate approximation for the energy released in a reactor as the result of a fission.

The Overall Energy Balance

We have not yet discussed the question of where the 200 MeV of energy released in fission "comes from." The answer is that it comes from the mass of the fissioned atom. Thus, if we measure the sum of the rest masses of all the products of a fission (fragments, neutrons, and β's) and subtract that sum from the mass of the original fissionable atom, we get a mass difference which, when multiplied by the square of the velocity of light, gives (after a change of units) 200 MeV. Hence the immediate result of a fission is the conversion of mass energy into the kinetic energy of the fission fragments and prompt neutrons and the radiation energy of γ rays and neutrinos. The fission fragments subsequently decay, further converting mass energy into γ rays and neutrinos and into the kinetic energy of released β rays and delayed neutrons. The neutrons, by direct collisions with nuclei, and the γ rays, by electromagnetic interactions, transmit their energy to charged particles. Thus, except for the neutrino energy, all the energy released as the result of a fission is transmitted to charged particles. These, because of the long-range Coulomb interaction with the nuclei making up the material medium, transmit their energy to other charged nuclei in a cascade effect, thus raising the average kinetic energy, or temperature, of the material comprising the medium.

The amount of energy obtained from the fissioning of a small amount of fissionable material is remarkable. A good figure to keep in mind in this regard is that a power level of one watt results from a fission rate of $\sim 3 \times 10^{10}$ fissions per second. In an absolute sense this is a sizeable fission rate. But, if one remembers that there are $\sim 6 \times 10^{23}$ atoms in 235 grams of U^{235}, it seems relatively small. In fact to sustain a power level of 4000 megawatts for a day requires the fissioning of only about 4 kilograms of fissionable material.

1.6 Nuclear Cross Sections

We have discussed a number of events (scattering, absorption, fission, etc.) that may occur when a neutron interacts with a nucleus. We must now make the discussion quantitative. That is, we must define numbers which specify the probability that, if a neutron, having a given kinetic energy, moves through a medium containing a given material, it will interact in a certain manner.

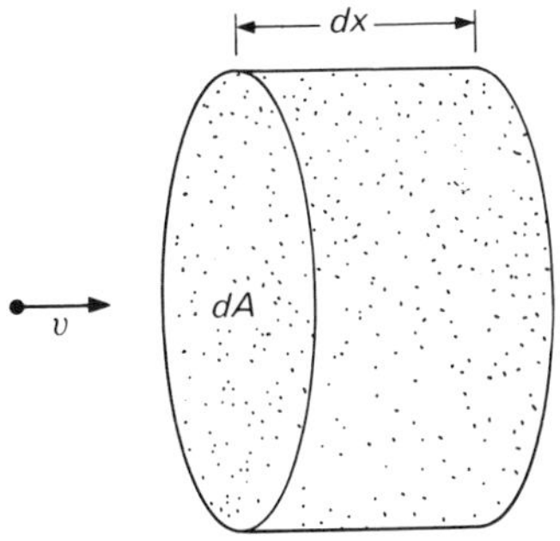

Figure 1.3 A cylinder of nuclei perpendicular to the direction of travel of a neutron of speed v.

It should first be noted that all we can do is to specify *probabilities*. Thus, if a neutron falls into a nucleus and excites it, the nucleus will de-excite in some definite and specific manner (i.e., via re-emission of a neutron, emission of a γ ray, fission, etc.). However we cannot predict exactly which means of de-excitation it will adopt for any particular case. All that we can do is to state that *on the average* it will de-excite in a particular fashion. This information does, however, permit us, given several thousand identical excitation events, to predict that the excited nucleus will de-excite by fission in, say, 70 percent of the cases, by γ-ray emission in, say, 13 percent of the cases, and by neutron re-emission in, say, 17 percent of the cases.

Definition of a Cross Section

To specify quantitatively the probabilities for the various things that can happen when a neutron interacts with a nucleus, physicists have introduced the concept of a *nuclear cross section*. This particular way of specifying the probability of an interaction comes from picturing the nucleus as presenting a cross-sectional area to a neutron traveling through a medium. To make things quantitative, consider a very small and very short cylinder of height dx and cross-sectional area dA (Figure 1.3), containing some pure material (for example U^{235}) having a concentration of n atoms/cc. (For solid U^{235} n would be $\sim 4 \times 10^{22}$ atoms/cc. In terms of mass density ρ, atomic weight A, and Avogadro's number $N = 0.6023 \times 10^{24}$ atoms/mole, $n = \rho N/A$.) Consider a neutron traveling in a direction perpendicular to the surface dA. Assume that dx is so small that nuclei within $dA\, dx$ do not "shadow" each other in the X direction. Then, on the average (i.e., if the experiment is repeated many times, the location at which the neutron crosses the surface dA being random), the probability that the neutron (considered as a "point" object) will "hit" a nucleus in traveling the distance dx through the cylinder is just the total cross-sectional area that the nuclei present to the neutron divided by dA. If the nuclei are pictured as having a cross-sectional area σ, the total cross-sectional area they present is the number of nuclei in $dx\, dA$ times σ. (Recall we have assumed that no nucleus shadows another in the X direction.) Thus:

Probability that a neutron will interact in going a distance dx

$$= \frac{\sigma n \, dx \, dA}{dA} = n\sigma \, dx. \tag{1.6.1}$$

The picture we have used to derive (1.6.1) leads to hopeless contradictions if taken literally. For instance it is known by measurement that the probability of interaction depends on the kinetic energy of the neutron. Thus a literal interpretation of our model would require that the nucleus adjust its "size" in accord with the energy of a neutron *about* to hit it. Nevertheless, if we shift our point of view, (1.6.1) is a very useful relationship. This is because the *left*-hand side (i.e., the probability of an interaction) *is* physically real and measurable. Thus we turn (1.6.1) around and use it, not to define the probability of an interaction, but rather to define the term *cross section*. The cross section is then not to be taken as a literal size of a nucleus, but only as a convenient way of expressing the probability of an interaction. Moreover, once we take this view, we can define cross sections for all possible nuclear processes (scattering, fission, etc.).

We are thus led to the following definition of a *microscopic cross section* for the interaction, by a specific process, between a nucleus and a passing neutron:

Probability that an interaction of type α will take place when a neutron having kinetic energy E moves a distance dx in a medium containing n^j atoms of isotope j per cc

$$= n^j \sigma_\alpha^j(E) \, dx. \tag{1.6.2}$$

This definition is still not entirely precise. The probability of interaction really depends on the *relative* kinetic energy between the neutron and the target nucleus. Thus (1.6.2) assumes either that the thermal motions of the target nuclei in $dA \, dx$ are negligibly slow in comparison with the speed of the neutron or that an average over these motions has been performed. For the formal definition of cross section we take the latter viewpoint.

Next we note that it is a good approximation to assume that the probability distribution of velocities of nuclei making up a given material depends only on the absolute temperature of the material. Thus, when we start with an interaction probability expressed as a function of the *relative* velocity between a neutron and a nucleus and then average over a distribution of nuclear velocities, we expect the result to depend only on the velocity (and hence kinetic energy) of the neutron relative to the center of mass of the material through which it is traveling and on the temperature of that material. Accordingly we shall think of the $\sigma_\alpha^j(E)$ defined by (1.6.2) as depending implicitly on the temperature of the isotope j and shall continue to look on E as the kinetic energy of the neutron relative to the laboratory system.

There are two situations in which the temperature dependence of $\sigma_\alpha^j(E)$ is important. One occurs when the neutron energy is in the so-called *thermal energy range* ($0 \leq E \lesssim 1$ eV) where the speed of the neutron is comparable to the thermal motion of the nuclei comprising the medium. The other occurs when the energy of the neutron corresponds to a "resonance energy" of the nucleus it strikes. We shall describe this latter situation qualitatively in Section 1.7 and more quantitatively in Chapter 5.

Units and Notation

A probability, being a ratio, is dimensionless. Thus, with n^j quoted in nuclei per cm^3 and dx in cm, the units of cross section as defined by (1.6.2) are cm^2. It is customary to use another unit, the *barn* (b), when quoting numerical values of cross sections determined by measurement or (quantum-mechanical) calculations. One barn is 10^{-24} cm^2. It is, of course, essential to convert from barns to cm^2 when cross sections are to be used for reactor calculations.

Cross sections for the various nuclear processes discussed in Sections 1.2 and 1.3 are distinguished by using particular letters for the general parameter α in (1.6.2). The processes we shall be concerned with and their corresponding "α symbols" are listed in Table 1.2. Thus $\sigma_f^j(E)$ is the microscopic fission cross section for a nucleus of isotope j interacting with a neutron having kinetic energy E.

We have already defined the quantity $\nu^j(E)$, the total number of neutrons emitted by fission of a nucleus of isotope j caused by a neutron of energy E. There are two other related parameters in common use. They are

$$\eta^j(E) = \frac{\nu^j(E)\sigma_f^j(E)}{\sigma_a^j(E)} \tag{1.6.3}$$

and

$$\alpha^j(E) = \frac{\sigma_a^j(E) - \sigma_f^j(E)}{\sigma_f^j(E)}. \tag{1.6.4}$$

Thus $\eta^j(E)$ (generally referred to simply as "the eta for isotope j at energy E") is the

Table 1.2 Significance of Subscripts

α Symbol	Process
s	elastic scattering
f	fission
a	total absorption (sum of fission, neutron-capture, and charged-particle processes)
i	inelastic scattering
t	total (sum of all interaction processes)

Table 1.3 Thermal (0.025 eV) Data for U^{233}, U^{235}, Natural Uranium, Pu^{239}, and Pu^{241}

	σ_a	σ_f	α	η	ν
U^{233}	579	531	0.0899	2.287	2.492
U^{235}	681	582	0.169	2.068	2.418
Natural uranium	7.59	4.19	0.811	1.335	2.418
Pu^{239}	1011	743	0.362	2.108	2.871
Pu^{241}	1377	1009	0.365	2.145	2.927

Source: BNL (1973).

total number of fission neutrons emitted per neutron of energy E absorbed in fissionable isotope j, and $\alpha^j(E)$ is the ratio of the cross section for neutron-absorption-not-leading-to-fission (i.e., *capture*) to the fission cross section for neutrons of energy E interacting with isotope j. Table 1.3 gives values for these quantities for $E = 0.025$ eV.

One other matter of notation needs to be mentioned. It is customary to identify the fissionable isotopes by assigning a number for the isotope symbol j. A two-digit number suffices. This is made up of the last digit of the atomic number and the last digit of the mass number. Thus $\sigma_s^{25}(E)$ represents the elastic-scattering cross section of ${}_{92}U^{235}$ for neutrons of energy E. Similarly $\eta^{41}(E)$ is the eta for ${}_{94}Pu^{241}$ at energy E. For the non-fissionable isotopes the chemical symbol is generally used for j (e.g., $\sigma_a^{B^{10}}(E)$). With no isotope identification the average cross section for the natural element is to be taken (e.g., $\sigma_s^{Zr}(E)$). This notation can be used with the fissionable isotopes themselves, so that we shall hereafter use U^{235} and U^{25}, for example, interchangeably.

1.7 Characteristics of Neutron Cross Sections

For reactor calculations the neutron energy range of interest extends to the top of the fission spectrum. (Equation (1.4.1) is a fit to the data; above a certain energy $\chi^j(E)$ should actually be zero rather than merely very, very small.) In practice the upper limit of interest is ~15 MeV. Even if (1.4.1) is taken as literally true, Figure 1.2 shows that very few neutrons are emitted with energies above this value.

The energy behavior of neutron cross sections in this 0 to 15 MeV range is all-important for the design of nuclear reactors. Unfortunately no simple rules exist for predicting this behavior. Nevertheless, since a few general trends can be pointed out, we shall discuss the "typical" behavior of the cross sections that are of greatest importance to reactors.

Behavior of Neutron-Absorption Cross Sections

Absorption cross sections for all elements tend to fall off as $1/\sqrt{E}$ at low (< 2 eV) energies. That is, if σ_a^j (0.025) is the value of the absorption cross section for isotope j

Table 1.4 Characteristics of Some Isotopes Important for Nuclear Reactors

Atomic Number	Isotope*	Atomic Abundance (%)	σ_a(2200 m/sec) (barns)
1	H^1	99.985	3.32×10^{-1}
	H^2	0.015	5.30×10^{-4}
3	Li^6	7.42	9.40×10^2
	Li^7	92.58	3.7×10^{-2}
5	B^{10}	19.6	3.84×10^3
	B^{11}	80.4	5.50×10^{-3}
6	C^{12}	98.89	3.40×10^{-3}
	C^{13}	1.11	9.00×10^{-4}
8	O(nat)	—	2.70×10^{-4}
11	Na^{23}	100	5.30×10^{-1}
13	Al^{27}	100	2.30×10^{-1}
24	Cr (nat)	—	3.10×10^0
25	Mn^{55}	100	1.33×10^1
26	Fe (nat)	—	2.55×10^0
27	Co^{59}	100	3.72×10^1
28	Ni (nat)	—	4.43×10^0
40	Zr (nat)	—	1.85×10^{-1}
48	Cd (nat)	—	2.45×10^3
54	Xe^{135}	9.17-hr half-life	2.65×10^6
62	Sm^{149}	13.83	4.10×10^4

* (nat) indicates the naturally occurring isotope.

for an incident neutron energy of 0.025 eV (corresponding to a speed of ~2200 m/sec), the corresponding cross section at other energies in the low energy range is given by

$$\sigma_a^j(E) = \sigma_a^j(0.025)\frac{\sqrt{0.025}}{\sqrt{E}} = \sigma_a^j(v = 2200 \text{ m/sec})\frac{2200}{v(E)}, \qquad (1.7.1)$$

where $v(E)$ is the speed of the neutron in meters/second corresponding to kinetic energy E. Such cross-sectional behavior is said to be "one-over-v" and the term "$1/v$ cross section" is common. Table 1.4 lists the values of σ_a^j (0.025) for a number of isotopes of interest for reactors. (See also Table 1.3.)

For most light elements the $1/v$ behavior persists up to high energies. However, for intermediate and heavy elements, the curve of $\sigma_a^j(E)$ versus E exhibits very high and very sharp peaks beginning in the energy range just above thermal. These peaks are associated with the existence of metastable energy levels in the excited nucleus. If the kinetic energy of the neutron about to strike the nucleus is such that, when added to the potential energy released when that neutron drops into the nucleus, the excitation energy of the nucleus corresponds to that of the metastable level, the formation of the compound nucleus is strongly favored. The situation is somewhat analogous to the process by which an electron is raised from its ground-state orbit to an orbit corresponding to an excited state. Absorption of a photon having an energy corresponding to the difference in energy between these two orbits is strongly favored. (Thus, if light of all frequencies is shone on an atom, the photons absorbed will be those having differences in energy close to the differences in energy between the electron orbits; light of other frequencies will be absorbed very little.)

In the nuclear case the peaks of the cross sections for resonances at low energy tend to be very high, and the resonance widths (i.e., the energy range about the peak energy within which the cross section exceeds half its peak value) are very narrow. For example Xe^{135}, which is created by β decay of I^{135}, an isotope belonging to one of the fission-fragment decay chains, has an absorption resonance with a peak cross section of about 3×10^6 b.

Figure 1.4 shows the total cross section for U^{28} as a function of energy in the range 1 to 10,000 eV. Most of the resonances shown are due to neutron absorption although the asymmetric dipping just beyond each resonance is a scattering effect). The figure

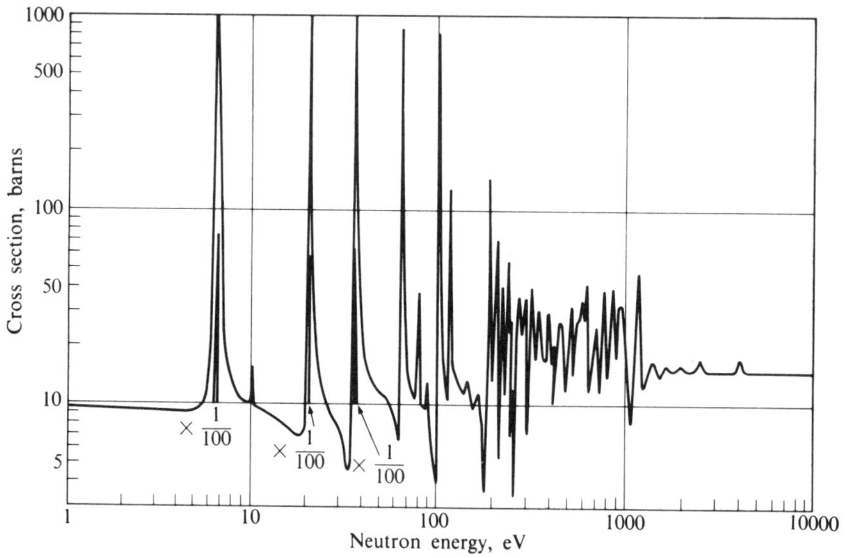

Figure 1.4 Total cross section of U^{28} in the range 1–10,000 eV. Source: BNL-325.

shows that the peak value of σ_t^{28}, corresponding to an incident-neutron energy of 6.67 eV, is ~7000 b. (due almost entirely to absorption). Actually this value is the result of measurements made at room temperature, so the curve shown for $\sigma_t^{28}(E)$ reflects the fact that the target nuclei were in thermal motion. Had the target nuclei been at rest, the peak value of σ_t^{28} for the 6.67-eV resonance would have been ~20,000 b, and the width of the resonance would have been 0.027 eV. Because of the thermal motion of the target nucleus, however, the energy range about 6.67 eV out of which neutrons are likely to be absorbed in U^{28} is actually broader than is implied by the 0.027-eV width, and the probability for absorbing neutrons having exactly 6.67 eV of energy is thus reduced. The net effect turns out to be that more neutrons in the reactor are in danger of being captured by the U^{28} nuclei, and the resonance is said to be *Doppler-broadened.* A quantitative treatment of the effect will be presented in Chapter 5. For the present we shall merely note that the "Doppler effect" is a very important automatic safety feature in a reactor. If for some reason the reactor temperature begins to rise (and hence the thermal motion of the constituent nuclei increases), the absorption rate of neutrons increases and the chain reaction consequently tends to shut down.

At neutron energies above a few hundred electron volts the resonance peaks in the curves of $\sigma_a^j(E)$ versus E for heavy elements become lower and broader. Moreover it becomes increasingly difficult to make precise measurements of $\sigma_a^j(E)$ in this energy range, and the experimental curves begin to take on a smooth appearance. Accordingly the energy range from a few hundred eV to several keV is known as the *unresolved* energy range.

Above the unresolved range absorption cross sections for both heavy and light elements exhibit broad low resonances and are in general small (a few barns in magnitude). At these high energies charged-particle reactions such as (n, p) (neutron in, proton out) begin to contribute to $\sigma_a^j(E)$.

Behavior of Neutron-Scattering Cross Sections

Scattering cross sections for most elements interacting with low-energy neutrons are elastic, constant as a function of energy, and small in magnitude (~10 barns). Important exceptions to this rule occur when the nucleus with which the neutron is interacting is bound chemically to other nuclei or is part of a crystalline structure. Under these conditions low-energy neutrons tend to interact with several nuclei at once. The scattering is then often inelastic, some of the incident kinetic energy of the neutron being retained by the molecule in the form of excited vibrational or rotational motions of its constituent atoms. One particularly important consequence of these effects for reactor analysis is that the scattering cross sections of hydrogen in H_2O and of deuterium in D_2O rise significantly at low incident-neutron energies rather than remaining constant.

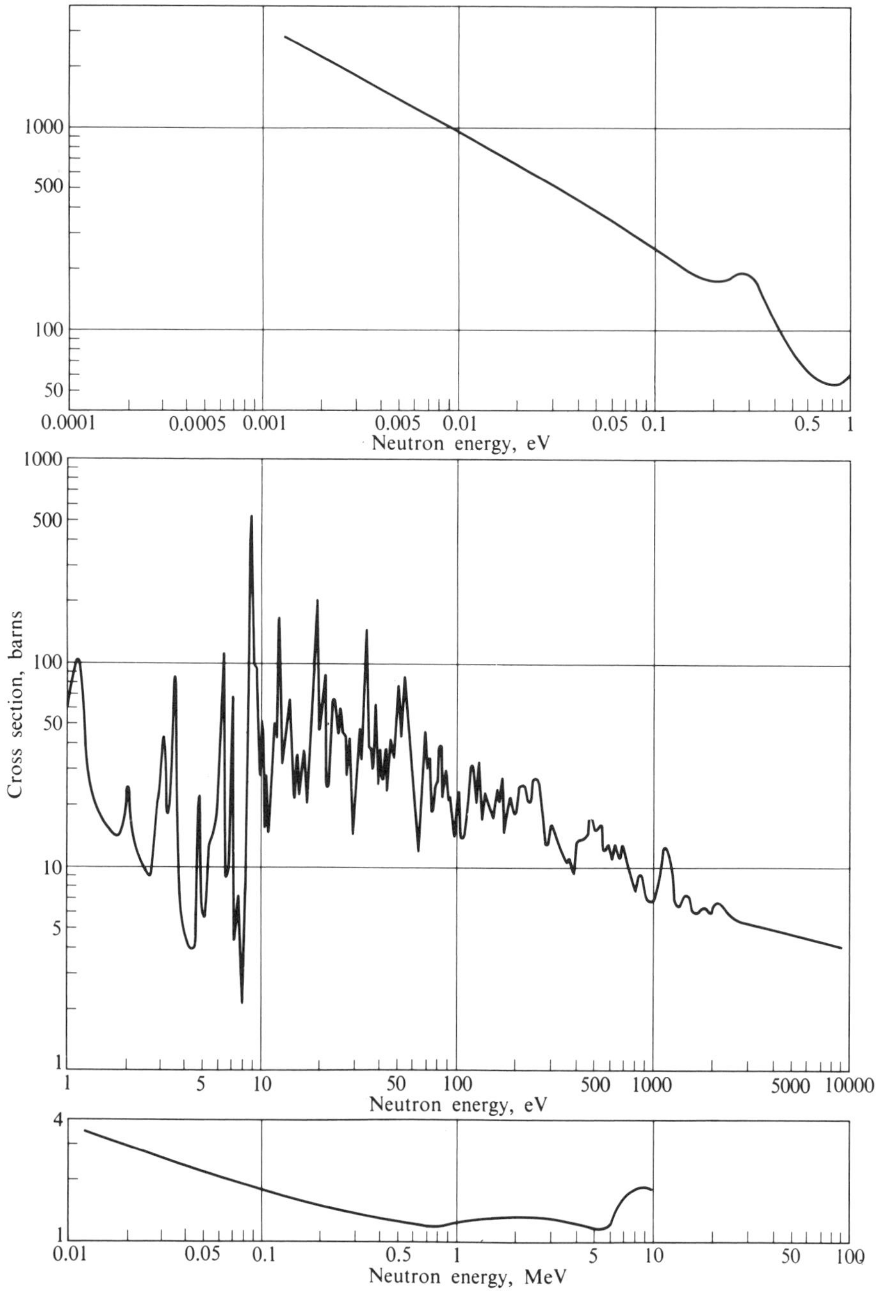

Figure 1.5 The fission cross section of U^{25}. Source: BNL-325.

Scattering cross sections for light and intermediate elements remain constant up to the MeV range, where broad and fairly low resonances can appear. For heavy elements scattering resonances can appear along with the absorption resonances at incident-neutron energies just above the thermal range.

Behavior of Fission Cross Sections

For fissile isotopes fission cross sections in the thermal range tend to have a $1/v$ shape and to be very large. Thus, at an incident neutron energy of 0.0253 eV, $\sigma_f^{25} = 582$ barns and $\sigma_f^{41} = 1009$ barns. This should be compared with the 0.0253-eV absorption cross section of hydrogen (0.332 barns), zirconium (0.18 barns), iron (2.53 barns), or U^{28} (2.73 barns). Thus, even in a mixture containing 200 atoms of U^{28} to one atom of U^{25}, a thermal neutron is more likely to cause fission in U^{25} than be absorbed in U^{28}. At higher energies the competition for neutrons is not nearly so one-sided.

Above the thermal range ($E > 1$ eV) large narrow resonances appear. These decrease in height and widen in the keV range (although part of this apparent effect is due to the difficulty of making precise measurements). In the high (keV to MeV) energy range the fission cross sections of the fissile nuclei level off at a few barns. Figure 1.5 shows the behavior of $\sigma_f^{25}(E)$ over the entire range of interest.

The fissionable isotopes (U^{28}, Pu^{40}, Th^{02}, etc.) all have energy thresholds around one or two MeV. Below these thresholds the cross sections for fission (except for containing very small resonances) are essentially zero. As shown in Figure 1.6 they rise very quickly above the thresholds but level off at a few barns.

Compilations of Cross Sections

A set of references in which nuclear cross-sectional data are tabulated appears at the

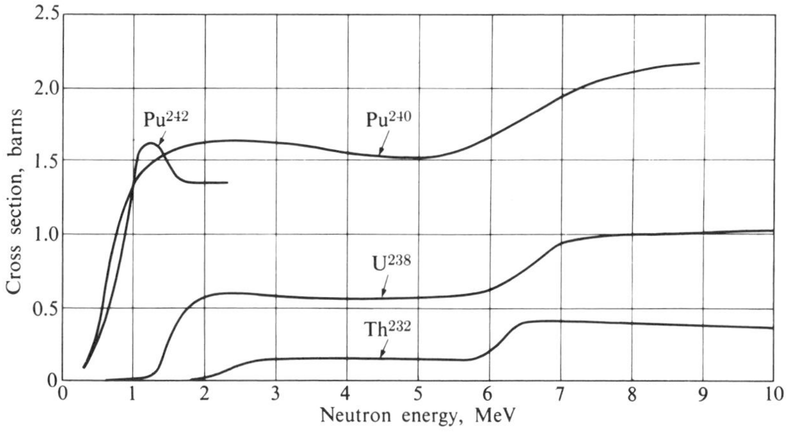

Figure 1.6 The fission cross sections of Th^{02}, U^{28}, Pu^{40}, and Pu^{42}. Source: BNL-325.

end of this chapter. Of these BNL-325 is probably the most instructive document to peruse in order to obtain a broad picture of the characteristics of neutron cross sections in the energy range of interest for reactor design.

A computer tabulation, called the *Evaluated Nuclear Data File* and referred to as ENDF-B, is available, although only on a limited basis at present. A rather elaborate effort has been undertaken (with the support of the United States Atomic Energy Commission) to compile ENDF-B, and new, improved versions appear periodically. Nuclear-data experts from many laboratories throughout the United States are asked to contribute various segments of the data, other experts are asked to judge the quality of these contributions, and a group at Brookhaven National Laboratory has been assigned the job of tabulating the data (on computer tape), creating computer programs for manipulating it, and making available tapes containing a single, current set of recommended data for all possible interactions between a neutron and any given isotope. The data are presented for a set of energy points, the number of points for a given interaction with a given nucleus being determined by the accuracy with which the data are known. (The rule is to provide sufficient data that approximating the energy behavior of a given cross section between data points by a straight-line interpolation will yield values that are within the experimental uncertainties of the measured data.) Thus ENDF-B is a file of "recommended" neutron-interaction data. Laboratories making use of the file are free to add other sets of data to their own private recommended set. The intent is that, whether or not they accept the current recommended values, many groups will use the ENDF-B format in tabulating basic neutron cross-sectional data. This makes exchange of such basic data between laboratories particularly simple. More important, it insures that, if any given laboratory improves the accuracy of a particular tabulation, that improved version will be created in a form that can immediately be exchanged with other ENDF-B users.

1.8 Nuclear Reactors

The qualitative discussion of nuclear physics presented in this chapter by no means covers the field in the depth needed to prepare libraries of cross sections for reactor-design purposes. On the other hand it should provide a sufficient familiarity with the basic definitions and concepts to serve as a starting point for the development of the field of reactor physics.

Before beginning this study, however, it may be helpful to discuss its final goal—reactors—in a qualitative fashion. We shall do this by first introducing some general terms and then describing various types of reactors from several different viewpoints.

General Considerations and Definitions

A reactor is an assembly of fissile material and other material (described below), put together in such a geometrical configuration and in such concentrations that a sustained neutron chain reaction can take place. A "sustained neutron chain reaction" is one in which the total number of free neutrons created per second by the fission process throughout the reactor exactly equals the total number of neutrons lost per second by absorption and by leakage out of the reactor. A reactor in this condition is said to be *critical*. A reactor in which the number of neutrons created per second exceeds those lost is said to be *supercritical*. One in which the number of neutrons lost per second exceeds those created is said to be *subcritical*.

Almost all reactors consist of a fuel-bearing region called the *core* and a surrounding non-fuel-bearing region called the *reflector*. In certain types of reactors there is also a region containing fertile material, called the *blanket* and usually located between the core and the reflector. (As such a reactor operates, neutrons will be absorbed in the blanket and fissile material will be produced. However, even when the fissile material builds up, this region is still called the blanket.)

Reactors can be classified in several ways:

1. By the energy range of the neutrons that cause most of the fissioning
2. By the material constituents of the reactor, including their physical state (solid, liquid, gas) and their purpose (fuel, structure, coolant, etc.)
3. By the purpose (or purposes) for which the reactor is to be used.

We shall discuss these classifications separately, although any particular reactor is usually characterized by a mixture of terms from the different classifications.

Classification of Reactors by the Energy Range in Which Fissions Occur

With respect to the neutron energy in which most of the fissions take place, the reactors being built in the world today are said to be *thermal* or *fast*. In a thermal reactor most of the fissions are caused by neutrons having an energy less than 1 eV. In a fast reactor the energy range in which most of the fissions take place is much wider, extending from ~ 100 keV to the top of the range of the fission spectrum (~ 15 MeV).

At present there seems to be no motivation for designing a reactor in which most of the fissions take place in the intermediate or resonance energy range (from 1 eV to 1 keV). The capture to fission ratio $\alpha^j(E)$ tends to be high in this range so that $\eta^j(E)$ is small: thus an insufficient number of extra neutrons (over the number needed to continue the chain reaction) become available for profitable use to justify the extra cost of building a reactor that operates in this energy range.

We shall see in Chapter 2 that when a neutron scatters from a light nucleus (H, D,

Be, C) it can transfer a sizeable amount of its energy to that nucleus—the lighter the nucleus, the more energy transferred. Hence the neutron will emerge from the scattering event with a greatly reduced energy. It follows that, if we wish to build a thermal reactor, we must include along with the fissionable material (called the *fuel*) a significant amount of one or more of the lightweight elements (called *moderators*). Thermal reactors, for example, will always be found to contain H_2O or D_2O or graphite or ZrH_2 or Be, etc.

Conversely, if we wish to build a fast reactor, we must *avoid* the presence of light elements. Thus fast reactors have no moderator (except accidentally). Moreover they must not be cooled by a material, such as liquid water, that contains hydrogen. Instead it is necessary to cool with a gas or a liquid metal such as sodium.

Classification of Reactors by Material Constituents

The materials in a reactor usually are present for a specific purpose, and a classification of reactors according to their material constituents frequently includes mention of what each of the materials is used for. Thus a reactor can contain fuel, fuel cladding, moderator, coolant, fertile material, structural material, and control material (neutron absorbers which can be inserted or withdrawn in order to adjust the critical condition of the reactor).

Fuels. The term "fuel" is a bit ambiguous since the actual fissile material is almost always mixed with fertile material or diluent, and it is this mixture that is referred to as "fuel." Thus we have fuels of 15 percent $Pu^{49}O_2$ in natural UO_2, or "slightly enriched" UO_2 (i.e., uranium oxide having a content of 2 or 3 percent U^{25} rather that the 0.7 percent in the natural ore), or a "highly enriched" uranium-zirconium alloy, the uranium being 93 percent U^{25}.

Cladding. The fuel is generally covered with a protective material or *cladding* in order to prevent corrosion and in order to prevent the radioactive fission fragments (created by fissions taking place on the fuel surface) from getting into the coolant or moderator. The fuel plus its cladding is called a *fuel element.* A typical fuel element for a light-water-moderated thermal reactor is a 0.4-inch-diameter cylinder, 10 feet long, containing 3 percent enriched UO_2, and clad with 0.025-inch-thick zirconium. For a fast reactor a typical fuel element (frequently called a *fuel pin*) is a cylinder 0.25 inches in diameter and 4 feet long, containing 15 percent $Pu^{49}O_2$ in natural UO_2 and clad with 0.015-inch-thick stainless steel.

The ideal fuel-cladding material is strong, highly resistant to corrosion by the coolant, a low absorber of neutrons, and cheap. The best compromise between these conflicting specifications appears today to be an alloy containing principally zirconium (used for water-cooled thermal reactors) and stainless steel (used for sodium-cooled fast reactors).

Moderators. The ideal moderator is also a material that can only be approximated in nature. It should be a cheap, dense, chemically stable material of very low atomic weight having a very low absorption cross section and a very high boiling point. For the first atomic-powered electricity plant built (Calder Hall in England), graphite was used. It is relatively cheap and has a very low absorption cross section (0.0034 barns at 0.025 eV), but it is not too good a moderator (carbon has atomic weight 12) and thus leads to reactors of large size. Most of the power-generating reactors in the United States are now moderated by light water, which also serves as a coolant. The hydrogen in light water is the best of all moderators, but $\sigma_a^H(0.025)$ is 0.332 barns and the material has a low boiling point so that the reactor must be contained in a "pressure vessel." The use of heavy water is favored in Canada. With heavy water the chief problem is cost ($\sim$ \$55 per pound).

Other materials light enough to be effective moderators for neutrons are He, Li, Be, and B. Helium and boron have never been used; the first is a gas, and the second has $\sigma_a^B(0.025) = 759$ barns. Lithium and beryllium have been used for special-purpose reactors and are the principal moderating materials in molten-salt reactors, but they are expensive and difficult to handle. Organic compounds having a high boiling point have also been investigated for use as reactor moderators (and coolants). However the problem of chemical stability at high temperatures in a radiation field is still considered severe for those compounds investigated to date.

Coolants. If a reactor operates in a steady-state fashion at a power level in excess of a few kilowatts, it is generally necessary to cool it to prevent overheating of the fuel elements. In the case of power reactors the heat removed by the coolant is, of course, used profitably. However in some cases (for example reactors designed as high-intensity neutron sources) this heat is just thrown away.

The fluid used to remove heat from a reactor is called the *coolant.* It is usually pumped through the reactor in the form of a liquid or a gas. The ideal coolant should thus have a high specific heat and boiling point (if a liquid) and should be easy to pump. It should be cheap, noncorrosive, and chemically stable with regard to both high temperatures and radiation. It should have a low neutron-absorption cross section, and, if it is made radioactive in passing through the core, this radioactivity should decay very quickly so that the coolant need not be shielded after it leaves the reactor. Finally, for fast reactors, the coolant should be a poor moderator. In light-water-moderated reactors the light water is also used as a coolant, and the whole reactor—fuel, moderator, and coolant—is contained in a single "pressure vessel" and maintained at pressures in the range 1000–2000 pounds per square inch. For heavy-water moderation a separately pressurized coolant system using either light or heavy water is employed. Graphite-moderated reactors also have a separate cooling system employing light water, carbon dioxide, or helium. The most common coolant for fast reactors is sodium, although both helium and steam-cooled systems have been investigated. (The moderating effect of sodium is small because of its relatively high atomic weight, and that of the the gases is small because of their exceedingly low densities at high temperature.)

Fertile Materials. The fertile materials used in reactors are Th^{02}, which yields U^{23}, and U^{28} (or natural U), which yields Pu^{49}. They are generally present in the form of their oxides. (The solid metal has been used, but it has a tendency to grow in size when subject to fission-fragment and high-energy-neutron bombardment. Also it interacts chemically in an almost explosive fashion if the cladding fails and the metal is placed in contact with hot water.) The fertile material is either mixed with the fissile material to form the overall fuel element or is separately clad and placed in a separate region of the reactor called the *blanket*.

Structural and Control Materials. It is frequently necessary to support the fuel elements inside a reactor so that they do not vibrate or so that the coolant flow is directed in some desired way. The ideal properties of the extra "structural material" used to provide such support are the same as those of the cladding, and, in fact, the same material is generally used for both structure and clad.

The ideal properties of control material are also the same as those of the clad, with one essential exception: For control material we want the neutron-absorption cross section to be as *large* as possible. The most common control materials are an alloy of B^{10} in stainless steel ($\sigma_a^{B^{10}}(0.025) = 3837$ b) and hafnium, a metal having about the same metallurgical properties as zirconium along with a thermal-absorption cross section of ~ 100 b and a great number of absorption resonances in the energy range just above thermal.

The control material is usually present in the form of long, small-diameter rods or thin blades that can move in and out of the reactor.

Classification of Reactors According to the Purposes for Which They Are Used

We shall classify reactors by their applications under the categories:

1. To make neutrons
2. To test reactor theory
3. To convert one material into another
4. To generate power for propulsion
5. To generate electrical power.

Rather than providing a thorough discussion of each category, we shall cite some typical examples.

Reactors to Make Neutrons. The research reactor at MIT provides a good example of a facility used to make neutrons. Figure 1.7 shows a cutaway view of this reactor and some of its associated equipment.

The reactor itself is composed of fuel elements which are boxes (2 ft × 3 in. × 3 in.)

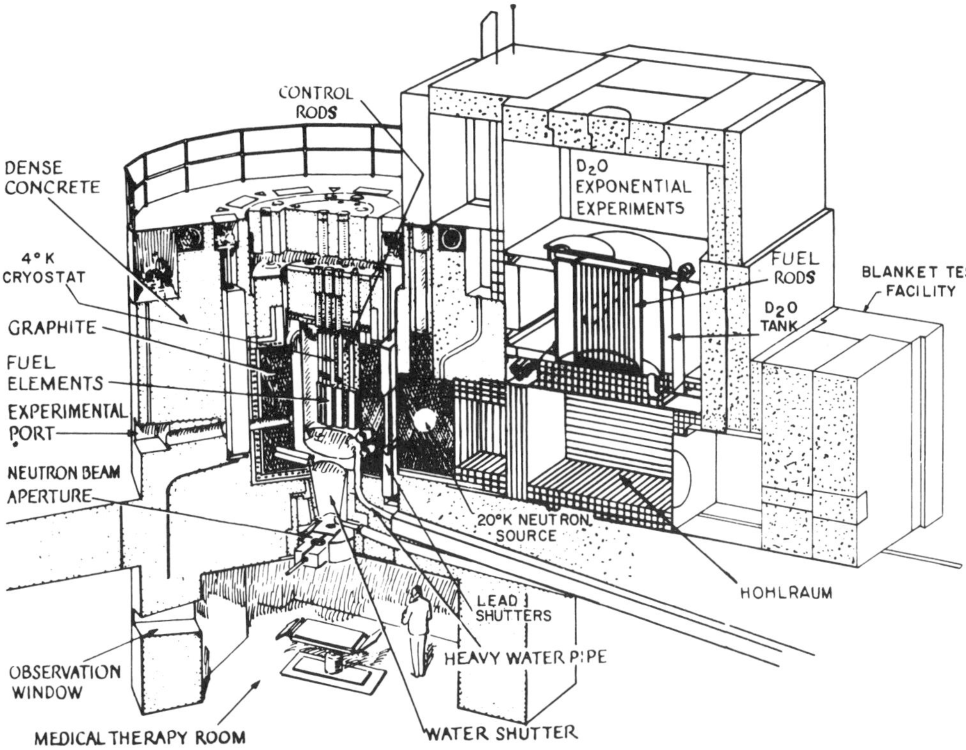

Figure 1.7 A view of the MIT research reactor showing major components and experimental facilities.

of aluminum-clad fuel plates separated by coolant channels, the fuel being 93 percent enriched uranium alloyed with aluminum. These fuel-element boxes along the reactor control rods are contained in a four-foot-diameter tank. Moderation and cooling is provided by D_2O pumped through and around the fuel boxes. A two-foot-thick graphite reflector surrounds the reactor tank. The operating power level is five megawatts.

Figure 1.7 also shows some of the facilities that use the neutrons leaking from the reactor. The largest of these is a hohlraum (cavity) into which low-energy neutrons leak (through a graphite column). Such neutrons then scatter from the walls of the hohlraum and thus supply an intense and extended isotropic source of low-energy neutrons. Shown above the hohlraum is a facility for studying the behavior of neutrons in a subcritical lattice of fuel rods, and shown to the right is a facility for studying the properties of fast-reactor blankets.

The medical-therapy room beneath the reactor can be seen clearly. However only one of some thirty neutron-beam ports piercing the concrete shield and graphite reflector is shown. Neutrons from these beam ports are used in conjunction with crystal spectrometers to investigate the interaction of neutrons with individual and chemically bound atoms.

Reactors to Test Reactor Theory. One of the facilities at the United States National Reactor Testing Site in Idaho (as well as a companion facility at the Argonne National Laboratory outside Chicago) has several zero-power fast reactors constructed of a core of U^{25} or Pu^{49}, and sodium, in stainless-steel cans surrounded by a simulated blanket of natural uranium and sodium (again in cans). Since they operate at only a few watts of fission power, the cooling requirements of these reactors are minimal. The relative proportions of the materials in the reactors can be adjusted, and different reactor configurations are constructed in order to validate theoretical predictions of neutron densities and of the reaction rates due to these densities.

If flexible reactors of this type are constructed so that they closely match the geometrical and material characteristics of some proposed design, they are called *mock-ups*. If, on the other hand, the geometry is kept very simple so that few approximations need be made in their theoretical analysis, they are said to be *clean critical reactors*. An additional term, *benchmark experiment*, is used to designate a cross between the clean critical and mock-up types of experiment. A benchmark experiment is supposed to be clean enough so that the theory can be tested unambiguously, yet close enough to some proposed design so that a model which successfully accounts for the benchmark behavior can be applied with some confidence to the actual design. There is some question of whether one can really have the best of both worlds.

Figure 1.8 shows a sketch of a reactor belonging to this class, the ZPR-6 facility. The two halves of the reactor are mounted on separate tables which can be pushed into contact to create a critical assembly. (There are also control rods available for fine adjustment and safety purposes.) The square grid visible on the face of the far section is where the drawers containing the full, fertile material and sodium wafers—all canned in stainless steel—are inserted. By altering the mixture of wafers in a given drawer and filling some drawers entirely with reflector material, experimenters can simulate many different kinds of reactor compositions and geometries.

Reactors to Convert One Material into Another. At the USAEC's Savannah River Facility in South Carolina there are a number of reactors used to convert one material into another—for example U^{28} into Pu^{49}, or Th^{02} into U^{23}, or Pu^{42} into the transplutonium elements Americium, Curium, Berkelium, and Californium. Every effort is made in designing these reactors to insure that all neutrons not needed to sustain the chain reaction are absorbed in fertile material. Thus the design consists of cylindrical fuel elements moderated and cooled by D_2O. The fuel elements themselves may be composed

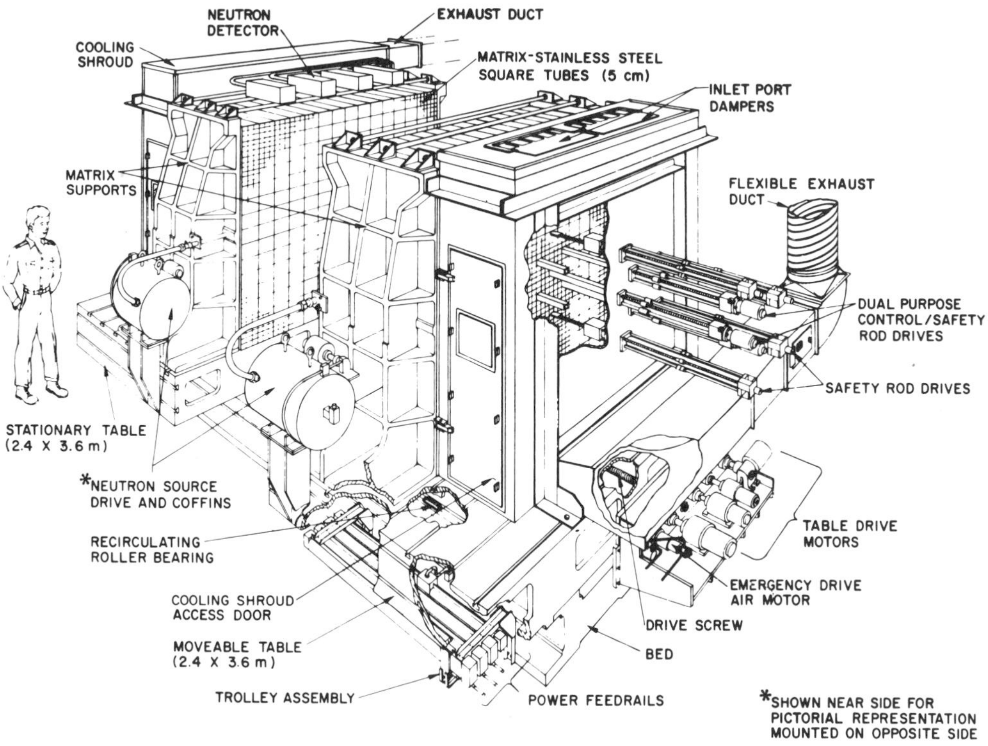

Figure 1.8 The Argonne fast critical facility (ZPR-6). Courtesy of Argonne National Laboratory.

of concentric shells or of a cluster of small cylindrical rods. The reactors are large, the D_2O tanks being in excess of 15 feet high and in excess of 16 feet in diameter, and the reactors operate at very high power. Figure 1.9 shows a lattice used to produce an extremely high neutron flux for the purpose of converting Pu^{42} into transplutonium isotopes. (The dark lattice positions were occupied by control rods and the lighter positions by fuel and target assemblies.) The cross section of a fuel element (two concentric shells of fuel with an inner and outer housing) is shown on the side. The element is about 3.5 inches in diameter, and the active part of the reactor lattice is approximately 6 feet high and 7 feet in diameter.

Reactors to Generate Power for Propulsion. In this category nuclear reactors for marine application provide the most common example. Highly enriched, light-water-moderated, cooled, and reflected cores with fuel elements consisting of a zirconium alloy have been used for this purpose. The hot coolant (H_2O at very high pressure) emerging from such a reactor flows through tubes of a boiler and is then pumped back through the core in

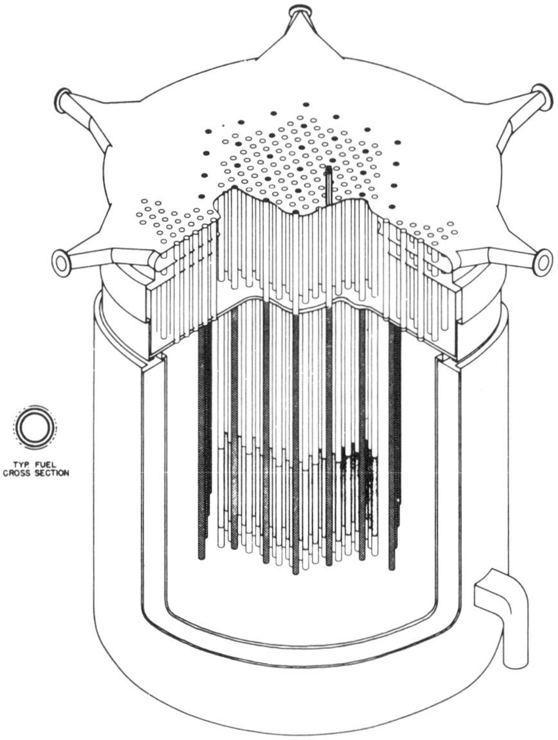

Figure 1.9 A high-flux demonstration lattice in a Savannah River reactor. Courtesy of E. I. Du Pont, De Nemours & Company.

a continuous closed-loop arrangement. Lower-pressure water on the other side of the boiler tubes is converted to steam which runs a turbine that, in turn, drives the ship's propeller.

The Nerva nuclear-powered rocket engine also illustrates the application of nuclear power to propulsion. Figure 1.10 shows a sketch of a proposed design. In this case hydrogen, originally in liquid form, is pumped into a highly enriched graphite-moderated core operating at extremely high power. The heat transferred to the hydrogen causes it to vaporize, expand, and blast out of the open nozzle at the bottom of the core, thus providing thrust.

Reactors to Generate Electrical Power. In the United States there are two classes of reactors that have been used most commonly to generate electric power. They are PWR's (Pressurized-Water Reactors) and BWR's (Boiling-Water Reactors). Both are slightly enriched, light-water-moderated, cooled, and reflected reactors making use of Zr-clad UO_2 fuel elements. A PWR is similar to the marine propulsion reactor just discussed,

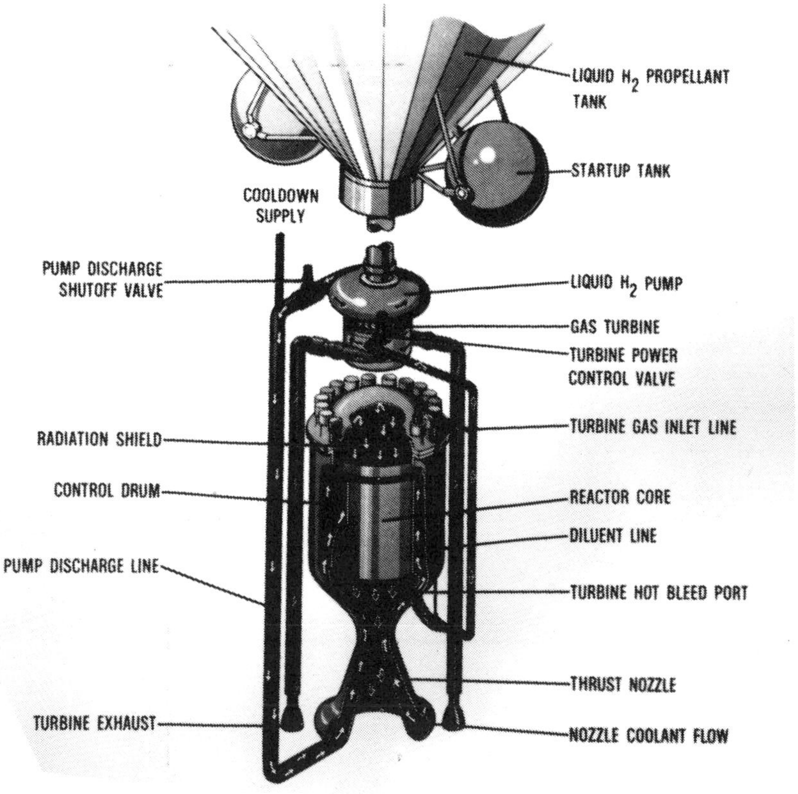

Figure 1.10 A schematic flow diagram of the XE nuclear-powered engine. Courtesy of Westinghouse Electric Corporation.

except that the steam turbine runs an electric generator rather than a propeller. In a BWR the moderator-coolant is allowed to boil in the core, and the generated steam drives the turbine directly. There is thus no intermediate heat exchanger.

A cutaway view of a PWR (the pressure vessel and its internals) is shown as Figure 1.11. The overall height of the reactor vessel and head is 44 ft. The height of the core is 12 ft and its diameter is $\sim$11 ft. The individual subassemblies making up the core have a cross-sectional area of 9 in. $\times$ 9 in. They are composed of 0.42-inch-diameter, zirconium-clad, slightly enriched UO_2 rods. The reactor is controlled by multipronged control rods that fit into guide tubes replacing certain fuel rods.

Both PWR's and BWR's are called *burner reactors* because they make no attempt to produce large amounts of plutonium but, instead, try to burn in place much of the plutonium they make.

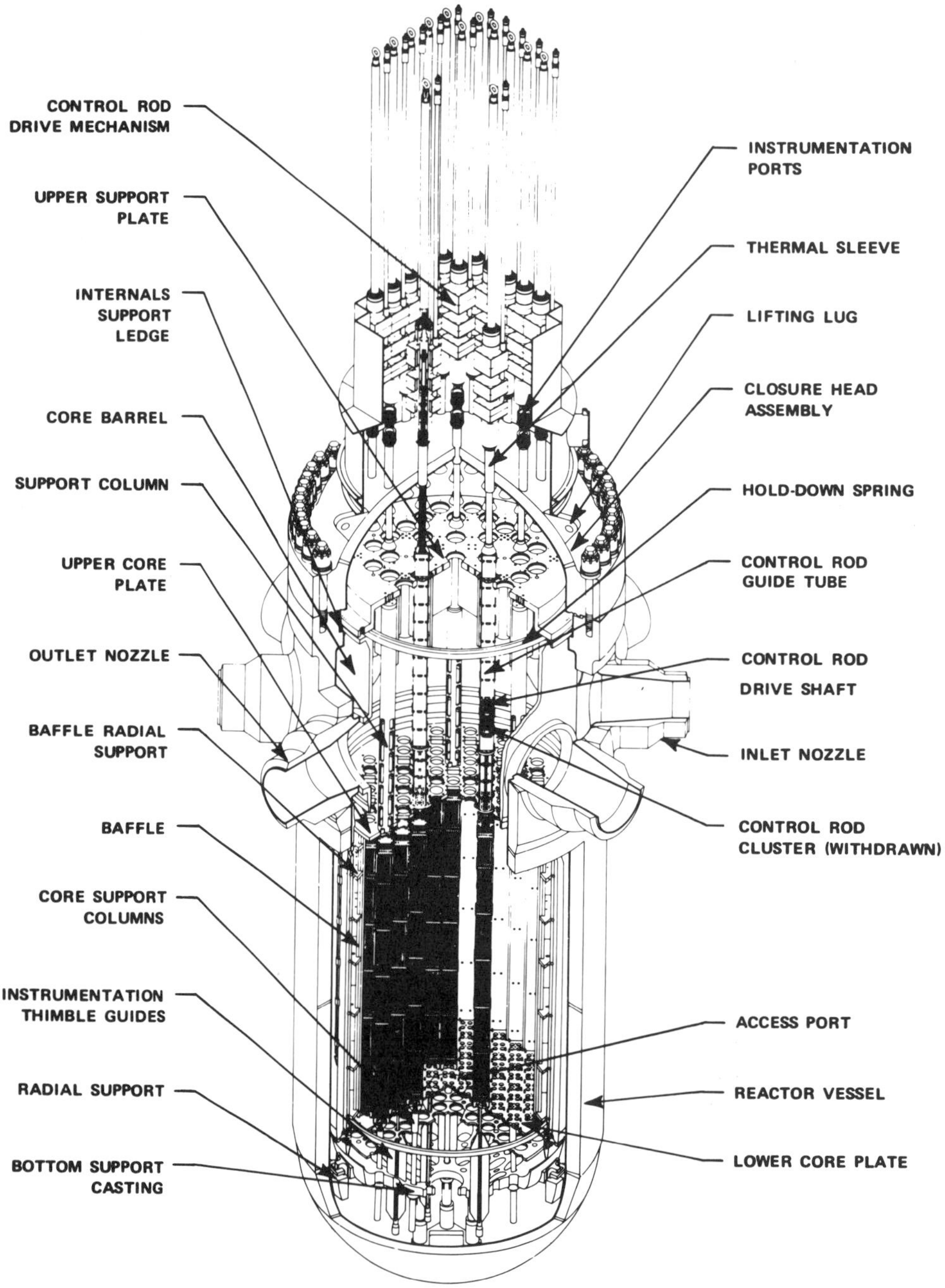

Figure 1.11 Cutaway of a typical pressurized-water reactor. Courtesy of Westinghouse Electric Corporation.

Another type of power reactor that shows long-range promise is the *breeder reactor*. The basic idea here, besides producing usable electrical power, is to make more fissionable material (Pu^{49} or Pu^{41}) than is destroyed. The favored design for this purpose is the fast reactor, fueled with stainless-steel-clad pins composed of $Pu^{49}O_2$ and natural uranium and cooled by liquid sodium. Although some Pu^{49} is created from the natural uranium that comprises 85 percent of the mixture in the fuel pins, most of the breeding (i.e., the creation of Pu^{49} from U^{28}) takes place in a sodium-cooled blanket of natural—or depleted—uranium that surrounds the core. A graphite reflector is usually placed outside this blanket to reflect as many neutrons as possible into the blanket area.

References

Evans, R. D., 1955. *The Atomic Nucleus* (New York: McGraw-Hill).

Kaplan, I., 1963. *Nuclear Physics*, 2nd ed. (Reading, Mass.: Addison-Wesley).

Keeping, G. R., 1965. *Physics of Nuclear Kinetics* (Reading, Mass.: Addison-Wesley).

Lamarsh, J. R., 1966. *Introduction to Nuclear Reactor Theory* (Reading, Mass.: Addison-Wesley), Chapters 1–4.

Nuclear Data

CINDA 69, An Index to the Literature on Microscopic Neutron Data, 1969. Edited by USAEC/DTIE, USSR/NDIC, ENFA/CCDN, and IAEA/NDU.

Drake, M. K., 1970. *Data Formats and Procedures for the ENDF Neutron Cross Section Library*, BNL–50274 (T–601).

Honeck, H. C., 1966. *ENDF 102*, BNL–50066 (T–467) (revised by S. Pearlstein, 1967).

Hughes, D. J., and Schwartz, R. B., 1958. *Neutron Cross Sections*, BNL–325, 2nd ed. (see also Supplement 2, Volumes 1 (1964), 2A, B, and C (1966), and 3 (1965); Volume 1 of the third edition was published in 1973).

Schmidt, J. J., 1962–1966. *Neutron Cross Sections for Fast Reactor Materials*, Part 1, *Evaluation*; Part 2, *Tables*; Part 3, *Graphs*, KFK–120 (EANDC–E–35 "U").

Schmidt, J. J., and Wall, O., 1968. *Tables of Evaluated Neutron Cross Sections for Fast Reactors*, KFK–750 (EANDC–E–88 "U"; EUR–37152).

Problems

1. The radius of the earth is ~ 6400 km. Roughly what would that radius be if the earth were "solid" nuclear matter?

2. What, according to (1.4.1), is the most probable energy of a neutron emitted in fission? What is its *average* energy?

3. What, according to (1.4.1), is the probability that a neutron born in fission will have an initial energy in the range 1–2 eV? What is the probability of its appearing with an energy of 1 eV?

4. Suppose a reactor operates at 2400 megawatts for one year and is then shut down. What would be the decay power in megawatts one year after the shutdown?

Suppose, instead, that all the energy released during the one year of power operation had been released as a burst at the beginning of that period. What would the decay power (of the debris) be two years after that burst?

5. Suppose a reactor experiences a power runaway lasting only a few milliseconds but releasing 2.4 million megawatt-seconds of energy. What will the decay heat in watts be one year after that burst?

6. If a reactor burns U^{25} at a power level of 4000 megawatts for one year, how many kilograms of mass are converted to energy?

7. What is the probability per centimeter of travel that a neutron having energy 0.025 eV and moving in pure Pu^{49} (which had a density of 19.6 g/cc) will be captured?

8. One might think that, for dx in (1.6.2) sufficiently large, the probability of an event could exceed unity. Why is this an invalid conclusion?

2 Reaction Rates

2.1 Introduction

In order to express mathematically the neutron-balance condition that will permit us to predict the characteristics of the neutron population throughout a reactor, we must first derive expressions for the rates at which various nuclear events will occur at any given location and involving neutrons of any given energy. The derivation of expressions for such reaction rates is the subject of the present chapter.

The description of reaction rates involves two stages of refinement. We shall first derive expressions specifying the rate at which any given type of reaction (absorption, fission, scattering, etc.) takes place between neutrons of a given energy and material of a given composition. Then, for the case of scattering events, we will need to know, not only the rate of occurrence, but also the energies and angular distribution of the scattered neutrons. Thus we shall define "differential scattering cross sections" that specify the rate at which neutrons having some particular initial energy E and traveling in some particular initial direction $\mathbf{\Omega}$ scatter into a small range of energies dE' about a new energy E' and a small range of directions $d\Omega'$ about a new direction $\mathbf{\Omega}'$.

There are a variety of mathematical forms in which differential scattering cross sections can be expressed, and we shall spend some time developing the relationships between these forms. Moreover, for the case of elastic scattering, the amount of energy lost by a neutron as the result of a scattering collision is rigidly related to the angle between its initial and final directions of travel. We shall develop this relationship and show how it permits us to express one form for a differential scattering cross section in terms of another.

2.2 Cross Sections for Neutron Interactions

Having introduced the definition (1.6.2) of a cross section—a quantity that specifies the probability of a given nuclear event when a *single* neutron moves a distance dx through a given region—we are now in a position to predict what will happen when a whole distribution of neutrons is present in the region. All that must be done is to focus attention on a beam of neutrons all having roughly the same energy and all traveling in roughly the same direction.

Derivation of the Basic Relationship

Recall that $N(\mathbf{r}, \mathbf{\Omega}, E, t)\, dV\, dE\, d\Omega$ is the number of neutrons that are, at time t, in dV about the point $\mathbf{r}$, in dE about the energy E, and in $d\Omega$ about the direction $\mathbf{\Omega}$. It follows (see Figure 2.1) that, since all these neutrons are traveling with very close to the same speed $v(E)$, and very close to the same direction $\mathbf{\Omega}$, the number that cross the right-hand surface dS of the cylinder in a small time dt is the number of this kind of neutron (i.e., "having" energy E and direction $\mathbf{\Omega}$) contained in a volume of height $dl = v(E)\, dt$ and

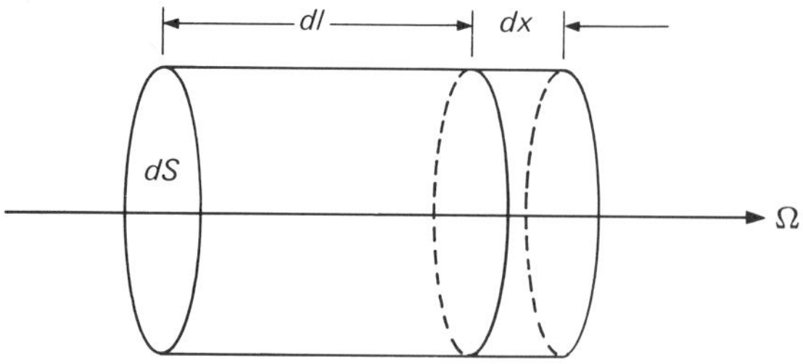

Figure 2.1 A portion of a beam of interacting neutrons.

base dS. That is:

Number of neutrons having energies in dE and directions in $d\Omega$ that cross dS in time dt $= v(E)N(\mathbf{r}, \mathbf{\Omega}, E, t)\, d\Omega\, dE\, dS\, dt.$ (2.2.1)

It follows that the number of this kind of neutron that cross the surface dS per second is $v(E)N(\mathbf{r}, \mathbf{\Omega}, E, t)\, d\Omega\, dE\, dS$.

Since $n^j(\mathbf{r})\sigma_\alpha^j(E)\, dx$ is the probability that a *single* neutron of this kind will cause interaction α with isotope j in moving a distance dx, the expected number of interactions of this kind per second in a volume $dS\, dx$ ($\equiv dV$) is just the number of neutrons per second that cross the surface dS perpendicular to dx times the probability $n^j(\mathbf{r})\sigma_\alpha^j(E)\, dx$, namely

$$[v(E)N(\mathbf{r}, \mathbf{\Omega}, E, t)\, d\Omega\, dE\, dS][n^j(\mathbf{r})\sigma_\alpha^j(E)\, dx] = n^j(\mathbf{r})\sigma_\alpha^j(E)v(E)N(\mathbf{r}, \mathbf{\Omega}, E, t)\, dV\, d\Omega\, dE. \quad (2.2.2)$$

We now make the fundamental assumption that the *actual* interaction rate equals this *expected* interaction rate. In other words we assume that there are so many interactions going on per second in a reactor that the fluctuations in reaction rates arising because of the random character of nuclear events cancel one another out. With 75 billion neutrons being born (and hence also being absorbed) per second in a reactor running at a power level of only one watt, this assumption seems plausible. Even if we consider only a small (but not infinitesimal) region of phase space (say a phase volume with $\Delta V = 1.0$ cc, $\Delta E = 1.0$ eV, and $\Delta\Omega = 1/1000$), the number of interactions of a particular type in any given second is likely to be very close to the number in any other second (provided, of course, that the reactor is operating at some fixed power level).

We are thus led to the following fundamental expression:

Number of type-α interactions per second between neutrons in $dV\, d\Omega\, dE$ and nuclei of isotope j (of concentration $n^j(\mathbf{r})$) at time t
$= n^j(\mathbf{r})\sigma_\alpha^j(E)v(E)N(\mathbf{r}, \mathbf{\Omega}, E, t)\, dV\, d\Omega\, dE.$ (2.2.3)

For complete generality we should also indicate that the number density $n^j(\mathbf{r})$ and the microscopic cross section $\sigma_\alpha^j(E)$ (because of the temperature effects mentioned earlier) can also depend on time. However, since in the initial development of reactor theory we shall be concerned with steady-state situations, we shall avoid this extra notational complexity. In fact, until we deal explicitly with time-dependent situations (Chapters 6 and 7), we shall suppress the explicit indication that $N(\mathbf{r}, \mathbf{\Omega}, E, t)$ can depend on time and write the neutron density in phase space as $N(\mathbf{r}, \mathbf{\Omega}, E)$.

Macroscopic Cross Sections and the Scalar Flux Density

Products of the type $n^j(\mathbf{r})\sigma_\alpha^j(E)$ appear in all expressions for reaction rates, and it is customary to represent them by a single symbol:

$$\Sigma_\alpha^j(\mathbf{r}, E) \equiv n^j(\mathbf{r})\sigma_\alpha^j(E), \tag{2.2.4}$$

where $\Sigma_\alpha^j(\mathbf{r}, E)$ is called the *macroscopic cross section* at $\mathbf{r}$ for interaction α between neutrons of energy E and nuclei of type j. Thus the macroscopic cross section is the sum of the microscopic cross sections of all the nuclei present in a cubic centimeter. It also has a more direct physical significance which we shall discuss presently. The units of the macroscopic cross section are cm^{-1}.

There may be several different isotopes present in the volume dV. For example, if dV were a region inside a fuel zone composed of natural UO_2 plus other isotopes formed during the prolonged operation of the reactor at high power, dV would contain U^{28}, U^{25}, O, U^{26}, Pu^{49}, Pu^{40}, Pu^{41}, fission products, plus other traces of uranium and plutonium isotopes. The *total* type-α interaction rate with neutrons in the beam $v(E)$ $N(\mathbf{r}, \mathbf{\Omega}, E)$ would then be a sum over all the isotopes j. We thus define

$$\Sigma_\alpha(\mathbf{r}, E) \equiv \sum_j \Sigma_\alpha^j(\mathbf{r}, E) = \sum_j n^j(\mathbf{r})\sigma_\alpha^j(E). \tag{2.2.5}$$

Finally we note that, since $\Sigma_\alpha(\mathbf{r}, E)$ is independent of the direction $\mathbf{\Omega}$ of the incident beam, we can find the *total* rate for type-α interactions in $dV\,dE$ by summing (i.e., integrating) over all the infinitesimal cones of direction $d\Omega$. We are then led to the result:

Number of type-α interactions per second between neutrons in the energy range dE and nuclei in the volume dV

$$= \left[\int_{\mathbf{\Omega}} \Sigma_\alpha(\mathbf{r}, E) v(E) N(\mathbf{r}, \mathbf{\Omega}, E)\, d\Omega\right] dV\, dE. \tag{2.2.6}$$

We now define the *scalar flux density* $\Phi(\mathbf{r}, E)$:

$$\Phi(\mathbf{r}, E) \equiv \int_{\mathbf{\Omega}} v(E) N(\mathbf{r}, \mathbf{\Omega}, E)\, d\Omega = v(E) \int_{\mathbf{\Omega}} N(\mathbf{r}, \mathbf{\Omega}, E)\, d\Omega. \tag{2.2.7}$$

It is clear from this definition that $\Phi(\mathbf{r}, E)$ is the product of the total number of neutrons (without regard to their direction of travel) per unit volume about $\mathbf{r}$ and per unit energy about E times their speed. When talking about $\Phi(\mathbf{r}, E)$, one frequently drops the word "scalar" and refers simply to the "flux density" or, more often, to the "flux." It is important to keep in mind though that $\Phi(\mathbf{r}, E)$ is still a density in energy. Its units are neutrons/cm^2/sec/eV.

The scalar flux density is an extremely important quantity for reactor physics. In fact a large part of the subject is concerned with ways of determining it. The reason for this is that, knowing $\Phi(\mathbf{r}, E)$ and the $\Sigma_\alpha(\mathbf{r}, E)$ (the latter from given nuclear data), we can compute any reaction rate anywhere in a reactor. That is, (2.2.6) may be written:

Number of type-α interactions per sec between neutrons in dE and nuclei in dV

$$= \Sigma_\alpha(\mathbf{r}, E)\Phi(\mathbf{r}, E)\, dV\, dE. \tag{2.2.8}$$

It follows that:

Total number of type-α interactions per cc per sec

$$= \int_0^\infty \Sigma_\alpha(\mathbf{r}, E)\Phi(\mathbf{r}, E)\, dE, \tag{2.2.9}$$

where for convenience we have set the limits of the energy integral as zero and infinity. (Since $\Phi(\mathbf{r}, E) = 0$ for $E \gtrsim 15$ MeV, we could equally well have made the upper limit 15 MeV.)

2.3 The Interaction of Neutron Beams with Matter

Neutrons that interact in any manner α in accordance with (2.2.9) cease to be members of the beam $v(E)N(\mathbf{r}, \mathbf{\Omega}, E)\, d\Omega\, dE$. It is a simple matter to determine the attenuation of the beam that results from such a loss. Having done this, we can then provide a physical interpretation of the macroscopic cross section and introduce the useful concept of a mean free path.

Attenuation of a Neutron Beam

Equation (2.2.3) can be used to find the attenuation of a beam of neutrons as it passes through a material medium, and from this result we can derive a physical interpretation of Σ_α.

To do this we first rewrite (2.2.3) summed over all isotopes j and specialized to subscript $\alpha = \mathrm{t}$ (designating the *total* interaction rate):

Total number of interactions per sec in $dV\, d\Omega\, dE$

$$= \Sigma_\mathrm{t}(\mathbf{r}, E)v(E)N(\mathbf{r}, \mathbf{\Omega}, E)\, dV\, d\Omega\, dE. \tag{2.3.1}$$

Now (2.2.1) states that the number of neutrons of this kind crossing the left-hand base dS of a cylinder of height dx per second is $v(E)N(\mathbf{r}, \mathbf{\Omega}, E)\, dS\, d\Omega\, dE$. Because of the neutrons removed in traveling the distance dx, the number of neutrons passing through the right-hand base dS per second is less than the number entering the cylinder $dS\, dx$. (See Figure 2.1.) To express this fact mathematically we say that the number leaving the cylinder per second is $v(E)[N(\mathbf{r}, \mathbf{\Omega}, E) + dN(\mathbf{r}, \mathbf{\Omega}, E)]\, dS\, d\Omega\, dE$, where $dN(\mathbf{r}, \mathbf{\Omega}, E)$, the "increase" in N in traversing dx, will of course turn out to be a negative number.

We can now get a differential equation that specifies the way the neutron density attenuates. We simply state mathematically that the difference between the number of neutrons entering and the number leaving $dS\, dx$ per second equals the number that interact per second in dx. (Remember, *any* interaction—absorption, scattering, fission—removes a neutron from the beam, and we are ignoring any neutrons which may be added to the beam in $dS\, dx$.) We thus have

$$v(E)N(\mathbf{r}, \mathbf{\Omega}, E)\, dS\, d\Omega\, dE - v(E)[N(\mathbf{r}, \mathbf{\Omega}, E) + dN(\mathbf{r}, \mathbf{\Omega}, E)]\, dS\, d\Omega\, dE$$
$$= \Sigma_t(\mathbf{r}, E)v(E)N(\mathbf{r}, \mathbf{\Omega}, E)\, dS\, dx\, d\Omega\, dE,$$

or, if we recognize that $\mathbf{\Omega}$ is the direction of dx,

$$\frac{1}{N(\mathbf{r}, \mathbf{\Omega}, E)} \frac{dN(\mathbf{r}, \mathbf{\Omega}, E)}{dx} = -\Sigma_t(\mathbf{r}, E) \text{ for } \mathbf{\Omega} \text{ in direction of increase in } x. \tag{2.3.2}$$

This result, interpreted physically, states that the fractional decrease of neutrons in the beam (dN/N) per centimeter of travel is $\Sigma_t(\mathbf{r}, E)$. Thus $\Sigma_t(\mathbf{r}, E)$ is the probability per centimeter of travel that a neutron will be removed from the beam. Moreover, since $\Sigma_t(\mathbf{r}, E)$ is a sum over particular interactions σ_α, and since, according to (2.2.5), each $\Sigma_\alpha(\mathbf{r}, E)$ is a sum over all isotopes j present, we are led to the following physical interpretation for the macroscopic cross section:

Probability per cm of travel that a neutron located at $\mathbf{r}$ and having kinetic energy E will undergo a type-α interaction with a nucleus of isotope j
$= \Sigma_\alpha^j(\mathbf{r}, E)$. (2.3.3)

It follows, since $v(E)\, dt$ is the distance a neutron of kinetic energy E moves in time dt, that $\Sigma_\alpha^j(\mathbf{r}, E)v(E)\, dt$ is the probable number of type-α interactions with isotope j this neutron will undergo during dt. Thus:

Probability per sec that a neutron located at $\mathbf{r}$ and having kinetic energy E will undergo a type-α interaction with a nucleus of isotope j
$= v(E)\Sigma_\alpha^j(\mathbf{r}, E)$. (2.3.4)

With these physical interpretations established, we now return to (2.3.2) and consider the special case of a homogeneous medium (in which $\Sigma_t(\mathbf{r}, E) = \Sigma_t(E)$ is independent of position). We can then integrate (2.3.2) between $x = 0$ and $x = X$. The result is

$$N(\mathbf{r}, \mathbf{\Omega}, E)|_{x=X} = N(\mathbf{r}, \mathbf{\Omega}, E)|_{x=0} \cdot \exp(-\Sigma_t(E)X), \tag{2.3.5}$$

where $\mathbf{\Omega}$ is in the direction of increase in x.

Thus a beam of neutrons traveling through a medium is attenuated exponentially. We shall use this result later on when we talk about neutrons "leaking" from one portion of a reactor to another.

It should be noticed that (2.3.5) is a steady-state relationship based on the assumption that the original beam $N(\mathbf{r}, \mathbf{\Omega}, E)|_{x=0}$ remains constant in time. It is not hard to show that a similar equation applies to a burst of neutrons starting in $d\Omega\, dE$ and in a range dx about $x = 0$ and traveling in the X direction.

Mean Free Paths

Equation (2.3.5) permits us to calculate the average distance a neutron travels in a homogeneous medium before it is removed from the beam. All we need do is find out how many neutrons are removed from the beam between x and $x + dx$ in a time dt and multiply that number by the distance x. This gives us the number of "neutron-centimeters" contributed by those neutrons removed in dx in a time dt. If we add up all such contributions, we get the total number of centimeters traveled by all the neutrons removed from the beam in the time dt. Dividing by the number of neutrons that start out in dt then gives us, for this steady-state situation, the average distance traveled by neutrons in the beam. This distance, denoted $\lambda_t(E)$, is called the *total mean free path* for neutrons of energy E in the medium.

To proceed quantitatively we note that the number of neutrons removed in time dt between x and $x + dx$ is, according to (2.3.1),

$$\Sigma_t(E)\, v(E)\, N(\mathbf{r}, \mathbf{\Omega}, E)|_x\, dS\, dx\, d\Omega\, dE\, dt.$$

Then, making use of (2.3.5) and (2.2.1), we obtain

$$\lambda_t(E) = \frac{[\int_0^\infty x\Sigma_t(E)v(E)N(\mathbf{r}, \mathbf{\Omega}, E)|_{x=0} \exp(-\Sigma_t(E)x)\, dx]\, dS\, d\Omega\, dE\, dt}{v(E)N(\mathbf{r}, \mathbf{\Omega}, E)|_{\alpha=0}\, dS\, d\Omega\, dE\, dt} = \frac{1}{\Sigma_t(E)}. \tag{2.3.6}$$

Thus the mean free path for removal of neutrons of energy E from a beam in a homogeneous medium is the reciprocal of the total macroscopic cross section for neutrons of that energy.

It is customary to extend this notion and talk about a mean free path for absorption ($\lambda_a(E) = 1/\Sigma_a(E)$), a mean free path for fission ($\lambda_f(E) = 1/\Sigma_f(E)$), etc. These quantities are the average distances neutrons of energy E would travel before removal from the beam if the process in question (absorption, fission, etc.) were the only one that removed neutrons from the beam.

2.4 Differential Scattering Cross Sections

In order to predict how the neutron population in a reactor behaves, we must know more than just the rate at which neutrons interact. We must also know the energies and directions of travel of the neutrons resulting from any fission or scattering event. Describing mathematically the characteristics of neutrons emitted by fission is a straightforward procedure which we shall develop in Chapter 3. Dealing with scattered neutrons, however, gives rise to many complications; the remainder of the present chapter will deal with the definitions and analysis that take care of these difficulties.

Double-Differential Scattering Cross Sections

What we seek ultimately is a quantitative expression for the rate at which the $N(\mathbf{r}, \boldsymbol{\Omega}, E)\, dV\, d\Omega\, dE$ neutrons in dV that are initially traveling with directions very close to the unit vector $\boldsymbol{\Omega}$ and have energies very close to E will scatter (within dV) into directions all nearly parallel to some different unit vector $\boldsymbol{\Omega}'$ and into a small energy interval dE' about some different energy E'.

We have already seen how to handle the scattering problem if our interest is only in the *total* scattering rate out of the beam (i.e., scattering into all possible directions and energies). Equation (2.2.3) summed over all isotopes j present in dV gives, for $\alpha = \mathrm{s}$, the total elastic-scattering rate from the beam and, for $\alpha = \mathrm{i}$, the total inelastic-scattering rate. Hence we have:

Number of scattering events (elastic plus inelastic) per sec between neutrons in $dV\, d\Omega\, dE$ and the material in dV

$$= \sum_j n^j(\mathbf{r})[\sigma_\mathrm{s}^j(E) + \sigma_\mathrm{i}^j(E)]v(E)N(\mathbf{r}, \boldsymbol{\Omega}, E)\, dV\, d\Omega\, dE$$

$$= [\Sigma_\mathrm{s}(\mathbf{r}, E) + \Sigma_\mathrm{i}(\mathbf{r}, E)]v(E)N(\mathbf{r}, \boldsymbol{\Omega}, E)\, dV\, d\Omega\, dE. \tag{2.4.1}$$

To specify the desired extra information, namely the scattering rate from this beam into $dE'\, d\Omega'$, we need to know, for each isotope j, the probabilities of elastic and inelastic scattering from $dE\, d\Omega$ into $dE'\, d\Omega'$. Accordingly we define double-differential scattering-probability densities by:

Probability that a neutron having energy E and traveling in direction $\boldsymbol{\Omega}$ will, as the result of an elastic ($\alpha = \mathrm{s}$) or an inelastic ($\alpha = \mathrm{i}$) scattering interaction with a nucleus of isotope j, emerge with energy in dE' and direction in $d\Omega'$

$$= f_\alpha^j(E \to E', \boldsymbol{\Omega} \to \boldsymbol{\Omega}')\, dE'\, d\Omega'. \tag{2.4.2}$$

Since the f_α^j are probability densities and the neutron must scatter into *some* energy

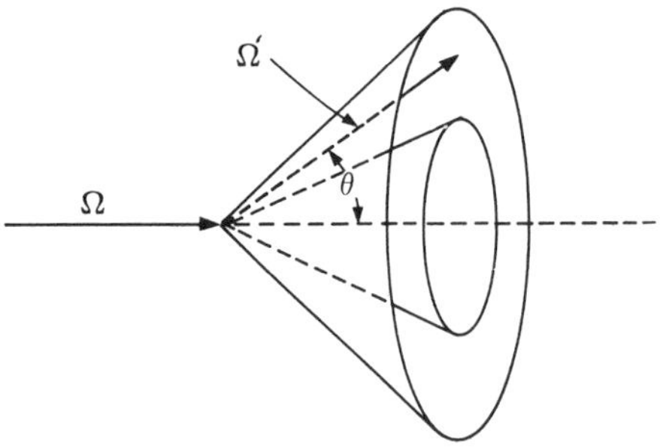

Figure 2.2 Directions $\mathbf{\Omega}'$ for which the scattering probability f_α^j is constant.

and direction, we get immediately

$$\int_0^\infty dE' \int_{\Omega'} d\Omega' f_\alpha^j(E \to E', \mathbf{\Omega} \to \mathbf{\Omega}') = 1 \qquad (\alpha = \text{s, i}). \tag{2.4.3}$$

A simplification of the dependence of the f_α^j's on $\mathbf{\Omega}$ and $\mathbf{\Omega}'$ is possible since in reactor theory one is almost always interested in interactions between randomly oriented particles. For random orientation and for a given initial and final energy, the probability of scattering from $\mathbf{\Omega}$ to $\mathbf{\Omega}'$ depends only on the "scattering angle" θ between the directions of $\mathbf{\Omega}$ and $\mathbf{\Omega}'$. Provided this angle (and the energies E and E') are fixed, f_α^j will be the same for all initial directions of motion $\mathbf{\Omega}$. Moreover, for $\mathbf{\Omega}$ fixed, it will be the same for all final directions $\mathbf{\Omega}'$ lying in the conical shell between θ and $\theta + d\theta$. (See Figure 2.2.) We can express the simplified functional dependence of the f_α^j in terms of $\mathbf{\Omega}$ and $\mathbf{\Omega}'$ by noting that the scalar product $\mathbf{\Omega}\cdot\mathbf{\Omega}'$ (recall that $\mathbf{\Omega}$ and $\mathbf{\Omega}'$ are unit vectors) equals the cosine of the scattering angle θ. We are thus led to define

$$\mu_0 \equiv \cos\theta = \mathbf{\Omega}\cdot\mathbf{\Omega}', \tag{2.4.4}$$

and we shall henceforth denote the simplified functional dependence of the f_α^j on $\mathbf{\Omega}$ and $\mathbf{\Omega}'$ by writing $f_\alpha^j(E \to E', \mathbf{\Omega}\cdot\mathbf{\Omega}')$ or, more simply, $f_\alpha^j(E \to E', \mu_0)$. Note that $\mathbf{\Omega}$, $\mathbf{\Omega}'$, θ, and μ_0 are all measured relative to the laboratory frame of reference in which the center of mass of the reactor is at rest.

Mathematically the simpler dependence of the f_α^j on $\mathbf{\Omega}$ and $\mathbf{\Omega}'$ implies that, when we find the actual mathematical formula for $f_\alpha^j(E \to E', \mathbf{\Omega} \to \mathbf{\Omega}')$, the vectors $\mathbf{\Omega}$ and $\mathbf{\Omega}'$ will never appear individually, but always in the combination $\mathbf{\Omega}\cdot\mathbf{\Omega}'$. We can use this fact to simplify the normalization equation (2.4.3). Specifically, since we know that the orientation of $\mathbf{\Omega}$ won't affect the result, we can measure $\mathbf{\Omega}'$ relative to $\mathbf{\Omega}$ and in that way simplify the integration over $\mathbf{\Omega}'$.

Figure 2.3 shows an axis system that will accomplish this purpose. The initial direc-

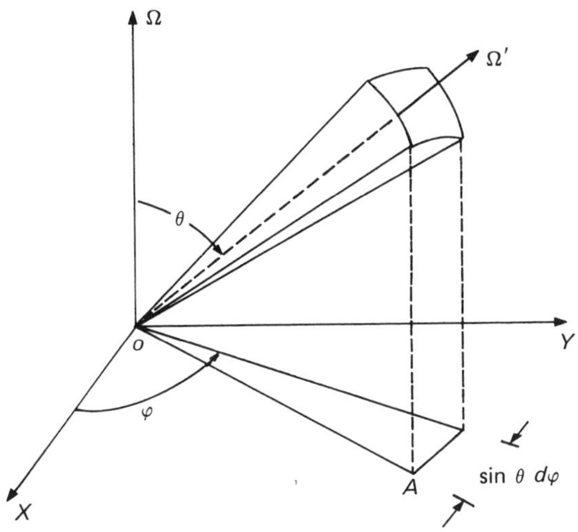

Figure 2.3 A convenient coordinate system for specifying scattering probabilities.

tion $\mathbf{\Omega}$ is taken as that of the Z axis, and the final direction $\mathbf{\Omega}'$ is specified by the angles θ and φ, where φ is the angle between the X axis and the projection OA of $\mathbf{\Omega}'$ onto the XY plane. We define the solid angle $d\Omega'$ to be the ratio of the differential area dS $(= \sin\theta\, d\varphi\, d\theta)$ of the piece of unit sphere immediately surrounding the tip of the vector $\mathbf{\Omega}'$ to the total area (4π) of the unit sphere.* Thus

$$d\Omega' = \frac{\sin\theta\, d\varphi\, d\theta}{4\pi} = \frac{-d\varphi\, d(\cos\theta)}{4\pi} = -\frac{d\varphi\, d\mu_0}{4\pi}. \tag{2.4.5}$$

With this definition

$$\int_{\Omega'} d\Omega' = \int_0^{\pi} d\theta \int_0^{2\pi} d\varphi(\sin\theta/4\pi) = 1. \tag{2.4.6}$$

Equation (2.4.3) can now be written

$$\int_0^{\infty} dE' \int_0^{2\pi} d\varphi \int_0^{\pi} d\theta(\sin\theta/4\pi) f_\alpha^j(E \to E', \mu_0) = 1$$

$$= \int_0^{\infty} dE' \int_{-1}^{1} \frac{d\mu_0}{2} f_\alpha^j(E \to E', \mu_0). \tag{2.4.7}$$

* Strictly speaking, to get a true solid angle $d\Omega'$, we should divide by the square of the radius (1) rather than by the surface area (4π) of the unit sphere. If we did this, however, $d\Omega'$ would be 4π times larger, and (2.4.3) would have to be multiplied by $1/4\pi$. No change in the physics would result, but many extra 4π's would appear in the reactor equations. Hence the definition (2.4.5) appears preferable.

Henceforth we shall assume that all double-differential scattering cross sections are normalized in this manner.

The f_α^j are probabilities that, *in a given scattering event*, a certain change in energy and direction will occur. We now combine the f_α^j with the σ_α^j and define a *double-differential scattering cross section*:

$$\sigma_\alpha^j(E \to E', \mu_0) \equiv \sigma_\alpha^j(E) f_\alpha^j(E \to E', \mu_0) \qquad (\alpha = \text{s, i}). \tag{2.4.8}$$

It follows immediately from (2.4.7) that

$$\int_0^\infty dE' \int_{-1}^1 \frac{d\mu_0}{2} \sigma_\alpha^j(E \to E', \mu_0) = \sigma_\alpha^j(E) \qquad (\alpha = \text{s, i}). \tag{2.4.9}$$

We can use (2.4.8) to write an equation analogous to (2.4.1) for scattering in dV from $d\Omega\, dE$ into $d\Omega'\, dE'$:

Rate at which neutrons in dV are scattered from $d\Omega\, dE$ into $d\Omega'\, dE'$

$$= \sum_j n^j(\mathbf{r})[\sigma_\text{s}^j(E \to E', \mathbf{\Omega}\cdot\mathbf{\Omega}') + \sigma_\text{i}^j(E \to E'), \mathbf{\Omega}\cdot\mathbf{\Omega}')]\, dE'\, d\Omega'[v(E)N(\mathbf{r}, \mathbf{\Omega}, E)]\, dV\, d\Omega\, dE. \tag{2.4.10}$$

If we integrate over all final values of φ, we then find:

Rate at which neutrons in dV are scattered from the beam in $d\Omega\, dE$ into the conical shell $(d\mu_0/2)\, dE'$

$$= \sum_j n^j(\mathbf{r})[\sigma_\text{s}^j(E \to E', \mu_0) + \sigma_\text{i}^j(E \to E', \mu_0)]\, dE' \frac{d\mu_0}{2} v(E)N(\mathbf{r}, \mathbf{\Omega}, E)\, dV\, d\Omega\, dE. \tag{2.4.11}$$

Double-differential cross sections have the units of barns per eV per "unit cosine."

Single-Differential Scattering Probabilities

For a great many applications it is not necessary to know both the final direction and the energy of scattered neutrons. Instead we can make use of probabilities that specify the fraction of neutrons scattered into a final energy interval dE' regardless of the final direction of travel, or ones that specify the fraction of neutrons scattered into a conical shell $d\mu_0/2$ regardless of the final energy. To define such probabilities we simply integrate the double-differential probability over one of its variables, thus obtaining:

$$f_\alpha^j(E, \mu_0) \frac{d\mu_0}{2} \equiv \left[\int_0^\infty dE' f_\alpha^j(E \to E', \mu_0)\right] \frac{d\mu_0}{2} \qquad (\alpha = \text{s, i}) \tag{2.4.12}$$

and

$$f_\alpha^j(E \to E')\, dE' \equiv \left[\int_{-1}^1 \frac{d\mu_0}{2} f_\alpha^j(E \to E', \mu_0)\right] dE' \qquad (\alpha = \text{s, i}). \tag{2.4.13}$$

Physically the first of these is probability that the path of the scattered neutron will make an angle θ with its original direction such that $\mu_0 = \cos\theta$ lies in some small range $d\mu_0$; the second is the probability that the energy of the scattered neutron will lie in dE'.

In writing (2.4.12) and (2.4.13), we have adopted a notational convention common among physicists but sometimes confusing to mathematicians. Namely, we have used function symbol $f_\alpha^j(\ \ )$ to refer to a *physical* quantity (a probability density) rather than to a specific mathematical formula. Thus $f_s^j(E, \mu_0)$ is a different mathematical function of E and μ_0 than $f_s^j(E \to E')$ is of E and E'. In fact the dimensions of the two are different. For this reason it might be preferable to use a different function symbol for one of the probability densities. We might, for example, replace $f_s^j(E \to E')$ in (2.4.13) by $g_s^j(E \to E')$. However, as long as the arguments of f_s^j are stated explicitly, the physics convention will lead to no trouble. Because of this, and because it is so common, we shall adopt it.

There is yet another differential scattering probability that will prove useful. This one is based on measuring the initial and final directions of travel of the scattered neutron relative to a set of axes at rest with respect to the center of mass of the system composed of the neutron and its target nucleus. Thus, if $\mathbf{\Omega}_c$ and $\mathbf{\Omega}_c'$ are these two directions, we define a double-differential scattering probability $f_\alpha^j(E \to E', \mathbf{\Omega}_c \to \mathbf{\Omega}_c')\, dE'\, d\Omega_c'$ completely analogous to (2.4.2) except that $\mathbf{\Omega}_c$, $\mathbf{\Omega}_c'$, and $d\Omega_c'$ are measured in the center-of-mass (CM) system. Then, with μ_c ($\equiv \mathbf{\Omega}_c \cdot \mathbf{\Omega}_c'$) being the cosine of the scattering angle as measured in the CM system, we define

$$
\begin{aligned}
f_\alpha^j(E, \mu_c)\frac{d\mu_c}{2} &\equiv \left[\int_0^\infty dE' \int_0^{2\pi} \frac{d\varphi}{2\pi} f_\alpha^j(E \to E', \mu_c)\right]\frac{d\mu_c}{2} \\
&= \left[\int_0^\infty dE'\, f_\alpha^j(E \to E', \mu_c)\right]\frac{d\mu_c}{2} \qquad (\alpha = \mathrm{s, i}).
\end{aligned}
\tag{2.4.14}
$$

Note that, although μ_c is the cosine of the scattering angle in the CM system, the initial energy E of the neutron is still to be measured in the lab system. Note also that we are again letting the function symbol $f_\alpha^j(\ \ )$ stand for a physical quantity and not a particular algebraic formula.

The reason for introducing $f_\alpha^j(E, \mu_c)$ is that this scattering probability is the one usually used to present the differential scattering data calculated or measured by nuclear physicists. The reason, in turn, why nuclear physicists favor this way of presenting data is that, for a great many isotopes and over a wide range of incident-neutron energies, the neutrons emitted from scattering events viewed in the CM system have no preferred directions of travel. The scattering is then said to be *isotropic in the center-of-mass system*. For such systems $f_\alpha^j(E, \mu_c)$ becomes a function of E only.

In complete analogy with (2.4.8) we can define microscopic differential scattering cross sections $\sigma_\alpha^j(E \to E')$, $\sigma_\alpha^j(E, \mu_0)$, and $\sigma_\alpha^j(E, \mu_c)$:

$$\sigma_\alpha^j(E \to E')\, dE' \equiv \sigma_\alpha^j(E) f_\alpha^j(E \to E')\, dE' \qquad (\alpha = \text{s, i}), \tag{2.4.15}$$

$$\sigma_\alpha^j(E, \mu_0)\frac{d\mu_0}{2} \equiv \sigma_\alpha^j(E) f_\alpha^j(E, \mu_0)\frac{d\mu_0}{2} \qquad (\alpha = \text{s, i}), \tag{2.4.16}$$

$$\sigma_\alpha^j(E, \mu_c)\frac{d\mu_c}{2} \equiv \sigma_\alpha^j(E) f_\alpha^j(E, \mu_c)\frac{d\mu_c}{2} \qquad (\alpha = \text{s, i}). \tag{2.4.17}$$

Finally we may define *macroscopic* differential scattering cross sections that include, for the mixture of isotopes, both elastic and inelastic scattering:

$$\Sigma_s(\mathbf{r}, E \to E') \equiv \sum_j n^j(\mathbf{r})[\sigma_s^j(E \to E') + \sigma_i^j(E \to E')], \tag{2.4.18}$$

$$\Sigma_s(\mathbf{r}, E, \mu_0) \equiv \sum_j n^j(\mathbf{r})[\sigma_s^j(E, \mu_0) + \sigma_i^j(E, \mu_0)]. \tag{2.4.19}$$

An analogous expression involving the $\sigma_\alpha^j(E, \mu_c)$ will not be defined, since μ_c, the cosine of the scattering angle measured in the CM system of a neutron and an isotope j that have relative energy E, would involve a different coordinate system for each isotope and each value of E.

Relationships between Differential Scattering Probabilities for Elastic Collisions

The job of finding numerical values for differential scattering cross sections is dependent on what kind of scattering (inelastic or elastic) is involved. In the case of inelastic scattering the energy E' of the scattered neutron is not rigidly related to the cosine of the scattering angles μ_0 or μ_c. Thus, for a given value of μ_0 or μ_c, the scattered neutrons can be in a whole *range* of energies. A combination of nuclear-physics measurements and theory supplies the numbers $f_\alpha^j(E \to E', \mu_0)$ or $f_\alpha^j(E \to E', \mu_c)$ in this case.

For elastic scattering the situation is different. As will be proven in the next section, there is a rigid relationship between the energy change $E \to E'$ and μ_0 or μ_c. In this case nuclear physics supplies only the total scattering cross section for the isotope and the probability that the scattering will be through an angle $\cos^{-1} \mu_c$. The energy change due to that scattering and the angle $\cos^{-1} \mu_0$ must be computed by the user through application of the laws of dynamics. The mathematical consequence of this rigid physical relationship is that, in the quantities $f_s^j(E \to E', \mu_0)$ (or $f_s^j(E \to E', \mu_c)$), μ_0 and E' (or μ_c and E') are not independent variables. Thus, for a given E and E', the scattering probability $f_s^j(E \to E', \mu_0)$ is nonzero for one and only one value of μ_0. Conversely, for a given value of E and μ_0, $f_s^j(E \to E', \mu_0)$ is nonzero for one and only one value of E'. By employing Dirac delta functions, we can write a function $f_s^j(E \to E', \mu_0)$ that has these properties, and we shall have occasion to do this in Chapter 9. However, for our present purposes, it will not be necessary to introduce this complexity.

Instead we need merely note that for elastic scattering there are precise and unique values of $d\mu_0$ and $d\mu_c$ corresponding to any given dE'. Thus, when dE', $d\mu_0$, and $d\mu_c$ are such corresponding intervals, measuring the probability that a neutron scattered at E will emerge with energy in dE' must yield the same number as measuring the probability that it will emerge in the conical shell $d\mu_0$ in the lab system or $d\mu_c$ in the CM system. In mathematical terms:

$$f_s^j(E \to E')\,dE' = f_s^j(E, \mu_0)\frac{d\mu_0}{2} = f_s^j(E, \mu_c)\frac{d\mu_c}{2} \tag{2.4.20}$$

or, from (2.4.15)–(2.4.17),

$$\sigma_s^j(E \to E')\,dE' = \sigma_s^j(E, \mu_0)\frac{d\mu_0}{2} = \sigma_s^j(E, \mu_c)\frac{d\mu_c}{2}. \tag{2.4.21}$$

Note that these relationships in no sense imply that the numbers $f_s^j(E \to E')$, $f_s^j(E, \mu_0)$, and $f_s^j(E, \mu_c)$ are equal.

Since E, E', and μ_0 are all measured in the laboratory system, they are the same for all isotopes j. Thus we can use the first equality in (2.4.21) to derive two mathematically dissimilar, but physically identical, expressions for the effect on a beam of neutrons of elastic scattering:

Rate at which neutrons in dV are scattered elastically from the beam $d\Omega\, dE$ into the conical shell $d\mu_0/2$

= Rate at which neutrons in dV are scattered elastically from the beam $d\Omega\, dE$ into the energy range dE' corresponding to this conical shell

$$= \sum_j \Sigma_s^j(E) f_s^j(E, \mu_0)\frac{d\mu_0}{2} v(E) N(\mathbf{r}, \mathbf{\Omega}, E)\, dV\, d\Omega\, dE \tag{2.4.22}$$

$$= \sum_j \Sigma_s^j(E) f_s^j(E \to E')\, dE'\, v(E) N(\mathbf{r}, \mathbf{\Omega}, E)\, dV\, d\Omega\, dE.$$

For some applications in reactor theory it will be convenient to use the first of these expressions; for others, the second.

In either case it will be necessary to know the relationship connecting μ_0, E', and E and thus the one connecting $f_s^j(E, \mu_0)$ and $f_s^j(E \to E')$. We shall now set about finding these relationships.

2.5 The Kinematics of Elastic Neutron Scattering

The prediction of the probable scattering angle for a given elastic-scattering event requires the use of quantum mechanics combined with measurements. However the relationships among μ_0, μ_c, E', and E can be simply derived from classical mechanics by invoking the laws of conservation of energy and of linear momentum. We start this

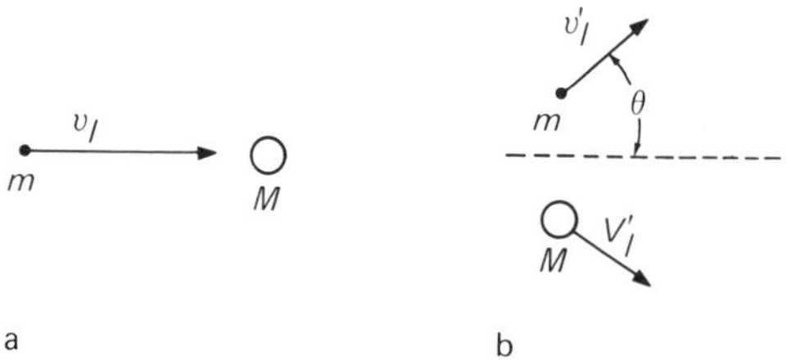

Figure 2.4 Trajectories of a neutron and its target nucleus (a) before and (b) after a scattering event as seen from the laboratory reference system.

derivation by analyzing the collision in a coordinate system in which the target nucleus is at rest. Since, *on the average*, the nuclei in a reactor are at rest relative to the center of mass of the reactor, we shall refer to this coordinate system as the laboratory system. We have already discussed the fact that the thermal motion of the nuclei relative to the center of mass of the reactor can be neglected, except for interactions involving neutrons whose energies are either very low or in the region where the cross sections have high and narrow resonances. Thus the results we shall obtain from consideration of a single collision between a neutron and a nucleus at rest will be applicable to a neutron slowing down inside a reactor at energies above the narrow-resonance region ($\sim$100 eV). Moreover, for interactions with light (nonresonance) materials, which we shall show are the best moderators, the results will be directly applicable down to about 1 eV.

As to notation, we shall use small letters m and v to refer to the mass and speed of the neutron and capital letters M and V to refer to the mass and speed of the target nucleus. A subscript l on v or V will indicate measurements relative to the laboratory system; and a prime on v or V will indicate measurements *after* the collision. With these conventions the trajectories of the neutron and target nucleus before and after a scattering collision are as shown in Figure 2.4.

Note that v_l, v_l', etc., are scalars (not vectors) and indicate "speeds" (not "velocities"). The directions of these speeds are as indicated in the figure.

An elastic collision is one in which the total kinetic energy of the particles before the collision (here $\frac{1}{2}mv_l^2$) equals the total kinetic energy of the particles after the collision ($\frac{1}{2}mv_l'^2 + \frac{1}{2}MV_l'^2$). For such a collision we seek the relationship between $\frac{1}{2}mv_l^2$, $\frac{1}{2}mv_l'^2$, and θ.

It turns out that the easiest way to get this relationship is to analyze the dynamics of the collision in the center-of-mass (CM) system. Proceeding in this manner also leads to several other very important results. Accordingly we shall switch our viewpoint to the CM system, defined by the requirement that the total linear momentum of the neutron and the nucleus measured relative to that system shall vanish. Thus, with subscript c

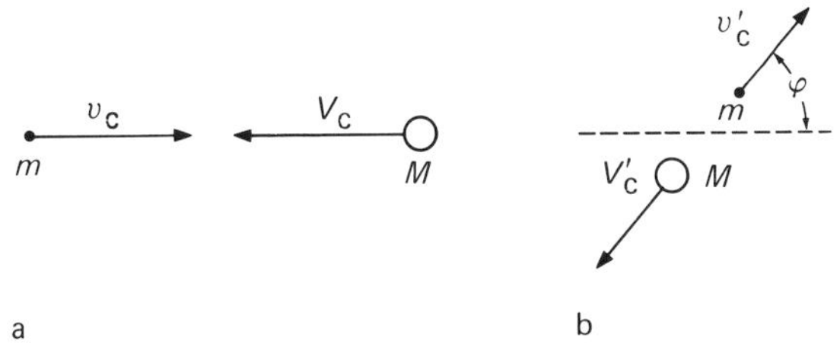

Figure 2.5 Trajectories of a neutron and its target nucleus (a) before and (b) after a scattering event as seen from the center-of-mass system.

indicating that the speeds v_c, v'_c, V_c, and V'_c are taken relative to the CM system, the picture of the collision is as shown in Figure 2.5. The collision angle relative to the CM system is designated as φ.

Using the requirement that the total linear momentum vanish and the law of conservation of momentum, we immediately find that

$$v_c m - V_c M = 0, \tag{2.5.1}$$
$$v'_c m - V'_c M = 0. \tag{2.5.2}$$

Because of the conservation of energy,

$$\tfrac{1}{2}mv_c^2 + \tfrac{1}{2}MV_c^2 = \tfrac{1}{2}mv_c'^2 + \tfrac{1}{2}MV_c'^2. \tag{2.5.3}$$

Using (2.5.1) and (2.5.2) to eliminate v_c and v'_c, we get

$$[\tfrac{1}{2}m(M/m)^2 + \tfrac{1}{2}M]V_c^2 = [\tfrac{1}{2}m(M/m)^2 + \tfrac{1}{2}M]V_c'^2. \tag{2.5.4}$$

Therefore

$$V_c = V'_c, \tag{2.5.5}$$

and insertion of this result into (2.5.1) and (2.5.2) yields

$$v_c = v'_c. \tag{2.5.6}$$

With these relationships established, let us relate the description of the scattering event in the CM system to the corresponding description in the laboratory system. To do this we first note that, since the target nucleus is at rest in the lab system (Figure 2.4a) and moving to the left (Figure 2.5a) with speed V_c in the CM system, the CM system itself must be moving to the right relative to the lab system with this same speed V_c. Thus, if we use V_{cm} to be the speed of the CM system relative to the laboratory system, the *magnitudes* of V_{cm} and V_c are the same. (Their *directions* are, of course, op-

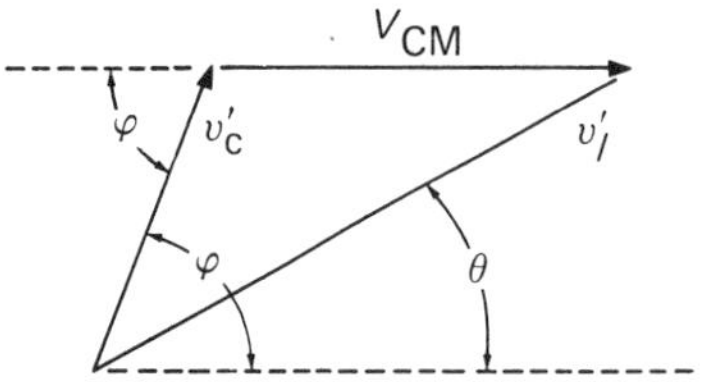

Figure 2.6 Vector relationship between the velocity of a neutron after a collision measured in the laboratory system and that measured in the CM system.

posite.) Hence

$$v_c = v_l - V_{cm} = v_l - V_c, \tag{2.5.7}$$

and, using (2.5.1), we find that

$$V_c = \frac{v_l m}{M + m}. \tag{2.5.8}$$

Moreover, if we create a composite of parts of Figures 2.4 and 2.5, we get the vector diagram shown in Figure 2.6. Figure 2.6 indicates (as it must) that the horizontal component ($v'_l \cos \varphi$) of the speed of the neutron after the collision measured in the lab system equals the horizontal component ($v'_c \cos \varphi$) of the final speed measured in the CM system plus the speed (V_{cm}) of the CM system relative to the lab; that is,

$$v'_l \cos \theta = v'_c \cos \varphi + V_{cm}. \tag{2.5.9}$$

It also indicates that the vertical components of these speeds ($v'_l \sin \theta$) and ($v'_c \sin \varphi$) are the same:

$$v'_l \sin \theta = v'_c \sin \varphi. \tag{2.5.10}$$

We are trying to get a relationship between the scattering angle and the kinetic energy of the neutron before (E) and after (E') the collision. To this end we apply the law of cosines to Figure 2.6, obtaining

$$v_l'^2 = v_c'^2 + V_c^2 + 2v'_c V_c \cos \varphi. \tag{2.5.11}$$

Making use of (2.5.2), (2.5.5), and (2.5.8), we can then show that

$$v_l'^2 = \left[\left(\frac{M}{m}\right)^2 + 1 + 2\frac{M}{m}\cos \varphi\right]\left(\frac{v_l m}{M + m}\right)^2. \tag{2.5.12}$$

If we now define the mass ratio

$$A \equiv \frac{M}{m} \tag{2.5.13}$$

and note that

$$\frac{E'}{E} = \frac{\frac{1}{2}mv_1'^2}{\frac{1}{2}mv_1^2} = \frac{v_1'^2}{v_1^2},$$

then

$$\frac{E'}{E} = \frac{A^2 + 1 + 2A \cos \varphi}{(1 + A)^2}. \tag{2.5.14}$$

Equation (2.5.4) is an extremely useful result. However it is a somewhat hybrid expression. It relates the neutron energies *in the lab system* before and after the collision to the cosine of the scattering angle *in the CM system.* To complete our original program we must relate the scattering angle in the lab system (θ) to that in the CM system (φ). To do this we merely divide (2.5.10) into (2.5.9), making use of (2.5.5) and (2.5.2). The result is

$$\cot \theta = \frac{\cos \varphi + (1/A)}{\sin \varphi}. \tag{2.5.15}$$

It follows from the trigonometry of a right triangle that, if we write $\cot \theta = N/D$,

$$\cos \theta = \frac{N}{\sqrt{(N^2 + D^2)}} = \frac{1 + A \cos \varphi}{\sqrt{(A^2 + 2A \cos \varphi + 1)}}. \tag{2.5.16}$$

This too is an extremely useful result.

It turns out that some of the subsequent algebra will be simplified if we define another quantity to replace the mass ratio A in (2.5.14). That quantity is

$$\alpha \equiv \left(\frac{A - 1}{A + 1}\right)^2, \tag{2.5.17}$$

in terms of which (2.5.14) may be rewritten as

$$\frac{E'}{E} = \frac{1 + \alpha}{2} + \frac{1 - \alpha}{2} \cos \varphi. \tag{2.5.18}$$

Equations (2.5.16) and (2.5.18) establish the fundamental relationships among the initial and final energies and the scattering angles for elastic collisions between neutrons and nuclei. Note that, although scattering angles for both the lab and CM system appear, the energies E and E' are measured relative to the lab system only. We shall not need to introduce a symbol designating the kinetic energy of the neutrons relative to the CM system. (Note that, if we did, one symbol would suffice, since (2.5.6) shows that $\frac{1}{2}mv_c^2 = \frac{1}{2}mv_c'^2$, i.e., that the kinetic energy of the neutron in the CM system is unaltered by the collision.) It should also be noted that, since, in the lab system, the target nucleus is initially at rest, E is also the kinetic energy of the neutron measured relative to the target nucleus. Thus it is the energy on which the nuclear cross sections $\sigma_\alpha^j(E)$ depend.

General Consequences of the Laws of Elastic Scattering

Equations (2.5.16), (2.5.17), and (2.5.18) permit us to make some general conclusions about the energy degradation of neutrons due to elastic scattering.

First of all, we note from (2.5.18) that, for $\varphi = 0$, E' and E are the same for all values of α. Thus "straight-ahead collisions" (permitted by the compound-nucleus model) result in no energy change.

When $\varphi = \pi$ radians, so that the scattering results in a complete reversal in the direction of neutron travel, the ratio E'/E takes on its minimum value, namely α. Thus we have

$$\begin{aligned}&\text{Minimum energy of neutron following an elastic collision}\\&= E'_{\text{min}} = \alpha E,\\&\text{Maximum energy loss due to an elastic collision}\\&= E - E'_{\text{min}} = E(1 - \alpha).\end{aligned} \tag{2.5.19}$$

Since α runs from zero for $A = 1$ (a collision with hydrogen) to near unity for A large, we have here a proof that the lighter the element, the more effective the moderator. In fact (2.5.19) shows that, for hydrogen ($\alpha = 0$), a neutron can lose all its energy in a single collision. In this case the neutron doesn't reverse its direction of travel after the collision; it simply stops.

It is clear from Figure 2.6 that the scattering angle (φ) in the CM system is always greater than that (θ) in the lab system. In fact, for hydrogen, (2.5.16) becomes

$$\cos\theta = \sqrt{\frac{1 + \cos\varphi}{2}}. \tag{2.5.20}$$

Thus, for hydrogen scattering, as φ goes from 0 to π, $\cos\theta$ goes from 1 to 0, and the maximum scattering angle in the laboratory system is $\pi/2$ radians.

Resultant Relationships between Differential Scattering Probabilities

Equations (2.5.16) and (2.5.18) provide the rigid relationships between μ_0, μ_c, and E' needed to express any one of the differential scattering probabilities $f_s^j(E \to E')$, $f_s^j(E, \mu_0)$, or $f_s^j(E, \mu_c)$ in terms of any other one. The most useful such relationship for present purposes is, from (2.4.20) and (2.5.18),

$$\begin{aligned}f_s^j(E \to E')\,dE' &= f_s^j(E, \mu_c)\frac{1}{2}\frac{d\mu_c}{dE'}\,dE'\\&= f_s^j(E, \mu_c)\frac{dE'}{E(1 - \alpha_j)} \qquad \text{for elastic scattering.}\end{aligned} \tag{2.5.21}$$

A subscript j has been added to α in (2.5.21) to indicate that the scattering in question is between a neutron and a nucleus of material j; in general the following definitions will prove useful:

$$\alpha_j \equiv \left(\frac{A_j - 1}{A_j + 1}\right)^2 \quad \text{and} \quad A_j \equiv \frac{M_j}{m}.$$

If a scattering event is isotropic in the CM system and if we imagine the scattering center to be surrounded by a sphere of unit radius so that the trajectories of the scattered neutrons lie along radii of that sphere, the probability that the scattered neutron will pass through any particular area on the surface of the unit sphere is just that area divided by 4π, the total area of the unit sphere. All areas of equal size on the surface of the sphere, no matter what their shape, have an equal probability that the scattered neutron will pass through them. In particular, if we consider, as in Figure 2.7, a circular ring of area $2\pi \sin \varphi \, d\varphi$ on the surface of the sphere, the probability that the scattered neutron will pass through that ring is

$$\frac{2\pi \sin \varphi \, d\varphi}{4\pi} = \frac{1}{2} \sin \varphi \, d\varphi = -\frac{1}{2} d(\cos \varphi) = \frac{-d\mu_c}{2},$$

where $d\mu_c$ is a negative number if $d\varphi$ is positive.

But this probability is, by (2.4.14) with $\alpha = \mathrm{s}$, $f_s^j(E, \mu_c) d\mu_c/2$, except that in (2.4.14) $d\mu_c$ is considered to be a positive number. Since there is no difference between integrating $-d\mu_c$ from $+1$ to -1 and integrating $+d\mu_c$ from -1 to $+1$, we adopt the latter convention when dealing with $d\mu_c$. We thus conclude that, for isotropic scattering in the CM system, $f_s^j(E, \mu_c) = 1$. It follows at once from (2.5.21) that

$$f_s^j(E \to E') \, dE' = \frac{dE'}{E(1 - \alpha_j)} \tag{2.5.22}$$

for elastic scattering isotropic in the CM system.

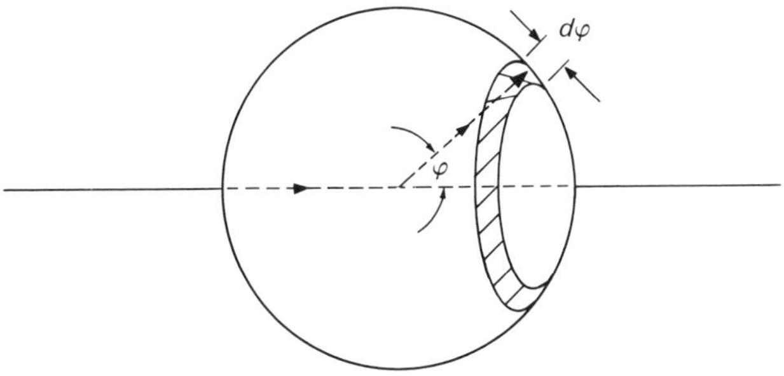

Figure 2.7 A scattered neutron passing through the surface of a unit sphere.

This equation is extremely useful. It shows that, for the isotropic-elastic-scattering case, the probability of scattering into any energy interval dE' in the accessible range of energies $E(1 - \alpha_j)$ is uniform, i.e., does not depend on E' and is, hence, equally likely for all equal intervals dE' in the range.

If the scattering is *not* isotropic in the CM system, the value of $f_s^j(E, \mu_c)$ in (2.5.21) for any E and μ_c must be supplied by nuclear physics. If the data *is* supplied in that form (rather than as tables of $f_s^j(E \to E')$), we may still deal with $f_s^j(E, \mu_c)$ mathematically as a function of E and E' by writing it as $f_s^j[E, \mu_c(E'/E)]$, where $\mu_c(E'/E)$ is found from (2.5.18). The procedure for finding a number for $f_s^j(E, \mu_c)$ when given values of E and E' is thus: First compute $\mu_c(E'/E)$, and then use nuclear-data tables for $f_s^j(E, \mu_c)$.

The Average Cosine of the Laboratory Scattering Angle and the Average Loss in the Logarithm of the Neutron Energy per Elastic-Scattering Event

The results obtained so far can be applied to determine two quantities that will be useful in later developments. The first of these is $\bar{\mu}_0^j$, the average value of the cosine μ_0 of the scattering angle θ measured in the laboratory system.

Even if the scattering is isotropic in the CM system (so that the average of the scattering angle φ measured in that system is $\pi/2$ and the average of cos φ is 0) the average value of θ will be less than $\pi/2$, and the scattering will be "forward scattering" in the lab system. This is clear from Figure 2.6, which shows that θ is always less than φ. In fact, when a neutron scatters from hydrogen, (2.5.20) shows that $\theta \leq \pi/2$ for all scattering events.

To find $\bar{\mu}_0^j$ we simply multiply each value of μ_0 by the fraction of scattered neutrons $f_s^j(E, \mu_0)\, d\mu_0/2$ that will have a trajectory lying in the conical shell between μ_0 and $\mu_0 + d\mu_0$ and then integrate over all possible values of μ_0. The result is

$$\bar{\mu}_0^j = \int_{-1}^{1} \mu_0 f_s^j(E, \mu_0) \frac{d\mu_0}{2}. \tag{2.5.23}$$

The easiest way to evaluate this integral is to make a change of variable in the integrand from μ_0 to μ_c. Equation (2.5.16) gives us μ_0 as a function of μ_c, and (2.4.20) shows that $f_s^j(E, \mu_0) = f_s^j(E, \mu_c)\, d\mu_c/d\mu_0$. Thus (2.5.23) becomes

$$\bar{\mu}_0^j = \int_{-1}^{1} \frac{1 + A_j\mu_c}{\sqrt{(A_j^2 + 2A_j\mu_c + 1)}} f_s^j(E, \mu_c) \frac{d\mu_c}{2}. \tag{2.5.24}$$

If the scattering is isotropic in the CM system, $f_\alpha(E, \mu_c)$ is again unity, and the integration of (2.5.24) yields

$$\bar{\mu}_0^j = \frac{2}{3A_j}. \tag{2.5.25}$$

This result shows that, for elastic scattering isotropic in the CM system, the average cosine of the scattering angle in the lab system is always positive. Thus the corresponding average value of θ (which we can define as $\bar{\theta}^j = \cos^{-1}\bar{\mu}_0^j$) will always lie in the first quadrant. Moreover the smaller the mass of the target nucleus, the larger is $\bar{\mu}_0^j$ and, hence, the smaller is $\bar{\theta}^j$ (i.e., the more "forward" is the scattering). This "forwardness" of the scattering process in the laboratory system has an effect on the average distance a neutron moves in slowing down. We shall consider this situation quantitatively in Chapter 4.

Another quantity that is useful in reactor analysis is the average loss in the logarithm of the energy during a collision. (Recall that, whenever we use a symbol, such as E, E', etc., to represent the kinetic energy of the neutron, we are referring to that energy relative to the lab system.) The symbol ξ is generally used to represent this average loss. For the case of elastic scattering isotropic in the CM system, (2.5.22) yields

$$\begin{aligned} \xi_j &= \int_{\alpha_j E}^{E} [\ln E - \ln E'] \frac{dE'}{E(1-\alpha_j)} \\ &= 1 + \frac{\alpha_j}{1-\alpha_j} \ln \alpha_j, \end{aligned} \tag{2.5.26}$$

where we are taking dE' as positive and are summing the losses in $\ln E$ starting with values of dE' about the lowest energy that the neutron can attain in a collision ($\alpha_j E$) and going up to values of dE' about the highest (E, the initial energy). (We could equally well sum from E down to $\alpha_j E$, but then the increments would be negative, $-dE'$, and the result would be the same.)

Making use of (2.5.17) and the Taylor expansion

$$\ln\left(1 \pm \frac{1}{A_j}\right) = \pm\frac{1}{A_j} - \frac{1}{2A_j^2} \pm \frac{1}{3A_j^3} \cdots,$$

we then find that

$$\xi_j = \frac{2}{A_j + \frac{2}{3}} + 0\left(\frac{1}{A_j^3}\right) \tag{2.5.27}$$

for elastic scattering isotropic in the CM system.

Lethargy

The logarithm of the neutron energy appears with sufficient frequency in reactor-physics calculations that in the early days of reactor theory, when calculations were done by hand, a special term, called the *lethargy*, was introduced to describe changes in this quantity. Its precise definition is

$$u \equiv \ln\frac{E_0}{E}, \tag{2.5.28}$$

where E is the energy of the neutron and E_0 is some arbitrary fixed energy usually taken as 10 or 15 MeV (effectively the highest energy of a neutron emitted in the fission process).

Thus, with E_0 being the highest energy of a neutron in a reactor, the lethargy starts at $u = 0$ (for $E = E_0$). Then, as the energy of neutrons decreases, their lethargy increases (hence the name "lethargy"). A neutron at rest ($E = 0$) has an infinite lethargy.

Since

$$\ln E - \ln E' = \ln E_0 - \ln E' - (\ln E_0 - \ln E) = \ln \frac{E_0}{E'} - \ln \frac{E_0}{E} = u' - u, \qquad (2.5.29)$$

we see that ξ_j is the average gain in the lethargy of a neutron scattered by nucleus j. Equations (2.5.26) and (2.5.27) then give exact and approximate expressions for this average lethargy gain when the scattering is elastic and isotropic in the CM system. With the gradual abandonment of analytical methods in favor of machine (numerical) methods for solving the equations of reactor theory, the use of lethargy in slowing-down calculations is becoming less common. In fact we shall work almost exclusively with energy as a variable. However it is important to be aware of the lethargy concept in order to follow many of the early papers in the field.

The quantity ξ_j serves as another measure of the slowing-down capability of isotope j—a measure that is independent of the initial energy of the scattered neutron. We see from (2.5.26) that the largest average lethargy gain is obtained when a neutron scatters from a hydrogen nucleus ($\alpha = 0$). For this case, since $\lim_{\alpha \to 0} \alpha \ln \alpha = 0$, $\xi_H = 1$. The approximation (2.5.27) is not adequate for this case; it gives $\xi_H = 1.2$. However, as A_j increases, it soon becomes adequate (it is only 3.3 percent in error for deuterium). For U^{28} it is good to better than one part in ten million and gives $\xi_{28} = 0.00838$.

Values of ξ_j can provide an estimate of the average number of collisions required to moderate a fission neutron. Thus, if the most probable energy of a neutron emitted in fission is 0.85 MeV, the total loss in the logarithm of the energy associated with slowing the neutron down to 0.025 eV is $[\ln(8.5 \times 10^5) - \ln 0.025] = \ln(3.4 \times 10^7) = 17.3$. Dividing this total loss by the average loss per collision gives the average number of collisions involved in the slowing-down process. Applying this procedure, we find that for collisions with hydrogen the average number of collisions to slow down is 17.3. For graphite ($\xi_C = 0.158$) it is 110, and for U^{28} it is 2065. It is clear that even with graphite the chance that a neutron will leak out of a given volume before it slows down to 0.025 eV is much greater than it is with hydrogen. As a result graphite-moderated thermal reactors must be made much larger than H_2O-moderated ones.

2.6 Summary

The various expressions we have defined and derived for reaction rates will permit us to write an equation specifying the balance between neutron creation and loss rates in a reactor. In so doing, it will be advantageous to employ a compact notation to describe the various nuclear processes, and it may prove helpful to summarize here as much of this notation as has been already introduced.

General Expression for Interaction with a Beam of Neutrons

We first derived expressions for interaction rates between neutrons in a beam and the medium through which the beam is traveling. Thus, from (2.2.3) and (2.2.4), we get:

Number of type-α interactions per sec between the material present in a volume dV at point $\mathbf{r}$ and neutrons with energies in a range dE about E traveling in a beam within a cone $d\Omega$ about a direction $\mathbf{\Omega}$

$$= \sum_j n^j(\mathbf{r})\sigma_\alpha^j(E)v(E)N(\mathbf{r}, \mathbf{\Omega}, E)\,dV\,d\Omega\,dE \tag{2.6.1}$$
$$= \Sigma_\alpha(\mathbf{r}, E)v(E)N(\mathbf{r}, \mathbf{\Omega}, E)\,dV\,d\Omega\,dE,$$

where the sum is over all isotopes j of concentration $n^j(\mathbf{r})$ in dV, $\sigma_\alpha^j(E)$ is the microscopic cross section for an interaction of type α between neutrons of kinetic energy E and nuclei of isotope j, and $v(E)$ is the speed of the neutrons in the beam $N(\mathbf{r}, \mathbf{\Omega}, E)\,dV\,d\Omega\,dE$.

The notation $\sigma_\alpha^j(E)$, indicating that microscopic cross sections are functions of the energy of the neutron measured in the laboratory system, implies that an average over thermal motions of the target nuclei has been made; hence the $\sigma_\alpha^j(E)$ depend in general on the local temperature of the medium.

Interactions with Neutrons without Regard for Their Directions of Travel

For a great many applications we shall not be concerned with the directions from which the interacting neutrons approach the target nuclei. Under these circumstances we can sum over all such directions and deal only with the scalar flux density $\Phi(\mathbf{r}, E)$ defined by (2.2.7). The result is:

Number of type-α interactions per sec between neutrons in dE and nuclei in dV, without regard to the initial direction of neutron travel

$$= \Sigma_\alpha(\mathbf{r}, E)\left[\int_{\mathbf{\Omega}} v(E)N(\mathbf{r}, \mathbf{\Omega}, E)\,d\Omega\right]dV\,dE \tag{2.6.2}$$
$$= \Sigma_\alpha(\mathbf{r}, E)\Phi(\mathbf{r}, E)\,dV\,dE.$$

Scattering from One Beam to Another

For scattering events it is necessary to introduce the probabilities $f_\alpha^j(E \to E', \mathbf{\Omega} \to \mathbf{\Omega}')\,dE'\,d\Omega'$ that, as the result of a scattering event, a neutron having an initial energy E

and direction of travel $\mathbf{\Omega}$ will scatter into dE' about E' and $d\Omega'$ about $\mathbf{\Omega}'$. For the purposes of reactor analysis we can always assume that these probabilities depend only on the scattering angle θ (or, equivalently, on its cosine $\mu_0 = \mathbf{\Omega}\cdot\mathbf{\Omega}'$) between the initial and final directions of motion of the neutron. The fundamental expression (2.4.10) for the rate at which neutrons in dV scatter from E and $\mathbf{\Omega}$ into dE' and $d\Omega'$ then results. By using (2.2.4), (2.4.5), and (2.4.7), this expression can be written in the somewhat simpler form:

Rate at which neutrons in dV are scattered from $dE\, d\Omega$ into $dE'\, d\Omega'$

$$= \sum_j [\Sigma_s^j(\mathbf{r}, E) f_s^j(E \to E', \mu_0) + \Sigma_i^j(E \to E', \mu_0)]\, dE' \frac{d\varphi\, d\mu_0}{4\pi} [v(E)N(\mathbf{r}, \mathbf{\Omega}, E)] dV\, dE\, d\Omega, \tag{2.6.3}$$

where the angle $\cos^{-1}\mu_0$, which, along with φ, specifies $d\Omega'$, is measured relative to the direction $\mathbf{\Omega}$ of initial travel.

Scattering from a Beam into an Energy Interval without Regard to Final Direction and into a Conical Shell of Directions without Regard to Final Energy

It is a simple matter to derive from (2.6.3) expressions for the rate of scattering from a beam into a final energy interval dE' without regard to the final directions of motion of the scattered neutrons or into a conical shell $d\mu_0/2$ without regard to final energies. To find the first expression we integrate (2.6.3) over all final directions $\mathbf{\Omega}'$ and make use of (2.4.13) to obtain:

Rate at which neutrons in dV are scattered from $dE\, d\Omega$ into dE', without regard to their final directions of motion

$$\begin{aligned} &= \left\{ \int_0^{2\pi} \frac{d\varphi}{2\pi} \int_{-1}^{1} \frac{d\mu_0}{2} \sum_j [\Sigma_s^j(\mathbf{r}, E) f_s^j(E \to E', \mu_0) + \Sigma_i^j(\mathbf{r}, E) f_i^j(E \to E', \mu_0)] \right\} dE' \\ &\quad \times [v(E)N(\mathbf{r}, \mathbf{\Omega}, E)]\, dV\, d\Omega\, dE \\ &= \sum_j [\Sigma_s^j(\mathbf{r}, E) f_s^j(E \to E') + \Sigma_i^j(\mathbf{r}, E) f_i^j(E \to E')]\, dE' [v(E)N(\mathbf{r}, \mathbf{\Omega}, E)]\, dV\, d\Omega\, dE. \end{aligned} \tag{2.6.4}$$

To find the second expression we integrate over all final energies and get:

Rate at which neutrons in dV are scattered from $dE\, d\Omega$ into $d\mu_0$, without regard to their final energies

$$\begin{aligned} &= \left\{ \int_0^{2\pi} \frac{d\varphi}{2\pi} \int_0^{\infty} dE' \sum_j [\Sigma_s^j(\mathbf{r}, E) f_s^j(E \to E', \mu_0) + \Sigma_i^j(\mathbf{r}, E) f_i^j(E \to E', \mu_0)] \right\} \\ &\quad \times \frac{d\mu_0}{2} [v(E)N(\mathbf{r}, \mathbf{\Omega}, E)]\, dV\, d\Omega\, dE \\ &= \sum_j [\Sigma_s^j(\mathbf{r}, E) f_s^j(E, \mu_0)] + \Sigma_i^j(\mathbf{r}, E) f_i^j(E, \mu_0)] \frac{d\mu_0}{2} [v(E)N(\mathbf{r}, \mathbf{\Omega}, E)]\, dV\, d\Omega\, dE. \end{aligned} \tag{2.6.5}$$

Equation (2.6.4) is the more useful of these results. We can simplify it by introducing the total macroscopic differential cross section $\Sigma_s(\mathbf{r}, E \to E')\, dE'$ for scattering from energy E to energy E' in dE' (2.4.18):

Rate at which neutrons in dV are scattered from $dE\, d\Omega$ into dE', without regard to their final directions of travel

$$= \Sigma_s(\mathbf{r}, E \to E')\, dE'[v(E)N(\mathbf{r}, \mathbf{\Omega}, E)]\, dV\, d\Omega\, dE. \tag{2.6.6}$$

Note that we have defined $\Sigma_s(\mathbf{r}, E \to E')$ to include both elastic and inelastic scattering.

For elastic scattering isotropic in the CM system, the $f_s^j(E \to E')$ in (2.6.4) may, according to (2.5.22), be replaced by the functions

$$f_s^j(E \to E') = \begin{cases} 0 \quad \text{for } E' > E, \\ \dfrac{1}{E(1-\alpha_j)} \qquad \text{for } E\alpha_j \le E' \le E, \\ 0 \quad \text{for } E' < E\alpha_j. \end{cases} \tag{2.6.7}$$

Note that, in making this substitution, we are neglecting any effects due to thermal motion of the isotopes j. Equation (2.5.22) was derived for isotope j initially at rest relative to the laboratory frame of reference. Thus it is valid only if the speed of the neutron is much greater than the speed of the target nuclei.

As a matter of notation we shall always understand that, in writing $f_s^j(E \to E')$ and all similar differential expressions as functions of energies E and E' measured relative to the laboratory system, an average over the thermal motion of the target nuclei has been made. Thus (2.6.7) is to be regarded as an approximation, valid for neutrons having energies greater than a few electron-volts. Room temperature corresponds to an average target energy of $\sim$0.025 eV so that the ratio of the speed of a neutron having energy E to the speed of a nucleus of mass ratio A_j is $\sim \sqrt{(EA_j/0.025)}$. For neutrons in the thermal energy range (2.6.7) is not valid. In fact, because of the thermal motion of the target nuclei, the $f_s^j(E \to E')$ will be nonzero for $E' > E$; thus thermal neutrons can be scattered "up" in energy.

Scattering from One Energy Interval to Another

Since $\Sigma_s(\mathbf{r}, E \to E')$ is not dependent on the initial direction of neutron travel, we may derive an even further "reduced interaction rate" by integrating (2.6.6) over all initial directions $\mathbf{\Omega}$ from which the neutrons enter dV. Doing this and making use of the definition (2.2.7) of the scalar flux yields:

Rate at which neutrons in dV are scattered from dE into dE', without regard to either their initial or final directions of travel

$$= [\Sigma_s(\mathbf{r}, E \to E')\, dE']\left[\int_{\mathbf{\Omega}} d\Omega\, v(E)N(\mathbf{r}, \mathbf{\Omega}, E)\right] dV\, dE \tag{2.6.8}$$

$$= \Sigma_s(\mathbf{r}, E \to E')\, dE'\, \Phi(\mathbf{r}, E)\, dV\, dE.$$

For the case of elastic scattering isotropic in the CM system this expression can be written in a more explicit form. Equation (2.6.7) for $f_s^j(E \to E')\,dE'$ again applies, and, for neutrons of energy greater than about 1 eV, we get:

Rate at which neutrons in dV are elastically scattered from dE into dE' isotropically in the CM system and without regard to either their initial of final directions of travel

$$= \sum_j \Sigma_s^j(\mathbf{r}, E) \frac{dE'}{E(1 - \alpha_j)} \Phi(\mathbf{r}, E)\, dV\, dE. \tag{2.6.9}$$

References

Goodjohn, A. J., and Pomraning, G. C., eds., 1966. *Reactor Physics in the Resonance and Thermal Regions*, Volume 1 (Cambridge, Mass.: The MIT Press).

Lamarsh, J. R., 1966. *Introduction to Nuclear Reactor Theory* (Reading, Mass.: Addison-Wesley), Chapter 2.

Meghreblian, R. V., and Holmes, D. K., 1960. *Reactor Analysis* (New York: McGraw-Hill), Chapters 1–3.

Parks, D. E., Nelkin, M. S., Beyster, J. R., and Wikner, N. F., 1970. *Slow Neutron Scattering and Thermalization* (New York: Benjamin).

Nuclear Data

Goldberg, M. D., May, V. M., and Stehn, J. R., 1962. *Neutron Angular Distributions*, BNL–400, 2nd ed., Volumes 1 and 2.

Problems

1. On the average, how many neutrons are born per second by fission in a reactor that is a cube ten feet on a side and that operates at 4000 megawatts?

2. Suppose that for $E = 0.0253$ eV and at a particular location $\mathbf{r}$, direction $\mathbf{\Omega}$, and time t, $N(\mathbf{r}, \mathbf{\Omega}, 0.0253, t)$ is known to be 10^4 neutrons per cm per eV per unit solid angle. How many neutrons per second at t are crossing a unit surface at $\mathbf{r}_1$ traveling in the direction $\mathbf{\Omega}_1$?

3. Suppose it is known that the neutrons in an energy range $\Delta E = 0.001$ eV containing $E = 0.025$ eV, in a volume interval ΔV of 1 cc containing point $\mathbf{r}$ and having directions in a cone $\Delta\Omega = .01$ surrounding the direction $\mathbf{\Omega}$ produce a power level of 10 microwatts. If the average U^{25} density near $\mathbf{r}$ is 0.235 gm/cc, what is the approximate value of $N(\mathbf{r}, \mathbf{\Omega}, 0.025)$? How many neutrons are contained in the element of phase space $\Delta V\, \Delta\Omega\, \Delta E$?

4. Show that, if a sample of neutrons $N(\mathbf{r}, \mathbf{\Omega}, E, t_0)\, dV\, d\Omega\, dE$ is known to be in the element of phase space $dV\, d\Omega\, dE$ at a time t_0 when the sample is located at $x = 0$, $y = y_0$, $z = z_0$ and $\mathbf{\Omega}$ is in the $+X$ direction, then the number of neutrons left in that sample at time t is given by (2.3.5) with $X = v(t - t_0)$, where v is the neutron speed corresponding to energy E.

5. What is the mean free path for absorption of neutrons moving at 2200 m/sec in a homogeneous

medium of natural uranium and H_2O for which the volume ratio of U to H_2O is 1: 1 at room temperature and pressure?

6. Suppose that during an inelastic-scattering collision the target nucleus absorbs an amount of energy Q. Show that (2.5.14) becomes

$$\frac{E'}{E} = \frac{A^2\tau^2 + 1 + 2A\tau\mu}{(1 + A)^2},$$

where

$$\tau \equiv \left(1 - \frac{1 + A}{EA} Q\right)^{1/2}.$$

7. For elastic scattering between a neutron and a nucleus that have a mass ratio A find the cosine of the scattering angle in the laboratory system ($\cos \theta$) as a function of the lethargy gain $U \equiv u' - u$ associated with the collision.

8. What is the probability that a neutron with an initial energy of 1 MeV, scattered elastically from hydrogen, will emerge from the scattering collision with an energy below 10 eV? Answer the same question if the collision is with a deuterium nucleus. (Both scattering events may be taken as isotropic in the center-of-mass system.)

9. Suppose that neutrons scatter from isotope j elastically and isotropically in the center-of-mass system. Derive an algebraic expression for $\sigma_s^j(E, \mu_0)$ in terms of $\sigma_s^j(E)$ and μ_0.

10. In terms of lethargy the differential elastic-scattering cross section $\sigma_s^j(u \to u')$ for isotope j is defined by

$$\sigma_s^j(u \to u')(-du') \equiv \sigma_s^j(E \to E')dE',$$

where u is the lethargy corresponding to energy E and u' is that corresponding to E'.

For scattering isotropic in the center-of-mass system express $\sigma_s^j(u \to u')$ in terms of $\sigma_s^j(u)$ ($\equiv \sigma_s^j(E)$) and the lethargy gain $U \equiv u' - u$.

11. Derive (2.5.27) from (2.5.26).

12. What is the ratio of the average speed of neutrons in thermal equilibrium with H_2O molecules at room temperature to that of the molecules? To have this same speed ratio relative to atomic hydrogen at 3000°K what must the energy of the neutrons be?

13. For an energy interval $\Delta E = .001$ eV containing $E = 0.025$ eV, a volume $\Delta V = 0.1$ cc containing $\mathbf{r}$, and a solid angular interval $\Delta\Omega = .01$ containing $\mathbf{\Omega}$, $N(\mathbf{r}, \mathbf{\Omega}, E) = 10^{10}$ neutrons per cc per unit solid angle per unit energy. If solid Pu^{49} is present in ΔV, how many fissions per second take place? How many fissions per second take place at point $\mathbf{r}$?

14. Consider a mixture of 10 percent (by volume) Pu^{49} and 90 percent graphite. Relevant densities are $\rho^{49} = 19.6$ g/cc and $\rho^{C} = 1.6$ g/cc, and the scattering cross sections (assumed constant in energy, and isotropic in the center-of-mass system) are $\sigma_s^{49} = 10$b and $\sigma_s^{C} = 5$b.

a. What is the probability that a neutron traveling in this mixture and having an initial energy 100 eV will scatter into the range 70–80 eV?

b. Suppose that within this material $\Phi(\mathbf{r}, E)$ is $10^{10}/E$ neutrons/cc/sec/eV. What is the rate at which neutrons will scatter from 100 eV into the range 70–80 eV?

c. Again with $\Phi(\mathbf{r}, E) = 10^{10}/E$ neutrons/cc/sec/eV, what is the rate at which neutrons will scatter from the range 100–110 eV into the range 70–80 eV?

15. Answer part c to Problem 14 for the initial and final energy ranges 100–100.1 eV, and 99.9–100 eV.

16. Answer part c to Problem 14 for the initial and final energy ranges 100–100.1 eV and 70–80 eV.

3 The Energy Distribution of Neutrons in an Infinite, Homogeneous, Critical Reactor

3.1 Introduction

Since the expression $\Sigma_\alpha(\mathbf{r}, E)\Phi(\mathbf{r}, E)\,dV\,dE$ of (2.6.2) gives the number of type-α interactions per second in the volume dV due to neutrons having energies in dE, a detailed knowledge of all the different nuclear events taking place throughout a reactor can be attained provided the scalar flux density $\Phi(\mathbf{r}, E)$ and the nuclear data needed to construct the $\Sigma_\alpha(\mathbf{r}, E)$ are available. Thus the computation of $\Phi(\mathbf{r}, E)$ is a problem of prime importance in reactor physics.

The rigorous way to determine the scalar flux density is first to find $v(E)N(\mathbf{r}, \mathbf{\Omega}, E)$ and then to integrate that quantity over all $\mathbf{\Omega}$ in accord with the definition (2.2.7). This transport-theory approach to the problem will be described in Chapters 8 and 9. It leads to very severe mathematical difficulties and to equations that are too cumbersome to solve for most reactor systems.

Fortunately an approximate equation for $\Phi(\mathbf{r}, E)$ itself can be derived and shown to be valid for a great many cases of practical interest. The approximate mathematical model leading to this equation is called *diffusion theory*. This model will be introduced and discussed in the next chapter (although it will not be derived as a systematic approximation to the transport equation until Chapter 9).

In the present chapter we shall be concerned exclusively with investigating the energy dependence of the scalar flux density in an infinite homogeneous medium containing just enough fissionable material to be exactly critical.

An infinite, homogeneous, critical medium is an idealization. By simply mixing all the materials intimately, we can in principle create a *homogeneous* medium (defined as one in which the number densities $n^j(\mathbf{r})$ are the same at all locations $\mathbf{r}$), but making the medium infinite is, of course, a practical impossibility. We can approximate infinity, however, by making the medium very large, or else we can, at least theoretically, create a finite region that has the properties of an infinite medium by surrounding it with a "perfect neutron mirror" that reflects all neutrons without changing their energies. (In fact no such perfect neutron mirror exists.)

Nevertheless the idealization is conceptually clear. Moreover analyzing the neutron behavior in an infinite homogeneous medium permits us to attack the problem of finding the energy dependence of $\Phi(\mathbf{r}, E)$ without at the same time having to determine its spatial dependence. The reason for this that, in an infinite, homogeneous, critical reactor, the steady-state scalar flux density $\Phi(\mathbf{r}, E)$ is the same at all locations. It is easy to prove this assertion by finding $\Phi(\mathbf{r}, E)$ in a *finite*, homogeneous, critical reactor (in which $\Phi(\mathbf{r}, E)$ *does* depend on $\mathbf{r}$) and then showing that the dependence on $\mathbf{r}$ in a fixed-sized, central region of that reactor vanishes as the size of the reactor is permitted to increase without limit (the critical condition being maintained by changing the fuel concentration).

Even without such a proof, though, it seems plausible that $\Phi(\mathbf{r}, E)$ would be spatially constant if the medium were infinite, since, with the statistically large number of neutrons involved, if there were some local region which had an above-average neutron concentration, there would be a net leakage out of that region until the local neutron density matched that of the neighboring regions—just as the molecules of a gas will diffuse until the density is, on the average, spatially uniform.

Accordingly we assume that $\Phi(\mathbf{r}, E)$ in an infinite, homogeneous, critical reactor containing a statistically large number of neutrons does not depend on position and can thus be written $\Phi(E)$.

The present chapter will be concerned first with deriving an integral equation whose solution will be $\Phi(E)$ for the infinite, homogeneous, critical medium. Then methods—both numerical and analytical—for solving that equation will be discussed, and the general physical characteristics of the solution will be examined. Finally some of the parameters traditionally used to analyze thermal reactors composed of a lattice of fuel rods will be defined and discussed.

3.2 The Fundamental Neutron-Balance Equation

To find the flux density $\Phi(E)$ we write in mathematical terms the physical statement that, in an infinite, homogeneous, critical reactor operating in a steady-state condition, the rate at which neutrons are produced in any small volume dV and any small energy range dE exactly equals the rate at which they are removed. (If these two rates were not equal, the neutron population in $dV\,dE$ would change, and we would thus not be in the assumed steady-state critical condition.)

Neutrons can be produced in $dV\,dE$ by fission and by scattering in from other energies.

The rate at which fissions occur in $dV\,dE'$ (where dE' is some energy interval other than dE) is given by (2.2.8) with $\alpha = \mathrm{f}$; the total number of neutrons produced by such a fission is, for isotope j, $\nu^j(E')$, and the fraction of those neutrons that is emitted in the energy range dE is $\chi^j(E)\,dE$ (1.4.1). Thus:

Total rate at which fission neutrons are produced in dV

$$= \left[\int_0^\infty \sum_j \nu^j \Sigma_\mathrm{f}^j(E')\Phi(E')\,dE'\right] dV.$$

Steady-state rate at which neutrons are produced by fission in $dV\,dE$

$$= \sum_j \chi^j(E)\,dE\left[\int_0^\infty \nu^j \Sigma_\mathrm{f}^j(E')\Phi(E')\,dE'\right] dV, \tag{3.2.1}$$

where we have omitted indication of any dependence on position for this case of an infinite homogeneous medium.

The rate at which neutrons are produced in $dV\,dE$ by scattering from dE' is the integral of (2.6.6) over all initial neutron energies.

Thus, reversing the symbols E' and E in that expression, we get:

Rate at which neutrons are produced in $dV\,dE$ by scattering from all other energies

$$= dE\left[\int_0^\infty \Sigma_s(E' \to E)\Phi(E')\,dE'\right]dV. \tag{3.2.2}$$

Finally we have from (2.2.8) with $\alpha = \mathrm{t}$:

Total rate at which neutrons are removed from $dV\,dE$ by all processes

$$= \Sigma_t(E)\Phi(E)\,dE\,dV. \tag{3.2.3}$$

Thus the steady-state balance equation is

$$\Sigma_t(E)\Phi(E)\,dE\,dV = \left\{\int_0^\infty dE'\left[\sum_j \chi^j(E)\nu^j\Sigma_f^j(E') + \Sigma_s(E' \to E)\right]\Phi(E')\right\}dE\,dV. \tag{3.2.4}$$

With the $dE\,dV$ omitted from both sides, this expression yields the following fundamental integral equation for $\Phi(E)$:

$$\Sigma_t(E)\Phi(E) = \int_0^\infty dE'\left[\sum_j \chi^j(E)\nu^j\Sigma_f^j(E') + \Sigma_s(E' \to E)\right]\Phi(E'). \tag{3.2.5}$$

3.3 Numerical Solution of the Neutron-Balance Equation

For all $E \geq 0$, we seek a real, nonnegative solution to (3.2.5). Since the $\chi^j(E)$ can be taken as zero for $E > 15$ MeV and since the probability that a neutron will scatter from an energy below 15 MeV to an energy above it is negligible, we expect on physical grounds that $\Phi(E) \approx 0$ for $E > 15$ MeV. We also expect that $\Phi(E) \to 0$ as $E \to 0$, since in that energy range absorption cross sections, being proportional to $1/v$, are very large, and the probability that the neutron will gain energy in a scattering collision (because of the thermal motion of the nuclei) is very high.

The Multigroup Approximation

Crude approximations to the solution of (3.2.5) that are valid over limited energy ranges can be determined analytically, and we shall investigate some of these shortly. However, because of the complicated energy dependence of the parameters $\Sigma_t(E)$, $\Sigma_s(E' \to E)$, etc., it is practically impossible to obtain an analytical solution for the general case. Nevertheless, by using digital computers, we can in a straightforward fashion obtain a solution to (3.2.5) having as high a degree of accuracy as we might desire—and certainly well within the accuracy with which the nuclear data is known at present. To obtain a solution of this nature we simply partition the energy range from $E = 0$ to $E = 15$ MeV

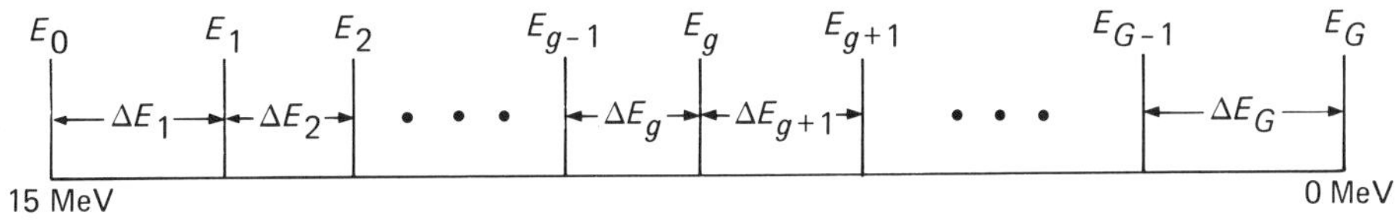

Figure 3.1 Multigroup partitioning of the energy range 0–15 MeV.

into a large number G of subintervals ΔE_g (not necessarily all the same size). We call the subintervals *energy groups* and label them (in accord with custom) *in increasing order from high to low energies* as in Figure 3.1. Thus we have

$$\Delta E_g = E_{g-1} - E_g \qquad (g = 1, 2, \ldots, G). \tag{3.3.1}$$

We next integrate (3.2.5) over any one of the energy groups ΔE_g, replacing the integral on the right by a sum of integrals over all the G energy groups. This gives

$$\begin{aligned}\int_{E_g}^{E_{g-1}} \Sigma_t(E)\Phi(E)\,dE = \sum_{g'=1}^{G} \Bigg[\sum_j \int_{E_g}^{E_{g-1}} \chi^j(E)\,dE \int_{E_{g'}}^{E_{g'-1}} \nu^j\Sigma_f^j(E')\Phi(E')\,dE' \\ + \int_{E_g}^{E_{g-1}} dE \int_{E_{g'}}^{E_{g'-1}} \Sigma_s(E' \to E)\Phi(E')\,dE' \Bigg] \qquad (g = 1, 2, \ldots, G).\end{aligned} \tag{3.3.2}$$

Now we formally define the following *group parameters*:

$$\begin{aligned}\Phi_g &\equiv \int_{E_g}^{E_{g-1}} \Phi(E)\,dE,\\ \chi_g^j &\equiv \int_{E_g}^{E_{g-1}} \chi^j(E)\,dE,\\ \Sigma_{tg} &\equiv \frac{\int_{E_g}^{E_{g-1}} \Sigma_t(E)\Phi(E)\,dE}{\Phi_g},\\ \nu^j\Sigma_{f\,g'}^j &\equiv \frac{\int_{E_{g'}}^{E_{g'-1}} \nu^j\Sigma_f^j(E')\Phi(E')\,dE'}{\Phi_{g'}},\\ \Sigma_{gg'} &\equiv \frac{\int_{E_g}^{E_{g-1}} dE \int_{E_{g'}}^{E_{g'-1}} \Sigma_s(E' \to E)\Phi(E')\,dE'}{\Phi_{g'}}.\end{aligned} \tag{3.3.3}$$

(Note the order of the subscripts in $\Sigma_{gg'}$: the scattering is *from* group g' *to* group g.)

Equation (3.3.2) then becomes

$$\Sigma_{tg}\Phi_g = \sum_{g'=1}^{G} \left[\sum_j \chi_g^j \nu^j \Sigma_{fg'}^j + \Sigma_{gg'} \right] \Phi_{g'} \qquad (g = 1, 2, \ldots, G). \tag{3.3.4}$$

The quantity Φ_g is called the *group flux*. Note that it is not a density in energy, and thus it has dimensions different from $\Phi(E)$. Physically Φ_g is the number of neutrons per unit volume in the energy range ΔE_g multiplied by their average speed.

The physical meaning of (3.3.4) is that the total rate at which neutrons are removed from energy group g (per unit volume in an infinite, homogeneous, critical reactor) equals the sum of the rate at which they appear in group g because of fissions in all groups plus the rate at which they appear because of scattering from all groups (including group g itself).

It is much easier to solve the algebraic equations (3.3.4) for the Φ_g than to solve the integral equation (3.2.5) for $\Phi(E)$. To determine the Σ_{tg}, χ_g^j, etc., appearing in (3.3.4) and defined by (3.3.3), however, we must know $\Phi(E)$. Thus we appear to be unable to take advantage of the simpler form of (3.3.4).

To overcome this difficulty we make the following approximation:

$$\Phi(E) = \text{constant} = C_g \quad \text{for } E_g < E \le E_{g-1} \qquad (g = 1, 2, \ldots, G). \tag{3.3.5}$$

In words, we approximate $\Phi(E)$ within each energy group ΔE_g by a constant. For G large and the locations of the partition energies carefully chosen, the approximation (3.3.5) is an excellent one. (Computer programs that can handle many thousand energy groups are currently in operation, although, for most applications, use of fewer than 100 groups provides satisfactory results.)

With $\Phi(E)$ constant within each ΔE_g, the flux-dependent "group cross sections" defined by (3.3.3) become

$$\begin{aligned}
\Sigma_{tg} &= \frac{\int_{E_g}^{E_{g-1}} \Sigma_t(E)\, dE}{\Delta E_g}, \\
\nu^j \Sigma_{fg'}^j &= \frac{\int_{E_{g'}}^{E_{g'-1}} \nu^j \Sigma_f^j(E')\, dE'}{\Delta E_{g'}}, \\
\Sigma_{gg'} &= \frac{\int_{E_g}^{E_{g-1}} dE \int_{E_{g'}}^{E_{g'-1}} \Sigma_s(E' \to E)\, dE'}{\Delta E_{g'}}.
\end{aligned} \tag{3.3.6}$$

These parameters, along with the χ_g^j in (3.3.4), can be obtained directly from the basic nuclear data and material concentrations in the medium. As a result we can solve (3.3.4) for the Φ_g without first having to know $\Phi(E)$. Moreover, insofar as the approximate group constants (3.3.6) are close to the exact group constants (3.3.3), we expect that these approximate constants, when used in the group equations (3.3.4), will yield values of Φ_g that are close to those of the exact Φ_g defined by (3.3.3).

Solution of the Multigroup Equations

To solve (3.3.4) on a computer in an efficient manner is a standard but nontrivial problem in numerical analysis. We shall not discuss it other than to say that it would take about 20 minutes on a CDC-6600 computer for $G = 100{,}000$.

We shall, however, look at some of the mathematical properties associated with

(3.3.4). Such an examination will illustrate the kind of numerical analysis frequently applied to the equations of reactor physics.

To proceed, we write down all the G equations (3.3.4) at once, using a matrix notation and, for simplicity, assuming that only one fissionable isotope is present:

$$\begin{bmatrix} \Sigma_{t1} - \chi_1 \nu\Sigma_{f1} - \Sigma_{11} & -\chi_1\nu\Sigma_{f2} - \Sigma_{12} & \cdots & -\chi_1\nu\Sigma_{fG} - \Sigma_{1G} \\ -\chi_2\nu\Sigma_{f1} - \Sigma_{21} & \Sigma_{t2} - \chi_2\nu\Sigma_{f2} - \Sigma_{22} & \cdots & -\chi_2\nu\Sigma_{fG} - \Sigma_{2G} \\ \vdots & \vdots & & \vdots \\ -\chi_G\nu\Sigma_{f1} - \Sigma_{G1} & -\chi_G\nu\Sigma_{f2} - \Sigma_{G2} & \cdots & \Sigma_{tG} - \chi_G\nu\Sigma_{fG} - \Sigma_{GG} \end{bmatrix} \begin{bmatrix} \Phi_1 \\ \Phi_2 \\ \vdots \\ \Phi_G \end{bmatrix} = 0. \tag{3.3.7}$$

Many of the terms formally appearing in these equations actually vanish. For example, since, outside the thermal range, there is no scattering from lower energy groups into higher groups, $\Sigma_{12}, \Sigma_{13}, \ldots, \Sigma_{1G}$, etc., are zero. Similarly, since no neutrons from fission are born in the thermal energy range, the χ_g for large values of g will vanish. We shall nevertheless retain the algebraic expressions for these terms in (3.3.4) and (3.3.7), since none of the mathematical properties we wish to discuss are affected by their presence.

The first thing we notice about (3.3.7) is that it is a *homogeneous* matrix equation (i.e., if the column vector $[\Phi] \equiv \text{Col}\{\Phi_1, \Phi_2, \ldots, \Phi_G\}$ is a solution, so is the vector $[c\Phi] \equiv \text{Col}\{c\Phi_1, c\Phi_2, \ldots, c\Phi_G\}$, where c is any constant). An immediate consequence of this is the mathematical fact that we can obtain a nontrivial solution of (3.3.7) (i.e., one other than $\Phi_1 = \Phi_2 = \cdots = \Phi_G = 0$) if and only if the determinant of the matrix multiplying $[\Phi]$ vanishes. Let us use the symbol $[H]$ to represent that matrix and the symbol $|H|$ to represent its determinant. Then the theorem just quoted becomes

$$[\Phi] \neq 0 \text{ if and only if } |H| = 0. \tag{3.3.8}$$

The value of $|H|$ will, of course, depend on the numbers $\Sigma_{tg}, \chi_g, \nu\Sigma_{fg}$, etc., that make up the elements of $[H]$, and it is highly unlikely that mixing materials together without paying special attention to the exact amounts of fuel, moderator, etc., will result in isotopic concentrations $n^j(\mathbf{r})$ such that $|H|$ will vanish. The mathematics is thus telling us that, if we just mix moderator and fuel in a random way, there will be no equilibrium solution for the flux in the reactor other than $[\Phi] = 0$. In other words the reactor will not be critical.

But (3.3.7) tells us much more. In fact it gives us a way of predicting the exact conditions under which the reactor *will* be critical. For example we can adjust the concentration of fissile material until $|H|$ *does* vanish. The resultant infinite mixture of homogeneous material will then be such that the critical neutron-balance condition will be met exactly. With $|H|$ zero it will also be possible to determine the (nonzero) flux vector $[\Phi]$ corresponding to this critical condition, so that subsequent computation of any reaction rate will be possible.

The Eigenvalue λ

It is easy to show mathematically that the procedure just described for determining critical concentrations will not yield a unique result. If the concentration n^k of some particular isotope k in the homogeneous mixture is treated as unknown (the concentrations of all other isotopes being fixed), each element in the matrix $[H]$ can be written as the sum of a fixed term plus a term proportional to n^k. For example Σ_{tg} becomes $\sum_{j \neq k} \Sigma_{tg}^j + n^k \sigma_{tg}^k$. As a result $[H]$ will be a polynomial containing powers of n^k up to $(n^k)^G$.

There will be, in general, G values of n^k that make this polynomial equal to zero. Thus, mathematically, there appear to be as many concentrations n^k that make the reactor critical as there are energy groups in the multigroup model.

Most of these values of n^k will be negative or complex or will lead to a corresponding flux vector $[\Phi]$ having some negative elements. Thus most of the n^k that cause $[H]$ to vanish will be unacceptable on physical grounds.

Fortunately it is possible to avoid these unphysical solutions altogether. There is an indirect, but systematic, procedure that yields only physically acceptable values of n^k. This procedure is based on the fact that, for a given *fixed* set of nuclear concentrations n^j, it would always be possible to make a reactor critical if we could alter the number of neutrons emitted per fission. Of course it is not possible physically to effect such an alteration in the ν^j's. Nevertheless we can represent the possibility mathematically and in that way cause the determinant $|H|$ to vanish. As we shall see, this possibility provides a systematic way of determining a set of physically real concentrations n^j that will make a critical system.

To set up a mathematical representation of a reactor in which the ν^j can be altered arbitrarily, we divide each $\nu\Sigma_{fg'}$ in (3.3.4) by a real, positive number λ. Equation (3.3.4) then becomes:

$$\Sigma_{tg}\Phi_g = \sum_{g'=1}^{G} \left[\sum_j \chi_g^j \frac{1}{\lambda} \nu^j \Sigma_{fg'}^j + \Sigma_{gg'} \right] \Phi_{g'} \qquad (g = 1, 2, \ldots, G). \tag{3.3.9}$$

Since division by λ has the mathematical effect of replacing each ν^j by ν^j/λ, we see on physical grounds that there always exists some unique real positive value of λ that will make the reactor critical (albeit artifically). Thus, if there is too much neutron production in the reactor because of an excess of fuel over the amount required for criticality, division by a value of λ greater than one will have the effect of reducing the ν^j's and thereby decreasing the neutron-production rate. Similarly, if there is too little fuel or too much absorber in the core to permit a sustained chain reaction, division by a value of λ less than one will have the effect of increasing the rate of neutron production. It is clear, then, that there is always some value of λ which, in a artificial way, will make the system critical, and hence for which $|H|$ will vanish and all the Φ_g will be greater than zero. If that value of λ happens to be one, we know that this mathematical solution also corresponds to a real, physical solution. Moreover, if the λ required to produce a positive solution of (3.3.9) is not one, its value gives us a quantitative notion of how far the system is from being physically critical.

The mathematical consequences of introducing the number λ become clearer if (3.3.9) is written in matrix form. Accordingly we define $G \times G$ matrices $[A]$ and $[M]$ whose gg' elements are

$$\begin{aligned} A_{gg'} &\equiv \Sigma_{tg}\delta_{gg'} - \Sigma_{gg'}, \\ M_{gg'} &\equiv \sum_j \chi_g^j \nu^j \Sigma_{fg'}^j, \end{aligned} \tag{3.3.10}$$

where $\delta_{gg'}$ is the Kronecker delta. Equation (3.3.9) may then be written

$$[A][\Phi] = [M][\Phi]/\lambda. \tag{3.3.11}$$

It is possible to prove that $[A]^{-1}$, the inverse of the matrix $[A]$, exists for any physically real set of cross sections and number densities (and, moreover, that every element of it is nonnegative). Thus (3.3.11) can be multiplied by $\lambda[A]^{-1}$ to give

$$[A]^{-1}[M][\Phi] = \lambda[\Phi]. \tag{3.3.12}$$

This result has the standard eigenvalue-equation form. The number λ is thus an eigenvalue of the matrix $[A]^{-1}[M]$, and a nontrivial solution for $[\Phi]$ will exist if and only if the determinant of $[A]^{-1}[M] - \lambda[I]$ vanishes ($[I]$ being the $G \times G$ unit diagonal matrix).

In general, just as with n^k, there will be G values of λ for which (3.3.12) will have a nontrivial solution. However, in this case, it is possible (for any physically realizable set of multigroup parameters) to prove the following mathematical statements:

1. There is a unique real, positive eigenvalue of (3.3.12) greater in magnitude than any other eigenvalue.
2. All the elements of the eigenvector corresponding to that eigenvalue are real and positive.
3. All other eigenvectors of (3.3.12) either have some elements that are zero or have elements that differ in sign from each other.

These properties guarantee that there always exists a solution to (3.3.12) corresponding to a real positive value of λ and that the eigenvector $[\Phi]$ corresponding to that eigenvalue is the only physically acceptable (all-positive) solution. Moreover the fact that the eigenvalue we seek is greater in magnitude than any other eigenvalue of (3.3.12) leads to a systematic procedure for solving (3.3.9) that is guaranteed to yield the desired eigenvalue and eigenvector.

If we make the simplification that the fission spectra (values of χ_g^j) are the same for all isotopes (i.e., $\chi_g^j = \chi_g$ for all j), the proofs of some of these mathematical assertions become quite straightforward. To see this, we define the column vectors

$$[\chi] \equiv \mathrm{Col}\{\chi_1, \chi_2, \ldots, \chi_G\},$$
$$[F] \equiv \mathrm{Col}\left\{\sum_j \nu^j \Sigma_{f1}^j, \sum_j \nu^j \Sigma_{f2}^j, \ldots, \sum_j \nu^j \Sigma_{fG}^j\right\}. \tag{3.3.13}$$

Then the elements $M_{gg'}$ of the matrix $[M]$ defined by (3.3.10) can be written

$$M_{gg'} = \chi_g \sum_j \nu^j \Sigma_{fg'}^j \tag{3.3.14}$$

and $[M]$ becomes

$$[M] = [\chi][F]^{\mathrm{T}}, \tag{3.3.15}$$

where $[F]^{\mathrm{T}}$ is a row vector, the transpose of $[F]$. The $G \times G$ matrix $[M]$ is now a *dyad* (a column vector multiplied into a row vector). (It is not to be confused with a quantity such as $[F]^{\mathrm{T}}[\chi]$, which, being a row multiplied into a column, is a number, the *scalar product*.)

With these definitions (3.3.12) becomes

$$[A]^{-1}[\chi][F]^{\mathrm{T}}[\Phi] = \lambda[\Phi]. \tag{3.3.16}$$

Multiplying this equation by the row vector $[F]^{\mathrm{T}}$ yields

$$([F]^{\mathrm{T}}[A]^{-1}[\chi])([F]^{\mathrm{T}}[\Phi]) = \lambda([F]^{\mathrm{T}}[\Phi]).$$

Since $[A]^{-1}[\chi]$ is a column vector and $[F]^{\mathrm{T}}$ is a row vector, the term $([F]^{\mathrm{T}}[A]^{-1}[\chi])$ is a scalar. Moreover, since, for physically real parameters, $[F]$ and $[\chi]$ have real nonnegative elements and $[A]^{-1}[\chi]$ can be proven to be real and positive, $([F]^{\mathrm{T}}[A]^{-1}[\chi])$ must be real and positive. Thus either the number $[F]^{\mathrm{T}}[\Phi]$ is zero or λ equals the unique positive real number $([F]^{\mathrm{T}}[A]^{-1}[\chi])$. This validates the first part of property one. In addition (3.3.16) shows that, for $[F]^{\mathrm{T}}[\Phi] \neq 0$, the vector $[\Phi]$ is proportional to a vector $[A]^{-1}[\chi]$ having all positive elements (property two).

Finally (3.3.16) shows that, if $[F]^{\mathrm{T}}[\Phi] = 0$, the eigenvalue λ is also zero. Thus the second part of property one is proven. In addition, since $[F]$ has only nonnegative elements and $[F]^{\mathrm{T}}[\Phi] = \sum_{g=1}^{G} F_g\Phi_g$, any vector $[\Phi]$ corresponding to $\lambda = 0$ must either have at least one element that is zero or have nonzero elements that differ in sign (property three).

Thus we have validated all three properties for the special case in which the fission spectra χ_g^j may be assumed the same for all fissionable isotopes present in the mixture.

It should be noted that finding the most positive value of λ and its corresponding positive flux vector will always yield a result, and that result will be unique. The same cannot be said if we search for one of the concentrations n^k. A physically acceptable solution may not exist (for example if the fissile and fertile material concentrations are fixed and we are trying to go critical by varying the moderator concentration). Equally disturbing is the fact that it is possible to get more than one value of n^k that will yield a physically acceptable solution. (If n^k is the concentration of a fissile isotope, there may be two values—one very large, corresponding to a fast reactor, and one very small, corresponding to a thermal reactor—both of which make the system critical.)

Because of all these potential difficulties associated with a direct attempt to find critical concentrations, it is preferable to proceed indirectly, making use of the attractive mathematical features of (3.3.11).

To find material concentrations corresponding to a critical state, then, we start by guessing at an initial set of n^j values and finding a λ such that, when it is used in (3.3.7), the determinant $|H|$ vanishes and a positive solution $[\Phi]$ results. Then we systematically adjust the concentrations at our disposal (generally the fuel concentration, or a control poison concentration) until, finally, a λ of one is obtained. The concentrations of material that lead to this particular λ correspond to a physically realizable critical state—or at least to one that would be critical if the nuclear data were all correct.

The Effective Multiplication Constant k_{eff}

It was mentioned above that the number λ corresponding to a positive solution for $[\Phi]$ provides a quantitative measure of how far an infinite homogeneous mixture of materials is from being a critical reactor capable of operating in a steady-state fashion. We can develop this idea further by showing that λ has an approximate physical interpretation. To do so we return to (3.3.9) and sum over all groups g, obtaining

$$\sum_{g=1}^{G} \Sigma_{tg}\Phi_g - \sum_{g=1}^{G} \sum_{g'=1}^{G} \Sigma_{gg'}\Phi_{g'} = \frac{1}{\lambda} \sum_{g'=1}^{G} \nu\Sigma_{fg'}\Phi_{g'}, \tag{3.3.17}$$

where we have written $\nu\Sigma_{fg'} = \sum_j \nu^j \Sigma_{fg'}^j$ and made use of the fact that

$$\sum_{g=1}^{G} \chi_g^j = 1, \tag{3.3.18}$$

an equation which states that the sum of the probabilities that fission neutrons will be born into the various energy ranges ΔE_g is unity.

The second term on the left of (3.3.17) permits a similar simplification. Recall that $\Sigma_{gg'}\Phi_{g'}$ is the rate (per unit volume) at which neutrons scatter out of group g' into group g. If we sum over all possible groups (including g) into which the neutron can scatter, we obtain the total scattering rate for group g'. This can be seen mathematically by summing the last line of (3.3.3) (multiplied by $\Phi_{g'}$) over all g. We get

$$\sum_{g=1}^{G} \Sigma_{gg'}\Phi_{g'} = \int_0^\infty dE \int_{E_{g'}}^{E_{g'-1}} \Sigma_s(E' \to E)\Phi(E')\, dE' = \Sigma_{sg'}\Phi_{g'}, \tag{3.3.19}$$

where $\Sigma_{sg'}$ is the total macroscopic group scattering cross section (elastic and inelastic) for group g'.

Since

$$\sum_{g'=1}^{G} \Sigma_{sg'}\Phi_{g'} = \sum_{g=1}^{G} \Sigma_{sg}\Phi_{g},$$

g and g' being dummy indices, (3.3.17) becomes

$$\sum_{g=1}^{G} (\Sigma_{tg} - \Sigma_{sg})\Phi_g = \frac{1}{\lambda}\sum_{g=1}^{G} \nu\Sigma_{fg}\Phi_g. \tag{3.3.20}$$

Then, with Σ_{ag} defined as the total macroscopic absorption cross section for group g, we obtain

$$\lambda = \frac{\sum_{g=1}^{G} \nu\Sigma_{fg}\Phi_g}{\sum_{g=1}^{G} \Sigma_{ag}\Phi_g}. \tag{3.3.21}$$

If concentrations of materials are such that the reactor is critical, we have already seen that λ will be unity. Now (3.3.21) tells us that λ is the ratio of the total rate of neutron production to the total rate of neutron absorption. For an infinite critical reactor this ratio is indeed unity.

For a noncritical reactor λ is most precisely thought of as that number by which the number of neutrons emitted per fission must be divided if the reactor is to be made critical in an artificial manner. If we write (3.3.21) in the form

$$1 = \frac{\sum_{g=1}^{G} \nu\Sigma_{fg}\Phi_g/\lambda}{\sum_{g=1}^{G} \Sigma_{ag}\Phi_g}, \tag{3.3.22}$$

this interpretation is apparent, for then (3.3.22) tells us that in an artificially critical reactor the ratio of the artificially altered neutron-production rate to the neutron-absorption rate is unity. We can then interpret (3.3.21) as saying that λ is the ratio of the *unaltered* fission rate to the absorption rate *in the artificially critical reactor.*

This latter interpretation is the common one. People think and speak of λ as the ratio of neutron-production rate to neutron-absorption rate characteristic of a particular assembly of materials. Thus, if λ is greater than unity, the ratio of production to absorption in such an assembly will exceed unity, and the reactor will be supercritical. It is important to recognize, however, that this ratio is not the physical one that would be observed if such a supercritical assembly were actually constructed and an initial population of neutrons were inserted. If that were done, there *would* be a ratio of production rate to absorption rate different from unity as the neutron population increased, and an equation similar to (3.3.21) would yield that ratio. However the group fluxes Φ_g actually present would differ somewhat from the solution to (3.3.9). In particular the physical fluxes in such an assembly would change with time.

Thus the value of λ corresponding to a physically acceptable $[\Phi]$ *is* a precisely defined ratio of neutron-production to neutron-absorption rates, but it is *not* a physically observable number. Rather it is a ratio associated with an assembly of materials made critical in a fictitious manner. Nevertheless it has an approximate physical significance, and, more important, it does provide a quantitative measure of how close an assembly of materials is to being critical. As such it is a very important and useful quantity.

In fact it is so commonly used that it has been given a name, the *effective multiplication factor*, and is often referred to by another symbol, k_{eff}, that reflects this name. Thus for a supercritical assembly $k_{\text{eff}} > 1$, for a critical assembly $k_{\text{eff}} = 1$, and for a subcritical assembly $k_{\text{eff}} < 1$.

The physical arguments that led us to conclude that there always exists a real positive λ that can make a system artificially critical can be extended in a straightforward manner to the case of a reactor of finite size from which some neutrons are lost by leakage and in which materials may be distributed in a heterogeneous manner. (We shall do this in Chapter 4.) Thus the concept of an effective multiplication constant is not restricted to the case of an infinite homogeneous medium. No matter how complicated the geometry, it is always possible to find a number k_{eff} which, when divided into the ν^j, will artificially cause a reactor to be critical. We plan, then, to generalize the definition of k_{eff} so that it will always be unity for a critical reactor. The effective multiplication constant will then become a property of the entire *reactor*. This being the case, it is convenient to define some quantity that characterizes the degree of criticality of a *material* alone. Accordingly we define a term

$$k_\infty(\text{for a given material}) \equiv k_{\text{eff}}(\text{for an infinite amount of that material}). \tag{3.3.23}$$

For the infinite homogeneous assembly we are examining in this chapter, there is no distinction between k_∞ and k_{eff}. For a reactor of finite size composed of several different material regions, however, the two will differ. There will be one value of k_{eff} for the whole reactor, but each material region will have its own value of k_∞. (The k_∞ of a region containing no fissionable material is zero; however the term is rarely applied to such composition.)

Reactivity

The effective multiplication constant is not the only term used to characterize quantitatively the degree of criticality of a reactor. Another quantity called the *reactivity* (ρ) is also very commonly used. Unfortunately there is no universally accepted definition of reactivity. Different authors define it in a number of different ways which lead to slightly different numerical values for a particular assembly. The definition to be introduced at this point is very commonly used in reactor-design calculations. An alternative, more general definition will be developed in Chapter 7. Thus, for the present, we define

$$\rho_\lambda \equiv \frac{\lambda - 1}{\lambda} = 1 - \frac{1}{k_{\text{eff}}}, \tag{3.3.24}$$

where λ is the number that yields a nontrivial solution of (3.3.9). Clearly $\rho_\lambda > 0$ for a supercritical reactor, $\rho_\lambda = 0$ for a critical reactor, and $\rho_\lambda < 0$ for a subcritical reactor.

Using (3.3.21), we find that

$$\rho_\lambda = \frac{\sum_{g=1}^{G} (\nu\Sigma_{\text{f}g} - \Sigma_{\text{a}g})\Phi_g}{\sum_{g=1}^{G} \nu\Sigma_{\text{f}g}\Phi_g}. \tag{3.3.25}$$

Thus, according to this definition, the reactivity is the ratio of the *net* production rate of neutrons (excess of those produced by fission over those absorbed) to the production rate due to fission—all this in a reactor made critical by artificially dividing the number of neutrons produced per fission by λ. (Note that the fission rates referred to in the definition are the *unaltered* ones; if we used $\lambda^{-1}\nu\Sigma_{\text{f}g}$ in (3.3.25), ρ_λ would always be zero.)

The reactivity, like the effective multiplication constant, is a property of the whole reactor—not just of a given material composition. Thus, when we generalize the definition of k_{eff} to apply to finite heterogeneous reactors, (3.3.24) will still apply, and a definition of the reactivity of the finite heterogeneous reactor will be immediate. There can be some confusion about this global character of reactivity in that people often refer to "the reactivity of a control rod." What is meant is the *change* in the ρ_λ of the *reactor* when that particular control rod is inserted. It is not meaningful to speak of the reactivity of a control rod considered as an isolated piece of material.

A Three-Group Example

Before leaving the mathematical problem of finding λ and the corresponding Φ_g by the energy-group method, it may be helpful to work out the algebra of an illustrative example.

To keep things as simple as possible we shall restrict attention to a three-group ($G = 3$) situation (even though an accurate spectrum computation requires the use of 30 or more groups). In addition we shall neglect all "up-scattering" (so that $\Sigma_{gg'} = 0$ for $g' > g$) and shall assume that there is only one fissionable isotope present and that no fission neutrons appear directly in the lowest energy group ($\chi_3 = 0$). Equation (3.3.7) then becomes

$$\begin{bmatrix} \Sigma_{t1} - \Sigma_{11} - \frac{1}{\lambda}\chi_1\nu\Sigma_{f1} & -\frac{1}{\lambda}\chi_1\nu\Sigma_{f2} & -\frac{1}{\lambda}\chi_1\nu\Sigma_{f3} \\ -\frac{1}{\lambda}\chi_2\nu\Sigma_{f1} - \Sigma_{21} & \Sigma_{t2} - \Sigma_{22} - \frac{1}{\lambda}\chi_2\nu\Sigma_{f2} & -\frac{1}{\lambda}\chi_2\nu\Sigma_{f3} \\ -\Sigma_{31} & -\Sigma_{32} & \Sigma_{t3} - \Sigma_{33} \end{bmatrix} \begin{bmatrix} \Phi_1 \\ \Phi_2 \\ \Phi_3 \end{bmatrix} = 0. \tag{3.3.26}$$

We now introduce another commonly used symbol, Σ_g, the total removal (absorption plus scattering-*out*) cross section for group g:

$$\Sigma_g \equiv \Sigma_{tg} - \Sigma_{gg} = \Sigma_{ag} + \Sigma_{sg} - \Sigma_{gg} = \Sigma_{ag} + \sum_{g' \neq g} \Sigma_{g'g}. \tag{3.3.27}$$

The critical condition for (3.3.26) then becomes

$$\begin{vmatrix} \Sigma_1 - \frac{1}{\lambda}\chi_1\nu\Sigma_{f1} & -\frac{1}{\lambda}\chi_1\nu\Sigma_{f2} & -\frac{1}{\lambda}\chi_1\nu\Sigma_{f3} \\ -\frac{1}{\lambda}\chi_2\nu\Sigma_{f1} - \Sigma_{21} & \Sigma_2 - \frac{1}{\lambda}\chi_2\Sigma_{f2} & -\frac{1}{\lambda}\chi_2\nu\Sigma_{f3} \\ -\Sigma_{31} & -\Sigma_{32} & \Sigma_3 \end{vmatrix} = 0. \tag{3.3.28}$$

If we multiply the second line of this determinant by χ_1, substract from the result the first line multiplied by χ_2, and then multiply the first line by λ (assumed nonzero), we obtain

$$\begin{vmatrix} \lambda\Sigma_1 - \chi_1\nu\Sigma_{f1} & -\chi_1\nu\Sigma_{f2} & -\chi_1\nu\Sigma_{f3} \\ -\chi_1\Sigma_{21} - \chi_2\Sigma_1 & \chi_1\Sigma_2 & 0 \\ -\Sigma_{31} & -\Sigma_{32} & \Sigma_3 \end{vmatrix} = 0. \tag{3.3.29}$$

It is clear from this result that the algebraic equation equivalent to (3.3.29) will be one of first order in λ and hence will yield only one value of that quantity. Expansion of the determinant shows that this value is

$$\lambda = \frac{(\chi_1\Sigma_{21} + \chi_2\Sigma_1)(\nu\Sigma_{f3}\Sigma_{32} + \nu\Sigma_{f2}\Sigma_3) + \chi_1\Sigma_2(\nu\Sigma_{f3}\Sigma_{31} + \nu\Sigma_{f1}\Sigma_3)}{\Sigma_1\Sigma_2\Sigma_3}. \tag{3.3.30}$$

Since the χ's and Σ's in this equation are real positive numbers, we have thus demonstrated mathematically for this case that there always exists a real positive value of λ that causes $|H|$ to vanish. Moreover we have found that there is only one such value of λ.

The other two eigenvalues of (3.3.26) are both $\lambda = 0$. The corresponding eigenvectors, Col $\{\Phi_1, \Phi_2, \Phi_3\}$, must be such that $\nu\Sigma_{f1}\Phi_1 + \nu\Sigma_{f2}\Phi_2 + \nu\Sigma_{f3}\Phi_3 = 0$, and thus must have elements that differ in sign.

Since (3.3.26) is a homogeneous equation, we can find the column vector $[\Phi]$ only to within a multiplicative constant; or, to put it another way, we are free to set any one of the group fluxes Φ_1, Φ_2, or Φ_3 equal to an arbitrary constant and find the other two fluxes in terms of that constant. Let us fix the element Φ_1. The last two algebraic equations implied by (3.3.26) then yield (for $\lambda \neq 0$)

$$\begin{aligned} &\left(\Sigma_2 - \frac{1}{\lambda}\chi_2\nu\Sigma_{f2}\right)\Phi_2 - \frac{1}{\lambda}\chi_2\nu\Sigma_{f3}\Phi_3 = \left(\frac{1}{\lambda}\chi_2\nu\Sigma_{f1} + \Sigma_{21}\right)\Phi_1, \\ &-\Sigma_{32}\Phi_2 + \Sigma_3\Phi_3 = \Sigma_{31}\Phi_1. \end{aligned} \tag{3.3.31}$$

Inverting the matrix that multiplies Col$\{\Phi_2, \Phi_3\}$ when (3.3.31) is cast in matrix form yields

$$\begin{bmatrix} \Phi_2 \\ \Phi_3 \end{bmatrix} = \frac{\Phi_1}{\Sigma_3(\Sigma_2 - \lambda^{-1}\chi_2\nu\Sigma_{f2}) - \lambda^{-1}\chi_2\Sigma_{32}\nu\Sigma_{f3}} \begin{bmatrix} \Sigma_3 & \frac{1}{\lambda}\chi_2\nu\Sigma_{f3} \\ \Sigma_{32} & \Sigma_2 - \frac{1}{\lambda}\chi_2\nu\Sigma_{f2} \end{bmatrix} \begin{bmatrix} \frac{1}{\lambda}\chi_2\nu\Sigma_{f1} + \Sigma_{21} \\ \Sigma_{31} \end{bmatrix}.$$

This equation, along with the value of λ given by (3.3.30), constitutes the solution of (3.3.26) that is of physical interest.

3.4 Analytical Computations of the Scalar Flux Density in an Infinite Homogeneous Medium

We have seen that an approximate solution (3.2.5) for $\Phi(E)$, the scalar flux density in an infinite homogeneous medium, can be obtained by making the single assumption (3.3.5) that $\Phi(E)$ can be represented by a series of constants $C_g(= \Phi_g/\Delta E_g)$ in energy ranges ΔE_g. If the total number of energy groups G is taken to be very large and the energy ranges ΔE_g of the group are all very small, this approximation is excellent. Moreover, if we extend the method by approximating $\Phi(E)$ by a series of functions linear within each energy group and continuous across the energy cut points $E_1, E_2, \ldots, E_{G-1}$, that is, if we take

$$\Phi(E) = \frac{E_{g-1} - E}{E_{g-1} - E_g}\Phi(E_g) + \frac{E - E_g}{E_{g-1} - E_g}\Phi(E_{g-1}) \quad \text{for } E_g \leq E \leq E_{g-1},$$

we can make the approximation even better and can, in fact, approximate the exact solution $\Phi(E)$ of (3.2.5) as closely as we wish. Thus it is legitimate to claim that, by solving a rather straightforward problem on a computer, we can determine the scalar flux density in an infinite, homogeneous, critical medium of arbitrary composition with an accuracy limited only by our knowledge of the concentrations of materials and the nuclear data. Moreover we can predict quantitatively (by finding the effective multiplication constant k_{eff} or the reactivity ρ_λ) how far a noncritical system is from being critical.

Unfortunately it is not practical to try using hand calculations to get a feeling for the numbers that result from such a computation. We have seen that even a simplified three-group solution of the critical equation (3.3.9) leads to complicated algebraic formulas. Accordingly, since it is important to acquire some feeling for the way $\Phi(E)$ behaves as a function of material composition and energy, we shall now return to the fundamental integral equation (3.2.5) and obtain some approximate analytical solutions for certain very simple situations. In doing so, we shall be attacking the slowing-down problem in the classical manner developed before the days of computers. Certain important quantities introduced in those days, and still quite useful, will be defined and discussed.

The first thing to note about (3.2.5) is that it is a homogeneous integral equation and (just as with the analogous energy-group equation (3.3.4) derived from it) is likely to have only the trivial solution $\Phi(E) = 0$ unless exactly the right amounts of fuel, moderator, and control poison are present in the medium.

However, also as with the energy-group equation, it is expected on physical grounds that a nontrivial solution can always be obtained by dividing the number of neutrons emitted per fission by some positive real number λ. Thus we rewrite (3.2.5) as

$$\Sigma_t(E)\Phi(E) = \int_0^\infty dE'\left[\frac{1}{\lambda}\sum_j \chi^j(E)\nu^j\Sigma_f^j(E') + \Sigma_s(E' \to E)\right]\Phi(E'). \qquad (3.4.1)$$

It will be convenient to investigate simple solutions for this equation in three ranges:

1. the fission-source range;
2. the intermediate "slowing-down" range; and
3. the thermal energy range.

Approximate Solutions for $\Phi(E)$ in the Fission-Source Energy Range

In the very high energy range (> 1 MeV), the fission-source term is significant. To simplify the algebra we assume the fission spectra $\chi^j(E)$ are the same for all j and write that term as $\chi(E)\int_0^\infty dE'\nu\Sigma_f(E')\Phi(E')$, which, since it involves fissions from the *whole* energy range, tends to be greater than $\int_0^\infty dE'\Sigma_s(E' \to E)\Phi(E')$. In view of this situation a first approximation for $\Phi(E)$ can be obtained by neglecting the latter term in (3.4.1).

The result is

$$\Phi(E) \approx \frac{\chi(E)}{\Sigma_t(E)} \int_0^\infty dE' \frac{1}{\lambda} \nu\Sigma_f(E')\Phi(E'), \quad E \text{ in the fission-source range.} \tag{3.4.2}$$

Thus, since $\int_0^\infty dE' \, \lambda^{-1}\nu\Sigma_f(E')\Phi(E')$ is just some constant—call it F—and since $\Sigma_t(E)$ is usually fairly constant in the fission-source range, the energy dependence of the flux in that range is, to a first approximation, proportional to that of the fission source.

This result can now be improved by using (3.4.2) as an approximation for $\Phi(E')$ in (3.4.1), yielding

$$\Sigma_t(E)\Phi(E) \approx \chi(E)F + \int_0^\infty dE' \, \Sigma_s(E' \to E) \frac{F\chi(E')}{\Sigma_t(E')}, \tag{3.4.3}$$

or

$$\begin{aligned} \Phi(E) &\approx \frac{\chi(E)F}{\Sigma_t(E)}\left[1 + \frac{1}{\chi(E)} \int_E^\infty dE' \Sigma_s(E' \to E) \frac{\chi(E')}{\Sigma_t(E')}\right] \\ &\equiv \frac{\chi(E)F}{\Sigma_t(E)}[1 + C(E)], \end{aligned} \tag{3.4.4}$$

where the lower limit on the integral has been changed from 0 to E since, in this energy range, $\Sigma_s(E' \to E) = 0$ for $E' < E$.

Equation (3.4.4) adds a correction term $C(E)$ to (3.4.2). We can thus acquire some notion of the validity of both (3.4.2) and (3.4.4) by estimating the size and energy dependence of this correction term. To illustrate the procedure, we assume that all scattering is elastic and isotropic in the center-of-mass system and that $\Sigma_s^j(E')$ and $\Sigma_t(E')$ are constant in this energy range. Then (2.6.8) and (2.6.9) yield for the correction term $C(E)$:

$$C(E) = \sum_j \frac{\Sigma_s^j}{\Sigma_t\chi(E)} \int_E^{E/\alpha_j} \frac{\chi(E')\,dE'}{E'(1 - \alpha_j)} = \frac{1}{\Sigma_t\chi(E)} \sum_j \frac{\Sigma_s^j}{1 - \alpha_j} \int_E^{E/\alpha_j} \frac{\chi(E')\,dE'}{E'}, \tag{3.4.5}$$

where the range of integration for each isotope j is over all energies from which an elastic scattering event at E' can place a neutron into the range dE about E. By using the expression (1.4.1) for $\chi(E)$, we can evaluate this expression for any desired mixture of isotopes. However, rather than going through such a procedure for any specific mixture, we shall consider only the limiting cases of hydrogen ($\alpha_j = 0$) and some heavy element ($\alpha_j \approx 1$). For $\alpha_j = 0$ we get

$$C(E) = \frac{\Sigma_s}{\Sigma_t} \frac{1}{\chi(E)} \int_E^\infty \frac{\chi(E')\,dE'}{E'}. \tag{3.4.6}$$

At high energies ($E > 0.85$ MeV) the function $[1/\chi(E)]\int_E^\infty[\chi(E')/E']\,dE'$ has a small value ($\lesssim 0.5$) that decreases as E increases. Thus (3.4.4) can be expected to provide a moderately good approximation for $\Phi(E)$, and even (3.4.2) might be satisfactory. At lower energies, however, $\chi(E)$ decreases rapidly with energy (see Figure 1.2) while $\int_E^\infty[\chi(E')/E']\,dE'$ approaches a constant value approximately equal to 0.5. Thus $C(E)$ becomes very large and even the approximation (3.4.4) is questionable.

For heavy elements the energy range E to E/α_j becomes small, and, as a rough approximation, we can replace $\chi(E')/E'$ by $\chi(E)/E$ in the integrand of (3.4.5) and obtain

$$C(E) = \frac{1}{\Sigma_t\chi(E)}\sum_j\frac{\Sigma_s^j}{1-\alpha_j}\int_E^{E/\alpha_j}\frac{\chi(E)dE'}{E} = \sum_j\frac{\Sigma_s^j}{\alpha_j\Sigma_t}. \tag{3.4.7}$$

This estimate shows that the correction factor $C(E)$ for heavy elements is a constant of order one. Consequently (3.4.4) and (3.4.2) both predict the same energy dependence for $\Phi(E)$, and we expect the approximation (3.4.2) to be fairly good.

The overall conclusion is that the energy dependence of the scalar flux density $\Phi(E)$ is roughly proportional to the fission spectrum $\chi(E)$ for energies above about 0.5 MeV. For mixtures containing hydrogen it is probably better to use (3.4.4) in this range. Below this range, particularly if light elements are present, both approximations are questionable.

One conclusion that follows immediately from these results is that it would be preferable, in computing the energy-group parameters (3.3.3) for $E \gtrsim 0.5$, to assume that $\Phi(E) \approx \chi(E)$ rather than using the histogram approximation (3.3.5). The former approach is equivalent to assuming that the "shape" (in energy) of $\Phi(E)$ is well approximated by *pieces* of the shape of $\chi(E)$ over *limited* energy ranges ΔE_g. This is a much milder assumption than requiring the two shapes to be similar over the *entire* fission energy range.

Approximate Solutions for $\Phi(E)$ in the Slowing-Down Energy Range

The slowing-down energy range is generally defined as extending from the energy below which few fission neutrons appear ($\sim$50,000 eV) to the top of the thermal energy range ($\sim$1 eV), below which up-scattering becomes important. Thus we assume that in the slowing-down energy range $\chi(E) = 0$ and $\Sigma_s(E' \to E) = 0$ for $E' < E$. The general equation (3.4.1) for slowing down then becomes

$$\Sigma_t(E)\Phi(E) = \int_E^\infty \Sigma_s(E' \to E)\Phi(E')\,dE', \quad E \text{ in the slowing-down range.} \tag{3.4.8}$$

All of the approximate analytic solutions that we shall obtain for this equation will be restricted to the case of elastic scattering isotropic in the center-of-mass system. Hence we immediately make use of (2.6.7) to simplify $\Sigma_s(E' \to E)$. This yields:

$$\Sigma_t(E)\Phi(E) = \sum_j \int_E^{E/\alpha_j} \frac{\Sigma_s^j(E')\Phi(E')\,dE'}{E'(1-\alpha_j)}, \tag{3.4.9}$$

where we have explicitly written the upper limits of the integrals as the highest energies (E/α_j) from which a neutron, scattered elastically and isotropically in the center-of-mass system from isotope j, can reach the energy E in a single collision. We shall find approximate solutions for this equation for five different situations:

a. moderation by hydrogen only;
b. no neutron absorption;
c. weak absorption;
d. absorption by narrow, widely spaced reasonances; and
e. moderation by heavy elements only.

Moderation by Hydrogen Only. If we take the mass of all isotopes j, except hydrogen, to be infinite, we see from the definition (2.5.17) of α_j that for hydrogen $\alpha_H = 0$, while for all other isotopes $\alpha_j = 1$. Hence, for these other isotopes, the lowest energy ($\alpha_j E$) attainable as the result of an elastic-scattering collision from initial energy E is E itself. That is, the scattering events lead to no energy loss. As a result the range of integration in (3.4.9) tends to zero, and we can legitimately assume that, throughout this vanishingly small range,

$$\frac{\Sigma_s^j(E')\Phi(E')}{E'} \to \frac{\Sigma_s^j(E)\Phi(E)}{E}.$$

Thus (3.4.9) becomes

$$\begin{aligned}\Sigma_t(E)\Phi(E) &= \int_E^{E/\alpha_H} \frac{\Sigma_s^H(E')\Phi(E')\,dE'}{E'(1-\alpha_H)} + \sum_{j\neq H} \int_E^{E/\alpha_j} \frac{\Sigma_s^j(E)\Phi(E)\,dE'}{E(1-\alpha_j)} \\ &= \int_E^{\infty} \frac{\Sigma_s^H(E')\Phi(E')\,dE'}{E'} + \sum_{j\neq H} \frac{1}{\alpha_j}\Sigma_s^j(E)\Phi(E),\end{aligned} \tag{3.4.10}$$

where $\Sigma_s^H(E')$ is the macroscopic scattering cross section for hydrogen.

If we now make use of the fact that $\Sigma_t(E) = \Sigma_a(E) + \Sigma_s^H(E) + \sum_{j\neq H}\Sigma_s^j(E)$ and go to the limit of $\alpha_j = 1$, we get

$$[\Sigma_a(E) + \Sigma_s^H(E)]\Phi(E) = \int_E^{\infty} \frac{\Sigma_s^H(E')\Phi(E')\,dE'}{E'}. \tag{3.4.11}$$

As might have been expected, the heavy-element scattering, which has been assumed not to degrade the neutron energy, has disappeared from both sides of the equation.

Equation (3.4.11) can be turned into a differential equation by the simple procedure of taking the derivative of both sides with respect to energy. The result is

$$\frac{d}{dE}\{[\Sigma_a(E) + \Sigma_s^H(E)]\Phi(E)\} = -\frac{\Sigma_s^H(E)\Phi(E)}{E}. \tag{3.4.12}$$

Dividing both sides by $[\Sigma_a(E) + \Sigma_s^H(E)]\Phi(E)$ and integrating from some lower energy E in the slowing-down region to some higher energy E_1 then yields

$$\ln\frac{[\Sigma_a(E_1) + \Sigma_s^H(E_1)]\Phi(E_1)}{[\Sigma_a(E) + \Sigma_s^H(E)]\Phi(E)} = -\int_E^{E_1}\frac{\Sigma_s^H(E')\,dE'}{[\Sigma_a(E') + \Sigma_s^H(E')]E'} \tag{3.4.13}$$

or

$$\begin{aligned}[\Sigma_a(E) + \Sigma_s^H(E)]\Phi(E) &= [\Sigma_a(E_1) + \Sigma_s^H(E_1)]\Phi(E_1)\exp\left(\int_E^{E_1}\frac{\Sigma_s^H(E')\,dE'}{[\Sigma_a(E') + \Sigma_s^H(E')]E'}\right)\\ &= [\Sigma_a(E_1) + \Sigma_s^H(E_1)]\Phi(E_1)\exp\left(\int_E^{E_1}\frac{dE'}{E'} - \int_E^{E_1}\frac{\Sigma_a(E')\,dE'}{[\Sigma_a(E') + \Sigma_s^H(E')]E'}\right).\end{aligned} \tag{3.4.14}$$

Since $\exp(\ln(E_1/E)) = E_1/E$, we obtain as the final result

$$\Phi(E) = \frac{[\Sigma_a(E_1) + \Sigma_s^H(E_1)]E_1\Phi(E_1)}{[\Sigma_a(E) + \Sigma_s^H(E)]E}\exp\left(-\int_E^{E_1}\frac{\Sigma_a(E')\,dE'}{[\Sigma_a(E') + \Sigma_s^H(E')]E'}\right), \tag{3.4.15}$$

under the assumption that the masses of all isotopes other than hydrogen are infinite. Thus, with hydrogen the only moderator, the scalar flux density, assumed to have a value $\Phi(E_1)$ at some initial high energy, rises roughly as E_1/E as the energy decreases. The overall $1/E$ behavior is modified by the presence of $[\Sigma_a(E) + \Sigma_s^H(E)]$ in the denominator and by the exponential term that accounts for the decrease in flux at lower energies due to neutron absorption in the range E to E_1.

This latter term leads to an interesting effect when the system contains materials having large narrow resonances. (See Figure 3.2.) Under such circumstances $\Sigma_a(E)$ will be very large for E in the range of a resonance, and the term $[\Sigma_a(E) + \Sigma_s^H(E)]$ will cause the flux to dip at energies corresponding to the resonance. At energies just below the resonance (where $\Sigma_a(E)$ once more becomes small) the flux $\Phi(E)$ will again rise. Physically such behavior is due to the fact that, in collisions with hydrogen, neutrons of initial energy E can emerge with any energy in the range 0 to E; hence there is little likelihood that they will emerge in the energy range corresponding to a given narrow resonance. Thus the neutron population at energies just below an absorption resonance will build up because of the appearance of neutrons scattered from energies above the range of the resonance.

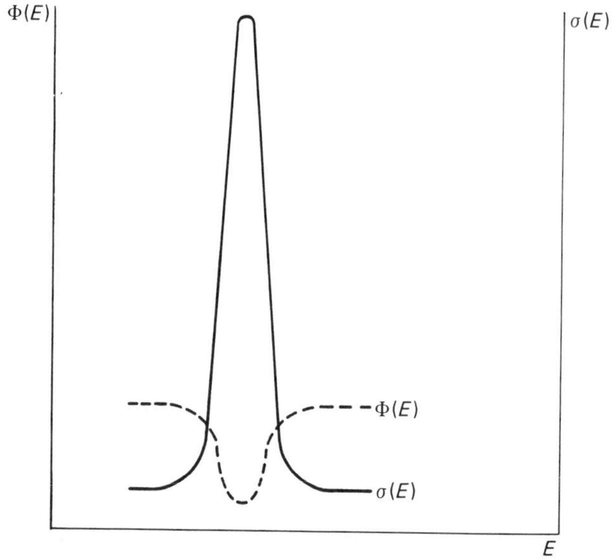

Figure 3.2 Behavior of the neutron flux density in the energy range about a large absorption resonance (in an infinite homogeneous medium, with hydrogen the only moderator).

Of course, since a resonance removes neutrons from the system, there will be *some* lasting effect on $\Phi(E)$. Mathematically this is accounted for through the exponential term

$$\exp\left(-\int_E^{E_1} \frac{\Sigma_a(E')dE'}{[\Sigma_a(E') + \Sigma_s^H(E')]E'}\right).$$

Even here, however, the magnitude of the effect is kept moderately small by the fact that $\Sigma_a(E')$ appears in the denominator as well as in the numerator of the integrand. To see this point more clearly, suppose that one of the isotopes in the reactor has a very large resonance absorption cross section, which we shall take to be infinite between E and $E + \Delta E$ and zero outside this range. (ΔE thus corresponds to the width of the resonance.) We have, then,

$$\exp\left(-\int_E^{E_1} \frac{\Sigma_a(E')\,dE'}{[\Sigma_a(E') + \Sigma_s^H(E')]E'}\right) = \exp\left(-\int_E^{E+\Delta E} \frac{dE'}{E'}\right). \tag{3.4.16}$$

Thus the attenuation in $\Phi(E)$ in the energy range E to $E + \Delta E$ is

$$\exp\left(-\ln\frac{E + \Delta E}{E}\right) = \frac{E}{E + \Delta E}.$$

For the first large absorption resonance at 6.67 eV in U^{28}, the unbroadened width of the resonances is 0.027 eV. The attenuation due to this resonance would be 6.67/6.697 ≈ 0.96. Hence only about 4 percent of the neutrons slowing down in a hydrogen medium would be captured in this resonance even if its absorption cross section were infinite. Again the physical reason for this is that only 4 percent of the neutrons slowing down ever acquire energies in this particular 0.027-eV energy range. (With $\Sigma_a \to \infty$ in that range, all of them that do appear in the range are absorbed.) This type of saturation effect is known as "energy self-shielding." It occurs whenever $\Sigma_a \gg \Sigma_s^H$.

We see then that the analytical solution (3.4.15) provides us with considerable insight into the behavior of $\Phi(E)$ in the slowing-down range. The fact that it accounts for the absorption effects of all isotopes in the mixture is its most valuable asset. Unfortunately it is of little quantitative value since the neglect of all heavy-element slowing down leads to significant errors for most compositions encountered in practical situations. Accordingly, to get some feeling for the effect of elastic heavy-element scattering, we turn to another approximation.

No Neutron Absorption. With $\Sigma_a(E) = 0$, Equation (3.4.9) becomes

$$\Sigma_s(E)\Phi(E) = \sum_j \int_E^{E/\alpha_j} \frac{\Sigma_s^j(E')\Phi(E')\,dE'}{E'(1-\alpha_j)} = \sum_j \Sigma_s^j(E)\Phi(E). \tag{3.4.17}$$

The trick of converting this integral equation into a differential equation analogous to (3.4.12) will not work here since the upper limit of the integrals for $j \neq \mathrm{H}$ are functions of E. However we can apply some of the information obtained in the last section to a situation where some isotope other than hydrogen is the only moderator. Specifically we can use (3.4.15) with $\Sigma_a(E) = 0$ to show that

$$\Phi(E) = \frac{\Sigma_s^H(E_1)E_1\Phi(E_1)}{\Sigma_s^H(E)E} \tag{3.4.18}$$

for hydrogen moderation and the assumption that all isotopes other than hydrogen have infinite mass.

This result will have to hold also for (3.4.17) if hydrogen is the only moderator. As a consequence it seems plausible to assume that in a mixture of isotopes j, only one of which ($j = k$) is a moderator (the others being assumed to have infinite mass),

$$\Phi(E) = \frac{C_k}{E\Sigma_s^k(E)}, \tag{3.4.19}$$

where C_k is a constant expected to depend on the properties of the moderating isotope. (For hydrogen (3.4.18) shows us that $C_k = C_H = \Sigma_s^H(E_1)E_1\Phi(E_1)$.)

To verify that (3.4.19) is a solution of (3.4.17), we substitute. The result is

$$\sum_j \Sigma_s^j(E)\Phi(E) = \sum_{j\neq k} \Sigma_s^j(E)\Phi(E) + \frac{C_k}{E}$$
$$= \sum_{j\neq k} \lim_{\alpha_j \to 1} \int_E^{E/\alpha_j} \frac{\Sigma_s^j(E')\Phi(E')\,dE'}{E'(1-\alpha_j)} + \int_E^{E/\alpha_k} \frac{C_k\,dE'}{(E')^2(1-\alpha_j)}. \tag{3.4.20}$$

The argument used to derive (3.4.10) again removes the nonmoderating, scattering term, so that we get

$$\frac{C_k}{E} = \frac{C_k}{1-\alpha_k}\left(-\frac{\alpha_k}{E} + \frac{1}{E}\right) = \frac{C_k}{E}. \tag{3.4.21}$$

Thus (3.4.19) is a possible solution for this case. We have not however, shown that it is the *only* possible solution. Further analysis would show that (3.4.19) is a valid solution at energies sufficiently far below the energy at which the fission-source term $\chi(E)\,dE \int_0^\infty \nu\Sigma_f(E')\Phi(E')\,dE'$ becomes negligible in comparison with the source of neutrons scattered into dE from higher energies. Thus (3.4.19) is a valid solution for this simplified nonabsorbing, single-scatterer case throughout the slowing-down region except at its upper range, close to energies at which $\chi(E)$ is still significant. (See Problem 13.)

It is possible to relate the constant C_k to a quantity of direct physical significance. This quantity is the *slowing-down density*, symbolized by $q(\mathbf{r}, E, t)$ and defined in general by:

Number of neutrons per sec in dV about $\mathbf{r}$ whose energies drop below E at time t
$= q(\mathbf{r}, E, t)\,dV.$ (3.4.22)

Note that, in contrast to $\Phi(\mathbf{r}, E)$, the slowing-down density is a density only in space and *not* in energy.

For elastic steady-state conditions in an infinite homogeneous medium and with scattering isotropic in the center-of-mass system, $q(\mathbf{r}, E, t)$ depends only on E, and a general relationship between $q(E)$ and $\Phi(E)$ can be obtained by direct calculation. Thus, referring to Figure 3.3, we first compute the rate at which neutrons scatter from dE' into dE''. Equation (2.6.9) shows that this is given by:

Rate of scattering from dE' to dE''

$$= \sum_j \Sigma_s^j(E')\frac{dE''}{E'(1-\alpha_j)}\Phi(E')\,dE'. \tag{3.4.23}$$

For fixed E' we thus get:

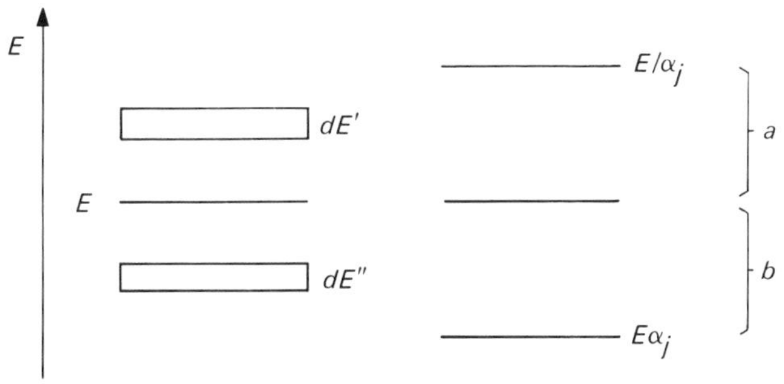

Figure 3.3 Energy intervals of interest in computing the slowing-down density: a is the range from which neutrons can slow down past E by colliding with isotope j; b is the range into which neutrons having an initial energy $\geq E$ can slow down by colliding with isotope j.

Rate at which neutrons scatter from dE' to some energy below E

$$= \sum_j \left[\int_{\alpha_j E'}^{E} \Sigma_s^j(E') \frac{dE''}{E'(1 - \alpha_j)} \right] \Phi(E')\, dE', \tag{3.4.24}$$

where the lower limit of integration is $\alpha_j E'$, the lowest energy attainable when a neutron of initial energy E' is scattered by isotope j.

Finally, by integrating over all energies E' in the range E to E/α_j from which scattered neutrons can fall below E, we get

$$q(E) = \sum_j \int_{E}^{E/\alpha_j} dE' \int_{\alpha_j E'}^{E} dE'' \frac{\Sigma_s^j(E')\Phi(E')}{E'(1 - \alpha_j)}. \tag{3.4.25}$$

This expression is valid for any mixture of isotopes, and it remains valid even if there is neutron absorption.

For the special case of moderation by only one isotope, we shall use the symbol q_k for $q(E)$ since all the neutrons that lose energy do so as a result of scattering interactions with isotope k. Moreover, since there is no absorption, q_k will turn out to be the same at all energies. To see this, and to find C_k in terms of q_k, we substitute (3.4.19) into (3.4.25), restricting the sum to the single term $j = k$. The result is

$$\begin{aligned} q_k &= \int_{E}^{E/\alpha_k} dE' \int_{\alpha_k E'}^{E} \frac{C_k dE''}{(E')^2(1 - \alpha_k)} \\ &= C_k\left[1 + \frac{\alpha_k \ln \alpha_k}{1 - \alpha_k}\right] = C_k \xi_k, \end{aligned} \tag{3.4.26}$$

where ξ_k, defined by (2.5.26), is the average loss in the logarithm of the energy per collision with isotope k.

Equation (3.4.19) then yields a very important relationship:

$$\Phi(E) = \frac{q_k}{E\xi_k \Sigma_s^k(E)}, \tag{3.4.27}$$

for an infinite homogeneous medium with no absorption, with moderation due to just one elastic scatterer, and with scattering isotropic in the center-of-mass system.

It appears to be impossible to generalize this expression rigorously to the case of many moderators since (3.4.19) cannot be simultaneously valid for many isotopes unless $C_k/\Sigma_s^k(E)$ is the same function of energy for all k. If, however, such behavior for the $\Sigma_s^k(E)$ is assumed, a straightforward generalization *is* possible and yields

$$\Phi(E) = \frac{q}{E\sum_j \xi_j \Sigma_s^j(E)} \tag{3.4.28}$$

for an infinite homogeneous medium with no absorption, with the energy behavior of $\Sigma_s^j(E)$ the same for all j and with scattering elastic and isotropic in the CM system.

(Here $q = \sum_j q_j$ is the total slowing-down density.)

If we define an average loss in the logarithm of the neutron energy for a composition of several isotopes by

$$\bar{\xi}(E) = \frac{\sum_j \xi_j \Sigma_s^j(E)}{\Sigma_s(E)}, \tag{3.4.29}$$

(3.4.28) may be written

$$\Phi(E) = \frac{q}{E\bar{\xi}(E)\Sigma_s(E)}. \tag{3.4.30}$$

Note that, since all the $\Sigma_s^j(E)$ have the same energy dependence, $\bar{\xi}(E)$ is actually a constant. However, since a later approximation will permit us to relax the condition that all the $\Sigma_s^j(E)$ must have the same energy dependence, we shall continue to indicate that $\bar{\xi}$ is a function of energy.

The simplest and most common way of meeting the requirement that the $\Sigma_s^j(E)$ have the same energy behavior is to approximate them by constants. When this is done, both Σ_s and $\bar{\xi}$ become independent of energy, and we get the result that, with no absorption and with constant scattering cross sections, the scalar flux density is proportional to $1/E$ in the slowing-down region.

This general $1/E$ behavior is characteristic of the flux in the slowing-down energy range. It arises basically from the kinematics of the scattering interaction. As we have seen, though, it gets distorted by the energy behavior of the scattering cross sections and by neutron-absorption processes. To obtain some further insight into this latter cause of flux distortion, we shall next modify (3.4.28) to account for absorption—first weak absorption, then absorption in isolated narrow resonances.

Weak Absorption. We have already noted that, with no neutron absorption, the slowing-down density q in an infinite homogeneous medium is a constant at all energies. Clearly, for a steady-state situation in a slowing-down energy range within which $\chi(E)$ is effectively zero, this constancy of q must be valid on physical grounds, even for arbitary behavior of the $\Sigma_s^j(E)$.

Just as clearly, if we include the effects of neutron absorption, the slowing-down density will become dependent on energy, decreasing as energy decreases. In order to account for this effect we express in mathematical terms the physical fact that the decrease in $q(E)$ between energies E and $E - dE$ equals in magnitude the absorption rate $\Sigma_a(E)\Phi(E)\,dE$ within this range. Thus

$$q(E) - [q(E) - dq(E)] = \Sigma_a(E)\Phi(E)\,dE \tag{3.4.31}$$

or

$$\frac{dq(E)}{dE} = \Sigma_a(E)\Phi(E). \tag{3.4.32}$$

To determine $q(E)$ and $\Phi(E)$ we need another equation relating them. On the grounds that the absorption is small we take (3.4.30) (derived by assuming $\Sigma_a = 0$) for this desired relationship. It is difficult to judge the validity of this approximation, but the following physical argument makes it plausible.

Note that (3.4.30) can be written

$$\Sigma_s(E)\Phi(E)\,dE = q\frac{d(\ln E)}{\bar{\xi}(E)}\,. \tag{3.4.33}$$

The left-hand side of this equation is the scattering rate in dE expressed in terms of the flux. On the right-hand side, since $\bar{\xi}(E)$ is approximately the average total loss in the logarithm of the neutron energy for collisions occurring at E and $d(\ln E)$ is the logarithmic energy width of the energy range in question, $d(\ln E)/\bar{\xi}(E)$ is approximately the fraction of those neutrons slowing down past E that lose energy (i.e., that scatter) in dE. Thus the right-hand side is also the scattering rate in dE. Moreover, since energy losses can occur only through scattering events, this mathematical expression ought to remain approximately valid even in the presence of absorption, when q becomes dependent on energy. Accordingly, since the left-hand side of (3.4.33) is a completely valid expression for the scattering rate in dE, with or without absorption, it seems reasonable to continue to use that equation even when some absorption occurs. We therefore combine (3.4.32) and (3.4.30), with q now assumed to depend on energy, to obtain

$$\frac{dq(E)}{dE} = \frac{\Sigma_a(E)q(E)}{E\bar{\xi}(E)\Sigma_s(E)}. \tag{3.4.34}$$

Dividing by $q(E)$ and integrating from E to E_1 yields

$$q(E) = q(E_1)\exp\left(-\int_E^{E_1} \frac{\Sigma_a(E')\,dE'}{E'\bar{\xi}(E')\Sigma_s(E')}\right). \tag{3.4.35}$$

Again making use of (3.4.30) to relate the slowing-down density to the flux, we get

$$\Phi(E) = \frac{E_1\bar{\xi}(E_1)\Sigma_s(E_1)\Phi(E_1)}{E\bar{\xi}(E)\Sigma_s(E)}\exp\left(-\int_E^{E_1} \frac{\Sigma_a(E')\,dE'}{E'\bar{\xi}(E')\Sigma_s(E')}\right) \tag{3.4.36}$$

for the case of weak absorption and with the assumption of identical energy dependences for all $\Sigma_s^j(E)$.

The result is quite similar to the corresponding one (3.4.15) for hydrogen moderation only. The chief difference lies in the fact that the absorption cross section appears only in the numerator of the integrand of the exponential attenuation factor. Thus the resonance energy self-shielding effects discussed in connection with the hydrogen moderator will not appear. This deficiency is, however, of little significance here since (3.4.36) is applicable only to weak absorbers.

Narrow, Widely Space Absorption Resonances. The expression (3.4.32) for dq/dE was derived on physical grounds and is valid for any behavior of the absorption cross section. However the argument used to justify (3.4.33) for weak absorption is no longer valid for energies in the range of large absorption resonances since, for an infinitesimal energy interval dE in a range with a high absorption resonance, the scattering-*in* rate given by the right-hand side of (3.4.33) no longer equals the scattering-out rate.

To treat this difficulty mathematically, we first note that, if the width of the resonance is small compared to the average energy lost when a neutron collides, most of the neutrons that fall into an interval dE within the width of the resonance have been scattered from an energy outside the range of the resonance. Thus, if we compute the average loss in the logarithm of the energy ($\bar{\xi}(E)$) using scattering cross sections that neglect resonance scattering, the ratio $d(\ln E)/\bar{\xi}(E)$ ought still to give approximately the fraction of neutrons slowing down that appear in the interval dE. (Such an assumption does minimal violence to the assumption of an effectively constant scattering cross section on which (3.4.33) is based.) Thus we argue that, with $\bar{\xi}(E)$ computed without taking scattering resonances into consideration, $q(E)d(\ln E)/\bar{\xi}(E)$ still represents the number of neutrons per cc per sec appearing in dE, even when dE is within the energy range of a resonance so that $q(E)$ may be changing fairly rapidly because of neutron absorption. We can then write a mathematical statement of the fact that the total number of interactions per second (absorptions plus scattering) in dE equals the total rate at which neutrons appear in dE. That is

$$[\Sigma_a(E) + \Sigma_s(E)]\Phi(E)\,dE = \frac{q(E)d(\ln E)}{\bar{\xi}(E)} = \frac{q(E)\,dE}{E\bar{\xi}(E)}. \tag{3.4.37}$$

Combining this result with (3.4.32) yields

$$\frac{dq(E)}{dE} = \frac{\Sigma_a(E)q(E)}{E\bar{\xi}(E)[\Sigma_a(E) + \Sigma_s(E)]}. \tag{3.4.38}$$

The solution of (3.4.38) is

$$q(E) = q(E_1)\exp\left(-\int_E^{E_1} \frac{\Sigma_a(E')\,dE'}{E'\bar{\xi}(E')[\Sigma_a(E') + \Sigma_s(E')]}\right), \tag{3.4.39}$$

where E_1 is an arbitrary reference energy just above the resolved-resonance range, or, with (3.4.37) used to eliminate $q(E)$,

$$\Phi(E) = \frac{[\Sigma_a(E_1) + \Sigma_s(E_1)]E_1\bar{\xi}(E_1)\Phi(E_1)}{[\Sigma_a(E) + \Sigma_s(E)]E\bar{\xi}(E)}\exp\left(-\int_E^{E_1} \frac{\Sigma_a(E')\,dE'}{E'\bar{\xi}(E')[\Sigma_a(E') + \Sigma_s(E')]}\right). \tag{3.4.40}$$

Note the great similarity between this result and (3.4.15) for hydrogen moderation. The previous discussion about energy self-shielding applies immediately. Moreover the fact that the two formulas are so similar provides some degree of assurance that the assumptions made in deriving (3.4.40) may be legitimate. Note also that, if $\Sigma_a(E')$ is sufficiently small that it can be neglected in comparison with $\Sigma_s(E)$ at all energies, (3.4.40) reduces to the weak-absorber result (3.4.36). Thus we may regard (3.4.36) as a special case of (3.4.40) and use the latter result for both weak and resonance absorbers.

Moderation by Heavy Elements: The Age Approximation. The trick of converting the integral equation (3.4.9) for the scalar flux density in the slowing-down energy range into a differential equation works only for hydrogen ($\alpha_H = 0$) since only then is the upper limit of the integral not a function of E. There is, however, another approach for converting to a differential equation. In its simplest form it is called the *age approximation* because, when the same idea is applied to the case of a *finite* homogeneous medium, it yields the "Fermi age equation," originally given its name because it has the same mathematical form as the time-dependent heat-transfer equation. (Fermi called the quantity analogous to time in the heat-transfer equation the neutron "age.") We shall define the neutron age precisely and discuss the age equation in Chapter 4; however the term "age approximation" is so common for the method we are about to develop that it seems appropriate to introduce it at this point.

We start by returning to (3.4.9), the basic equation for the slowing-down region in an infinite homogeneous medium:

$$\Sigma_t(E)\Phi(E) = \sum_j \int_E^{E/\alpha_j} \frac{\Sigma_s^j(E')\Phi(E')\,dE'}{E'(1-\alpha_j)}, \tag{3.4.9}$$

for elastic scattering isotropic in the center-of-mass system.

The difficulty in dealing with this equation arises from the fact that the energy dependence of $\Phi(E')$ and hence of $\Sigma_s^j(E')\Phi(E')$ is not known a priori. However, from the results developed so far (see (3.4.15), (3.4.28), and (3.4.40)), we expect that, over ranges where the absorption cross section is a fairly smooth function of energy, $E'\Sigma_s^j(E')\Phi(E')$ will be a slowly varying function of energy. In fact, for $\Sigma_a(E') = 0$, all the equations just cited show that $E'\Sigma_s^j(E')\Phi(E')$ is a constant that is independent of E'. Thus, for energies E' fairly close to E, we should be able to approximate $E'\Sigma_s^j(E')\Phi(E')$ by using only a few terms of the Taylor-series expansion about the energy E:

$$E'\Sigma_s^j(E')\Phi(E') = E\Sigma_s^j(E)\Phi(E) + \left\{\frac{d}{dE}[E\Sigma_s^j(E)\Phi(E)]\right\}(E'-E) + \frac{1}{2}\left\{\frac{d^2}{dE^2}[E\Sigma_s^j(E)\Phi(E)]\right\}(E'-E)^2 + \cdots. \tag{3.4.41}$$

If a series expansion is to be employed, it is desirable that the coefficients multiplying $(E'-E)^n$ $(n = 0, 1, 2, \ldots)$ all have the same dimension. Otherwise the relative sizes of the coefficients will depend on the units used. The series (3.4.41) is deficient in this respect. Since $E\Sigma_s^j(E)\Phi(E)$ has the units $\text{cm}^{-3}\,\text{sec}^{-1}$ the coefficient multiplying $(E'-E)^n$ has the units $\text{cm}^{-3}\,\text{sec}^{-1}\,(\text{energy})^{-n}$. To avoid this inconvenience it is necessary to expand in a dimensionless parameter rather than in powers of $(E'-E)$. For this reason, in developing approximations of this type it is customary to expand in powers of $(\ln E' - \ln E)$, which is equivalent to $\ln(E'/E)$.

Accordingly we shall get around the difficulty of not knowing the energy dependence of $\Phi(E')$ in (3.4.9) by using the expansion

$$E'\Sigma_s^j(E')\Phi(E') = E\Sigma_s^j(E)\Phi(E) + \left\{\frac{d}{d(\ln E)}[E\Sigma_s^j(E)\Phi(E)]\right\}(\ln E' - \ln E) + \frac{1}{2}\left\{\frac{d^2}{d(\ln E)^2}[E\Sigma_s^j(E)\Phi(E)]\right\}(\ln E' - \ln E)^2 + \cdots. \tag{3.4.42}$$

The "age approximation" results when only the first two terms of this series are retained. We anticipate, then, that the approximation will be valid when $\Sigma_a(E)$ is a fairly smooth function of energy and when the total energy range $[(E/\alpha_j) - E]$ throughout which the expansion must be applied is small. Thus we do not expect the age approximation to be valid in the range of large narrow resonances or when $[(E/\alpha_j) - E]$ is large (as it would be in the case of hydrogen or deuterium moderators).

Inserting the first two terms of (3.4.42) into the integrands of (3.4.9) yields

$$\begin{aligned}\Sigma_t(E)\Phi(E) &= \sum_j \int_E^{E/\alpha_j} \frac{dE'}{(E')^2(1-\alpha_j)}\left\{E\Sigma_s^j(E)\Phi(E) + \ln(E'/E)\frac{d}{d(\ln E)}[E\Sigma_s^j(E)\Phi(E)]\right\}\\ &= \sum_j \left\{\Sigma_s^j(E)\Phi(E) + \frac{1}{E}\left(1 + \frac{\alpha_j \ln \alpha_j}{1-\alpha_j}\right)\frac{d}{d(\ln E)}[E\Sigma_s^j(E)\Phi(E)]\right\}. \qquad (3.4.43)\end{aligned}$$

The definitions (2.5.26) and (3.4.29) lead to the final result

$$\Sigma_t(E)\Phi(E) = \Sigma_s(E)\Phi(E) + \frac{d}{dE}[E\xi(E)\,\Sigma_s(E)\Phi(E)], \qquad (3.4.44)$$

where, since the age approximation does not require that all the $\Sigma_s^j(E)$ have the same energy dependence, the indication that ξ depends on E is now necessary. Thus making the age approximation has permitted us to convert the integral equation (3.4.9) into a simple differential equation. Integration of this latter result then gives

$$\Phi(E) = \frac{E_1\xi(E_1)\Sigma_s(E_1)\Phi(E_1)}{E\xi(E)\Sigma_s(E)} \exp\left(-\int_E^{E_1} \frac{\Sigma_a(E')\,dE'}{E'\xi(E')\Sigma_s(E')}\right) \qquad (3.4.45)$$

for the case of no resonance absorption and with no hydrogen present.

This formula for the scalar flux density in the slowing-down region is identical with (3.4.36). The difference between the two lies in the different restrictions needed to justify the derivations. To obtain (3.4.36) we had to assume that the absorption was weak and that the dependence on energy of all the $\Sigma_s^j(E)$ was the same; there was, however, no restriction to heavy-element scattering. The justification of (3.4.45) also requires that the absorption be weak—or, more precisely, that $\Sigma_a(E)$ be a smooth function of energy. However it puts no requirement (other than smoothness) on the $\Sigma_s^j(E)$. On the other hand the derivation of (3.4.45) makes it questionable whether the formula is valid for light elements, particularly hydrogen.

All of this illustrates a situation that is encountered fairly frequently: the same result may be obtained by several different approaches based on rather different assumptions. Actually such a situation is a happy one. The more ways there are for looking at a given approximate result, the more understanding can be developed about its range of validity. In the present instance our use of a variety of starting assumptions has led to the conclusion that (3.4.40), with resonance scattering ignored in the computation of $\bar{\xi}(E)$, should be a valid approximation whether or not there is resonance scattering, whether or not there is hydrogen moderation, and whether or not all isotopic scattering cross sections have the same energy dependence.

Approximate Solutions for $\Phi(E)$ in the Thermal Energy Range

The thermal energy range ($E \leq 1$ eV) is characterized by the fact that, because of the thermal motion of the target nuclei, a colliding neutron can gain energy. Thus the elastic and inelastic scattering probabilities $f_s^j(E' \to E)\,dE$ and $f_i^j(E' \to E)\,dE$ defined by (2.4.13) do not become zero for $E' < E$. The calculation of these probabilities involves an integration over the thermal motions of the nuclei making up the medium, and even for such a simple (and unrealistic) case as moderation by a monatomic hydrogen gas, the resulting expressions are extremely complicated. This complexity leads to no significant problems in computing the multigroup cross sections $\Sigma_{gg'}$ in accord with (3.3.6). However it makes finding an analytical expression for $\Phi(E)$ an involved mathematical procedure. Because of this situation, we shall not attempt to solve (3.4.1) for $\Phi(E)$ in the thermal energy range even in an approximate manner. Instead we shall discuss the energy distribution of the neutron flux density in that range on the basis of more general physical considerations.

The argument is a simple one. If there were no absorption of neutrons in the infinite homogeneous medium under consideration and if there were no production of neutrons due to fission, any neutron population present would neither decay nor increase. Instead the neutrons would come into thermal equilibrium with the material of the medium. They would behave like a very rare gas and would thus have a Maxwellian distribution of velocities given by

$$n(v)\,dv \propto v^2 \exp\left(-\frac{1}{2}\frac{mv^2}{kT}\right)dv, \tag{3.4.46}$$

where m and v are the mass and speed of the neutron $n(v)\,dv$ is the number of neutrons per unit volume having speeds between v and $v + dv$, T is the absolute temperature of the medium, and k is the Boltzmann constant.

The quantity $n(v)\,dv$ is also equal to the number of neutrons per cc having energies between E and $E + dE$, where $E = \frac{1}{2}mv^2$ is the kinetic energy corresponding to speed v. Thus, if we define $n(E)$, a number density per unit *energy* such that $n(E)\,dE = n(v)\,dv$, we get

$$n(E)\,dE \propto v^2 \exp\left(-\frac{1}{2}\frac{mv^2}{kT}\right)dv. \tag{3.4.47}$$

(Note that we are again using the physics convention that $n(\ \ )$ stands for a physical quantity and not a mathematical function; thus $n(E)$ and $n(v)$ have different units.)

Since the scalar flux density is $\Phi(E) = v(E)n(E)$ and $E \propto v^2$, we then find that

$$\Phi(E)\,dE \propto E\exp\left(\frac{-E}{kT}\right)dE \quad \text{if} \quad \Sigma_a(E) = 0. \tag{3.4.48}$$

This result, being based on a statistical argument, is independent of the nature of the scattering probabilities. In fact one of the requirements imposed on physically acceptable expressions for the scattering probabilities f_s^j and f_i^j is that, when they are used in the integral equation

$$\Sigma_s(E)\Phi(E) = \sum_j \int_0^\infty [\Sigma_s^j(E')f_s^j(E' \to E) + \Sigma_i^j(E')f_i^j(E' \to E)]\Phi(E')\,dE', \tag{3.4.49}$$

the solution of this equation must be the Maxwellian flux spectrum (3.4.48).

The Maxwellian distribution function $E\exp(-E/kT)$ is one of the curves shown on Figure 3.4. It becomes vanishingly small for $E \gg kT$ and has a maximum at the energy

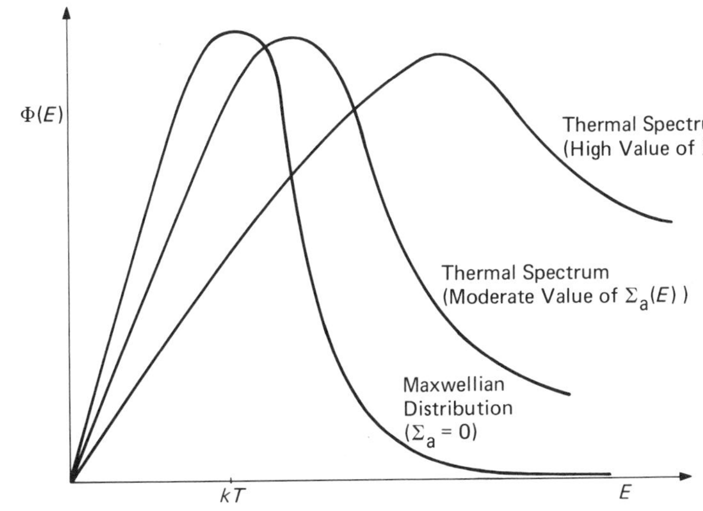

Figure 3.4 Typical behavior of flux density versus neutron energy in the thermal energy region.

$E = kT$, which is called the *most probable energy* of the Maxwellian distribution. Use of the constants $k = 8.62 \times 10^{-5}$ eV/°K; $m = 1.67 \times 10^{-24}$ gm leads to the result that, for a medium having a temperature of 20°C (293°K or 68°F), the most probable energy of a Maxwellian neutron spectrum is 0.025 eV; the corresponding speed is 2.2×10^5 cm/sec.

The distribution (3.4.48) is a completely acceptable solution for $\Phi(E)$ in an infinite source-free medium in which there is no neutron absorption. It is not, strictly speaking, an exact solution for the physical problem since it is necessary to make a number of assumptions about the system—among them that the neutrons do not interact with one another and that their energies form a continuum—in order to derive the Maxwellian velocity distribution (3.4.46) in the first place. Nevertheless, since these necessary conditions are closely approximated for the situation at hand, the solution is completely acceptable.

When absorption and fission are present, however, the basis for assuming a Maxwellian distribution is no longer valid. The exact natures of the scattering probabilities $f_s^j(E' \to E)$ and $f_i^j(E' \to E)$ become important, and (3.4.1) must actually be solved in order to determine $\Phi(E)$. With the equations cast in the energy-group form (3.3.9), this solution is quite straightforward. Fairly accurate expressions for the scattering probabilities (often called *scattering kernels*) can be found using quantum-mechanical methods, and, from these, the necessary group-to-group scattering cross sections $\Sigma_{gg'}$ can be found by applying (3.3.6) or (3.3.3), with the Maxwellian spectrum used to approximate $\Phi(E')$ in the expression for $\Sigma_{gg'}$. It has been found that use of 30 or 40 energy groups in the range 0–1 eV yields solutions for the group equations that are very little improved in accuracy by increasing the number of groups.

When solutions of this nature are determined, it is found—as would be expected—that, for values of the absorption cross section $\Sigma_a(E)$ small in comparison with $\Sigma_s(E)$, the dependence of the flux on energy in the thermal region approaches the Maxwellian distribution (3.4.48). Figure 3.4 shows several spectra corresponding to different values of the absorption cross section. The most noticeable differences in the curves come in the higher energy range of the thermal region (0.1–1 eV). Here the Maxwellian distribution goes to zero as $\exp(-E/kT)$, a factor which for $kT = 0.025$ eV is of order 10^{-18} at $E = 1$ eV. The correct solution, with slowing-down source and absorption present, falls off approximately as $1/E$ at energies in the higher portion of the thermal region. This is, of course, to be expected since we have already shown that, with no up-scattering and at energies below the fission-source region, $\Phi(E)$ takes on a $1/E$ behavior.

This qualitative discussion suggests that a rough approximation to $\Phi(E)$ in the thermal range can be obtained by assuming it to have a Maxwellian shape $KE\exp(-E/kT)$, where K is a constant, up to some energy cut point $E_c \approx 0.1$ eV, and to be proportional to $1/E$ throughout the remainder of the thermal range. To connect these two segments together and to connect the thermal range to the slowing-down range, we simply require that $\Phi(E)$ be a continuous function of energy and that neutron balance be maintained. Thus we require that the flux be continuous in energy at $E = E_2 \approx 1$ eV and at $E = E_c$ and that the total absorption rate in the energy range 0–E_2 equal the slowing-down density $q(E_2)$. If we normalize so that the $1/E$ part of the flux is $E_2\Phi(E_2)/E$, the requirement of continuity at $E = E_c$ yields

$$KE_c \exp\left(\frac{-E_c}{kT}\right) = \frac{E_2\Phi(E_2)}{E_c}, \tag{3.4.50}$$

so that, from (3.4.30) and (3.4.49),

$$\begin{aligned}\int_0^{E_c} K\Sigma_a(E)E \exp\left(\frac{-E}{kT}\right)dE + \int_{E_c}^{E_2} \frac{\Sigma_a(E)\, E_2\Phi(E_2)}{E}\, dE &= q(E_2)\\ &= \xi(E_2)\, \Sigma_s(E_2)\Phi(E_2)\, E_2\\ &= E_2\Phi(E_2)\left[\int_0^{E_c} \frac{\Sigma_a(E)E}{E_c^2} \exp\left(\frac{E_c - E}{kT}\right)dE + \int_{E_c}^{E_2} \frac{\Sigma_a(E)}{E}\, dE\right].\end{aligned} \tag{3.4.51}$$

Thus

$$\int_0^{E_c} \frac{\Sigma_a(E)E}{\xi(E_2)\Sigma_s(E_2)E_c^2} \exp\left(\frac{E_c - E}{kT}\right)dE + \int_{E_c}^{E_2} \frac{\Sigma_a(E)\, dE}{\xi(E_2)\Sigma_s(E_2)E} = 1. \tag{3.4.52}$$

The energy cut point E_c therefore depends on T, $\Sigma_a(E)$, and $\xi(E_2)\Sigma_s(E_2)$ through a transcendental equation.

The approximate value of $\Phi(E)$ that results from this procedure is then

$$\Phi(E) = \begin{cases} \dfrac{E_2 E\Phi(E_2)}{E_c^2} \exp\left(\dfrac{E_c - E}{kT}\right) & \text{for} \quad 0 \le E \le E_c, \\ \dfrac{E_2\Phi(E_2)}{E} & \text{for} \quad E_c < E \le E_2 \approx 1 \text{ eV}. \end{cases} \tag{3.4.53}$$

The slope of this flux shape will be discontinuous at $E = E_c$, but as a rough approximation the formula is adequate. It should be noted, however, that whenever absorption is sizeable or whenever there are absorption resonances in the thermal energy range, it is unlikely that (3.4.53) will give accurate values of thermal neutron activation rates. Under these circumstances there may, in fact, be no value of E_c in the range 0–E_2 that yields a solution for (3.4.52).

Summary of Analytical Approximations for $\Phi(E)$ in an Infinite Homogeneous Reactor
Combining (3.4.4), (3.4.40), and (3.4.53) yields an approximate analytical result for $\Phi(E)$ for an infinite homogeneous medium:

$$\Phi(E) \approx \begin{cases} \dfrac{\chi(E)F}{\Sigma_t(E)}\left[1 + \dfrac{1}{\chi(E)}\displaystyle\int_E^\infty dE'\Sigma_s(E' \to E)\frac{\chi(E')}{\Sigma_t(E')}\right] & \text{for } E_1 \le E \le E_0, \\ \dfrac{[\Sigma_a(E_1) + \Sigma_s(E_1)]E_1\xi(E_1)\Phi(E_1)}{[\Sigma_a(E) + \Sigma_s(E)]E\xi(E)} \exp\left(-\displaystyle\int_E^{E_1} \frac{\Sigma_a(E')dE'}{E'\xi(E')[\Sigma_a(E') + \Sigma_s(E')]}\right) & \\ \qquad \text{for } E_2 \le E \le E_1, & \\ \dfrac{E_2\Phi(E_2)}{E} \quad \text{for } E_c \le E \le E_2, & \\ \dfrac{E_2\Phi(E_2)}{E_c^2} E \exp\left(\dfrac{E_c - E}{kT}\right) \quad \text{for } 0 \le E \le E_c, & \end{cases} \tag{3.4.54}$$

where $E_0 \approx 1.5 \times 10^7$ eV, $E_1 \approx 1.0 \times 10^5$ eV, $E_2 \approx 1$ eV, E_c is given implicitly by (3.4.52), $\Phi(E_2)$ is given in terms of $\Phi(E_1)$ by the second equation in (3.4.54), and $\Phi(E_1)$ is given in terms of F by the first equation in (3.4.54).

To get a rough quantitative idea of this overall shape we arbitrarily set $F/\Sigma_t(E) = 1$ neutron/cm^2/sec for E equal to the energy E_1. At E_1 (Figure 1.2) $\chi(E)$ has a value of ~ 0.1 (eV)$^{-1}$, and, if we assume that the term $\int_E^\infty dE'\, \Sigma_s(E' \to E)\chi(E')/\Sigma_t(E')$ has the value 0.5 at E_1, we get $\Phi(E_1) = 0.6$ neutrons/cm^2/sec/eV.

If it were not for absorption, $\Phi(E)$ would rise by roughly a factor of 10^5 as the energy decreases from 10^5 to 1 eV. Absorption of neutrons can, however, cause a significant amount of attenuation throughout this range.

The exponential factor on the second line of (3.4.54) is close to unity if $\Sigma_a(E')$ is insignificant throughout the slowing-down range, but it can become small if $\Sigma_a(E')$ is sizeable. Moreover it can become *very* small if $\xi(E')$, the average loss in the logarithm of the energy per collision, is small (i.e., if there are no significant concentrations of light elements in the medium). To see this point more quantitatively we can estimate $\Phi(E_1)$ under the simplifying assumptions that $\Sigma_a(E)$, $\Sigma_s(E)$, and hence $\xi(E)$, are constants in the slowing-down range. Under these assumptions (3.4.54) yields

$$\begin{aligned} \Phi(E) &= \Phi(E_1)\frac{E_1}{E}\exp\left(-\frac{\Sigma_a}{\xi(\Sigma_a + \Sigma_s)}\ln\left(\frac{E_1}{E}\right)\right) \\ &= \Phi(E_1)\left(\frac{E_1}{E}\right)^{1-\Sigma_a/[\xi(\Sigma_a+\Sigma_s)]} \qquad \text{for } E_2 \le E \le E_1. \end{aligned} \tag{3.4.55}$$

If $\Sigma_a/(\Sigma_a + \Sigma_s) = 0.1$ and $\xi = 0.5$ (as it might be in a uranium–water mixture), we get $\Phi(E_2) = (E_1/E_2)^{0.8}\,\Phi(E_1) = 10^4\Phi(E_1)$. Thus the flux density at 1 eV in a light-water-moderated reactor, although a factor of ten smaller than what it would be with $\Sigma_a = 0$, is still much larger than at 10^5 eV. If, on the other hand, $\bar{\xi} = 0.05$ (as it might be in a sodium-cooled fast reactor), $\Phi(E_2) = 10^{-5}\Phi(E_1)$. Thus in a fast reactor the flux at 1 eV is negligibly small.

To relate the peak flux at $E = kT$ in the thermal range to $\Phi(E_2)$ we must first solve (3.4.52) for E_c, the energy cut point between the Maxwellian and $1/E$ portions of the assumed shape in the thermal range. Rather than facing this troublesome algebraic problem, we can make an estimate by simply assuming that a typical cut point would occur at $E_c = 0.1$ eV. Then (3.4.54) gives

$$\Phi(kT) = \frac{kT\Phi(E_2)}{0.01}\exp\left(\frac{0.1 - kT}{kT}\right). \tag{3.4.56}$$

For $kT = 0.025$ eV, this expression gives $\Phi(kT) = 50\Phi(E_2)$, and for $kT = 0.050$ eV ($T = 586°\text{K} \approx 595°\text{F}$), $\Phi(kT) = 13.6\Phi(E_2)$.

Figure 3.5 shows curves of the thermal portion of (3.4.54) for several values of E_c and a fixed value of the slowing-down density $q(E_2)$. Since the medium is infinite, the rate of neutron absorption in the range $0 \le E \le E_2$ must equal $q(E_2)$ (see (3.4.51)). Hence large values of $\Sigma_a(E)$ require that the function $\Phi(E)$ be smaller (so that $\int_0^{E_2}\Sigma_a(E)\Phi(E)\,dE$ can still equal $q(E_2)$). Figure 3.5 thus implies that E_c and $\Phi(kT)$ will be smaller for a heavily absorbing material.

We also expect from (3.4.51) that the cut point E_c will be at a lower energy if $\bar{\xi}(E_2)\Sigma_s(E_2)$ decreases. In fact, if $\bar{\xi}(E_2)\Sigma_s(E_2)$ is very small and if $\Sigma_a(E)$ is very large, there may be *no* value of E_c between 0 and 1 eV that will yield a solution of (3.4.52). When this happens, the representation of $\Phi(E)$ in the thermal range by a Maxwellian shape followed by a $1/E$ tail breaks down completely.

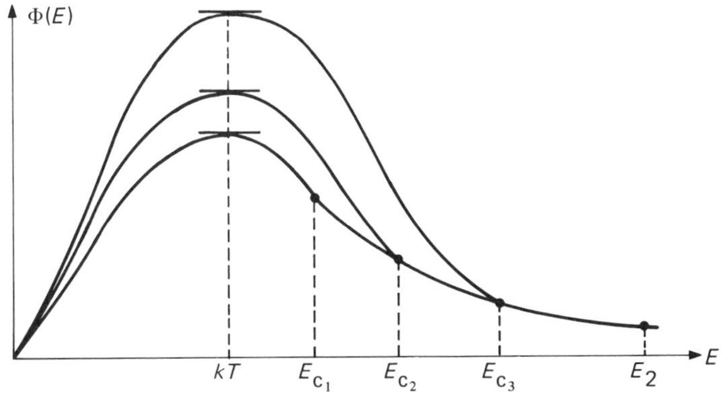

Figure 3.5 The thermal flux approximated by a Maxwellian distribution joined to a $1/E$ tail at various cut points E_c.

3.5 The Four-Factor Formula

In Section 3.3 we introduced the number λ (also called k_{eff}) which, when divided into ν, the number of neutrons emitted per fission, guarantees a nontrivial solution for the group flux equations in an infinite homogeneous medium. First we argued physically that a real positive value of λ always exists if the reactor contains some fissionable material. Then we showed mathematically that λ is, not only real and positive, but also unique. We further showed in (3.3.21) that λ can be interpreted physically as the ratio of the production rate to the absorption rate for neutrons in the infinite medium in the presence of a neutron flux corresponding to the fictitiously critical state (i.e., the solution of (3.3.9)). Finally we pointed out that the definition of λ can readily be extended to a reactor of finite size composed of many different material regions. Thus the effective multiplication constant is a property of the entire reactor. The term k_∞ was defined by (3.3.23) for a given material composition as the k_{eff} for an infinite amount of that material. We shall now consider methods for determining this important property of a given multiplying composition.

In order to compute k_∞ for a given mixture of materials we need to know detailed nuclear data over the entire energy range from 0 to 15 MeV so that we may determine the group parameters Σ_{fg}, $\Sigma_{gg'}$, etc. If the number of groups is at all sizeable, we need a computing machine.

When reactor theory was first being developed, the nuclear data was very poorly known and computing machines were very rare and, by today's standards, very primitive. Thus other methods had to be devised for determining whether a given material composition would have a multiplication constant k_∞ greater or less than one. For the thermal reactors designed at that time a very effective way of overcoming these difficulties was invented. It had the double advantage of being simple mathematically and of requiring as input the ratios of certain reaction rates averaged over large ranges of energy. These ratios could be determined fairly accurately by direct experiment. Thus the need for detailed cross-sectional values at all energies was circumvented.

This simplified method is used today only for large D_2O- or graphite-moderated thermal reactors. It is quite inappropriate for fast reactors, and it has been extensively modified for application to slightly enriched PWR's and BWR's. Nevertheless the reaction-rate ratios that form such an important part of the method are still measured as a means of validating the accuracy of detailed nuclear data. Thus it is important to know what the terms mean, even though they may not today be used directly in reactor analysis.

It is possible to develop the simple theory without approximation from (3.3.21), and we shall do essentially that. However it will make the notation simpler if we work with integrals rather than sums. Hence our first step will be to derive the analogue of (3.3.27) for the continuous-energy case. This can be done by simply integrating (3.4.1) over the entire energy range of interest. The result is

$$k_\infty = \frac{\int_0^\infty dE \int_0^\infty dE' \sum_j \chi^j(E) \nu^j \Sigma_f^j(E')\Phi(E')}{\int_0^\infty \Sigma_t(E)\Phi(E)\, dE - \int_0^\infty dE \int_0^\infty dE' \Sigma_s(E' \to E)\Phi(E')} = \frac{\int_0^\infty \nu\Sigma_f(E)\Phi(E)\, dE}{\int_0^\infty \Sigma_a(E)\Phi(E)\, dE}, \tag{3.5.1}$$

the second expression arising from the following facts:

$$\int_0^\infty dE \chi^j(E) = 1, \qquad \nu\Sigma_f(E) \equiv \sum_j \nu^j \Sigma_f^j(E),$$

$$\int_0^\infty dE\, \Sigma_s(E' \to E) = \Sigma_s(E'), \qquad \Sigma_a(E) = \Sigma_t(E) - \Sigma_s(E).$$

Equation (3.5.1) is the continuous-energy analogue of (3.3.27), and thus k_∞ can still be interpreted as discussed earlier. It is the ratio of the neutron-production rate to the neutron-absorption rate in a given infinite homogeneous medium in the presence of a flux $\Phi(E)$ that is the everywhere-positive solution of (3.4.1).

If we could somehow create the flux $\Phi(E)$ and measure the two reaction rates, we would be able to determine k_∞ by simply taking their ratio. We can almost do this by building a reactor and arranging things so that it is critical with a region of the material in question in the center of it. The flux at the center of the zone of material being examined would turn out to be fairly close to the desired $\Phi(E)$, and the fact that we are taking a *ratio* in (3.5.1) with $\Sigma_f(E)$ and $\Sigma_a(E)$ having roughly the same energy dependence would further decrease errors. Thus k_∞ could be inferred fairly accurately if it were possible to measure the neutron-production rate and the neutron-absorption rate in the material of interest.

It turns out that no simple, direct method for measuring these two rates is available. However it is possible to break up the ratio of total production to absorption into four factors, each of which is itself a ratio of reaction rates that can be measured with a fair degree of precision. We shall first write out this factorization in an exact manner and then discuss each of the four factors.

In introducing the four-factor formula at this stage, we shall be forced to treat qualitatively one extremely important fact, namely, that the formula was first introduced, not for *homogeneous* infinite media, but rather for infinite lattices of fuel rods distributed periodically in a moderator. A square array of cylindrical natural-uranium rods with a diameter of about one inch and embedded in graphite with, say, eight inches between fuel-rod centers provides an example of such a lattice. Even though we have not yet discussed the spatial dependence of $\Phi(\mathbf{r}, E)$ in a reactor, it is intuitively clear that, in an infinite array of such rods, the spatial shape of the flux will be periodic. Specifically $\Phi(\mathbf{r}, E)$ for any given square "cell" (fuel rod and its associated moderator) will be the same as that for any other cell. Thus, if we know $\Phi(\mathbf{r}, E)$ for one cell, we know it for all. (Just as, if we know $\Phi(E)$ at one *point* in an infinite homogeneous medium, we know it for all points.)

The ratio of the total production rate to the total absorption rate of neutrons for a single cell (in which $\Phi(\mathbf{r}, E)$ is that flux which would be present if the infinite lattice of cells could be made critical by dividing ν by λ) can easily be expressed as a generalization of (3.5.1). In keeping with custom, we shall continue to use the symbol k_∞ for this ratio and refer to k_∞ as the *infinite-medium multiplication constant of the lattice* or simply the *k-infinity of the lattice*. Accordingly we define

$$k_\infty \equiv \frac{\int_{\text{cell}} dV \int_0^\infty dE\, \nu\Sigma_f(\mathbf{r}, E)\Phi(\mathbf{r}, E)}{\int_{\text{cell}} dV \int_0^\infty dE\, \Sigma_a(\mathbf{r}, E)\Phi(\mathbf{r}, E)}, \tag{3.5.2}$$

where now $\nu\Sigma_f(\mathbf{r}, E)$, $\Sigma_a(\mathbf{r}, E)$, and $\Phi(\mathbf{r}, E)$ all depend on position, and the volume integral is over a single cell of the infinite array. Since $\Sigma_f(\mathbf{r}, E)$ is zero except in the fuel rod, the integral in the numerator could equally well be taken over just the fuel rod.

We now rewrite the ratio on the right-hand side of (3.5.2) as the product of four ratios as follows:

$$\begin{aligned} k_\infty &\equiv \frac{\int_{\text{fuel}} dV \int_0^\infty dE\, \nu\Sigma_f(\mathbf{r}, E)\Phi(\mathbf{r}, E)}{\int_{\text{cell}} dV \int_0^\infty dE\, \Sigma_a(\mathbf{r}, E)\Phi(\mathbf{r}, E)} \\ &= \frac{\int_{\text{fuel}} dV \int_0^\infty dE\, \nu\Sigma_f(\mathbf{r}, E)\Phi(\mathbf{r}, E)}{\int_{\text{fuel}} dV \int_0^{E_2} dE\, \nu\Sigma_f(\mathbf{r}, E)\Phi(\mathbf{r}, E)} \cdot \frac{\int_{\text{fuel}} dV \int_0^{E_2} dE\, \nu\Sigma_f(\mathbf{r}, E)\Phi(\mathbf{r}, E)}{\int_{\text{fuel}} dV \int_0^{E_2} dE\, \Sigma_a^{\text{fuel}}(\mathbf{r}, E)\Phi(\mathbf{r}, E)} \\ &\quad \cdot \frac{\int_{\text{fuel}} dV \int_0^{E_2} dE\, \Sigma_a^{\text{fuel}}(\mathbf{r}, E)\Phi(\mathbf{r}, E)}{\int_{\text{cell}} dV \int_0^{E_2} dE\, \Sigma_a(\mathbf{r}, E)\Phi(\mathbf{r}, E)} \cdot \frac{\int_{\text{cell}} dV \int_0^{E_2} dE\, \Sigma_a(\mathbf{r}, E)\Phi(\mathbf{r}, E)}{\int_{\text{cell}} dV \int_0^\infty dE\, \Sigma_a(\mathbf{r}, E)\Phi(\mathbf{r}, E)}, \end{aligned} \tag{3.5.3}$$

where E_2 is a cutoff energy (~ 1.0 eV) separating the thermal energy region (in which up-scattering is important) from the slowing-down region.

Each of the four ratios into which we have factored k_∞ has a name, and we shall discuss them one by one, indicating in a rough way how they can be measured. The definitions chosen guarantee that the product of the four factors is k_∞. There are, however, a variety of alternative, slightly different, definitions of the factors, some of which yield k_∞ only approximately when multiplied together, but which correspond to quantities easier to measure. We shall not discuss these alternate ways of factoring k_∞.

The Fast-Fission Factor ε

The first of the four factors is called the *fast-fission factor*, represented by the symbol

ε and defined as follows:

$$\varepsilon \equiv \frac{\int_{\text{fuel}} dV \int_0^\infty dE\, \nu\Sigma_f(\mathbf{r}, E)\Phi(\mathbf{r}, E)}{\int_{\text{fuel}} dV \int_0^{E_2} dE\, \nu\Sigma_f(\mathbf{r}, E)\Phi(\mathbf{r}, E)}. \tag{3.5.4}$$

It is thus the ratio of the total rate of production of fission neutrons in the lattice to the rate of production due to fissions in the thermal energy range. It is clear from the definition that ε depends directly only on the $\nu^j\sigma_f^j(E)n^j(\mathbf{r}, E)$ of the various fissionable isotopes contained in the fuel rod.

There is, of course, an important *indirect* dependence on the other constituents of the medium through the fact that $\Phi(\mathbf{r}, E)$ within the fuel rod depends on them. Nevertheless, if a good approximation to $\Phi(\mathbf{r}, E)$ can be constructed experimentally, ε can be measured by finding the ratio of total to thermal fission-neutron production rates in a small sample of fuel (for example a foil) irradiated by the flux. The actual fuel-rod or lattice geometry need not be simulated. Moreover, if we assume that ν^j is a constant and the same for all the fissionable isotopes in the fuel, or if we make a theoretical correction to the data to account for variations in the ν^j, a measurement of the ratio of total to thermal fission rates,

$$\frac{\int_{\text{fuel}} dV \int_0^\infty dE\, \Sigma_f(\mathbf{r}, E)\Phi(\mathbf{r}, E)}{\int_{\text{fuel}} dV \int_0^{E_2} dE\, \Sigma_f(\mathbf{r}, E)\Phi(\mathbf{r}, E)},$$

will yield a number for ε.

It is this latter ratio that is usually determined in order to infer ε. Such a determination can be made by irradiating a bare foil in $\Phi(\mathbf{r}, E)$ for a fixed time and then measuring the level of radioactivity (β's or γ's emitted per second) caused by decay of the radioactive fragments created by fissions during the irradiation period. If corrections are made for the fact that the mix of fission products (and hence the decay activity) may be different for different fissionable isotopes (U^{25}, Pu^{49}, U^{28}, etc.), a number proportional to $\int_{\text{fuel}} dV \int_0^\infty dE\, \Sigma_f(\mathbf{r}, E)\Phi(\mathbf{r}, E)$ can be found. Then, if a similar foil is covered with cadmium (which has an absorption cross section—a resonance—that rises from essentially 0 to $\sim$7500 b at $E = 0.5$ eV and remains high throughout the thermal region), irradiated, and counted in the same way, a number proportional to $\int_{\text{fuel}} dV \int_{E_{\text{Cd}}}^\infty dE\, \Sigma_f(\mathbf{r}, E)\Phi(\mathbf{r}, E)$ is measured, the energy range below E_{Cd} being excluded from the integral since neutrons with energies in that range are absorbed in the Cd cover and never reach the foil. The energy E_{Cd} is called the *cadmium cutoff.* If we then take $E_{\text{Cd}} = 0.5$ eV to be E_2 (or make a theoretical correction for the difference), the fast-fission factor is found experimentally from

$$\varepsilon \approx \frac{B}{B - C}, \tag{3.5.5}$$

where B is the fission-product activity of bare foil irradiated in $\Phi(\mathbf{r}, E)$ and C is the fission-product activity of a Cd-covered foil irradiated in $\Phi(\mathbf{r}, E)$. Thus, provided we can obtain a neutron flux close to $\Phi(\mathbf{r}, E)$, we can infer a value for ε experimentally without having to know any of the cross sections involved.

The most accurate way to obtain a flux close to the desired $\Phi(\mathbf{r}, E)$ is to construct a portion of the infinite lattice being investigated and arrange for it to be part of a critical reactor. Then the fuel foils can be taken as actual slices of the fuel rod, and the flux to which they are exposed will be an excellent approximation to $\Phi(\mathbf{r}, E)$. If such an elaborate "inner-lattice" experiment cannot be performed, irradiation of a foil in some other lattice similar to the one for which a value of ε is desired, or even irradiation in a neutron flux generated by neutrons arising from a charged-particle reaction (say a (D-D) reaction), will provide approximate measurements of ε.

In a natural-uranium-fueled thermal reactor most of the fissions occur in the thermal energy range, so the fast-fission factor is only about 1.05. Moreover most of fissioning in the energy range E_2–∞ comes from fissions in U^{28} induced by neutrons having energies above the U^{28} fission threshold. Of course high-energy neutrons fission in U^{25} also, but the number ratio n^{28}/n^{25} is about 140 for natural uranium while, above the U^{28} threshold, the fission cross sections of the two isotopes are about the same size. Thus in that energy range U^{28} fission is far more probable. As a result ε is often thought of as a measure of the augmentation of the fission rate caused by fast-neutron fission of U^{28}.

The Eta Factor

The second of the four factors into which (3.5.3) is split is the ratio η, called simply the *eta of the fuel.* It is the ratio of the rate at which fission neutrons are created by thermal fission to the rate at which thermal neutrons are absorbed in the fuel; it is thus defined by

$$\eta \equiv \frac{\int_{\text{fuel}} dV \int_0^{E_2} dE\, \nu\Sigma_f(\mathbf{r}, E)\Phi(\mathbf{r}, E)}{\int_{\text{fuel}} dV \int_0^{E_2} dE\, \Sigma_a^{\text{fuel}}(\mathbf{r}, E)\Phi(\mathbf{r}, E)}. \tag{3.5.6}$$

This quantity is a generalization of the energy-dependent $\eta^j(E)$ defined by (1.6.3) as $\nu^j(E)\sigma_f^j(E)$, the cross section for production of fission neutrons for isotope j at energy E, divided by $\sigma_a^j(E)$. The relationship between the $\eta^j(E)$ and the η of (3.5.6) is given by

$$\eta \equiv \frac{\int_{\text{fuel}} dV \int_0^{E_2} dE \sum_j n^j(\mathbf{r})\sigma_a^j(E)\Phi(\mathbf{r}, E)\eta^j(E)}{\int_{\text{fuel}} dV \int_0^{E_2} dE\, \Sigma_a^{\text{fuel}}(\mathbf{r}, E)\Phi(\mathbf{r}, E)}. \tag{3.5.7}$$

It is important to remember that the "fuel" generally refers to a mixture of fissile material with fertile material or diluent. Specifically, for the natural-uranium reactors for which the four-factor formula was invented, Σ_a^{fuel} includes absorption in U^{28} and (after the reactor has operated) absorption in any fission products or plutonium isotopes that have been formed. Any new fissile isotopes formed from fertile material during operation of the reactor must also be included in the sum in the numerator of (3.5.7). Thus the η for a lattice making up a power-producing reactor will change during the reactor's lifetime. For example, in a natural-uranium fuel element exposed to a Maxwellian spectrum, the value of η at the beginning of life will be about 1.34. This value may increase slightly in the early stages of life as U^{28} is converted to Pu^{49}. ($\nu^{49}\sigma_f^{49}(E)$ is much greater than $\nu^{25}\sigma_f^{25}(E)$ in the thermal range.) Eventually, however, the continual increase in absorption cross section associated with the creation of fission products will cause η to decrease in size.

If all the cross sections required to determine η from (3.5.6) had the same dependence on energy, η itself would be independent of the energy shape of $\Phi(\mathbf{r}, E)$. This condition is attained only approximately since, for example, $\sigma_a^{28}(E) \propto 1/v$ in the thermal range, while $\sigma_a^{25}(E)$ and $\sigma_f^{25}(E)$ fall somewhat faster than $1/v$. Nevertheless the energy dependence of the relevant cross sections is sufficiently similar that η is moderately insensitive to variations in the energy shape of $\Phi(\mathbf{r}, E)$. Thus a value of η measured in any thermal-reactor spectrum yields an acceptable first approximation for most natural-uranium lattices. Thus, like ε, η depends primarily on the isotopic composition of the fuel. Other materials in the lattice have an effect on η through $\Phi(\mathbf{r}, E)$, but this effect is not great.

Because of this relative insensitivity to details of the thermal spectra, η is often calculated directly from the nuclear data and some approximation to $\Phi(\mathbf{r}, E)$. There are, however, methods for estimating it directly from experiment. For example one may use the following method, called the *manganese-bath technique.*

In this scheme a small sample of fuel is placed in the center of a straight hollow tube running through a tank of water in which some soluble manganese salt is dissolved. (See Figure 3.6.) A beam of thermal neutrons is sent through the tube, and the attenuation of the beam due to absorption in the fuel sample is measured. For thermal neutrons absorption does, in fact, account for most of the attenuation, so that, if a small correction for scattering is made, it is possible to find a numerical value for the number of neutrons absorbed per second in the sample (i.e., for $\int_{\text{fuel}} dV \int_0^{E_2} dE\, \Sigma_a^{\text{fuel}}(\mathbf{r}, E)\Phi(\mathbf{r}, E)$, where the flux is that of the beam, which, if the beam is taken from a reactor, will have an energy shape that is approximately Maxwellian).

Most of the neutrons emitted by fission in the sample will be absorbed in the manganese bath by either the manganese ($\sigma_a^{\text{Mn}}(0.025) = 13$ b) or by the H_2O ($\sigma_a^{H_2O}(0.025)$

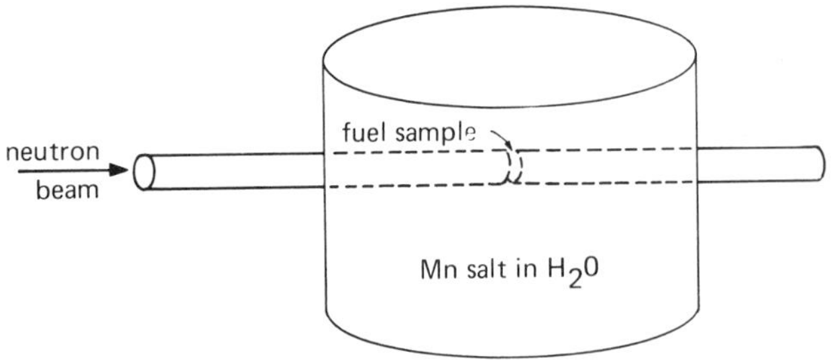

Figure 3.6 The manganese-bath experiment for measuring eta.

= 0.664 b). (A correction can be made to account for the small number of fission neutrons that escape through the ends of the tube.) Thus a measurement of the rate at which neutrons are absorbed in the bath gives the rate at which they are being admitted by fissioning in the sample; that is, it gives $\int_{\text{fuel}} dV \int_0^{E_2} dE\, \Sigma_f(\mathbf{r}, E)\Phi(\mathbf{r}, E)$. The number being absorbed in the bath can be found by stirring the bath continually and then measuring the manganese activity (having a 2.6-hour half-life) in a sample of the solution of known volume. Multiplication by the ratio of the total volume to the volume of the sample and correction back to the time at which the sample was removed from the bath gives the total activity level of the manganese in the bath. For a sufficiently long fuel-sample irradiation time an equilibrium condition will be obtained such that the number of disintegrations of manganese per cc per second equals the number of neutron captures in manganese per cc per second. Thus we can determine the rate of capture of neutrons in manganese, and, by knowing the concentration of manganese in the water and the relative capture cross sections of Mn and H_2O, we can then infer the total capture rate in the bath. As a result both the numerator and denominator of (3.5.6) are measured, and η can be inferred from their ratio.

The Thermal Utilization *f*

The third of the four factors into which we split (3.5.3) is called the *thermal utilization.* It is represented by the symbol f and defined as follows:

$$f \equiv \frac{\int_{\text{fuel}} dV \int_0^{E_2} dE\, \Sigma_a^{\text{fuel}}(\mathbf{r}, E)\Phi(\mathbf{r}, E)}{\int_{\text{cell}} dV \int_0^{E_2} dE\, \Sigma_a(\mathbf{r}, E)\Phi(\mathbf{r}, E)}. \tag{3.5.8}$$

The thermal utilization is the ratio of the rate at which thermal neutrons are absorbed in fuel to the total rate at which all thermal neutrons are absorbed. It is thus the fraction of the thermal neutrons absorbed in the cell that are absorbed in fuel.

For computations of f the spatial shape of the flux is much more important than it is for computations of ε and η. This is because the flux throughout the whole cell appears in the denominator, while only the flux throughout the fuel appears in the numerator. Thus the ratio of the level of the thermal flux outside the fuel to the level inside matters. For this reason measurements of f must be made in the actual geometry of the cell of interest.

In a natural-uranium lattice it is a fairly good approximation to assume that the energy behavior of $\Phi(\mathbf{r}, E)$ is the same at all spatial points $\mathbf{r}$ throughout the cell. That is, we write

$$\Phi(\mathbf{r}, E) \approx R(\mathbf{r})\psi(E). \tag{3.5.9}$$

The flux $\Phi(\mathbf{r}, E)$ is then said to be "separable."

If this "separability assumption" is legitimate, the exact energy dependence of the energy part $\psi(E)$ does not affect the value of f significantly, since the energy dependence of all the cross sections involved is approximately $1/v$. (If all cross sections *were* $1/v$, and if the flux $\Phi(\mathbf{r}, E)$ were separable, f, like η, would be completely independent of $\psi(E)$.) Thus the value of the thermal utilization depends primarily on the relative values of microscopic cross sections at some fixed energy (e.g., 0.025 eV) and the shape of the thermal flux throughout the cell.

Measurements of f are generally based on determination of this shape. Tiny foils are irradiated at various positions throughout the cell, and the shape $R(\mathbf{r})$ is inferred from the relative activity levels of these foils. Then (3.5.9) is used to transform (3.5.8) into

$$\frac{1}{f} = 1 + \frac{\int_{(\text{cell}-\text{fuel})} dV \int_0^{E_2} dE\, \Sigma_a(\mathbf{r}, E) R(\mathbf{r})\psi(E)}{\int_{\text{fuel}} dV \int_0^{E_2} dE\, \Sigma_a^{\text{fuel}}(\mathbf{r}, E) R(\mathbf{r})\psi(E)}, \tag{3.5.10}$$

where $\Sigma_a(\mathbf{r}, E)$ will change in a stepwise fashion as $\mathbf{r}$ goes from one region of the cell to the next.

With the best estimate of $\psi(E)$ used along with the best estimates of the nuclear cross sections $\sigma_a^j(E)$ for isotopes j present in the cell, measurement of $\int_i R(\mathbf{r})\, dV_i$ in the various regions i (fuel, clad, coolant, moderator) will yield an approximate value of f. The approximation can be improved by making various theoretical and experimental corrections.

Values of the thermal utilization for natural-uranium lattices usually run around 0.90. To obtain a lattice having a large infinite-medium multiplication factor, one would like f to be as large (as close to unity) as possible. Thus one would like the absorption cross sections and volumes of the nonfuel materials and the values of the flux level $R(\mathbf{r})$ *in* those nonfuel regions to be as *small* as possible. Unfortunately the material constituents of a lattice are determined primarily by other considerations, and changes in the lattice geometry that will increase f usually cause a decrease in one of the other factors. Thus obtaining a maximum value for k_∞ for a lattice involves the balancing of a number of factors.

The Resonance-Escape Probability p

The last of the four factors defined by (3.5.3) is called the *resonance-escape probability*. It is denoted by the symbol p and defined as follows:

$$p \equiv \frac{\int_{\text{cell}} dV \int_0^{E_2} dE\, \Sigma_a(\mathbf{r}, E)\Phi(\mathbf{r}, E)}{\int_{\text{cell}} dV \int_0^\infty dE\, \Sigma_a(\mathbf{r}, E)\Phi(\mathbf{r}, E)}$$
$$= 1 - \frac{\int_{\text{cell}} dV \int_{E_2}^\infty dE\, \Sigma_a(\mathbf{r}, E)\Phi(\mathbf{r}, E)}{\int_{\text{cell}} dV \int_0^\infty dE\, \Sigma_a(\mathbf{r}, E)\Phi(\mathbf{r}, E)}. \tag{3.5.11}$$

It is the ratio of the rate of thermal-neutron absorption throughout the cell to the rate of absorption at all energies throughout the cell. It is thus the fraction of all neutrons absorbed in the cell that are absorbed thermally. Hence it is the probability that a neutron will *escape* capture at energies *above* thermal. Finally, since, in natural or slightly enriched uranium lattices, most nonthermal capture is in the resonance part of the slowing-down region, it is the "resonance-escape probability."

An entirely equivalent definition of p can be given in terms of the slowing-down density. Thus, if we apply the definition (3.4.22) of $q(\mathbf{r}, E)\, dV$, the number of neutrons per second in volume dV containing point $\mathbf{r}$ whose energy drops below the energy E, then the rate at which neutrons are absorbed in the thermal energy range equals $\int_{\text{cell}} q(\mathbf{r}, E_2)\, dV$, the total number of neutrons slowing down past the cut point E_2. It follows that, in an infinite lattice where the total rate at which neutrons are absorbed equals the total rate at which they are produced by fission,

$$p = \frac{\int_{\text{cell}} q(\mathbf{r}, E_2)\, dV}{\int_{\text{cell}} dV \int_0^\infty dE\, \nu\Sigma_f(\mathbf{r}, E)\Phi(\mathbf{r}, E)}. \tag{3.5.12}$$

For a *homogeneous* mixture of reactor materials this expression permits us to derive a very simple and fairly accurate explicit formula for p applicable when the resonance absorber in the mixture has narrow, widely spaced resonances. Under these conditions (3.4.39) for $q(E)$ applies. Moreover, if we take E_1 to be an energy just below the point where $\chi(E)$ becomes negligibly small ($\sim 10^5$ eV) and if we neglect any neutron absorption *above* E_1, we see that $q(E_1) = \int_0^\infty dE\, \nu\Sigma_f(\mathbf{r}, E)\Phi(\mathbf{r}, E)$; that is, the number of neutrons per cc per sec slowing down past E_1 equals the rate at which neutrons are created by the fission process. It follows then from (3.5.12) and (3.4.39) that, for a homogeneous medium,

$$p = \frac{\int_{\text{cell}} q(E_2)\, dV}{\int_{\text{cell}} q(E_1)\, dV} = \frac{q(E_2)}{q(E_1)} = \exp\left(-\int_{E_2}^{E_1} \frac{\Sigma_a(E')\, dE'}{E'\bar{\xi}(E')[\Sigma_a(E') + \Sigma_s(E')]}\right). \tag{3.5.13}$$

If this formula is applied to a *homogenized* cell containing materials of the kind and proportions used for natural-uranium-fueled reactors, values of p of about 0.7 are obtained. The energy self-shielding, which we discussed in connection with hydrogen moderation, decreases the resonance absorption to a value lower than what we might expect from consideration of the magnitude of the resonance cross sections alone. Nevertheless around 30 percent of the neutrons will be absorbed while slowing down in such a medium, and for this reason it turns out to be impossible to create homogeneous mixtures fueled by natural uranium that have a value of k_∞ greater than unity. Thus even an infinite homogeneous assembly of natural uranium and moderating material cannot be made to go critical. The neutron leakage associated with a *finite* assembly makes matters even worse.

The way around this difficulty—and the primary reason why all natural-uranium-fueled reactors are composed of lattices of cells—is to *lump* the uranium and moderator in separate regions. If this is done, a neutron that happens to acquire, in the slowing-down process, an energy corresponding to the peak of a resonance (and, hence, that almost certainly would be absorbed in a homogeneous mixture) may very well scatter from another moderating nucleus without getting near the lumps of fuel. If this does happen, the neutron will have its energy degraded further and will thus escape capture in the resonance. For example the macroscopic scattering cross section for graphite in the slowing-down region is about 0.4 cm^{-1}. The mean free path for scattering $1/\Sigma_s$ is therefore about 2.5 cm, and, in a lattice of natural-uranium rods laid out parallel to one another in a square array with their centers eight inches (20.3 cm) apart, only those neutrons that happen to slow down near a fuel rod have a significant chance of being captured, even if their energies correspond to that of a resonance.

This phenomenon, which protects neutrons from being captured in lumped fuel, is called *spatial self-shielding*. For maximum protection we would like the size of the lumps and the overall ratio of the number of moderating atoms to fuel atoms to be as large as possible. However both these conditions adversely affect the value of the thermal utilization. Hence designing a natural-uranium lattice so that k_∞ is a maximum is an optimization process. The fast-fission factor and eta are little affected by the moderator-to-fuel volume ratio. However both the thermal utilization and the resonance-escape probability are quite sensitive to this ratio, and a balance between the two must be struck. In an optimized natural-uranium lattice both f and p are about 0.9.

It is very difficult to compute accurately the resonance-escape probability for a heterogeneous lattice cell. (See Chapter 5.) In the energy range of a given resonance the flux density $\Phi(\mathbf{r}, E)$ dips significantly in both energy and position. Thus the assumption of separability in (3.5.9) is very poor, and a detailed space-energy solution must be made.

Exact calculations are possible, but they require application of elaborate computing-machine methods that are difficult to program. Fortunately there do exist a number of simple algebraic prescriptions that are based on somewhat unconvincing theoretical arguments but have been tested (and empirically corrected) by comparison with measurement and more exact theory. We shall discuss some of those formulas in Chapter 5. They relate p to the fuel-rod size and the moderator-to-fuel volume ratio and are quite reliable over the range for which they have been validated.

Because the resonance-escape probability depends directly on the fuel-rod diameter and the moderator-to-fuel volume ratio, it must be measured in the lattice of interest. Even then, however, the measurements are rather indirect. One method is based on measuring the Np^{39} activity of bare and cadmium-covered foils having the same diameter as, and being an integral part of, a fuel rod. (Note that $U^{28} + n \rightarrow U^{29} \xrightarrow{\beta^-} Np^{39} \xrightarrow{\beta^-} Pu^{49}$.) Detection of the equilibrium Np^{39} concentration by measurement of a particular decay γ yields a number that can be related to the rate at which neutrons are being captured in U^{28}. Thus the ratio

$$\frac{\int_{\text{fuel}} dV \int_{E_{\text{Cd}}}^{\infty} dE\, \Sigma_{\text{a}}^{28}(\mathbf{r}, E)\Phi(\mathbf{r}, E)}{\int_{\text{fuel}} dV \int_{0}^{E_{\text{Cd}}} dE\, \Sigma_{\text{a}}^{28}(\mathbf{r}, E)\Phi(\mathbf{r}, E) + \int_{\text{fuel}} dV \int_{E_{\text{Cd}}}^{\infty} dE\, \Sigma_{\text{a}}^{28}(\mathbf{r}, E)\Phi(\mathbf{r}, E)}$$

can be obtained. Then, if the ratio

$$\frac{\int_{\text{fuel}} dV \int_{0}^{E_{\text{Cd}}} dE\, \Sigma_{\text{a}}^{28}(\mathbf{r}, E)\Phi(\mathbf{r}, E)}{\int_{\text{cell}} dV \int_{0}^{E_{\text{Cd}}} dE\, \Sigma_{\text{a}}(\mathbf{r}, E)\Phi(\mathbf{r}, E)}$$

can be measured or computed, if

$$\int_{\text{cell}} dV \int_{E_{\text{Cd}}}^{\infty} dE\, [\Sigma_{\text{a}}(\mathbf{r}, E) - \Sigma_{\text{a}}^{28}(\mathbf{r}, E)]\Phi(\mathbf{r}, E)$$

is assumed negligible, and if E_2 is taken to be E_{Cd}, a value of p can be inferred from (3.5.11).

References

Clark, M., Jr., and Hansen, K. F., 1964. *Numerical Methods of Reactor Analysis* (New York: Academic Press).

Lamarsh, J. R., 1966. *Introduction to Nuclear Reactor Theory* (Reading, Mass.: Addison-Wesley), Chapter 6.

Meghreblian, R. V., and Holmes, D. K., 1960. *Reactor Analysis* (New York: McGraw-Hill), Chapter 4.

Problems

1. For the homogeneous plutonium-graphite mixture specified in Problem 14, Chapter 2, find $\Sigma_{gg'}$ for $E_g = 70$ eV, $E_{g-1} = 80$ eV, $E_{g'} = 100$ eV, and $E_{g'-1} = 110$ eV. Also find $\Sigma_{g'g}$.

2. If neutrons were produced uniformly in an infinite homogeneous medium, not by fission, but rather by some external sources (for example a uniformly distributed mixture of radium and beryllium) and this "external" source emitted S_g neutrons/cc/sec into energy group g, (3.3.9) would become

$$\sum_{g'} [\Sigma_{tg}\delta_{gg'} - \Sigma_{gg'}]\Phi_{g'} = S_g$$

or, in matrix form,

$$[A][\Phi] = [S].$$

Thus $[A]^{-1}[S]$ can be interpreted as the column vector of group fluxes due to a source $[S]$.

Use this result to interpret $[A]^{-1}[\chi]$ and thus give a physical interpretation of the nonzero value of λ proved to exist when $[\chi] = [\chi^j]$ for all j.

3. Find two independent eigenvectors for (3.3.26) corresponding to the eigenvalue $\lambda = 0$. Show that any other vector corresponding to eigenvalue $\lambda = 0$ can be expressed as a linear combination of these two.

4. Validate (3.3.30).

5. For the *general* two-group case (Σ_{21}, Σ_{12}, χ_1^j, and χ_2^j *all* nonzero) derive an algebraic formula for λ analogous to (3.3.30). Prove that, for physically real group parameters (Σ's, χ's > 0; $\Sigma_1 > \Sigma_{12}$; etc.), there exists a most positive real value of λ.

6. For E_1 below the fission-source range and hydrogen the only moderator show that

$$q(E) = q(E_1) \exp\left(-\int_E^{E_1} \frac{\Sigma_a(E')dE'}{[\Sigma_a(E') + \Sigma_s^H(E')]E'}\right).$$

Derive (3.4.15) directly from this result.

7. Consider a solution of boric acid (H_3BO_3) in water such that the ratio of the molecules of H_3BO_3 per cc of solution to the molecules of H_2O per cc of solution is 0.01. The absorption cross sections of the constituents of this mixture are all $1/v$, having values at 0.025 eV of $\sigma_a^{B(nat)}(0.025) = 760$b; $\sigma_a^H(0.025) = 0.315$b; $\sigma_a^O(0.025) \approx 0$. Also, in the range 1 eV to 10 eV, the scattering cross sections may be taken as constant with $\sigma_s^B = 4$b; $\sigma_s^H = 20$b; $\sigma_s^O = 4.2$b.

Find the ratio of the slowing-down density in this mixture at 1 eV to that at 10 eV under the following conditions:

a. Assuming hydrogen is the only moderator.
b. Accounting for the moderation of all the constituents.
c. Using the narrow-resonance expression (3.4.39) and the approximation $1/(1 + \Sigma_a/\Sigma_s) \approx 1 - \Sigma_a/\Sigma_s$.

8. Suppose that, at $E = 100$ eV, U^{28} has an absorption resonance that can be approximated as a square with a width of 1 eV ($\sigma_a^{28} \gg \sigma_s^{28}$; 99.5 eV $\leq E \leq$ 100.5 eV; $\sigma_a^{28} = 0$ for several eV's on either side of the range).

What would be the ratio $q(99.5)/q(100.5)$ in a mixture of U^{28} and H_2O so dilute that all moderation could be considered as due to the H_2O, yet such that $\Sigma_a \gg \Sigma_s$? (Use $\sigma_s^O = 4$b; $\sigma_s^H = 20$b).

What would that ratio be in solid U^{28}?

9. Write (3.4.27) and (3.4.40) with energy replaced by lethargy.

10. Write the expansion (3.4.42) in terms of the lethargy u corresponding to E and the lethargy gain U ($\equiv u - u'$), where u' corresponds to E'. Use this result and (3.4.9) to derive (3.4.45) in lethargy form.

11. Show that kT is the most probable energy of a Maxwellian distribution. What is the corresponding *average* energy?

12. Suppose the macroscopic absorption cross section for a homogeneous mixture of materials is known to vary with neutron velocity as $1/v^2$ in the thermal energy range ($0 \leq E \leq E_2 = 1$ eV) and that $\xi(E_2)\Sigma_s(E_2) = 0.2\ \text{cm}^{-1}$. What must be the value of $\Sigma_a(E)$ at $E = 0.025$ eV if the cut point E_c, at which the Maxwellian and $1/E$ shapes are joined in the approximate spectrum (3.4.53), is to be $E_c = 0.05$ eV? Compare this value of $\Sigma_a(0.025)$ with that for pure H_2O at room temperature and pressure.

13. It has been pointed out that (3.4.19) is valid only for energies well below the energies at which source neutrons appear. The other extreme is to consider (for an infinite homogeneous medium containing a simple, nonabsorbing isotope that scatters elastically and isotropically in the center-of-mass system) the flux $\Phi(E)$ at energies close to the energy E_0 at which a source emits neutrons at a rate q_0 per unit volume per second.

Mathematical difficulties arise if we try to express the fact that *all* the neutrons in this source are assumed to appear with exactly the energy E_0. An entirely equivalent physical picture is to let the "monochromatic" source neutrons make a single collision and then to regard the neutrons emerging from this single collision as a source extended in energy. Thus (2.5.22) shows that the case of q_0 neutrons appearing per unit volume per second with exactly the energy E_0 is equivalent (in this infinite, purely scattering medium) to a source $q_0 dE/[E_0(1 - \alpha)]$ appearing per unit volume per second within dE in the energy range αE_0–E_0. The flux due to this first collision source then obeys a modification of (3.4.17), namely,

$$\Sigma_s(E)\Phi(E) = \int_E^{E_0} \frac{\Sigma_s(E')\Phi(E')}{E'(1-\alpha)} + \frac{q_0}{E_0(1-\alpha)} \quad \text{for} \quad \alpha E_0 \leq E < E_0.$$

a. Show that a solution of this equation in the range $\alpha E_0 \leq E < E_0$ is

$$\Phi(E) = \frac{q_0}{\Sigma_s(E)E_0(1-\alpha)}\left(\frac{E_0}{E}\right)^{1/(1-\alpha)}.$$

b. Show further that, in the range $\alpha^2 E_0 \leq E < \alpha E_0$ where there is no direct source so that (3.4.17) written for a simple isotope is valid (but where $\Phi(E)$ in the range αE_0–E/α is known from the result just found),

$$\Phi(E) = \frac{q_0}{\Sigma_s(E)E_0(1-\alpha)}\left(\frac{\alpha E_0}{E}\right)^{1/(1-\alpha)}\left\{\alpha^{-1/(1-\alpha)} - 1 - \ln\left[\left(\frac{\alpha E_0}{E}\right)^{1/(1-\alpha)}\right]\right\} \quad \text{for} \quad \alpha^2 E_0 \leq E < \alpha E_0.$$

c. Sketch $\Phi(E)$ in the range $\alpha^2 E_0$–E_0.

Hint: In both energy ranges the relevant integral equations for $\Phi(E)$ involve integrals that have a constant as one limit. Thus the procedure used to solve (3.4.11) is possible. The constant of integration for the range αE_0–E_0 can be found from the relevant integral equation by taking the limit as $E \rightarrow E_0$. For the range $\alpha^2 E_0$–αE_0 the integration constant can be found by requiring that $\Sigma_s(\alpha E_0^-)\,\Phi(\alpha E_0^-)\,dE_0^-$, the rate at which neutrons scatter out of an interval dE_0^- just below αE_0, equal the rate at which they scatter into that interval from the range αE_0–E_0.

4 Diffusion Theory

4.1 Introduction

Real reactors are finite in size and almost always contain a number of different material compositions. As a result the scalar flux density, which is a function only of energy in an infinite, homogeneous, critical reactor, becomes a function, $\Phi(\mathbf{r}, E)$, of both energy and position. In this chapter we shall discuss methods for determining $\Phi(\mathbf{r}, E)$ according to the *diffusion-theory approximation*. The conditions under which this approximation is valid are met for a number of practical reactors. Moreover there are systematic procedures (to be discussed in Chapters 5 and 10) for extending the applicability of diffusion theory to situations where transport theory would otherwise have to be applied. Such extensions make it possible to apply diffusion theory to the analysis of almost all reactor designs. The method is thus of fundamental importance for reactor analysis.

4.2 The Net Current Density

To begin the development of diffusion theory we introduce a new concept—that of the net current density. We shall first define it physically and then show how it can be defined mathematically in terms of $N(\mathbf{r}, \mathbf{\Omega}, E)$, the neutron density in phase space.

To define the net current density physically consider a small surface dS located at some position $\mathbf{r}$ in the medium of interest.

As indicated in Figure 4.1, neutrons traveling in random directions are continually passing through dS. Let us consider a subset of those neutrons, namely all of them that pass through dS in one second having a kinetic energy between E and $E + dE$ (but still moving in random directions). If we arbitrarily call one side of dS the (+) side and the other side the (−) side, we can talk about the number of neutrons in dE crossing from the (−) to the (+) side in one second and the number crossing from the (+) to the (−) side in that same second. These two numbers may happen to be the same, but in general they will not be and there will be a *net* number of neutrons crossing dS, either from (+) to (−) or from (−) to (+). For example, if 1200 neutrons with energies in the range dE cross dS from (−) to (+) in one second and 1000 cross from (+) to (−), a *net* of 200 neutrons in the range dE cross dS in that second from the (−) to the (+) side.

The net number in dE crossing dS per second will in general depend on the orientation of dS. Moreover, except for the case in which the net number crossing is zero for *all* orientations (as it would be in an infinite, homogeneous, critical reactor), there is one unique orientation of dS such that the net number crossing from the (−) to (+) side is a maximum as compared with all other orientations of dS at that point. With this physical picture as background we then define the *net current density* $\mathbf{J}(\mathbf{r}, E)$ by:

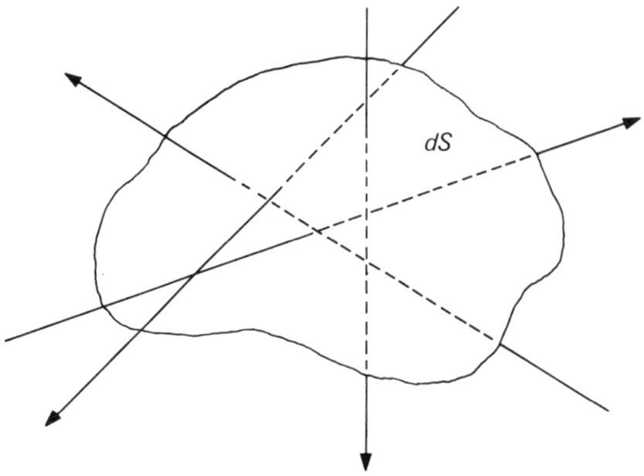

Figure 4.1 Neutrons contributing to the net current density.

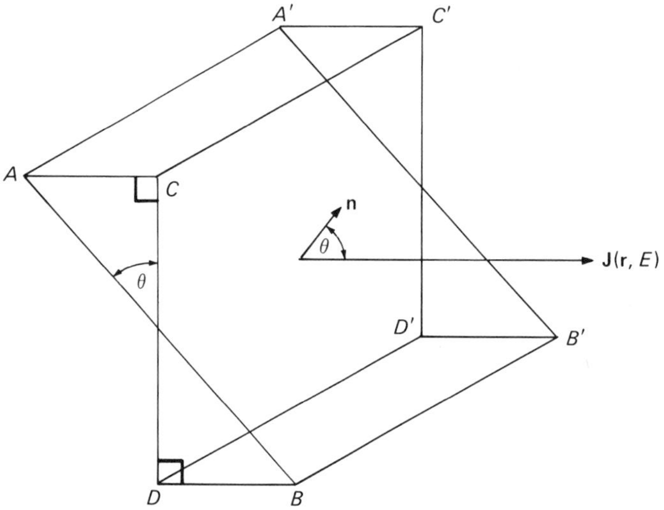

Figure 4.2 A vector decomposition of the net current density.

The maximum over all orientations of a unit surface at $\mathbf{r}$ of the net number of neutrons with energies between E and $E + dE$ crossing that unit surface per second, the direction of the vector being the direction of this maximum net flow

$$\equiv \mathbf{J}(\mathbf{r}, E)\, dE. \tag{4.2.1}$$

The units of $\mathbf{J}(\mathbf{r}, E)$ are neutrons per unit energy per cm^2 per sec.

It is important to recognize that the *direction* of $\mathbf{J}$ is *not* the direction of some specific group of neutrons. It is instead the direction of the *net flow*. In fact it can happen that no neutrons making up $\mathbf{J}$ are actually moving in the direction of $\mathbf{J}$. For example, if there are two beams of neutrons of equal strength crossing at $\mathbf{r}$, one moving in the $+X$ direction, specified by unit vector $\mathbf{i}$, and the other in the $+Y$ direction, specified by unit vector $\mathbf{j}$, the direction of the *net* current $\mathbf{J}$ will be along the 45° line between the X and Y axes, that is, the direction will be that of the vector sum of $\mathbf{i}$ and $\mathbf{j}$.

Once we know $\mathbf{J}(\mathbf{r}, E)$, which is the net current through one *particular* unit surface (namely the one oriented so that this net flow is a maximum), we can find the net current through any surface at $\mathbf{r}$. Thus, if we specify the orientations of surfaces dS by unit vectors $\mathbf{n}$ normal to dS, we have

Net current (neutrons per unit energy per sec) through $\mathbf{n}\, dS$

$$= \mathbf{J}(\mathbf{r}, E)\cdot\mathbf{n}\, dS. \tag{4.2.2}$$

To see the reason for this consider Figure 4.2. Let dS be a square, the length of one side being $(AB) = (A'A)$ so that $dS = (AB)(A'A)$. Since $\mathbf{J}$ is the net flow in the direction for which that flow is a maximum, there is no *net* flow in directions perpendicular to $\mathbf{J}$. (If there were, the vector sum of this net flow and $\mathbf{J}$ would produce a net flow larger than $\mathbf{J}$, so that $\mathbf{J}$ would not, in fact, be the direction of maximum net flow.) Thus the net flow through $(AB)(A'A)$ is the same as the net flow through the rectangle $(CD)(C'C) = (CD)(A'A)$ perpendicular to $\mathbf{J}$, that is, $(CD)(A'A)J(\mathbf{r}, E)\, dE$, where $J(\mathbf{r}, E)$ is the magnitude of $\mathbf{J}(\mathbf{r}, E)$. But

$$(CD)(A'A)J(\mathbf{r}, E)\, dE = (AB)(A'A) \cos\theta\, J(\mathbf{r}, E)\, dE = (AB)(A'A)\mathbf{n}\cdot\mathbf{J}(\mathbf{r}, E)\, dE. \tag{4.2.3}$$

This justifies (4.2.2).

We can use this method of breaking up a neutron current into components to relate $\mathbf{J}(\mathbf{r}, E)$ to $N(\mathbf{r}, \mathbf{\Omega}, E)$. To do this we first reintroduce (2.2.1) in an equivalent "per second per unit surface" form:

Number of neutrons having energies in dE and directions in $d\Omega$ that cross a unit surface per second

$$= v(E)N(\mathbf{r}, \mathbf{\Omega}, E)\, d\Omega\, dE. \tag{4.2.4}$$

We derived this equation by observing that all neutrons traveling in direction $\mathbf{\Omega}$ with a speed $v(E)$ and contained in a cylinder of base dS and height $v(E)$ cross dS in one second. By the same argument that led to (4.2.3) we can obtain expressions for the number of neutrons in this beam that cross unit surfaces perpendicular to the X, Y, and Z directions. These partial currents are proportional to the cosines of the angles between $\mathbf{\Omega}$, the direction of the beam, and $\mathbf{i}$, $\mathbf{j}$, and $\mathbf{k}$, the unit vectors representing the directions of the X, Y, and Z axes; that is,

Neutrons per second of the type $N(\mathbf{r}, \mathbf{\Omega}, E)$ passing through a unit surface perpendicular to the X axis

$$= [v(E)\, N(\mathbf{r}, \mathbf{\Omega}, E)\, d\Omega\, dE]\mathbf{\Omega}\cdot\mathbf{i}, \tag{4.2.5}$$

with similar expressions for the Y and Z directions.

If there are neutron beams at point $\mathbf{r}$ traveling in all directions, we may compute the net number of neutrons in dE going through a unit surface perpendicular to any particular axis by adding up all the contributions from different directions $\mathbf{\Omega}$. For example:

Net current of neutrons in dE through a unit surface in the X direction

$$= \left[\int_{\mathbf{\Omega}} d\Omega\, \mathbf{\Omega}\cdot\mathbf{i}\, v(E)\, N(\mathbf{r}, \mathbf{\Omega}, E)\right] dE. \tag{4.2.6}$$

This number will be positive or negative depending on the sign of the integral over the beam directions. A positive sign implies that the net flow is in the $+\mathbf{i}$ direction; a negative sign implies net flow in the $-\mathbf{i}$ direction.

Now it follows from (4.2.2) that the component of the net current of neutrons in dE across a unit surface in the X direction is $\mathbf{J}(\mathbf{r}, E)\cdot\mathbf{i} \equiv J_x(\mathbf{r}, E)\, dE$. Thus, from (4.2.6) with $\Omega_x \equiv \mathbf{\Omega}\cdot\mathbf{i}$,

$$J_x(\mathbf{r}, E) = \int_{\mathbf{\Omega}} d\Omega\, \Omega_x v(E) N(\mathbf{r}, \mathbf{\Omega}, E). \tag{4.2.7}$$

Repeating this argument for the Y and Z directions, we find that

$$\begin{aligned} \mathbf{J}(\mathbf{r}, E) &\equiv \mathbf{i}J_x(\mathbf{r}, E) + \mathbf{j}J_y(\mathbf{r}, E) + \mathbf{k}J_z(\mathbf{r}, E) \\ &= \int_{\mathbf{\Omega}} d\Omega(\mathbf{i}\Omega_x + \mathbf{j}\Omega_y + \mathbf{k}\Omega_z) v(E) N(\mathbf{r}, \mathbf{\Omega}, E) \\ &= \int_{\mathbf{\Omega}} d\Omega\, \mathbf{\Omega}\, v(E) N(\mathbf{r}, \mathbf{\Omega}, E). \end{aligned} \tag{4.2.8}$$

Equation (4.2.8) is the mathematical relationship between $\mathbf{J}$ and N that follows from the definition (4.2.1). Note the similarity to the relationship (2.2.7) between the scalar

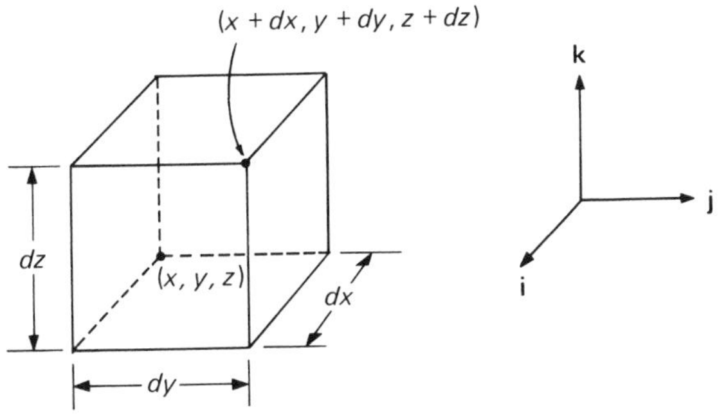

Figure 4.3 A volume element for the computation of the net neutron leakage.

flux density and N:

$$\Phi(\mathbf{r}, E) \equiv \int_{\Omega} d\Omega\ v(E)N(\mathbf{r}, \mathbf{\Omega}, E). \tag{2.2.7}$$

4.3 The Net Neutron Loss per Unit Volume Due to Leakage

The net current density can be used to obtain an expression for the net loss of neutrons in an energy interval dE due to leakage out of a volume element dV. To do so we take $dV = dx\, dy\, dz$ to be a small rectangular parallelopiped as in Figure 4.3.

Equation (4.2.2) states that the net rate at which neutrons in the energy range dE leave dV through the surface of area $dy\, dz$ at a distance x from the YZ plane is $\mathbf{J}(x, y, z, E)\cdot(-\mathbf{i})\, dy\, dz\, dE$, where the minus sign on the unit vector $\mathbf{i}$ is due to the fact that the direction leaving $dx\, dy\, dz$ on the back side of the parallelopiped is $-\mathbf{i}$.

Similarly the net rate of which neutrons pass through the surface $dy\, dz$ at $x + dx$, the front side of the box, is $\mathbf{J}(x + dx, y, z, E)\cdot\mathbf{i}\, dy\, dz\, dE$, where we take y and z to be the same values that appear in $\mathbf{J}(x, y, z, E)$.

It follows that the total rate at which neutrons are lost to dV by net leakage across surfaces perpendicular to the X axis is the difference of these two terms, which for convenience we write as

$$\left[\frac{\mathbf{J}(x + dx, y, z, E) - \mathbf{J}(x, y, z, E)}{dx}\right]\cdot\mathbf{i}\, dx\, dy\, dz\, dE. \tag{4.3.1}$$

Thus, in the limit as $dx \to 0$, the net rate of leakage out of dV across surfaces perpendicular to the X axis is

$$\left[\left(\mathbf{i}\frac{\partial}{\partial x}\right)\cdot\mathbf{J}(x, y, z, E)\right] dx\, dy\, dz\, dE,$$

where we have formally associated the derivative operation with the unit vector **i**.

Repeating this analysis for the Y and Z directions, we get

Total rate at which neutrons with energies in the range E to $E + dE$ leak out of $dV = dx\,dy\,dz$

$$= \left[\left(\mathbf{i}\frac{\partial}{\partial x}\right)\cdot\mathbf{J}(x, y, z, E) + \left(\mathbf{j}\frac{\partial}{\partial y}\right)\cdot\mathbf{J}(x, y, z, E) + \left(\mathbf{k}\frac{\partial}{\partial z}\right)\cdot\mathbf{J}(x, y, z, E)\right] dx\,dy\,dz\,dE$$
$$= \nabla\cdot\mathbf{J}(\mathbf{r}, E)\,dV\,dE. \tag{4.3.2}$$

Thus the leakage rate per unit volume per unit energy is $\nabla\cdot\mathbf{J}$, the divergence of the net current density.

4.4 A Fundamental Neutron-Balance Condition

For a reactor in a steady-state critical condition the rate at which neutrons leak out of $dV\,dE$ must equal the rate at which they are added to dE within dV (by fission and scattering-in) less the rate at which they are removed by interaction with the material in dV (i.e., by absorption and scattering-out). We can express this result quantitatively. The leakage rate is given by (4.3.2), the production ratio due to fission and scattering-in by (3.2.1) and (3.2.2), and the total interaction rate by (3.2.3). The physical statement that leakage equals production less removal then becomes

$$\nabla\cdot\mathbf{J}(\mathbf{r}, E)\,dV\,dE = \sum_j \chi^j(E)\,dV\,dE\int_0^\infty \nu^j\Sigma_f^j(\mathbf{r}, E')\Phi(\mathbf{r}, E')\,dE'$$
$$+ dV\,dE\int_0^\infty \Sigma_s(\mathbf{r}, E'\to E)\Phi(\mathbf{r}, E')\,dE' - \Sigma_t(\mathbf{r}, E)\Phi(\mathbf{r}, E)\,dV\,dE \tag{4.4.1}$$

or

$$\nabla\cdot\mathbf{J}(\mathbf{r}, E) + \Sigma_t(\mathbf{r}, E)\Phi(\mathbf{r}, E) = \int_0^\infty \left[\sum_j \chi^j(E)\nu^j\Sigma_f^j(\mathbf{r}, E') + \Sigma_s(\mathbf{r}, E'\to E)\right]\Phi(\mathbf{r}, E')\,dE', \tag{4.4.2}$$

where, for this general, heterogeneous, finite assembly, the cross sections, flux density, and current density are all expressed as functions of both energy and position.

Equation (4.4.2) is rigorous for the external-source-free, steady-state condition. We have made no approximations in deriving it other than those inherent in the basic expressions for reaction rates. Thus, if we could obtain $\Phi(\mathbf{r}, E)$ from (4.4.2) without approximation, it would be possible to compute the power level at all locations in a critical reactor with an accuracy limited only by the reliability of the nuclear data. Unfortunately, since the equation contains $\mathbf{J}$ as well as Φ, it cannot be solved for either of these functions without making use of some second equation relating them. If we knew the density in phase space $N(\mathbf{r}, \boldsymbol{\Omega}, E)$, there would of course be no difficulty, since the definition (2.2.7) of $\Phi(\mathbf{r}, E)$ would give us that quantity immediately, and there would be no need to solve (4.4.2). But knowing $N(\mathbf{r}, \boldsymbol{\Omega}, E)$ implies that we have solved the Boltzmann transport equation, and that is much more difficult than solving (4.4.2). To proceed along the present lines we thus seek some auxiliary relationship between the net current and the scalar flux.

4.5 Fick's Law

As might be expected, the best that can be done to relate $\mathbf{J}$ to Φ without actually finding a solution for $N(\mathbf{r}, \boldsymbol{\Omega}, E)$ is an approximation. Fortunately there turns out to be a rather accurate approximation that is quite satisfactory for many reactor-design calculations. It is called *Fick's Law*, and it relates $\mathbf{J}(\mathbf{r}, E)$ to $\nabla\Phi(\mathbf{r}, E)$ by the equation

$$\mathbf{J}(\mathbf{r}, E) = -D(\mathbf{r}, E)\nabla\Phi(\mathbf{r}, E), \tag{4.5.1}$$

where $D(\mathbf{r}, E)$ is called the *diffusion constant.*

This relationship between the net current and the gradient of the scalar flux seems plausible if we recall from (2.2.7) that $\Phi(\mathbf{r}, E)$ is $v(E)\, n(\mathbf{r}, E)$, where $n(\mathbf{r}, E)\, dE$ is the number of neutrons in dE per unit volume. Thus (4.5.1) states that the net current across a surface of neutrons with energies in dE is proportional to the rate of decrease of the density of neutrons in dE across that surface. Moreover the direction of the net current is the direction in which $n(\mathbf{r}, E)$ is decreasing at its maximum rate. (Recall that, mathematically, the gradient of a function is a vector pointing in the direction of the maximum rate of increase of that function; thus $-\nabla n(\mathbf{r}, E)$ points in the direction of the maximum rate of *decrease* of $n(\mathbf{r}, E)$.) The neutron population thus tends to drift from a region of high concentration to one of low concentration, like gas diffusing through a porous plug, and this drift gives rise to a net current.

The Validity of Fick's Law

Fick's Law can be derived in several ways, each of which results in a slightly different value for $D(\mathbf{r}, E)$ in terms of the cross sections of the medium. The most accurate, straightforward, and convincing of these derivations is based on finding an approximate expression for $N(\mathbf{r}, \boldsymbol{\Omega}, E)$ and then, in effect, relating the net current to the gradient of the flux by computing them each directly from the approximate N, making use of the definitions $\mathbf{J} \equiv \int \boldsymbol{\Omega} v N\, d\Omega$ and $\Phi \equiv \int v N\, d\Omega$. This method is the one most frequently used today. It does not lead, without further approximation, to a simple formula for the diffusion constant, since it proceeds indirectly by finding an approximate value for $N(\mathbf{r}, \boldsymbol{\Omega}, E)$. Nevertheless numerical values of $D(\mathbf{r}, E)$ emerge that reflect such important physical phenomena as inelastic scattering, anisotropic scattering, and—most important —neutron moderation.

We shall defer the detailed analysis leading to Fick's Law until Chapter 9, since it will require first setting up the transport equation for $N(\mathbf{r}, \boldsymbol{\Omega}, E)$. Here we shall simply describe the approximation for $N(\mathbf{r}, \boldsymbol{\Omega}, E)$ on which the method is based in order to give some insight into the conditions under which (4.5.1) should be valid.

This approximation is the assumption that $N(\mathbf{r}, \boldsymbol{\Omega}, E)$ has the form

$$\begin{aligned} N(\mathbf{r}, \boldsymbol{\Omega}, E) &= F(\mathbf{r}, E) + \boldsymbol{\Omega}\cdot\mathbf{V}(\mathbf{r}, E) \\ &= F(\mathbf{r}, E) + \Omega_x V_x(\mathbf{r}, E) + \Omega_y V_y(\mathbf{r}, E) + \Omega_z V_z(\mathbf{r}, E), \end{aligned} \tag{4.5.2}$$

where $F(\mathbf{r}, E)$ and $\mathbf{V}(\mathbf{r}, E)$ are scalar and vector functions of position and energy (but *not* of direction $\boldsymbol{\Omega}$), and where $\Omega_x = \boldsymbol{\Omega}\cdot\mathbf{i}$, $\Omega_y = \boldsymbol{\Omega}\cdot\mathbf{j}$, and $\Omega_z = \boldsymbol{\Omega}\cdot\mathbf{k}$.

To understand the nature of this approximation, suppose first that all three components of $\mathbf{V}$ are zero: (4.5.2) then reduces to $N(\mathbf{r}, \boldsymbol{\Omega}, E) = F(\mathbf{r}, E)$. Thus the number of neutrons in $dV\, dE\, d\Omega$ are the same for all directions $\boldsymbol{\Omega}$ (i.e., $N(\mathbf{r}, \boldsymbol{\Omega}_1, E) = N(\mathbf{r}, \boldsymbol{\Omega}_2, E)$ for any $\boldsymbol{\Omega}_1$ and $\boldsymbol{\Omega}_2$). The neutron density in phase space at $\mathbf{r}$ and E is thus *isotropic* in the directions of neutron travel, and a polar plot of $N(\mathbf{r}, \boldsymbol{\Omega}, E)$ for the magnitude of $N(\mathbf{r}, \boldsymbol{\Omega}, E)$ in various directions yields a sphere (see Figure 4.4a).

If, instead, F and V_x are positive (with $V_x < F$) while $V_y = V_z = 0$, a polar plot of $N(\mathbf{r}, \boldsymbol{\Omega}, E)$ looks like Figure 4.4b, where the lengths of the two vectors represent the *magnitudes* of $N(\mathbf{r}, \boldsymbol{\Omega}, E)$ in the directions of the arrows. In this case there is thus a net drift of neutrons in the $+X$ direction. Similarly, finite values of V_y and V_z imply net drifts in the Y and Z directions. The neutron distribution described by (4.5.2) with $\mathbf{V} \neq 0$ is thus said to be *linearly anisotropic*.

The absolute value of $\mathbf{V}$ must be less than or equal to F; otherwise $N(\mathbf{r}, \boldsymbol{\Omega}, E)$ will take on unphysical (negative) values for certain directions. Thus the magnitude of the components $V_x(\mathbf{r}, E)$, $V_y(\mathbf{r}, E)$, and $V_z(\mathbf{r}, E)$ can never exceed $F(\mathbf{r}, E)$, and (4.5.2) cannot come close to representing a "beam" of neutrons (in which $N(\mathbf{r}, \boldsymbol{\Omega}, E)$ is zero except for a small cone of directions around some fixed value of $\boldsymbol{\Omega}$).

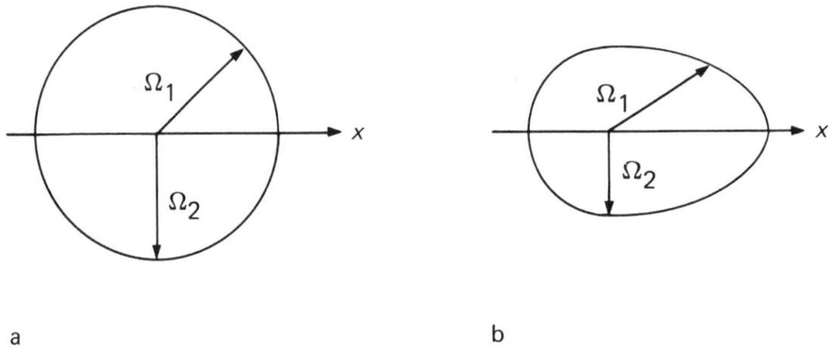

Figure 4.4 Polar plots of the neutron density in phase space.

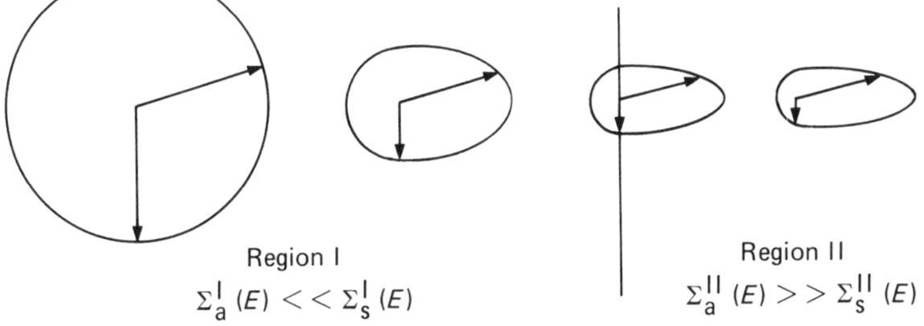

Figure 4.5 Polar plots of $N(\mathbf{r}, \mathbf{\Omega}, E)$ near an interface between a heavily and a lightly absorbing medium.

Thus Fick's Law is expected to be a poor approximation whenever the angular distribution of the neutrons is large in one preferred direction. This can happen near an interface between two media if either the scattering cross section or the absorption cross section changes abruptly.

As an example consider $N(\mathbf{r}, \mathbf{\Omega}, E)$ for $\mathbf{r}$ values near the interface in Figure 4.5. If $\Sigma_a^{II}(E) \gg \Sigma_s^{II}(E)$, $\Sigma_a^{I}(E) \ll \Sigma_s^{I}(E)$, and $\Sigma_a^{II}(E) \gg \Sigma_a^{I}(E)$, neutrons in Region II will be strongly absorbed. Hence few neutrons will enter I from II, whereas many will enter II from I. Thus, near the interface, $N(\mathbf{r}, \mathbf{\Omega}, E)$ will be large for directions $\mathbf{\Omega}$ pointing from I to II but small for directions pointing from II to I. Equation (4.5.2) cannot describe such a distribution at all well, and Fick's Law will be a poor approximation. Moreover it will continue to be a poor approximation at points further in the interior of Region II, since, with $\Sigma_a^{II}(E) \gg \Sigma_s^{II}(E)$, $N(\mathbf{r}, \mathbf{\Omega}, E)$ will continue to decrease very fast as $\mathbf{r}$ moves away from the interface. Thus the current of neutrons moving to the right will continue to be much greater than the current moving to the left (although, because of strong absorption, both currents will become very small).

To the left of the interface, however, the dependence of $N(\mathbf{r}, \mathbf{\Omega}, E)$ on $\mathbf{\Omega}$ should soon become much more smooth. With $\Sigma_s^{I}(E) \gg \Sigma_a^{I}(E)$, beam-like distributions of neutrons cannot persist. Because of scattering, the neutrons will behave like a gas diffusing through a porous plug. There may be a net drift of neutrons, but their distribution in direction can be well represented by (4.5.2). Thus, a few scattering mean free paths beyond the interface, Fick's Law is a good approximation.

A very similar argument shows that Fick's Law breaks down close to the interface if Region II is outside the reactor, where Σ_a^{II} and Σ_s^{II} are very small and are usually set equal to zero.

We conclude then that (4.5.2) should be valid a few scattering mean free paths within regions in which $\Sigma_s(\mathbf{r}, E)$ and $\Sigma_a(\mathbf{r}, E)$ do not vary rapidly as a function of position and in which $\Sigma_s(\mathbf{r}, E) \gg \Sigma_a(\mathbf{r}, E)$. It is expected to break down near the outside surface of a reactor and near (or within) strong absorbers.

Approximate Expressions for the Diffusion Constant

One useful formula for $D(\mathbf{r}, E)$ falls out of the derivation of Fick's Law based on (4.5.2), provided some extra assumptions are made. It is

$$D(\mathbf{r}, E) = \frac{1}{3[\Sigma_t(\mathbf{r}, E) - \bar{\mu}_0 \Sigma_s(\mathbf{r}, E)]}, \tag{4.5.3}$$

where $\bar{\mu}_0$ is the cosine of the scattering angle in the laboratory system averaged over all the isotopes present. Specifically we have

$$\bar{\mu}_0 \equiv \frac{\sum_j \bar{\mu}_0^j \Sigma_s^j(\mathbf{r}, E)}{\Sigma_s(\mathbf{r}, E)}, \tag{4.5.4}$$

where $\bar{\mu}_0^j$ is the average cosine of the scattering angle for isotope j, as defined by (2.5.23).

The quantity $\Sigma_t(\mathbf{r}, E) - \bar{\mu}_0 \Sigma_s(\mathbf{r}, E)$ is frequently called the *transport cross section* $\Sigma_{tr}(\mathbf{r}, E)$ for neutrons of energy E. Its inverse is called the *transport mean free path* $\lambda_{tr}(\mathbf{r}, E)$ for neutrons of energy E. Thus we have

$$D(\mathbf{r}, E) = \frac{1}{3\Sigma_{tr}(\mathbf{r}, E)} = \frac{\lambda_{tr}(\mathbf{r}, E)}{3}. \tag{4.5.5}$$

Equation (4.5.3) for $D(\mathbf{r}, E)$ does not include certain effects due to the slowing down of neutrons, but it is adequate for rough calculations.

4.6 The Energy-Dependent Diffusion Equation

Inserting the approximation (4.5.1) (Fick's Law) into the exact equation (4.4.2) relating $\mathbf{J}$ to Φ gives us an approximate equation for the scalar flux density alone:

$$-\nabla \cdot D(\mathbf{r}, E)\nabla\Phi(\mathbf{r}, E) + \Sigma_t(\mathbf{r}, E)\Phi(\mathbf{r}, E)$$
$$= \int_0^\infty \left[\sum_j \chi^j(E)\nu^j\Sigma_f^j(\mathbf{r}, E') + \Sigma_s(\mathbf{r}, E' \to E)\right]\Phi(\mathbf{r}, E')dE'. \tag{4.6.1}$$

Equation (4.6.1) is called the *continuous-energy diffusion equation.* Since the only assumption made in deriving it is the validity of Fick's Law, it can be expected to predict accurate values for $\Phi(\mathbf{r}, E)$ within weakly absorbing media, several scattering mean free paths from an external boundary or a strong neutron absorber. Moreover, if a reactor is made up of a number of regions (for example a nonfueled reflector surrounding several homogeneous fuel zones in which the enrichments of the fuel are different from one another), we expect that, if the diffusion equation is valid within each of the zones individually, it will be valid throughout the entire reactor. When (4.6.1) is applied throughout the reactor, we say that we are "representing the flux by the diffusion-theory model."

Conditions Imposed on $\Phi(\mathbf{r}, E)$ within a Reactor

The continuous-energy diffusion-theory equation is an integrodifferential equation in the three space coordinates and the energy. The coefficients (D, Σ_t, $\nu\Sigma_f$, etc.) are, in general, functions of position and energy and may be discontinuous in both these variables. Possible discontinuities in the energy pose no essential mathematical problems, however, since no derivatives with respect to energy appear and since integration of discontinuous functions is straightforward.

The situation with respect to the spatial dependence is more complicated. Equation (4.6.1) is a second-order partial-differential equation in the position coordinates, and, if it is to be meaningful, it is necessary that $\Phi(\mathbf{r}, E)$ and the three components of the vector $D(\mathbf{r}, E)\ \nabla\Phi(\mathbf{r}, E)$ be differentiable. (Otherwise $\nabla\Phi$ and $\nabla \cdot D\nabla\Phi$ would not be defined.)

We get around these potential difficulties by imposing on the desired solutions for (4.6.1) internal boundary conditions suggested by the physical meanings of the scalar flux and the net current. Thus, since $\Phi(\mathbf{r}, E) = v(E)\ n(\mathbf{r}, E)$, and since the neutron concentration $n(\mathbf{r}, E)$ must, on physical grounds, be a continuous function of position, we require that $\Phi(\mathbf{r}, E)$ be a continuous function of $\mathbf{r}$ within the reactor. We also require on physical grounds that the component $\mathbf{n}\cdot\mathbf{J}(\mathbf{r}, E)$ of the net current across an internal surface separating two different materials ($\mathbf{n}$ being the normal to that surface) be continuous. Physically this last requirement states that the net number of neutrons starting across the "minus" side of an internal surface equals the net number emerging from the "plus" side. If Fick's Law is assumed valid, this condition is equivalent to requiring that $-\mathbf{n}\cdot D(\mathbf{r}, E)\ \nabla\Phi(\mathbf{r}, E)$ be continuous in the normal direction $\mathbf{n}$ at any point $\mathbf{r}$ on a surface separating two different compositions within the reactor. Since Fick's Law is the basis for the diffusion-theory model, we then impose this condition in solving the diffusion equation (4.6.1). It follows from this condition that, since $D(\mathbf{r}, E)$ will generally change discontinuously between one composition and another, the normal component of the gradient, $\mathbf{n}\cdot\nabla\Phi(\mathbf{r}, E)$ (as distinct from $D(\mathbf{r}, E)\mathbf{n}\cdot\nabla\Phi(\mathbf{r}, E)$), will *not* be continuous across an interface between two different material compositions. Note also that *within* a uniform, homogeneous medium *any* internal surface element dS can be viewed as separating two materials that happen to be the same. Thus $D(E)\mathbf{n}\cdot\nabla\Phi(\mathbf{r}, E)$—and, as a result, $\mathbf{n}\cdot\nabla\Phi(\mathbf{r}, E)$—will be continuous across any such surface, no matter how it is oriented.

The current-continuity condition leads to mathematical difficulties if it is applied at "corner points" (for example, in two dimensions, at the four corners of a square core surrounded by a reflector). Strict application of the condition at such a point would lead to the conclusion that the components of $\nabla\Phi$ must there be either zero or infinite (although Φ itself could remain finite). Generally one avoids this difficulty by ignoring it and never attempting to make $\mathbf{n}\cdot D\nabla\Phi$ continuous "at" such a point. This is done on the physical grounds that one should only speak about a current across a finite surface element (and not "at" a point). This works in practice since, as we shall see, (4.6.1) is always solved, for complex geometrical situations, by a finite-difference approximation to the differential equation, and the application of current continuity "at" corner points is not necessary in the derivation of such difference equations.

External Boundary Conditions

The conditions we have imposed on Φ and $D\nabla\Phi$ insure that the continuous-energy diffusion equation is well defined within a reactor. However, before we can face the problem of solving this equation, it is necessary to specify boundary conditions on the outer surface of the reactor.

The definition of the "outer surface" of a reactor is itself somewhat arbitrary. We want it to be a surface such that any neutrons passing through it from the reactor side never get back into the reactor. If reactors were imbedded in a vacuum or in a nonscattering medium (and had convex surfaces), the definition of the outer surface would cause no problems: it would simply be the interface between the reactor and the vacuum or pure absorber. Any neutrons traveling out through this surface would never return. In fact, however, reactors are always surrounded by media that scatter back some of the emerging neutrons. Faced with this theoretical difficulty, we simply define the outer surface to be the smallest surface surrounding the reactor such that neutrons which *return* through that surface, having emerged from it at some prior time, produce a "negligible" effect on the criticality of the reactor and on the flux shape within it. Since almost all reactors are purposefully reflected by a region of nonfissionable material, it is usually satisfactory to define the surface of the reactor as the outer surface of this reflector. The magnitude of $\Phi(\mathbf{r}, E)$ is generally very small on this surface and continues to decrease exponentially in the material beyond it. Hence the effect on criticality and flux shape of any neutrons scattered back from that material is negligible. As a result the most common boundary condition placed on $\Phi(\mathbf{r}, E)$ is:

$$\Phi(\mathbf{r}, E) = 0 \text{ for } \mathbf{r} \text{ on the outer surface of the reflector.} \tag{4.6.2}$$

The best way to test the appropriateness of this condition is to solve (4.6.1) twice, first with the condition (4.6.2) and then with the condition that $\Phi(\mathbf{r}, E) = 0$ for $\mathbf{r}$ on some surface *beyond* the outer surface of the reflector. If including in the calculation whatever additional material is between these two surfaces produces a negligible change in the solution to (4.6.1), the assumption that $\Phi(\mathbf{r}, E)$ goes to zero on the first surface is satisfactory.

It follows immediately from these considerations that another procedure which should lead to a boundary condition fully as satisfactory as (4.6.2) is to assume that the reflector material extends to infinity and that $\Phi(\mathbf{r}, E)$ vanishes there. Use of this second boundary condition somewhat simplifies the mathematics of solving (4.6.1) analytically.

There is a third boundary condition applicable to (4.6.1) for those rare cases in which the reactor is unreflected. In such a case neutron production due to fissions can take place right up to the outer boundary. There is thus a considerable amount of neutron leakage through this boundary into the surrounding vacuum, and it is a somewhat questionable procedure to set $\Phi(\mathbf{r}, E)$ to zero there.

In order to obtain a more justifiable boundary condition for this situation, we first take note of the exact physical condition that $N(\mathbf{r}, \mathbf{\Omega}, E)$ ought to obey on such a boundary. This condition is quite simple: $N(\mathbf{r}, \mathbf{\Omega}, E)$ must be zero on the outer surface for all directions $\mathbf{\Omega}$ pointed *into* the reactor. Mathematically, if $\mathbf{n}$ is an outward-pointing normal at the outer surface, we have

$$N(\mathbf{r}, \mathbf{\Omega}, E) = 0 \text{ for } \mathbf{r} \text{ on outer surface and } \mathbf{\Omega}\cdot\mathbf{n} < 0. \tag{4.6.3}$$

In order to obtain from this exact condition one which applies to the $\Phi(\mathbf{r}, E)$ appearing in the approximate equation (4.6.1), we make use of the basic approximation (4.5.2) concerning the functional form of $N(\mathbf{r}, \mathbf{\Omega}, E)$, which underlies Fick's Law—and hence (4.6.1). By substituting the approximation for $N(\mathbf{r}, \mathbf{\Omega}, E)$ given by (4.5.2) into the definitions (4.2.8) for $\mathbf{J}(\mathbf{r}, E)$ and (2.2.7) for $\Phi(\mathbf{r}, E)$, we can easily show that the values of $\mathbf{J}$ and Φ consistent with (4.5.2) are $\mathbf{J} = v\mathbf{V}/3$, $\Phi = vF$. Thus (4.5.2) becomes

$$\begin{aligned} v(E)\, N(\mathbf{r}, \mathbf{\Omega}, E) &= \Phi(\mathbf{r}, E) + 3\mathbf{\Omega}\cdot\mathbf{J}(\mathbf{r}, E) \\ &= \Phi(\mathbf{r}, E) + 3[\Omega_x J_x(\mathbf{r}, E) + \Omega_y J_y(\mathbf{r}, E) + \Omega_z J_z(\mathbf{r}, E)]. \end{aligned} \tag{4.6.4}$$

It is fairly clear that, with this simplified functional form assumed, $N(\mathbf{r}, \mathbf{\Omega}, E)$ cannot be zero on the surface for *all* directions $\mathbf{\Omega}$ pointing into the reactor. The mathematical reason for this is that requiring there to be no returning neutron beam for any *particular* direction yields an algebraic equation relating $\Phi(\mathbf{r}, E)$ to the components of $\mathbf{J}(\mathbf{r}, E)$ on the surface; imposing this requirement for five or more different directions of reentry thus yields an overdetermined set of equations for Φ, J_x, J_y, and J_z, the only solution to which is $\Phi = J_x = J_y = J_z = 0$, a condition which is clearly unreasonable on physical grounds.

We circumvent these difficulties by requiring only that the *net incoming current* through each square centimeter of surface area be zero. According to (4.2.5) the rate at which neutrons in the beam $N(\mathbf{r}, \mathbf{\Omega}, E)d\Omega\, dE$ cross a unit surface with normal $\mathbf{n}$ is $v\mathbf{n}\cdot\mathbf{\Omega}N(\mathbf{r}, \mathbf{\Omega}, E)d\Omega\, dE$. Thus, if we add together all *inward*-directed currents on a surface whose *outward*-directed normal is $\mathbf{n}$ and require that this sum (integral) be zero, we get the boundary condition

$$0 = \int_{\mathbf{n}\cdot\mathbf{\Omega}<0} d\Omega\, \mathbf{n}\cdot\mathbf{\Omega}\, v(E)\, N(\mathbf{r}, \mathbf{\Omega}, E) \text{ for } \mathbf{r} \text{ on the outer surface,} \tag{4.6.5}$$

where the notation $\mathbf{n}\cdot\mathbf{\Omega} < 0$ means that the integral over $\mathbf{\Omega}$ is to be taken only over those directions for which $\mathbf{n}\cdot\mathbf{\Omega} < 0$, $\mathbf{n}$ being the outward-directed normal to the surface.

It is clear that setting $N(\mathbf{r}, \mathbf{\Omega}, E)$ equal to zero for *all* $\mathbf{\Omega}$ such that $\mathbf{n}\cdot\mathbf{\Omega} < 0$ will satisfy the condition (4.6.5). Thus the condition is consistent with the true physical boundary condition. However (4.6.5) can also be satisfied by many other functions $N(\mathbf{r}, \mathbf{\Omega}, E)$ that are not zero for all inward-directed $\mathbf{\Omega}$. In particular the form (4.6.4) can be made to satisfy (4.6.5); and, by forcing this condition, a relationship between Φ and the components of $\mathbf{J}$ on the outer surface can be obtained.

To simplify the algebra we shall carry through the details for only one surface of a reactor having the shape of a rectangular parallelopiped. If we choose this surface as the one with outward normal pointing in the $+Z$ direction (i.e., take $\mathbf{n}$ in (4.6.5) equal to the unit vector $\mathbf{k}$), we have, from Figure 2.3,

$$\begin{aligned} \mathbf{\Omega}\cdot\mathbf{n} &= \cos\theta, \\ \mathbf{\Omega} &= \mathbf{i}\sin\theta\cos\chi + \mathbf{j}\sin\theta\sin\chi + \mathbf{k}\cos\theta, \\ d\Omega &= \frac{1}{4\pi}\sin\theta\, d\theta\, d\chi. \end{aligned} \tag{4.6.6}$$

Substitution of (4.6.4) into (4.6.5) then yields

$$\begin{aligned} 0 = \int_0^{2\pi}\frac{d\chi}{2\pi}\int_{\pi/2}^{\pi}\frac{d\theta}{2}\sin\theta\cos\theta\,\{\Phi(\mathbf{r}, E) + 3[\sin\theta\cos\chi\, J_x(\mathbf{r}, E) + \sin\theta\sin\chi\, J_y(\mathbf{r}, E) \\ + \cos\theta\, J_z(\mathbf{r}, E)]\} \text{ on the } z^+ \text{ surface,} \end{aligned} \tag{4.6.7}$$

or, with $\mu \equiv \cos\theta$,

$$0 = \int_0^{-1} -\frac{d\mu}{2}[\mu\Phi(\mathbf{r}, E) + 3\mu^2 J_z(\mathbf{r}, E)] = -\frac{1}{4}\Phi(\mathbf{r}, E) + \frac{1}{2}J_z(\mathbf{r}, E). \tag{4.6.8}$$

It is consistent with the approximation (4.6.4) to use Fick's Law to relate $J_z(\mathbf{r}, E)$ to $\Phi(\mathbf{r}, E)$. When we do this, (4.6.8) becomes

$$\Phi(\mathbf{r}, E)|_{z^+} = -2D(\mathbf{r}, E)\left.\frac{\partial\Phi(\mathbf{r}, E)}{\partial z}\right|_{z^+}, \tag{4.6.9}$$

the subscript indicating that quantities are to be evaluated at (x, y, z^+).

Analogous results can be derived for the other surfaces of the reactor. The general expression is

$$\Phi(\mathbf{r}, E)|_{S_0} = -2D(\mathbf{r}, E)\mathbf{n}\cdot\nabla\Phi(\mathbf{r}, E)|_{S_0} \text{ for } \mathbf{n} \text{ normal to the outer surface } S_0. \tag{4.6.10}$$

This boundary condition is a bit awkward to apply. However we can derive from it an alternate condition which is almost equivalent. To illustrate, we again restrict attention to the z^+ surface. Since $\Phi(\mathbf{r}, E)$ is positive on that surface, (4.6.9) shows that the corresponding $\partial\Phi/\partial z|_{z^+}$ must be negative. A sketch of Φ versus z near the surface is shown in Figure 4.6a.

If the plot of Φ versus z is extended as a straight line with its slope equal to the value on the surface, Φ will go to zero at some point beyond z^+. Figure 4.6b shows this behavior with the distance between the physical surface and the point at which $\Phi(\mathbf{r}, E)$ extrapolates to zero represented by the symbol δ. Since a straight-line extrapolation with a slope equal to that at the surface has been made, we find that

$$\frac{0 - \Phi|_{z^+}}{\delta} = \left.\frac{\partial\Phi}{\partial z}\right|_{z^+}. \tag{4.6.11}$$

Then from (4.6.9) and (4.5.5) we obtain the result

$$\delta = 2D(\mathbf{r}, E)|_{z^+} = 0.67\lambda_{\mathrm{tr}}(\mathbf{r}, E)|_{z^+}. \tag{4.6.12}$$

The quantity δ is known as the *extrapolation distance*. It is important to remember that it has *mathematical* rather than *physical* significance. The *physical* value of the scalar flux doesn't fall to zero δ centimeters after entering a vacuum. In fact the neutrons making up this flux continue moving out to infinity once they leak into the vacuum. The extrapolation distance is thus an entirely artificial quantity: *if* the slope of the scalar flux density remained constant at distances into the vacuum beyond the physical surface, *then* $\Phi(\mathbf{r}, E)$ would go to zero δ centimeters from that physical surface. It follows that,

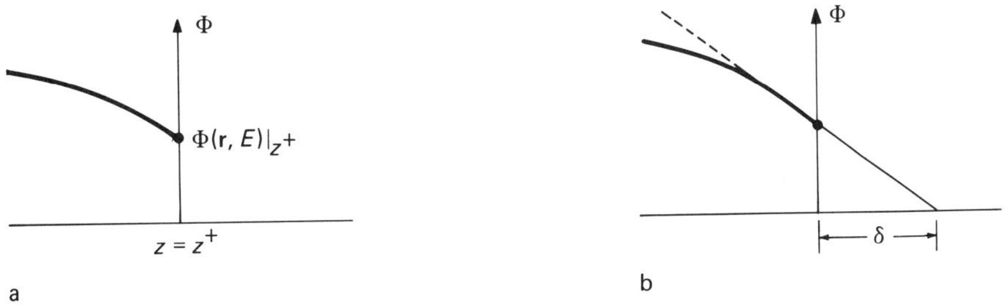

Figure 4.6 The flux shape and extrapolation distance near the outer surface of a reactor embedded in a vacuum.

if we artificially extend the physical reactor out a distance δ beyond its physical surface and there impose the boundary condition $\Phi = 0$, the resultant solution for the diffusion equation (4.6.1) will come very close to obeying (4.6.10) at the *physical* surface, and thus will be an acceptable solution to the physical problem.

Thus we may replace the condition on the logarithmic derivative $(1/\Phi)(\partial\Phi/\partial z)|_{z^+}$ at the physical surface of the reactor by the condition that $\Phi(\mathbf{r}, E)$ be taken as zero on the outer surface of a reactor whose outermost composition is extended a distance 0.67 λ_{tr} beyond the actual physical surface. This latter condition is sometimes more convenient to deal with mathematically. The two conditions are not *exactly* equivalent since we do not continue any curvature that $\Phi(\mathbf{r}, E)$ may have into the extrapolation zone. But they are so close that there is little to choose between them, particularly in view of the approximate nature of the expression (4.6.4) for $N(\mathbf{r}, \mathbf{\Omega}, E)$, which is the basis of this whole development of diffusion theory.

The extrapolation distance is rarely larger than a centimeter or two, and thus, for bare reactors having overall dimensions of a meter or more, it is generally satisfactory to take as the boundary condition on $\Phi(\mathbf{r}, E)$ in (4.6.1) that Φ vanishes on the actual physical boundary. Only for small bare cores is it necessary to use the boundary condition (4.6.10), or its near equivalent that $\Phi(\mathbf{r}, E)$ be zero a distance δ beyond the physical surface of the reactor.

Thus the special boundary conditions for surfaces adjacent to a vacuum are rarely employed. Simply requiring $\Phi(\mathbf{r}, E)$ to vanish on the outer surface of the reactor is almost always sufficient.

4.7 Accurate Solution of the Diffusion-Theory Equation for Bare, Homogeneous Cores

The continuous-energy diffusion equation (4.6.1) is considerably easier to deal with than is the Boltzmann transport equation (the equation for $N(\mathbf{r}, \mathbf{\Omega}, E)$). Nevertheless it is still a very difficult equation to solve even for the simplest reactor. We shall approach the problem by considering cases of increasing complexity, starting with a description of methods suitable for bare, homogeneous reactors and going on to indicate what refinements—or, more correctly, what completely new methods—must be employed to deal with the complex multiregional cores almost always encountered in practice.

By definition a homogeneous reactor is one in which the concentrations of materials $n^j(\mathbf{r})$ are uniform and hence independent of $\mathbf{r}$. It follows that all the macroscopic cross sections are also independent of $\mathbf{r}$, and (4.6.1) therefore reduces to

$$-D(E)\nabla^2\Phi(\mathbf{r}, E) + \Sigma_t(E)\Phi(\mathbf{r}, E) = \int_0^\infty \left[\frac{1}{\lambda}\chi(E)\nu\Sigma_f(E') + \Sigma_s(E' \to E)\right]\Phi(\mathbf{r}, E')\,dE', \quad (4.7.1)$$

where, to reduce detail, we assume that $\chi(E)$ is the same for all isotopes j.

We can find a solution to this equation by writing $\Phi(\mathbf{r}, E)$ as a separable function:

$$\Phi(\mathbf{r}, E) = R(\mathbf{r})\psi(E). \tag{4.7.2}$$

Substituting this assumed form for $\Phi(\mathbf{r}, E)$ into (4.7.1) yields

$$-D(E)\psi(E)\frac{\nabla^2 R(\mathbf{r})}{R(\mathbf{r})} + \Sigma_t(E)\psi(E) = \int_0^\infty \left[\frac{1}{\lambda}\chi(E)\nu\Sigma_f(E') + \Sigma_s(E' \to E)\right]\psi(E')\,dE', \tag{4.7.3}$$

or

$$-\frac{\nabla^2 R(\mathbf{r})}{R(\mathbf{r})} = -\frac{\Sigma_t(E)}{D(E)} + \frac{1}{D(E)\psi(E)}\int_0^\infty \left[\frac{1}{\lambda}\chi(E)\nu\Sigma_f(E') + \Sigma_s(E' \to E)\right]\psi(E')\,dE'. \tag{4.7.4}$$

This result states that a function of $\mathbf{r}$ alone (the left-hand side of (4.7.4)) equals a function of E alone (the right-hand side of (4.7.4)), a situation which can be true only if both functions equal the same constant. We call this constant $B_{\mathbf{r}}^2$ and expect, on physical grounds, that $B_{\mathbf{r}}^2$ will be real (although it may be negative). It follows that the assumption (4.7.2) can be valid only if

$$\nabla^2 R(\mathbf{r}) = -B_{\mathbf{r}}^2 R(\mathbf{r}). \tag{4.7.5}$$

Accordingly we seek a function $R(\mathbf{r})$ that obeys (4.7.5). Moreover, if (4.7.2) is to be a solution, $R(\mathbf{r})\psi(E)$ must meet both the condition that $\Phi(\mathbf{r}, E)$ and $\mathbf{n}\cdot D(E)\nabla\Phi(\mathbf{r}, E)$ be continuous at all interior points for all directions $\mathbf{n}$ and the boundary condition (4.6.2).

The Infinite Slab Reactor

To illustrate how such an acceptable solution can be found we first consider the simple case of a one-dimensional slab reactor, assuming that the reactor extends to $\pm\infty$ in the Y and Z directions but is of thickness L_x in the X direction. Then the physically acceptable value of $R(\mathbf{r}) = R(x, y, z)$ becomes independent of the variables y and z and can therefore be written $R(x)$, so that (4.7.5) reduces to

$$\nabla^2 R(\mathbf{r}) = \frac{\partial^2 R(x)}{\partial x^2} + \frac{\partial^2 R(x)}{\partial y^2} + \frac{\partial^2 R(x)}{\partial z^2} = \frac{\partial^2 R(x)}{\partial x^2} = -B_x^2 R(x). \tag{4.7.6}$$

The general solution for $R(x)$ is

$$R(x) = C_1 \sin B_x x + C_2 \cos B_x x, \tag{4.7.7}$$

where C_1 and C_2 are constants to be determined from other considerations.

Although (4.7.7) is a solution of (4.7.5) for the case at hand, we cannot accept it without checking to see if it meets the boundary conditions.

With respect to the internal conditions it is satisfactory. The spatial continuity properties of $R(x)\psi(E)$ are determined entirely by $R(x)$, which is continuous. Moreover the gradient of $R(x)$ is $\mathbf{i}\partial(\sin B_x x)/\partial x = \mathbf{i}B_x \cos B_x x$, and, since $D(E)$ does not depend on x in this case, $\mathbf{n}\cdot\mathbf{i}D(E)\partial R(x)/\partial x$ is continuous in the direction of $\mathbf{n}$.

The external boundary conditions are a different matter. The form (4.7.2), when inserted into (4.6.9), implies that

$$\frac{\partial}{\partial x}\ln R(x)\bigg|_{x^+} = \frac{-1}{2D(E)}.$$

But this can be true only if we replace $D(E)$ by some constant average value. Requiring that $R(x)$ vanish at a distance δ beyond the physical boundary of the reactor leads to a similar problem since δ is also a function of energy (see (4.6.12)). Thus, rather than abandoning the separable form (4.7.2) or using an average value of $D(E)$, we shall apply the simpler, zero-flux condition (4.6.2). To investigate this last condition explicitly let us assume that the left-hand side of the reactor is at $x = 0$ and the right-hand side is at $x = L_x$. Thus $R(0) = C_2$ for any value of B_x, and, hence, C_2 must be zero.

We next want $\sin B_x x = 0$ for $x = 0$ and for $x = L_x$. The condition at $x = 0$ is met for all values of B_x. At $x = L_x$, however, $\sin B_x L_x$ will be zero only if

$$B_x \equiv B_{xn} = \frac{n\pi}{L_x} \qquad (n = 0, \pm 1, \pm 2, \ldots). \tag{4.7.8}$$

There are thus an infinite number of values of $B_x (B_{x1}, B_{x2}, B_{x3}$, etc.) which yield solutions that vanish at $x = 0$ and $x = L_x$. Since $n = 0$ leads to the trivial solution $R(x) = 0$ everywhere, and solutions for negative n differ only in the sign of C_1, we need consider only positive n. To pick from these remaining solutions the one appropriate to the physical problem we impose another condition of acceptability on the solution $\Phi(\mathbf{r}, E)$: we require, on physical grounds, that

$$\Phi(\mathbf{r}, E) \geq 0 \quad \text{for all } E \text{ and all } \mathbf{r} \text{ within the reactor.} \tag{4.7.9}$$

Since $\sin(n\pi x/L_x)$ is negative somewhere in the range $0 < x < L_x$ for all values of n except $n = 1$, this condition eliminates all of the B_{xn} except B_{x1}. (Note that condition (4.7.9) along with the requirement $R(0) = R(L_x) = 0$ also eliminates the possibility of B^2 being negative.) We then have, for this slab reactor, that

$$R(x) = C_1 \sin B_{x1} x = C_1 \sin\frac{\pi x}{L_x}. \tag{4.7.10}$$

To complete the solution for $\Phi(x, E)$ we must find $\psi(E)$. This can be done by inserting (4.7.10) into (4.7.3):

$$[D(E)B_{x1}^2 + \Sigma_t(E)]\psi(E) = \int_0^\infty \left[\frac{1}{\lambda}\chi(E)\nu\Sigma_f(E') + \Sigma_s(E' \to E)\right]\psi(E')\,dE'. \tag{4.7.11}$$

Except for the added "leakage" term $D(E)\,B_{x1}^2$ and the neglect of possible differences in the $\chi^j(E)$, (4.7.11) is identical with the infinite-homogeneous-medium equation (3.2.4). Thus all the theory we developed in Sections 3.3 and 3.4 for dealing with that equation is immediately applicable. We simply add the term $D(E)B_{x1}^2$ to $\Sigma_t(E)$ in all the former results.

The term $D(E)B_{x1}^2$ has a physical interpretation based on the fact that $C_1 D(E)B_{x1}^2 \sin B_{x1}x\ \psi(E)\ dV\ dE$ is the number of neutrons in the energy range dE that leak out of the volume dV about point x per second. Our results thus show that, in a homogeneous, bare, slab reactor, the leakage rate per unit volume per unit flux is $D(E)B_{x1}^2$, the same at all locations in the reactor. Hence introducing the leakage term $D(E)B_{x1}^2$ in the infinite-medium equation (3.2.4) produces the same effect on the neutron economy as would an increase in the absorption cross section. For a large reactor, moreover, this effect is quite small. For example, if L_x, the thickness of the reactor, is 314 cm (about 10 ft), $B_{x1}^2 = (\pi/314)^2 \approx 10^{-4}\ \text{cm}^{-2}$. Thus, for $D(E) = 2$ cm (a typical value), $D(E)B_{x1} \approx 0.0002\,\text{cm}^{-1}$, a small number compared to the value of $\Sigma_a(E)$, which generally lies in the range 0.01 to 0.1 cm^{-1}. Thus, for a large slab, the spectrum $\psi(E)$ is little affected by neutron leakage and will be quite similar to $\Phi(E)$, the spectrum in an infinite medium of the same material.

Our previous results then show that, with B_{x1} known from the geometry of the system, the most accurate way to determine $\psi(E)$ and thus complete the problem of the bare slab, is to split the energy range into many groups and solve the resultant algebraic equations for the multigroup fluxes. To do this we define a group diffusion constant D_g by

$$D_g \equiv \frac{\int_{E_g}^{E_{g-1}} D(E)\psi(E)\,dE}{\int_{E_g}^{E_{g-1}} \psi(E)\,dE} \tag{4.7.12}$$

and then in (3.3.4) add terms $D_g B_{x1}^2$ to all the Σ_{tg}. If we represent the group flux analogous to the Φ_g of (3.3.3) by ψ_g, the energy-group equivalent of (4.7.11) becomes

$$(D_g B_{x1}^2 + \Sigma_{tg})\psi_g = \sum_{g'=1}^{G} \left[\frac{1}{\lambda}\chi_g \nu\Sigma_{fg'} + \Sigma_{gg'}\right]\psi_{g'} \qquad (g = 1, 2, \ldots, G). \tag{4.7.13}$$

Solution of these equations gives us an excellent approximation for the energy part $\psi(E)$ of the solution $\Phi(\mathbf{r}, E) = R(\mathbf{r})\psi(E)$. It also gives us the value of λ (the "eigenvalue") required if this solution is to be positive for all energy groups. If λ turns out to be unity, the theory is predicting that the particular slab reactor being analyzed will be critical. If λ is *not* unity, we can determine what must be done to *make* the system critical. For example, if solution of the critical determinant associated with (4.7.13) yields a value of λ greater than unity, we know from our earlier discussion that the reactor will be supercritical. We may then search for the critical conditions by decreasing the fuel loading, increasing a control poison, or increasing the leakage out of the reactor (i.e., decreasing L_x and hence increasing $D(E)B_{x1}^2$).

Just as in the infinite-medium case, a real positive value of λ, leading to a positive vector $[\psi] \equiv \text{Col}\{\psi_1, \psi_2, \ldots, \psi_G\}$ whose components satisfy (4.7.13), always exists. Thus we can always find an excellent approximation to $\psi(E)$; and a complete solution of the diffusion equations for the slab reactor is then given by

$$\Phi(x, E) = C_1\psi(E) \sin B_{x1}x = C_1\psi(E) \sin \frac{\pi x}{L_x}. \tag{4.7.14}$$

The normalization constant C_1 in (4.7.14) is usually expressed in terms of the power level per unit volume of the reactor. If this quantity, for which we shall use the symbol $\bar{P}$, is given in watts per unit volume, we have (with a power level of one watt corresponding to 3×10^{10} fissions per second)

$$\bar{P} = \frac{1}{3 \times 10^{10}} \frac{\int_{\text{reactor}} dV \int_0^\infty dE\, \Sigma_f(\mathbf{r}, E)\Phi(\mathbf{r}, E)}{\int_{\text{reactor}} dV}. \tag{4.7.15}$$

Inserting (4.7.14) into this expression, we get for the bare slab core

$$C_1 = \frac{3 \times 10^{10}\, \bar{P}\pi/2}{\int_0^\infty \Sigma_f(E)\psi(E)\, dE}. \tag{4.7.16}$$

Often there is interest only in the *relative* flux from point to point or energy to energy. If this is the case, any convenient arbitrary value can be set for C_1. Notice that this freedom to set any value for C_1 implies that a reactor can be critical at *any* power level (until feedback effects change local atomic number densities or temperature-averaged cross sections). Criticality is a *balance condition.* A reactor will be critical if the total number of neutrons created per second by fission equals the total number destroyed (by absorption or leakage out of the reactor). The criticality condition tells us nothing about the *level* of these production and destruction processes.

Relationship between k_{eff} and k_∞: The Nonleakage Probability

Because of the similar mathematical structures of the finite-medium equations (4.7.11) and (4.7.13) and their infinite-medium counterparts (3.2.5) and (3.3.9), all of the analytical approximations discussed in connection with the infinite-medium case are immediately applicable to the case of the finite homogeneous reactor. One important example of this occurs if we repeat the procedure used to derive the expression (3.3.21) for k_{eff}. Summing (4.7.13) over all energy yields

$$\lambda \equiv k_{\text{eff}} = \frac{\sum_{g=1}^G \nu\Sigma_{fg}\psi_g}{\sum_{g=1}^G [D_g B_{x1}^2 + \Sigma_{ag}]\psi_g}, \tag{4.7.17}$$

which can be rewritten

$$k_{\text{eff}} = \frac{1}{1 + [\sum_{g=1}^{G} D_g\psi_g / \sum_{g=1}^{G} \Sigma_{ag}\psi_g] B_{x1}^2} \frac{\sum_{g=1}^{G} \nu\Sigma_{fg}\psi_g}{\sum_{g=1}^{G} \Sigma_{ag}\psi_g}. \tag{4.7.18}$$

Comparison with (3.3.21) and application of (3.3.23) shows that, except for the slight difference between the Φ_g and ψ_g (the energy dependence of the fluxes in the infinite and finite media), the last term in (4.7.18) is what we have defined as k_∞. The leakage effect (i.e., the part involving B_{x1}^2) is contained in the first term. In fact it is customary to call this first term the *nonleakage probability*. Accordingly we have the result that, for a homogeneous slab reactor, k_{eff} is closely approximated by the product of the nonleakage probability and the infinite-medium multiplication constant for the material making up the slab:

$$k_{\text{eff}} = (1 - P_{\text{L}})k_\infty, \tag{4.7.19}$$

where we define the *leakage probability* P_{L} for the slab by

$$P_{\text{L}} \equiv \frac{[\int_0^\infty D(E)\psi(E)\,dE / \int_0^\infty \Sigma_{\text{a}}(E)\psi(E)\,dE] B_{x1}^2}{1 + [\int_0^\infty D(E)\psi(E)\,dE / \int_0^\infty \Sigma_{\text{a}}(E)\psi(E)\,dE] B_{x1}^2}. \tag{4.7.20}$$

As would be expected, P_{L} increases as $B_{x1}(\equiv \pi/L_x)$ increases (i.e., as the core thickness L_x *decreases*). It also increases as $\Sigma_{\text{a}}(E)$ decreases and as $D(E)$ increases (and thus as $\Sigma_{\text{tr}}(E)$ decreases). Decreases in either $\Sigma_{\text{a}}(E)$ or $\Sigma_{\text{tr}}(E)$ permit neutrons to move further between interactions, and this leads to a greater probability that they will leak out of the reactor before any interaction can occur.

Extension to Other Bare-Core Geometries

The extension of the results we have developed for the homogeneous slab reactor to other geometries is straightforward. It is only necessary to find a real, positive B_{r}^2 for (4.7.5) such that the corresponding $R(\mathbf{r})$ vanishes on the outer boundary of the reactor and is nonnegative within the reactor, and to substitute this value of B_{r}^2 for the B_{x1}^2 in the expressions developed for the slab case. There is a mathematical theorem which guarantees that such a solution always exists, although the problem of finding this solution can be extremely difficult for complicated geometrical shapes. For rectangular parallelopipeds, spheres, and cylinders, however, it is quite simple. If, for example, the reactor has the shape of a rectangular parallelopiped extending from 0 to L_x in the X direction, from 0 to L_y in the Y direction, and from 0 to L_z in the Z direction, then

$$R(\mathbf{r}) = R(x, y, z) = C \sin\frac{\pi x}{L_x} \sin\frac{\pi y}{L_y} \sin\frac{\pi z}{L_z}. \tag{4.7.21}$$

Substituting this solution into (4.7.5) yields

$$\begin{aligned}\nabla^2 R(\mathbf{r}) &= \left(\frac{\partial^2}{\partial x^2} + \frac{\partial^2}{\partial y^2} + \frac{\partial^2}{\partial z^2}\right) C \sin\frac{\pi x}{L_x} \sin\frac{\pi y}{L_y} \sin\frac{\pi z}{L_z} \\ &= -C\left[\left(\frac{\pi}{L_x}\right)^2 + \left(\frac{\pi}{L_y}\right)^2 + \left(\frac{\pi}{L_z}\right)^2\right] \sin\frac{\pi x}{L_x} \sin\frac{\pi y}{L_y} \sin\frac{\pi z}{L_z} \\ &= -\left[\left(\frac{\pi}{L_x}\right)^2 + \left(\frac{\pi}{L_y}\right)^2 + \left(\frac{\pi}{L_z}\right)^2\right] R(\mathbf{r}) \\ &= -B_{\mathbf{r}}^2 R(\mathbf{r}).\end{aligned} \tag{4.7.22}$$

Expressions for the Laplacian ∇^2, the solutions, and the corresponding values of $B_{\mathbf{r}}^2$ for rectangular parallelopipeds, spheres, and cylinders are given in Table 4.1. The number $B_{\mathbf{r}}^2$ is called the *geometrical buckling*. Its units are cm^{-2}. As we have seen, it provides a measure of the amount of leakage out of the reactor.

Examination of (4.7.4) and (4.7.5) shows that the spectrum $\psi(E)$ in a bare, homogeneous reactor depends only on the total geometrical buckling and not on the particular shape of the reactor. A sphere and a cylinder having the same geometrical buckling will have the same neutron energy distribution, and the neutrons in such cores will have the same leakage probability.

We see from Table 4.1 that, in the case of the sphere, the everywhere-positive solution $R(\mathbf{r})$ $(= R(r, \theta, \varphi))$ depends only on the radial position r and not on the polar and azimuthal angles. Similarly the solution for cylinders $(R(\mathbf{r}) = R(\rho, \varphi, z))$ depends only on ρ and z. (In the cylindrical case $J_0(2.405\rho/R)$ is the zeroth-order "ordinary Bessel function of the first kind," a function which is finite at $\rho = 0$ and becomes zero when its argument is 2.405; i.e., $J_0(2.405) = 0$.)

Just as in the case of the slab reactor, so there are for all geometries other values of $B_{\mathbf{r}}^2$ and corresponding solutions $R(\mathbf{r})$ that meet the internal continuity and external boundary conditions. Here too, however, these are found to yield flux shapes that are negative for certain regions within the core; hence they are not acceptable physical solutions.

Summary

To sum up the situation for the finite homogeneous reactor: the solution to the continuous-energy diffusion equation (4.7.1) for a bare, homogeneous reactor is

$$\Phi(\mathbf{r}, E) = R(\mathbf{r})\psi(E), \tag{4.7.2}$$

where $R(\mathbf{r})$ is the one everywhere-positive solution of

$$\nabla^2 R(\mathbf{r}) = -B_{\mathbf{r}}^2 R(\mathbf{r}), \tag{4.7.5}$$

$R(\mathbf{r})$ being the *shape* of the flux and $B_{\mathbf{r}}^2$, the geometrical buckling (often simply called

Table 4.1 Expressions for the Laplacian, Flux Shape, and Geometrical Buckling of Bare, Homogeneous Reactors Having Simple Geometries

Geometry	∇^2	$R(\mathbf{r})$	$B_{\mathbf{r}}^2$
Rectangular parallelopiped of dimensions $L_x \times L_y \times L_z$	$\frac{\partial^2}{\partial x^2} + \frac{\partial^2}{\partial y^2} + \frac{\partial^2}{\partial z^2}$	$C \sin \frac{\pi x}{L_x} \sin \frac{\pi y}{L_y} \sin \frac{\pi z}{L_z}$	$\left[\left(\frac{\pi}{L_x}\right)^2 + \left(\frac{\pi}{L_y}\right)^2 + \left(\frac{\pi}{L_z}\right)^2\right]$
Sphere of radius R 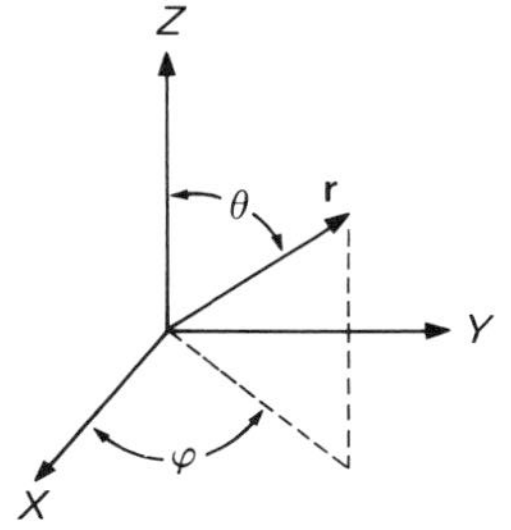	$\frac{1}{r^2 \sin\theta}\left[\sin\theta \frac{\partial}{\partial r}\left(r^2 \frac{\partial}{\partial r}\right) + \frac{\partial}{\partial\theta}\left(\sin\theta \frac{\partial}{\partial\theta}\right) + \frac{1}{\sin\theta}\frac{\partial^2}{\partial\varphi^2}\right]$	$\frac{C \sin \frac{\pi r}{R}}{r}$	$\left(\frac{\pi}{R}\right)^2$
Cylinder of height L_z and radius R 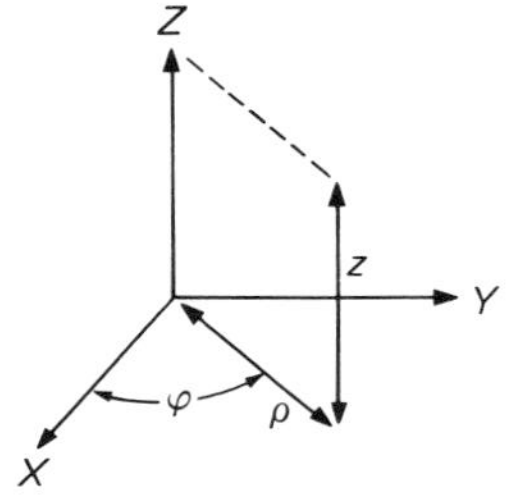	$\frac{1}{\rho}\left[\frac{\partial}{\partial\rho}\left(\rho \frac{\partial}{\partial\rho}\right) + \frac{1}{\rho}\frac{\partial^2}{\partial\varphi^2} + \rho \frac{\partial^2}{\partial z^2}\right]$	$C J_0\left(\frac{2.405\,\rho}{R}\right) \sin \frac{\pi}{L_z}$	$\left[\left(\frac{2.405}{R}\right)^2 + \left(\frac{\pi}{L_z}\right)^2\right]$

the buckling), being a number that depends on the geometry and dimensions of the reactor, and where the spectrum function $\psi(E)$ is the solution of

$$[D(E)B_r^2 + \Sigma_t(E)]\psi(E) = \int_0^\infty \left[\frac{1}{\lambda}\chi(E)\nu\Sigma_f(E') + \Sigma_s(E' \to E)\right]\psi(E'). \tag{4.7.23}$$

A sequence of steps for calculating the criticality of a bare core is as follows:

1. From the dimensions and geometry (assuming it is simple) obtain B_r^2 and $R(\mathbf{r})$ using the formulas given in Table 4.1. (Note that for the bare core the spatial shape of the flux depends only on the geometry and not on the material constituents.)
2. Knowing B_r^2, solve (4.7.11) (usually cast in the energy-group form (4.7.13)) for λ and $\psi(E)$.
3. If $\lambda \neq 1$, so that the reactor is not predicted to be critical, alter whatever parameter is adjustable (fuel loading, control-poison concentrations, size—i.e., B_r^2) until $\lambda = 1$.

For a large, bare, homogeneous reactor for which the leakage probability P_L defined by (4.7.20) is less than about 5 percent, the accuracy with which k_{eff} (i.e., λ) can be computed by this procedure (assuming cross sections are known) is better than a few tenths of a percent. Thus, for the large, bare, homogeneous core, errors in predicting criticality due to nuclear data will almost always outweigh errors inherent in the use of the diffusion-theory approximation.

4.8 Age Theory

Just as in the infinite-medium case, there are approximate analytical ways of dealing with the continuous-energy diffusion equation for a bare, homogeneous reactor. The most famous of these is called *age theory*. We have already touched on the approximation for the infinite homogeneous medium. In this section we shall extend it to the case of the bare, homogeneous reactor. The result will be a simple, approximate expression for k_{eff} in terms of k_∞, B_r^2 (the geometrical buckling of the bare core), and two new quantities called the *Fermi age* and the *thermal diffusion length*. The great advantage of introducing the age is that it can be computed directly from cross-sectional data without first having to determine the neutron spectrum $\psi(E)$ by elaborate multigroup methods. Moreover both the age and the thermal diffusion length can be measured directly by integral experiments that do not require any information about the cross sections of the core materials. Since the four factors making up k_∞ can also be measured in this integral manner, age theory permits a prediction of criticality that does not require the availability of detailed nuclear data or of a large computing machine.

The Fermi Age Equation

As was shown in Section 3.4 (see (3.4.42)), the essence of the age approximation is to deal with the scattering-in integrals $\int_E^{E/\alpha_j}[\Sigma_s^j(E')\Phi(E')]/[E'(1-\alpha_j)]\,dE'$ of (3.4.9) by expanding $E'\Sigma_s^j(E')\Phi(E')$ in a power series about the final energy E. If we make the assumption that scattering is elastic and isotropic in the center-of-mass system, these same integrals will appear on the right-hand side of the diffusion equation (4.7.1) for the bare, homogeneous reactor. Since the age approximation involves only the scattering integrals, the former results (3.4.44) can be applied at once to the scattering-in integral $\int_0^\infty \Sigma_s(E' \to E)\Phi(\mathbf{r}, E)\,dE'$ of (4.7.1). The result is

$$-D(E)\nabla^2\Phi(\mathbf{r}, E) + \Sigma_a(E)\Phi(\mathbf{r}, E) = \int_0^\infty \frac{1}{\lambda}\chi(E)\nu\Sigma_f(E')\Phi(\mathbf{r}, E')\,dE' + \frac{\partial}{\partial E}[E\xi(E)\Sigma_s(E)\Phi(\mathbf{r}, E)], \tag{4.8.1}$$

where we have again replaced $\Sigma_t(E) - \Sigma_s(E)$ by $\Sigma_a(E)$.

We expect this approximation to be adequate for a homogeneous reactor in which scattering is elastic and isotropic in the center-of-mass system, in which $\Sigma_a(E)$ in the slowing-down region is small compared with $\Sigma_s(E)$ (or, if not, is smoothly varying), and in which $\xi(E)$ is small (no significant scattering from hydrogen).

If, in keeping with these approximations, we assume that, at each point in the reactor, the flux and slowing-down density are related by the infinite-medium, no-absorption relationship derived earlier,

$$\Phi(\mathbf{r}, E) \approx \frac{q(\mathbf{r}, E)}{E\xi(E)\Sigma_s(E)} \tag{3.4.30}$$

(where the fact that the slowing-down density depends on position is explicitly indicated), then we can obtain, from (4.8.1),

$$-\nabla^2 q(\mathbf{r}, E) + \frac{\Sigma_a(E)}{D(E)} q(\mathbf{r}, E) = \frac{E\xi(E)\Sigma_s(E)}{D(E)}\frac{\partial q(\mathbf{r}, E)}{\partial E} + \frac{E\xi(E)\Sigma_s(E)}{D(E)}\int_0^\infty \frac{1}{\lambda}\chi(E)\nu\Sigma_f(E')\Phi(\mathbf{r}, E')\,dE'. \tag{4.8.2}$$

In order to simplify this equation we will make a change of variable from energy E to a quantity $\tau(E)$, which we call the *age of the neutrons at energy* E and which we define by

$$d\tau(E) \equiv -\frac{D(E)}{E\xi(E)\Sigma_s(E)}\,dE = -\frac{\lambda_{tr}(E)}{3E\xi(E)\Sigma_s(E)}\,dE, \tag{4.8.3}$$

where the second relationship comes from applying (4.5.5).

Setting $\tau(E_0) = 0$, where E_0 is the energy at the top of the fission spectrum ($\sim$15 MeV) and integrating $-d\tau(E)$ from E to E_0 yields

$$\tau(E) = \int_{E}^{E_0} \frac{\lambda_{tr}(E')\,dE'}{3E'\bar{\xi}(E')\Sigma_s(E')}. \tag{4.8.4}$$

We can use (4.8.3) to show that

$$\frac{\partial q(\mathbf{r}, E)}{\partial E} = \frac{\partial q(\mathbf{r}, E)}{\partial \tau}\frac{d\tau}{dE} = -\frac{D(E)}{E\bar{\xi}(E)\Sigma_s(E)}\frac{\partial q(\mathbf{r}, E)}{\partial \tau}, \tag{4.8.5}$$

and (4.8.2) can then be rewritten as

$$-\nabla^2 q(\mathbf{r}, \tau) + \frac{\Sigma_a(\tau)}{D(\tau)} q(\mathbf{r}, \tau) = -\frac{\partial q(\mathbf{r}, \tau)}{\partial \tau} + Q_f(\mathbf{r}, \tau), \tag{4.8.6}$$

where we have indicated that the quantities q, Σ_a, D, and Q_f can be thought of as depending on the age $\tau(E)$. Thus we speak of $q(\mathbf{r}, \tau)$ as the number of neutrons per unit volume at $\mathbf{r}$ slowing down "past τ," i.e., past an energy corresponding to the age $\tau(E)$. In (4.8.6) we have also introduced the definition

$$\begin{aligned} Q_f(\mathbf{r}, \tau) &\equiv \frac{E\bar{\xi}(E)\Sigma_s(E)}{D(E)} \int_0^\infty \frac{1}{\lambda}\chi(E)\nu\Sigma_f(E')\Phi(\mathbf{r}, E')\,dE' \\ &= \frac{E\bar{\xi}(E)\Sigma_s(E)}{D(E)}\chi(\tau)\left(\frac{-d\tau}{dE}\right)\frac{1}{\lambda}\int_0^\infty \nu\Sigma_f(E')\Phi(\mathbf{r}, E')\,dE' \\ &= \chi(\tau)\frac{1}{\lambda}\int_0^\infty \nu\Sigma_f(E')\Phi(\mathbf{r}, E')\,dE', \end{aligned} \tag{4.8.7}$$

where $\chi(\tau)\,d\tau \equiv -\chi(E)\,dE$ and we have used (4.8.3) to get $d\tau/dE$.

Equation (4.8.6) can be simplified one step further by making the transformation

$$\begin{aligned} q(\mathbf{r}, \tau) &\equiv q'(\mathbf{r}, \tau) \exp\left(-\int_0^\tau \frac{\Sigma_a(\tau')}{D(\tau')}\,d\tau'\right) \\ &= q'(\mathbf{r}, \tau) \exp\left(-\int_E^{E_0} \frac{\Sigma_a(E')\,dE'}{E'\bar{\xi}(E')\Sigma_s(E')}\right) \\ &\equiv q'(\mathbf{r}, \tau)p(E), \end{aligned} \tag{4.8.8}$$

where comparison with (3.5.13) shows that, for this low-absorption case ($\Sigma_a(E') \ll \Sigma_s(E')$) and for absorption in the fission energy range $E_1 - E_0$ neglected, $p(E)$ is an approximation to the resonance-escape probability at energy E.

Since

$$\frac{\partial q(\mathbf{r}, \tau)}{\partial \tau} = p(E)\left[\frac{\partial q'(\mathbf{r}, \tau)}{\partial \tau} - \frac{\Sigma_a(\tau)}{D(\tau)} q'(\mathbf{r}, \tau)\right],$$

we can simplify (4.8.6) to

$$\nabla^2 q'(\mathbf{r}, \tau) + \frac{Q_f(\mathbf{r}, \tau)}{p(E)} = \frac{\partial q'(\mathbf{r}, \tau)}{\partial \tau}. \tag{4.8.9}$$

Comparison of this result with (4.8.6) shows that $q'(\mathbf{r}, \tau)$ can be interpreted physically as the slowing-down density in a core having a source $Q_f/p(E)$ and an absorption cross section $\Sigma_a = 0$.

Equation (4.8.9) is called the *Fermi age equation.* The word "age" comes from the fact that the age equation has the same mathematical form as the time-dependent heat-conduction equation, with τ in (4.8.9) taking the place of "time" in the latter equation. The age, however, is not a "time" at all. It has the dimensions cm^2.

Physical Significance of the Fermi Age

It can be shown by solving (4.8.9) that for a "point source" of neutrons at energy E_0 in an infinite homogeneous medium (i.e., for the spatially and energetically extended source $Q_f(\mathbf{r}, \tau(E))/p(E)$ in (4.8.9) replaced by a source located at a point $\mathbf{r} = 0$ and having a single energy $E_0 = 15$ MeV) $\tau(E)$ is one-sixth the average squared distance a neutron moves from $\mathbf{r} = 0$ in the course of slowing down from E_0 to E. In symbols this is

$$\begin{aligned}\tau(E) &= \frac{1}{6}\overline{r^2(E)} \\ &\equiv \frac{\int_0^\infty r^2 q(\mathbf{r}, \tau) 4\pi r^2 \, dr}{6\int_0^\infty q(\mathbf{r}, \tau) 4\pi r^2 \, dr} \\ &= \frac{\int_0^\infty r^4 q'(\mathbf{r}, \tau) \, dr}{6\int_0^\infty r^2 q'(\mathbf{r}, \tau) \, dr}.\end{aligned} \tag{4.8.10}$$

Thus the age is a measure of how far a neutron moves in slowing down in an infinite homogeneous medium. Because the age can be interpreted in this physical manner, it can, in principle, be measured directly as an integral parameter. An isotropic point source emitting neutrons of energy E_0 is placed at $\mathbf{r} = 0$ in the medium, and the activity induced in a detector sensitive to neutrons of energy E is measured at various locations $\mathbf{r} = \mathbf{r}_1, \mathbf{r}_2, \mathbf{r}_3$, etc. (Indium, which has a large absorption resonance at 1.4 eV, is generally used.) Since the detector activity is, by (2.6.2), proportional to $\Phi(\mathbf{r}, E)$ and since (3.4.30) and (4.8.8) show that, spatially, $q(\mathbf{r}, E) \approx \Phi(\mathbf{r}, E)$, the ratio of the integrals in (4.8.10) can be inferred from the data, and $\tau(E_2)$ can thus be determined without a detailed knowledge of the cross sections of the medium.

It should be noted, however, that in making such a measurement it is necessary to prevent any fission events from occurring. These would give rise to an extended source

Q_f in (4.8.7), and it would then be impossible to distinguish between neutrons slowing down from this source and those slowing down from the point source. The best way to avoid these unwanted fissions is to remove all fissionable material from the mixture for which the age is being measured. Such material is heavy and thus contributes little to $\xi(E')\,\Sigma_s(E')$ in (4.8.4) (see (3.4.29) and (2.5.27)). Moreover it is often relatively dilute. Thus the age in a mixture of materials from which the fissionable material has been removed is close to that in the same mixture with the fissionable material present.

The Age Criticality Formula

Since the separability assumption (4.7.2), and hence the relationship $\nabla^2\Phi(\mathbf{r}, E) = -B_\mathbf{r}^2\Phi(\mathbf{r}, E)$, is valid in general for a bare, homogeneous reactor, (3.4.30) and (4.8.8) show that $\nabla^2 q'(\mathbf{r}, \tau) = -B_\mathbf{r}^2 q'(\mathbf{r}, \tau)$, where $B_\mathbf{r}^2$ is the geometrical buckling of the system. It follows immediately from (4.8.9) that

$$-B_\mathbf{r}^2 q'(\mathbf{r}, \tau) + \frac{Q_f(\mathbf{r}, \tau)}{p(E)} = \frac{\partial q'(\mathbf{r}, \tau)}{\partial \tau}, \tag{4.8.11}$$

the solution of which is

$$\begin{aligned} q'(\mathbf{r}, \tau) &= \exp(-B_\mathbf{r}^2\tau)\int_0^\tau d\tau' \exp(B_\mathbf{r}^2\tau')\frac{Q_f(\mathbf{r}, \tau')}{p(\tau')} \\ &= \exp(-B_\mathbf{r}^2\tau)\int_0^\tau d\tau' \exp(B_\mathbf{r}^2\tau')\frac{\chi(\tau')}{p(\tau')}\int_0^\infty \frac{\nu\Sigma_f(E')}{\lambda}\Phi(\mathbf{r}, E')\, dE', \end{aligned} \tag{4.8.12}$$

where we have used (4.8.7) to eliminate $Q_f(\mathbf{r}, \tau')$

Now, if we further assume that, in the fission energy range where $\chi(\tau')$ is significant, τ' is small enough (i.e., E is close enough to E_0) that we can set $\exp(B_\mathbf{r}^2\tau') = 1$ and the absorption is small enough that $p(\tau') \approx 1$, we may write approximately

$$\int_0^\tau d\tau' \exp(B_\mathbf{r}^2\tau')\frac{\chi(\tau')}{p(\tau')} = 1 \text{ for } \tau \text{ sufficiently large that it is in the range where } \chi(\tau) \approx 0. \tag{4.8.13}$$

It is then possible to obtain a simple formula for $q(\mathbf{r}, \tau_2)$, the number of neutrons per second per unit volume at $\mathbf{r}$ slowing down past τ_2, the age corresponding to the energy at the top of the thermal range. This formula, which comes from (4.8.8), (4.8.12), and (4.8.13), is

$$q(\mathbf{r}, \tau_2) = p(E_2)q'(\mathbf{r}, \tau_2) = p(E_2)\exp(-B_\mathbf{r}^2\tau_2)\int_0^\infty \frac{\nu\Sigma_f(E')}{\lambda}\Phi(\mathbf{r}, E')\, dE'. \tag{4.8.14}$$

We use this result to get an approximate criticality equation for a bare, homogeneous thermal reactor. This is done by expressing mathematically the physical fact that the number of neutrons slowing down into the thermal energy range per unit volume equals the sum of the number absorbed in that volume plus the number that leak out of it. Mathematically this statement is

$$\begin{aligned} q(\mathbf{r}, \tau_2) &= \int_0^{E_2} [D(E)B_r^2 + \Sigma_a(E)]\Phi(\mathbf{r}, E)\, dE \\ &= \left[\frac{\int_0^{E_2} D(E)\psi(E)\, dE}{\int_0^{E_2} \Sigma_a(E)\psi(E)\, dE} B_r^2 + 1\right] \int_0^{E_2} \Sigma_a(E)R(\mathbf{r})\psi(E)\, dE, \end{aligned} \tag{4.8.15}$$

where we have made use of (4.7.2) and have separated out the buckling-dependent term.

We now define a quantity L^2, where L is called the *thermal diffusion length*:

$$L^2 \equiv \frac{\int_0^{E_2} D(E)\psi(E)\, dE}{\int_0^{E_2} \Sigma_a(E)\psi(E)\, dE}. \tag{4.8.16}$$

Substituting (4.8.15) into (4.8.14) then yields

$$\lambda = \frac{\exp(-B_r^2\tau_2)}{1 + L^2B_r^2}\, p(E_2)\, \frac{\int_0^{\infty} \nu\Sigma_f(E)\psi(E)\, dE}{\int_0^{E_2} \Sigma_a(E)\psi(E)\, dE}, \tag{4.8.17}$$

or, using the definitions of the four factors from (3.5.3), (3.5.4), (3.5.6), and (3.5.8),

$$\lambda = k_{\text{eff}} = \frac{\exp(-B_r^2\tau_2)}{1 + L^2B_r^2}\, p(E_2)\, \varepsilon\eta f = \frac{k_\infty \exp(-B_r^2\tau_2)}{1 + L^2B_r^2}. \tag{4.8.18}$$

Equation (4.8.18) is the criticality equation of Fermi age theory. The principal difference between this and the exact result (4.7.19) is the way in which the nonleakage probability is expressed. The exact result (4.7.20) requires that we know $D(E)$ and $\Sigma_a(E)$ and be able to compute $\psi(E)$. In the Fermi expression, however, the nonleakage probability is $\exp(-B_r^2\tau_2)/(1 + L^2B_r^2)$, and both τ_2 and L^2 can be determined experimentally. (The latter can be found, it turns out, by measuring the spatial attenuation of the thermal flux due to a plane source of thermal neutrons.) Thus a detailed knowledge of the energy behavior of cross sections and a sophisticated computer are not needed to determine the nonleakage probability in the age approximation. This fact explains why the Fermi approach was so important in the early days of reactor analysis.

The great shortcoming of age theory (aside from the fact that it does not apply to H_2O-moderated cores) is that, strictly speaking, it is applicable only to bare, homogeneous reactors. This is because, if the diffusion constant or scattering cross section are position-dependent, the steps leading to the basic equation (4.8.9) are invalid. For reactors composed of a uniform lattice of cells, prescriptions have been devised for circumventing this difficulty. For a reactor composed of several zones of material having different slowing-down properties, however, no effective way of extending the method has been found. Thus, for fast reactors and for reflected and multizone loaded cores, age theory is not a useful analytical tool.

4.9 The Diffusion Equation for Multiregion Reactors: General Considerations

The last two sections provide practical methods for dealing with the energy-group diffusion equation for a bare, homogeneous reactor. The essential mathematical characteristic that leads to tractable equations is the separability of $\Phi(\mathbf{r}, E)$ into spatial ($R(\mathbf{r})$) and energy-dependent ($\psi(E)$) parts that are connected only by the fact that the geometrical buckling $B_{\mathbf{r}}^2$ appears as a parameter in the equation for $\psi(E)$. Solution of (4.7.5) for $R(\mathbf{r})$ by analytical methods is then straightforward for simple geometries. Moreover, with $B_{\mathbf{r}}^2$ known from this solution, $\psi(E)$ and k_{eff} may be determined with great accuracy and small cost by solving the multigroup equations (4.7.13) or, in many cases of interest, by performing hand calculations based on measured integral parameters.

Unfortunately completely homogeneous reactors are so rare that they may be regarded as mathematical abstractions. Even a homogeneous solution of uranyl nitrate in water placed in a thin-shelled spherical container and suspended three hundred feet in the air is not a completely homogeneous reactor. The container will act as a partial reflector, and calculations of the U^{25} concentration needed to go critical will be incorrect unless the neutron scattering and absorption in the thin shell are taken into account. Thus in practice the parameters $D(\mathbf{r}, E)$, $\Sigma_t(\mathbf{r}, E)$, etc., appearing in the continuous-energy diffusion equation (4.6.1) are, at best, independent of position only in each of the several regions that make up the reactor (core, reflector, blanket, etc.). In order to attack problems of practical interest, different methods must be developed.

Accordingly we shall next consider the problem of solving the energy-dependent diffusion equation for a reactor composed of a number of large, homogeneous, material regions. There are a great many reactors that can be well represented by this mathematical model:

First there are those reactors that actually are composed of homogeneous material regions—for instance a solution of uranyl nitrate in an aluminum container surrounded by a graphite reflector.

Second there are those reactors which are "isotopically homogeneous" even though they are composed of regions that are heterogeneous with respect to their material compositions. An example is a reactor in which the fuel is a dilute alloy of U^{25} in thin aluminum plates separated by thin coolant channels. If the mean free paths for scattering and absorption are several times larger than the thickness of the fuel plates and coolant channels, the probabilities of what happens to neutrons in this fuel assembly will differ very little from the corresponding probabilities for a homogeneous mixture of the material making up the plates and channels. Under these conditions one can simply treat the fuel assembly as a homogeneous region.

Finally there are reactors containing regions (for example slightly enriched fuel rods or plates interspersed with coolant channels) which are isotopically heterogeneous but for which "equivalent" (or "effective"), homogenized diffusion-theory parameters can be found. These equivalent constants are defined so that, ideally, when they are used in a diffusion-theory calculation for some particular region of the reactor, the predicted *average* reaction rates for that region will be correct. In this same sense equivalent diffusion-theory parameters can be found for homogeneous material regions, such as control rods, within which the basic assumption (Fick's Law) underlying diffusion theory is invalid. The determination of equivalent, homogenized diffusion-theory parameters will be discussed at length in Chapters 5 and 10. Until then we shall merely assume that some method of homogenization has been applied so that the reactor may be regarded as composed of a number of isotopically homogeneous regions, the diffusion-theory parameters in (4.6.1) being constant for all $\mathbf{r}$ within each such region.

Thus we seek a solution to (4.6.1) for the special but very practical case in which the equation has the form (4.7.1) (i.e., for which $D(E)$, $\Sigma_t(E)$, etc., are independent of position within each of several different regions making up the reactor). In order to represent this behavior of the diffusion-theory parameters we rewrite (4.6.1) as

$$-\nabla \cdot D^k(E)\nabla\Phi(\mathbf{r}, E) + \Sigma_t^k(E)\Phi(\mathbf{r}, E) = \int_0^\infty \left[\frac{1}{\lambda}\chi^k(E)\nu\Sigma_f^k(E') + \Sigma_s^k(E' \to E)\right]\Phi(\mathbf{r}, E')\, dE' \quad (k = 1, 2, 3, \ldots, K), \tag{4.9.1}$$

where superscript k indicates a particular region of the reactor (core, reflector, etc.). The parameters $D^k(E)$, $\Sigma_t^k(E)$, etc., will, in general, change from region to region.

4.10 One-Group Theory

The separability trick used to solve the energy-dependent diffusion equation for the bare-core case cannot be applied directly to the solution of (4.9.1), since the form $\Phi(\mathbf{r}, E) = R(\mathbf{r})\psi(E)$ cannot be made to fit this equation. (If it could, $\psi(E)$, substituted into the right-hand side of (4.7.4), would have to yield the same number for K sets of parameters $D^k(E)$, $\Sigma_t^k(E)$, etc.)

Nevertheless, a few mean paths inside a given region k, one might expect to find a solution which could be fairly well represented by

$$\Phi(\mathbf{r}, E) \approx R^k(\mathbf{r})\psi^k(E), \tag{4.10.1}$$

where $\psi^k(E)$ is a spectrum function characteristic of the materials present in region k. This fact is the basis of an approximation called *one-group diffusion theory.*

The essential strategy of one-group diffusion theory is: assume that an equation of the form (4.10.1) is valid within each region k; find an appropriate spectrum $\psi^k(E)$ for each region; then connect the $R^k(\mathbf{r})$ for each region to the corresponding spatial functions of neighboring regions.

The Spectrum Function for Material Region k

The reason for expecting the form (4.10.1) to be valid a few mean free paths inside region k is that any arbitrary distribution of neutrons that diffuse across a boundary into region k will interact with the materials in that region within a short distance of the surface. The cross sections for the fissioning, scattering, or absorption that occur in these interactions will be those of region k, and the neutrons that emerge from these events will have an energy distribution more characteristic of the material in region k than of the material in the region from which they entered k. After several such interactions within k the neutrons "forget" where their ancestors came from and behave with respect to absorption, scattering, fission, *and leakage* as if they were located in a (finite) homogeneous reactor composed *entirely* of region-k material. Under these conditions the leakage term $-\nabla \cdot D^k(E)\nabla\Phi(\mathbf{r}, E)$ can be closely approximated by $D^k(E)(B_m^k)^2\Phi(\mathbf{r}, E)$, the form it has in a bare, homogeneous reactor. In this case, however, the number $(B_m^k)^2$ is found not from the geometry of the region, but rather from the condition that, since region k is part of a critical reactor, λ in (4.9.1) should be unity. Thus we find $(B_m^k)^2$ by writing (4.9.1) as

$$[D^k(E)(B_m^k)^2 + \Sigma_t^k(E)]\Phi(\mathbf{r}, E) = \int_0^\infty [\chi^k(E)\nu\Sigma_f^k(E') + \Sigma_s^k(E' \to E)]\Phi(\mathbf{r}, E')\, dE'$$

for $\mathbf{r}$ in the interior of region k, (4.10.2)

where $(B_m^k)^2$ is that number which gives a positive solution to (4.10.2) when λ of (4.9.1) is taken as unity.

The number $(B_m^k)^2$ is characteristic of the mixture of materials making up region k and, for that reason, is called the *materials buckling.* Mathematically it is an eigenvalue of the homogeneous equation (4.10.2). As can be seen from that equation, its effect on neutron balance is to regulate the leakage rate so that the portion of the reactor in the interior of region k behaves in a sustained, critical fashion. If the k_∞ of that region exceeds unity, the number of neutrons removed from $dE\, dV$ per second by out-leakage, scattering, and absorption will equal the number introduced into $dE\, dV$ by fissioning and scattering from other energies; while, if the k_∞ is less than unity, the number removed by scattering and absorption will equal the number introduced by fissioning, scattering, and leakage *into* the region. Thus, if the k_∞ for region-k material is less than unity,

$(B_m^k)^2$ will be a negative number so that the extra neutrons needed for overall balance will be supplied by *negative* leakage, i.e., by net leakage *into* dV. In reflector material, for which $\nu\Sigma_f(E) = 0$, $(B_m^k)^2$ will always be negative.

We conjecture then—and mathematical arguments supporting this conjecture will be presented in Chapters 8 and 9—that, several mean free paths within each region of a reactor composed of a number of homogeneous regions, the solution for the energy part of $\Phi(\mathbf{r}, E)$ in (4.9.1) will be very close to a solution of (4.10.2). If this is indeed the case, then spatial and energy behavior are separable in the interior of the individual regions, and we can find the spectrum functions $\psi^k(E)$ for such regions by simply substituting (4.10.1) into (4.10.2) and dividing by $R^k(\mathbf{r})$. The result is

$$[D^k(E)(B_m^k)^2 + \Sigma_t^k(E)]\psi^k(E) = \int_0^\infty [\chi^k(E)\nu\Sigma_f^k(E') + \Sigma_s^k(E' \to E)]\psi^k(E')\,dE'. \tag{4.10.3}$$

The solution of (4.10.3) is accomplished by the same energy-group or approximate analytical methods outlined earlier in connection with (3.3.9) and (4.7.13). In principle we simply vary $(B_m^k)^2$ until the critical determinant analogous to (3.3.7) vanishes and the associated $\psi^k(E)$ is positive. In practice more approximate ways of finding $\psi^k(E)$ are used. These will be described in Chapter 9.

It is important to remember that $(B_m^k)^2$, unlike the $B_\mathbf{r}^2$ appearing, through (4.7.5), in (4.7.4), has nothing to do with the geometry of a reactor. It never appears explicitly in the spatial part of the solution for $\Phi(\mathbf{r}, E)$; rather its effect is felt indirectly through the spectrum function $\psi^k(E)$. Once $\psi^k(E)$ is found, $(B_m^k)^2$ is no longer needed.

To summarize: for each material composition we can find a spectrum function $\psi^k(E)$ such that, if this composition forms part of a critical reactor, separable functions of the form $R^k(\mathbf{r})\psi^k(E)$ will be solutions to the diffusion equation (4.9.1) several mean free paths into the interior of the composition.

This part of the theory of the multiregion core is fairly rigorous. It is when we try to connect the solutions in the various regions together that cruder approximations become necessary. The basic problem is that the separable forms $R^k(\mathbf{r})\psi^k(E)$ $(k = 1, 2, \ldots, K)$, which are good approximations to $\Phi(\mathbf{r}, E)$ in the interior portions of the regions, *cannot* be valid near interfaces between regions, since, if superscripts k and l designate two adjacent regions of different composition so that $\psi^k(E)$ and $\psi^l(E)$ are different functions of E, then, if $\mathbf{r}_c$ is any point on the interface between these regions, $R^k(\mathbf{r}_c)\psi^k(E)$ cannot equal $R^l(\mathbf{r}_c)\psi^l(E)$ for *all* energies. (If it did, then $\psi^k(E)$ would have to have the *same* energy shape as $\psi^l(E)$.) Thus assuming that the separable form $R^k(\mathbf{r})\psi^k(E)$ is valid *throughout* each region k makes it impossible to meet the continuity-of-flux boundary condition at the interface between regions.

Physically, then, we expect that there are "transition regions" near the interfaces between material compositions in a multiregion reactor where $\Phi(\mathbf{r}, E)$ is not separable. Nevertheless, since the separable form is valid in the interior of each material composition, it is reasonable to impose it on $\Phi(\mathbf{r}, E)$ *throughout* each composition and then try to find some *approximate* boundary conditions at interfaces which will permit us to connect each $R^k(\mathbf{r})$ to its neighbors.

An Equation for the Spatial Function $R(\mathbf{r})$

One-group theory is the simplest scheme for providing a connection between the $R^k(\mathbf{r})$. In it the difficulties about the continuity conditions at *each* energy are circumvented by imposing the much weaker boundary condition that the integrals over all energies of $\Phi(\mathbf{r}, E)$ and $\mathbf{n}\cdot\nabla\Phi(\mathbf{r}, E)$ be continuous. This boundary condition is entirely consistent with the continuity of $\Phi(\mathbf{r}, E)$ and $\mathbf{n}\cdot\nabla\Phi(\mathbf{r}, E)$ at all energies since, if there were continuity at each energy, the integrals over all energy would certainly be continuous. (The reverse implication cannot, of course, be drawn.)

To be specific, one-group theory is based on the following assumptions:

1. $\Phi(\mathbf{r}, E) = R^k(\mathbf{r})\psi^k(E)$ *throughout* each composition k, $\psi^k(E)$ being found by solving (4.10.3).
2. If k and l indicate any two adjacent compositions and $\mathbf{r}_c$ is a point on the interface, we require that

$$\int_0^\infty dE\, R^k(\mathbf{r}_c)\psi^k(E) = \int_0^\infty dE\, R^l(\mathbf{r}_c)\psi^l(E)$$

and

$$\int_0^\infty dE\, \mathbf{n}_k\cdot D^k(E)\nabla[R^k(\mathbf{r}_c)\psi^k(E)] = \int_0^\infty dE\, \mathbf{n}_k\cdot D^l(E)\nabla[R^l(\mathbf{r}_c)\psi^l(E)].$$

To find the functions $R^k(\mathbf{r})$ we substitute the assumed form $R^k(\mathbf{r})\psi^k(E)$ into (4.9.1) and integrate the result over all energies. It will simplify this process to normalize the solutions $\psi^k(E)$ of (4.10.3) in such a way that

$$\int_0^\infty \psi^k(E)\, dE = 1 \qquad (k = 1, 2, \ldots, K). \tag{4.10.4}$$

When this is done, assumption 2 becomes

$$R^k(\mathbf{r}_c) = R^l(\mathbf{r}_c) \tag{4.10.5}$$

and

$$\left[\int_0^\infty D^k(E)\psi^k(E)\, dE\right]\mathbf{n}_k\cdot\nabla R^k(\mathbf{r}_c) = \left[\int_0^\infty D^l(E)\psi^l(E)\, dE\right]\mathbf{n}_k\cdot\nabla R^l(\mathbf{r}_c). \tag{4.10.6}$$

Then, if (4.10.1) is substituted into (4.9.1) and the result is integrated over all energies, we get

$$
\begin{aligned}
&-\nabla\cdot\left[\int_0^\infty D^k(E)\psi^k(E)\,dE\right]\nabla R^k(\mathbf{r}) + \left[\int_0^\infty \Sigma_t^k(E)\psi^k(E)\,dE\right]R^k(\mathbf{r}) \\
&= \int_0^\infty \frac{1}{\lambda}\chi^k(E)\,dE\left[\int_0^\infty \nu\Sigma_f^k(E')\psi^k(E')\,dE'\right]R^k(\mathbf{r}) + \left[\int_0^\infty dE\int_0^\infty dE'\Sigma_s^k(E'\to E)\psi^k(E')\right]R^k(\mathbf{r}) \\
&= \frac{1}{\lambda}\left[\int_0^\infty \nu\Sigma_f^k(E')\psi^k(E')\,dE'\right]R^k(\mathbf{r}) + \left[\int_0^\infty \Sigma_s^k(E')\psi^k(E')\,dE'\right]R^k(\mathbf{r}).
\end{aligned}
\tag{4.10.7}
$$

To simplify (4.10.7) we define a set of *one-group cross sections* by

$$
\begin{aligned}
D^k &\equiv \int_0^\infty D^k(E)\psi^k(E)\,dE, \\
\Sigma_t^k &\equiv \int_0^\infty \Sigma_t^k(E)\psi^k(E)\,dE, \\
\nu\Sigma_f^k &\equiv \int_0^\infty \nu\Sigma_f^k(E)\psi^k(E)\,dE, \\
\Sigma_s^k &\equiv \int_0^\infty \Sigma_s^k(E)\psi^k(E)\,dE, \\
\Sigma^k &\equiv \Sigma_t^k - \Sigma_s^k = \Sigma_a^k.
\end{aligned}
\tag{4.10.8}
$$

These numbers are independent of both space and energy. They may be computed for any region k once we have determined the nuclear concentrations n^j and the microscopic cross sections $\sigma^j(E)$ of the material in region k and have obtained the spectrum functions $\psi^k(E)$ by solving (4.10.3), either by the energy-group method or by some of the approximate schemes discussed in Section 3.4.

When the definitions (4.10.8) are substituted into (4.10.7), the latter becomes

$$
-\nabla\cdot D^k\nabla R^k(\mathbf{r}) + \Sigma^k R^k(\mathbf{r}) = \frac{1}{\lambda}\nu\Sigma_f^k R^k(\mathbf{r}),
\tag{4.10.9}
$$

and the internal boundary conditions (4.10.5) and (4.10.6) become

$$
\begin{aligned}
R^k(\mathbf{r}_c) &= R^l(\mathbf{r}_c), \\
\mathbf{n}_k\cdot D^k\nabla R^k(\mathbf{r}_c) &= \mathbf{n}_k\cdot D^l\nabla R^l(\mathbf{r}_c).
\end{aligned}
\tag{4.10.10}
$$

The condition that the flux is to vanish on the external surface of the reactor is met by requiring that the appropriate $R^k(\mathbf{r})$ vanish for any $\mathbf{r}$ on that surface.

Finally, since $R^k(\mathbf{r})$ is everywhere continuous throughout the reactor, the superscript k is redundant. The point $\mathbf{r}$ automatically specifies the region k. Accordingly we shall replace all the $R^k(\mathbf{r})$ by the single function $\Phi(\mathbf{r})$, the *one-group scalar flux*. (Note that $\Phi(\mathbf{r})$ is *not* a flux per unit energy; its units are (speed) $\times$ (neutrons/cc) $=$ neutrons per square cm per sec.)

With this change in notation the one-group diffusion equation for a reactor composed of several homogeneous regions becomes

$$-\nabla \cdot D^k \nabla \Phi(\mathbf{r}) + \Sigma^k \Phi(\mathbf{r}) = \frac{1}{\lambda} \nu \Sigma_f^k \Phi(\mathbf{r}), \tag{4.10.11}$$

with the following boundary conditions:

1. $\Phi(\mathbf{r})$ and the normal component $\mathbf{n} \cdot D\nabla\Phi(\mathbf{r})$ are continuous across interfaces between different compositions;
2. $\Phi(\mathbf{r}) = 0$ on the outer boundary of the reactor.

Equation (4.10.11) is sometimes called the *one-velocity diffusion equation* since it can also be derived by assuming that all the neutrons are traveling at a single speed. The trouble with this approach is that it provides no systematic procedure for finding the one-group constants D^k, Σ^k, and $\nu\Sigma_f^k$, and makes it hard to account for such effects as fast fission, resonance capture, and fast-neutron leakage. It also makes one-group theory appear much more crude than it actually is. According to the view we have adopted, the one-group theory actually provides a very detailed energy-dependent flux, namely $\psi^k(E)\Phi(\mathbf{r})$, in each region. The essential approximation consists of a very crude treatment of the "transition zones" between the different material compositions. In these zones the true scalar flux density $\Phi(\mathbf{r}, E)$ actually changes gradually from the $\psi^k(E)$ appropriate to the interior parts of one region to that appropriate to the interior parts of the next. We replace this gradual change by an abrupt one and in so doing make an error in the net leakage rate from one region to the next. Nevertheless one-group theory is surprisingly accurate and is, in fact, the basis of several design procedures in use today.

Numerical Solution of the One-Group Diffusion Equation

The standard way to solve the one-group diffusion equation is to transform it from a differential equation into a finite-difference equation and solve the latter on a digital computer. The essential step in this procedure is to replace the continuous function $\Phi(\mathbf{r})$ by a sequence of discrete (average) values over small but finite intervals.

For example, in the interior of a region k of a one-dimensional reactor, (4.10.11) becomes

$$-D^k \frac{d^2\Phi(x)}{dx^2} + \Sigma^k \Phi(x) = \frac{1}{\lambda} \nu \Sigma_f^k \Phi(x). \tag{4.10.12}$$

If we partition the X axis into a series of equal intervals of length h and define $\Phi^{(n)}(n = 0, 1, 2, \ldots, N)$ to be the values of $\Phi(x)$ at the points separating these intervals (see Figure 4.7), then we may approximate $d\Phi(x)/dx$ at the midpoint of the interval

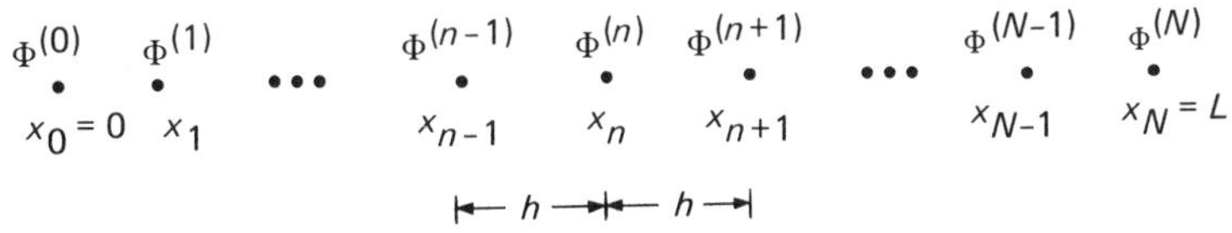

Figure 4.7 A partitioning of the X axis to be used in forming difference equations.

between points x_n and x_{n+1} by

$$\frac{d\Phi(x)}{dx} \approx \frac{\Phi^{(n+1)} - \Phi^{(n)}}{h}. \tag{4.10.13}$$

Similarly, at the midpoint between x_{n-1} and x_n, we have

$$\frac{d\Phi(x)}{dx} \approx \frac{\Phi^{(n)} - \Phi^{(n-1)}}{h}. \tag{4.10.14}$$

We may then approximate $d^2\Phi(x)/dx^2$ at point x_n by

$$\frac{d^2\Phi(x)}{dx^2} \approx \frac{1}{h}\left[\frac{\Phi^{(n+1)} - \Phi^{(n)}}{h} - \frac{\Phi^{(n)} - \Phi^{(n-1)}}{h}\right] = \frac{\Phi^{(n+1)} - 2\Phi^{(n)} + \Phi^{(n-1)}}{h^2}. \tag{4.10.15}$$

Thus, at mesh point x_n within region k, the continuous-energy one-group diffusion equation may be approximated by

$$-D^k \frac{\Phi^{(n+1)} - 2\Phi^{(n)} + \Phi^{(n-1)}}{h^2} + \Sigma^k\Phi^{(n)} = \frac{1}{\lambda}\nu\Sigma_f^k\Phi^{(n)}, \tag{4.10.16}$$

and, if x_n is at the interface between regions k and $k + 1$, it is easy to show that this generalizes to

$$-\frac{1}{h}\left[D^{k+1}\frac{\Phi^{(n+1)} - \Phi^{(n)}}{h} - D^k\frac{\Phi^{(n)} - \Phi^{(n-1)}}{h}\right] + \frac{1}{2}(\Sigma^k + \Sigma^{k+1})\Phi^{(n)} = \frac{1}{\lambda}\frac{1}{2}(\nu\Sigma_f^k + \nu\Sigma_f^{k+1})\Phi^{(n)}. \tag{4.10.17}$$

Thus, for *each* point x_n $(n = 1, \ldots, N - 1)$, we can derive from (4.10.12) a difference equation relating the flux at x_n to those at x_{n-1} and x_{n+1}, giving a total of $N - 1$ equations. Moreover the boundary conditions on the outer surface require that $\Phi^{(0)} = \Phi^{(N)} = 0$, so we have exactly $N - 1$ unknown fluxes. Thus we have $N - 1$ homogeneous equations in $N - 1$ unknowns. These will have a nontrivial solution if and only if the determinant of the coefficients of the unknowns vanish, and the requirement that this happen yields an equation from which we can find the eigenvalue λ. As always, if λ is one and if the corresponding interior fluxes $\Phi^{(1)}, \Phi^{(2)}, \ldots, \Phi^{(N-1)}$ are positive, the reactor is predicted to be critical.

There are mathematical theorems that guarantee the existence of a unique, real, positive eigenvalue λ leading to everywhere-positive interior fluxes for the problem. Moreover there are systematic iterative procedures that are guaranteed to converge to this desired solution. Thus determination of the most positive λ and its corresponding physically acceptable flux is a straightforward procedure. Just as in the infinite-medium case, we can then vary the fuel or control-poison concentration (or, in this instance, the physical dimensions of the reactor) until λ equals unity. If, however, this procedure leads to an altered composition for any region k, we must be careful to redetermine the spectrum function ψ^k and the corresponding one-group parameters for that region.

The finite-difference method can be generalized to two and three dimensions in a clear-cut manner. Also it is possible to derive difference equations in which the mesh intervals h are unequal. Thus h can be made small where $\Phi(x)$ is changing rapidly (so that the accuracy of (4.10.14) and (4.10.15) is improved) and large where it is changing slowly. In this way improved accuracy can be obtained without unduly increasing the number of simultaneous algebraic equations that must be solved.

One extremely important characteristic of the finite-difference method is that it is directly applicable to the situation in which the material properties of the reactor vary continuously as a function of position. We shall not discuss here the derivation of the one-group equation for this situation, but, as might be expected, the net effect on (4.10.11) is that the parameters D^k, Σ^k, and $\nu\Sigma_f^k$ become dependent on position. Under these circumstances solution of (4.10.11) by conventional analytical methods becomes virtually impossible. The extension of the finite-difference approach is, however, uncomplicated: one simply replaces the position-dependent values of the one-group parameters by their average values over each mesh interval and uses (4.10.17) at every point, rather than just at the interfaces between regions. In other words the number of compositions K becomes equal to the number of mesh intervals N.

Analytical Solutions of the One-Group Diffusion Equation

The finite-difference method for solving the diffusion equation leads to no simple algebraic formulas for the flux shape or eigenvalue. Thus the most straightforward way to determine how changes in nuclear composition or geometry affect the characteristics of the solution is to examine a series of test cases. This procedure is somewhat clumsy, however, and it would be preferable to have analytical formulas from which the sensitivity of the flux shape and eigenvalue might be estimated by inspection. Thus, even with large computers available, there is still motivation for trying to determine solutions for (4.10.11) analytically. (Prior to the advent of computers the approach was, of course, almost mandatory.)

Unfortunately, closed-form, analytical solutions of the one-group diffusion equation are possible only for essentially one-dimensional cases. Moreover, as a practical matter, solutions are restricted to cases involving only two or three regions. Nevertheless the physical insight to be gained by studying the properties of a closed-form algebraic solution makes it worthwhile to examine the methods usually applied to the problem.

The Two-Region Slab, Infinite in the Y and Z Directions. We shall first investigate the two-composition slab reactor, infinite in the Y and Z directions. Moreover we shall assume that the reactor is symmetric and consists of a central core with two identical reflectors on either side, so that $\nu\Sigma_f = 0$ in the reflector regions. If we set up the origin of the X axis at the center plane of the slab and let $x = \pm x_1$ be the positions of the interfaces between the core and its reflector and $x = \pm x_2$ be the outer boundaries of the reactor, (4.10.12) becomes

$$\begin{aligned}
&-D^{(1)}\frac{d^2}{dx^2}\Phi(x) + \Sigma^{(1)}\Phi(x) = \frac{1}{\lambda}\nu\Sigma_f^{(1)}\Phi(x) \quad \text{for } -x_1 \le x \le x_1, \\
&-D^{(2)}\frac{d^2}{dx^2}\Phi(x) + \Sigma^{(2)}\Phi(x) = 0 \quad \text{for } -x_2 \le x \le -x_1 \text{ and } x_1 \le x \le x_2, \\
&\Phi(\pm x_1)^- = \Phi(\pm x_1)^+, \\
&D^{(1)}\frac{d}{dx}\Phi(\pm x_1)^- = D^{(2)}\frac{d}{dx}\Phi(\pm x_1)^+, \\
&\Phi(\pm x_2) = 0,
\end{aligned} \tag{4.10.18}$$

where superscripts (1) and (2) refer to core and reflector material and superscripts $-$ and $+$ refer to the core and reflector sides of the interfaces at $\pm x_1$.

Since the slab reactor under consideration is symmetrical about its center plane, we need only solve (4.10.18) in the range $0 \le x \le x_2$, imposing as a boundary condition at $x = 0$ the relationship $d\Phi(x)/dx|_{x=0} = 0$, a condition which states, physically, that the net current in the X direction at $x = 0$ is zero and, mathematically, that the solution must be symmetric about $x = 0$. This latter interpretation is the one we shall use. Equations (4.10.18) may then be written

$$\begin{aligned}
&-\frac{d^2}{dx^2}\Phi(x) = \frac{\lambda^{-1}\nu\Sigma_f^{(1)} - \Sigma^{(1)}}{D^{(1)}}\Phi(x) \equiv \kappa^2\Phi(x) \quad \text{for } 0 \le x \le x_1, \\
&\qquad\qquad \text{where } \kappa \equiv \left[\frac{\lambda^{-1}\nu\Sigma_f^{(1)} - \Sigma^{(1)}}{D^{(1)}}\right]^{1/2}, \\
&\frac{d^2}{dx^2}\Phi(x) = \frac{\Sigma^{(2)}}{D^{(2)}}\Phi(x) \equiv \frac{\Phi(x)}{L^2} \quad \text{for } x_1 \le x \le x_2, \text{ where } L \equiv \left[\frac{D^{(2)}}{\Sigma^{(2)}}\right]^{1/2}, \\
&\left.\frac{d}{dx}\Phi(x)\right|_{x=0} = 0, \\
&\Phi(x_1)^- = \Phi(x_1)^+, \\
&D^{(1)}\frac{d}{dx}\Phi(x_1)^- = D^{(2)}\frac{d}{dx}\Phi(x_1)^+, \\
&\Phi(x_2) = 0.
\end{aligned} \tag{4.10.19}$$

These equations are second-order; therefore the general solutions in the core and the reflector will each have two undetermined constants. Such general solutions are

$$\begin{aligned}\Phi(x) &= C_1 \sin \kappa x + C_2 \cos \kappa x \quad \text{for } 0 \le x \le x_1,\\ \Phi(x) &= C_3 \sinh \frac{x}{L} + C_4 \cosh \frac{x}{L} \quad \text{for } x_1 \le x \le x_2.\end{aligned} \tag{4.10.20}$$

At $x = 0$

$$\left.\frac{d\Phi(x)}{dx}\right|_{x=0} = \left.(C_1 \kappa \cos \kappa x - C_2 \kappa \sin \kappa x)\right|_{x=0} = \kappa C_1. \tag{4.10.21}$$

Therefore, if $d\Phi(x)/dx|_{x=0}$ is to vanish, C_1 must be zero. At $x = x_2$ we want the flux to vanish. To apply this boundary condition it is convenient to use a different form for $\Phi(x)$ in the reflector region. This form is

$$\Phi(x) = C_5 \sinh\left(C_6 - \frac{x}{L}\right) = C_5 \sinh C_6 \cosh \frac{x}{L} - C_5 \cosh C_6 \sinh \frac{x}{L}, \tag{4.10.22}$$

which would be identical to the second equation in (4.10.20) if we set $C_5 \sinh C_6 \equiv C_3$ and $-C_5 \cosh C_6 \equiv C_4$. If, now, $\Phi(x_2)$ is to vanish, we must have $\sinh(C_6 - (x_2/L)) = 0$; and this can happen only if $C_6 = x_2/L$.

Thus the solutions (4.10.20) and (4.10.22) become

$$\Phi(x) = \begin{cases} C_2 \cos \kappa x & \text{for } 0 \le x \le x_1,\\ C_5 \sinh\left(\dfrac{x_2 - x}{L}\right) & \text{for } x_1 \le x \le x_2.\end{cases} \tag{4.10.23}$$

Now, applying the internal boundary condition at x_1, we get

$$\begin{aligned} C_2 \cos \kappa x_1 &= C_5 \sinh\left(\frac{x_2 - x_1}{L}\right),\\ -C_2 D^{(1)} \kappa \sin \kappa x_1 &= -C_5 D^{(2)} \left(\frac{1}{L}\right) \cosh\left(\frac{x_2 - x_1}{L}\right).\end{aligned} \tag{4.10.24}$$

Dividing the first of these equations into the second yields

$$D^{(1)} \kappa \tan \kappa x_1 = \frac{D^{(2)}}{L} \coth \frac{x_2 - x_1}{L} \tag{4.10.25}$$

or

$$\begin{aligned} &[D^{(1)}(\lambda^{-1} \nu \Sigma_f^{(1)} - \Sigma^{(1)})]^{1/2} \tan \left[\frac{\lambda^{-1} \nu \Sigma_f^{(1)} - \Sigma^{(1)}}{D^{(1)}}\right]^{1/2} x_1\\ &= (D^{(2)} \Sigma^{(2)})^{1/2} \coth \left[\frac{\Sigma^{(2)}}{D^{(2)}}\right]^{1/2} (x_2 - x_1).\end{aligned} \tag{4.10.26}$$

For a given set of compositions (1) and (2), and for fixed dimensions x_1 and x_2, the transcendental equation (4.10.26) determines an infinite number of discrete values of λ. The largest such value will be the only one for which $\Phi(x)$ is everywhere nonnegative. It will thus be the k_{eff} of the reactor. For this reason (4.10.26) is called the *criticality equation for the two-region slab*. Once the largest value of λ is found, we can determine κ and L and then use (4.10.24) to determine one of the constants (C_2, C_5) in terms of the other. The remaining constant can either be set arbitrarily or be determined in terms of the average power per unit volume by applying (4.7.15). Thus we have a complete analytical solution for $\Phi(x)$ throughout the slab reactor, and, in the course of determining it, we have found the critical eigenvalue k_{eff}.

If we are searching for the predicted physically critical condition, it is necessary to vary whatever parameters are available for adjustment until the calculated value of λ is unity. If the dimensions of the reactor are being varied to coincide with a predicted "critical size," no recomputation of the one-group parameters appearing in (4.10.26) is necessary. However, if criticality is being obtained by varying the concentrations of various materials, a new spectrum calculation (solution of (4.10.3)) must be performed each time in order to obtain the $\psi^k(E)$ needed to compute the one-group parameters (4.10.8) used in (4.10.26).

The Two-Region Slab, Finite in the Y and Z Directions. If, for a given x, the material properties of the reactor are the same for all values of y and z, the result we have just obtained can easily be generalized to a reactor that is finite in the Y and Z directions. This extension is possible because, under the conditions just postulated, the reactor is a bare, homogeneous slab as far as the Y and Z dimensions go. Thus, just as with the bare, rectangular-parallelopiped reactor (see (4.7.21)), we expect to be able to find a spatially separable solution for the one-group equation (4.10.11).

Specifically, if the Y dimension of the reactor extends from $y = 0$ to $y = L_y$ and if the Z dimension extends from $z = 0$ to $z = L_z$, we expect the function

$$\Phi(\mathbf{r}) = X(x) \sin \frac{\pi y}{L_y} \sin \frac{\pi z}{L_z} \equiv X(x) \sin B_{y1} y \sin B_{z1} z \tag{4.10.27}$$

to provide a solution of (4.10.11). This function is zero for $y = 0$, $y = L_y$, $z = 0$, and $z = L_z$ and positive for all internal y and z values if $X(x) > 0$. For any fixed value of x it is continuous in y and z with continuous values of $D^k \partial\Phi/\partial y$ and $D^k \partial\Phi/\partial z$. Thus (4.10.27) meets the internal and external boundary conditions.

To see if (4.10.27) can be made to satisfy the differential equation, we substitute it into (4.10.11). The result is

$$-D^k\left(\frac{\partial^2}{\partial x^2} + \frac{\partial^2}{\partial y^2} + \frac{\partial^2}{\partial z^2}\right)X(x) \sin B_{y1} y \sin B_{z1} z + \left(\Sigma^k - \frac{1}{\lambda}\nu\Sigma_{\text{f}}^k\right)X(x) \sin B_{y1} y \sin B_{z1} z = 0 \tag{4.10.28}$$

or

$$-D^k \sin B_{y1}y \sin B_{z1}z \frac{d^2}{dx^2} X(x) + D^k(B_{y1}^2 + B_{z1}^2)X(x) \sin B_{y1}y \sin B_{z1}z$$

$$+ \left(\Sigma^k - \frac{1}{\lambda}\nu\Sigma_{\rm f}^k\right)X(x) \sin B_{y1}y \sin B_{z1}z = 0. \tag{4.10.29}$$

Dividing (4.10.29) by $\sin B_{y1}y \sin B_{z1}z$ then yields

$$-D^k \frac{d^2X(x)}{dx^2} + (D^kB_{y1}^2 + D^kB_{z1}^2 + \Sigma^k)X(x) = \frac{1}{\lambda}\Sigma_{\rm f}^k X(x). \tag{4.10.30}$$

This equation is identical in form with (4.10.12), the corresponding one-group diffusion equation for a slab reactor infinite in the transverse directions. The only difference is that the one-group absorption cross section Σ^k is augmented by the Y and Z leakage terms $D^kB_{y1}^2$ and $D^kB_{z1}^2$, which, as we have seen, produce the same effect on the overall neutron economy of a bare slab as an absorption cross section. Thus all the analysis leading to the criticality condition (4.10.26) is immediately applicable provided all Σ^k are replaced by $\Sigma^k + D^k(B_{y1}^2 + B_{z1}^2)$ and κ and L in (4.10.19) are altered accordingly.

The Two-Region Cylindrical Reactor. As a final example of an analytical solution of the one-group diffusion equation we consider a cylindrical reactor composed of a central cylinder of one fuel-bearing material embedded in an annulus of a second fuel-bearing material. As shown in the Figure 4.8, the assembly is unreflected.

We assume that multigroup spectra have been determined for the different (nuclearly homogeneous) materials in regions (1) and (2) by solving (4.10.3) and that the one-group cross sections of (4.10.8) have been obtained by averaging over these spectra. The resulting form of the one-group equation (4.10.11) for the cylindrical reactor is thus

$$D^k\left(\frac{1}{r}\frac{\partial}{\partial r}r\frac{\partial}{\partial r} + \frac{\partial^2}{\partial z^2}\right)\Phi(r, z) + \left(\frac{1}{\lambda}\nu\Sigma_{\rm f}^k - \Sigma_{\rm a}^k\right)\Phi(r, z) = 0, \tag{4.10.31}$$

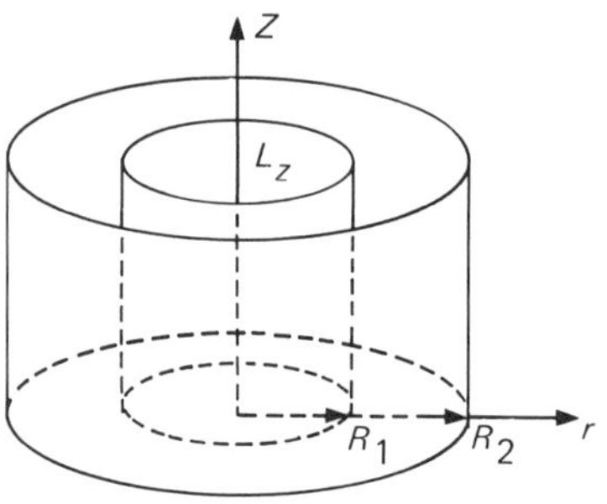

Figure 4.8 A two-region cylindrical reactor.

with the continuity and boundary conditions

$$\Phi(R_1, z)|^- = \Phi(R_1, z)|^+,$$
$$D^1 \frac{\partial\Phi(R_1, z)}{\partial r}\bigg|^- = D^2 \frac{\partial\Phi(R_1, z)}{\partial r}\bigg|^+, \tag{4.10.32}$$
$$\Phi(r, z) = 0 \quad \text{for } z = 0,\ z = L_z, \text{ or } r = R_2.$$

The one-group parameters D^k, $\nu\Sigma_f^k$, and Σ_a^k are all uniform in the Z direction. Thus we may assume again here that Φ is separable:

$$\Phi(r, z) = R(r)Z(z). \tag{4.10.33}$$

Substituting this form into (4.10.31), dividing by $D^k R(r)Z(z)$, and rearranging gives

$$\frac{1}{R(r)}\frac{1}{r}\frac{d}{dr}r\frac{d}{dr}R(r) + \frac{\lambda^{-1}\nu\Sigma_f^k - \Sigma_a^k}{D^k} = -\frac{1}{Z(z)}\frac{d^2}{dz^2}Z(z) \qquad (k = 1, 2) \tag{4.10.34}$$

and, since a function of r alone now equals a function of z alone, both functions must equal the same constant. Moreover, since the right-hand side of (4.10.34) is the same for both values of k, this "separation constant" must be independent of k. We may thus write

$$\frac{d^2Z(z)}{dz^2} = -B_z^2 Z(z), \tag{4.10.35}$$

where B_z^2 is the separation constant.

Note that we can proceed in this manner only because the reactor is uniform in the Z direction for *all* r. If this were not so, the D^k, $\nu\Sigma_f^k$, and Σ_a^k would be functions of z, and the left-hand side of (4.10.34) would no longer be a function of r alone.

The general solution of (4.10.35) is

$$Z(z) = A_1 \sin B_z z + A_2 \cos B_z z. \tag{4.10.36}$$

However, since the external boundary conditions and the requirements for a physical solution demand that $Z(0) = Z(L_z) = 0$ and $Z(z) > 0$ for $0 < z < L_z$, we have

$$A_2 = 0,$$
$$B_{zn}L_z = n\pi, \tag{4.10.37}$$
$$n = 1.$$

Thus

$$B_{z1} = \frac{\pi}{L_z} \tag{4.10.38}$$

and (4.10.36) becomes

$$Z(z) = A_1 \sin\frac{\pi z}{L_z}, \tag{4.10.39}$$

where the constant A_1 will be determined later as part of an overall normalization constant.

The choice $n = 1$ in (4.10.38) is made so that $Z(z)$ will be positive for all z within the reactor. A linear combination $\sum_n \alpha_n \sin B_{zn} z$ of terms having different values of n could be made positive, but this combination would no longer be a solution of (4.10.35).

Substitution of (4.10.39) into (4.10.34) and rearrangement yields

$$\frac{1}{r}\frac{d}{dr}r\frac{d}{dr}R(r) + \frac{\lambda^{-1}\nu\Sigma_{\text{f}}^k - \Sigma_{\text{a}}^k - D^k B_{z1}^2}{D^k}R(r) = 0 \qquad (k = 1, 2). \tag{4.10.40}$$

If we define

$$(\kappa^k)^2 \equiv \frac{\lambda^{-1}\nu\Sigma_{\text{f}}^k - \Sigma_{\text{a}}^k - D^k B_{z1}^2}{D^k} \qquad (k = 1, 2), \tag{4.10.41}$$

then (4.10.40) can be written

$$\frac{1}{r}\frac{d}{dr}r\frac{d}{dr}R(r) + (\kappa^k)^2 R(r) = 0. \tag{4.10.42}$$

Note that κ^k can be either real or pure imaginary depending on the magnitudes of the reactor parameters and the axial height (i.e., B_{z1}^2). The analysis which follows will be valid for either case; however we shall introduce functions more appropriate for real values of κ^k. We thus write, for the general solution of (4.10.42) in region k,

$$R(r) = C_1^k J_0(\kappa^k r) + C_2^k Y_0(\kappa^k r) \qquad (k = 1, 2), \tag{4.10.43}$$

where J_0 and Y_0 are the zeroth-order Bessel functions of the first and second kind.

Since $Y_0(\kappa^k r) \to -\infty$ as $r \to 0$, we must set $C_2^{(1)}$ equal to zero in region (1). Moreover application of the external boundary condition requiring $R(r)$ to vanish at $r = R_2$ shows us that in region (2) the solution must be of the form

$$R(r) = C^{(2)}[J_0(\kappa^{(2)}r)Y_0(\kappa^{(2)}R_2) - J_0(\kappa^{(2)}R_2)Y_0(\kappa^{(2)}r)], \tag{4.10.44}$$

where the relationship of $C^{(2)}$ to $C_1^{(2)}$ and $C_2^{(2)}$ can be found by comparison with (4.10.43).

The continuity conditions at $r = R_1$ now require that

$$\begin{aligned} C^{(1)}J_0(\kappa^{(1)}R_1) &= C^{(2)}[J_0(\kappa^{(2)}R_1)Y_0(\kappa^{(2)}R_2) - J_0(\kappa^{(2)}R_2)Y_0(\kappa^{(2)}R_1)], \\ C^{(1)}D^{(1)}\kappa^{(1)}J_1(\kappa^{(1)}R_1) &= C^{(2)}D^{(2)}\kappa^{(2)}[J_1(\kappa^{(2)}R_1)Y_0(\kappa^{(2)}R_2) - J_0(\kappa^{(2)}R_2)Y_1(\kappa^{(2)}R_1)]. \end{aligned} \tag{4.10.45}$$

In these equations the subscript 1 has been dropped from $C_1^{(1)}$, J_1 and Y_1 are first-order Bessel functions of the first and second kind, and we have used the relationships

$$\frac{d}{d\xi} J_0(\xi) = -J_1(\xi), \quad \frac{d}{d\xi} Y_0(\xi) = -Y_1(\xi). \tag{4.10.46}$$

Equations (4.10.45) are linear and homogeneous in $C^{(1)}$ and $C^{(2)}$. Therefore they will have a solution other than the trivial one $C^{(1)} = C^{(2)} = 0$ if and only if the determinant of the coefficients of the C^k vanishes. The equation obtained by setting the determinant to zero is the critical equation for the two-region cylinder. However, as in the slab case (4.10.26), the critical equation may be obtained more directly by dividing the second equation in (4.10.45) into the first. Thus there will be a mathematical solution to the problem of the one-group, two-region cylindrical reactor if and only if

$$\frac{J_0(\kappa^{(1)}R_1)}{D^{(1)}\kappa^{(1)}J_1(\kappa^{(1)}R_1)} = \frac{J_0(\kappa^{(2)}R_1)Y_0(\kappa^{(2)}R_2) - J_0(\kappa^{(2)}R_2)Y_0(\kappa^{(2)}R_1)}{D^{(2)}\kappa^{(2)}[J_1(\kappa^{(2)}R_1)Y_0(\kappa^{(2)}R_2) - J_0(\kappa^{(2)}R_2)Y_1(\kappa^{(2)}R_1)]}. \tag{4.10.47}$$

Any physical conditions that lead to values of $\kappa^{(1)}$, $\kappa^{(2)}$, $D^{(1)}$, $D^{(2)}$, R_1, and R_2 such that this equation is satisfied (with $\lambda = 1$) represent a critical condition according to the one-group model. Alternatively, the geometry and material compositions may be fixed and (4.10.47) used to determine the value of λ, or k_{eff}, of the system according to the one-group model. In either case the algebra is tedious since the reactor parameters to be varied so that (4.10.47) is satisfied appear in the arguments of Bessel functions.

It is now possible to construct an overall solution $\Phi(r, z)$ for the initial partial-differential equation (4.10.31). Thus, combining (4.10.33), (4.10.39), and (4.10.43) yields, for $r \leq R_1$,

$$\Phi(r, z) = R(r)Z(z) = A_1 C^{(1)} J_0(\kappa^{(1)} r) \sin\frac{\pi z}{L_z} \quad \text{for } 0 \leq r \leq R_1, \quad 0 \leq z \leq L_z. \tag{4.10.48}$$

Similarly, for $r \geq R_1$, (4.10.33), (4.10.39), (4.10.44), and the first equation in (4.10.45) give

$$\begin{aligned}\Phi(r, z) &= A_1 C^{(2)}[J_0(\kappa^{(2)}r)Y_0(\kappa^{(2)}R_2) - J_0(\kappa^{(2)}R_2)Y_0(\kappa^{(2)}r)] \sin\frac{\pi z}{L_z} \\ &= A_1 C^{(1)} \frac{J_0(\kappa^{(1)}R_1)[J_0(\kappa^{(2)}r)Y_0(\kappa^{(2)}R_2) - J_0(\kappa^{(2)}R_2)Y_0(\kappa^{(2)}r)]}{J_0(\kappa^{(2)}R_1)Y_0(\kappa^{(2)}R_2) - J_0(\kappa^{(2)}R_2)Y_0(\kappa^{(2)}R_1)} \sin\frac{\pi z}{L_z}\end{aligned} \tag{4.10.49}$$

for $R_1 \leq r \leq R_2, \quad 0 \leq z \leq L_z$.

To construct the complete one-group estimate of the scalar flux density $\Phi(r, z, E)$ we simply multiply by the multigroup spectrum functions appropriate to the two radial regions. If, for convenience, we express these as continuous (rather than multigroup) functions of energy, $\psi^{(1)}(E)$ and $\psi^{(2)}(E)$, and if we replace the constant $A_1C^{(1)}$ by the single symbol C, we have the overall one-group solution for the two-region cylinder

$$\Phi(r, z, E) = \begin{cases} C\psi^{(1)}(E)J_0(\kappa^{(1)}r)\sin\dfrac{\pi z}{L_z} \quad \text{for } 0 \le r \le R_1, \quad 0 \le z \le L_z, \\ C\psi^{(2)}(E)\dfrac{J_0(\kappa^{(1)}R_1)[J_0(\kappa^{(2)}r)Y_0(\kappa^{(2)}R_2) - J_0(\kappa^{(2)}R_2)Y_0(\kappa^{(2)}r)]}{J_0(\kappa^{(2)}R_1)Y_0(\kappa^{(2)}R_2) - J_0(\kappa^{(2)}R_2)Y_0(\kappa^{(2)}R_1)}\sin\dfrac{\pi z}{L_z} \\ \qquad \text{for } R_1 \le r \le R_2, \quad 0 \le z \le L_z. \end{cases} \tag{4.10.50}$$

The integration constant C is arbitrary as far as the solution of the homogeneous differential equation (4.10.31) is concerned. We may determine it so that the fission rate corresponds to the average power level $\bar{P}$ of the reactor in watts per cc per second. Equation (4.7.15) accomplishes this normalization; Written out for the case at hand it becomes

$$\begin{aligned} \bar{P} &= \frac{1}{3 \times 10^{10} \times \pi R_2^2 L_z}\int_0^\infty dE \int_0^{R_2} dr \int_0^{L_z} dz\, \Sigma_f(r, z, E)\Phi(r, z, E) \\ &= \frac{1}{3 \times 10^{10} \times \pi R_2^2 L_z}\int_0^{L_z} dz \left[\int_0^{R_1} \Sigma_f^{(1)}\Phi(r, z)\, dr + \int_{R_1}^{R_2} \Sigma_f^{(2)}\Phi(r, z)\, dr\right], \end{aligned} \tag{4.10.51}$$

where $\Phi(r, z, E)$ is given by (4.10.50) and $\Phi(r, z)$ by (4.10.49), $\Sigma_f^{(1)}$ and $\Sigma_f^{(2)}$ being the one-group macroscopic fission cross sections for regions (1) and (2).

Expressing the complete solution (4.10.50) as a function of **r** and E makes it clear that a one-group approximation is actually far more general than a "one-speed" approximation in which all the neutrons are assumed to have a single constant velocity.

Limiting and Equivalent Forms of the Critical Equation

There are two limiting conditions under which we expect the critical equation (4.10.47) to take a much simpler form. These occur when the two-region radial problem reduces to a one-region problem, that is, when $R_1 \to R_2$ or when $(D^{(2)},\kappa^{(2)}) \to (D^{(1)},\kappa^{(1)})$. Examination shows that, as $R_1 \to R_2$, the numerator of the right-hand side of (4.10.47) approaches zero while the denominator remains finite. Thus the criticality condition (4.10.47) becomes, as $R_1 \to R_2$,

$$J_0(\kappa^{(1)}R_2) = 0. \tag{4.10.52}$$

Moreover it is obvious that (4.10.52) is sufficient to meet condition (4.10.47) as $(D^{(2)}, \kappa^{(2)}) \to (D^{(1)}, \kappa^{(1)})$. (The second terms in both the numerator and denominator of (4.10.47) become zero for $J_0(\kappa^{(2)}R_2) = 0$.) A closer look shows that the condition (4.10.52) is also necessary for this case.

Thus both limiting conditions are—as they must be—the same. Moreover, if we want $R(\mathbf{r})$ to be positive for all r in the range $0 \leq r \leq R_2$, we must select $\kappa^{(1)}R_2$ to be the smallest number such that (4.10.52) is obeyed. Since $\xi \approx 2.405$ is the smallest number such that $J_0(\xi) = 0$, (4.10.52) implies

$$\kappa^{(1)}R_2 = 2.405. \tag{4.10.53}$$

From (4.10.41) and (4.10.38), this limiting criticality condition becomes

$$\left(\frac{2.405}{R_2}\right)^2 + \left(\frac{\pi}{L_z}\right)^2 = \frac{\lambda^{-1}\nu\Sigma_f - \Sigma_a}{D} \tag{4.10.54}$$

or

$$\lambda = \frac{\nu\Sigma_f}{D[(2.405/R_2)^2 + (\pi/L_z)^2] + \Sigma_a} = \frac{\nu\Sigma_f}{DB_r^2 + \Sigma_a}, \tag{4.10.55}$$

where we have dropped superscripts on the reactor parameters since there is only one material composition in this limiting case.

Equation (4.10.55) is the one-group criticality condition for the bare cylindrical reactor. Note that it is a specific case of the multigroup, bare-core formula (4.7.17) with the total number G of groups taken as one and B_{x1}^2 replaced by the total geometrical buckling (from Table 4.1) of the finite cylinder.

Variant forms of the critical equation (4.10.47) appear whenever $\kappa^{(1)}$ or $\kappa^{(2)}$ are pure imaginary. This can happen if $\lambda^{-1}\nu\Sigma_f^{(1)}$ or $\lambda^{-1}\nu\Sigma_f^{(2)}$ is sufficiently small or zero (as in a nonmultiplying material). The critical condition (4.10.47) is still valid for this case. For applications, however, it is more convenient to replace the Bessel functions of pure imaginary argument by the corresponding modified Bessel functions of real argument. Specifically we have

$$\begin{aligned} J_0(i\xi) &= I_0(\xi), \\ J_1(i\xi) &= iI_1(\xi), \\ Y_0(i\xi) &= iI_0(\xi) - \frac{2}{\pi}K_0(\xi), \\ Y_1(i\xi) &= -I_1(\xi) + \frac{2i}{\pi}K_1(\xi), \end{aligned} \tag{4.10.56}$$

where $I_0(\xi)$, $K_0(\xi)$, $I_1(\xi)$, and $K_1(\xi)$ are modified Bessel functions of the first and second kind of order zero and one.

Thus, if, for example, $\kappa^{(2)}$ is pure imaginary and we define

$$[k^{(2)}]^2 = -[\kappa^{(2)}]^2 = \frac{\Sigma_a^{(2)} + D^{(2)}B_z^2 - \lambda^{-1}\nu\Sigma_f^{(2)}}{D^{(2)}}, \tag{4.10.57}$$

(4.10.47) becomes

$$\frac{J_0(\kappa^{(1)}R_1)}{D^{(1)}\kappa^{(1)}J_1(\kappa^{(1)}R_1)}$$
$$= \frac{I_0(k^{(2)}R_1)[iI_0(k^{(2)}R_2) - (2/\pi)K_0(k^{(2)}R_2)] - I_0(k^{(2)}R_2)[iI_0(k^{(2)}R_1) - (2/\pi)K_0(k^{(2)}R_1)]}{iD^{(2)}k^{(2)}\{iI_1(k^{(2)}R_1)[iI_0(k^{(2)}R_2) - (2/\pi)K_0(k^{(2)}R_2)] - I_0(k^{(2)}R_2)[-I_1(k^{(2)}R_1) + (2i/\pi)K_1(k^{(2)}R_1)]\}}$$
$$= \frac{-I_0(k^{(2)}R_1)K_0(k^{(2)}R_2) + I_0(k^{(2)}R_2)K_0(k^{(2)}R_1)}{D^{(2)}k^{(2)}[I_1(k^{(2)}R_1)K_0(k^{(2)}R_2) + I_0(k^{(2)}R_2)K_1(k^{(2)}R_1)]}. \qquad (4.10.58)$$

This same result can, of course, be obtained directly if we recognize that $I_0(k^{(2)}r)$ and $K_0(k^{(2)}r)$ are the two particular solutions of (4.10.42) when $ik^{(2)}$ replaces $\kappa^{(2)}$ (in accord with (4.10.57)), so that (4.10.44) is replaced by

$$R(\mathbf{r}) = C^{(2)}[I_0(k^{(2)}r)K_0(k^{(2)}R_2) - I_0(k^{(2)}R_2)K_0(k^{(2)}r)]. \qquad (4.10.59)$$

Applying the continuity conditions, as in (4.10.45), leads again to the result (4.10.58).

Analogous bare-core or equivalent forms for the criticality equation can be obtained by starting with the two-region-slab formula (4.10.26).

4.11 Two-Group Theory

The basic approximation which leads to one-group theory is that $\Phi(\mathbf{r}, E)$ has a separable energy dependence $\psi^k(E)$ throughout each given material composition k. We have seen that this approximation is very questionable at interfaces between regions k and l, since the form of $\psi^k(E)$ will not in general match that of $\psi^l(E)$ at such a boundary. We tried to minimize this problem by matching $R^k(\mathbf{r}_c)\psi^k(E)$ and $R^l(\mathbf{r}_c)\psi^l(E)$ in an energy-integral sense. But if $\psi^k(E)$ and $\psi^l(E)$ are quite different (as will be in the case if k is core material and l is reflector material), there are bound to be significant mismatches over particular subranges of the overall energy range 0 to ∞.

For example, at an interface between core and reflector material, the net leakage of high-energy neutrons, which are originally created by fission in the core, is from the core to the reflector, whereas the net leakage of low-energy neutrons, created in abundance by the superior moderating power of the reflector material, is in the opposite direction. A one-group model cannot describe this process. Depending on the sign of $\mathbf{n}_k \cdot D^k \nabla R^k(\mathbf{r})$ $(= \mathbf{n}_k \cdot D^l \nabla R^l(\mathbf{r}))$, either a net number of neutrons having the energy distribution $\psi^k(E)$ leak from k to l per second or a net number having the distribution $\psi^l(E)$ leak from l to k.

Two-group theory represents an attempt to improve the accuracy with which the flux can be described near such interfaces. The basic idea is to split the spectrum functions $\psi^k(E)$ into two parts, $\psi_1^k(E)$ (for energy group one, extending from a "cut-point" energy E_c to ∞) and $\psi_2^k(E)$ (for energy group two, extending from 0 to E_c) and to associate separate spatial functions $R_1^k(\mathbf{r})$ and $R_2^k(\mathbf{r})$ with the neutrons belonging to each of these two groups. Continuity of flux and current across an interface is then required in an integral sense, individually, for the ranges 0 to E_c and E_c to ∞. Thus the two-group model permits a net leakage rate of group-one neutrons in one direction across an interface and a net leakage rate of group-two neutrons in the opposite direction. In the interior portions of a given region, where it is expected that the separable form $\Phi(\mathbf{r}, E) = R^k(\mathbf{r})\psi^k(E)$ will be valid over the *entire* energy range, the two-group approximation will yield the result that $R_1^k(\mathbf{r})$ and $R_2^k(\mathbf{r})$ have the same spatial shape, so that $R_1^k(\mathbf{r})\psi_1^k(E)$ and $R_2^k(\mathbf{r})\psi_2^k(E)$ for the two energy ranges fit together to form a single function of the form $R^k(\mathbf{r})\psi^k(E)$.

Derivation of the Two-Group Equations

Mathematically the assumptions of two-group theory may be summarized as follows:

1. The scalar flux function may be written, for $\mathbf{r}$ in region k, as

$$\Phi(\mathbf{r}, E) = \begin{cases} R_1^k(\mathbf{r})\psi_1^k(E) & \text{for } E_c < E < \infty, \\ R_2^k(\mathbf{r})\,\psi_2^k(E) & \text{for } E \le E_c, \end{cases} \tag{4.11.1}$$

where $\psi_1^k(E)$ and $\psi_2^k(E)$ are parts of $\psi^k(E)$ found by solving (4.10.3).

2. Boundary conditions, for k and l indicating any two adjacent compositions and $\mathbf{r}_c$ being a point on the interface separating them, require that

$$\begin{aligned}
\int_{E_c}^{\infty} dE R_1^k(\mathbf{r}_c)\psi_1^k(E) &= \int_{E_c}^{\infty} dE R_1^l(\mathbf{r}_c)\psi_1^l(E), \\
\int_{0}^{E_c} dE R_2^k(\mathbf{r}_c)\psi_2^k(E) &= \int_{0}^{E_c} dE R_2^l(\mathbf{r}_c)\psi_2^l(E), \\
\int_{E_c}^{\infty} dE\, \mathbf{n}_k \cdot D^k(E)\nabla[R_1^k(\mathbf{r}_c)\psi_1^k(E)] &= \int_{E_c}^{\infty} dE\, \mathbf{n}_k \cdot D^l(E)\nabla[R_1^l(\mathbf{r}_c)\psi_1^l(E)], \\
\int_{0}^{E_c} dE\, \mathbf{n}_k \cdot D^k(E)\nabla[R_2^k(\mathbf{r}_c)\psi_2^k(E)] &= \int_{0}^{E_c} dE\, \mathbf{n}_k \cdot D^l(E)\nabla[R_2^l(\mathbf{r}_c)\psi_2^l(E)].
\end{aligned} \tag{4.11.2}$$

As with the one group case, it simplifies the algebra if, after having found the *shape* in energy of $\psi_1^k(E)$ and $\psi_2^k(E)$ by solving (4.10.3) for the material of region k, we renormalize these two segments of $\psi^k(E)$ so that

$$\begin{aligned}
\int_{E_c}^{\infty} \psi_1^k(E)\, dE &= 1 \qquad (k = 1, 2, \ldots, K), \\
\int_{0}^{E_c} \psi_2^k(E)\, dE &= 1 \qquad (k = 1, 2, \ldots, K).
\end{aligned} \tag{4.11.3}$$

Then the continuity equations (4.11.2) show that $R_1^k(\mathbf{r}_c) = R_1^l(\mathbf{r}_c)$ and $R_2^k(\mathbf{r}_c) = R_2^l(\mathbf{r}_c)$ for points $\mathbf{r}_c$ on the interface between regions k and l. Thus, as in the one-group case, the superscript k on the functions R are superfluous, and we shall change notation by replacing the various $R_1^k(\mathbf{r})$ by the single, everywhere-continuous function $\Phi_1(\mathbf{r})$, and similarly replacing the $R_2^k(\mathbf{r})$ by $\Phi_2(\mathbf{r})$. The functions $\Phi_1(\mathbf{r})$ and $\Phi_2(\mathbf{r})$ are called the *two-group fluxes.* Note that they are *not* fluxes per unit energy (in the way that $\Phi(\mathbf{r}, E) = v(E)n(\mathbf{r}, E)$ is). Physically $\Phi_1(\mathbf{r})$ is the two-group approximation to the number of neutrons per cc having energies in the range E_c to ∞ ($\int_{E_c}^{\infty} n(\mathbf{r}, E)\, dE$) multiplied by the the average "fast" speed

$$v_1 = \frac{\int_{E_c}^{\infty} v(E)n(\mathbf{r}, E)\, dE}{\int_{E_c}^{\infty} n(\mathbf{r}, E)\, dE},$$

and $\Phi_2(\mathbf{r})$ is the two-group approximation to the number in the range 0 to E_c multiplied by the average thermal speed

$$v_2 = \frac{\int_0^{E_c} v(E)n(\mathbf{r}, E)\, dE}{\int_0^{E_c} n(\mathbf{r}, E)\, dE}.$$

(These interpretations are somewhat ambiguous since (4.11.1), on which they are based, is an approximation which cannot be rigorously correct.)

To find differential equations for the two-group fluxes $\Phi_1(\mathbf{r})$ and $\Phi_2(\mathbf{r})$ we substitute the approximation (4.11.1) into (4.9.1). This will give us different results for $E > E_c$ and $E < E_c$:

$$\begin{aligned} &-D^k(E)\psi_1^k(E)\nabla^2\Phi_1(\mathbf{r}) + \Sigma_t^k(E)\psi_1^k(E)\Phi_1(\mathbf{r}) \\ &= \int_{E_c}^{\infty} \left[\frac{1}{\lambda}\chi^k(E)\nu\Sigma_f^k(E')\psi_1^k(E') + \Sigma_s^k(E' \to E)\psi_1^k(E')\right] dE'\, \Phi_1(\mathbf{r}) \\ &\quad + \int_0^{E_c} \left[\frac{1}{\lambda}\chi^k(E)\nu\Sigma_f(E')\psi_2^k(E')\right] dE'\Phi_2(\mathbf{r}) \quad \text{for } E > E_c \end{aligned} \tag{4.11.4}$$

and

$$\begin{aligned} &-D^k(E)\psi_2^k(E)\nabla^2\Phi_2(\mathbf{r}) + \Sigma_t^k(E)\psi_2^k(E)\Phi_2(\mathbf{r}) \\ &= \int_{E_c}^{\infty} \left[\frac{1}{\lambda}\chi^k(E)\nu\Sigma_f^k(E')\psi_1^k(E') + \Sigma_s^k(E' \to E)\psi_1^k(E')\right] dE'\, \Phi_1(\mathbf{r}) \\ &\quad + \int_0^{E_c} \left[\frac{1}{\lambda}\chi^k(E)\nu\Sigma_f^k(E')\psi_2^k(E') + \Sigma_s^k(E' \to E)\psi_2^k(E')\right] dE'\, \Phi_2(\mathbf{r}) \quad \text{for } E \leq E_c, \end{aligned} \tag{4.11.5}$$

where on physical grounds we have omitted from (4.11.4) the scattering from $E' < E_c$ to $E > E_c$. (For thermal reactors the cut point E_c will always be such that $\chi^k(E)$ is zero for $E < E_c$; hence no fission terms will appear in (4.11.5). We retain them for possible application to fast reactors, for which E_c will be much higher.)

There is no solution $(\Phi_1(\mathbf{r}), \Phi_2(\mathbf{r}))$ that will satisfy (4.11.4) and (4.11.5) at all energies since the form (4.11.1) is not sufficiently general. However we can force equality of the right- and left-hand sides in an integral sense and in that way find equations which, when solved, will give us $\Phi_1(\mathbf{r})$ and $\Phi_2(\mathbf{r})$. Accordingly we shall integrate (4.11.4) from E_c to ∞ and (4.11.5) from 0 to E_c and require that the resultant equations be valid at all $\mathbf{r}$. To simplify the result we first define a set of "two-group constants":

$$\begin{aligned}
&D_1^k \equiv \int_{E_c}^{\infty} D(E)\psi_1^k(E)\,dE, \qquad D_2^k \equiv \int_0^{E_c} D(E)\psi_2^k(E)\,dE,\\
&\Sigma_{t1}^k \equiv \int_{E_c}^{\infty} \Sigma_t^k(E)\psi_1^k(E)\,dE, \qquad \Sigma_{t2}^k \equiv \int_0^{E_c} \Sigma_t^k(E)\psi_2^k(E)\,dE,\\
&\chi_1^k \equiv \int_{E_c}^{\infty} \chi^k(E)\,dE, \qquad \chi_2^k \equiv \int_0^{E_c} \chi^k(E)\,dE,\\
&\nu\Sigma_{f1}^k \equiv \int_{E_c}^{\infty} \nu\Sigma_f^k(E)\psi_1^k(E)\,dE, \qquad \nu\Sigma_{f2}^k \equiv \int_0^{E_c} \nu\Sigma_f^k(E)\psi_2^k(E)\,dE,\\
&\Sigma_{11}^k \equiv \int_{E_c}^{\infty} dE \int_{E_c}^{\infty} dE'\Sigma_s^k(E' \to E)\psi_1^k(E'), \qquad \Sigma_{22}^k \equiv \int_0^{E_c} dE \int_0^{E_c} dE'\Sigma_s^k(E' \to E)\psi_2^k(E'),\\
&\Sigma_{21}^k \equiv \int_0^{E_c} dE \int_{E_c}^{\infty} dE'\,\Sigma_s^k(E' \to E)\psi_1^k(E'), \qquad \Sigma_1^k \equiv \Sigma_{t1}^k - \Sigma_{11}^k, \qquad \Sigma_2^k \equiv \Sigma_{t2}^k - \Sigma_{22}^k.
\end{aligned} \tag{4.11.6}$$

Using these definitions, integrating (4.11.4) from E_c to ∞, and integrating (4.11.5) from 0 to E_c, we get

$$\begin{aligned}
-D_1^k\nabla^2\Phi_1(\mathbf{r}) + \Sigma_1^k\Phi_1(\mathbf{r}) &= \frac{1}{\lambda}\chi_1^k[\nu\Sigma_{f1}^k\Phi_1(\mathbf{r}) + \nu\Sigma_{f2}^k\Phi_2(\mathbf{r})],\\
-D_2^k\nabla^2\Phi_2(\mathbf{r}) + \Sigma_2^k\Phi_2(\mathbf{r}) &= \frac{1}{\lambda}\chi_2^k[\nu\Sigma_{f1}^k\Phi_1(\mathbf{r}) + \nu\Sigma_{f2}^k\Phi_2(\mathbf{r})] + \Sigma_{21}^k\Phi_1(\mathbf{r}).
\end{aligned} \tag{4.11.7}$$

In view of (4.11.3) and (4.11.6), the boundary conditions (4.11.2) become:

1. $\Phi_1(\mathbf{r})$, $\Phi_2(\mathbf{r})$ must be continuous everywhere,
2. $\mathbf{n}\cdot D_1\nabla\Phi_1(\mathbf{r})$ and $\mathbf{n}\cdot D_2\nabla\Phi_2(\mathbf{r})$ must be continuous across interfaces separating different material compositions, (4.11.8)
3. $\Phi_1(\mathbf{r}) = \Phi_2(\mathbf{r}) = 0$ on the outer boundary of the reactor.

Equations (4.11.7) are the "two-group diffusion equations." They are the standard workhorses of thermal-reactor design. It can be proved that a unique, positive solution corresponding to a most-positive real eigenvalue λ always exists for the two-group equations. Thus a physically acceptable solution for the group fluxes Φ_1 and Φ_2 can always be found, and from this the two-group approximation $(\Phi_1(\mathbf{r})\psi_1^k(E);\ \Phi_2(\mathbf{r})\psi_2^k(E))$ for the scalar flux $\Phi(\mathbf{r}, E)$ can be constructed throughout each material composition k.

Solution of the Two-Group Equations by Finite-Difference Methods

The standard procedure for solving the partial-differential equations (4.11.7) is first to convert them into a set of algebraic equations by finite-difference methods and then to solve this latter set using a digital computer. This method is a straightforward extension of the scheme described in Section 4.10 for the one-dimensional, one-group case. We shall illustrate its application in the case of two-dimensional (XY) geometry.

The basic step in converting the continuous, position-dependent equations of (4.11.7) into finite-difference form for an XY geometry is to superimpose an XY mesh grid over the space occupied by the reactor and then to replace the computation of the $\Phi_g(\mathbf{r})$ at all points $\mathbf{r}$ by a computation of the point values of the $\Phi_g(\mathbf{r})$ at the intersections of mesh lines.

To proceed we consider a typical interior mesh point and its nearest neighbors, as illustrated in Figure 4.9.

Note that, if x_i and y_j represent the spatial coordinates of the point (i, j), the "mesh intervals" $h_x^{(i)}$ and $h_y^{(j)}$, which need not be constant in size, are defined as

$$h_x^{(i)} \equiv x_{i+1} - x_i,$$
$$h_y^{(j)} \equiv y_{j+1} - y_j.$$

We shall refer to the area $h_x^{(i)} h_y^{(j)}$ as a "mesh rectangle."

In two-dimensional XY geometry, the two-group equations to be solved are, from (4.11.7),

$$\begin{aligned} &-\frac{\partial}{\partial x}\left[D_1(x, y)\frac{\partial}{\partial x}\Phi_1(x, y)\right] - \frac{\partial}{\partial y}\left[D_1(x, y)\frac{\partial}{\partial y}\Phi_1(x, y)\right] + \Sigma_1(x, y)\Phi_1(x, y) \\ &= \frac{1}{\lambda}[\nu\Sigma_{f1}(x, y)\Phi_1(x, y) + \nu\Sigma_{f2}(x, y)\Phi_2(x, y)], \\ &-\frac{\partial}{\partial x}\left[D_2(x, y)\frac{\partial}{\partial x}\Phi_2(x, y)\right] - \frac{\partial}{\partial y}\left[D_2(x, y)\frac{\partial}{\partial y}\Phi_2(x, y)\right] + \Sigma_2(x, y)\Phi_2(x, y) \\ &= \Sigma_{21}(x, y)\Phi_1(x, y), \end{aligned} \tag{4.11.9}$$

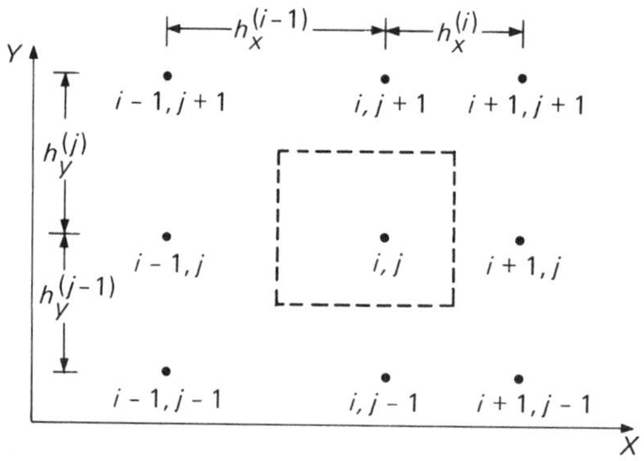

Figure 4.9 A mesh layout in XY geometry.

where, for greater generality, we have assumed that the group parameters may vary arbitrarily with x and y, but where, to simplify the algebra, we have set $\chi_2 = 0$ and $\chi_1 = 1$.

The next step in deriving the finite-difference form of these equations is to replace the position-dependent reactor parameters by their average values over the mesh rectangles. Thus, over the mesh rectangle $h_x^{(i)}h_y^{(j)}$ to the right and above mesh point (i, j) on the figure, we make the replacements

$$\begin{aligned} &D_g(x, y) \to D_g^{(i,j)}, \\ &\Sigma_g(x, y) \to \Sigma_g^{(i,j)}, \\ &\nu\Sigma_{fg}(x, y) \to \nu\Sigma_{fg}^{(i,j)}, \\ &\Sigma_{21}(x, y) \to \Sigma_{21}^{(i,j)}. \end{aligned} \tag{4.11.10}$$

Referring to Figure 4.9, we then integrate (4.11.9) over the area within the dotted lines connecting the centers of the four mesh rectangles. If the coordinates of these center points are labeled $(x_{i-\frac{1}{2}}, y_{j-\frac{1}{2}})$, $(x_{i+\frac{1}{2}}, y_{j-\frac{1}{2}})$, etc., the result of this operation is (for the group-two equation)

$$\begin{aligned} &-\int_{y_{j-½}}^{y_{j+½}} dy \left[D_2(x_{i+\frac{1}{2}}, y)\frac{\partial}{\partial x}\Phi_2(x, y)\bigg|_{x_{i+½}} - D_2(x_{i-\frac{1}{2}}, y)\frac{\partial}{\partial x}\Phi_2(x, y)\bigg|_{x_{i-½}} \right] \\ &-\int_{x_{i-½}}^{x_{i+½}} dx \left[D_2(x, y_{j+\frac{1}{2}})\frac{\partial}{\partial y}\Phi_2(x, y)\bigg|_{y_{j+½}} - D_2(x, y_{j-\frac{1}{2}})\frac{\partial}{\partial y}\Phi_2(x, y)\bigg|_{y_{j-½}} \right] \\ &+\int_{x_{i-½}}^{x_{i+½}} \int_{y_{j-½}}^{y_{j+½}} [\Sigma_2(x, y)\Phi_2(x, y) - \Sigma_{21}(x, y)\Phi_1(x, y)] = 0, \end{aligned} \tag{4.11.11}$$

with an analogous result for the group-one equation.

We next make the approximations

$$\begin{aligned} &\frac{\partial}{\partial x}\Phi_g(x, y)\bigg|_{x_{i+½}} \approx \frac{\Phi_g^{(i+1,j)} - \Phi_g^{(i,j)}}{h_x^{(i)}} \quad \text{for } y_{j-\frac{1}{2}} \le y \le y_{j+\frac{1}{2}}, \\ &\frac{\partial}{\partial y}\Phi_g(x, y)\bigg|_{y_{j+½}} \approx \frac{\Phi_g^{(i,j+1)} - \Phi_g^{(i,j)}}{h_y^{(j)}} \quad \text{for } x_{i-\frac{1}{2}} \le x \le x_{i+\frac{1}{2}}, \\ &\Phi_g(x, y) \approx \Phi_g^{(i,j)} \quad \text{for } x_{i-\frac{1}{2}} \le x \le x_{i+\frac{1}{2}}, \quad y_{j-\frac{1}{2}} \le y \le y_{j+\frac{1}{2}}, \end{aligned} \tag{4.11.12}$$

where $\Phi_g^{(i,j)}$ is the value of the flux for group g at mesh point (i, j).

When we replace the position-dependent group parameters in (4.11.11) by the average piecewise-constant parameters (4.11.10), the former equation becomes

$$\begin{aligned} &-\left[\frac{1}{2}h_y^{(j-1)}D_2^{(i,j-1)} + \frac{1}{2}h_y^{(j)}D_2^{(i,j)}\right]\frac{\Phi_2^{(i+1,j)} - \Phi_2^{(i,j)}}{h_x^{(i)}} \\ &+\left[\frac{1}{2}h_y^{(j-1)}D_2^{(i-1,j-1)} + \frac{1}{2}h_y^{(j)}D_2^{(i-1,j)}\right]\frac{\Phi_2^{(i,j)} - \Phi_2^{(i-1,j)}}{h_x^{(i-1)}} \end{aligned}$$

(equation continued)

$$
\begin{aligned}
&-\left[\frac{1}{2}h_x^{(i-1)}D_2^{(i-1,j)} + \frac{1}{2}h_x^{(i)}D_2^{(i,j)}\right]\frac{\Phi_2^{(i,j+1)} - \Phi_2^{(i,j)}}{h_y^{(j)}} \\
&+\left[\frac{1}{2}h_x^{(i-1)}D_2^{(i-1,j-1)} + \frac{1}{2}h_x^{(i)}D_2^{(i,j-1)}\right]\frac{\Phi_2^{(i,j)} - \Phi_2^{(i,j-1)}}{h_y^{(j-1)}} \\
&+\frac{1}{4}[\Sigma_2^{(i-1,j-1)}h_x^{(i-1)}h_y^{(j-1)} + \Sigma_2^{(i,j-1)}h_x^{(i)}h_y^{(j-1)} + \Sigma_2^{(i,j)}h_x^{(i)}h_y^{(j)} + \Sigma_2^{(i-1,j)}h_x^{(i-1)}h_y^{(j)}]\Phi_2^{(i,j)} \\
&-\frac{1}{4}[\Sigma_{21}^{(i-1,j-1)}h_x^{(i-1)}h_y^{(j-1)} + \Sigma_{21}^{(i,j-1)}h_x^{(i)}h_y^{(j-1)} + \Sigma_{21}^{(i,j)}h_x^{(i)}h_y^{(j)} + \Sigma_{21}^{(i-1,j)}h_x^{(i-1)}h_y^{(j)}]\Phi_1^{(i,j)} = 0.
\end{aligned}
\tag{4.11.13}
$$

This equation and the corresponding difference equation for the first energy group relate the fluxes $\Phi_1^{(i,j)}$ and $\Phi_2^{(i,j)}$ at point (i, j) to those at the four nearest-neighbor points. For each group there are as many of these equations as there are internal mesh points, and since, for a boundary condition of zero flux on the external boundary, the fluxes at mesh points on that boundary are set to zero, the number of unknown fluxes is also equal to the number of internal mesh points. Thus, if there are I internal mesh points in the X direction and J in the Y direction, (4.11.13) and the corresponding difference equation for group one, when written for all internal mesh points, represent a system of $2IJ$ homogeneous, linear, algebraic equations in $2IJ$ unknowns. These equations will have a solution if and only if the $(2IJ) \times (2IJ)$ determinant of the coefficients of the $\Phi_g^{(i,j)}$ vanishes. That value of λ which makes that determinant vanish and at the same time yields a solution for the point fluxes that is positive at all internal mesh points for both groups is the k_{eff} of the reactor, and the corresponding $\Phi_g^{(i,j)}$ are the fluxes we seek. It is possible to prove that a unique solution to this very sizeable algebraic problem always exists for any physically realistic set of reactor parameters and for any finite number of mesh points. Computer programs exist that will solve (4.11.13) for over a million mesh points.

The accuracy with which the fluxes $\Phi_g^{(i,j)}$ obtained by finite-difference methods match the corresponding $\Phi_g(x_i, y_j)$ that would be obtained if it were possible to solve (4.11.9) analytically depends on the validity of the approximations (4.11.12). If the $\Phi_g(x, y)$ vary wildly with position, small mesh spacings will be required. Fortunately the proper mesh sizes can be determined empirically, since it can be proved that the solution of the difference equations converges to a unique limit as the $h_x^{(i)}$ and $h_y^{(j)}$ are decreased. Thus, if halving the mesh intervals produces a negligible change in the flux values, we can be sure that the original mesh intervals were adequately small.

By starting with approximations more elaborate than (4.11.12), we can derive difference equations that will yield more accurate values of the $\Phi_g^{(i,j)}$ for a given mesh layout. However these "higher-order difference equations" will be more complicated than (4.11.13).

For example the fluxes $\Phi_2^{(i+1,j+1)}$, $\Phi_2^{(i+1,j-1)}$, $\Phi_2^{(i-1,j+1)}$, and $\Phi_2^{(i-1,j-1)}$ may also appear in the higher-order difference equation analogous to (4.11.13), so that each $\Phi_2^{(i,j)}$ will be connected to eight (rather than four) nearest neighbors. Devising a set of difference equations that are the best compromise between accuracy and ease of solution is a problem in numerical analysis that is of major importance for reactor design.

An Analytical Solution of the Two-Group Equations

As in the one-group case, there is a strong motivation for obtaining closed-form analytical solutions for the two-group diffusion equations. Such solutions would permit ready insight into how changes in cross sections or material compositions affect the criticality and flux-shape characteristics of the reactor. Unfortunately it is possible to obtain such closed-form solutions only for essentially one-dimensional reactors for which the reactor parameters are constant in each region. Nevertheless, since many real reactors can be approximated by such a model, the analytical approach is worth pursuing as long as the resultant expressions do not become so complicated that they are no longer an aid to physical understanding.

Accordingly let us now consider the analytical solution for a two-group model of a reactor consisting of two nuclearly homogeneous materials assembled into a rectangular parallelopiped. We shall take the material properties to be uniform in the Y and Z directions for any fixed value of x, but in the X direction we shall assume they are assembled in the symmetric form shown in Figure 4.10. Material (1) is present in the region $(-X_1 \le x \le X_1, -Y_1 \le y \le Y_1, -Z_1 \le z \le Z_1)$ and material (2) in the regions $(X_1 \le x \le X_1 + X_2, -Y_1 \le y \le Y_1, -Z_1 \le z \le Z_1)$ and $(-(X_1 + X_2) \le x \le -X_1, -Y_1 \le y \le Y_1, -Z_1 \le z \le Z_1)$.

Let us again assume that we have performed auxiliary multigroup computations in order to find normalized spectrum functions $\psi_g^k(E)$ for the two materials ($k = 1, 2$) in

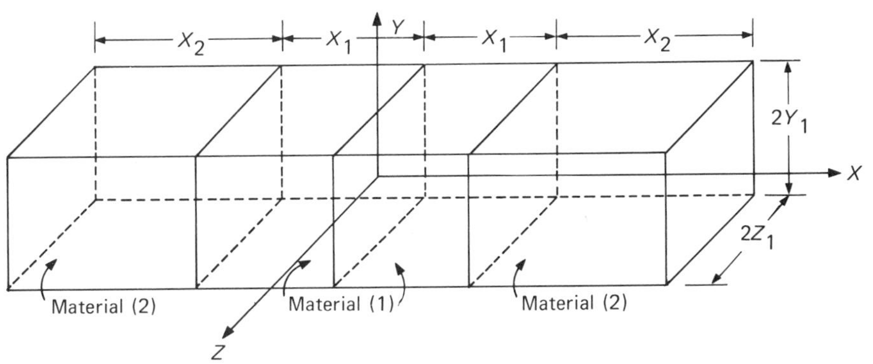

Figure 4.10 A three-region reactor.

the two energy ranges ($g = 1, 2$), and that we have used these four spectrum functions to determine, by (4.11.6), two-group parameters for the two material regions. In terms of these parameters the two-group equations (4.11.7) become, for the case at hand,

$$\begin{aligned} -D_1^k\nabla^2\Phi_1(x, y, z) + \left(\Sigma_1^k - \frac{1}{\lambda}\chi_1\nu\Sigma_{f1}^k\right)\Phi_1(x, y, z) - \frac{1}{\lambda}\chi_1\nu\Sigma_{f2}^k\Phi_2(x, y, z) &= 0, \\ -D_2^k\nabla^2\Phi_2(x, y, z) + \left(\Sigma_2^k - \frac{1}{\lambda}\chi_2\nu\Sigma_{f2}^k\right)\Phi_2(x, y, z) - \left(\frac{1}{\lambda}\chi_2\nu\Sigma_{f1}^k + \Sigma_{21}^k\right)\Phi_1(x, y, z) &= 0. \end{aligned} \tag{4.11.14}$$

Since the reactor is, at any x, homogeneous in the Z and Y directions, we can immediately separate the variables. Accordingly, motivated by the analysis of Section 4.10, we write

$$\Phi_g(x, y, z) = X_g(x)\cos\frac{\pi y}{2Y_1}\cos\frac{\pi z}{2Z_1} \qquad (g = 1, 2). \tag{4.11.15}$$

Note that $\cos(\pi y/2Y_1)\cos(\pi z/2Z_1)$ vanishes at $y = \pm Y_1$ and $z = Z_1$ and is everywhere positive for y and z values within those limits.

If we now define

$$B_{yz}^2 \equiv \left(\frac{\pi}{2Y_1}\right)^2 + \left(\frac{\pi}{2Z_1}\right)^2, \tag{4.11.16}$$

substitution of (4.11.15) into (4.11.14) yields

$$\begin{aligned} -D_1^k\frac{d^2X_1(x)}{dx^2} + \left(D_1^kB_{yz}^2 + \Sigma_1^k - \frac{1}{\lambda}\chi_1\nu\Sigma_{f1}^k\right)X_1(x) - \frac{1}{\lambda}\chi_1\nu\Sigma_{f2}^kX_2(x) &= 0, \\ -D_2^k\frac{d^2X_2(x)}{dx^2} + \left(D_2^kB_{yz}^2 + \Sigma_2^k - \frac{1}{\lambda}\chi_2\nu\Sigma_{f2}^k\right)X_2(x) - \left(\frac{1}{\lambda}\chi_2\nu\Sigma_{f1}^k + \Sigma_{21}^k\right)X_1 &= 0. \end{aligned} \tag{4.11.17}$$

Thus the partial-differential equations (4.11.14) in x, y, and z have been reduced to regular-differential equations in x alone.

In order to solve (4.11.17) we first note that we could eliminate X_1 and d^2X_1/dx^2 from the first equation by using the second equation and the second equation differentiated twice. The result would be a single *fourth-order* equation in X_2. It follows from this that, for each region k, there will be four linearly independent, particular solutions for X_2, and, since the second equation in (4.11.17) gives us X_1 in terms of X_2, there will be four *corresponding* linearly independent solutions for X_1. Any one of these four pairs of corresponding, particular solutions (X_1, X_2) will satisfy the equations (4.11.17). If linear combinations of such solutions for each region k can be made to satisfy the continuity and boundary conditions, the problem will be solved. We shall therefore search for pairs of particular solutions to (4.11.17) with the expectation of finding four such pairs in each region k.

To motivate the next step we note that, if D_1^k were zero, we could use the first equation in (4.11.17) to eliminate X_1 from the second equation; the second for any given region k would then have the form

$$\frac{d^2X_2}{dx^2} + B^2X_2 = 0. \tag{4.11.18}$$

Equation (4.11.18) has two particular solutions (sin Bx and cos Bx), and the corresponding particular solutions for X_1 could be found immediately from the first equation (D_1^k still being zero). They would also obey (4.11.18), since the coefficients in (4.11.17) are all independent of x for a given region k. These circumstances suggest that, in the case of interest where D_1^k is *not* zero, we may still be able to find particular solutions for X_2 in each region k that satisfy an equation of the form (4.11.18). In fact, if, for the *pair* of equations (4.11.17), we can find *two* values of B^2 and thus *four* linearly independent solutions obeying (4.11.18), these will be the solutions we seek. Accordingly we substitute (4.11.18) and an analogous expression for $X_1(x)$ involving the *same* B^2 into (4.11.17) and look for two values of B^2 each of which causes particular solutions of (4.11.18) also to be particular solutions of (4.11.17). The substitution yields

$$\begin{aligned}&\left(D_1^kB^2 + D_1^kB_{yz}^2 + \Sigma_1^k - \frac{1}{\lambda}\chi_1\nu\Sigma_{f1}^k\right)X_1(x) - \frac{1}{\lambda}\chi_1\nu\Sigma_{f2}^kX_2(x) = 0,\\ &-\left(\frac{1}{\lambda}\chi_2\nu\Sigma_{f1}^k + \Sigma_{21}^k\right)X_1(x) + \left(D_2^kB^2 + D_2^kB_{yz}^2 + \Sigma_2^k - \frac{1}{\lambda}\chi_2\nu\Sigma_{f2}^k\right)X_2(x) = 0.\end{aligned} \tag{4.11.19}$$

These two homogeneous algebraic equations will have a nontrivial solution if and only if the determinant of the coefficients of $X_1(x)$ and $X_2(x)$ vanishes, that is, if and only if

$$\begin{vmatrix} D_1^kB^2 + D_1^kB_{yz}^2 + \Sigma_1^k - \frac{1}{\lambda}\chi_1\nu\Sigma_{f1}^k & -\frac{1}{\lambda}\chi_1\nu\Sigma_{f2}^k \\ -\frac{1}{\lambda}\chi_2\nu\Sigma_{f1}^k - \Sigma_{21}^k & D_2^kB^2 + D_2^kB_{yz}^2 + \Sigma_2^k - \frac{1}{\lambda}\chi_2\nu\Sigma_{f2}^k \end{vmatrix} = 0. \tag{4.11.20}$$

With μ_k^2 and $-\nu_k^2$ defined to be the two values of B^2 that satisfy this condition, we find that

$$\begin{aligned}\mu_k^2 = -B_{yz}^2 &- \frac{1}{2}\left(\frac{\Sigma_2^k - \lambda^{-1}\chi_2\nu\Sigma_{f2}^k}{D_2^k} + \frac{\Sigma_1^k - \lambda^{-1}\chi_1\nu\Sigma_{f1}^k}{D_1^k}\right)\\ &+ \left[\left(\frac{\Sigma_2^k - \lambda^{-1}\chi_2\nu\Sigma_{f2}^k}{2D_2^k} - \frac{\Sigma_1^k - \lambda^{-1}\chi_1\nu\Sigma_{f1}^k}{2D_1^k}\right)^2 + \frac{\lambda^{-1}\chi_1\nu\Sigma_{f2}^k(\lambda^{-1}\chi_2\nu\Sigma_{f1}^k + \Sigma_{21}^k)}{D_1^k D_2^k}\right]^{1/2},\\ \nu_k^2 = +B_{yz}^2 &+ \frac{1}{2}\left(\frac{\Sigma_2^k - \lambda^{-1}\chi_2\nu\Sigma_{f2}^k}{D_2^k} + \frac{\Sigma_1^k - \lambda^{-1}\chi_1\nu\Sigma_{f1}^k}{D_1^k}\right)\\ &+ \left[\left(\frac{\Sigma_2^k - \lambda^{-1}\chi_2\nu\Sigma_{f2}^k}{2D_2^k} - \frac{\Sigma_1^k - \lambda^{-1}\chi_1\nu\Sigma_{f1}^k}{2D_1^k}\right)^2 + \frac{\lambda^{-1}\chi_1\nu\Sigma_{f2}^k(\lambda^{-1}\chi_2\nu\Sigma_{f1}^k + \Sigma_{21}^k)}{D_1^kD_2^k}\right]^{1/2}.\end{aligned} \tag{4.11.21}$$

The form of this result shows that both values of B^2 for a given region are real and that one value $(-\nu_k^2)$ is always negative, whereas the other value (μ_k^2) can be negative or positive depending on the relative magnitudes of the two-group parameters. Note that for $\Sigma_{f1}^k = \Sigma_{f2}^k = 0$ the roots become

$$\mu_k^2 = -B_{yz}^2 - \frac{\Sigma_1^k}{D_1^k},$$
$$\nu_k^2 = B_{yz}^2 + \frac{\Sigma_2^k}{D_2^k}. \qquad (4.11.22)$$

That is, both values of B^2 are negative for a nonmultiplying material.

Let us now return to the solution of (4.11.17). We have found two values of B^2, namely μ_k^2 and $-\nu_k^2$, such that a particular solution of (4.11.18) will also be a particular solution of (4.11.17). Writing (4.11.18) for these two values of B^2 yields

$$\frac{d^2X_2(x)}{dx^2} + \mu_k^2 X_2(x) = 0 \quad \text{and} \quad \frac{d^2X_2(x)}{dx^2} - \nu_k^2 X_2(x) = 0, \qquad (4.11.23)$$

so that satisfactory particular solutions for $X_2(x)$ are

$$X_2(x) = \sin \mu_k x, \quad \cos \mu_k x, \quad \sinh \nu_k x, \quad \cosh \nu_k x. \qquad (4.11.24)$$

To find the particular solutions $X_1(x)$ corresponding to (4.11.24) we simply use the second equation in (4.11.19) to get the ratios $X_1(x)/X_2(x)$ for the two values of B^2. We symbolize these ratios by s_k and t_k:

$$s_k \equiv \frac{D_2^k(\mu_k^2 + B_{yz}^2) + \Sigma_2^k - \lambda^{-1}\chi_2 \nu\Sigma_{f2}^k}{\lambda^{-1}\chi_2\nu\Sigma_{f1}^k + \Sigma_{21}^k}, \qquad (4.11.25)$$

$$t_k \equiv \frac{D_2^k(-\nu_k^2 + B_{yz}^2) + \Sigma_2^k - \lambda^{-1}\chi_2\nu\Sigma_{f2}^k}{\lambda^{-1}\chi_2\nu\Sigma_{f1}^k + \Sigma_{21}^k}. \qquad (4.11.26)$$

Then the particular solutions $X_1(x)$ corresponding to (4.11.24) are

$$X_1(x) = s_k \sin \mu_k x, \quad s_k \cos \mu_k x, \quad t_k \sinh \nu_k x, \quad t_k \cosh \nu_k x. \qquad (4.11.27)$$

Note that, as they must, these particular solutions also obey (4.11.23).

To get the general solution to (4.11.17) in each region k we simply form linear combinations of the particular solutions (4.11.24) and (4.11.27) and then determine the coefficients of combination so that the boundary and continuity conditions are met.

For the problem at hand, instead of imposing external boundary conditions $X_g(x) = 0$ at $x = \pm(X_1 + X_2)$, we can take advantage of the symmetry of the situation and set $dX_g(x)/dx = 0$ at $x = 0$ and $X_g(x) = 0$ at $x = X_1 + X_2$. Linear combinations of the particular solutions (4.11.24) and (4.11.27) that satisfy these conditions are

$$\left.\begin{aligned} X_1(x) &= C_1^{(1)} s_1 \cos \mu_1 x + C_2^{(1)} t_1 \cosh \nu_1 x \\ X_2(x) &= C_1^{(1)} \cos \mu_1 x + C_2^{(1)} \cosh \nu_1 x \end{aligned}\right\} \text{for } 0 \le x \le X_1 \tag{4.11.28}$$

and

$$\left.\begin{aligned} X_1(x) &= C_1^{(2)} s_2 \sin \mu_2(X_1 + X_2 - x) + C_2^{(2)} t_2 \sinh \nu_2(X_1 + X_2 - x) \\ X_2(x) &= C_1^{(2)} \sin \mu_2(X_1 + X_2 - x) + C_2^{(2)} \sinh \nu_2(X_1 + X_2 - x) \end{aligned}\right\} \text{for } X_1 \le x \le (X_1 + X_2). \tag{4.11.29}$$

Imposing continuity of flux and current at $x = X_1$ then yields

$$\begin{aligned} &C_1^{(1)} s_1 \cos \mu_1 X_1 + C_2^{(1)} t_1 \cosh \nu_1 X_1 = C_1^{(2)} s_2 \sin \mu_2 X_2 + C_2^{(2)} t_2 \sinh \nu_2 X_2, \\ &C_1^{(1)} s_1 D_1^{(1)} \mu_1 \sin \mu_1 X_1 - C_2^{(1)} t_1 D_1^{(1)} \nu_1 \sinh \nu_1 X_1 = C_1^{(2)} s_2 D_1^{(2)} \mu_2 \cos \mu_2 X_2 \\ &\qquad + C_2^{(2)} t_2 D_1^{(2)} \nu_2 \cosh \nu_2 X_2, \\ &C_1^{(1)} \cos \mu_1 X_1 + C_2^{(1)} \cosh \nu_1 X_1 = C_1^{(2)} \sin \mu_2 X_2 + C_2^{(2)} \sinh \nu_2 X_2, \\ &C_1^{(1)} D_2^{(1)} \mu_1 \sin \mu_1 X_1 - C_2^{(1)} D_2^{(1)} \nu_1 \sinh \nu_1 X_1 = C_1^{(2)} D_2^{(2)} \mu_2 \cos \mu_2 X_2 + C_2^{(2)} D_2^{(2)} \nu_2 \cosh \nu_2 X_2. \end{aligned} \tag{4.11.30}$$

These equations will have a nontrivial solution if and only if the determinant of the coefficient of the C's vanishes. Thus the two-group criticality condition for the parallelopiped reactor being analyzed is

$$\begin{vmatrix} s_1 \cos \mu_1 X_1 & t_1 \cosh \nu_1 X_1 & -s_2 \sin \mu_2 X_2 & -t_2 \sinh \nu_2 X_2 \\ s_1 D_1^{(1)} \mu_1 \sin \mu_1 X_1 & -t_1 D_1^{(1)} \nu_1 \sinh \nu_1 X_1 & -s_2 D_1^{(2)} \mu_2 \cos \mu_2 X_2 & -t_2 D_1^{(2)} \nu_2 \cosh \nu_2 X_2 \\ \cos \mu_1 X_1 & \cosh \nu_1 X_1 & -\sin \mu_2 X_2 & -\sinh \nu_2 X_2 \\ D_2^{(1)} \mu_1 \sin \mu_1 X_1 & -D_2^{(1)} \nu_1 \sinh \nu_1 X_1 & -D_2^{(2)} \mu_2 \cos \mu_2 X_2 & -D_2^{(2)} \nu_2 \cosh \nu_2 X_2 \end{vmatrix} = 0. \tag{4.11.31}$$

Any combination of material properties and geometric sizes which, for $\lambda = 1$, causes this determinant to vanish and results in positive solutions to (4.11.28) and (4.11.29) constitutes a critical condition according to the two-group theory. Conversely, for fixed material compositions and dimensions, that value of λ which causes (4.11.31) to vanish and which results in solutions $\Phi_1(x, y, z)$ and $\Phi_2(x, y, z)$ that are everywhere positive is the k_{eff} of the reactor assembly.

Once (4.11.31) is satisfied, we can return to (4.11.30) and, using any three of the four equations, determine three of the C's in terms of the fourth. (The fact that any three will do follows from the vanishing of the determinant (4.11.31).) The solutions $X_1(x)$ and $X_2(x)$ in both regions can then be found to within a multiplicative constant. Knowing these, we can form the $\Phi_g(x, y, z)$ for both regions from (4.11.15). Finally, using the spectrum functions $\psi_g^k(E)$ for the two groups and two compositions, we can form two-group estimates of the scalar flux density $\Phi(\mathbf{r}, E)$ for any energy and at any point in the reactor. For example, at a point (x, y, z) located in the region $(-X_1 \le x \le X_1, -Y_1 \le y \le Y_1, \quad -Z_1 \le z \le Z_1)$, the two-group prediction of the scalar flux density is

$$\Phi(x, y, z, E) = \begin{cases} C_2^{(1)}\psi_1^{(1)}(E)\left(\dfrac{C_1^{(1)}}{C_2^{(1)}} s_1 \cos \mu_1 x + t_1 \cosh \nu_1 x\right) \cos \dfrac{\pi y}{2Y_1} \cos \dfrac{\pi z}{2Z_1} \\ \qquad \text{for } E_c \le E < \infty, \\ C_2^{(1)}\psi_2^{(1)}(E)\left(\dfrac{C_1^{(1)}}{C_2^{(1)}} \cos \mu_1 x + \cosh \nu_1 x\right) \cos \dfrac{\pi y}{2Y_1} \cos \dfrac{\pi z}{2Z_1} \\ \qquad \text{for } 0 \le E \le E_c, \end{cases} \tag{4.11.32}$$

where the ratio $C_1^{(1)}/C_2^{(1)}$ can be obtained by simultaneously solving any three of the equations in (4.11.30), and the constant C_2 can be found from the average power level, as in (4.10.51).

4.12 Some Applications of Analytical Solutions of the Two-Group Equations

Analytical solutions of the group diffusion equations serve a variety of purposes. For very simple bare-core geometrical situations they provide algebraic formulas relating the k_{eff} of a reactor to the two-group parameters. Moreover many of the terms in such formulas can be interpreted physically so that the effect on various reaction rates of altering cross sections or nuclear concentrations can be readily judged. For more complex geometries the criticality equation and analytical solutions for the flux (for example (4.11.31) and (4.11.32)) become so cumbersome that examining them produces little physical insight. Nevertheless they are useful in assessing the validity of certain approximations (for example in suggesting when a one-group model will be adequate, or when replacing a reflector of finite thickness by one extending to infinity is legitimate, or when a certain interpretation of an experiment is valid). Moreover examination of analytical results for multiregion reactors brings to light certain simplifying approximations that permit us to define an "equivalent bare core" to which the simple algebraic criticality formulas can again be applied.

In this section we shall use the results of the two-group analysis just developed for the three-region slab reactor (Figure 4.10) to illustrate all these applications.

Two-Group Bare-Core Formulas

We shall first specialize the two-group, three-region results to the limiting case of a bare, homogeneous reactor. We could of course analyze this case directly starting with (4.7.18) for the two-group situation. However there is some value in showing that multiregion results reduce in the proper fashion.

In the limit where there is only one composition in the X direction, the criticality condition (4.11.31) reduces to a much simpler form; it is not hard to show that:

1. for the thickness $X_1 = 0$, (4.11.31) implies $\cos \mu_2 X_2 = 0$,
2. for the thickness $X_2 = 0$, (4.11.31) implies $\cos \mu_1 X_1 = 0$, (4.12.1)
3. for identical properties in all X regions, (4.11.31) implies $\cos \mu(X_1 + X_2) = 0$,

where we have dropped the subscript in the last case since $\mu_1 = \mu_2$. (Note that these three conditions are, in fact, identical.) Thus, for a homogeneous reactor of half-thickness X_1,

$$\mu X_1 = (2n + 1)\frac{\pi}{2} \qquad (n = 0, 1, 2, \ldots), \tag{4.12.2}$$

where μ^2 is the more positive root of (4.11.20) for the material making up the reactor. Moreover the first equation in (4.11.30) with $X_2 = 0$ now shows that $C_2^{(1)}$ must vanish. Thus (4.11.32) for the two-group fluxes becomes

$$\Phi(x, y, z, E) = \begin{cases} C_1 s\psi_1(E) \cos \mu x \cos \dfrac{\pi y}{2Y_1} \cos \dfrac{\pi z}{2Z_1} \text{ for } E_c \leq E < \infty, \\ C_1\psi_2(E) \cos \mu x \cos \dfrac{\pi y}{2Y_1} \cos \dfrac{\pi z}{2Z_1} \text{ for } 0 \leq E \leq E_c, \end{cases} \tag{4.12.3}$$

where superscripts and subscripts indicating composition have been dropped.

Since $\Phi(x, y, z, E)$ is to be everywhere positive, n in (4.12.2) must be zero. Thus the criticality condition for the bare, homogeneous, parallelopiped reactor is (4.12.2) with $n = 0$. If we then use $\mu^2 = (\pi/2X_1)^2$ for the value of B^2 in (4.11.19) and introduce the definition of the geometrical buckling B_r^2,

$$B_r^2 \equiv \left(\frac{\pi}{2X_1}\right)^2 + \left(\frac{\pi}{2Y_1}\right)^2 + \left(\frac{\pi}{2Z_1}\right)^2, \tag{4.12.4}$$

we obtain (with the help of (4.11.16) and dropping superscripts k)

$$(D_1B_r^2 + \Sigma_1)(D_2B_r^2 + \Sigma_2) - \frac{1}{\lambda}[(D_1B_r^2 + \Sigma_1)\chi_2\nu\Sigma_{f2} + (D_2B_r^2 + \Sigma_2)\chi_1\nu\Sigma_{f1}]$$
$$- \frac{1}{\lambda}\Sigma_{21}\chi_1\nu\Sigma_{f2} = 0 \tag{4.12.5}$$

or

$$\lambda \equiv k_{\text{eff}} = \frac{\Sigma_{21}\chi_1\nu\Sigma_{f2}}{(D_1B_r^2 + \Sigma_1)(D_2B_r^2 + \Sigma_2)} + \frac{\chi_2\nu\Sigma_{f2}}{(D_2B_r^2 + \Sigma_2)} + \frac{\chi_1\nu\Sigma_{f1}}{(D_1B_r^2 + \Sigma_1)}. \tag{4.12.6}$$

Obvious simplifications result if either χ_2 or $\nu\Sigma_{f1}$ is zero. Since two-group theory is generally applied to thermal reactors, the term in χ_2 rarely appears.

Equation (4.12.6) is called the *two-group, homogeneous, bare-core criticality formula.* The same result can, of course, be obtained directly by substituting $\nabla^2\Phi_g = -B_r^2\Phi_g$ into (4.11.14) written for the bare, homogeneous reactor. The formula is extremely useful for estimating critical conditions for reasonably uniform reactors. Moreover the various terms of the expression can be interpreted as probabilities that the neutrons in the reactor will interact in particular ways, and we can thus obtain some insight into the problem of altering the reactor design to produce desired effects.

To see how the probabilities emerge we rewrite (4.12.6) in terms of the ratios of reaction rates:

$$\begin{aligned} k_{\text{eff}} = &\left[1 - \frac{\int D_1 B_r^2 \Phi_1\, dV}{\int (D_1 B_r^2 + \Sigma_1)\Phi_1\, dV}\right]\left[1 - \frac{\int D_2 B_r^2 \Phi_2\, dV}{\int (D_2 B_r^2 + \Sigma_2)\Phi_2\, dV}\right]\left[\frac{\int \Sigma_{21}\Phi_1\, dV}{\int \Sigma_1 \Phi_1\, dV}\right]\chi_1 \nu \\ &\times \left[\frac{\int \Sigma_{f2}\Phi_2\, dV}{\int \Sigma_{a2}^{\text{fuel}}\Phi_2\, dV}\right]\left[\frac{\int \Sigma_{a2}^{\text{fuel}}\Phi_2\, dV}{\int \Sigma_2 \Phi_2\, dV}\right] + \left[1 - \frac{\int D_2 B_r^2 \Phi_2\, dV}{\int (D_2 B_r^2 + \Sigma_2)\Phi_2\, dV}\right]\chi_2 \nu \left[\frac{\int \Sigma_{f2}\Phi_2\, dV}{\int \Sigma_{a2}^{\text{fuel}}\Phi_2\, dV}\right] \\ &\times \left[\frac{\int \Sigma_{a2}^{\text{fuel}}\Phi_2\, dV}{\int \Sigma_2 \Phi_2\, dV}\right] + \left[1 - \frac{\int D_1 B_r^2 \Phi_1\, dV}{\int (D_1 B_r^2 + \Sigma_1)\Phi_1\, dV}\right]\chi_1 \nu \left[\frac{\int \Sigma_{f1}\Phi_1\, dV}{\int \Sigma_1 \Phi_1\, dV}\right], \end{aligned} \tag{4.12.7}$$

where the volume integrals are over the entire (bare, homogeneous) reactor. (To see that (4.12.7) and (4.12.6) are the same, cancel the volume integrals $\int \Phi_g\, dV$ and simplify the result.)

Since $\int D_g B_r^2 \Phi_g(\mathbf{r})\, dV$ is the rate at which group-g neutrons leak out of the reactor and $\int (D_g B_r^2 + \Sigma_g)\Phi_g\, dV$ is the total rate of "destruction" (i.e., leakage plus absorption plus transfer to another group) of group-g neutrons in the reactor, their ratio is the probability that a group-g neutron will be "destroyed" by leakage. Thus we are led to generalize (4.7.20) by defining a nonleakage probability for group-g neutrons as

$$P_{Lg} \equiv 1 - \frac{\int D_g B_r^2 \Phi_g\, dV}{\int (D_g B_r^2 + \Sigma_g)\Phi_g\, dV} = \frac{1}{1 + (D_g/\Sigma_g)B_r^2}. \tag{4.12.8}$$

Similarly $\int \Sigma_{21}\Phi_1\, dV$ is the rate at which group-one neutrons become group-two neutrons, and $\int \Sigma_1 \Phi_1\, dV$ is the sum of this rate plus the rate at which group-one neutrons are absorbed. Their ratio is thus the fraction of group-one neutrons that interact (without leaking) in such a way that they escape from group one. Thus this ratio can be interpreted as the two-group expression for the resonance-escape probability p. Specifically, we set

$$p \equiv \frac{\int \Sigma_{21}\Phi_1\, dV}{\int \Sigma_1 \Phi_1\, dV} = \frac{\Sigma_{21}}{\Sigma_1}. \tag{4.12.9}$$

Note that this formula differs slightly from what we would obtain if the earlier definition (3.5.11) of p were reduced directly to the two-group form for a homogeneous mixture. The difference is small and vanishes as $B_r^2 \to 0$. The situation is an example of one of the alternate, slightly different definitions of the four factors mentioned in Section 3.5.

The remaining ratios in (4.12.7) can be interpreted as follows:

$$f \equiv \frac{\int \Sigma_{a2}^{\text{fuel}} \Phi_2 \, dV}{\int \Sigma_2 \Phi_2 \, dV} = \frac{\Sigma_{a2}^{\text{fuel}}}{\Sigma_2} = \frac{\text{rate of absorption of group-two neutrons in fuel}}{\text{total rate of absorption of group-two neutrons}}, \tag{4.12.10}$$

where f is the thermal utilization (this definition is identical with that given by the homogeneous two-group reduction of (3.5.10));

$$\frac{\int \Sigma_{f2} \Phi_2 \, dV}{\int \Sigma_{a2}^{\text{fuel}} \Phi_2 \, dV} = \frac{\Sigma_{f2}}{\Sigma_{a2}^{\text{fuel}}} = \frac{\text{rate of fission due to group-two neutrons}}{\text{rate of absorption of group-two neutrons in fuel}}; \tag{4.12.11}$$

$$\eta \equiv \frac{\nu \int \Sigma_{f2} \Phi_2 \, dV}{\int \Sigma_{a2}^{\text{fuel}} \Phi_2 \, dV} = \frac{\nu \Sigma_{f2}}{\Sigma_{a2}^{\text{fuel}}}, \tag{4.12.12}$$

the total number of neutrons produced by fission per group-two neutron absorbed in fuel (this too agrees with the homogeneous two-group counterpart of (3.5.7)); and finally

$$\frac{\int \Sigma_{f1} \Phi_1 \, dV}{\int \Sigma_1 \Phi_1 \, dV} = \frac{\Sigma_{f1}}{\Sigma_1} \tag{4.12.13}$$

is the probability that, if a neutron interacts in group one (without leaking), it will undergo a fission interaction.

Using some of these definitions, we rewrite (4.12.7) once again as

$$k_{\text{eff}} = \nu \left\{ \chi_1 \left[P_{L1} p P_{L2} f \frac{\Sigma_{f2}}{\Sigma_{a2}^{\text{fuel}}} + P_{L1} \frac{\Sigma_{f1}}{\Sigma_1} \right] + \chi_2 P_{L2} f \frac{\Sigma_{f2}}{\Sigma_{a2}^{\text{fuel}}} \right\}. \tag{4.12.14}$$

To interpret this result physically we suppose that a fission has occurred in the reactor. This produces a total of ν neutrons, a fraction χ_1 of which appear in group one and a fraction χ_2 in group two. Of the $\nu\chi_1$ neutrons appearing in group one $P_{L1}p$ escape both capture and leakage in group one and slow down to become group-two neutrons. Of these a fraction $P_{L1}f$ are absorbed in the fuel, and of these a fraction $\Sigma_{f2}/\Sigma_{a2}^{\text{fuel}}$ cause fission. Thus $\nu\chi_1 P_{L1} p P_{L2} f(\Sigma_{f2}/\Sigma_{a2}^{\text{fuel}})$ is the number of "second-generation" fissions due to those "first-generation" neutrons which appear in group one, slow down, and then cause another fission. Similarly $\nu\chi_1 P_{L1}(\Sigma_{f1}/\Sigma_1)$ is the number of fissions which appear in group one and cause a fission in that same group. Finally $\nu\chi_2 P_{L2} f(\Sigma_{f2}/\Sigma_{a2}^{\text{fuel}})$ is the number of first-generation neutrons which appear directly in group two, escape leakage, are absorbed in fuel, and cause a fission in group two.

We conclude that k_{eff} may be interpreted approximately as the number of "second-generation" fissions due to a single "first-generation" fission. This interpretation is, of course, consistent with the fact that k_{eff} is greater than unity for a supercritical reactor, equal to unity for a critical reactor, and less than unity for a subcritical reactor. However, as was discussed in Section 3.3, this interpretation is somewhat artificial since a reactor with $k_{\text{eff}} \neq 1$ will be in a transient condition.

There is another algebraic form of the two-group, homogeneous, bare-core criticality formula which is very common for thermal reactors (for which $\chi_1 = 1$ and $\chi_2 = 0$). To derive this we use the second equation in (4.11.14) for the bare, homogeneous reactor ($\nabla^2\Phi_2 = -B_r^2\Phi_2$) to obtain

$$\frac{\int \Sigma_{21}\Phi_1 \, dV}{\int (D_2 B_r^2 + \Sigma_2)\Phi_2 \, dV} = 1. \tag{4.12.15}$$

Multiplying the last term of (4.12.6) by this factor and adding to the first term yields (with $\chi_1 = 1$; $\chi_2 = 0$)

$$k_{\text{eff}} = \frac{\Sigma_{21}(\nu\Sigma_{f2}/\Sigma_{a2}^{\text{fuel}})[\int(\nu\Sigma_{f2}\Phi_2 + \nu\Sigma_{f1}\Phi_1)\, dV/\int \nu\Sigma_{f2}\Phi_2 \, dV]}{(D_1 B_r^2 + \Sigma_1)(D_2 B_r^2 + \Sigma_2)}. \tag{4.12.16}$$

We now introduce the definitions

$$\varepsilon \equiv \frac{\text{total fission rate}}{\text{thermal fission rate}} = \frac{\int(\nu\Sigma_{f2}\Phi_2 + \nu\Sigma_{f1}\Phi_1)\, dV}{\int \nu\Sigma_{f2}\Phi_2 \, dV}, \tag{4.12.17}$$

$$L_g^2 \equiv \frac{D_g}{\Sigma_g} \qquad (g = 1, 2). \tag{4.12.18}$$

The quantity ε is the *fast-fission factor* (note the agreement with (3.5.4)); L_2^2 is the *thermal diffusion length*; and L_1^2 is sometimes called the *two-group age*. These definitions along with (4.12.9) and (4.12.12) permit us to rewrite (4.12.16) as

$$k_{\text{eff}} = \frac{p\eta f\varepsilon}{(1 + L_1^2 B_r^2)(1 + L_2^2 B_r^2)}. \tag{4.12.19}$$

The numerator of this expression is the infinite-medium multiplication factor k_∞:

$$k_\infty \equiv f\eta\varepsilon p. \tag{4.12.20}$$

Note that k_∞ appears not to depend on leakage. Actually there is a slight dependence since the two-group cross sections used in the definitions of the four factors are based on a multigroup spectrum which, in turn, depends on a materials buckling. This dependence is small. Nevertheless the subscript ∞ on k_∞ is a bit misleading since it implies a quantity appropriate to an infinite medium with $B_r^2 = 0$.

Using the definition (4.12.20), we can then write

$$k_{\text{eff}} = \frac{k_\infty}{(1 + L_1^2 B_r^2)(1 + L_2^2 B_r^2)}. \tag{4.12.21}$$

Finally, for a large reactor (X_1, Y_1, and Z_1 large), (4.12.4) shows that B_r^2 will be small. If then we note that, for small $L_1^2 B_r^2$,

$$\frac{1}{1 + L_1^2 B_r^2} \approx 1 - L_1^2 B_r^2 \approx \exp(-L_1^2 B_r^2), \tag{4.12.22}$$

we may rewrite (4.12.21) in the "Fermi-age form" (see (4.8.18))

$$k_{\text{eff}} = \frac{k_\infty \exp(-L_1^2 B_r^2)}{1 + L_2^2 B_r^2}. \tag{4.12.23}$$

All these definitions are important because they permit us to make contact between few-group theory and the older "lattice-theory" methods of reactor analysis. It is most important to remember, however, that the simple formulas for k_{eff} have been derived for a *bare, homogeneous reactor*. The geometrical buckling B_r^2 is defined as the number that yields an everywhere-positive solution of (4.7.5) which vanishes on the outer boundary. (Equation (4.12.4) specifies such a value of B_r^2 for the parrallelopiped geometry.) The derivation of (4.7.5), however, is based on the separability assumption (4.7.2), and that assumption is only valid for a completely homogeneous reactor. Hence, for a reflected reactor or for a reactor constructed of several different multiplying materials, (4.12.6), (4.12.14), (4.12.21), and (4.12.23) cannot be used unless we can find some alternate scheme for determining the number B_r^2. We shall see below that for large homogeneous cores reflected by a nonmultiplying material there are ways to estimate an "equivalent" B_r^2. However, if the core is small, is not uniform in composition, or has an internal reflector, or if the external "reflector" is multiplying (as with a breeder blanket), the bare-core formulas are generally useless.

Determining When a Reflector Is Effectively Infinite

One useful application of the analytical result (4.11.31) is to provide a quantitative prediction of when a reflector may be considered infinitely thick so that addition of further reflector material will have only a negligible effect on criticality. To make this prediction we simply rearrange the critical determinant (4.11.31) so that the sensitivity of k_{eff} to changes in the thickness X_2 of the reflector becomes apparent.

The first step of this rearrangement is to note that there is generally no fissionable material in a reflector. Equation (4.11.22) shows that in this case μ_2^2 will be negative. Accordingly, to avoid complex terms in the critical determinant (4.11.31) we make the transformation

$$\mu_2 \equiv i\kappa_2, \tag{4.12.24}$$

so that κ_2 is real; the trigonometric functions $\sin \mu_2 X_2$ and $\cos \mu_2 X_2$ in (4.11.31) then become

$$\begin{aligned} \sin \mu_2 X_2 &= \sin i\kappa_2 X_2 = i \sinh \kappa_2 X_2, \\ \cos \mu_2 X_2 &= \cos i\kappa_2 X_2 = \cosh \kappa_2 X_2. \end{aligned} \tag{4.12.25}$$

It also follows from (4.11.22) and (4.11.26) that, for $\nu\Sigma_{f1}^{(2)} = \nu\Sigma_{f2}^{(2)} = 0$, t_2 vanishes. As a result the critical determinant (4.11.31) becomes

$$\begin{vmatrix} s_1 \cos \mu_1 X_1 & t_1 \cosh \nu_1 X_1 & -s_2 i \sinh \kappa_2 X_2 & 0 \\ s_1 D_1^{(1)} \mu_1 \sin \mu_1 X_1 & -t_1 D_1^{(1)} \nu_1 \sinh \nu_1 X_1 & -s_2 D_1^{(2)} i\kappa_2 \cosh \kappa_2 X_2 & 0 \\ \cos \mu_1 X_1 & \cosh \nu_1 X_1 & -i \sinh \kappa_2 X_2 & -\sinh \nu_2 X_2 \\ D_2^{(1)} \mu_1 \sin \mu_1 X_1 & -D_2^{(1)} \nu_1 \sinh \nu_1 X_1 & -D_2^{(2)} i\kappa_2 \cosh \kappa_2 X_2 & -D_2^{(2)} \nu_2 \cosh \nu_2 X_2 \end{vmatrix} = 0. \tag{4.12.26}$$

The relationship between λ and the reactor parameters implied by this result will be unaltered if we divide the third column by $-i \cosh \kappa_2 X_2$ and the fourth column by $-\cosh \nu_2 X_2$. Doing so, we obtain

$$\begin{vmatrix} s_1 \cos \mu_1 X_1 & t_1 \cosh \nu_1 X_1 & s_2 \tanh \kappa_2 X_2 & 0 \\ s_1 D_1^{(1)} \mu_1 \sin \mu_1 X_1 & -t_1 D_1^{(1)} \nu_1 \sinh \nu_1 X_1 & s_2 D_1^{(2)} \kappa_2 X_2 & 0 \\ \cos \mu_1 X_1 & \cosh \nu_1 X_1 & \tanh \kappa_2 X_2 & \tanh \nu_2 X_2 \\ D_2^{(1)} \mu_1 \sin \mu_1 X_1 & -D_2^{(1)} \nu_1 \sinh \nu_1 X_1 & D_2^{(2)} \kappa_2 & D_2^{(2)} \nu_2 \end{vmatrix} = 0. \tag{4.12.27}$$

Since the hyperbolic tangent approaches unity for large values of its argument, we expect very little change in the value of the critical determinant in the form (4.12.27) as X_2 increases beyond the point where $\kappa_2 X_2$ and $\nu_2 X_2$ exceed, say, 4.0 ($\tanh 4.0 = 0.99933$). Thus reflector thicknesses X_2 leading to values of $\kappa_2 X_2$ and $\nu_2 X_2$ of this size or greater are effectively infinite.

Judging When a One-Group Model Will Be Adequate

The essential assumption on which a one-group model is based is that the scalar flux density $\Phi(\mathbf{r}, E)$ may be expressed as a separable product $R^k(\mathbf{r})\psi^k(E)$ of position and energy within each material composition. The two-group model provides an approximate test of this assumption.

The two-group analogue of the separability assumption (4.10.1) is

$$\Phi(\mathbf{r}, E) = \begin{cases} \psi_1^k(E) R^k(\mathbf{r}) & \text{for } E_c \le E < \infty, \\ \psi_2^k(E) R^k(\mathbf{r}) & \text{for } 0 \le E \le E_c. \end{cases} \tag{4.12.28}$$

Equation (4.11.32), however, shows that this form for $\Phi(\mathbf{r}, E)$ is possible only if $C_2^{(1)} = 0$

(the bare-core case) or if $s_1 = t_1$, a condition that examination of (4.11.25), (4.11.26), and (4.11.21) shows to be impossible. Nevertheless (4.11.32) also shows that the separability condition may be valid throughout most of the region under some circumstances. For situations where this is so, we expect the one-group model to be an adequate approximation.

The conditions under which separability will be almost valid occur when one of the two spatial terms, $\cos \mu_1 x$ or $\cosh \nu_1 x$, dominates throughout most of the region $-X_1 \leq x \leq X_1$. Now, since it can be shown that $t_1 \leq 0$, dominance of the term in $\cosh \nu_1 x$ implies a physically unacceptable zero or negative ratio of the group-one to the group-two flux. Thus, if any term is to dominate, it must be the term in $\cos \mu_1 x$. This will happen if $C_1^{(1)} \gg C_2^{(1)}$, for then the term in $\cosh \nu_1 x$ will be important only for $x \approx X_1$, where $\cosh \nu_1 x$ becomes large and $\cos \mu_1 x$ becomes small.

Typical values of μ_1 and ν_1 for a thermal reactor are $\mu_1 \approx 0.02\ \text{cm}^{-1}$, $\nu_1 \approx 0.2\ \text{cm}^{-1}$; s_1 is always positive and of order unity; t_1 is always negative and generally somewhat larger in magnitude than s_1. Thus, although the particular ratio of $C_1^{(1)}$ to $C_2^{(1)}$ will depend on the properties and thickness of the external portion of the reactor, we know that, for values of X_1 such that $\mu_1 X_1$ is close to $\pi/2$, the term $\cosh \nu_1 X_1$ will be approximately $\cosh 10(\pi/2) \approx 3.3 \times 10^6$ so that $C_1^{(1)}/C_2^{(1)}$ will have to be several times 3.3×10^6. (Otherwise $(C_1^{(1)}/C_2^{(1)})s_1 \cos \mu_1 X_1 + t_1 \cosh \nu_1 X_1$ will be negative.) It follows that for $x \ll X_1$, the term in $\cosh \nu_1 x$ will be negligible in comparison with the term in $\cos \mu_1 x$.

The situation is shown graphically in Figure 4.11. In the region $0 \leq x \leq X'$ the ratio

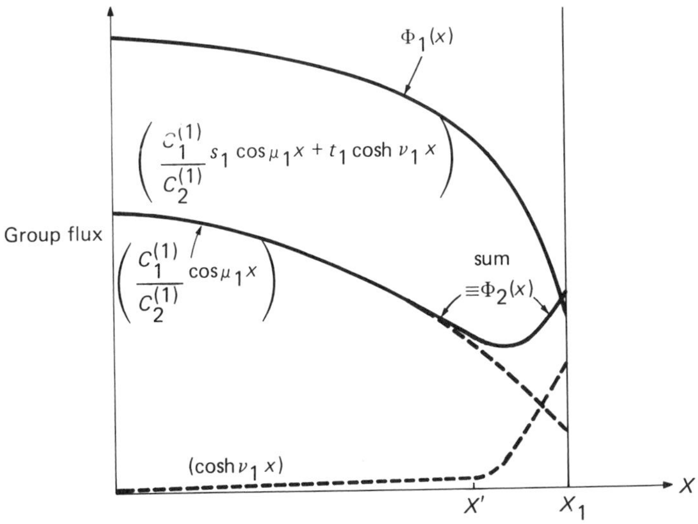

Figure 4.11 Two-group fluxes showing the zone of separability and the zone of nonseparability.

of $\Phi_1(x)$ to $\Phi_2(x)$ is almost constant (and equal to s_1). In the region $X' \le x \le X_1$ the ratio is far from constant. A one-group model would assume that $\Phi_1(x)/\Phi_2(x)$ is constant throughout the *entire* region $0 \le x \le X_1$. Thus, insofar as $(X_1 - X')$ is a small fraction of X_1, we expect the one-group picture to be adequate for the region $0 \le x \le X_1$. (Whether it is adequate for the whole reactor will depend on an analogous analysis for the region $X_1 \le x \le (X_1 + X_2)$.)

The essential part of this argument involves the comparison between $\mu_1 X_1$ and $\nu_1 X_1$, and carrying out an analogous analysis for the region $X_1 \le x \le (X_1 + X_2)$ would show that $\mu_2 X_2$ and $\nu_2 X_2$ are what matter in this range. Thus the analytical approach allows us to make some judgment about the adequacy of a one-group analysis from a knowledge simply of the μ_k^2 and ν_k^2 defined by (4.11.21).

Reflector Savings

In the region $0 \le x \le X'$ in Figure 4.11, where the curves of Φ_1 and Φ_2 have the same shape, the solution is said to be "asymptotic"; the ratio Φ_1/Φ_2 is constant there and Φ_1 and Φ_2 are said to have their "asymptotic shapes." It is not difficult to show that the total curvature $\mu^2 + B_{yz}^2$ associated with this asymptotic shape in a critical reactor equals the materials buckling $(B_m^k)^2$ of (4.10.3) for the material in the region $0 \le x \le X_1$. Moreover comparison of (4.10.3) with (4.7.23) shows that, for a critical, bare core, the geometrical buckling B_r^2 equals the materials buckling.

It follows that the total curvature of the asymptotic flux in the interior of the central zone of the parallelopiped reactor of Figure 4.10 is the same as would be found in a critical, *bare* reactor composed of material (1). Thus, provided the flux in the central part of a multizone reactor is asymptotic, it behaves, in that region, just as it would if the entire reactor were made of this central material. As a consequence, if the asymptotic flux shape in the center of the reactor is extrapolated (maintaining its curvature) until it goes to zero, the point at which that happens will specify the critical size of a bare reactor made entirely of the central material.

For the case at hand, (4.11.32) shows that the flux is asymptotic in the Y and Z directions at all points. (The ratio Φ_1/Φ_2 is independent of y and z at all points.) With respect to the X direction, Figure 4.11 shows that the asymptotic flux is proportional to $\cos \mu_1 x$ and extrapolates to zero at some point beyond X_1. We designate the distance from X_1 to that point as S_x and call this distance the *reflector savings*, since, in the presence of the reflector the critical size of the reactor decreases by an amount S_x from what it would have to be if it were to be critical without the reflector.

We define the reflector savings mathematically (provided the reactor is critical and μ_1 is real) as that distance S_x such that

$$\cos \mu_1 (X_1 + S_x) = 0. \tag{4.12.29}$$

Thus

$$S_x = \frac{\pi}{2\mu_1} - X_1, \tag{4.12.30}$$

where μ_1 is given by (4.11.21) with $\lambda = 1$ and the reactor parameters and geometry are such that the critical condition (4.11.31) is fulfilled.

It has been found that, for a given homogeneous, nonmultiplying reflector material sufficiently thick that it may be treated as an infinite reflector, the reflector savings for a given *type* of core material (U^{25}-Zr-H_2O; natural U–graphite; etc.) is only slightly dependent on the detailed composition of the core. Thus, for a highly enriched, light-water-moderated core composed of zirconium-clad fuel elements, the reflector savings of a light-water reflector at room temperature is about 7 cm, and this number stays nearly constant over a significant range of metal-to-water ratios and sizes of the core; in fact it is little different for a slightly enriched U^{25}-U^{28} core or for stainless-steel clad material instead of zirconium. (Note, however, that it is several times larger than the extrapolation distance δ of (4.6.12). The latter, which can be regarded as the reflector savings of a vacuum, would be about 2 cm.)

Making the assumption that the reflector savings is constant permits us to determine criticality directly from the bare-core formula (4.12.6), rather than by solving the four-by-four determinant (4.11.31). We simply add the reflector savings to the physical size of the core region and apply our analysis to a bare, homogeneous core of this extended size. The reflected core is thus replaced by an "equivalent bare core" having an extended size. All the bare-core formulas apply immediately if the geometrical buckling is taken to be

$$B_{\mathrm{r}}^2 = \left(\frac{\pi}{2X_1 + 2S_x}\right)^2 + \left(\frac{\pi}{2Y_1}\right)^2 + \left(\frac{\pi}{2Z_1}\right)^2, \tag{4.12.31}$$

where we have added $2S_x$ to the physical size $2X_1$ in the X direction since there are reflectors on both sides (and where we have neglected the extrapolation distances in the Y and Z directions).

We can extend the use of reflector savings to the other two dimensions and in that way analyze a reactor that is reflected on all sides. Similarly, if the reactor is reflected in two of its dimensions but is truly multizonal in the third (for example, if both material compositions in the X dimension are multiplying), we can add reflector savings to the two reflected sides (thereby finding an equivalent value of B_{yz}^2) and use the one-dimensional, multiregional analysis of (4.11.31) to account for flux behavior in the third direction. Thus we can reduce what is really a three-dimensional problem (and hence intractable analytically) to one that is, mathematically, one-dimensional.

It is important to keep in mind that the reflector-savings technique may not always work. For the method to be valid it is essential that there be a zone of significant size in the central part of the core region where the flux has its asymptotic shape, that is, where "transient terms" such as cosh $\nu_1 x$ and t_1 cosh $\nu_1 x$ in (4.11.32) are negligible in comparison with the "fundamental" ($\cos \mu_1 x$) terms. It is this fundamental shape that we extend to zero to define the reflector savings. Hence it is this fundamental shape that we recover when we analyze the equivalent bare core. Unless the true shape is very close to the fundamental throughout some portion of the core, our analysis will be in error. It follows that the reflector-savings procedure should not be used if the pertinent core dimension ($2X_1$ in the case of S_x) is not large compared to the mean free paths of neutrons traveling in the core material. The scheme will also fail if the outer region $X_1 \leq x \leq (X_1 + X_2)$ is multiplying to an extent that there is a net current of both group-one and group-two neutrons *into* the central zone. For this to happen μ_1^2 would have to be negative so that, with $\mu_1 = i\kappa_1$ (κ real), $\cos \mu_1 x$ in (4.11.32) would become $\cosh \kappa_1 x$. Thus, even if there were an asymptotic region where the $\cosh \kappa_1 x$ term exceeded greatly the cosh $\nu_1 x$ term, the asymptotic flux shape in that region would never extrapolate to zero, but would instead become infinite for large x. Generally speaking the reflector-savings method for analyzing a reflected reactor by bare-core formulas is useful only when the reflector is nonmultiplying.

The Two-Group Analysis of Exponential Experiments

One of the most useful techniques for measuring the nuclear properties of reactor materials is called the *exponential experiment*. As a last illustration of the utility of having an analytical solution for the two-group diffusion equations, we shall consider the analysis of such experiments. As we shall see, two-group theory is not recommended for detailed analysis of the experiment. However it does appear able to provide a good indication of when a meaningful measurement can be made.

An exponential experiment consists of placing a subcritical amount of the material to be tested next to an intense source of neutrons such as that provided by an operating critical reactor. Measurements of the asymptotic flux shape in the material can then be analyzed to provide a value of the materials buckling that can be compared with theory (using, for example, (4.10.3)). Such measurements have been a particularly useful means of investigating the properties of slightly enriched and natural-uranium lattices. They have the significant practical advantages of requiring less fissionable material than would be needed to construct a critical assembly and of not requiring the extra safety features that must be incorporated into an assembly that is capable of becoming critical.

In most exponential experiments a reactor provides the source of neutrons needed to drive the assembly. Generally the neutrons emitted by the reactor are first thermalized by being made to diffuse through a region that is a good moderator but a weak absorber (for example several feet of graphite—called a *thermal column*). This procedure has the advantage of attenuating the flux to a magnitude sufficiently large for the experiment at hand but not so great as to provide shielding problems. (Thus neutrons can be used from a general-purpose, high-power reactor designed to produce neutron sources of varying intensity for many different purposes.) In addition use of a thermal column cuts out the direct, fast-neutron source emanating from the driver core and thus permits an asymptotic spectrum that is uniquely appropriate to the test material to be established a fairly short distance inside the test assembly.

To analyze the exponential experiment with a thermal column between the test assembly and the driver reactor would require that we deal analytically with at least three different material regions. However, since we are primarily interested here in using the experiment to illustrate some of the properties of the general solution to the two-material, rectangular-parallelopiped reactor, we shall assume there is no intervening thermalizing region. Accordingly we now assume that material (2) in the zones between $-(X_1 + X_2)$ and $-X_1$ and between X_1 and $X_1 + X_2$ constitutes two identical exponential experiments of a certain test material and inquire how we can infer the materials buckling of the test material from measurements of gross flux shapes.

In principle the procedure is quite simple. The shape of the two-group fluxes within material (2) in the X direction is given by (4.11.29), the ratio $C_1^{(2)}/C_2^{(2)}$ being found from (4.11.30) under the condition that (4.11.31) is obeyed. If that shape is measured by activating and counting interactions in small foils placed at various locations along the X axis, it should in principle be possible to infer from it values of μ_2 and ν_2. Comparison of these experimentally inferred values with the theoretical expressions derived from (4.11.31) (λ being equal to unity) serves as a check on the theory.

There are practical difficulties with this direct approach. First, the $C^{(2)}$'s themselves depend on μ_2 and ν_2 as well as on the $D^{(2)}$'s, s_2, and t_2. Thus the formula being fit to the data is a very complicated one. Second, even if we assume we know the $D^{(2)}$'s and all parameters except μ_2 and ν_2 in the expressions (4.11.25) and (4.11.26) for s_2 and t_2, we must perform a two-parameter fit to the data. Thus to infer accurate values for μ_2 and ν_2 will require flux-shape measurements of extreme accuracy. Finally, since the fit depends on computed values of two-group parameters and on an algebraic expression for the flux shape derived from the two-group model, it appears that the numbers μ_2 and ν_2 inferred from the measurements in this manner would be as much a mathematical property of the model as a physical property of the test material. It is a property having the latter nature that we really want to infer from the experiment.

To circumvent these difficulties we shall invoke (but not prove) an important property of the solution to the continuous-energy diffusion equation (4.6.1):

Asymptotically far from boundaries of a *homogeneous*, multiplying, but nonsupercritical medium, the spatial shape of the scalar flux density $\Phi(\mathbf{r}, E)$ approaches a solution such that, for all energies, $\nabla^2\Phi = -B_m^2\Phi$, where B_m^2 is the materials buckling yielding a solution of (4.10.3) that is positive for all energies.

We shall call this characteristic of the solution to (4.6.1) the *asymptotic-behavior property.*

Two examples of the asymptotic-behavior property have been encountered already. We saw that in a bare core $\Phi(\mathbf{r}, E)$ obeys the condition everywhere (see (4.7.3) and (4.7.5)) provided B_m^2 is the *geometrical buckling.* We also encountered the asymptotic-behavior property in our discussion of the conditions under which a one-group approximation is valid. What mattered there was that the part of the solution (4.11.32) involving $\cosh \nu_1 x$ become negligibly small, so that the remaining part would obey

$$\nabla^2\Phi = -\left[\mu_1^2 + \left(\frac{\pi}{2Y_1}\right)^2 + \left(\frac{\pi}{2Z_1}\right)^2\right]\Phi.$$

For a large enough central region we saw that this condition would be fulfilled in a central, asymptotic region $0 \leq x \leq X'$. Thus our analysis illustrates the property and shows that the two-group expression for the materials buckling is

$$B_m^2 = \mu^2 + \left(\frac{\pi}{2Y_1}\right)^2 + \left(\frac{\pi}{2Z_1}\right)^2, \tag{4.12.32}$$

with μ defined by (4.11.21).

We now adopt the viewpoint that an exponential experiment can be analyzed to provide a value for the materials buckling of a given material. All that needs be done is to measure the flux shapes in the X, Y, and Z directions in a central portion of a large exponential assembly and to infer from these shapes the total curvature (second derivative) of $\Phi(\mathbf{r}, E)$, which the asymptotic-behavior property assures us will be constant in that region and equal to the materials buckling. The measurement thus yields a physical property of a material that is not dependent on a mathematical model or on cross-sectional values.

The two-group analysis can now be used to estimate how large the exponential assembly must be for the flux in its central portion to have an asymptotic shape.

For the two-material example we have been analyzing, there is no reflector in the Y and Z directions. Thus the two-group model shows that the flux curvature in those directions is constant everywhere and equal to $(\pi/2Y_1)^2 + (\pi/2Z_1)^2$. In the X direction the flux shape in the region $X_1 \leq x \leq (X_1 + X_2)$ is given by (4.11.29), and we see from this expression that it will be asymptotic wherever the terms involving $\sinh \nu_2(X_1 + X_2 - x)$ are small compared to those involving $\sin \mu_2(X_1 + X_2 - x)$.

It will be easier to get a qualitative picture of the flux shapes (4.11.29) if we redefine the constants $C_1^{(2)}$ and $C_2^{(2)}$. However it will be convenient to start by making the transformation

$$\mu_2 \equiv i\kappa_2 \tag{4.12.33}$$

so that the two-group expression (4.12.32) for the materials buckling becomes

$$B_m^2 = -\kappa_2^2 + \left(\frac{\pi}{2Y_1}\right)^2 + \left(\frac{\pi}{2Z_1}\right)^2. \tag{4.12.34}$$

We make this transformation because, although exponential experiments are performed for materials having both positive and negative values of B_m^2, the transverse dimensions Y_1 and Z_1 are generally chosen so that μ_2^2 is negative. This is done to avoid the danger that the exponential region, which is usually constructed with no provision made for control rods, might go critical on its own even with the central driving reactor shut down. Thus, in the exponential region, μ_2 is pure imaginary, and it is convenient to replace it with the real parameter κ_2. Since $\sin i\kappa_2(X_1 + X_2 - x) = i \sinh \kappa_2(X_1 + X_2 - x)$, (4.11.29) becomes

$$\begin{aligned} X_1(x) &= iC_1^{(2)}s_2 \sinh \kappa_2(X_1 + X_2 - x) + C_2^{(2)}t_2 \sinh \nu_2(X_1 + X_2 - x), \\ X_2(x) &= iC_1^{(2)} \sinh \kappa_2(X_1 + X_2 - x) + C_2^{(2)} \sinh \nu_2(X_1 + X_2 - x). \end{aligned} \tag{4.12.35}$$

To make it easier to get a feeling for the shapes of $X_1(x)$ and $X_2(x)$ implied by this result we now introduce constants $C_1^{(3)}$ and $C_2^{(3)}$ defined by

$$\begin{aligned} iC_1^{(2)} &= C_1^{(3)}/\cosh \kappa_2 X_2, \\ C_2^{(2)} &= C_2^{(3)}/\cosh \nu_2 X_2. \end{aligned} \tag{4.12.36}$$

Then (4.12.35) becomes

$$\begin{aligned} X_1(x) &= C_1^{(3)}s_2 \frac{\sinh \kappa_2(X_1 + X_2 - x)}{\cosh \kappa_2 X_2} + C_2^{(3)}t_2 \frac{\sinh \nu_2(X_1 + X_2 - x)}{\cosh \nu_2 X_2}, \\ X_2(x) &= C_1^{(3)} \frac{\sinh \kappa_2(X_1 + X_2 - x)}{\cosh \kappa_2 X_2} + C_2^{(3)} \frac{\sinh \nu_2(X_1 + X_2 - x)}{\cosh \nu_2 X_2}, \end{aligned} \tag{4.12.37}$$

or, if we express the hyperbolic functions as exponentials,

$$\begin{aligned} X_1(x) = {} & C_1^{(3)}s_2 \frac{\exp[-\kappa_2(x - X_1)] - \exp[-\kappa_2(X_1 + 2X_2 - x)]}{1 + \exp(-2\kappa_2 X_2)} \\ & + C_2^{(3)}t_2 \frac{\exp[-\nu_2(x - X_1)] - \exp[-\nu_2(X_1 + 2X_2 - x)]}{1 + \exp(-2\nu_2 X_2)}, \\ X_2(x) = {} & C_1^{(3)} \frac{\exp[-\kappa_2(x - X_1)] - \exp[-\kappa_2(X_1 + 2X_2 - x)]}{1 + \exp(-2\kappa_2 X_2)} \\ & + C_2^{(3)} \frac{\exp[-\nu_2(x - X_1)] - \exp[-\nu_2(X_1 + 2X_2 - x)]}{1 + \exp(-2\nu_2 X_2)}. \end{aligned} \tag{4.12.38}$$

The range of x of interest lies between X_1 and $X_1 + X_2$. Thus $x - X_1$ takes on values between 0 and X_2, and $X_1 + 2X_2 - x$ takes on values between $2X_2$ and X_2. Moreover ν_2 is generally much greater than κ_2. Thus, for $C_1^{(3)}$ roughly equal in magnitude to $C_2^{(3)}$ and X_2 sufficiently large, there is a range of positions near $x = X_1 + \frac{1}{2}X_2$ where (4.12.38) becomes approximately

$$\begin{aligned} X_1(x) &\approx C_1^{(3)} s_2 \frac{\exp[-\kappa_2(x - X_1)]}{1 + \exp(-2\kappa_2 X_2)}, \\ X_2(x) &\approx C_1^{(3)} \frac{\exp[-\kappa_2(x - X_1)]}{1 + \exp(-2\kappa_2 X_2)}, \end{aligned} \tag{4.12.39}$$

In this range two-group theory predicts that the flux will have its asymptotic shape and, moreover, that that shape will be exponentially decaying. (Hence the name "exponential experiment.") Computing the actual two-group values for $C_1^{(3)}/C_2^{(3)}$, κ_2, and ν_2 will quantitatively define the limits of this zone according to the two-group model and thus provide a good estimate of the range in which a measurement of the total flux curvature will directly yield a value of the materials buckling.

4.13 Multigroup Diffusion Theory

The essential assumption (4.10.1) of one-group theory is that there is a rigid spectrum function $\psi^k(E)$ $(0 \leq E < \infty)$ present throughout each material composition k. Similarly the essential assumption (4.11.1) of two-group theory is that there are two rigid spectrum functions, $\psi_1^k(E)$ for the range $E \geq E_c$ and $\psi_2^k(E)$ for the range $E < E_c$, present throughout each composition k. Clearly, by dividing the asymptotic spectrum (solution of (4.10.3) for material k) into more segments and associating a different spatial shape with each segment, we can derive energy-group models that are better able to match the true energy shape of the flux at interfaces between material regions, where the energy dependence of $\Phi(\mathbf{r}, E)$ is not close to any asymptotic spectrum. This representation of the energy dependence, since it is made up of rigid pieces of the asymptotic spectrum, will be discontinuous at the energy cut points between groups. Nevertheless the overall representation of $\Phi(\mathbf{r}, E)$ at such points will be better. This point is illustrated qualitatively by Figure 4.12, which shows that, while the true energy-dependent flux $\Phi(\mathbf{r}_0, E)$ at some point $\mathbf{r}_0$ might not be matched by the asymptotic shape $\psi^k(E)$ for the material in which $\mathbf{r}_0$ is embedded, it can be well approximated by three pieces of that asymptotic flux even though the result is discontinuous in energy at 10^{-1} and 10^5 eV.

It has been found that a model using three or four groups provides a very good representation of a thermal reactor. This is because most of the important neutron interactions take place at energies below 1 eV. For fast reactors, however, twenty or thirty groups are often necessary.

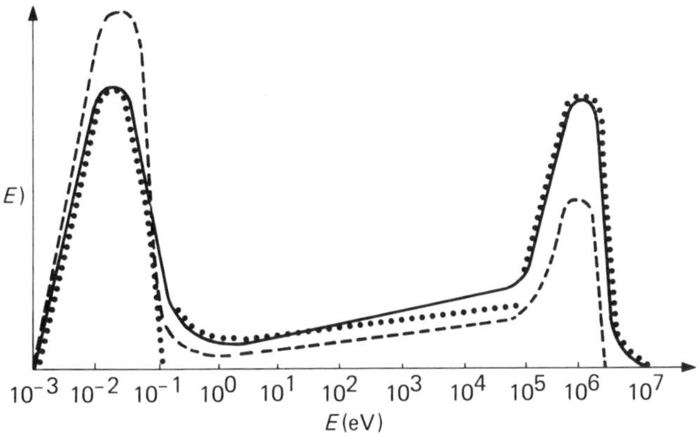

Figure 4.12 A three-group approximation to $\Phi(\mathbf{r}_0, E)$. The solid line is the true flux at $\mathbf{r}_0$; the dashed line is the asymptotic spectrum $\psi(E)$; and the dotted line is the three-group approximation to $\Phi(\mathbf{r}_0, E)$ using pieces of $\psi(E)$ separately normalized over the ranges 0–10^{-1} eV, 10^{-1}–10^5 eV, and 10^5–10^7 eV.

As the number of groups in the energy-group model increases, the individual segments into which the energy range 0–15 MeV is partitioned become increasingly small, and the behavior of the portions of $\psi^k(E)$ within a given group becomes increasingly uniform. In fact, in the limit where the number of energy groups for the spatial problem equals the number used to solve the spectrum equation (4.10.3), there is no longer any reason to perform separate spectrum calculations. Instead, as with (4.10.3), we may assume that the energy shape of $\Phi(\mathbf{r}, E)$ within each of a very large number of groups is flat (as in (3.3.5)) or (in view of the discussion of approximate flux shapes in Chapter 3) is proportional to the fission spectrum $\chi(E)$ in the high-energy range, to $1/E$ in the slowing-down range, and to a Maxwellian distribution in the thermal range. We shall call the group diffusion equation containing group parameters obtained by averaging over portions of spectra having an *arbitrarily* assumed shape, *multigroup diffusion equations.* Group equations, such as (4.11.7), for which the parameters are found by averaging over spectra $\psi^k(E)$ determined for each material composition by a separate, infinite-medium, multigroup calculation will be called *few-group diffusion equations.* Note that, according to this convention, group parameters obtained by averaging over a detailed spectrum shape specific to the material at hand are "few-group constants" even though the number of groups may be 30 to 100. With the number of "few-groups" that large, however, the convention is generally abandoned and the numerical model is referred to as a multigroup scheme.

In the next section an energy-group formalism sufficiently general in form to be used for both multigroup and few-group models will be developed. We shall then specialize this to the multigroup case.

General Energy-Group Diffusion Equations

To derive energy-group theory in a general way we first partition the reactor into a number of spatial regions (quantities associated with the kth spatial region being labeled with superscript k) and partition the energy range into a number of energy groups, $\Delta E_1 = E_0 - E_1, \Delta E_2 = E_2 - E_1, \ldots, \Delta E_g = E_{g-1} - E_g, \ldots$, and $\Delta E_G = E_{G-1} - E_G$, where E_0 is the highest energy of interest to reactor problems ($E_0 \approx 15$ MeV) and $E_G = 0$. The essential approximation is that the flux density $\Phi(\mathbf{r}, E)$ for $\mathbf{r}$ in region k and E in energy group ΔE_g may be written

$$\Phi(\mathbf{r}, E) = \Phi_g(\mathbf{r})\psi_g^k(E), \tag{4.13.1}$$

where $\psi_g^k(E)$ is a "spectrum function" for each energy group g in region k normalized so that

$$\int_{E_g}^{E_{g-1}} \psi^k(E)\, dE = 1 \qquad (g = 1, 2, \ldots, G), \tag{4.13.2}$$

and $\Phi_g(\mathbf{r})$ is the "scalar flux for energy group g," found by solving the group equations (4.13.7), derived below. Note that, because of (4.13.1) and (4.13.2), we have

$$\Phi_g(\mathbf{r}) \approx \int_{E_g}^{E_{g-1}} \Phi(\mathbf{r}, E)\, dE. \tag{4.13.3}$$

Thus $\Phi_g(\mathbf{r})$ has the dimensions neutrons/cm^2/sec and, unlike $\Phi(\mathbf{r}, E)$, is a flux rather than a flux density. ($\Phi(\mathbf{r}, E)$ has the units neutrons/cm^2/sec/eV.)

Assuming, for the moment, that the $\psi_g^k(E)$ are known, we find coupled partial-differential equations for the $\Phi_g(\mathbf{r})$ by simply substituting the approximation (4.13.1) into (4.6.1) and integrating the resulting equation G different times over the G energy ranges ΔE_g. (Note that we are now permitting $D(\mathbf{r}, E)$, $\Sigma_t(\mathbf{r}, E)$, etc., to be arbitrary functions of position.)

The first step in this process, the substitution of (4.13.1) into (4.6.1), yields, for $\mathbf{r}$ in region k and E in ΔE_g,

$$\begin{aligned} &-\nabla \cdot D(\mathbf{r}, E)\nabla[\psi_g^k(E)\Phi_g(\mathbf{r})] + \Sigma_t(\mathbf{r}, E)\psi_g^k(E)\Phi_g(\mathbf{r}) \\ &\approx \sum_{g'=1}^{G} \int_{E_{g'}}^{E_{g'-1}} \left[\frac{1}{\lambda}\sum_j \chi^j(E)\nu^j\Sigma_f^j(\mathbf{r}, E') + \Sigma_s(\mathbf{r}, E' \to E)\right]\psi_{g'}^k(E')\Phi_{g'}(\mathbf{r})\, dE'. \end{aligned} \tag{4.13.4}$$

Since $\psi_g^k(E)\Phi_g(\mathbf{r})$ is only an approximation to $\Phi(\mathbf{r}, E)$, the right- and left-hand sides of this equation cannot be exactly equal for all E and $\mathbf{r}$. (Otherwise the approximation (4.13.1) would be exact.) As before, however, we require that (4.13.4) be true in "an integral sense." That is, for each energy group we require that $\Phi_g(\mathbf{r})$ be such that the *integral* of the left-hand side of (4.13.4) over that group be *exactly* equal to the *integral* of the right-hand side. Thus we require that, for **r** in region k,

$$-\nabla\cdot\left[\int_{\Delta E_g} D(\mathbf{r}, E)\psi_g^k(E)\,dE\right]\nabla\Phi_g(\mathbf{r}) + \left[\int_{\Delta E_g} \Sigma_t(\mathbf{r}, E)\psi_g^k(E)\,dE\right]\Phi_g(\mathbf{r})$$
$$= \sum_{g'=1}^{G}\left\{\frac{1}{\lambda}\sum_j \int_{\Delta E_g} \chi^j(E)\,dE\left[\int_{\Delta E_{g'}} \nu^j\Sigma_f^j(\mathbf{r}, E')\psi_g^k(\mathbf{r}, E')\,dE'\right]\Phi_{g'}(\mathbf{r})\right. \tag{4.13.5}$$
$$\left. + \left[\int_{\Delta E_g} dE \int_{\Delta E_{g'}} dE'\,\Sigma_s(\mathbf{r}, E' \to E)\psi_{g'}^k(E')\right]\Phi_{g'}(\mathbf{r})\right\} \qquad (g = 1, 2, \ldots, G).$$

Now we define the "group parameters":

$$D_g^k(\mathbf{r}) \equiv \int_{\Delta E_g} D(\mathbf{r}, E)\psi_g^k(E)\,dE,$$
$$\Sigma_{tg}^k(\mathbf{r}) \equiv \int_{\Delta E_g} \Sigma_t(\mathbf{r}, E)\psi_g^k(E)\,dE,$$
$$\chi_g^j \equiv \int_{\Delta E_g} \chi^j(E)\,dE, \tag{4.13.6}$$
$$\nu^j\Sigma_{fg}^{jk}(\mathbf{r}) \equiv \int_{\Delta E_g} \nu^j\Sigma_f^j(\mathbf{r}, E)\psi_g^k(E)\,dE,$$
$$\Sigma_{gg'}^k(\mathbf{r}) \equiv \int_{\Delta E_{g'}} dE \int_{\Delta E_{g'}} dE'\Sigma_s(\mathbf{r}, E' \to E)\psi_{g'}^k(E').$$

Since it is assumed that we know the $\psi_g^k(E)$ and that $D(\mathbf{r}, E)$, $\Sigma_t(\mathbf{r}, E)$, etc., are known from the nuclear data and the isotopic concentrations of materials throughout the reactor, all these group parameters are known. The plethora of subscripts and superscripts attached to them is unfortunate; however it seems unavoidable. One simplification that has been made is to omit a possible superscript k from the $\Phi_g(\mathbf{r})$. That is, we take advantage of the fact that the region k is implied by the location **r**. This notation is in keeping with our definition of the $\Phi_g(\mathbf{r})$ as being solutions of (4.13.5).

Substitution of (4.13.6) into (4.13.5) yields the general form of the energy-group diffusion equations:

$$-\nabla\cdot[D_g^k(\mathbf{r})\nabla\Phi_g(\mathbf{r})] + \Sigma_{tg}^k(\mathbf{r})\Phi_g(\mathbf{r}) = \sum_{g'=1}^{G}\left\{\frac{1}{\lambda}\sum_j \chi^j\nu^j\Sigma_{fg}^{jk}(\mathbf{r}) + \Sigma_{gg'}^k(\mathbf{r})\right\}\Phi_{g'}(\mathbf{r}). \tag{4.13.7}$$

The continuity conditions on the $\Phi_g(\mathbf{r})$ are obtained from those imposed on $\Phi(\mathbf{r}, E)$ in a manner similar to that used in the one- and two-group cases; namely we require continuity of the integrals over each energy group ΔE_g of the approximate flux density $\Phi_g(\mathbf{r})\psi_g^k(E)$ and of the component of the approximate net current density $D(\mathbf{r}, E)\nabla[\Phi_g(\mathbf{r})\,\psi_g^k(E)]$ normal to internal surfaces at a material discontinuity (with the surface normal specified by the unit vector **n**). Because of (4.13.2) and (4.13.6) this requirement implies

$$\begin{aligned}&\text{the } \Phi_g(\mathbf{r}) \text{ are continuous everywhere;}\\ &\mathbf{n}\cdot D_g^k(\mathbf{r})\nabla\Phi_g(\mathbf{r}) \text{ is continuous across surfaces at material discontinuities.}\end{aligned} \tag{4.13.8}$$

Note that, across internal surfaces separating two different materials, the $D_g^k(\mathbf{r})$ will in general be discontinuous. The second condition in (4.13.8) then implies that the normal components of the gradients, $\mathbf{n}\cdot\nabla\Phi_g(\mathbf{r})$, will also be discontinuous across these interfaces.

Physically the conditions (4.13.8) imply that the velocity-weighted neutron density *integrated over the group* is everywhere continuous and that the net number of neutrons *belonging to a given group* and leaving one region across a given unit surface is the same as the net number entering the adjacent region through that surface. Because we are approximating $\Phi(\mathbf{r}, E)$ by $\Phi_g(\mathbf{r})\psi_g^k(E)$, where the $\psi_g^k(E)$ can change from one region k to the next, we cannot have continuity of flux and net current at *each* energy E. Instead we require by (4.13.8) that there be continuity for the *sum* of all neutrons having energies lying in a given group.

Boundary conditions for the $\Phi_g(\mathbf{r})$ on the external surface of the reactor can be found by substituting (4.13.1) into (4.6.10) and integrating over each of the G energy groups. However, just as with $\Phi(\mathbf{r}, E)$, the most common boundary condition—one which is almost always sufficiently accurate—is to require

$$\Phi_g(\mathbf{r}) = 0 \qquad (g = 1, 2, \ldots, G) \qquad \text{for } \mathbf{r} \text{ on the outer surface of the reactor.} \tag{4.13.9}$$

If there are internal planes about which the material properties of the reactor are symmetrically distributed, the critical flux shape on one side of such planes will be a mirror image of that on the other side. Under these conditions the group equations (4.13.7) need be solved in only the smallest portion of the reactor bounded by planes of symmetry, the net currents perpendicular to such planes ($-\mathbf{n}\cdot D_g(\mathbf{r})\ \nabla\Phi_g(\mathbf{r})$ for $\mathbf{r}$ in a plane of symmetry) being set equal to zero for each group.

Multigroup Theory

The point has already been made that the essential distinction between a multigroup and a few-group approximation lies in the manner in which the spectrum functions $\psi_g^k(E)$ are determined. For a few-group computation a separate spectrum must be calculated (for example by a multigroup solution of the position-independent equation (4.10.3)) for each composition k. For a multigroup model, the spectra $\psi_g(E)$ are assumed known beforehand and are taken to be the same for all material compositions.

Actually this distinction is not made universally. Some authors use *only* the term "multigroup theory" and mean by it any approximation in which the total number of energy groups exceeds one. Moreover the definition of a multigroup procedure about to be given is not in strict accord with practice for some of the more sophisticated techniques being used for current reactor design. Nevertheless, to emphasize the fact that there are two distinctly different ways by which the spectrum functions $\psi_g^k(E)$ are obtained, it helps to introduce precise (but not universally accepted) definitions for the two approximations.

Accordingly we shall continue to call a multigroup theory one in which:

1. For a given energy group g the same $\psi_g(E)$ is used throughout the entire reactor (i.e., superscript k is dropped).
2. The "energy shape" of each $\psi_g(E)$ (i.e., a plot of $\psi_g(E)$ versus E in the range of ΔE_g) is taken to be the same for all reactors to which (4.13.7) is applied.

Once the shapes $\psi_g(E)$ are chosen, it is possible to compute and tabulate *microscopic* multigroup cross sections $\sigma_{\alpha g}^j$ for reaction α of isotope j in energy group g. The corresponding macroscopic multigroup cross sections can then be written as a sum of microscopic cross sections multiplied by local number densities. We have, for example,

$$\Sigma_{tg}(\mathbf{r}) = \sum_j n^j(\mathbf{r})\sigma_{tg}^j. \tag{4.13.10}$$

Thus all the multigroup parameters, $D_g(\mathbf{r})$, $\Sigma_{tg}(\mathbf{r})$, etc., in (4.13.7) can be obtained from a knowledge of the local concentrations of isotopes $n^j(\mathbf{r})$, the microscopic cross sections for the various processes in each energy group being permanently tabulated as a "cross section library." The accuracy with which the solution of (4.13.7) obtained using the resultant multigroup parameters matches the exact solution of (4.6.1) is limited only by the number of energy groups that it is practical to employ.

Limitations of the Multigroup Procedure

In principle the multigroup diffusion-theory model is capable of predicting neutron behavior throughout a reactor with an accuracy limited only by the validity of the diffusion-theory approximation and by the quality of the basic nuclear data. In practice it is also limited by the expense of solving the multigroup diffusion equations. As with the one- and two-group models, the only practical method for solving the spatial part of the equations for a multidimensional, multicompositional reactor is to convert the differential equations (4.13.7) into some kind of difference form. The ultimate set of equations is thus algebraic. However the number of unknowns is roughly equal to the number of spatial mesh points multiplied by the number of energy groups, and, moreover, the equations containing these unknowns are coupled together. It is not today realistic to attempt a solution for a problem involving, say, a hundred thousand spatial mesh points and two thousand energy groups in three dimensions.

We have seen that the few-group approximation greatly reduces the number of unknowns by partially decoupling the energy solution from the spatial solution. Rather than, in effect, computing 2000-group spectra at each of 100,000 mesh points (2×10^8 simultaneous unknowns), we instead determine a greatly reduced number of spectra (perhaps ten for an undepleted reactor) appropriate to each of a number of compositions. In a few-group approach there are thus a number of 2000-group, position-independent calculations to be performed, followed by a 100,000-point spatial calculation for perhaps two or three energy groups (assuming the reactor is a thermal one). This problem (2 or 3×10^5 simultaneous unknowns) is still extremely formidable, but it can be done.

We shall return in Chapter 11 to the investigation of methods for reducing further the cost of performing accurate and detailed diffusion-theory calculations. However there is a problem of greater priority: we must develop a scheme for homogenizing the effects of small heterogeneous regions such as fuel elements, their associated clad, coolant, and possibly moderator. Otherwise we shall not be justified in using the diffusion-theory approximation in the first place.

The next chapter will deal with some aspects of this problem. Specifically we shall see how to homogenize heterogeneous effects and reduce the number of energy groups needed to describe the resolved-resonance region. The result will be a set of 30- to 50-group, "equivalent, homogenized" cross sections that can be used in multigroup or few-group diffusion-theory analysis of fast or thermal reactors in which the fuel is distributed in heterogeneous cells.

References

Lamarsh, J. R., 1966. *Introduction to Nuclear Reactor Theory* (Reading, Mass.: Addison-Wesley), Chapters 5, 6, 8, 9, 10.

Meghreblian, R. V., and Holmes, D. K., 1960. *Reactor Analysis* (New York: McGraw-Hill), Chapters 5, 6, 8.

Problems

1. Derive the expressions for the scalar flux density $\Phi(\mathbf{r}, E)$ and the net current density $\mathbf{J}(\mathbf{r}, E)$ implied by (4.5.2).

2. Assuming that $V_y = V_z = 0$, that $F(\mathbf{r}, E) = 1$, and that $N(\mathbf{r}, \mathbf{\Omega}, E)$ is to be nonnegative for all values of $\mathbf{\Omega}$, make a polar plot of $N(\mathbf{r}, \mathbf{\Omega}, E)$ that represents the closest approximation to a beam of neutrons traveling in the $+X$ direction obtainable from the form (4.5.2).

3. Consider a slab of material of thickness $L = Z_2 - Z_1$ and infinite in the X and Y directions, and suppose the material is homogeneous and a "pure absorber" ($\Sigma_t(\mathbf{r}, E) = \Sigma_a(E)$; $\Sigma_s(\mathbf{r}, E) = \Sigma_f(\mathbf{r}, E) = 0$). Suppose further that, over one face of the slab (at $z = Z_1$), the neutron density

in phase space $N(x, y, Z_1, \mathbf{\Omega}, E)$ is independent of x and y and isotropic (i.e., independent of $\mathbf{\Omega}$) for $\mathbf{\Omega}$ directed into the slab.

a. Derive an expression for $N(x, y, Z_2, \mathbf{\Omega}, E)$ in terms of $N(x, y, Z_1, \mathbf{\Omega}, E)$.

b. If $L = 10$ cm and $\Sigma_a(E) = 0.2\ \text{cm}^{-1}$, find the ratio of $N(x, y, Z_2, \mathbf{\Omega}, E)$ for $\mathbf{\Omega}$ making an angle of 60° with the Z axis to the value for $\mathbf{\Omega}$ parallel to the Z axis.

4. Derive the fact that the leakage density per unit volume is $\nabla \cdot \mathbf{J}(\mathbf{r}, E)$ directly from the meaning of $\mathbf{J}(\mathbf{r}, E)$ and the divergence theorem.

5. Assuming that $D(E)$ may be replaced by an average, $\bar{D} = 2$ cm, at all energies, use all three of the boundary conditions (4.6.2), (4.6.9), and $\Phi(L_x + \delta) = 0$ to find $\bar{D}B_x^2$ for two homogeneous slab reactors, the first one 10 π cm thick and the second one 100 π cm thick.

6. Derive (4.8.10).

Hint: Take the Laplace transform of the age equation for $\tau > 0$. Then solve for the spatial solution of the transformed $q(\mathbf{r}, \tau)$, finding the integration constants from the fact that, in an absorber-free medium (such as is obtained if the source is divided by $p(E)$), the integral of the slowing-down density over all of space for any value of τ is equal to its initial value.

7. The age of a neutron slowing down from 1 MeV to 1 eV in light water is $\sim$ 25 cm². Assuming that the ratio of Σ_{tr}^{U} for pure uranium to $\Sigma_{tr}^{H_2O}$ for pure H_2O is independent of energy and equal to one in this range and that ξ^U may be set to zero, compute the corresponding age in a homogeneous mixture of U and H_2O, the volume ratio of uranium to water being 0.8.

8. The age to the indium resonance energy of fission neutrons in light water is $\sim$ 26 cm². What is the corresponding age in saturated steam at 200°C, for which the density is 7.84 kg/m³?

9. Consider a dilute, homogeneous mixture of U^{25} and light water having a k_∞ value of 1.5 and with $\varepsilon = p = 1$. Assume that $\eta^{25}(E)$ equals 2.08, a constant in the thermal energy region and that, for pure H_2O, the thermal diffusion length is 2.85 cm and $1/\Sigma_{tr}^{H_2O}(E)$ averaged over the thermal energy range is 0.48 cm. Suppose that the U^{25} is so dilute that the age to thermal in the mixture is 26 cm², that of pure H_2O, and the density of H_2O in the mixture may be taken as that of pure H_2O. Estimate by age theory the minimum volume of this material required to go critical. If the average value of $\sigma_a^{25}(E)$ in the thermal region is 600 b, what is the critical mass of U^{25} contained in this volume?

10. Redo Problem 9 with D_2O in place of H_2O. Take $L_{D_2O} = 170$ cm; $\langle 1/\Sigma_{tr}^{D_2O}(E)\rangle_{av} = 2.61$ cm; and the age of fission neutrons to thermal in $D_2O = 131$ cm².

11. The first of the internal continuity conditions in (4.10.10) is based on the assumption that the integrated scalar flux density $\int_0^\infty dE\, \Phi(\mathbf{r}_c, E)$ is continuous at interface points $\mathbf{r}_c$. It is just as plausible to require that the integrated neutron density $\int_0^\infty dE \Phi(\mathbf{r}_c, E)/v(E)$ be continuous. If we do so and approximate $\Phi(\mathbf{r}, E)$ for $\mathbf{r}$ in region k by $\Phi(\mathbf{r}, E) \approx N^k(\mathbf{r})v(E)\psi^k(E)$, the first of the conditions in (4.10.10) becomes $N^k(\mathbf{r}_c) = N^l(\mathbf{r}_c)$.

a. Show by derivation that, within each region k, $N^k(\mathbf{r})$ obeys an equation identical in form with (4.10.9) but with different definitions for D^k, Σ^k, and Σ_f^k.

b. If continuity of the normal component of the net neutron current is still the desired condition, what does the second condition in (4.10.10) become when expressed in terms of $N^k(\mathbf{r}_c)$ and $N^l(\mathbf{r}_c)$?

12. Generalize the difference equation (4.10.16) to the case where the parameters D, Σ, and $v\Sigma_f$, as well as the size of the mesh interval, may change from one interval to the next.

13. For the general conditions posed in Problem 12, derive difference equations relating the *average* flux $\bar{\Phi}^{(n)}$ within h_n ($\equiv x_{n+1} - x_n$) to $\bar{\Phi}^{(n+1)}$ and $\bar{\Phi}^{(n-1)}$. Assume that at x_n the net

neutron current may be approximated by both

$$-D_{n-1}\frac{\Phi^{(n)}-\overline{\Phi}^{(n-1)}}{(h_{n-1}/2)} \text{ and } -D_n\frac{\overline{\Phi}^{(n)}-\Phi^{(n)}}{(h_n/2)},$$

where D_{n-1} and D_n are the one-group diffusion constants in k_{n-1} and k_n, and $\Phi^{(n)}$ is the one-group flux *at* x_n.

14. A solution of the difference equations (4.10.16) consists of the value of λ and the corresponding values of the fluxes $\Phi^{(n)}$ ($n = 0, 1, 2, \ldots, N$) at all the mesh points. Consider the case of a homogeneous slab with three internal mesh points (and $\Phi^{(0)} = \Phi^{(4)} = 0$).

a. Find all the solutions of (4.10.16). (In all cases normalize so that $\Phi^{(1)} = 1.0$.)

b. For $D = 2$ cm, $h = 10$ cm, $\Sigma = 0.06$ cm^{-1}, and the reactor critical, find the magnitude of the eigenvalues λ.

c. Find the corresponding eigenvalues for a core of this size and composition predicted from an analytical solution of (4.10.12) (assuming $\Phi(x_0) = \Phi(x_4) = 0$). Show that, for this homogeneous case, the ratios $\Phi(x_2)/\Phi(x_1)$ determined from the analytical solutions equal the $\Phi^{(2)}/\Phi^{(1)}$ ratios found by the finite-difference method.

15. Consider two mathematical abstractions of bare, homogeneous reactors, the first an infinitely long cylinder of radius R and the second a slab of total thickness Z in the Z direction, infinite in extent in the X and Y directions. Suppose that the density of the material making up the two reactors is decreased in a uniform way so that R for the first reactor and Z for the second increase. With any possible changes in the spectrum function neglected, what does one-group diffusion theory predict will be the change in k_{eff} for the two reactors due to these expansions?

16. In spherical geometry the one-group diffusion equation for a bare, homogeneous sphere is

$$D\frac{1}{r^2}\frac{d}{dr}\left[r^2\frac{d\Phi(r)}{dr}\right] + (\nu\Sigma_f - \Sigma_a)\Phi(r) = 0.$$

Under the assumption that $\Phi(r)$ is zero on the outer surface of the sphere, calculate the critical mass of a sphere of pure U^{25}. Assume the following characteristics for U^{25}: density, 19.1 g/cc; number density, $.048 \times 10^{24}$ atoms/cc; one-group parameters: $\sigma_s = 8$b, $\sigma_t = 4$b, $\sigma_a = 3.85$b, $\nu = 2.6$, $\bar{\mu}_0 = 0$.

17. Suppose that a slab reactor extending from $x = 0$ to $x = L$ is symmetric about the center plane $x = L/2$ and that the one-group parameters vary linearly, so that, for $0 \leq x \leq L/2$,

$$D(x) = D_0(1 + \alpha x),$$
$$\Sigma(x) = \Sigma_0(1 + \alpha x),$$
$$\nu\Sigma_f(x) = \nu\Sigma_{f0}(1 + \alpha x).$$

Find an analytical formula analogous to (4.10.26) for this case.

Hint: There is a transformation of variables that will convert the one-group diffusion equation for $\Phi(x)$ with variable parameters $D(x)$, $\Sigma(x)$, and $\nu\Sigma_f(x)$ into Bessel's equation (4.10.42).

18. One method of estimating the extent to which a given region of a reactor may be considered isotopically homogeneous is to examine the difference between the average group constants determined under the assumption that the material is completely homogeneous and those determined by the weighting process using the detailed flux shape actually present in the heterogeneous geometry. For a mixture of thin-slab fuel plates separated by thin-slab water channels, the equation for the neutron flux in any given energy group of a multigroup scheme can be written as

$$-D_g^k\frac{d^2\Phi_g(x)}{dx^2} + \Sigma_{tg}^k\Phi_g(x) = \sum_{g'=1}^{G}\left[\frac{1}{\lambda}\chi_g\nu\Sigma_{fg'}^k + \Sigma_{gg'}^k\right]\Phi_{g'}(x),$$

where $\nu\Sigma_{fg'}^k$ will be zero unless region k is a fuel plate and $\Sigma_{gg'}^k$ will be much smaller in the fuel plate than in the water channel. For energy groups below the fission source range, χ_g will be zero; the neutrons appearing in group g will thus be due solely to scattering from other energies and they will therefore be significant primarily in the water channel. Under these circumstances we can get an estimate of the flux shape $\Phi_g(x)$ across a fuel channel and its neighboring coolant channel by assuming that $\sum_{g' \neq g} \Sigma_{gg'}^k \Phi_{g'}(x) = S$, where S is a constant that is spatially flat in the coolant channels and zero in the fuel plates. If $k = (1)$ signifies fuel-plate material and $k = 2$ signifies coolant water, we then have

$$-D^{(1)} \frac{d^2\Phi(x)}{dx^2} + \Sigma^{(1)}\Phi(x) = 0 \quad \text{for } x \text{ in the fuel plate,}$$

$$-D^{(2)} \frac{d^2\Phi(x)}{dx^2} + \Sigma^{(2)}\Phi(x) = S \quad \text{for } x \text{ in the water channel,}$$

where we have used the definition $\Sigma_g^k \equiv \Sigma_{tg}^k - \Sigma_{gg}^k$ and then dropped all subscripts g.

If we impose boundary conditions of zero net current ($-Dd\Phi/dx = 0$, and thus $d\Phi/dx = 0$) at the center lines of the fuel plate and water channel (so-called "cell" boundary conditions), we can determine $\Phi(x)$ and then the flux-weighted average value $\bar{\Sigma}$ of the Σ^k for the cell. Specifically, if we place the origin of an axis system at the interface between the fuel plate and the water channel (with the fuel plate to the left of this interface), we have

$$\bar{\Sigma} = \frac{\int_{-t_1}^{0} \Sigma^{(1)}\Phi(x)dx + \int_0^{t_2} \Sigma^{(2)}\Phi(x)\,dx}{\int_{-t_1}^{0} \Phi(x)dx + \int_0^{t_2} \Phi(x)dx},$$

where t_1 and t_2 are the half-thicknesses of the fuel plate and water channel respectively.

a. Show that

$$\bar{\Sigma} = \frac{t_2}{\dfrac{t_2}{\Sigma^{(2)}} + \dfrac{\dfrac{D^{(1)}\kappa^{(1)}}{\kappa^{(2)}}\left(\dfrac{1}{\Sigma^{(1)}} - \dfrac{1}{\Sigma^{(2)}}\right) \sinh \kappa^{(1)}t_1 \sinh \kappa^{(2)}t_2}{D^{(2)}\kappa^{(2)} \cosh \kappa^{(1)}t_1 \sinh \kappa^{(2)}t_2 + D^{(1)}\kappa^{(1)} \sinh \kappa^{(1)}t_1 \cosh \kappa^{(2)}t_2}},$$

where

$$\kappa^{(1)} = (\Sigma^{(1)}/D^{(1)})^{1/2} \quad \kappa^{(2)} = (\Sigma^{(2)}/D^{(2)})^{1/2}.$$

b. Use the result of part a to show that, for small values of $\kappa^{(2)}t_2$ and $\kappa^{(1)}t_1$, $\bar{\Sigma}$ goes over to the completely homogeneous expression

$$\bar{\Sigma} \approx \frac{\Sigma^{(1)}t_1 + \Sigma^{(2)}t_2}{t_1 + t_2}.$$

19. Find, from (4.10.26), the one-group criticality equation for a slab reactor imbedded in an infinitely thick reflector.

20. Derive a criticality equation, analogous to (4.10.26) and involving *real* functions, for the case where $\nu\Sigma_f^{(1)} = 0$ and $\nu\Sigma_f^{(2)} > 0$.

21. Consider a reactor having a central, homogeneous fuel region of size $L_y = L_z = 100$ cm and extending from $x = -50$ cm to $x = +50$ cm in the X direction. Let two of the three one-group parameters characterizing this region be $D^{(1)} = 2.0$ cm and $\Sigma^{(1)} = 0.08$ cm^{-1}. Suppose the reactor is reflected on both sides in the X direction, the thickness of the reflectors being 20 cm (L_y and L_z again being 100 cm). Let the one-group parameters characterizing the reflector be $D^{(2)} = 1.0$ cm, $\Sigma^{(2)} = 0.03$ cm^{-1}, and $\nu\Sigma_f^{(2)} = 0$.

If the reactor fuel is U^{25}, the one-group cross section $\sigma_f^{25}(\equiv \int_0^\infty \sigma(E)_f^{25}\psi(E)dE)$ being 500 barns, how many kilograms are required for criticality?

22. Find the one-group criticality equation for a reactor consisting of a homogeneous sphere of fuel material surrounded by a homogeneous spherical-shell reflector.

Hint: Find the differential equation obeyed by $r\Phi(r)$.

23. Find one-group criticality conditions analogous to (4.10.47) and complete one-group solutions analogous to (4.10.50) for the two-region, finite cylinder under the conditions that:
a. $\kappa^{(1)} = 0, \quad \kappa^{(2)} > 0.$
b. $\kappa^{(2)} = 0, \quad \kappa^{(1)} > 0.$
The κ^k referred to are those defined by (4.10.41).

24. Suppose that, for two-group theory, we were to require continuity of the neutron concentrations $\int_{E_c}^{\infty} dE\, \Phi(\mathbf{r}, E)/v(E)$ and $\int_0^{E_c} dE\, \Phi(\mathbf{r}, E)/v(E)$ at interfaces between homogeneous material regions. Equation (4.11.1) would then be replaced by

$$\Phi(\mathbf{r}, E) = \begin{cases} N_1(\mathbf{r})\, v(E)\, \psi_1^k(E) & \text{for } E_c < E < \infty, \\ N_2(\mathbf{r})\, v(E)\, \psi_2^k(E) & \text{for } E \leq E_c. \end{cases}$$

(The normalization (4.11.3) could be retained.)

By doing this and by defining two-group parameters analogous to (4.11.6), derive two-group equations and continuity conditions analogous to (4.11.7) and (4.11.8) for $N_1(\mathbf{r})$ and $N_2(\mathbf{r})$.

25. Derive a finite-difference equation analogous to (4.11.13) relating the *average* group-two flux

$$\frac{1}{h_x^{(i)} h_y^{(j)}} \int_{x_i}^{x_{i+1}} dx \int_{y_j}^{y_{j+1}} dy\, \Phi_2(x, y)$$

in an interior mesh rectangle (i, j) to the average fluxes in its four nearest neighbors. Use the method suggested in Problem 13 to approximate the average currents across the surfaces of the mesh rectangle.

26. The general solution to (4.11.17) in region (1) of Figure 4.10 is

$$X_1(x) = C_1^{(1)} s_1 \cos \mu_1 x + C_2^{(1)} t_1 \cosh \nu_1 x + C_3^{(1)} s_1 \sin \mu_1 x + C_4^{(1)} t_1 \sinh \nu_1 x,$$
$$X_2(x) = C_1^{(1)} \cos \mu_1 x + C_2^{(1)} \cosh \nu_1 x + C_3^{(1)} \sin \mu_1 x + C_4^{(1)} \sinh \nu_1 x.$$

The derivatives dX_1/dx and dX_2/dx will vanish at $x = 0$ if $C_3^{(1)} = C_4^{(1)} = 0$; and (4.11.28) is based on this fact.

However this is not the only condition that will cause these derivatives to vanish. If material (1) has certain very unique characteristics leading to a very special relationship among the D's and Σ's, the derivatives will again vanish. Find this condition and show that, when it obtains, the solution in region (1) has the form

$$X_1(x) = a_1 \cosh \nu_1 x,$$

$$X_2(x) = a_2 \cosh \nu_1 x - a_1 \frac{\Sigma_{21}}{2\nu_1 D_2} x \sinh \nu_1 x,$$

where $\nu_1^2 = B_{yz}^2 + (\Sigma_2^{(1)}/D_2^{(1)})$ and where a_1 and a_2 are constants that can be determined by applying the continuity conditions at $x = X_1$.

27. One of the standard diffusion-theory reactor models devised to save computing costs for thermal-reactor problems is "one-and-a-half-group theory." It is based on the observation that the thermal diffusion constant is a small number, so that, in the interior of a reactor described by two-group theory, $D_2 \nabla^2 \Phi_2$ is small compared to $\Sigma_2 \Phi_2$. Accordingly in 1.5-group theory the group-two diffusion constants D_2^k for the various reactor regions k in a two-group model are set to zero, and the continuity condition on $\Phi_2(x, y, z)$ is relaxed. From (4.11.7) with $\chi_2 = 0$ derive the criticality condition according to the 1.5-group model for a bare, homogeneous reactor. For the slab geometry case sketch what $\Phi_1(x)$ and $\Phi_2(x)$ will look like across an interface between two different media.

28. We have defined the materials buckling as being the value of B^2 which yields a solution of (4.10.3) in multigroup form such that all the $\Phi_g(\mathbf{r})$ have the same sign.

Show, for the two-group model with a single fissionable isotope present in the mixture, $\chi_2 = 0$, and $\nu\Sigma_{f1} = 0$, that there exists such a value of B^2 and that there exists another value of B^2 such that Φ_1 and Φ_2 have different signs.

29. Prove the first and third assertions of (4.12.1).

30. Derive the two-group criticality equation (4.12.6) for a bare, homogeneous core directly from (4.7.3) written in the form

$$[D(E)B_r^2 + \Sigma_t(E)]R(\mathbf{r})\psi(E) = \int_0^\infty \left[\frac{1}{\lambda}\chi(E)\nu\Sigma_f(E') + \Sigma_s(E' \to E)R(\mathbf{r})\right]\psi(E')\,dE'.$$

Hint: Use (4.11.14) to determine the ratio $\Phi_2(\mathbf{r})/\Phi_1(\mathbf{r})$.

31. In this problem use the following microscopic parameters for pure U^{25} (density = 19 g/cc) with the energy cut point at 1.0 eV:

$\nu\sigma_{f1} = 5\text{b}$,
$\sigma_{tr1} = 10\text{b}$,
$\nu\sigma_{f2} = 1200\text{b}$,
$\sigma_1(\equiv \sigma_{t1} - \sigma_{s1}) = 3\text{b}$.

a. Assume that for pure U^{25} the resonance-escape probability as defined by (4.12.9) is $p = 10^{-7}$. What does the two-group, bare-core formula predict for the critical mass of a cube of pure U^{25}?

b. For this cubic reactor k_{eff} (or λ) is unity. Find the next largest value of λ that yields a nontrivial solution of the two-group diffusion equations for this reactor.

c. Sketch (without doing the actual calculations) how you would estimate by the two-group formula the minimum critical mass of a homogeneous mixture of U^{25} and H_2O having the geometrical shape of a rectangular parallelopiped. (Assume that $p = 0.9$ for all U^{25}–H_2O mixtures.)

32. Suppose the parallelopiped reactor depicted in Figure 4.10 has a height $2Y_1$ and width $2Z_1$ both equal to one meter. Assuming that the solution for k_{eff} of the critical determinant (4.12.27) will be changed negligibly if values of the hyperbolic tangents exceeding 0.999 are replaced by 1.0, find the minimum thickness X_2 of a reflector composed of (a) D_2O and (b) H_2O such that that reflector is effectively infinite. (Take the two-group age and thermal diffusion length in D_2O to be 111 cm² and 170 cm and the corresponding quantities for H_2O to be 26 cm² and 2.85 cm.)

33. The critical radius of a certain bare spherical reactor is 31.42 cm, and it is composed of homogeneous material having a two-group age of 50 cm² and a thermal diffusion length of 1.5 cm. If the reflector savings of a light-water reflector is 7 cm, estimate the k_{eff} of the sphere immersed in water.

5 The Determination of Energy-Group Constants in the Presence of Resonance Absorbers

5.1 Introduction

For the basic assumptions of multigroup theory to be valid in the presence of resonance absorbers, it may be necessary to deal with a large number (thousands) of energy groups. Thus, even for a completely homogeneous reactor in which diffusion theory and the few-group assumption of an asymptotic spectrum are legitimate (for example a reactor in which the fuel is dissolved in the moderator-coolant), the determination of the few-group parameters can be relatively expensive.

For the much more common case of a reactor composed of lattices of clad fuel rods, the situation is considerably worse. The rods are so small that the spectrum never becomes asymptotic within them. Moreover, for the case of moderation by light water, there is so little spacing between the rods that the spectrum also fails to become asymptotic in the moderator. It would appear necessary, then, to go to a full multigroup space-energy model to represent such a reactor. Unfortunately even this is not a sufficiently accurate procedure, since, at any energy corresponding to an absorption resonance, the cross section of the fuel is so large (thousands of barns) that Fick's Law (4.5.1), the basic approximation of diffusion theory, becomes invalid. Thus, if we want to analyze a system such as a pressurized-water reactor in a direct, brute-force manner, we must determine a multigroup solution (involving thousands of groups) of some approximation to the transport equation that is more accurate than diffusion theory. Such an approach is not impossible if only gross parameters such as k_{eff} are desired. But it is expensive and, today, rarely taken. Instead we shall develop a technique called *cell theory*, which will permit us to find "equivalent *homogenized*" few-group constants, spatially constant over a "cell" (the region occupied by a fuel rod and its associated clad and moderator-coolant). These homogenized few-group D's and Σ's will be equivalent in the sense that (ideally) they will reproduce the *average* reaction rates throughout the cell. Since most reactors are composed of "fuel assemblies" made up of a number of cells that are identical at the beginning of core-life and that do not differ very much from one another at the end of life, the homogenization procedure will recreate for us the conditions under which the assumptions of few-group diffusion theory are valid. Thus we shall again be able, at a reasonable cost, to carry out a meaningful, detailed analysis of a geometrically complex reactor system.

The most common cell encountered in the analysis of reactors consists of a clad cylindrical rod of reactor fuel surrounded by moderator-coolant zones. Examples are:

1. one-inch-diameter, aluminum-clad, natural-uranium metal rods surrounded by thin cylindrical annuli through which coolant (H_2O or gas) flows, the rods and coolant annuli being embedded in graphite in a regular square or hexagonal array;

2. clusters of one-inch-diameter, aluminum-clad, natural-uranium rods (with 19 or 37

rods per cluster) closely spaced in a cylindrical aluminum tube through which coolant (H_2O or D_2O) flows, the reactor being composed of a regular array of such tubes surrounded by D_2O moderator;

3. half-inch-diameter, slightly-enriched-UO_2 rods, clad with Zr or stainless steel, packed in a regular square or hexagonal array, and cooled with either liquid or boiling H_2O (see Figure 1.11);

4. quarter-inch-diameter, PuO_2-depleted uranium pins, clad with stainless steel, packed in a regular array, and cooled with liquid sodium.

The process of finding equivalent homogenized constants for these cells may be split into a number of more or less distinct parts, each of which can be attacked with various degrees of precision. Thus it will be convenient to examine the following aspects of the overall problem individually:

1. determination of resonance cross sections and expressions for obtaining Doppler-broadened values;
2. determination of neutron spectra for completely homogeneous mixtures of resonance absorbing materials and moderators;
3. formal analysis leading to expressions for effective homogenized constants in terms of "escape probabilities" that predict the likelihood that a neutron born in one material region will next interact in some other material region;
4. methods for determining escape probabilities;
5. corrections to elementary formulas to account for heterogeneous lattice properties;
6. extension of the methods developed for analysis in the resonance range to problems in other energy ranges.

5.2 Doppler Broadening of Resonance Cross Sections

The Breit-Wigner Formula

Figure 1.4 is a plot of the total cross section of U^{28} versus energy. The large, narrow resonances shown are typical of those that cause the trouble mentioned in the previous section.

The resonance absorption cross section shown in this figure can be represented mathematically by the single-level Breit-Wigner formula for zero-angular-momentum neutrons:

$$\sigma_a(E_r) = \frac{\sigma_0[E_0/E_r]^{1/2}[\Gamma(E_0)/\Gamma(E_r)]^2}{1 + [(E_r - E_0)/(\Gamma(E_r)/2)]^2}, \tag{5.2.1}$$

where E_r is the relative kinetic energy between the neutron and the nucleus, E_0 is the value of E_r corresponding to the resonance peak, σ_0 is $\sigma_a(E_0)$, and $\Gamma(E)$ is the total resonance width, the sum of the "radiation width" Γ_γ (a constant) and the neutron width $\Gamma_n(E)$ given by

$$\Gamma_n(E) = \Gamma_n(E_0)\left[\frac{E}{E_0}\right]^{1/2}. \tag{5.2.2}$$

Note that, since E_0 is usually large compared to $\Gamma(E_r)$, the cross section for $E_r \ll E_0$ becomes

$$\sigma_a(E_r) \approx \sigma_0\left[\frac{\Gamma(E_0)}{2E_0}\right]^2\left[\frac{E_0}{E_r}\right]^{1/2} \approx \frac{1}{v_r}, \tag{5.2.3}$$

where v_r is the relative speed of the neutron. Thus although, for the first (6.67 eV) resonance of U^{28}, σ_0 is about 20,000 b, the amount that this resonance contributes to the total absorption cross section at .0667 eV ($\Gamma(E_0)$ being only 0.027 eV) is approximately $20{,}000\,(.027/13.34)^2\sqrt{100} = 0.82$ b.

Doppler-Broadened Expressions

Because $\sigma_a(E)$ is sizeable for only a small energy range about E_0, it is not legitimate in (5.2.3) to replace the relative kinetic energy E_r between the neutron and the nucleus by the kinetic energy E of the neutron relative to the laboratory system. Instead we must account for the thermal motion of the target nuclei. For nuclei in a solid material this motion is quite complicated. However it has been found legitimate in most cases of interest to assume that the thermal motion of the target nuclei is the same as it would be if these nuclei composed a dilute gas. In other words we assume that the distribution of kinetic energies of the target nuclei is Maxwellian:

$$n^j(E')\,dE' = 2n^j\left[\frac{E'}{(kT)^3\pi}\right]^{1/2}\exp\left(\frac{-E'}{kT}\right)dE', \tag{5.2.4}$$

where $n^j(E')\,dE'$ is the number of nuclei per unit volume having kinetic energies (in the laboratory system) between E' and $E' + dE'$, n^j is the total number of nuclei of the jth kind per unit volume, T is the absolute temperature of the material, and k is the Boltzmann constant (8.6×10^{-3} eV/°K). (Recall that this formula yields a "most probable" energy of kT for the nuclei and an "average" energy of $\frac{3}{2}kT$.)

If now we express relative kinetic energy E_r between the neutron and the target in terms of the kinetic energies of the neutron (E) and the target (E') relative to the laboratory system and then average over the energy distribution (5.2.4), it will be possible to obtain an expression for $\sigma_a(E, T)$, the cross section for a neutron interacting with resonance absorbers that have a Maxwellian distribution of kinetic energies parameterized by temperature T. This cross section is

$$\sigma_a(E, T) = \frac{\sigma_0}{2}\left[\frac{E_0}{E}\frac{A}{\pi kTE}\right]^{1/2} \int_0^\infty dE_r \frac{\Gamma^2(E_0)}{\Gamma^2(E_r) + 4(E_r - E_0)^2} \times\left[\exp\left(-\frac{A}{kT}(\sqrt{E_r} - \sqrt{E})^2\right) - \exp\left(-\frac{A}{kT}(\sqrt{E_r} + \sqrt{E})^2\right)\right], \tag{5.2.5}$$

where A is the ratio of the mass of the target to the mass of the neutron.

Averaging over the thermal motion of the target nuclei has the effect of broadening the resonance. For example the peak value of the 6.67 eV resonance in U^{28} is reduced from 20,000 b at 0°K to 7000 b at 300°K (about room temperature). Thus it is quite important to use (5.2.5) rather than (5.2.1) in reactor calculations.

For narrow resonances ($\Gamma(E_0)$ small) above the thermal range ($E_0 > 1$ eV) it is possible to simplify (5.2.5). This simplification is based on the fact that, for these conditions, because of the exponentials (with $kT < 0.1$ eV and $A > 100$), the integrand is very small except for $\sqrt{E_r}$ close to $\sqrt{E}$ and, because of the factor $\Gamma^2(E_0)/[\Gamma^2(E_r) + 4(E_r - E_0)^2]$, it is also very small except for E_r close to E_0. Thus we may approximate as follows:

$$\begin{aligned} &\Gamma(E_r) \to \Gamma(E_0), \\ &\sqrt{E_r} = \sqrt{[E + (E_r - E)]} \to \left(1 + \frac{E_r - E}{2E}\right)\sqrt{E}, \\ &\left[\frac{A}{\pi kTE}\right]^{1/2} \to \left[\frac{A}{\pi kTE_0}\right]^{1/2}, \\ &\int_0^\infty dE_r(\ ) \to \int_{-\infty}^\infty dE_r(\ ), \\ &\exp\left(-\frac{A}{kT}(\sqrt{E_r} + \sqrt{E})^2\right) \to 0. \end{aligned} \tag{5.2.6}$$

As a result, (5.2.5) becomes

$$\sigma_a(E, T) = \sigma_0\left[\frac{E_0}{E}\right]^{1/2} \psi(\xi, x), \tag{5.2.7}$$

where

$$\psi(\xi, x) = \frac{\xi}{2\sqrt{\pi}} \int_{-\infty}^\infty \frac{dy}{1 + y^2} \exp\left(-\frac{1}{4}\xi^2(x - y)^2\right), \tag{5.2.8}$$

with

$$\begin{aligned} x &\equiv \frac{2(E - E_0)}{\Gamma(E_0)}, \\ \xi &\equiv \frac{\Gamma(E_0)}{2}\left[\frac{A}{E_0 kT}\right]^{1/2}. \end{aligned} \tag{5.2.9}$$

Economical methods of determining the function $\psi(\xi, x)$ have been devised so that finding $\sigma_a(E, T)$ on a computer is straightforward.

It is important to recognize that (5.2.7) is valid only in the vicinity of one particular resonance. Any given isotope may have many resonances, and (5.2.7) (with different values of σ_0, E_0, and $\Gamma(E_0)$) will apply to each of them. Thus a general expression for the cross section of a resonance absorber is a sum of terms like (5.2.7). In such a general expression it is convenient to represent the $(E)^{-1/2}$ parts of all the resonances as a separate, smooth term. The more exact expression (5.2.5) shows that each resonance contributes to this term. (Also see (5.2.3).) Hence, since $\psi(\xi, x)$ in (5.2.7) is significant only when $E \approx E_0$, we can replace (5.2.7) by

$$\sigma_a(E, T) = \sigma_0\psi(\xi, x) + \sigma_{1/v}(E_0)\left[\frac{E_0}{E}\right]^{1/2}, \tag{5.2.10}$$

where $\sigma_{1/v}(E_0)$ is the value of the $1/v$ part of the cross section at $E = E_0$. Note that, since the constant $\sigma_{1/v}(E_0)\sqrt{E_0} = \sigma_{1/v}(E)\sqrt{E}$, it could equally well be written $\sigma_{1/v}(0.025)\sqrt{0.025}$, where $\sigma_{1/v}(0.025)$ is that part of the total $1/v$ absorption cross section at 0.025 eV due to the resonance peak at E_0.

If then we use σ_i, ξ_i, and x_i to represent the parameters σ_0, ξ, and x associated with the ith resonance, we have the complete expression for $\sigma_a(E, T)$ in the energy range within which the single-level formula is appropriate:

$$\sigma_a(E, T) = \sum_i \sigma_i\psi(\xi_i, x_i) + \sigma(0.025)\left[\frac{0.025}{E}\right]^{1/2}, \tag{5.2.11}$$

where $\sigma(0.025)$ is the total contribution of *all* the $1/v$ parts of the resonances at 0.025 eV. We expect that, at any given energy E, only a few terms in the sum of the resonance contributions will be important, all the other $\psi(\xi_i, x_i)$ being negligibly small.

In the case of scattering resonances the expressions for Doppler-broadened cross sections are more involved. This is because, in addition to resonance scattering due to the formation and decay of a compound nucleus, there is also potential scattering due to the interaction of the neutron with the nucleus as a whole. Moreover the neutron wave function due to the potential scattering interferes with that due to the resonance scattering, and this will introduce a third term in the overall expression for the scattering cross section.

A general expression for the scattering cross section in the resonance region is quite involved even without averaging over the nuclear motions. However, at low energies (but above thermal) and for isolated resonances, an approximate expression analogous to the Doppler-broadened absorption cross section (5.2.7) is

$$\sigma_s(E, T) = \frac{\sigma_0\Gamma_n(E_0)}{\Gamma_\gamma}\psi(\xi, x) + \frac{\sigma_0\Gamma(E_0)}{\lambda_0\Gamma_\gamma}R\chi(\xi, x) + 4\pi R^2, \tag{5.2.12}$$

where $\lambda\!\!\!{}^{-}_0 = \hbar/\sqrt{(2\mu E_0)}$, μ being the reduced mass of the neutron plus target, R is the nuclear radius, and

$$\chi(\xi, x) = \frac{\xi}{\sqrt{\pi}} \int_{-\infty}^{\infty} \frac{y dy}{1 + y^2} \exp\left(\frac{1}{4}\xi^2(x - y)^2\right). \tag{5.2.13}$$

For values of x such that E is far from E_0 (see (5.2.9)), both $\psi(\xi, x)$ and $\chi(\xi, x)$ become small, and $\sigma_s(E, T)$ takes on the constant, temperature-independent value $4\pi R^2$ (pure potential scattering). For E near E_0 ($x \approx 0$), however, the first two terms of (5.2.12) dominate, and $\sigma_s(E, T)$ takes on the resonance shape of $\sigma_a(E, T)$ modified by the term in $\chi(\xi, x)$. This latter term, since it is an odd function of x, can cause the overall shape to dip below the constant $4\pi R^2$ for E values just below E_0.

As in the absorption case, (5.2.12) applies to a single resonance. The appropriate expression over the whole energy range is

$$\sigma_s(E, T) = \sum_i \left[\frac{\sigma_i \Gamma_{in}(E_i)}{\Gamma_{i\gamma}} \psi(\xi_i, x_i) + \frac{\sigma_i \Gamma_i(E_i)}{\lambda\!\!\!{}^{-}_i \Gamma_{i\gamma}} R\chi(\xi_i, x_i)\right] + 4\pi R^2. \tag{5.2.14}$$

Limitations on Accuracy

The formulas (5.2.11) and (5.2.14) are quite adequate for widely spaced absorption and scattering resonances in the energy range above thermal. For broad resonances at high energies Doppler broadening is not important, and unbroadened expressions such as (5.2.1) can be used directly. At very low energies more exact expressions such as (5.2.5) are available.

It must be remembered, however, that all these formulas are based on a single-level Breit-Wigner shape. For some types of decay in some nuclei more than one level is involved, and more complicated "multilevel" formulas must be invoked to fit the measured data. Unfortunately low-lying fission resonances are of this nature; the shape of their cross sections cannot be fit by a single-level formula, and attempts to force a fit may be risky if one tries to account for Doppler broadening in the usual way.

5.3 Determination of Group Parameters for Homogeneous Media Composed of Mixtures of Resonance and Nonresonance Materials

There are really two kinds of difficulty associated with finding group parameters for reactor regions made up of cells containing resonance absorbing material. The first of these arises from the heterogeneous nature of the cell, and the second is due to the violent changes in cross section exhibited by resonance materials. Since methods for finding group parameters that account for the resonance nature of cross sections are applicable to both homogeneous and heterogeneous geometries, we shall deal with the second problem first. Specifically we shall look for ways to find few-group parameters for a homogeneous mixture of resonance scatterers and absorbers. It will be sufficient to discuss in detail the computation of only two parameters of this type, namely the absorption cross section and the group-to-group transfer cross section. Accordingly we state the problem in mathematical terms by saying that we seek values of

$$\Sigma_{ag} \equiv \int_{\Delta E_g} \Sigma_a(E)\psi_g(E)\,dE = \sum_j n^j \int_{\Delta E_g} \sigma_a^j(E)\psi_g(E)\,dE \tag{5.3.1}$$

and

$$\begin{aligned}\Sigma_{gg'} &\equiv \int_{\Delta E_g} dE \int_{\Delta E_{g'}} dE'\,\Sigma_s(E' \to E)\psi_{g'}(E') \\ &= \sum_j n^j \int_{\Delta E_g} dE \int_{\Delta E_{g'}} dE'\,\sigma_s^j(E' \to E)\psi_{g'}(E'),\end{aligned} \tag{5.3.2}$$

where superscripts j represent different isotopes in the mixture and where $\psi_g(E)$ within each energy range ΔE_g has the *shape* of $\psi(E)$, the fundamental solution of

$$[D(E)B_m^2 + \Sigma_t(E)]\psi(E) = \int_0^\infty dE'\left[\sum_j \chi^j(E)\nu\Sigma_f^j(E') + \Sigma_s(E' \to E)\right]\psi(E') \tag{5.3.3}$$

for the homogeneous mixture. Each $\psi_g(E)$ is normalized by

$$\int_{\Delta E_g} \psi_g(E)\,dE = 1 \qquad (g = 1, 2, \ldots, G). \tag{5.3.4}$$

An Ultrafine-Group Solution

The most straightforward way to solve this problem is to use a very large number (>1000) of "ultrafine" groups to obtain a very accurate approximation to $\psi(E)$ and then to use this approximation in (5.3.1) and (5.3.2). Thus we assume that, for E in ΔE_n, $\psi(E)$ in (5.3.3) may be written

$$\psi(E) \approx \frac{\Phi_n}{\Delta E_n}, \tag{5.3.5}$$

so that, with

$$\begin{aligned}D_n &\equiv \frac{1}{\Delta E_n}\int_{\Delta E_n} D(E)\,dE, \\ \nu\Sigma_{fn}^j &\equiv \frac{1}{\Delta E_n}\int_{\Delta E_n} \nu\Sigma_f^j(E)\,dE, \\ \Sigma_{tn} &\equiv \frac{1}{\Delta E_n}\int_{\Delta E_n} \Sigma_t(E)\,dE, \\ \Sigma_{nn'} &\equiv \int_{\Delta E_n} \frac{dE}{\Delta E_{n'}}\int_{\Delta E_{n'}} \Sigma_s(E' \to E)\,dE', \\ \chi_n^j &\equiv \int_{\Delta E_n} \chi^j(E)\,dE,\end{aligned} \tag{5.3.6}$$

(5.3.3) becomes

$$[D_n B_m^2 + \Sigma_{tn}]\Phi_n = \sum_{n'=1}^{N}\left[\sum_j \chi_n^j \nu\Sigma_{fn'}^j + \Sigma_{nn'}\right]\Phi_{n'} \qquad (n = 1, 2, \ldots, N). \tag{5.3.7}$$

When this equation is solved for the physically acceptable (i.e., positive) set of Φ_n values, the few-group parameters (5.3.1) and (5.3.2) can be found from

$$\Sigma_{ag} = \frac{\sum_{n\subset g}\Sigma_{an}\Phi_n}{\sum_{n\subset g}\Phi_n} = \sum_j n^{(j)}\left\{\frac{\sum_{n\subset g}\sigma_{an}^j\Phi_n}{\sum_{n\subset g}\Phi_n}\right\} \qquad (g = 1, 2, \ldots, G) \tag{5.3.8}$$

and

$$\Sigma_{gg'} = \sum_j n^j \frac{\sum_{n\subset g}\sum_{n'\subset g'}\sigma_{nn'}^j\Phi_{n'}}{\sum_{n'\subset g'}\Phi_{n'}} \qquad (g, g' = 1, 2, \ldots, G), \tag{5.3.9}$$

where the symbol $\sum_{n\subset g}$ indicates a sum over all ultrafine groups n having energy intervals ΔE_n lying in the larger energy interval ΔE_g.

The only drawback to this procedure is cost. For a fast reactor a thousand or so groups may be enough to give an accurate representation of $\psi(E)$. But in a thermal reactor, because of the great importance of the very high and narrow resonances at energies in the range 1 eV to 1000 eV, it may be necessary to use as many as 100,000 groups. Having to perform such calculations for each different material composition in the reactor adds substantially to the cost of design computations. The situation is particularly serious when nonuniform depletion causes material compositions that are initially identical to become different.

A Multigroup Library Formed from the Ultrafine-Group Solution

It was pointed out in Section 4.13 that, in a multigroup method where the $\psi_g(E)$ are assumed to be known functions of energy in the intervals ΔE_g, independent of the composition of the medium, a library of multigroup microscopic cross sections can be determined, stored, and used to construct macroscopic multigroup cross sections as needed. (See (4.13.10).) An analogous procedure can be used to reduce the cost of an ultrafine-group solution.

What is done is to construct a multigroup library of microscopic parameters from a single ultrafine computation for a particular material composition in the range of interest. Equations (5.3.8) and (5.3.9) show that the appropriate expressions for the absorption and group-transfer parameters are

$$\begin{aligned}\bar{\sigma}_{ag}^j &\equiv \frac{\sum_{n\subset g}\sigma_{an}^j\Phi_n}{\sum_{n\subset g}\Phi_n},\\ \bar{\sigma}_{gg'}^j &\equiv \frac{\sum_{n\subset g}\sum_{n'\subset g'}\sigma_{nn'}^j\Phi_{n'}}{\sum_{n'\subset g'}\Phi_{n'}},\end{aligned} \tag{5.3.10}$$

where the number of ultrafine groups may be several thousand, and the number of "multigroup" groups may be 30–100.

We can, in a similar fashion, find averaged microscopic cross sections for each element of other processes ($\bar{\sigma}^j_{tg}$, $\overline{\nu^j\sigma^j_{fg}}$, $\bar{\sigma}^j_{trg}$—recall that $D_g = [3\sum_j n^j \sigma^j_{trg}]^{-1}$) and thus form a complete library consisting of group-averaged, microscopic cross sections for each isotope in the mixture.

Then, for any set of nuclear concentrations fairly close to those for which the ultrafine spectrum was found, we can determine approximate values for multigroup macroscopic cross sections by using the formulas

$$\Sigma_{ag} = \sum_j n^j \bar{\sigma}^j_{ag} \tag{5.3.11}$$

and

$$\Sigma_{gg'} = \sum_j n^j \bar{\sigma}^j_{gg'}. \tag{5.3.12}$$

Since the Φ_n used to perform the averaging for (5.3.10) are dependent on the ultrafine macroscopic parameters Σ_{tn}, D_n, $\nu^j\Sigma^j_{fn}$, etc., and hence on the number densities n^j of the isotopes making up the particular mixture for which the Φ_n were found, the averaged cross sections $\bar{\sigma}^j_{ag}$, etc., are really not unique parameters associated with isotope j, but also depend on the relative amounts of all other isotopes making up the mixture. However it has been found that this sensitivity is, for many situations, not very great and that, if the number of groups g is moderately large (30–100), the $\bar{\sigma}^j_{ag}$ for isotope j over a moderate range of compositions about some average are quite close to one another. Thus, just as in the standard multigroup procedure, we can form a library of multigroup cross sections and use them in the standard way to compute few-group parameters. For example, after doing a single 2000-group calculation (i.e., solving (5.3.7) for the Φ_n), we can construct a library of 50-group, isotopic, microscopic cross sections using (5.3.10). Then, for a range of material compositions similar in character to the composition for which we performed the ultrafine computation, we may form a set of 50-group parameters from (5.3.11) and (5.3.12) and again use (5.3.7) to find a 50-group spectrum and, hence, construct, say, 4-group parameters.

This two-stage reduction to obtain few-group parameters takes into account in a detailed way (through the ultrafine-group calculation) changes in the spectrum due to resonances. Yet, because only one ultrafine calculation is required for a given range of material compositions, it is not inordinately expensive. The method works because a fairly large number of groups (30–100) is retained in the first step of the reduction. Thus it is not too important that the gross, *overall* energy shape of the basic ultrafine spectrum may not, for some compositions in the range, match the correct ultrafine spectrum. We require only that the shapes match over 30 to 100 *limited* energy ranges. Differences in overall gross shape are accounted for by the fact that the reduction to few groups is done for each particular composition individually, as in the conventional few-group procedure.

5.4 Analytical Procedures for Homogeneous Mixtures

The two-stage process just described is used primarily for the analysis of fast reactors, for which the resonances of importance are fairly broad; thus an ultrafine calculation involving, say, a thousand groups is sufficient to account for the details of the shape of the resonance cross sections. For thermal reactors the resonances of importance are at a lower energy and are thus much higher and much narrower. To describe such violently erratic behavior requires the use of many thousands of groups in the ultrafine calculation. Thus there is a real advantage to solving the integral equation (5.3.3) for $\psi(E)$ analytically when $\Sigma_t(E)$ has a resonance structure. Even a solution accurate over a limited range would be useful, since it would allow us to determine accurate group parameters for, say, a 50-group structure, and we could then find few-group parameters in the usual manner (using (5.3.8) and (5.3.9), for example).

Since the most urgent need for an analytical approach is in the energy range of the resolved resonances, we shall restrict our attention to the solution of (5.3.3) in that range. Two advantages immediately accrue: (1) we may ignore in (5.3.3) any direct source from fission (i.e., we may set $\chi^j(E)$ to zero); and (2) we may assume the scattering is elastic and isotropic in the center-of-mass system. As a result of (2), the expression for $\Sigma_s(E' \to E)$ becomes

$$\Sigma_s(E' \to E)\,dE = \sum_j \Sigma_s^j(E' \to E)\,dE,$$

$$\Sigma_s^j(E' \to E)\,dE = \begin{cases} \Sigma_s^j(E')\dfrac{dE}{E'(1-\alpha_j)} & \text{for } \alpha_j E' \le E \le E', \\ 0 & \text{for } E > E' \text{ or } E < \alpha_j E'. \end{cases} \tag{5.4.1}$$

Here $\alpha_j \equiv [(A_j - 1)/(A_j + 1)]^2$, A_j being the ratio of the mass of isotope j to the mass of the neutron (see (2.6.7)).

In view of these assumptions, (5.3.3) for the spectrum function in the homogeneous medium reduces to

$$[D(E)B_m^2 + \Sigma_t(E)]\psi(E) = \sum_j \int_E^{E/\alpha_j} \Sigma_s^j(E')\frac{\psi(E')dE'}{E'(1-\alpha_j)} \tag{5.4.2}$$

for E in the resolved-resonance range.

For material compositions of interest the materials buckling B_m^2 is generally small in magnitude ($|B_m^2| < 10^{-2}$ cm^{-2}). Moreover $D(E)$, being inversely proportional to $3\Sigma_{tr}(E)$, is usually less than 1 cm and becomes extremely small at energies corresponding to either a scattering or an absorbing resonance. Thus $D(E)B_m^2$ is generally negligible in comparison with $\Sigma_t(E)$, and neglecting it in (5.4.2) affects $\psi(E)$ insignificantly. Accordingly we can determine acceptably accurate spectrum functions in the resolved-resonance energy range by solving

$$\Sigma_t(E)\psi(E) = \sum_j \int_E^{E/\alpha_j} \Sigma_s^j(E') \frac{\psi(E')\,dE'}{E'(1-\alpha_j)}. \tag{5.4.3}$$

This equation is identical with (3.4.9) for the scalar flux density in an infinite homogeneous medium. Thus much of the analysis developed in Section 3.4 is immediately applicable.

Hydrogen the Only Moderator

In Chapter 3 we saw that there is one (and only one) special situation in which a closed-form solution can be obtained for (5.4.3). This is when the only moderator is hydrogen, that is, when we neglect any moderation due to other isotopes present. With α_j for hydrogen set equal to zero and with the neglect of moderation by other isotopes implying that their $\alpha_j \to 1$, we found that that (5.4.3) becomes

$$[\Sigma_a(E) + \Sigma_s^H(E)]\psi(E) = \int_E^{\infty} \Sigma_s^H(E') \frac{\psi(E')\,dE'}{E'}, \tag{5.4.4}$$

where $\Sigma_s^H(E)$ is the macroscopic scattering cross section for hydrogen and $\Sigma_a(E)$ is the total macroscopic absorption cross section.

Differentiating (5.4.4) with respect to E and reintegrating (see (3.4.12)–(3.4.15)) yields

$$\psi(E) = \frac{[\Sigma_a(E_1) + \Sigma_s^H(E_1)]E_1\psi(E_1)}{[\Sigma_a(E) + \Sigma_s^H(E)]E} \exp\left(-\int_E^{E_1} \frac{\Sigma_a(E')dE'}{E'[\Sigma_a(E') + \Sigma_s^H(E')]}\right), \tag{5.4.5}$$

where E_1 is some fixed energy at the top of the resolved-resonance region. We have discussed already in Section 3.4 the characteristic dip and recovery of $\psi(E)$ over an energy range that includes a large absorption resonance (see Figure 3.2).

The energy shape $\psi(E)$ may be normalized in accord with (5.3.4), and group parameters may be obtained for any desired group structure by applying the definitions (5.3.1), etc. Numerical integrations will be necessary at this point. However the fact that we no longer need to find a several-thousand-group solution to (5.3.7) greatly reduces the cost.

For the Σ_{ag} and $\Sigma_{gg'}$ further analytical simplifications are possible. Suppose that C_g

$(g = 1, 2, \ldots, G)$ are a set of normalized constants such that, with (5.4.5) written as

$$\psi_g(E) = \frac{C_g}{E[\Sigma_a(E) + \Sigma_s^H(E)]} \exp\left(-\int_E^{E_1} \frac{\Sigma_a(E')dE'}{E'[\Sigma_a(E') + \Sigma_s^H(E')]}\right), \tag{5.4.6}$$

the integrals $\int_{\Delta E_g} \psi_g(E)dE$ are all unity.

We now introduce a quantity $p(E)$ with the following definition:

$$p(E) \equiv \exp\left(-\int_E^{E_1} \frac{\Sigma_a(E')dE'}{E'[\Sigma_a(E') + \Sigma_s^H(E')]}\right). \tag{5.4.7}$$

Comparison with (3.5.13) shows that $p(E)$ is the resonance-escape probability for neutrons moderated by hydrogen ($\bar{\xi} = 1$) in the energy range E to E_1. Differentiation yields

$$\frac{dp(E)}{dE} = \frac{\Sigma_a(E)p(E)}{E[\Sigma_a(E) + \Sigma_s^H(E)]}. \tag{5.4.8}$$

Thus, from (5.3.1), (5.4.6), and (5.4.7), we have

$$\Sigma_{ag} = \int_{E_g}^{E_{g-1}} \frac{\Sigma_a(E)C_g p(E)\,dE}{E[\Sigma_a(E) + \Sigma_s^H(E)]} = C_g[p(E_{g-1}) - p(E_g)]. \tag{5.4.9}$$

Similarly, from (5.3.2) (with the symbols g and g' for convenience interchanged),

$$\begin{aligned}
\Sigma_{g'g} &= \int_{E_{g'}}^{E_{g'-1}} dE' \int_{E_g}^{E_{g-1}} dE \frac{C_g \Sigma_s^H(E)p(E)}{E^2[\Sigma_a(E) + \Sigma_s^H(E)]} \\
&= \Delta E_{g'} C_g \int_{E_g}^{E_{g-1}} \left(\frac{p(E)}{E^2} - \frac{\Sigma_a(E)p(E)}{E^2[\Sigma_a(E) + \Sigma_s^H(E)]}\right) dE \\
&= \Delta E_{g'} C_g \int_{E_g}^{E_{g-1}} \left(\frac{p(E)}{E^2} - \frac{1}{E}\frac{dp(E)}{dE}\right) dE \\
&= \Delta E_{g'} C_g \int_{E_g}^{E_{g-1}} -\frac{d}{dE}\left(\frac{p(E)}{E}\right) dE \\
&= \Delta E_{g'} C_g \left[\frac{p(E_g)}{E_g} - \frac{p(E_{g-1})}{E_{g-1}}\right]. \tag{5.4.10}
\end{aligned}$$

Thus to find the Σ_{ag} and $\Sigma_{g'g}$ we need only determine the $p(E_g)$ and C_g by numerical evaluation of the integrals in (5.4.7) and (5.3.4).

The group parameters Σ_{ag} and $\Sigma_{g'g}$ can be used to make contact with the classical expressions for the resonance-escape probability and slowing-down density in homogeneous, hydrogen-moderated material. Thus, if we sum $\Sigma_{g'g}$ over all groups g' into which the neutron might be slowed down, we obtain the cross section specifying the rate at which neutrons entering group g will slow down past its lower energy boundary. Other neutrons may also slow down past E_g by jumping over group g; but if group g is taken to be large, this number, even for hydrogen scattering, will be small. Thus, since

$$\sum_{g'=g+1}^{G} \Delta E_{g'} = E_g,$$

summing (5.4.10) from $g' = g + 1$ to G yields:

Cross section for a neutron in group g to slow down past E_g

$$= C_g\left[p(E_g) - \frac{E_g}{E_{g-1}} p(E_{g-1})\right]. \tag{5.4.11}$$

If ΔE_g is large so that E_g/E_{g-1} is small compared to unity, very few neutrons will jump over group g, and we have that $q(E_g)$, the slowing-down density at E_g, is proportional to $p(E_g)$.

A similar argument shows that the probability that a neutron in group g will escape capture is

$$\frac{\sum_{g'=g+1}^{G} \Sigma_{g'g}\Phi_g}{\sum_{g'=g+1}^{G} (\Sigma_{g'g} + \Sigma_{ag})\Phi_g} = \frac{p(E_g) - (E_g/E_{g-1})p(E_{g-1})}{[1 - (E_g/E_{g-1})]\,p(E_{g-1})}. \tag{5.4.12}$$

If ΔE_g is taken to be the whole resonance region, so that $E_{g-1} = E_1$ and $E_g/E_{g-1} \ll 1$, we see that the escape probability from the resonance region, defined by the left-hand side of (5.4.12), reduces to

$$\frac{p(E_g)}{p(E_{g-1})} = \exp\left(-\int_{E_g}^{E_{g-1}} \frac{\Sigma_a(E')\,dE'}{E'[\Sigma_a(E') + \Sigma_s^H(E')]}\right). \tag{5.4.13}$$

Basic Approximations Made When Moderation by All Isotopes Is Considered

For moderators other than hydrogen the α_j are nonzero, and the upper limits of the integrals in (5.4.3) depend on E. As a consequence the strategy of solving (5.4.3) by differentiating with respect to E no longer works, and it is necessary to turn to approximate schemes for solving the equation. It is most convenient in developing such approximations to deal with the slowing-down density $q(E)$ for the homogeneous medium. Accordingly we shall first review certain basic equations satisfied by this quantity that were developed in Chapter 3.

Recall that $q(E)$ is *defined* as the number of neutrons per cm^3 per sec in the homogeneous medium whose energies drop below the energy E (see (3.4.22)). For a pure scattering medium in which scattering is elastic and isotropic in the center-of-mass system and in which all the scattering cross sections $\sigma_s^j(E)$ have the same dependence on energy (a condition most easily met if the $\sigma_s^j(E)$ are all constant), the flux density and slowing-down density are related at energies below the fission source range by

$$\psi(E) = \frac{q(E)}{E\bar{\xi}(E)\Sigma_s(E)}, \tag{5.4.14}$$

where $\bar{\xi}(E)$ is the average loss in the logarithm of the energy per collision in the mixture (see (3.4.29) and (3.4.30)).

With no absorption (and below the fission source range), q is a constant. With absorption it becomes energy-dependent and obeys (3.4.32):

$$\frac{dq(E)}{dE} = \Sigma_a(E)\,\psi(E). \tag{5.4.15}$$

The Practical Width

The analytical method we shall develop makes use of the fact that the absorption cross section $\Sigma_a(E)$ of the mixture is very small except in the energy ranges immediately surrounding the peaks of the resonance. Thus outside those ranges (5.4.14) will be a good approximation. For a given resonance of a given isotope j, we shall use the term "practical width," represented by Γ_p, to designate the range about the energy of the peak throughout which $\Sigma_a^j(E)$, the macroscopic absorption cross section for resonance absorber j, is significant. The decision of just how small $\Sigma_a^j(E)$ must become before it is "insignificant" is, of course, arbitrary. However, since the corresponding microscopic cross section $\sigma_a^j(E)$ falls off sharply on both sides of a resonance peak, we need not be too concerned about this point. Accordingly we shall arbitrarily define Γ_p, the practical width of an absorption resonance, to be that range about the peak energy within which $\Sigma_a^j(E)$ $(=n^j\,\sigma_a^j(E)$, with $\sigma_a^j(E)$ given by (5.2.7)) exceeds the total macroscopic scattering cross section $\Sigma_s(E)$ for the material. (Note that $\Sigma_s(E)$ includes the resonance and potential scattering of isotope j [see (5.2.14)] as well as the potential scattering of all the other isotopes in the mixture.) The practical width of the resonance is then an energy range about the peak of the resonance within which there is a greater probability that the neutron will be captured than that it will be scattered. For resonances which are primarily absorbing ($\Gamma_\gamma \gg \Gamma_n$) the practical width generally exceeds the total width Γ by an order or magnitude or so. Moreover, since its definition involves the magnitude of the cross section out on the *wings* of the resonance, the practical width is not too sensitive to changes in temperature. It should be noted, however, that for resonances which are primarily scattering resonances ($\Gamma_n \gg \Gamma_\gamma$), it may happen that $\Sigma_a^j(E)$ never exceeds $\Sigma_s^j(E)$, so that the practical width, according to the definition just given, becomes zero. This situation is quite in keeping with the notation that the practical width is an energy range wherein the probability of capture exceeds that of scattering: for a purely scattering resonance such a range does not exist.

The Narrow-Resonance Approximation

Let us now return to the solution of (5.4.3) for an energy range in which some of the isotopes in the mixture are resonance absorbers with large, widely spaced resonances. We first of all rewrite (5.4.3) as

$$\Sigma_t(E)\psi(E)\,dE = \sum_j \left[\int_E^{E/\alpha_j} \Sigma_s^j(E') \frac{\psi(E')\,dE'}{E'(1 - \alpha_j)} \right] dE \tag{5.4.16}$$

and note that the left-hand side represents the total rate per unit volume at which neutrons leave the energy interval dE by being absorbed in that interval or by being scattered out of it, while the right-hand side represents the rate per unit volume at which neutrons appear in dE by virtue of being elastically scattered in from higher energies. Suppose that E is an energy in the range of the practical width of some absorption resonance, and consider the situation in which this width is narrow compared to all the energy ranges $(E/\alpha_j) - E$ within which neutrons have had their last collisions. Then, since $\Sigma_a(E)$ is close to zero in the range between the practical width in question and that associated with the next highest resonance and since we are assuming the resonances to be widely spaced, $\psi(E')$ in the integrals can be represented approximately by the pure-scatterer result (5.4.14). Thus we may write

$$\Sigma_t(E)\psi(E) \approx \sum_j \int_E^{E/\alpha_j} \frac{\Sigma_s^{pj}(E')q_0\,dE'}{(E')^2 \bar{\xi}^p(E')\Sigma_s^p(E')(1 - \alpha_j)}, \tag{5.4.17}$$

where q_0 is the (constant) slowing-down density above the resonance and $\Sigma_s^{pj}(E')$, $\bar{\xi}^p(E')$, and $\Sigma_s^p(E')$ contain only potential-scattering cross sections. Since these cross sections (and hence $\bar{\xi}^p$) are essentially constant, (5.4.17) becomes

$$\Sigma_t(E)\psi(E) \approx \frac{q_0}{E\bar{\xi}^p} \tag{5.4.18}$$

for widely spaced resonances and for $(E/\alpha_j) - E \gg \Gamma_p$ (all j).

This approximation is called the *narrow-resonance* (NR) *approximation*. Its use greatly simplifies the computation of resonance capture rates for a heterogeneous lattice, and we shall employ it when we get to that stage of our development.

A more accurate expression for $\psi(E)$ in the energy range of widely spaced resonances has already been derived in Section 3.4. Since the argument is short and important we shall repeat it here. The improvement in the accuracy of (5.4.18) is effected by accounting approximately for the fact that a small amount of the scattering-in source (that due to glancing collisions and hence involving a very small energy loss) fails to reach the lower portion of the practical width since the neutrons involved are absorbed in the upper range of Γ_p.

To account for this correction we note that, with no absorptions, $\Sigma_s(E)\psi(E)dE$ can be interpreted both as the number scattering into dE and the number scattering out. Thus (5.4.14) shows that, if $\Sigma_a(E) = 0$,

$$\frac{q_0\, dE}{E\bar{\xi}(E)} = \text{number of neutrons per unit volume scattered into } dE \text{ per second.} \tag{5.4.19}$$

Since q_0 is the slowing-down density past E (a constant for $\Sigma_a(E) = 0$) and $\bar{\xi}(E)$ is the average loss in the logarithm of the energy due to a scattering, we can picture (5.4.19) as describing a continuous process in which the fraction of those neutrons slowing down past E per second that appear in dE is $(d \ln E)/\bar{\xi}(E)$. It is as if the slowing-down neutrons were dispersed uniformly in a logarithmic energy range $\bar{\xi}(E)$. In view of this interpretation it seems a reasonable approximation to assume that, when there *is* absorption (and, hence, when q is no longer a constant but, instead, changes in accord with (5.4.15)), (5.4.19), with q_0 replaced by $q(E)$, will still be valid throughout the range corresponding to the practical width of the resonance. Moreover, since we are still assuming that Γ_p is small in comparison with all the ranges $(E/\alpha_j) - E$, we shall continue to employ a constant $\bar{\xi}^p$ based only on potential-scattering cross sections.

If, then, we replace the scattering-in source on the right-hand side of (5.4.16) by (5.4.19), with q_0 replaced by $q(E)$, and make use of the exact equation (5.4.15) to eliminate $\psi(E)$ on the left-hand side of (5.4.16), we get

$$\frac{dq(E)}{dE} = \frac{\Sigma_a(E)}{\Sigma_t(E)}\frac{q(E)}{E\bar{\xi}^p}. \tag{5.4.20}$$

Integration between E and E_1, some fixed energy immediately above the resolved-resonance range, yields

$$q(E) = q(E_1)\exp\left(-\int_E^{E_1}\frac{\Sigma_a(E')dE'}{E'\Sigma_t(E')\bar{\xi}^p}\right). \tag{5.4.21}$$

Taking the derivative with respect to E and again applying (5.4.15) gives us

$$\Sigma_t(E)\psi(E) = \frac{q(E_1)}{E\bar{\xi}^p}\exp\left(-\int_E^{E_1}\frac{\Sigma_a(E')dE'}{E'\Sigma_t(E')\bar{\xi}^p}\right) \tag{5.4.22}$$

for widely spaced resonances and for $(E/\alpha_j) - E \gg \Gamma_p$ (all j).

Comparison of this more exact result with (5.4.18) shows that the two differ by the exponential factor, which in (5.4.22) accounts for the decrease in $q(E)$ due to neutrons absorbed in resonances in the range E–E_1. Note that, according to (5.4.14), at an energy E_1 above the resonances, where the absorption cross section is essentially zero,

$$q(E_1) = E_1 \bar{\xi}^p \Sigma_s(E_1) \psi(E_1).$$

Thus the approximation (5.4.21) reduces to the exact result (5.4.5) when hydrogen is the only moderator (and hence $\bar{\xi}^p = 1$), while the more gross approximation (5.4.18) does not.

Nevertheless (5.4.18) provides a fairly accurate estimate of $\psi(E)$ in the range of a given resonance. This is because, except for the very-lowest-lying resonances, $\int_E^{E_1} (\Sigma_a(E')dE'/E'\Sigma_t(E')\bar{\xi}^p)$ is small across the energy range Γ_{p_i} of the ith resonance in which $\Sigma_a(E')$ is large. (A rough estimate of its size can be obtained by assuming $\Sigma_a(E')/\Sigma_t(E')\bar{\xi}^p$ to be unity within the energy range Γ_{p_i} and zero outside. Then

$$\int_E^{E_1} \frac{\Sigma_a(E')\, dE'}{E'\Sigma_t(E')\bar{\xi}} \approx \ln\left(\frac{E_i + \frac{1}{2}\Gamma_p}{E_i - \frac{1}{2}\Gamma_p}\right),$$

where E_i is the energy corresponding to the peak of the ith resonance.)

Since, as E decreases across the range Γ_{p_i}, the value of $\int_E^{E_1} (\Sigma_a(E')dE'/E'\Sigma_t(E')\bar{\xi}^p)$ increases only a small amount, the exponential $\exp(-\int_E^{E_1} (\Sigma_a(E')dE'/E'\Sigma_t(E')\bar{\xi}^p))$ decreases by only a small amount. On the other hand the flux density $\psi(E)$, being proportional to $1/\Sigma_t(E)$, dips by a factor of hundreds to thousands at the peak energy E_i.

At energies E below a resonance, the resonance-absorption part of $\Sigma_t(E)$ disappears, and $\int_E^{E_1} (\Sigma_a(E')dE'/E'\Sigma_t(E')\bar{\xi}^p)$ remains constant since $\Sigma_a(E')$ is zero over the extra range being added to the limits of integration. Thus $\psi(E)$ is again proportional to $1/E$. In fact (5.4.14) is again valid, except that the constant slowing-down density q must be replaced by the new constant $q \exp(-\int_{\Delta E_i} (\Sigma_a(E')dE'/E'\Sigma_t(E')\bar{\xi}^p))$, where the integration is over that range ΔE_i in which, because of the resonance at E_i, $\Sigma_a(E')$ is significant. Thus, at energies below the resonance, the slowing-down density is decreased by the exponential factor. It follows that

$$p_i \equiv \exp\left(-\int_{\Delta E_i} \frac{\Sigma_a(E')dE'}{E'\Sigma_t(E')\bar{\xi}^p}\right)$$

is the probability of escaping capture in resonance i, a result consistent with the definition of $p(E)$ for hydrogen moderation (5.4.7) and the resonance-escape probability defined by (3.5.13).

Figure 5.1 illustrates the behavior of the flux density $\psi(E)$ and the interaction rate $\Sigma_t(E)\psi(E)$ in an energy range E_g–E_{g-1} including two absorption resonances.

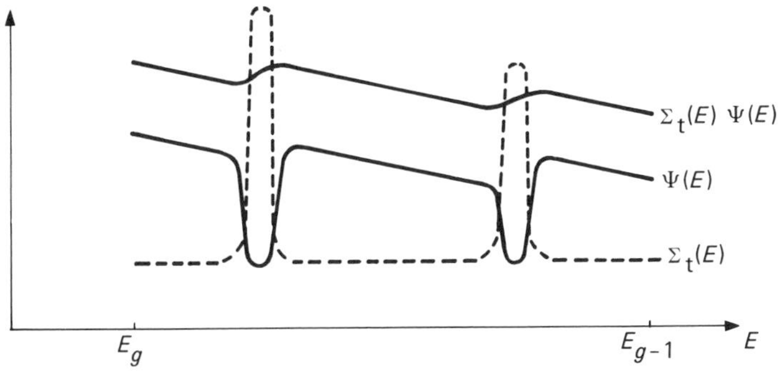

Figure 5.1 The scalar flux density and collision density versus energy for a homogeneous material (in the resolved-resonance range).

Clearly $\psi(E)$ behaves much more drastically that $\Sigma_t(E)\psi(E)$ (called the *collision density*). In fact, if we plotted $E\Sigma_t(E)\psi(E)$, it would be flat except in the ranges of resonance peaks. Even there it would vary slowly.

Because (5.4.22) accounts for the decrease in $E\Sigma_t(E)\psi(E)$ due to neutron captures in resonances, we may use it to estimate the spectrum function throughout an energy group containing many resonances. (Even with 50 energy groups in the range 1–10^7 eV, there may be eight or more resolved resonances in a given group.) Thus, if we are interested in a group g extending from E_g to E_{g-1}, we may choose E_1 in (5.4.22) to be E_{g-1} (or some slightly higher energy if E_{g-1} happens to fall in the middle of a resonance). Then, with a normalization constant C_g defined so that

$$\int_{E_g}^{E_{g-1}} C_g\psi(E)\,dE = \int_{E_g}^{E_{g-1}} \frac{C_g q(E_{g-1})}{E\xi^{\mathrm{p}}\Sigma_t(E)} \exp\left(-\int_E^{E_{g-1}} \frac{\Sigma_a(E')dE'}{E'\Sigma_t(E')\xi^{\mathrm{p}}}\right)dE \equiv \int_{E_g}^{E_{g-1}} \psi_g(E)dE = 1, \tag{5.4.23}$$

equations such as (5.3.1) and (5.3.2) give the group constants immediately.

The Wide-Resonance Approximation

For certain resonances at energies near thermal, the energy lost by a neutron making a scattering collision with the resonance absorber is less than the practical width of the resonance. For example Γ_p for the 6.67 eV resonance of U^{28} is approximately 0.7 eV. However the maximum energy loss $E(1-\alpha_{28})$ a neutron can sustain in an elastic collision with U^{28} at 6.67 eV is $6.67(1-.983) = 0.11$ eV. Under these circumstances the scattering-in source due to U^{28} on the right-hand side of (5.4.16) comes from an energy range E–$E/.983$, very close to E. Moreover the U^{28}-scattering part of $\Sigma_t(E)\psi(E)dE$ on the left-hand side doesn't remove neutrons from the range Γ_p, but instead only transfers them to a slightly lower part of that range. Thus neither scattering into dE nor scattering out of dE due to U^{28} in the range Γ_p does much to change the probabilities that neutrons within Γ_p will be absorbed or scattered out by collisions with nonresonance scatterers. Hence it seems reasonable to neglect the U^{28} scattering on both sides of (5.4.16).

We can implement this argument analytically by noting that, if α_{res} is the value of α_j in (5.4.1) for the resonance absorbing isotope and if $\Gamma_p \gg E_i(1 - 1/\alpha_{res})$, E_i being the energy corresponding to the peak of a resonance, the integral

$$\int_E^{E/\alpha_{res}} \Sigma_s^{res}(E') \frac{\psi(E')dE'}{E'(1 - \alpha_{res})}$$

on the right-hand side of (5.4.16) can be evaluated approximately by taking the limit as $\alpha_{res} \to 1$. The result is

$$\int_E^{E/\alpha_{res}} \Sigma_s^{res}(E') \frac{\psi(E')dE'}{E'(1 - \alpha_{res})} \approx \frac{\Sigma_s^{res}(E)\psi(E)}{E} \lim_{\alpha_{res} \to 1} \frac{(E/\alpha_{res}) - E}{1 - \alpha_{res}} = \Sigma_s^{res}(E)\psi(E), \tag{5.4.24}$$

where $\Sigma_s^{res}(E)$ is the scattering cross section of the resonance absorbing material. Equation (5.4.16) then becomes

$$[\Sigma_t(E) - \Sigma_s^{res}(E)]\psi(E) = \sum_{j \neq j_{res}} \int_E^{E/\alpha_j} \Sigma_s^j(E') \frac{\psi(E')dE'}{E'(1 - \alpha_j)} \quad \text{for } \Gamma_p \gg \left(\frac{1}{\alpha_{res}} - 1\right)E. \tag{5.4.25}$$

We shall call a resonance for which (5.4.24) is valid a *wide resonance* and shall refer to the approximation (5.4.24) as the *wide-resonance* (WR) *approximation*. This name has been used in the literature, but it is not the most common. Originally the approximation was referred to as the *narrow-resonance-infinite-absorber* (NRIA) *approximation* and subsequently as the *narrow-resonance-infinite-mass* (NRIM) *approximation*. WR is shorter.

Since "wide resonances" are still narrow with respect to the average energy lost in collisions with *other* materials in the mixture, all the theory developed for the NR approximation is immediately applicable. In particular (5.4.22) for $\psi(E)$ is again valid provided we exclude $\Sigma_s^{res}(E)$ from $\Sigma_t(E)$.

We thus have an approximate analytical method for determining group constants for energy groups containing widely spaced resonances that are narrow with respect to energy losses due to collisions with the moderator and either narrow or wide with respect to energy losses due to collisions with the resonance absorber itself. If these approximations are inadequate, an iteration procedure is possible. For example the first approximation for $\psi(E)$ as found from (5.4.22) can be used for $\psi(E')$ on the right-hand side of (5.4.16). Alternately (5.4.16) can be solved numerically throughout the energy ranges of interest. With all these possibilities we are justified in feeling that the determination of multigroup constants in the resolved-resonance region for a homogeneous mixture of materials is a tractable problem.

5.5 Determination of Group Parameters for Media Composed of a Number of Identical Cells

In almost all operating power reactors the fuel is not homogeneously mixed with moderator; it is, rather, present in the form of clad plates, pins, or rods. For fast reactors or thermal reactors fueled by highly enriched material the reason for this is primarily mechanical. There are advantages from the points of view of overall safety, shielding, and maintenance to containing the fuel in solid, clad elements. In order to do this and still provide a large amount of heat-transfer surface area and maintain an operating temperature throughout the fuel well below its melting point, elements are made in the form of plates or thin rods, their dimensions being determined as a compromise between high surface area, low center temperatures, mechanical integrity, and manufacturing economics. (The former two considerations favor thin elements; the latter two, thick ones.)

For slightly enriched and natural-uranium-fueled reactors nuclear considerations also enter into the selection of the fuel-element size. In fact they are overriding. The reason for this is that resonance capture in fertile material like U^{28} or Th^{02} will greatly reduce the k_{eff} of the system unless special precautions are taken to minimize such capture. The way to effect such a minimization is to lump the fuel so that those neutrons which acquire energies in a range near the peak of a given resonance during the slowing-down process will be more likely to scatter again before getting near a nucleus of resonance absorbing material. The lumping thus collects the fertile material into local regions so that it is protected from most of the neutrons that acquire energies in ranges near the resonance peaks. As mentioned in Section 3.5, this protection mechanism is called *spatial self-shielding*. Note that it operates in addition to the "energy self-shielding" discussed in Section 3.4.

To protect neutrons by spatial self-shielding from resonance capture one would like the lumps of resonance material to be as large as possible. However, when the fuel material is slightly enriched uranium, the U^{25} present in the fuel is also self-shielded, and in the thermal energy range neutron capture in the moderator relative to the U^{25} will be enhanced. (The thermal utilization will decrease.) Thus, for a fuel of given enrichment and a given type of moderator, there is an optimum size and spacing for the fuel elements.

Analyzing the effect on neutron reaction rates of lumping strongly absorbing material is an extremely complex problem. All the difficulties of dealing with cross sections that are violently changing functions of energy must be faced along with spatial-heterogeneity problems for which diffusion theory is not valid.

The analytical procedures for attacking this problem are based on the fact, established in Sections 5.3 and 5.4, that we *can* handle *homogeneous* mixtures containing resonance absorbing materials. Accordingly the key to analyzing heterogeneous lattices of fuel elements is to convert them mathematically into "equivalent homogeneous" regions by defining equivalent homogenized group cross sections. We shall see how this can be done in a formal manner and then make some simplifying assumptions that will permit a practical attack on the problem.

The development will be restricted to the energy range of epithermal resolved resonances ($1 \text{ eV} \leq E \leq 1000 \text{ eV}$) and to the simple geometrical situation of a large assembly consisting of a uniform lattice of unclad cylindrical fuel rods. The method can be extended to other energy ranges and to more realistic geometries (clad fuel). However the resolved-resonance range is the most common one to which it is applied, and the basic theoretical ideas are more readily understood if we omit the extra geometrical complexity associated with the presence of cladding. The cells to be analyzed are then as shown in Figure 5.2, where the first cell is part of a square lattice and the second is part of a hexagonal lattice. These cells might form part of a lattice of fuel rods such as the one shown in Figure 1.9.

Formal Analysis of the Problem

Ideally we are seeking "equivalent" group parameters $\nu\Sigma_{fg}$, D_g, and $\Sigma_{gg'}$ that are constant over the volume occupied by any given cell making up the reactor and that, when used in a group diffusion-theory calculation for the whole reactor, will reproduce the same *average* reaction rates over a given cell as would be determined if it were possible to solve the energy-dependent transport equation for the reactor with the heterogeneous geometrical characteristics of all cells treated explicitly.

It is important to note that, if we could actually find such exact parameters, their values would depend, not only on the exact composition of the cell, but also on its location within the reactor. Thus two identical cells of a given cluster, one next to a reflector

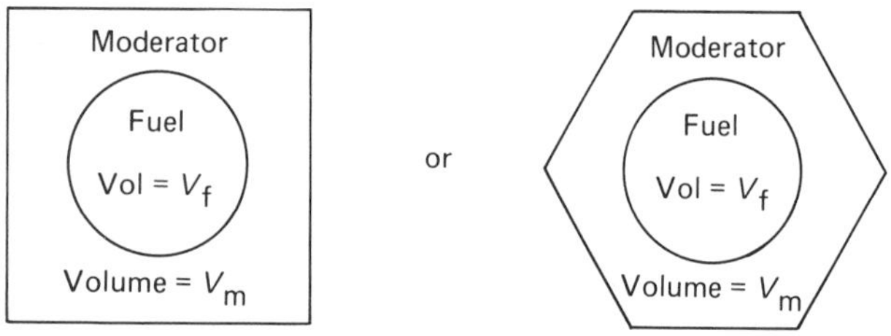

Figure 5.2 Fuel-element cells.

and the other in the interior of the cluster, would undoubtedly have different equivalent homogenized group parameters.

The first assumption we shall make in attacking the problem is to ignore this difference. In fact we shall attempt to find equivalent group parameters for a cell assumed to be part of an infinite, exactly critical array of identical cells, so that the component of the net current vector $\mathbf{J}(\mathbf{r}, E)$ perpendicular to the outer surface of the cell vanishes. In mathematical terms, if $\mathbf{n}$ is a unit vector normal to the surface, we assume that

$$\mathbf{n}\cdot\mathbf{J}(\mathbf{r}, E) = 0 \tag{5.5.1}$$

for all points $\mathbf{r}$ on the surface of the cell.

Once we make this assumption, a rigorously accurate definition of equivalent homogenized group constants follows from it. We simply integrate the fundamental, exact equation (4.4.2) over the volume V_c of the cell and split the energy dependence into groups. The result is

$$\begin{aligned}&\int_{V_c} dV \int_{\Delta E_g} dE\,\nabla\cdot\mathbf{J}(\mathbf{r}, E) + \int_{V_c} dV \int_{\Delta E_g} dE\,\Sigma_t(\mathbf{r}, E)\Phi(\mathbf{r}, E)\\ &= \frac{1}{\lambda}\sum_j \int_{\Delta E_g} dE\,\chi^j(E) \sum_{g'=1}^{G} \int_{\Delta E_{g'}} dE' \int_{V_c} dV\,\nu^j\Sigma_f^j(\mathbf{r}, E')\Phi(\mathbf{r}, E')\\ &\quad + \sum_{g'=1}^{G} \int_{\Delta E_g} dE \int_{\Delta E_{g'}} dE' \int_{V_c} dV\,\Sigma_s(\mathbf{r}, E'\to E)\Phi(\mathbf{r}, E').\end{aligned} \tag{5.5.2}$$

Now, by Gauss's Law and (5.5.1),

$$\int_{V_c} dV\,\nabla\cdot\mathbf{J}(\mathbf{r}, E) = \int_{S_c} \mathbf{J}(\mathbf{r}, E)\cdot\mathbf{n}\,dS = 0. \tag{5.5.3}$$

Then, if we define equivalent homogenized constants and fluxes for the cell by

$$\begin{aligned}\bar{\Sigma}_{tg} &\equiv \frac{\int_{V_c} dV \int_{\Delta E_g} dE\,\Sigma_t(\mathbf{r}, E)\Phi(\mathbf{r}, E)}{\int_{V_c} dV \int_{\Delta E_g} dE\,\Phi(\mathbf{r}, E)},\\ \overline{\nu\Sigma}_{fg}^{j} &\equiv \frac{\int_{V_c} dV \int_{\Delta E_g} dE\,\nu^j\Sigma_f^j(\mathbf{r}, E)\Phi(\mathbf{r}, E)}{\int_{V_c} dV \int_{\Delta E_g} dE\,\Phi(\mathbf{r}, E)},\\ \bar{\Sigma}_{gg'} &\equiv \frac{\int_{V_c} dV \int_{\Delta E_g} dE \int_{\Delta E_{g'}} dE'\,\Sigma_s(\mathbf{r}, E'\to E)\Phi(\mathbf{r}, E)}{\int_{V_c} dV \int_{\Delta E_{g'}} dE'\,\Phi(\mathbf{r}, E')},\\ \bar{\Phi}_g &\equiv \frac{1}{V_c}\int_{V_c} dV \int_{\Delta E_g} dE\,\Phi(\mathbf{r}, E),\end{aligned} \tag{5.5.4}$$

we may write homogenized group equations for the cell as

$$\bar{\Sigma}_{tg}\bar{\Phi}_g = \sum_{g'=1}^{G}\left[\frac{1}{\lambda}\sum_j \chi_g^j\overline{\nu\Sigma}_{fg'}^{j} + \bar{\Sigma}_{gg'}\right]\bar{\Phi}_{g'} \qquad (g = 1, 2, \ldots, G). \tag{5.5.5}$$

By comparing this equation with (5.5.2), we see that the solution for the homogenized group fluxes will lead to values of $\bar{\Phi}_g V_c$ identical with $\int_{V_c} dV \int_{\Delta E_g} dE \Phi(\mathbf{r}, E)$. Thus, insofar as (5.5.1) is a valid assumption, the terms $\bar{\Sigma}_{tg}\bar{\Phi}_g V_c$, etc., found by using $\bar{\Sigma}_{tg}$ as defined by (5.5.4) and $\bar{\Phi}_g$ as found by solving (5.5.5), will be identical with the exact reaction rates integrated over the cell and energy group ($\int_{V_c} dV \int_{\Delta E_g} dE \Sigma_t(\mathbf{r}, E)\Phi(\mathbf{r}, E)$, etc.).

This procedure is entirely formal in that we cannot solve (5.5.5) without knowing the parameters (5.5.4), and to find these exactly we must first determine $\Phi(\mathbf{r}, E)$ by performing a detailed solution of the transport equation. Nevertheless the formal procedure indicates clearly a goal: we must try to find good approximations to the $\bar{\Sigma}_{tg}$, $\overline{\nu\Sigma^j_{fg}}$, and $\bar{\Sigma}_{gg'}$. The closer we get to the ideal values (5.5.4), the more accurately will the solution $(\bar{\Phi}_1, \bar{\Phi}_2, \ldots, \bar{\Phi}_G)$ of (5.5.5) approximate the integrals $\int_{V_c} dV \int_{\Delta E_g} dE \Phi(\mathbf{r}, E)/V_c$ $(g = 1, 2, \ldots, G)$, and the closer will $\bar{\Sigma}_{tg}\bar{\Phi}_g V_c$, $\overline{\nu\Sigma^j_{fg}}\bar{\Phi}_g$, etc., be to the true integrated reaction rates over the cell volume V_c and the energy range ΔE.

Specialization to Two-Region Cells

So far the development has been general and limited only by the validity of approximation (5.5.1). If we now specialize to cells composed of two homogeneous regions, such as those sketched in Figure 5.2, we see that the expression for $\bar{\Sigma}_{tg}$ becomes

$$\bar{\Sigma}_{tg} = \frac{\int_{\Delta E_g} dE[\Sigma_{t(m)}(E)\int_{V_m}\Phi(\mathbf{r}, E)\,dV + \Sigma_{t(f)}(E)\int_{V_f}\Phi(\mathbf{r}, E)\,dV]}{\int_{\Delta E_g} dE[\int_{V_m}\Phi(\mathbf{r}, E)\,dV + \int_{V_f}\Phi(\mathbf{r}, E)dV]}, \tag{5.5.6}$$

where

$$\Sigma_{t(m)}(E) \equiv \Sigma_t(\mathbf{r}, E) \text{ for } \mathbf{r} \text{ in } V_m,$$
$$\Sigma_{t(f)}(E) \equiv \Sigma_t(\mathbf{r}, E) \text{ for } \mathbf{r} \text{ in } V_f,$$

with completely analogous expressions for the $\overline{\nu\Sigma^j_{fg}}$ and $\bar{\Sigma}_{gg'}$.

Hence, with spatially averaged fluxes $\bar{\Phi}^{(m)}(E)$ and $\bar{\Phi}^{(f)}(E)$, defined by

$$\bar{\Phi}^{(m)}(E) \equiv \frac{\int_{V_m}\Phi(\mathbf{r}, E)\,dV}{V_m},$$
$$\bar{\Phi}^{(f)}(E) \equiv \frac{\int_{V_f}\Phi(\mathbf{r}, E)\,dV}{V_f}, \tag{5.5.7}$$

(5.5.1) can be written

$$\bar{\Sigma}_{tg} = \frac{\int_{\Delta E_g}[\Sigma_{t(m)}(E)\bar{\Phi}^{(m)}(E)V_m + \Sigma_{t(f)}(E)\bar{\Phi}^{(f)}V_f]\,dE}{\int_{\Delta E_g}[\bar{\Phi}^{(m)}(E)V_m + \bar{\Phi}^{(f)}(E)V_f]\,dE}. \tag{5.5.8}$$

Thus the problem of determining rigorous values of the $\bar{\Sigma}_{tg}$, etc., requires only that we determine rigorous values of the spatially averaged fluxes $\bar{\Phi}^{(m)}(E)$ and $\bar{\Phi}^{(f)}(E)$. We shall attack that problem by first deriving an equation whose rigorous solution would yield $\overline{\Phi}^{(m)}(E)$ and $\overline{\Phi}^{(f)}(E)$, and then indicating how to go about getting an approximate solution to that equation.

We begin with (4.4.2), which, when specialized to the resolved-resonance region ($\chi^j(E) = 0$), becomes

$$\begin{aligned} \nabla\cdot\mathbf{J}(\mathbf{r}, E) + \Sigma_{t(m)}(E)\Phi(\mathbf{r}, E) &= \int_0^\infty \Sigma_{s(m)}(E' \to E)\Phi(\mathbf{r}, E')\,dE' \text{ for } \mathbf{r} \text{ in } V_m,\ E < E_1, \\ \nabla\cdot\mathbf{J}(\mathbf{r}, E) + \Sigma_{t(f)}(E)\overline{\Phi}(\mathbf{r}, E) &= \int_0^\infty \Sigma_{s(f)}(E' \to E)\Phi(\mathbf{r}, E)\,dE' \text{ for } \mathbf{r} \text{ in } V_f,\ E < E_1, \end{aligned} \tag{5.5.9}$$

where E_1 is an energy at the top of the resolved-resonance range.

Integration of the first of these equations over V_m and of the second over V_f and application of Gauss's theorem and the definitions in (5.5.7) yields

$$\begin{aligned} \int_{S_m} \mathbf{J}(\mathbf{r}, E)\cdot\mathbf{n}_m\,dS + \Sigma_{t(m)}(E)V_m\overline{\Phi}^{(m)}(E) &= V_m \int_0^\infty \Sigma_{s(m)}(E' \to E)\overline{\Phi}^{(m)}(E')\,dE', \\ \int_{S_f} \mathbf{J}(\mathbf{r}, E)\cdot\mathbf{n}_f\,dS + \Sigma_{t(f)}(E)V_f\overline{\Phi}^{(f)}(E) &= V_f \int_0^\infty \Sigma_{s(f)}(E' \to E)\overline{\Phi}^{(f)}(E')\,dE', \end{aligned} \tag{5.5.10}$$

where $\mathbf{n}_m$ and $\mathbf{n}_f$ are unit vectors drawn *outward*, perpendicular respectively to the surfaces S_m and S_f of V_m (the moderator) and V_f (the fuel).

Escape Probabilities

The moderator volume V_m has two surfaces, one the outer surface of the cell and the other the common surface S_f between the moderator and the fuel rod. Because of the cell boundary condition (5.5.1), the portion of $\int_{S_m}\mathbf{J}(\mathbf{r}, E)\cdot\mathbf{n}_m dS$ over the outer surface of the cell vanishes. Then, since $\mathbf{n}_m$ at a given point on the inner surface of V_m equals $-\mathbf{n}_f$ on the outer surface of V_f, we have

$$\int_{S_m} \mathbf{J}(\mathbf{r}, E)\cdot\mathbf{n}_m\,dS = -\int_{S_f} \mathbf{J}(\mathbf{r}, E)\cdot\mathbf{n}_f\,dS. \tag{5.5.11}$$

Physically this equation states that, there being no net leakage across the outer boundary of the cell, the net leakage out of V_m is equal and opposite to the net leakage out of V_f.

If we could express these two leakages in terms of $\overline{\Phi}^{(m)}(E)$ and $\overline{\Phi}^{(f)}(E)$, we would be able to convert (5.5.10) into a pair of integral equations whose solution would give us $\overline{\Phi}^{(m)}(E)$ and $\overline{\Phi}^{(f)}(E)$; this would permit us to find the rigorous homogenized cell constants by applying formulas such as (5.5.8). To accomplish this goal we introduce the two quantities $P_m(E)$ and $P_f(E)$, which we shall later show are the probabilities that "average" neutrons of energy E will escape from V_m and V_f respectively. (What is meant by "average" and "escape" will become clear shortly.)

To express the net current $\int_{S_f} \mathbf{J}(\mathbf{r}, E)\cdot\mathbf{n}_f dS$ in terms of such leakage probabilities we rewrite (5.5.9) in a different, entirely equivalent form. Specifically we think of $\Phi(\mathbf{r}, E)$ and $\mathbf{J}(\mathbf{r}, E)$ on the left-hand side of (5.5.9)—and *only* on the left-hand side—as split into two parts

$$\begin{aligned}\Phi(\mathbf{r}, E) &= \Phi_f(\mathbf{r}, E) + \Phi_m(\mathbf{r}, E),\\ \mathbf{J}(\mathbf{r}, E) &= \mathbf{J}_f(\mathbf{r}, E) + \mathbf{J}_m(\mathbf{r}, E),\end{aligned} \tag{5.5.12}$$

where the functions $\Phi_f(\mathbf{r}, E)$ and $\mathbf{J}_f(\mathbf{r}, E)$ are the flux and net current densities *throughout the lattice* due to sources $\int_0^\infty \Sigma_{s(f)}(E' \to E)\Phi(\mathbf{r}, E')dE'$ in the fuel rods alone, while $\Phi_m(\mathbf{r}, E)$ and $\mathbf{J}_m(\mathbf{r}, E)$ are the flux and net current densities throughout the lattice due to sources $\int_0^\infty \Sigma_{s(m)}(E' \to E)\Phi(\mathbf{r}, E')dE'$ in the moderator regions alone. These partial fluxes and currents thus obey the equations

$$\begin{aligned}\nabla\cdot\mathbf{J}_f(\mathbf{r}, E) + \Sigma_t(\mathbf{r}, E)\Phi_f(\mathbf{r}, E) &= \int_0^\infty dE'\,\Sigma_{s(f)}(E' \to E)\Phi(\mathbf{r}, E')H^{(f)},\\ \nabla\cdot\mathbf{J}_m(\mathbf{r}, E) + \Sigma_t(\mathbf{r}, E)\Phi_m(\mathbf{r}, E) &= \int_0^\infty dE'\Sigma_{s(m)}(E' \to E)\Phi(\mathbf{r}, E)H^{(m)},\end{aligned} \tag{5.5.13}$$

where $H^{(f)}$ is a function which is unity in the fuel rod and zero in the moderator and $H^{(m)} = 1 - H^{(f)}$.

Note that $\Phi(\mathbf{r}, E')$ in the integrals on the right-hand side of (5.5.13) is the total flux $\Phi_f + \Phi_m$. Note also that adding the equations in (5.5.13) together and applying (5.5.12) yields (5.5.9).

We now integrate (5.5.13) first over V_f and then over V_m and apply Gauss's Law and the definitions in (5.5.7). The result is

$$\begin{aligned}&\int_{S_f} \mathbf{J}_f(\mathbf{r}, E)\cdot\mathbf{n}_f\, dS + \Sigma_{t(f)}(E)\overline{\Phi}_f^{(f)}(E)V_f = V_f\int_0^\infty dE'\,\Sigma_{s(f)}(E' \to E)\overline{\Phi}^{(f)}(E'),\\ &\int_{S_f} \mathbf{J}_m(\mathbf{r}, E)\cdot\mathbf{n}_f\, dS + \Sigma_{t(f)}(E)\overline{\Phi}_m^{(f)}(E)V_f = 0,\\ &\int_{S_m} \mathbf{J}_f(\mathbf{r}, E)\cdot\mathbf{n}_m\, dS + \Sigma_{t(m)}(E)\overline{\Phi}_f^{(m)}(E)V_m = 0,\\ &\int_{S_m} \mathbf{J}_m(\mathbf{r}, E)\cdot\mathbf{n}_m\, dS + \Sigma_{t(m)}(E)\overline{\Phi}_m^{(m)}(E)V_m = V_m\int_0^\infty dE'\,\Sigma_{s(m)}(E' \to E)\overline{\Phi}^{(m)}(E'),\end{aligned} \tag{5.5.14}$$

where $\overline{\Phi}_m^{(f)}$, for example, is the average flux in the fuel due to the source in the moderator.

Since the fictitious physical situation that (5.5.13) describes still has lattice symmetry, the integrals over S_m are zero for that part of S_m which is an interface between cells and, hence, are finite only over the moderator-fuel interface S_f. Thus, since $\mathbf{n}_m = -\mathbf{n}_f$ on S_f, the last two equations in (5.5.14) become

$$
\begin{aligned}
&-\int_{S_f} \mathbf{J}_f(\mathbf{r}, E)\cdot\mathbf{n}_f\, dS + \Sigma_{t(m)}(E)\overline{\Phi}_f^{(m)}(E)V_m = 0,\\
&-\int_{S_f} \mathbf{J}_m(\mathbf{r}, E)\cdot\mathbf{n}_f\, dS + \Sigma_{t(m)}(E)\overline{\Phi}_m^{(m)}(E)V_m = V_m\int_0^\infty dE'\Sigma_{s(m)}(E'\to E)\overline{\Phi}^{(m)}(E').
\end{aligned}
\tag{5.5.15}
$$

We now define escape probabilities $P_f(E)$ and $P_m(E)$ by

$$
\begin{aligned}
&\int_{S_f} \mathbf{J}_f(\mathbf{r}, E)\cdot\mathbf{n}_f\, dS \equiv P_f(E)V_f\int_0^\infty dE'\,\Sigma_{s(f)}(E'\to E)\overline{\Phi}^{(f)}(E'),\\
&\int_{S_m} \mathbf{J}_m(\mathbf{r}, E)\cdot\mathbf{n}_m dS = -\int_{S_f} \mathbf{J}_m(\mathbf{r}, E)\cdot\mathbf{n}_f dS \equiv P_m(E)V_m\int_0^\infty dE'\,\Sigma_{s(m)}(E'\to E)\overline{\Phi}^{(m)}(E').
\end{aligned}
\tag{5.5.16}
$$

In view of the second equation in (5.5.14) and the first in (5.5.15), entirely equivalent definitions of $P_f(E)$ and $P_m(E)$ are

$$
\begin{aligned}
P_f(E) &\equiv \frac{\Sigma_{t(m)}(E)V_m\overline{\Phi}_f^{(m)}(E)}{V_f\int_0^\infty dE'\,\Sigma_{s(f)}(E'\to E)\overline{\Phi}^{(f)}(E')},\\
P_m(E) &\equiv \frac{\Sigma_{t(f)}(E)V_f\overline{\Phi}_m^{(f)}(E)}{V_m\int_0^\infty dE'\,\Sigma_{s(m)}(E'\to E)\overline{\Phi}^{(m)}(E')}.
\end{aligned}
\tag{5.5.17}
$$

These latter definitions give us direct physical interpretations of $P_f(E)$ and $P_m(E)$: $P_f(E)$ is the probability that neutrons scattered into the energy interval dE within the fuel rod will leave that energy interval by interactions taking place in the moderator; $P_m(E)$ is the probability that neutrons born in dE in the moderator will leave dE by interacting with nuclei in a fuel rod. They are thus "escape probabilities" for "average" neutrons in the fuel and moderator. Note that, in general, the sum of P_f and P_m is *not* unity; the two probabilities pertain to different initial samples of neutrons.

Subtracting the second equation in (5.5.16) from the first gives us the expression we have been seeking:

$$
\begin{aligned}
&\int_{S_f} [\mathbf{J}_f(\mathbf{r}, E) + \mathbf{J}_m(\mathbf{r}, E)]\cdot\mathbf{n}_f\, dS = \int_{S_f} \mathbf{J}(\mathbf{r}, E)\cdot\mathbf{n}_f\, dS\\
&= P_f(E)V_f\int_0^\infty dE'\,\Sigma_{s(f)}(E'\to E)\overline{\Phi}^{(f)}(E') - P_m(E)V_m\int_0^\infty dE'\,\Sigma_{s(m)}(E'\to E)\overline{\Phi}^{(m)}(E').
\end{aligned}
\tag{5.5.18}
$$

Finally, inserting this result into (5.5.10) and applying (5.5.11) leads in a formally rigorous fashion to two coupled integral equations for $\overline{\Phi}^{(m)}(E)$ and $\overline{\Phi}^{(f)}(E)$:

$$\begin{aligned}
\Sigma_{t(m)}(E)\overline{\Phi}^{(m)}(E) &= [1 - P_m(E)]\int_0^\infty \Sigma_{s(m)}(E' \to E)\overline{\Phi}^{(m)}(E')\,dE' \\
&\quad + P_f(E)\frac{V_f}{V_m}\int_0^\infty \Sigma_{s(f)}(E' \to E)\overline{\Phi}^{(f)}(E')\,dE', \\
\Sigma_{t(f)}(E)\overline{\Phi}^{(f)}(E) &= [1 - P_f(E)]\int_0^\infty \Sigma_{s(f)}(E' \to E)\overline{\Phi}^{(f)}(E')dE' \\
&\quad + P_m(E)\frac{V_m}{V_f}\int_0^\infty \Sigma_{s(m)}(E' \to E)\overline{\Phi}^{(m)}(E')\,dE'.
\end{aligned} \tag{5.5.19}$$

There is a straightforward physical interpretation of these equations: if the first is multiplied by $V_m dE$, it states that the total interaction rate with moderator of neutrons having energies in the range dE equals the sum of (1) their birth rate in the moderator multiplied by the probability $1 - P_m(E)$ that neutrons born in the moderator will *not* have their next interaction in fuel (and, therefore, *will* have it in the moderator) and (2) their birth rate in fuel multiplied by the probability $P_f(E)$ that neutrons born in fuel *will* have their next interaction in moderator. The second equation in (5.5.19) has an analogous interpretation.

As we shall see shortly, the solution of (5.5.19), given that $P_m(E)$ and $P_f(E)$ are known, is relatively straightforward. Thus we have lumped all the difficulty in determining equivalent homogenized group constants for a cell into the problem of finding the escape probabilities $P_m(E)$ and $P_f(E)$. If these are known, solution of (5.5.19) will yield $\overline{\Phi}^{(m)}(E)$ and $\overline{\Phi}^{(f)}(E)$, whose substitution into (5.5.8) (and the analogous equations for the $\overline{\nu\Sigma^j_{fg}}$, $\overline{\Sigma}_{gg'}$, and $\overline{\Sigma}_{trg}$) will yield numerical values for the homogenized group parameters.

The only formal assumption made in the development so far is that there is no net current across the outer boundary of the cell. For a cell in the interior of a cluster this assumption is likely to lead to negligible errors in the group parameters. However, for a cell at an interface between two grossly different material media (for example at an interface between the core and the reflector), some correction to the zero-leakage cell parameters is probably advisable if a high degree of accuracy is desired. One simple, approximate way to make a correction is to alter the escape probabilities for such a cell. We shall indicate how this can be done when we consider methods for finding these probabilities.

To complete the discussion of how to determine equivalent homogenized energy-group parameters for a fuel-moderator cell we must develop practical methods for finding the escape probabilities and for solving (5.5.19) once these are known. We shall examine the second, much simpler problem first.

Collision Densities

Let us assume for the time being that we know the escape probabilities $P_{\mathrm{m}}(E)$ and $P_{\mathrm{f}}(E)$ and wish to solve (5.5.19) for the average flux densities $\bar{\Phi}^{(\mathrm{m})}(E)$ and $\bar{\Phi}^{(\mathrm{f})}(E)$. We expect that these flux densities—particularly $\bar{\Phi}^{(\mathrm{f})}(E)$—will dip suddenly and significantly near energies corresponding to a resonance. For this reason it will be numerically simpler to reformulate (5.5.19) in terms of two related quantities, the *collision densities* $F^{(\mathrm{m})}(E)$ and $F^{(\mathrm{f})}(E)$ defined by

$$F^{(i)}(E) \equiv \Sigma_{\mathrm{t}(i)}(E)\bar{\Phi}^{(i)}(E) \qquad (i = \mathrm{m}, \mathrm{f}). \tag{5.5.20}$$

Physically $F^{(i)}(E)$ is the average total number of interactions per unit volume per unit energy per second in region (i). Since $\bar{\Phi}^{(i)}(E)$ will dip when $\Sigma_{\mathrm{t}(i)}(E)$ peaks, we expect $F^{(i)}(E)$ to be a much smoother function of energy than either of its factors. Examination of (5.4.18) for the homogeneous mixture confirms this expectation. (See also Figure 5.1.)

Carrying out this switch from flux densities to collision densities in (5.5.19) and writing the scattering integrals as sums over all isotopes j yields

$$\begin{aligned} F^{(\mathrm{m})}(E) &= [1 - P_{\mathrm{m}}(E)] \sum_j \int_E^{E/\alpha_j} \frac{\Sigma^j_{\mathrm{s(m)}}(E')F^{(\mathrm{m})}(E')\,dE'}{E'(1-\alpha_j)\Sigma_{\mathrm{t(m)}}(E')} \\ &\quad + P_{\mathrm{f}}(E)\frac{V_{\mathrm{f}}}{V_{\mathrm{m}}} \sum_j \int_E^{E/\alpha_j} \frac{\Sigma_{\mathrm{s(f)}}(E')F^{(\mathrm{f})}(E')\,dE'}{E'(1-\alpha_j)\Sigma_{\mathrm{t(f)}}(E')}, \\ F^{(\mathrm{f})}(E) &= [1 - P_{\mathrm{f}}(E)] \sum_j \int_E^{E/\alpha_j} \frac{\Sigma^j_{\mathrm{s(f)}}(E')F^{(\mathrm{f})}(E')\,dE'}{E'(1-\alpha_j)\Sigma_{\mathrm{t(f)}}(E')} \\ &\quad + P_{\mathrm{m}}(E)\frac{V_{\mathrm{m}}}{V_{\mathrm{f}}} \sum_j \int_E^{E/\alpha_j} \frac{\Sigma^j_{\mathrm{s(m)}}(E')F^{(\mathrm{m})}(E')\,dE'}{E'(1-\alpha_j)\Sigma_{\mathrm{t(m)}}(E')}. \end{aligned} \tag{5.5.21}$$

There are many ways to solve these two coupled integral equations. We shall sketch the details of two of them and indicate the basic idea behind a third class of approximations.

Multigroup Procedures for Finding Collision Probabilities

Since the $F^{(i)}(E)$ are much smoother functions than the $\bar{\Phi}^{(i)}(E)$, it is practical to solve (5.5.21) by direct numerical integration. All that is needed is some expression for the $F^{(i)}(E)$ at the top of the resolved-resonance range. The assumption most commonly made to obtain such an expression is that, in that energy range, there being no sharp resonances and little absorption, the cell is isotopically homogeneous. This is the same as assuming that the flux is spatially flat across the cell, i.e., that $\bar{\Phi}^{(\mathrm{m})}(E) = \bar{\Phi}^{(\mathrm{f})}(E) = \bar{\Phi}(\mathbf{r}, E)$. As a result the energy dependence of both $\bar{\Phi}^{(\mathrm{m})}(E)$ and $\bar{\Phi}^{(\mathrm{f})}(E)$ is given by the expression (5.4.14) for a homogeneous, nonabsorbing medium. Thus, for the case at hand, since the $\Sigma^j_{\mathrm{s}(i)}(E)$ are cross sections for the isotope j in region (i), we have for the energy range immediately above E_1, the top of the resolved-resonance region,

$$\bar{\Phi}^{(m)}(E) = \bar{\Phi}^{(f)}(E) \approx \frac{q(V_m + V_f)}{E[V_m\xi_m\Sigma_{s(m)} + V_f\xi_f\Sigma_{s(f)}]}, \tag{5.5.22}$$

so that (5.5.20) yields, for $E \geq E_1$,

$$F^{(m)}(E) = \frac{q\Sigma_{t(m)}(V_m + V_f)}{E[V_m\xi_m\Sigma_{s(m)} + V_f\xi_f\Sigma_{s(f)}]},$$
$$F^{(f)}(E) = \frac{q\Sigma_{t(f)}(V_m + V_f)}{E[V_m\xi_m\Sigma_{s(m)} + V_f\xi_f\Sigma_{s(f)}]}. \tag{5.5.23}$$

With these starting values for the $F^{(i)}(E)$, (5.5.21) may be broken into many energy groups and integrated numerically to find values of the $F^{(i)}(E)$ at energies below E_1. This procedure is straightforward but a bit expensive.

Use of the NR and WR Approximations to Find Collision Densities

By making the narrow- and wide-resonance approximations, we can reduce the equations in (5.5.21) to a simple algebraic set. To effect such a reduction for a given energy range ΔE_g we simply distinguish the nuclei having resonances with a practical width less than $(E/\alpha_j) - E$ from those having a practical width greater than $(E/\alpha_j) - E$. For the former class we make the narrow-resonance approximation that the $F^{(i)}(E')$ $(\equiv \Sigma_{t(i)}(E')\bar{\Phi}^{(i)}(E'))$ in the integrands on the right-hand side of (5.5.21) may be replaced by $C_g\Sigma_{t(i)}(E')/E'$, where C_g is some constant; for the latter class we assume $\alpha_j \to 1$ (see (5.4.24)). The NR approximation is based on the assumption that the flux density between the widely spaced resonances recovers the asymptotic $1/E$ shape given, in both fuel and moderator, by (5.5.22) with $C_g = q(V_m + V_f)/[V_m\xi_m^p\Sigma_{s(m)} + V_f\xi_f^p\Sigma_{s(f)}]$ (see (5.4.18)). Strictly speaking the normalization constant C_g should be reduced from one resonance to the next to reflect the dropping of q, the slowing-down density, due to neutron capture in the resonance (see Figure 5.1). However, if ΔE_g contains only a few resonances, assuming a single value of q, and hence of C_g, for the whole group leads to very little error.

If we make these approximations and split the sums on the right-hand side of (5.5.21) into sums over narrow resonances (with summation index n) and sums over wide resonances (with summation index w), and if we further assume that the moderating materials in region (m) will always be either potential scatterers or NR scatterers, we obtain

$$\begin{aligned} F^{(m)}(E) &\approx [1 - P_m(E)]\sum_n \int_E^{E/\alpha_n} \frac{\Sigma^n_{s(m)}C_g\,dE'}{(E')^2(1-\alpha_n)} + P_f(E)\frac{V_f}{V_m}\Bigg[\sum_n \int_E^{E/\alpha_n} \frac{\Sigma^n_{s(f)}C_g\,dE'}{(E')^2(1-\alpha_n)} \\ &\quad + \lim_{\alpha_w \to 1}\sum_w \int_E^{E/\alpha_w} \frac{\Sigma^w_{s(f)}(E')F^{(f)}(E')\,dE'}{E'(1-\alpha_w)\Sigma_{t(f)}(E')}\Bigg] \\ &= [1 - P_m(E)]\frac{\Sigma^p_{s(m)}C_g}{E} + P_f(E)\frac{V_f}{V_m}\left[\frac{\Sigma^p_{s(f)}C_g}{E} + \frac{\Sigma^{res}_{s(f)}(E)}{\Sigma_{t(f)}(E)}F^{(f)}(E)\right], \end{aligned} \tag{5.5.24}$$

where $\Sigma_{s(m)}^{p}$ is the total *potential* scattering cross section of the narrow-resonance material in the moderator region ($\Sigma_{s(m)}^{p} = \sum_n \Sigma_{s(m)}^{n}$, assumed constant in energy) and $\Sigma_{s(f)}^{res}(E)$ is the *total* scattering cross section (potential *plus* resonance) of the wide-resonance material in the fuel ($\Sigma_{s(f)}^{res}(E) = \sum_w \Sigma_{s(f)}^{w}(E)$).

In a similar fashion we obtain from the second equation in (5.5.21) the approximate result

$$F^{(f)}(E) \approx [1 - P_f(E)]\left[\frac{\Sigma_{s(f)}^{p} C_g}{E} + \frac{\Sigma_{s(f)}^{res}(E)}{\Sigma_{t(f)}(E)} F^{(f)}(E)\right] + P_m(E)\frac{V_m}{V_f}\frac{\Sigma_{s(m)}^{p} C_g}{E} \tag{5.5.25}$$

or

$$F^{(f)}(E) \approx \frac{[1 - P_f(E)]\Sigma_{s(f)}^{p} + P_m(E)(V_m/V_f)\Sigma_{s(m)}^{p}}{1 - \{[1 - P_f(E)]\Sigma_{s(f)}^{res}/\Sigma_{t(f)}(E)\}}\frac{C_g}{E}. \tag{5.5.26}$$

Substitution of this expression into (5.5.24) yields the corresponding formula for $F^{(m)}(E)$. Equation (5.5.8), along with the definition of the collision density (5.5.20), then permits us to compute $\bar{\Sigma}_{tg}$ and, in an analogous way, the other group parameters of (5.5.4). Note that the normalization constant C_g cancels in the expressions for the group constants.

One characteristic of the method just described requires care: some resonances of a given isotope may be narrow while others may be wide. For example at low energies the energy interval $(E/\alpha_{28}) - E$ for U^{28} tends to be small compared to the practical width of its absorption resonances. Hence resonances in this range are "wide." However at high energies the situation is reversed, and all U^{28} resonances are narrow. Thus, for some energy ranges ΔE_g, the fuel material may have no wide resonances, whereas for other ranges all its resonances may be wide. This complication is readily taken care of if we interpret $\Sigma_{s(f)}^{res}(E)$ in (5.5.25) as being the total scattering cross section for those isotopes that *have* wide resonances at energy E. If all resonances are narrow, then $\Sigma_{s(f)}^{res}(E)$ will be zero and (5.5.26) will be simpler.

Iterative Methods for Determining Collision Densities

If the approximations implicit in (5.5.24) and (5.5.26) are too gross, (5.5.21) may be solved in an iterative fashion based on simplifying only *some* of the scattering-in integrals. For example the narrow-resonance approximation may be an excellent one in the moderator. If so, we may use it to simplify the last term of the second equation in (5.5.21) and then solve that second equation for $F^{(f)}(E)$ by some numerical method. An improved value of $F^{(m)}(E)$ can then be obtained from the first equation.

5.6 Methods for Determining Escape Probabilities

The formal procedure we have developed up to this point permits a very accurate determination of the equivalent homogenized group parameters (5.5.4), provided the $P_i(E)$

are known. The only essential approximation is that the net neutron current normal to the surface of the cell is zero. We have avoided having to find values of $\mathbf{J}(\mathbf{r}, E)$ and $\Phi(\mathbf{r}, E)$ that satisfy (5.5.9) and have reduced the computational problem to that of solving equations in energy alone, such as (5.5.19). Moreover we have seen that there are simple approximate schemes for solving (5.5.19) which are expected to be quite accurate for most cases of interest. Of course all this formal manipulation will be of little value unless we can obtain accurate expressions for the $P_i(E)$ without having to find $\mathbf{J}(\mathbf{r}, E)$ and $\Phi(\mathbf{r}, E)$. We now examine how this can be done.

Equations (5.5.16) and (5.5.17) give equivalent mathematical definitions of $P_{\mathrm{f}}(E)$ and $P_{\mathrm{m}}(E)$. We used the former definition in the analysis leading to (5.5.19). However the latter will be more convenient for investigating practical ways to compute the $P_i(E)$. Moreover the essential approximation we are going to make will be clearer if we replace the $\bar{\Phi}^{(i)}(E)$, $\bar{\Phi}_{\mathrm{f}}^{(i)}(E)$, and $\bar{\Phi}_{\mathrm{m}}^{(i)}(E)$ in (5.5.17) by their definitions. Doing so, we obtain

$$P_{\mathrm{f}}(E) \equiv \frac{\int_{V_{\mathrm{m}}} \Sigma_{\mathrm{t(m)}}(E)\Phi_{\mathrm{f}}(\mathbf{r}, E)\, dV}{\int_0^\infty dE' \int_{V_{\mathrm{f}}} \Sigma_{\mathrm{s(f)}}(E' \to E)\Phi(\mathbf{r}, E')\, dV},$$
$$P_{\mathrm{m}}(E) \equiv \frac{\int_{V_{\mathrm{f}}} \Sigma_{\mathrm{t(f)}}(E)\Phi_{\mathrm{m}}(\mathbf{r}, E)\, dV}{\int_0^\infty dE' \int_{V_{\mathrm{m}}} \Sigma_{\mathrm{s(m)}}(E' \to E)\Phi(\mathbf{r}, E')\, dV}. \tag{5.6.1}$$

Recall that $\Phi_{\mathrm{f}}(\mathbf{r}, E)$, the solution of the first equation in (5.5.13), is a flux, defined for *all* points (in both fuel and moderator), but arising from sources $\int_0^\infty dE'\, \Sigma_{\mathrm{s(f)}}(E' \to E)$ $\Phi(\mathbf{r}, E')H^{(\mathrm{f})}$ in the fuel rods alone, these sources being due to scattering from higher energies E' of neutrons belonging to the total flux $\Phi(\mathbf{r}, E')$ (the *sum* of $\Phi_{\mathrm{f}}(\mathbf{r}, E')$ and $\Phi_{\mathrm{m}}(\mathbf{r}, E')$). Similarly the flux $\Phi_{\mathrm{m}}(\mathbf{r}, E)$ is due to sources in the moderator alone, depending again on the total flux $\Phi(\mathbf{r}, E)$. To determine $\Phi_{\mathrm{f}}(\mathbf{r}, E)$ and $\Phi_{\mathrm{m}}(\mathbf{r}, E)$ exactly would require that we solve two transport-type equations at all energies and at all positions throughout the cell. In order to avoid this extremely difficult problem we make an essential assumption: we assume that the sources $\int_0^\infty dE'\, \Sigma_{\mathrm{s(f)}}(E' \to E)\Phi(\mathbf{r}, E')H^{(\mathrm{f})}$ and $\int_0^\infty dE'\, \Sigma_{\mathrm{s(m)}}(E' \to E)$ $\Phi(\mathbf{r}, E')H^{(\mathrm{m})}$, where nonzero, are spatially flat, i.e., independent of $\mathbf{r}$. Mathematically this assumption can be expressed by

$$\int_0^\infty dE'\, \Sigma_{\mathrm{s(f)}}(E' \to E)\Phi(\mathbf{r}, E')H^{(\mathrm{f})} = S_{\mathrm{f}}(E)H^{(\mathrm{f})},$$
$$\int_0^\infty dE'\, \Sigma_{\mathrm{s(m)}}(E' \to E)\Phi(\mathbf{r}, E')H^{(\mathrm{m})} = S_{\mathrm{m}}(E)H^{(\mathrm{m})}, \tag{5.6.2}$$

where S_{f} and S_{m} are spatially constant. Then (5.5.13) becomes

$$\nabla \cdot \mathbf{J}_{\mathrm{f}}(\mathbf{r}, E) + \Sigma_{\mathrm{t}}(\mathbf{r}, E)\Phi_{\mathrm{f}}(\mathbf{r}, E) = S_{\mathrm{f}}(E)H^{(\mathrm{f})},$$
$$\nabla \cdot \mathbf{J}_{\mathrm{m}}(\mathbf{r}, E) + \Sigma_{\mathrm{t}}(\mathbf{r}, E)\Phi_{\mathrm{m}}(\mathbf{r}, E) = S_{\mathrm{m}}(E)H^{(\mathrm{m})}, \tag{5.6.3}$$

where $\mathbf{J} \cdot \mathbf{n} = 0$ on the outer surface of the cell.

Notice that the magnitudes of $S_f(E)$ and $S_m(E)$ are actually of no concern to us. We are interested in the $P_i(E)$, and (5.6.1) and (5.6.2) show that these are defined by ratios $\Phi_f(E)/S_f(E)$ and $\Phi_m(E)/S_m(E)$, which can be seen by (5.6.3) to be independent of $S_f(E)$ and $S_m(E)$. As a result we need make no assumption about the energy dependence of $\Phi(\mathbf{r}, E')$ or the $\Sigma_{s(i)}(E' \to E)$ in the source integrals (5.6.2). The assumption (5.6.2) concerning these integrals decouples the energy dependence of the partial fluxes. Thus, rather than having to solve a set of equations (5.5.13) in which the energy and space behavior must be accounted for simultaneously, we need only solve a set of individual spatial equations (5.6.3) at different energies. This latter problem is still very difficult. It is, however, simpler than the former one. Moreover assuming the sources are flat will permit us to develop a completely general relationship between $P_f(E)$ and $P_m(E)$, as well as some analytical expressions for them that will lead to simple algebraic formulas adequate for most design purposes.

The justification for assuming that the sources in the fuel rods and moderator are flat is the same one we used in deriving the NR approximation: most of the neutrons appearing in an energy range dE as result of scattering from higher energies come from energy ranges corresponding to valleys between the resonance peaks. In such regions the scattering cross section is small (~ 10 b) and the absorption cross section is essentially zero; the total mean free path is thus a few centimeters. As a result the flux shape, even though it dips severely at energies corresponding to absorption resonances, is fairly flat over most of the energy range of the integrands in (5.5.13). Accordingly we assume that the integrals themselves are independent of $\mathbf{r}$.

A Reciprocity Relationship between the Escape Probabilities $P_f(E)$ and $P_m(E)$

Equations (5.6.3) concern neutrons in an infinitesimal energy interval about dE. Any neutron born into such an interval will travel in a straight line until it is removed by an absorption or scattering interaction. A useful relationship follows from this result. Consider a beam source of neutrons $S(\mathbf{r}_f, \mathbf{\Omega}, E)dV_f\, dE\, d\Omega$ in dV_f about point $\mathbf{r}_f$ traveling in a cone of directions $d\Omega$ about the unit vector $\mathbf{\Omega}$ pointing from $\mathbf{r}_f$ to another point $\mathbf{r}_m$ in dV_m. The attenuation of that beam between $\mathbf{r}_f$ and $\mathbf{r}_m$ is $\exp(-\int_{\mathbf{r}_f}^{\mathbf{r}_m} \Sigma_t(\mathbf{r}, E)dl)$, where the integration is along the straight line connecting $\mathbf{r}_f$ and $\mathbf{r}_m$, and $\Sigma_t(\mathbf{r}, E)$ may vary with position $\mathbf{r}$ in a completely general way. Clearly this attenuation factor will be the same as that associated with a source $S(\mathbf{r}_m, -\mathbf{\Omega}, E)dV_m\, dE\, d\Omega$ starting in dV_m and traveling in the $-\mathbf{\Omega}$ direction. We can express this reciprocal relationship mathematically by the following definition:

Flux in $dV_m\, dE$ due to an isotropic source producing one neutron per unit volume per second at $\mathbf{r}_f$ and E

$$\equiv \Phi(\mathbf{r}_m; \mathbf{r}_f; E)dV_m\, dE. \tag{5.6.4}$$

Because scattering removes neutrons from dE, only those neutrons in the isotropic source headed initially in the direction $\mathbf{r}_m - \mathbf{r}_f$ have any chance of contributing to $\Phi(\mathbf{r}_m; \mathbf{r}_f; E)$. However this situation is no way invalidates the definition.

If we consider a volume dV_f surrounding $\mathbf{r}_f$, we have from (5.6.4) that the flux in $dV_m\, dE$ due to the unit source in that volume is $\Phi(\mathbf{r}_m; \mathbf{r}_f; E)dV_m\, dE\, dV_f$. Conversely the flux in $dV_f\, dE$ due to the source in dV_m is $\Phi(\mathbf{r}_f; \mathbf{r}_m; E)dV_f\, dE\, dV_m$. Now, if we take the two volumes dV_f and dV_m to be equal, there will be just as many source neutrons leaving dV_f and contributing to the flux in dV_m as there will be source neutrons leaving dV_m and contributing to the flux in dV_f. Since the attenuations of the two beams are identical, we must have

$$\Phi(\mathbf{r}_m; \mathbf{r}_f; E)dV_m\, dE\, dV_f = \Phi(\mathbf{r}_f; \mathbf{r}_m; E)dV_f\, dE\, dV_m, \qquad dV_f = dV_m,$$

so that

$$\Phi(\mathbf{r}_m; \mathbf{r}_f; E) = \Phi(\mathbf{r}_f; \mathbf{r}_m; E). \tag{5.6.5}$$

This relationship is valid for any geometrical configuration of materials between the two points $\mathbf{r}_f$ and $\mathbf{r}_m$. Moreover it can be shown to be true for the idealized situation in which neutron scattering doesn't change the energy of the scattered neutrons and, hence, in which neutrons starting in *any* direction from dV_m have a finite probability of contributing to the flux in dV_f. In this more general form, (5.6.5) is known as the *reciprocity theorem.* For the even more general case of scattering with energy exchange, an expression analogous to (5.6.5), but with the source neutrons at an energy different from E, can be defined (see Chapter 8). The corresponding reciprocity theorem, however, involves a quantity called *neutron importance.* We shall have occasion to deal with neutron importance in connection with reactor kinetics, but we shall not pursue the matter further at present.

Before we apply (5.6.5) to obtain a relationship between $P_f(E)$ and $P_m(E)$, we must examine more closely the physical significance of these escape probabilities. To do so we incorporate the flat-source approximation into the definitions (5.6.1) by substituting (5.6.2) for the denominator:

$$P_f(E) = \frac{\int_{V_m} \Sigma_{t(m)}(E)\Phi_f(\mathbf{r}, E)\, dV}{\int_{V_f} S_f H^{(f)}\, dV},$$
$$P_m(E) = \frac{\int_{V_f} \Sigma_{t(f)}(E)\Phi_m(\mathbf{r}, E)\, dV}{\int_{V_m} S_m H^{(m)}\, dV}. \tag{5.6.6}$$

Given the flat-source assumption, the partial fluxes $\Phi_f(\mathbf{r}, E)$ and $\Phi_m(\mathbf{r}, E)$ are defined to be solutions of (5.6.3) subject to the boundary condition (5.5.1) of zero net leakage in a direction perpendicular to the outer surface of the cell. It is important to recognize that this boundary condition is legitimate only because the cell of interest is assumed to be one of an infinite array of cells for which the high-energy neutron source is spatially flat. Thus the neutrons that make up $\Phi_f(\mathbf{r}, E)$ and $\Phi_m(\mathbf{r}, E)$ throughout a given cell do not come from the sources $S_f H^{(f)}$ or $S_m H^{(m)}$ of that cell alone. Strictly speaking *all* the cells in the lattice contribute to these fluxes. (There could not be a zero *net* current perpendicular to the outer surface of the cell unless neutrons born in other cells crossed that surface.) Thus, if we label quantities associated with different cells in a lattice by superscripts and consider the flux $\Phi_f(\mathbf{r}, E)$ in regions $V_f^{(1)}$ and $V_m^{(1)}$ of cell (1), we have that this flux is due, not only to the source $S_f H^{(f)}$ in $V_f^{(1)}$, but also to the sources $S_f H^{(f)}$ in $V_f^{(i)}$ $(i = 2, 3, \ldots)$, the fuel volumes of all the other cells of the lattice. Now, if we consider two adjacent cells (1) and (2), as in Figure 5.3, and the two symmetrically located points $\mathbf{r}^{(1)}$ and $\mathbf{r}^{(2)}$, it is clear that the contribution to $\Phi_f(\mathbf{r}, E)$ at $\mathbf{r}^{(1)}$ due to the source throughout $V_f^{(2)}$ is the same as the contribution to $\Phi_f(\mathbf{r}, E)$ at $\mathbf{r}^{(2)}$ due to the source throughout $V_f^{(1)}$. Consequently the contribution to $\int_{V_m^{(1)}} \Phi_f(\mathbf{r}, E)dV$ from the source in $V_f^{(2)}$ is the same as the contribution to $\int_{V_m^{(2)}} \Phi_f(\mathbf{r}, E)dV$ from the source in $V_f^{(1)}$. It follows that we may determine the total integral $\int_{V_m^{(1)}} \Phi_f(\mathbf{r}, E)dV$ in two ways: (1) we may find $\Phi_f(\mathbf{r}, E)$ by solving the first equation in (5.6.3) for a single cell with zero-net-current boundary conditions (thus implying that the cell in question is imbedded in an infinite array of cells all having sources $S_f H^{(f)}$ in the volumes $V_f^{(i)}$), or (2) we may solve for $\Phi_f^{(1)}(\mathbf{r}, E)$, the flux throughout the *entire* infinite array of cells (with the boundary condition $\Phi_f^{(1)}(\mathbf{r}, E) = 0$ at infinity), for the case of a source $S_f H^{(f)}$ *present in* $V_f^{(1)}$ *alone* and then perform the integration $\sum_{i=1}^{\infty} \int_{V_m^{(i)}} \Phi^{(1)}(\mathbf{r}, E)dV$ over the *entire* (infinite) lattice moderator or fuel volume.

An analogous argument applies to the determination of integrals involving $\Phi_m^{(1)}(\mathbf{r}, E)$, the flux throughout the lattice due to a source $S_m H^{(m)}$ in the moderator volume $V_m^{(1)}$ of cell one alone. (Note that $\Phi_f^{(1)}$ and $\Phi_m^{(1)}$ differ from Φ_f and Φ_m.)

In view of these alternate ways of determining the flux integrals we may rewrite (5.6.6)

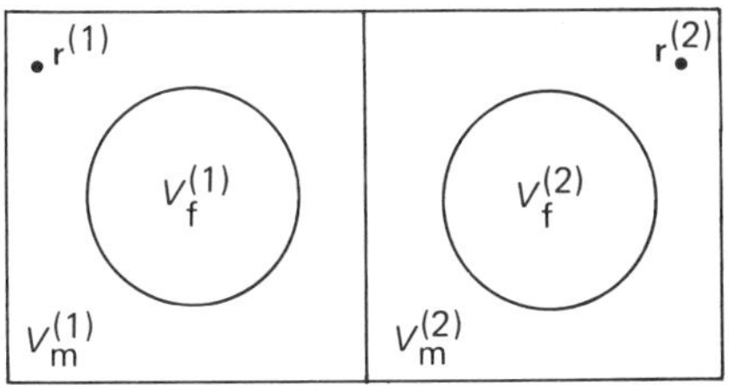

Figure 5.3 Adjacent cells used to derive relationships between escape probabilities.

for the escape probabilities $P_f(E)$ and $P_m(E)$ as

$$P_f(E) = \frac{\sum_{i=1}^{\infty} \int_{V_m^{(i)}} \Sigma_{t(m)}(E)\Phi_f^{(1)}(\mathbf{r}, E)\, dV}{\int_{V_f^{(1)}} S_f H^{(f)}\, dV},$$
$$P_m(E) = \frac{\sum_{i=1}^{\infty} \int_{V_f^{(i)}} \Sigma_{t(f)}(E)\Phi_m^{(1)}(\mathbf{r}, E)\, dV}{\int_{V_m^{(1)}} S_m H^{(m)}\, dV}, \qquad (5.6.7)$$

where the integrals involving $\Phi_f^{(1)}(\mathbf{r}, E)$ and $\Phi_m^{(1)}(\mathbf{r}, E)$ in the numerators extend over the entire infinite array of cells, while the sources in the denominators extend only over the moderator or fuel portions of a single central cell designated by superscript (1).

The physical interpretation of (5.6.7) is that $P_f(E)$ is the probability that an "average" neutron born in the energy interval dE in a given fuel rod will interact next with moderating material somewhere in the lattice (but not necessarily in the cell where the neutron was born). Similarly $P_m(E)$, according to (5.6.7), is the probability that an average neutron born in the moderator volume $V_m^{(1)}$ associated with one given fuel rod will next interact with fuel material (anywhere in the lattice).

These interpretations are to be contrasted with those implied by (5.6.6), namely that $P_f(E)$ is the probability that neutrons produced in *any* fuel rod throughout the lattice will next interact with the moderator *in* $V_m^{(1)}$ and that $P_m(E)$ is the probability that neutrons produced in *any* moderating material will next interact with the fuel *in* $V_f^{(1)}$. If, now, we use the interpretation of (5.6.7) for $P_f(E)$ and that of (5.6.6) for $P_m(E)$, and apply the reciprocity relation (5.6.5), we can relate $P_f(E)$ to $P_m(E)$. Thus the definition (5.6.4) gives us:

Flux in $dV_m^{(i)}\, dE$ due to a source of strength S_f in the fuel rod of cell (1)

$$= dV_m^{(i)}\, dE \int_{V_f^{(1)}} S_f \Phi(\mathbf{r}_m^{(i)}; \mathbf{r}_f^{(1)}; E)\, dV_f^{(1)}. \qquad (5.6.8)$$

Flux in $dV_f^{(1)}\, dE$ due to sources of strength S_m throughout the moderator region of the entire lattice

$$= dV_f^{(1)}\, dE \sum_{i=1}^{\infty} \int_{V_m^{(i)}} S_m \Phi(\mathbf{r}_f^{(1)}; \mathbf{r}_m^{(i)}; E) dV_m^{(i)}.$$

As a result we have

$$dE \sum_{i=1}^{\infty} \int_{V_m^{(i)}} \Sigma_{t(m)}(E)\Phi_f^{(1)}(\mathbf{r}, E) dV_m^{(1)} = dE \sum_{i=1}^{\infty} \Sigma_{t(m)}(E) \int_{V_m^{(i)}} dV_m^{(i)} \int_{V_f^{(1)}} S_f \Phi(\mathbf{r}_m^{(i)}; \mathbf{r}_f^{(1)}; E)\, dV_f^1,$$
$$dE \int_{V_f^{(1)}} \Sigma_{t(f)}(E)\Phi_m(\mathbf{r}, E)\, dV_f^{(i)} = dE\, \Sigma_{t(f)}(E) \int_{V_f^{(1)}} dV_f^{(1)} \sum_{i=1}^{\infty} \int_{V_m^{(i)}} S_m \Phi(\mathbf{r}_f^{(1)}; \mathbf{r}_m^{(i)}; E)\, dV_m^{(i)}.$$
$$(5.6.9)$$

Thus, by the first equation in (5.6.7) and the second in (5.6.6),

$$dE\, P_{\rm f}(E) S_{\rm f} V_{\rm f} = dE\, \Sigma_{\rm t(m)}(E) S_{\rm f} \sum_{i=1}^{\infty} \int_{V_{\rm m}^{(i)}} dV_{\rm m}^{(i)} \int_{V_{\rm f}^{(1)}} dV_{\rm f}^{(1)}\, \Phi(\mathbf{r}_{\rm m}^{(i)}; \mathbf{r}_{\rm f}^{(1)}; E),$$
$$dE\, P_{\rm m}(E) S_{\rm m} V_{\rm m} = dE\, \Sigma_{\rm t(f)}(E) S_{\rm m} \sum_{i=1}^{\infty} \int_{V_{\rm m}^{(i)}} dV_{\rm m}^{(i)} \int_{V_{\rm f}^{(1)}} dV_{\rm f}^{(1)}\, \Phi(\mathbf{r}_{\rm f}^{(1)}; \mathbf{r}_{\rm m}^{(i)}; E). \quad (5.6.10)$$

It follows then from (5.6.5) that

$$P_{\rm f}(E) V_{\rm f} \Sigma_{\rm t(f)}(E) = P_{\rm m}(E) V_{\rm m} \Sigma_{\rm t(m)}(E). \quad (5.6.11)$$

This general relationship between $P_{\rm f}(E)$ and $P_{\rm m}(E)$ for an array of two-region cells permits us to eliminate one of the two escape probabilities that appear in equations such as (5.5.21), (5.5.24), and (5.5.26) for the spectra $F^{(\rm f)}(E)$ and $F^{(\rm m)}(E)$ in the fuel rod and moderator. For example (5.5.26) becomes

$$F^{(\rm f)}(E) \approx \frac{[1 - P_{\rm f}(E)]\Sigma_{\rm s(f)}^{\rm p} + P_{\rm f}(E)[\Sigma_{\rm t(f)}(E)/\Sigma_{\rm t(m)}(E)]\Sigma_{\rm s(m)}^{\rm p}}{1 - \{[1 - P_{\rm f}(E)]\Sigma_{\rm s(f)}^{\rm res}(E)/\Sigma_{\rm t(f)}(E)\}} \frac{C_g}{E}. \quad (5.6.12)$$

The only additional assumption made to achieve this result is that the sources of neutrons are spatially flat within the fuel and within the moderator.

Calculation of $P_{\rm f}(E)$ for an Isolated Fuel Rod

The problem of an analytic determination of equivalent homogenized group parameters for a two-region cell is now reduced to that of computing the escape probability from the fuel rod. We shall first attack this problem for the limiting case of an isolated fuel rod and then outline an approximate method that accounts for the effect on $P_{\rm f}(E)$ of neighboring fuel rods.

The reason why the computation of $P_{\rm f}(E)$ is relatively simple for an isolated fuel rod is that the probability that a neutron born in this rod will have its next interaction with moderator material is simply the probability that the neutron will escape from the rod before interacting with fuel. This latter probability is just the rate at which neutrons in dE stream out of the fuel divided by the rate at which they are born in dE. For simple geometries, as we shall now show, these two rates can be computed exactly, and an analytical expression for $P_{\rm f}(E)$ can be obtained.

Consider a point $\mathbf{r}$ within a volume element dV inside a fuel rod of volume $V_{\rm f}$ having a convex shape, and specify a direction of travel by the unit vector $\mathbf{\Omega}$, as shown in Figure 5.4. For a spatially flat, isotropic source of S neutrons per unit volume per unit energy per unit solid angle per second, we have:

Number of neutrons emitted per second into $dV\, dE\, d\Omega$ about $\mathbf{r}$, E, and $\mathbf{\Omega}$
$= S\, dV\, dE\, d\Omega, \quad (5.6.13)$

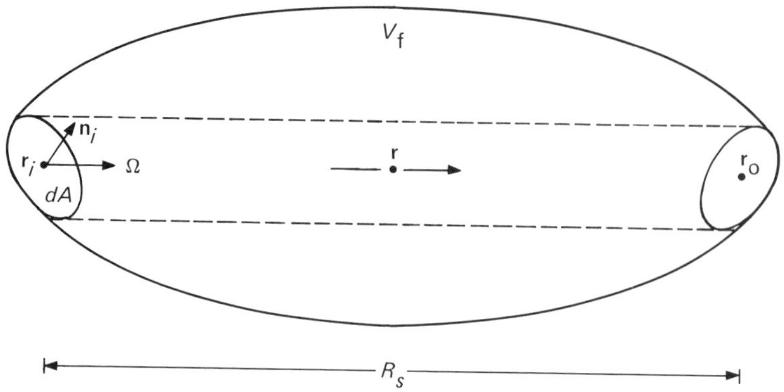

Figure 5.4 The escape of neutrons from a convex body.

where, to conform to convention, we shall here normalize so that $\int d\mathbf{\Omega} = 4\pi$.

If the total interaction cross section for the medium is $\Sigma_{t(f)}(E)$, the fraction of these neutrons reaching $\mathbf{r}_0$ on the surface of V_f is $\exp(-\Sigma_{t(f)}(E)|\mathbf{r} - \mathbf{r}_0|)$. Thus for a steady-state situation:

Number of those neutrons born in $dV\,dE\,d\mathbf{\Omega}$ about $\mathbf{r}$, E, and $\mathbf{\Omega}$ that leave the medium per second

$$= S\exp(-\Sigma_{t(f)}R)dV\,dE\,d\mathbf{\Omega}, \tag{5.6.14}$$

where $R \equiv |\mathbf{r} - \mathbf{r}_0|$.

Now, if we construct a thin tube extending in the direction $\mathbf{\Omega}$ from $\mathbf{r}_i$ to $\mathbf{r}_0$ and let dA be the amount of the surface area of V_f that this tube intersects, we see that the cross-sectional area of the tube *perpendicular* to $\mathbf{\Omega}$ is $\mathbf{n}_i \cdot \mathbf{\Omega}\,dA$, where $\mathbf{n}_i$ is the inward-directed normal to dA. Thus any small volume dV of the tube can be represented by

$$dV = \mathbf{n}_i \cdot \mathbf{\Omega}\,dA\,dR, \tag{5.6.15}$$

where dR is a small element of distance about $\mathbf{r}$ parallel to $\mathbf{\Omega}$. If we substitute this result into (5.6.14) and integrate from $\mathbf{r}_i$ to $\mathbf{r}_0$, we obtain:

Number of neutrons born in $R_s\,\mathbf{n}_i \cdot \mathbf{\Omega}\,dA\,dE\,d\mathbf{\Omega}$ that leave V_f per second

$$\begin{aligned} &= S\,\mathbf{n}_i \cdot \mathbf{\Omega}\,dA\,dE\,d\mathbf{\Omega} \int_0^{R_s} \exp(-\Sigma_{t(f)}(E)R)\,dR \\ &= \mathbf{n}_i \cdot \mathbf{\Omega} \frac{S\,dE\,dA\,d\mathbf{\Omega}}{\Sigma_{t(f)}(E)} [1 - \exp(-\Sigma_{t(f)}(E)R_s)], \end{aligned} \tag{5.6.16}$$

where $R_s \equiv |\mathbf{r}_i - \mathbf{r}_0|$.

To obtain the total rate at which neutrons born in V_f escape without interaction we first sum over all directions $\mathbf{\Omega}$ for which $\mathbf{n}_i \cdot \mathbf{\Omega} > 0$ and then sum over all surface elements dA; thus:

Rate at which neutrons born in dE within V_f escape without interacting

$$= \frac{S\,dE}{\Sigma_{t(f)}(E)} \int_A dA \int_{\mathbf{\Omega}\cdot\mathbf{n}_i>0} d\Omega\{\mathbf{n}_i \cdot \mathbf{\Omega}[1 - \exp(-\Sigma_{t(f)}(E)R_s)]\}. \tag{5.6.17}$$

Since the total rate at which neutrons are born in dE within V_f is

$$dE \int_{V_f} dV \int d\Omega\, S = 4\pi V_f S\, dE, \tag{5.6.18}$$

we have (using a superscript zero to indicate the result is for an isolated rod)

$$P_f^0(E) = \frac{1}{4\pi V_f \Sigma_{t(f)}(E)} \int_A dA \int_{\mathbf{\Omega}\cdot\mathbf{n}_i>0} d\Omega\, \mathbf{n}_i \cdot \mathbf{\Omega}[1 - \exp(-\Sigma_{t(f)}(E)R_s)]. \tag{5.6.19}$$

The integral on the right-hand side of this equation is rather difficult to determine since R_s is a function of $\mathbf{\Omega}$. It can be evaluated by a procedure, due to Dirac, called the *cord-length method.* The name comes from the fact that the integral can be viewed as being proportional to the average length of a distribution of cords R_s entering the surface of V_f with an isotropic distribution of directions and altered in length by the fact that $\Sigma_{t(f)}(E)$ is nonzero. We shall not go through any details of the method except to state that the "average cord length" $\bar{R}_s$ is defined to be the average value of R_s over the distribution of cord lengths for the case of no attenuation ($\Sigma_{t(f)}(E) = 0$). For any solid convex body it can be shown that

$$\bar{R}_s = \frac{4V}{A}, \tag{5.6.20}$$

where V is the volume of the body and A its surface area.

The cord-length method yields explicit expressions for $P_f(E)$ in terms of tabulated functions (exponentials, exponential integrals, and Bessel functions) for spheres, slabs, and cylinders. However, except for the sphere case, these are so complicated that much simpler, empirical fits to the exact expressions are used for practical calculations. The simplest of these is called the *rational approximation.* It is due to Wigner and states that, for an isolated rod,

$$P_f^0(E) \approx \frac{1}{1 + \Sigma_{t(f)}(E)\bar{R}_{s(f)}}, \tag{5.6.21}$$

where $\bar{R}_{s(f)}$ is the average cord length for the fuel rod. This simple formula is correct in the limits of very small and very large values of $\Sigma_{t(f)}(E)\bar{R}_{s(f)}$. However it always underestimates $P_f^0(E)$.

An improved approximation for $P_f^0(E)$ for cylinders, due to Sauer, has been derived by obtaining an empirical fit to the cord-length distribution function and using this in the integral yielding $P_f^0(E)$. This formula is

$$P_f^0(E) = \frac{1}{\Sigma_{t(f)}(E)\bar{R}_{s(f)}}\left[1 - \frac{1}{1 + (\Sigma_{t(f)}(E)\bar{R}_{s(f)}/4.58)^{4.58}}\right]. \tag{5.6.22}$$

Substitution of this expression into (5.6.12) provides us with a good analytical approximation for the collision density $F^{(f)}(E)$ inside an isolated fuel rod. Moreover for a lattice of widely spaced fuel rods (such as in a D_2O- or graphite-moderated, natural-uranium reactor) we may assume that almost all the neutrons escaping from a given fuel rod have their next interaction with a nucleus in the moderator. Thus the value of $P_f(E)$ for the lattice will be very close to that for the isolated rod, and we may solve (5.5.24) and (5.5.26), using (5.6.21) along with the reciprocity relationship (5.6.11), to obtain average flux spectra $\bar{\Phi}^{(f)}(E)$ and $\bar{\Phi}^{(m)}(E)$ for the fuel and moderator regions of the cell.

Calculation of Escape Probabilities for Close-Packed Lattices

When the fuel rods in a lattice are close together, a neutron that escapes from one particular rod may very well not have its next interaction in moderator, but may instead interact with a nucleus in some neighboring fuel rod. Under these circumstances $P_f(E)$ as defined by (5.5.16) or (5.5.17) will be less than (5.6.19), the escape probability from an isolated fuel rod.

The most accurate scheme for computing $P_f(E)$ under these circumstances is the so-called Monte Carlo method (see Chapter 8). As the name suggests, this is a statistical sampling procedure in which several thousand case histories are followed from the birth of a neutron in dE (at some random location in the fuel rod and traveling in some random direction) until that neutron leaves dE by interacting with either a fuel or a moderator nucleus somewhere in the lattice. The escape probability $P_f(E)$ is then just the fraction of neutrons in the sample that leave dE by interaction with moderator nuclei. Sampling a few thousand case histories usually provides a very accurate value of $P_f(E)$.

Determination of $P_f(E)$ by the Monte Carlo procedure takes only a minute or so on a large digital computer. Even for a fixed-lattice geometry, however, $P_f(E)$ depends on both $\Sigma_{t(f)}(E)$ and $\Sigma_{t(m)}(E)$, and the first of these cross sections can vary drastically with energy. To find $P_f(E)$ at many energies it is therefore necessary to construct a two-dimensional table of Monte Carlo results corresponding to sets of values of $\Sigma_{t(f)}$ and $\Sigma_{t(m)}$. For a given energy (and, hence, for particular values of $\Sigma_{t(f)}(E)$ and $\Sigma_{t(m)}(E)$) $P_f(E)$ may then be found by interpolation. This process is not too expensive on the modern generation of computing machines. However, on earlier machines and before various tricks for speeding up Monte Carlo calculations were discovered, it was costly, and those working in this area were therefore led to develop analytical methods. For many cases the accuracy of these methods compares favorably with Monte Carlo results, and the cost in running time is much less. Moreover, by using Monte Carlo techniques to determine certain fitting parameters that appear in the analytical formulation, we can, for a limited range of lattice parameters, obtain expressions for $P_{\mathrm{f}}(E)$ that combine the accuracy of the Monte Carlo method with the economy of an analytical procedure.

In order to find an analytical expression for $P_{\mathrm{f}}(E)$ we shall first extend the rigorous expression (5.6.19), for an isolated fuel rod, to the case of a fuel rod in a regular lattice, and then make a fundamental simplifying assumption.

Recall that expression (5.6.19) for $P_{\mathrm{f}}(E)$ was obtained by dividing the rate (5.6.18) at which neutrons are born in dE within the fuel-rod volume V_{f} into the rate (5.6.17) at which neutrons born in dE within V_{f} leave V_{f} without interacting. In order to determine $P_{\mathrm{f}}(E)$ for a rod in a lattice we must correct this result for the fact that not all neutrons escaping from V_{f} will leave dE by interacting with moderator material. They may instead enter another fuel rod and interact there. To account for this possible interaction consider a beam of neutrons some of which pass through several cells of the lattice.

As shown in Figure 5.5, neutrons in this beam pass alternately through fuel material and moderator material. With subscript (i) indicating fuel material (f) or moderator material (m) and superscript (j) indicating the sequence of distances traversed, let $R_{\mathrm{s}(i)}^{(j)}$ stand for the successive distances that a neutron in the beam will traverse as it passes through alternate regions of fuel and moderator. The $R_{\mathrm{s}(i)}^{(j)}$ are very complicated functions of the location of the surface element dA and the direction $\mathbf{\Omega}$—particularly since the

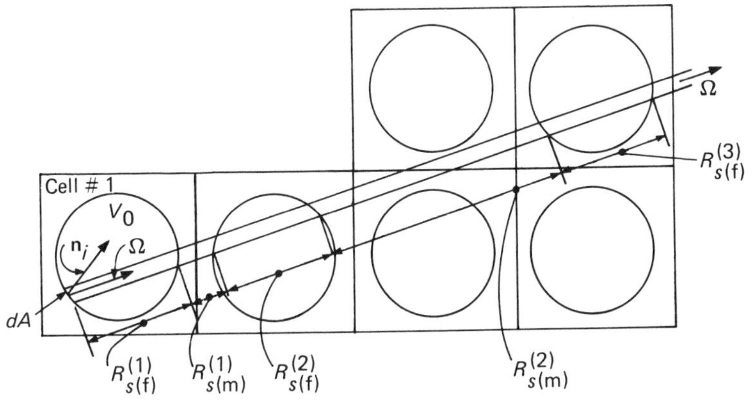

Figure 5.5 Interaction path of neutrons in a lattice.

beam need not lie in the plane of the figure. However we shall disregard this fact for the moment and write an exact expression for $P_f(E)$ in terms of them.

To simplify notation we define

$$G_i^{(j)}(E) \equiv \exp(-\Sigma_{t(i)}(E)R_{s(i)}^{(j)}) \qquad (i = \text{f, m}). \tag{5.6.23}$$

Thus $G_i^{(j)}(E)$ is the fraction of neutrons in the beam that traverse the distance $R_{s(i)}^{(j)}$ without interacting, and $1 - G_i^{(j)}(E)$ is the fraction that do interact in this distance.

From (5.6.16) we have (suppressing, for convenience, the E dependence of the G's) that $\mathbf{n}_i\cdot\mathbf{\Omega}\, S\, dE\, dA(1 - G_f^{(1)})/\Sigma_{t(f)}(E)$ is the number of neutrons born in $R_{s(f)}^{(1)}\mathbf{n}_i\cdot\mathbf{\Omega}\, dA\, dE\, d\Omega$ that leave the fuel region $V_f^{(1)}$ of cell (1) per second. Of these, a fraction $1 - G_m^{(1)}$ interact with moderator in traversing the distance $R_{s(m)}^{(1)}$. A fraction $G_m^{(1)}$ do not interact in traversing this distance, and of these a fraction $G_f^{(2)}(1 - G_m^{(2)})$ survive the passage through $R_{s(f)}^{(2)}$ but then interact with the moderator in traversing the distance $R_{s(m)}^{(2)}$. Thus an additional fraction $G_m^{(1)}G_f^{(2)}(1 - G_m^{(2)})$ of neutrons in the beam leaving the fuel region of cell (1) interact with moderator in traversing $R_{s(m)}^{(2)}$.

Continuing to follow this process, we obtain finally:

$$\begin{aligned}&\text{Rate at which neutrons born in } R_{s(f)}^{(1)}\mathbf{n}_i\cdot\mathbf{\Omega}\, dA\, dE\, d\Omega \text{ interact with moderator nuclei}\\ &= \mathbf{n}_i\cdot\mathbf{\Omega}\frac{S\, dE\, dA}{\Sigma_{t(f)}(E)}(1 - G_f^{(1)})[1 - G_m^{(1)} + G_m^{(1)}G_f^{(2)}(1 - G_m^{(2)})\\ &\quad + G_m^{(1)}G_f^{(2)}G_m^{(2)}G_f^{(3)}(1 - G_m^{(3)}) + \cdots]d\Omega.\end{aligned} \tag{5.6.24}$$

Integration of this result over all inward-pointing directions $\mathbf{\Omega}$ and over the fuel-rod surface of cell (1) gives us the total rate at which neutrons born in V_f of cell (1) will interact with moderator. Since the total rate of birth of such neutrons is still given by (5.6.18), we have for the escape probability for a neutron born in dE within a fuel rod that is part of a lattice:

$$\begin{aligned}P_f(E) &= \frac{1}{4\pi V_f\Sigma_{t(f)}(E)}\int_A dA\int_{\mathbf{\Omega}\cdot\mathbf{n}_i>0} d\Omega\, \mathbf{n}_i\cdot\mathbf{\Omega}(1 - G_f^{(1)})\\ &\quad\times[1 - G_m^{(1)} + G_m^{(1)}G_f^{(2)}(1 - G_m^{(2)}) + G_m^{(1)}G_f^{(2)}G_m^{(2)}G_f^{(3)}(1 - G_m^{(3)}) + \cdots].\end{aligned} \tag{5.6.25}$$

This expression for $P_f(E)$ is exact. Unfortunately, except for a lattice of slabs, it is virtually impossible to perform the indicated integrations. There is, however, an approximation suggested by Bell that makes the integration tractable. To see the motivation behind this approximation we return to the case of the isolated rod (5.6.19), to which (5.6.25) reduces (as it should) when $G_m^{(1)}$ (the probability that a neutron will survive interaction and enter the fuel rod nearest to the one in which it was born) approaches zero.

The term $\exp(-\Sigma_{t(f)}(E)R_s)$ in the integrand of (5.6.19) is the probability that a neutron will not interact in traversing the distance R_s. Thus the integral in (5.6.19) appears to involve an average of this probability over all cord lengths R_s that correspond to different locations for dA and different directions for $\mathbf{\Omega}$. Development of the cord-length method would show that this is indeed a correct interpretation, provided we introduce a normalization constant K so that, when $\Sigma_{t(f)}(E) \to 0$ and the probability of noninteraction is unity, the integral will yield this probability, i.e., so that

$$\lim_{\Sigma_{t(f)} \to 0} \frac{1}{K} \int_A dA \int_{\mathbf{\Omega} \cdot \mathbf{n}_i > 0} d\Omega\, \mathbf{n}_i \cdot \mathbf{\Omega} \exp(-\Sigma_{t(f)}(E) R_s) = 1. \tag{5.6.26}$$

The appropriate value of the normalization constant is clearly

$$K = \int_A dA \int_{\mathbf{\Omega} \cdot \mathbf{n}_i > 0} d\Omega\, \mathbf{n}_i \cdot \mathbf{\Omega}. \tag{5.6.27}$$

Equation (5.6.19) may then be rewritten

$$P_f^0(E) = \frac{K}{4\pi V_f \Sigma_{t(f)}(E)} \left[1 - \frac{1}{K} \int_A dA \int_{\mathbf{\Omega} \cdot \mathbf{n}_i > 0} d\Omega\, \mathbf{n}_i \cdot \mathbf{\Omega} \exp(-\Sigma_{t(f)}(E) R_s) \right], \tag{5.6.28}$$

where the double integral

$$\frac{1}{K} \int_A dA \int_{\mathbf{\Omega} \cdot \mathbf{n}_i > 0} d\Omega\, \mathbf{n}_i \cdot \mathbf{\Omega} \exp(-\Sigma_{t(f)}(E) R_s)$$

can now be interpreted as an average over cord length.

To evaluate K we first fix the location of the surface element dA and describe $d\Omega$ with respect to an axis system having $\mathbf{n}_i$ as its polar axis. In terms of the coordinates shown in Figure 5.6 we have

$$d\Omega = \frac{(\rho \sin\theta\, d\varphi)(\rho\, d\theta)}{\rho^2} = -d\mu\, d\varphi,$$

$$\mu \equiv \cos\theta = \mathbf{\Omega} \cdot \mathbf{n}_i.$$

Thus

$$K = \int_A dA \int_{\mathbf{\Omega} \cdot \mathbf{n}_i > 0} d\Omega\, \mathbf{n}_i \cdot \mathbf{\Omega} = -\int_A dA \int_0^{2\pi} d\varphi \int_1^0 \mu\, d\mu = \int_A \pi dA. \tag{5.6.29}$$

Since this result is true for all locations of dA on the surface of V_f, we obtain the simple result

$$K = \pi A. \tag{5.6.30}$$

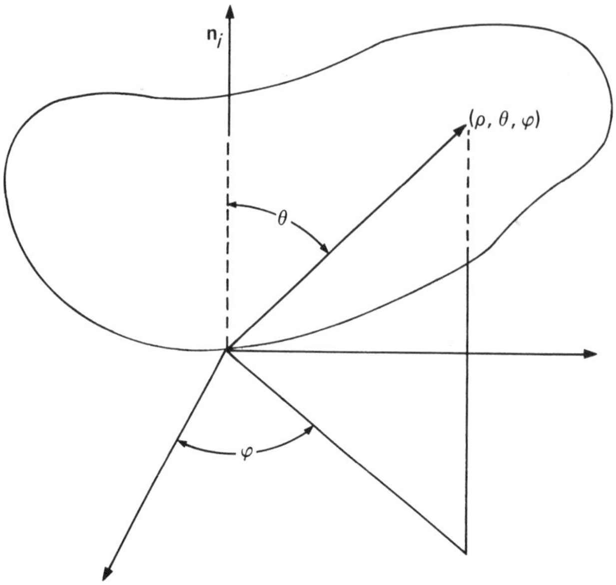

Figure 5.6 An axis system used in the evaluation of the normalization constant K.

Accordingly we define a "cord average" of $G_i^{(j)} \equiv \exp(-\Sigma_{t(i)}(E)R_{s(i)}^{(j)})$:

$$\langle G_i^{(j)} \rangle \equiv \frac{1}{\pi A}\int_A dA \int_{\mathbf{\Omega}\cdot\mathbf{n}_i>0} d\Omega\, \mathbf{n}_i \cdot \mathbf{\Omega}\, G_i^{(j)}. \tag{5.6.31}$$

Finally, by making use of the expression (5.6.20) for the average cord length $\bar{R}_{s(f)}$ for the fuel rod, we see that (5.6.28) becomes

$$P_f^0(E) = \frac{1}{\bar{R}_{s(f)}\Sigma_{t(f)}(E)}[1 - \langle G_f^{(1)} \rangle]. \tag{5.6.32}$$

If we now return to (5.6.25) and let angle brackets $\langle\ \ \rangle$ stand for the "cord-averaging operator" $(1/\pi A)\int_A dA \int_{\mathbf{\Omega}\cdot\mathbf{n}_i>0} d\Omega\, \mathbf{n}_i \cdot \mathbf{\Omega}[\ \ ]$, we may rewrite the expression for $P_f(E)$ as

$$P_f(E) = \frac{1}{\bar{R}_{s(f)}\Sigma_{t(f)}(E)}\langle(1 - G_f^{(1)})[1 - G_m^{(1)} + G_m^{(1)}G_f^{(2)}(1 - G_m^{(2)}) + G_m^{(1)}G_f^{(2)}G_m^{(2)}G_f^{(3)}(1 - G_m^{(3)}) + \cdots]\rangle. \tag{5.6.33}$$

The Bell approximation for this complicated average of the product of the attenuation factors $G_i^{(j)}$ is made by assuming the average of the product to be the product of the averages; (5.6.33) then becomes

$$P_f(E) \approx \frac{1}{\bar{R}_{s(f)}\Sigma_{t(f)}(E)}\{[1 - \langle G_f\rangle][1 - \langle G_m\rangle + \langle G_m\rangle\langle G_f\rangle(1 - \langle G_m\rangle)$$
$$+ (\langle G_m\rangle\langle G_f\rangle)^2(1 - \langle G_m\rangle) + \cdots]\}$$
$$= \frac{[1 - \langle G_f\rangle][1 - \langle G_m\rangle]}{\bar{R}_{s(f)}\Sigma_{t(f)}(E)}[1 + \langle G_m\rangle\langle G_f\rangle + (\langle G_m\rangle\langle G_f\rangle)^2 + \cdots]$$
$$= \frac{[1 - \langle G_f\rangle][1 - \langle G_m\rangle]}{\bar{R}_{s(f)}\Sigma_{t(f)}(E)[1 - \langle G_f\rangle\langle G_m\rangle]}, \tag{5.6.34}$$

where, since all the $\langle G_f^{(j)}\rangle$ $(j = 1, 2, \ldots)$ are the same and all the $\langle G_m^{(j)}\rangle$ $(j = 1, 2, \ldots)$ are the same, we have dropped the superscripts.

The problem of computing $P_f(E)$ now reduces to finding $\langle G_f\rangle$ and $\langle G_m\rangle$, the average transmission probabilities for the fuel-rod and moderator regions.

The first of these is easily determined. We simply use (5.6.32) to eliminate $\langle G_f\rangle$. The result is

$$P_f(E) = \frac{P_f^0(E)[1 - \langle G_m(E)\rangle]}{1 - [1 - \bar{R}_{s(f)}\Sigma_{t(f)}(E)P_f^0(E)]\langle G_m(E)\rangle}, \tag{5.6.35}$$

where $P_f^0(E)$ can be found using the Sauer approximation (5.6.22) and we have reintroduced the energy dependence of $G_m(E)$.

The Dancoff Factor

The average transmission probability $\langle G_m(E)\rangle$ for the moderator part of the cell is somewhat more difficult to estimate. The first approximation method, developed by Dancoff and Ginsburg, was based on a geometrical shadowing picture, with the shadowing fuel rods assumed to be black. As a result the quantity $\langle G_m(E)\rangle$ is frequently called the *Dancoff factor*, represented symbolically by the letter c, in many papers. The Dancoff factor can be used to account in an approximate fashion for the fact that the zero-current approximation (5.5.1), fundamental to the present development, may be quite poor for cells near the boundary between two clusters of different material composition.

A very simple approximation for the Dancoff factor was proposed by Bell, namely to assume that the escape probability $P_m^0(E)$ from an isolated column of moderator associated with a cluster of four fuel rods (see Figure 5.5) embedded in an infinite sea of fuel material is given by the rational approximation

$$P_m(E) \approx \frac{1}{1 + \bar{R}_{s(m)}\Sigma_{t(m)}(E)}, \tag{5.6.36}$$

where $\bar{R}_{s(m)}$ and $\Sigma_{t(m)}$ are the average cord length and total macroscopic cross section for the moderator. If we further assume, in analogy with (5.6.32), that $P_m(E)$ is related to

$\langle G_m(E)\rangle$ by

$$P_m(E) \approx \frac{1}{\bar{R}_{s(m)}\Sigma_{t(m)}(E)}[1 - \langle G_m(E)\rangle] \tag{5.6.37}$$

and equate these two expressions for $P_m(E)$, we obtain the approximation

$$\langle G_m(E)\rangle \approx \frac{1}{1 + \bar{R}_{s(m)}\Sigma_{t(m)}(E)}. \tag{5.6.38}$$

The formula found for $P_f(E)$ by substituting this result into (5.6.35) provides results quite close to those determined by Monte Carlo methods. (It is ultimately this agreement that justifies the Bell approximation (5.6.34).) Moreover results of even greater accuracy can be obtained by using a somewhat more accurate empirical formula for $\langle G_m(E)\rangle$ based on comparisons with Monte Carlo results.

5.7 Summary and Connection with Elementary Theory

Review of the Method

We thus come to the end of a long development, whose object has been the determination of analytical methods for finding equivalent homogenized energy-group parameters in the resonance energy range for a cell composed of a fuel rod and its associated moderator. The following summary of the principal assumptions and formulas making up the overall procedure may help the reader to distinguish the forest from the trees.

To determine the equivalent homogenized group parameters of (5.5.4) for a cell we proceed as follows:

1. Specialize to the case of a two-region cell. Make the assumption (5.5.1) concerning cell boundary conditions, and, without further approximation, rewrite the expressions for the group parameters in terms of the average flux densities $\bar{\Phi}^{(f)}(E)$ and $\bar{\Phi}^{(m)}(E)$ in the fuel and in the moderator ((5.5.8) and analogous equations for the other group parameters).

2. To obtain equations for the average flux densities split the source of neutrons in a given energy interval dE into its component in the moderator and its component in the fuel. This splitting (which involves no approximation) permits us to write equations ((5.5.12) and (5.5.13)) for the fluxes and currents $\Phi_f(\mathbf{r}, E)$ and $\mathbf{J}_f(\mathbf{r}, E)$ due to the source in the fuel alone and the fluxes and currents $\Phi_m(\mathbf{r}, E)$ and $\mathbf{J}_m(\mathbf{r}, E)$ due to the source in the moderator alone. In terms of these, we define the "escape probabilities" $P_f(E)$ and $P_m(E)$ of (5.5.16) and (5.5.17), $P_f(E)$ being the probability that an average neutron first appearing in dE in the fuel will have its next collision (either scattering or absorbing) in the moderator, and $P_m(E)$ being the probability that a neutron born in the moderator

will next collide in fuel. The escape probabilities then permit us to write (5.5.21), two coupled integral equations for the spectra $\overline{\Phi}^{(f)}(E)$ and $\overline{\Phi}^{(m)}(E)$. These may be solved numerically, or we may make the narrow- and wide-resonance approximations to convert the coupled integral equations into the coupled algebraic equations (5.5.24) and (5.5.26).

The only essential approximations so far are the restriction to a two-region cell and the imposition of cell boundary conditions. To find the $P_i(E)$ we must make one more essential approximation.

3. To determine the $P_i(E)$ we assume that the source of neutrons in dE is flat within the fuel rod and flat within the moderator. This assumption permits us to rewrite the expressions for the $P_i(E)$ in the simpler form (5.6.6). It also permits us to find the exact relationship (5.6.11) between $P_f(E)$ and $P_m(E)$, so that further effort can be focused solely on finding $P_f(E)$, the escape probability from the fuel.

4. For an isolated rod, $P_f^0(E)$ is just the probability that an average neutron born in the rod will get out, and the exact integral (5.6.19) for $P_f^0(E)$ can be obtained by averaging over cord lengths R_s. However the evaluation of this integral is so complicated that approximate formulas for $P_f^0(E)$ must be sought. The simplest of these is Wigner's rational approximation (5.6.21). A more accurate formula, however, is (5.6.22), due to Sauer.

5. For a rod that is part of a lattice, Monte Carlo methods can be used, or else the formally exact cord-length method can be extended to give (5.6.25). However (5.6.25) is completely intractable unless we make the Bell assumption that the average of a product of attenuation factors $G_i^{(j)}(E)$ (defined in (5.6.23) and appearing in (5.6.33)) is equal to the product of the averages. If this assumption is made, the expression for $P_f(E)$, the escape probability from a fuel rod in a lattice, is greatly simplified to (5.6.34), which depends only on the average attenuation factors $\langle G_f(E)\rangle$ for the rod and $\langle G_m(E)\rangle$ (called the Dancoff factor) for the moderator. The former can be expressed in terms of the escape probability for an isolated rod (5.6.32), and the latter can be determined, as in (5.6.38), either by assuming that the rational approximation (5.6.36) holds for the moderator portion of a cell or by fitting to Monte Carlo results. In either case the overall expressions for $P_f(E)$, and thus for the average fluxes $\overline{\Phi}^{(f)}(E)$ and $\overline{\Phi}^{(m)}(E)$, are quite accurate.

Application and Validation of the Method

The procedure that we have developed for analyzing resonance-energy neutron behavior in cells has been aimed at determining the equivalent homogenized energy-group constants (5.5.4). We can either use these cell-averaged constants directly in the group diffusion equations (4.13.7), or else we can use them to find asymptotic spectra for the various assemblies of cells making up the reactor (for example by solving (4.10.3)) and then obtain few-group constants by averaging over these spectra (as, for example, was done in (4.13.6)). The latter practice is the more common one since solving multigroup, multi-dimensional diffusion equations is very expensive.

We expect the overall method to be quite accurate since the approximations made to determine the spectra $\overline{\Phi}^{(i)}(E)$ are quite mild and since we are only interested in the energy *shape* of these spectra over *limited* ranges (the ΔE_g).

The energy-group parameters obtained in this manner can also be used to calculate various experimentally measurable properties of a cell in a lattice and can thus provide a check on the adequacy of the theoretical methods and of the basic nuclear data used in the calculations. For example an exponential experiment consisting of a uniform lattice of slightly-enriched-uranium fuel rods can be constructed, and the ratio of the capture rate in U^{28} of neutrons above the cadmium cutoff to the capture rate in U^{28} of neutrons of *all* energies can be measured and compared with theoretical predictions. Moreover, by performing several other experiments and making some small theoretical corrections to the data, we can infer from this ratio a value of the resonance-escape probability p for the cell. This quantity can also be compared with predictions of the theory. The comparison is somewhat indirect; however the resonance-escape probability is so fundamental in the elementary theoretical methods of analyzing reactors that it is still used as a basis of comparison between theory and experiment, even though today it rarely enters directly into design computations.

A validation of the theoretical method and basic data can be made more readily if small corrections are made to the results of lattice measurements in order to annul the effect of gross neutron leakage out of the lattice. If this is done, it is possible to infer a value of p for a cell with a zero-net-current boundary condition, and the methods we have developed to find $\overline{\Phi}^{(\mathrm{f})}(E)$ and $\overline{\Phi}^{(\mathrm{m})}(E)$ can then be tested without going through the intermediate stage of determining energy-group parameters. Instead an approximate value of p can be computed analytically and compared with experiment. We shall outline how this is done in order to make contact with the extensive theoretical work in the literature and to introduce several parameters and procedures used in the field.

A Definition of Resonance-Escape Probability for a Cell

Several slightly different definitions have already been given for the resonance-escape probability. (See (3.5.11), (3.5.13), and (5.4.13).) The reason for the variety is theoretical convenience: certain formulas are simpler when particular small effects (for example neutron capture in the moderator) are included in the definition of the resonance-escape probability; others are simpler when they are excluded. It is, of course, very important to adopt consistent definitions when comparisons between calculations or between calculation and experiment are made.

In extending the definition of resonance-escape probability to a heterogeneous cell, we shall adopt a definition well suited to the collision-theory methods we have been developing and to comparisons with measurements. To be specific we shall define it as the probability that neutrons originating from a spatially flat source in a cell (which is part of an infinite lattice) and slowing down past E_1, an energy at the top of the resonance region, will escape capture in fertile material (U^{28} or Th^{02}) in the energy range E_c–E_1, where E_c is the thermal cutoff energy ($E_c \approx 1.0$ eV).

Note that the definition involves capture in the fertile material alone. Thus a neutron captured in the moderator (for example by an atom of control poison) in the range E_c–E_1 or a neutron captured by U^{25} in the fuel has *escaped* resonance capture according to this definition.

For the particular case of the two-region cell we have been analyzing, we shall assume that the fertile material is U^{28}; we then have:

Capture rate in U^{28} in the cell over the resonance range

$$= V_f \int_{E_c}^{E_1} \Sigma_a^{28}(E)\bar{\Phi}^{(f)}(E)\,dE, \tag{5.7.1}$$

where $\bar{\Phi}^{(f)}(E)$ is the average flux density in the fuel rod given by (5.5.7) and $\Sigma_a^{28}(E) \equiv n^{28}\sigma_a^{28}(E)$, with n^{28} being the number density of a U^{28} atoms *in the fuel*. If we normalize $\bar{\Phi}^{(f)}(E)$ so that at E_1, the top of the resonance range, $\bar{\Phi}^{(f)}(E_1)$ corresponds to a slowing-down density $q(E_1)$ of one neutron per unit volume per second, we have from (5.5.22) that

$$\bar{\Phi}^{(f)}(E_1) = \frac{V_m + V_f}{E_1[V_m\xi_m\Sigma_{s(m)} + V_f\xi_f\Sigma_{s(f)}]} \equiv \frac{1}{E_1\overline{\xi\Sigma_s}}, \tag{5.7.2}$$

where the last term defines $\overline{\xi\Sigma_s}$.

Then, since $\bar{\Phi}^{(f)}(E)$ is a flux resulting from $(V_f + V_m)q(E_1)$ neutrons slowing down in the cell past E_1 per second, and $q(E_1)$ is unity, the fraction of those neutrons that escape capture in U^{28} is

$$p = 1 - \frac{V_f}{V_f + V_m}\int_{E_c}^{E_1} \Sigma_a^{28}(E)\bar{\Phi}^{(f)}(E)\,dE. \tag{5.7.3}$$

Thus p depends directly on a spectrum-weighted average of the absorption cross section for U^{28} in the resonance energy range and indirectly on the cross sections and concentrations of the other materials making up the cell insofar as they affect $\bar{\Phi}^{(f)}(E)$.

The Resonance Integral

To simplify the notation it is customary to define a "resonance integral" of U^{28} in the cell:

$$\bar{\sigma}^{28} \equiv \left(\frac{\overline{\xi\Sigma_s}}{q(E_1)}\right)\int_{E_c}^{E_1} \sigma_a^{28}(E)\bar{\Phi}^{(f)}(E)\,dE, \tag{5.7.4}$$

the division by $q(E_1)$ $(=1)$ being so that the units of $\bar{\sigma}^{28}$ are those of $\sigma_a^{28}(E)$ and the multiplication by $\overline{\overline{\xi\Sigma_s}}$ being so that, for the case of infinite dilution ($V_f \to 0$), when (with $\overline{\overline{\xi\Sigma_s}}$ constant) $\overline{\Phi}^{(f)}(E) = q(E_1)/(E\,\overline{\overline{\xi\Sigma_s}})$ over the *entire* energy range,

$$\bar{\sigma}^{28} = \int_{E_c}^{E_1} \sigma_a^{28}(E)\frac{dE}{E}. \tag{5.7.5}$$

Thus, for the infinitely dilute case, the resonance integral is totally independent of the concentrations or cross sections of other materials making up the cell.

Equation (5.7.3) expressed in terms of the resonance integral becomes

$$p = 1 - \frac{V_f}{(V_f + V_m)}\frac{n^{28}\bar{\sigma}^{28}}{\overline{\overline{\xi\Sigma_s}}}, \tag{5.7.6}$$

where n^{28} represents the number of nuclei of U^{28} per unit volume of the fuel rod and $q(E_1)$ $(=1)$ has been suppressed.

Since the resonance integral is defined as an integral over the entire range E_c–E_1, approximating $\overline{\Phi}^{(f)}(E)$ by the NR and WR formula (5.6.12) over this range will yield an overestimate of $\bar{\sigma}^{28}$ (and a corresponding underestimate of p). The reason for this is that the NR and WR approximations fail to account for the step decreases in the $1/E$ portion of the flux that occur at each resonance because of the decrease in $q(E)$ due to the neutrons absorbed in the resonance (see Figure 5.1). In order to avoid this error in computing p analytically, it is necessary to deal with each resonance individually. We accordingly define a resonance integral $\bar{\sigma}_i^{28}$ for the ith resonance by

$$\bar{\sigma}_i^{28} \equiv \overline{\overline{\xi\Sigma_s}}(E_i)\int_{\Delta E_i} \sigma_a^{28}(E)\overline{\Phi}^{(f)}(E)\,dE, \qquad \overline{\Phi}^{(f)}\left(\frac{E_i + E_{i-1}}{2}\right) = \frac{2}{(E_i + E_{i-1})\overline{\overline{\xi\Sigma_s}}(E_i)}, \tag{5.7.7}$$

where ΔE_i is an energy interval including the resonance in question and we have normalized $\overline{\Phi}^{(f)}(E)$ in that interval so that the slowing-down density at an energy $\frac{1}{2}(E_i + E_{i-1})$, midway between the ith and $(i-1)$th resonances, is unity. (A systematic way to choose ΔE_i is to let it extend from $E_i - \frac{1}{2}(E_i - E_{i+1})$ to $E_i + \frac{1}{2}(E_{i-1} - E_i)$, where $E_{i-1} > E_i > E_{i+1}$ are the energies corresponding to the peaks of the $(i-1)$th, ith, and $(i+1)$th resonance.) We also account in the formula for the possibility that the value of $\overline{\overline{\xi\Sigma_s}}(E_i)$ may change slightly from one range ΔE_i to the next by employing $\Sigma_{s(m)}(E_i)$ and $\Sigma_{s(f)}(E_i)$ for $\Sigma_{s(m)}$ and $\Sigma_{s(f)}$ in the definition (5.7.2) of $\overline{\overline{\xi\Sigma_s}}(E_i)$. Note that, because of the renormalization of $\overline{\Phi}^{(f)}(E)$ for each energy range ΔE_i, $\bar{\sigma}^{28} \neq \sum_i \bar{\sigma}_i^{28}$.

The resonance-escape probability for the ith resonance is then

$$p_i \equiv 1 - \frac{V_f}{(V_f + V_m)}\frac{n^{28}\bar{\sigma}_i^{28}}{\overline{\overline{\xi\Sigma_s}}(E_i)}. \tag{5.7.8}$$

Physically, because of the normalization of $\overline{\Phi}^{(f)}(E)$ in (5.7.7), p_i may be interpreted as the probability that neutrons slowing down past the upper limit of ΔE_i will escape capture in U^{28} in the energy range ΔE_i. Thus the total resonance-escape probability is a product over all resonances:

$$p = \prod_i p_i = \prod_i \left(1 - \frac{V_f}{(V_f + V_m)} \frac{n^{28}\bar{\sigma}_i^{28}}{\xi\Sigma_s(E_i)}\right). \tag{5.7.9}$$

Moreover, since the capture probabilities $1 - p_i$ for the individual resonances are small, we may approximate

$$p_i \approx \exp\left(-\frac{V_f}{(V_f + V_m)} \frac{n^{28}\bar{\sigma}_i^{28}}{\xi\Sigma_s(E_i)}\right), \tag{5.7.10}$$

so that

$$p \approx \exp\left(-\frac{V_f n^{28}}{(V_f + V_m)} \sum_i \frac{\bar{\sigma}_i^{28}}{\xi\Sigma_s(E_i)}\right), \tag{5.7.11}$$

where the sum is over all the resonances in the range E_c–E_1.

Now, if we use the NR and WR approximations (5.6.12) for $\bar{\Phi}^{(f)}(E)$ in (5.7.7), we get

$$\bar{\sigma}_i^{28} \approx \int_{\Delta E_i} \frac{[1 - P_f(E)]\Sigma_{s(f)}^{p} + P_f(E)[\Sigma_{t(f)}(E)/\Sigma_{t(m)}(E)]\Sigma_{s(m)}^{p}}{\Sigma_{t(f)}(E) - [1 - P_f(E)]\Sigma_{s(f)}^{res}(E)} \frac{\sigma_a^{28}(E)\, dE}{E}, \tag{5.7.12}$$

where, in accord with the discussion in Section 5.5 and the normalization of $\bar{\Phi}^{(f)}(E)$ in (5.7.7), we have taken the C_g in (5.6.12) for each individual interval ΔE_i to be $1/\xi\Sigma_s(E_i)$ and where we understand the escape probability $P_f(E)$ to be that for a fuel rod in a lattice. In applying (5.7.12), we must remember that, if the resonance in ΔE_i is narrow, $\Sigma_{s(f)}^{res}(E)$ should be set to zero (although $\Sigma_{s(f)}^{p}$ should still include that of the U^{28}). On the other hand, if the resonance is wide, $\Sigma_{s(f)}^{res}(E)$ must be retained.

Substitution of (5.7.12) into (5.7.11) provides an approximate algebraic expression for the resonance-escape probability of a two-region cell. To check this against results obtained earlier for the homogeneous case we note that, in the homogeneous limit as V_f and V_m both go to zero (but with the ratio V_f/V_m kept constant), the spatial shape of the flux density $\Phi(\mathbf{r}, E)$ becomes flat, so that $\bar{\Phi}_f^{(f)}(E) \approx \bar{\Phi}_f^{(m)}(E)$ at all energies, and the rigorous equation

$$\int_0^\infty dE' \int_{V_f + V_m} dV\, \Sigma_{s(f)}(E' \to E)\Phi(\mathbf{r}, E')H^{(f)} = \int_{V_f + V_m} \Sigma_t(\mathbf{r}, E)\Phi_f(\mathbf{r}, E)\, dV \tag{5.7.13}$$

(derivable directly from (5.5.13) by application of Gauss's Law) becomes

$$\begin{aligned}\int_0^\infty dE' \int_{V_f} dV\, \Sigma_{s(f)}(E' \to E)\Phi(\mathbf{r}, E') &= V_m\Sigma_{t(m)}(E)\bar{\Phi}_f^{(m)}(E) + V_f\Sigma_{t(f)}(E)\bar{\Phi}_f^{(f)}(E)\\ &= [V_m\Sigma_{t(m)}(E) + V_f\Sigma_{t(f)}(E)]\bar{\Phi}_f^{(f)}(E).\end{aligned} \tag{5.7.14}$$

As a result (5.6.1) shows that, in the homogeneous limit,

$$P_{\mathrm{f}}(E) = \frac{V_{\mathrm{m}}\Sigma_{\mathrm{t(m)}}(E)}{V_{\mathrm{m}}\Sigma_{\mathrm{t(m)}}(E) + V_{\mathrm{f}}\Sigma_{\mathrm{t(f)}}(E)}. \tag{5.7.15}$$

We see that this is indeed the probability that, in a homogeneous mixture of fuel and moderator, the next interaction will be with the moderator.

Equation (5.7.12) then simplifies to

$$\bar{\sigma}_i^{28} \approx \int_{\Delta E_i} \frac{V_{\mathrm{f}}\Sigma_{\mathrm{s(f)}}^{\mathrm{p}}(E_i) + V_{\mathrm{m}}\Sigma_{\mathrm{s(m)}}^{\mathrm{p}}(E_i)}{V_{\mathrm{m}}\Sigma_{\mathrm{t(m)}}(E) + V_{\mathrm{f}}[\Sigma_{\mathrm{t(f)}}(E) - \Sigma_{\mathrm{s(f)}}^{\mathrm{res}}(E)]} \frac{\sigma_{\mathrm{a}}^{28}(E)\,dE}{E}. \tag{5.7.16}$$

If, now, we introduce $\bar{\xi}(E_i)$, an average loss in the logarithm of the energy for the homogeneous mixture in the energy range ΔE_i, with the definition

$$\begin{aligned}\bar{\xi}(E_i) &\equiv \frac{V_{\mathrm{f}} + V_{\mathrm{m}}}{V_{\mathrm{f}}\Sigma_{\mathrm{s(f)}}^{\mathrm{p}}(E_i) + V_{\mathrm{m}}\Sigma_{\mathrm{s(m)}}^{\mathrm{p}}(E_i)} \bar{\xi}\bar{\Sigma}_{\mathrm{s}}(E_i) \\ &= \frac{V_{\mathrm{f}}\bar{\xi}_{\mathrm{f}}\Sigma_{\mathrm{s(f)}}^{\mathrm{p}}(E_i) + V_{\mathrm{m}}\bar{\xi}_{\mathrm{m}}\Sigma_{\mathrm{s(m)}}^{\mathrm{p}}(E_i)}{V_{\mathrm{f}}\Sigma_{\mathrm{s(f)}}^{\mathrm{p}}(E_i) + V_{\mathrm{m}}\Sigma_{\mathrm{s(m)}}^{\mathrm{p}}(E_i)},\end{aligned} \tag{5.7.17}$$

equation (5.7.10) for the resonance-escape probability of the ith resonance in the homogeneous mixture becomes

$$p_i \approx \exp\left(-\int_{\Delta E_i} \frac{V_{\mathrm{f}}\Sigma_{\mathrm{a}}^{28}(E)\,dE}{\{V_{\mathrm{m}}\Sigma_{\mathrm{t(m)}}(E) + V_{\mathrm{f}}[\Sigma_{\mathrm{t(f)}}(E) - \Sigma_{\mathrm{s(f)}}^{\mathrm{res}}(E)]\}} \frac{1}{\bar{\xi}(E_i)E}\right). \tag{5.7.18}$$

(Recall that Σ_{a}^{28}, $\Sigma_{\mathrm{t(f)}}$, and $\Sigma_{\mathrm{s(f)}}^{\mathrm{res}}$ are per unit volume of fuel and $\Sigma_{\mathrm{t(m)}}$ is per unit volume of moderator.)

It follows from this result and (5.7.9) that the total resonance-escape probability in the homogeneous limit is given approximately by a formula identical with (5.7.18), except that the integration is extended over the whole range from E_{c} to E_1. We note again that, for narrow resonances, $\Sigma_{\mathrm{s(f)}}^{\mathrm{res}}(E)$ is omitted from (5.7.18) and the potential scattering of U^{28} is included in the definition (5.7.17) of $\bar{\xi}(E_i)$. When this is done, (5.7.18) reduces exactly to the homogeneous, narrow-resonance result derived in Section 5.4.

Equivalence Relationships

It is possible to establish an approximate equivalence between the value of $\bar{\sigma}_i^{28}$ for a homogeneous and that for a heterogeneous distribution of fuel and moderator. Thus, if we make use of the approximation (5.6.35) for the escape probability from the fuel rod in a lattice and, in addition, express the average attenuation factor $\langle G_1(E)\rangle$ in the moderator by the rational approximation (5.6.38), then (5.6.35) reduces to

$$P_{\mathrm{f}}(E) \approx \frac{1}{1 + \{\bar{R}_{\mathrm{s(f)}} + [\bar{R}_{\mathrm{s(f)}}/\bar{R}_{\mathrm{s(m)}}\Sigma_{\mathrm{t(m)}}(E)]\}\Sigma_{\mathrm{t(f)}}(E)}. \tag{5.7.19}$$

Substituting this expression for $P_{\mathrm{f}}(E)$ into (5.7.12) and rearranging yields

$$\bar{\sigma}_i^{28} \approx \int_{\Delta E_i} \frac{V_{\mathrm{f}}\Sigma^{\mathrm{p}}_{\mathrm{s(f)}}(E) + \left[\dfrac{V_{\mathrm{f}}}{\bar{R}_{\mathrm{s(f)}}\Sigma_{\mathrm{t(m)}}(E) + (\bar{R}_{\mathrm{s(f)}}/\bar{R}_{\mathrm{s(m)}})}\right]\Sigma^{\mathrm{p}}_{\mathrm{s(m)}}(E)}{\left[\dfrac{V_{\mathrm{f}}}{\bar{R}_{\mathrm{s(f)}}\Sigma_{\mathrm{t(m)}}(E) + (\bar{R}_{\mathrm{s(f)}}/\bar{R}_{\mathrm{s(m)}})}\right]\Sigma_{\mathrm{t(m)}} + V_{\mathrm{f}}[\Sigma_{\mathrm{t(f)}}(E) - \Sigma^{\mathrm{res}}_{\mathrm{s(f)}}(E)]} \frac{\sigma_{\mathrm{a}}^{28}(E)\,dE}{E}. \quad (5.7.20)$$

Comparison with the value of $\bar{\sigma}_i^{28}$ in the homogeneous limit (5.7.16) suggests that, if the total cross section in the moderator is independent of energy, we may define an "effective moderator volume" for the cell by

$$V_{\mathrm{m}}^{*} \equiv \frac{V_{\mathrm{f}}}{\bar{R}_{\mathrm{s(f)}}\Sigma_{\mathrm{t(m)}} + (\bar{R}_{\mathrm{s(f)}}/\bar{R}_{\mathrm{s(m)}})}. \quad (5.7.21)$$

Then, in a heterogeneous cell having a fuel volume V_{f} and a moderator volume V_{m}, the resonance integral for the ith resonance will be the same as that in a homogeneous mixture of the same fuel and moderator material for which the ratio of the volume occupied by moderator to that occupied by fuel is $V_{\mathrm{m}}^{*}: V_{\mathrm{f}}$. It follows immediately that, because of (5.7.11), there is a corresponding equivalence principle between resonance-escape probabilities of homogeneous mixtures and lattices.

If a rod is isolated ($\bar{R}_{\mathrm{s(m)}} \to \infty$) and if the moderator is nonabsorbing, so that $\Sigma_{\mathrm{t(m)}} = \Sigma^{\mathrm{p}}_{\mathrm{s(m)}}$, the equivalence we have just derived can be expressed in terms of an "effective scattering cross section" of the mixture per absorber atom. That is, the numerator and denominator of (5.7.20) may be divided by $n^{\mathrm{f}} V_{\mathrm{f}}$ where n^{f} is the number of atoms of fuel element material per unit volume. Then, with σ_{s}^{*}, the "effective scattering cross section per absorber atom," defined by

$$\sigma_{\mathrm{s}}^{*} \equiv \frac{1}{n^{\mathrm{f}}\bar{R}_{\mathrm{s(f)}}}, \quad (5.7.22)$$

(5.7.20) becomes

$$\bar{\sigma}_i^{28} = \int_{\Delta E_i} \frac{\sigma^{\mathrm{p}}_{\mathrm{s(f)}}(E) + \sigma_{\mathrm{s}}^{*}}{\sigma_{\mathrm{s}}^{*} + \sigma_{\mathrm{t(f)}}(E) - \sigma^{\mathrm{res}}_{\mathrm{s(f)}}(E)} \frac{\sigma_{\mathrm{a}}^{28}(E)\,dE}{E}. \quad (5.7.23)$$

Thus the isolated fuel element in an infinite volume of moderator may be regarded as a single, homogeneous material having an altered scattering cross section.

Stating the equivalence in this way is an alternate to the slightly more general approach we have developed. Both schemes, of course, depend on the validity of the simple rational approximations for $P_{\mathrm{f}}(E)$ and $\langle G_{\mathrm{m}}(E)\rangle$. Hence we do not expect the correspondence predicted by the equivalence principle to be exact. Nevertheless the principle provides us with a very useful way of correlating different calculations and measurements of isolated fuel rods, homogeneous mixtures, and lattices.

For example it has been found that measurements of the total resonance integral $\bar{\sigma}^{28}$ for isolated uranium-metal and UO_2 fuel rods can be fit quite well to the empirical formulas (expressing $\bar{\sigma}^{28}$ in barns)

$$\begin{aligned} \bar{\sigma}^{28} &= 2.95 + 25.8\sqrt{(A/M)} \quad \text{for U metal,} \\ \bar{\sigma}^{28} &= 4.45 + 25.6\sqrt{(A/M)} \quad \text{for } UO_2, \end{aligned} \tag{5.7.24}$$

where A is the surface area per unit length and M the mass per unit length of the rod. (These expressions have the $1/v$ part of the absorption subtracted out.)

To find from these correlations $\bar{\sigma}^{28}$ for a rod in a lattice of fuel and moderator with volumes V_{f} and V_{m}, we proceed as follows:

1. Find $\bar{R}_{s(\mathrm{f})}$ and $\bar{R}_{s(\mathrm{m})}$ for the fuel and moderator portions of the cell in the lattice.
2. Find V_{m}^* from (5.7.21).
3. For this value of V_{m}^* and using (5.7.21) again, this time with $\bar{R}_{s(\mathrm{m})} \to \infty$, find $\bar{R}_{\mathrm{s(f)}}^{\mathrm{eff}}$ for an equivalent isolated rod of volume V_{f}. (Note that V_{m}^* is finite for an isolated rod even though $V_{\mathrm{m}} \to \infty$.)
4. From (5.6.20) in the form $\bar{R}_{\mathrm{s(f)}}^{\mathrm{eff}} = 4V_{\mathrm{f}}^{\mathrm{eff}}/A^{\mathrm{eff}}$, obtain the effective surface area and mass of the equivalent isolated rod.
5. Use this result in (5.7.24) to compute the resonance integral for this equivalent isolated rod. (Note that $A^{\mathrm{eff}}/M = 4/(R_{\mathrm{s(f)}}^{\mathrm{eff}}\rho_{\mathrm{f}})$, where ρ_{f} is the density of the fuel.)

According to the equivalence principle, the resonance integral for this equivalent rod is the same as that for the actual rod in the lattice. Equation (5.7.6) gives us at once the corresponding resonance-escape probability for the lattice.

5.8 Determination of Energy-Group Constants for Other Geometrical Arrangements and Other Energy Ranges

The methods developed for the two-region cell in the resolved-resonance range can be extended to more complex geometries (clad rods, for example) and applied to energy ranges where the resonances are unresolved or are broad and overlapping. In this way spatially averaged spectra suitable for finding equivalent homogeneized group constants for the cell in the fast and thermal energy ranges can be obtained. In addition analytical procedures can be derived for determining such traditional quantities as the fast-fission factor and the thermal utilization, and just as with the resonance-escupe probability p, the approximate theory and basic nuclear data can be checked against the values of these parameters inferred from experiment.

Such procedures, based on the use of escape probabilities, are called *collision-theory methods*. The essential characteristic that makes them efficient is the determination of the escape probabilities as a separate calculation. Isolating this part of the problem requires, in turn, that the sources of neutrons at each energy within a given region be spatially flat (see (5.6.2)). This assumption uncouples the space and energy parts of the overall problem. Unless it is valid, collision theory is unlikely to be accurate. The flat-source assumption is *least* likely to be true when the absorption cross sections of the materials making up the cell are large and fairly smooth functions of energy. Fortunately energy-group methods for solving the transport equation do not require the use of very many groups for this case and hence can be used to obtain accurate values for $\Phi(\mathbf{r}, E)$ throughout the cell when the flat-source approximation breaks down.

The treatment of cell heterogeneity in this chapter has involved another assumption that is very convenient but not essential. This is the zero-net-leakage assumption (5.5.1). Depending on the circumstances, alternate conditions may be more appropriate. Thus, for the cells associated with fuel rods next to the blade of a control rod, a more accurate way of eliminating $\mathbf{J}(\mathbf{r}, E)$ at the edge of the cell is a condition such as

$$\int_{S_r} [\mathbf{n}\cdot\mathbf{J}(\mathbf{r}, E)]\, dS = \frac{\alpha S_r}{V_m}\int_{V_m} \Phi(\mathbf{r}, E)\, dV, \tag{5.8.1}$$

where S_r is the part of the cell surface adjacent to the control blade and α is a constant determined by some separate calculation in which the control blade is represented explicitly. Similarly, if the cells in a cluster are not all identical (for example if slightly-enriched-U^{25} fuel rods are seeded with plutonium fuel rods), it is possible to make use of the fact that the net current out of one surface of a cell equals that into the surface of the adjacent cell and thus to couple the various cells in an accurate manner. Such a procedure leads to much more severe numerical problems since it results in coupled equations of the type (5.5.21)—one pair for each cell in the cluster—containing explicit terms representing currents across the boundaries of the cell. It also complicates the computation of the escape probabilities. The overall equations, however, are still simpler than a set of multigroup transport equations.

References

Bell, G. I., 1959. "A Simple Treatment for Effective Resonance Absorption Cross Sections in Dense Lattices," *Nuclear Science and Engineering* **5**: 138–139.

Case, K. M., DeHoffman, F., and Placzek, G., 1953. *Introduction to the Theory of Neutron Diffusion* (Washington, D. C.: U. S. Government Printing Office).

Chernick, J., 1955. *Proceedings of the International Conference on the Peaceful Uses of Atomic Energy*, Volume 5, Paper 603.

Chernick, J., and Vernon, R., 1958. "Some Refinements in the Calculation of Resonance Integrals," *Nuclear Science and Engineering* **4**: 649–672.

Dresner, L., 1960. *Resonance Absorption in Nuclear Reactors* (London and New York: Pergamon Press).

Goodjohn, A. J., and Pomraning, G. C., eds., 1966. *Reactor Physics in the Resonance and Thermal Regions*, Volume 2 (Cambridge, Mass.: The MIT Press).

Hinman, G. W., et al., 1963. "Accurate Doppler Broadened Absorption," *Nuclear Science and Engineering* **16**: 202–207.

Lamarsh, J. R., 1966. *Introduction to Nuclear Theory* (Reading, Mass.: Addison-Wesley), Chapters 2, 7.

Numerical Reactor Calculations, 1972. Proceedings of a seminar held in Vienna, Austria, 17–21 January 1972, Vienna IAEA, ST1/Pub/307, Sessions III and IV.

Rothenstein, W., 1960. "Collision Probabilities and Resonance Integrals for Lattices, "*Nuclear Science and Engineering* **7**: 162–171.

Solbrig, A. W., Jr., 1961. "Doppler Effect in Neutron Absorption Resonances," *American Journal of Physics* **29**: 257–261.

Problems

1. The Breit-Wigner single-level expressions for $\sigma_a(E)$ and $\sigma_s(E)$ in the neighborhood of a resonance may be written

$$\sigma_a(E) = \frac{\sigma_0}{1+x^2}, \quad \text{where } x \equiv \frac{2}{\Gamma}(E - E_0),$$

$$\sigma_s(E) = \frac{\sigma_0 \Gamma_n}{\Gamma_\gamma}\frac{1}{1+x^2} + \frac{2\sigma_0 \Gamma R}{\lambda\!\!\!^{-}_0 \Gamma_\gamma}\frac{x}{1+x^2} + 4\pi R^2, \quad \text{where } \lambda\!\!\!^{-}_0 \equiv \frac{\hbar}{\sqrt{(2\mu E_0)}}.$$

For the first resolved resonance in U^{28}, $\Gamma_n = 0.0015$ eV, $\Gamma_\gamma = 0.026$ eV, $\sigma_0 = 23{,}000$ b, and $E_0 = 6.67$ eV. Find the practical width at 0°K for this resonance in a mixture for which the ratio of the concentration of hydrogen atoms to that of U^{28} atoms is 10: 1. (Assume that the potential-scattering cross section for U^{28} is a constant 8.3 b and that that for hydrogen is 20 b. Take $\mu = 1.66 \times 10^{-24}$ gm, $\hbar = 1.05 \times 10^{-27}$ erg-sec, and 1 eV $= 1.6 \times 10^{-12}$ erg.)

2. Consider a homogeneous system in which there is no leakage and in which hydrogen is the only moderator. If we consider the whole energy range from 1 eV to the top of the fission source (15 MeV), above which $\chi(E)$, the fission source, and $\psi(E)$, the scalar flux density, vanish, the appropriate equation to determine $\psi(E)$ is

$$[\Sigma_a(E) + \Sigma_s^H(E)]\psi(E) \equiv F(E) = \int_E^\infty \Sigma_s^H(E')\frac{\psi(E')\,dE'}{E'} + \chi(E).$$

a. Show that, in this range,

$$F(E) = \chi(E) + \frac{1}{E}\int_E^\infty \frac{\chi(E')\Sigma_s^H(E')dE'}{\Sigma_a(E') + \Sigma_s^H(E')}\exp\left(-\int_E^{E'} \frac{\Sigma_a(E'')dE''}{E''[\Sigma_a(E'') + \Sigma_s^H(E'')]}\right).$$

b. If E_1 is an energy such that $\chi(E) = 0$ for $E \le E_1$, show that, for $E < E_1$, this result reduces exactly to (5.4.5).

3. Derive (5.2.7) from (5.2.5).

4. Consider a homogeneous mixture of hydrogen and a heavy isotope. In the energy range E_g–E_{g-1}, suppose that the microscopic absorption cross section of the heavy isotope can be represented by two square resonances as in Figure 5.7. Suppose that, within the resonance, the macroscopic abosrption cross sections are $\Sigma_{a1} = \Sigma_{a2} = 200$ cm^{-1} and the resonance scattering cross sections are $\Sigma_{s1}^{res} = \Sigma_{s2}^{res} = 20$ cm^{-1}. Outside the resonance, assume $\Sigma_a = 0$, and let the

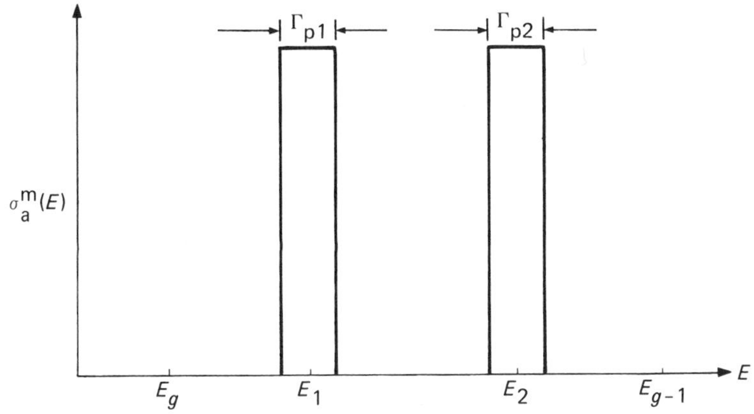

Figure 5.7

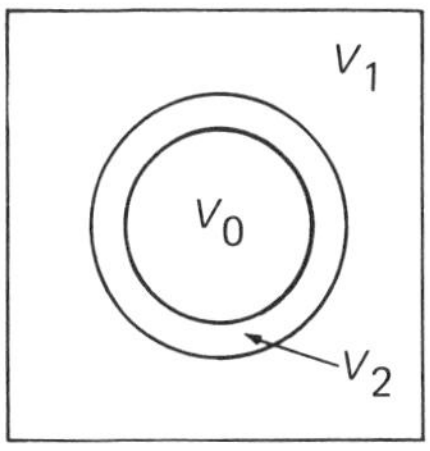

Figure 5.8

macroscopic potential-scattering cross sections for hydrogen (H) and the heavy absorber (m) be $\Sigma_{sH}^{p} = 0.5\ \text{cm}^{-1}$, $\Sigma_{sm}^{p} = 0.2\ \text{cm}^{-1}$. Take $\xi^{H} = 1$, $\xi^{m} = 0.01$, and let the practical widths be $\Gamma_{p1} = \Gamma_{p2} = 1$ eV.

If $E_g = 5$ eV, $E_{g-1} = 20$ eV, and the energies at the centers of the resonances are $E_1 = 10$ eV and $E_2 = 15$ eV, find the absorption cross section Σ_{ag} for a group extending from E_g to E_{g-1} using:

a. the NR expression (5.4.18);
b. the corrected expression (5.4.22) for $\psi(E)$;
c. the WR approximation.

Which approximation can be exprected to be most accurate in this case?

5. Show that, under the assumptions of no absorption, elastic scattering isotropic in the center-of-mass system, scattering cross sections independent of energy, and (5.6.11) valid, a solution to (5.5.19) is $\overline{\Phi}^{(m)}(E) = \overline{\Phi}^{(f)}(E) = C/E$.

Does this result imply that $\Phi(\mathbf{r}, E)$ throughout the cell is spatially flat?

6. For a cell composed of three regions, fuel (V_0), cladding (V_2), and moderator (V_1), as in Figure 5.8, (5.5.10) becomes

$$\int_{S_{12}} \mathbf{J}(\mathbf{r}, E)\cdot\mathbf{n}_{12}\, dS + \Sigma_{t(1)}(E) V_1 \overline{\Phi}^{(1)}(E) = \bar{Q}_1(E) \quad \text{for } \mathbf{r} \text{ in } V_1,$$

$$\int_{S_{12}} \mathbf{J}(\mathbf{r}, E)\cdot\mathbf{n}_{21}\, dS + \int_{S_{02}} \mathbf{J}(\mathbf{r}, E)\cdot\mathbf{n}_{20}\, dS + \Sigma_{t(2)}(E) V_2 \overline{\Phi}^{(2)}(E) = \bar{Q}_2(E) \quad \text{for } \mathbf{r} \text{ in } V_2,$$

$$\int_{S_{02}} \mathbf{J}(\mathbf{r}, E)\cdot\mathbf{n}_{02}\, dS + \Sigma_{t(0)}(E) V_0 \overline{\Phi}^{(0)}(E) = \bar{Q}_0(E) \quad \text{for } \mathbf{r} \text{ in } V_0,$$

where S_{ij} is the surface common to regions i and j, $\mathbf{n}_{ij}$ is the normal extending in the direction *from* region i *to* region j (so that $\mathbf{n}_{ij} = -\mathbf{n}_{ji}$), and the $\bar{Q}_i(E)$ are spatially integrated sources in region i (see (5.5.10)).

To replace the surface integrals with expressions involving the average fluxes $\bar{\Phi}^{(i)}(E)$, we may again split $\Phi(\mathbf{r}, E)$ and $\mathbf{J}(\mathbf{r}, E)$ into partial contributions, as in (5.5.12):

$$\Phi(\mathbf{r}, E) \equiv \Phi_0(\mathbf{r}, E) + \Phi_1(\mathbf{r}, E) + \Phi_2(\mathbf{r}, E),$$
$$\mathbf{J}(\mathbf{r}, E) \equiv \mathbf{J}_0(\mathbf{r}, E) + \mathbf{J}_1(\mathbf{r}, E) + \mathbf{J}_2(\mathbf{r}, E),$$

where the $\Phi_i(\mathbf{r}, E)$ and $\mathbf{J}_i(\mathbf{r}, E)$ obey

$$\nabla \cdot \mathbf{J}_i(\mathbf{r}, E) + \Sigma_t(\mathbf{r}, E)\Phi_i(\mathbf{r}, E) = Q_i(\mathbf{r}, E)H^{(i)} \qquad (i = 0, 1, 2),$$

analogous to (5.5.13), with the boundary condition that the normal component of $\mathbf{J}(\mathbf{r}, E)$ on the *outer* boundary of V_1 vanishes. Here

$$H^{(i)} = \begin{cases} 1 \text{ for } \mathbf{r} \text{ in } (i), \\ 0 \text{ for } \mathbf{r} \text{ not in } (i). \end{cases}$$

Show that, if we define six "escape probabilities" by

$$P_{ij}(E) \equiv \frac{\Sigma_{t(j)}(E)\, V_j \bar{\Phi}_i^{(j)}(E)}{\bar{Q}_i(E)} \qquad (i, j = 0, 1, 2),$$

where

$$\bar{\Phi}_i^{(j)}(E) \equiv \frac{1}{V_j}\int_{V_j} \Phi_i(\mathbf{r}, E)\, dV \quad \text{and} \quad \bar{Q}_i(E) \equiv \int_{V_i} Q_i(\mathbf{r}, E) H^{(i)}\, dV,$$

then

$$\int_{S_{02}} \mathbf{J}(\mathbf{r}, E)\cdot \mathbf{n}_{02}\, dS = \bar{Q}_0(E)[P_{01}(E) + P_{02}(E)] - \bar{Q}_1(E)P_{10}(E) - \bar{Q}_2(E)P_{20}(E).$$

Note that this result, along with an analogous expression for $\int_{S_{12}} \mathbf{J}(\mathbf{r}, E)\cdot \mathbf{n}_{12} dS$ will permit us, in a formal fashion, to eliminate the surface integrals from the first set of equations in this problem and derive equations analogous to (5.5.19).

7. Consider a slab cell composed of fuel and moderator, as shown in Figure 5.9, and assume that the net leakage of neutrons is zero at the outer surfaces of the cell, that is, at $X = \pm(\Delta_f + \Delta_m)$. Assume that, at some energy E, the diffusion-theory model is valid.

a. At that energy, and using the flat-source approximation (5.6.2), find analytical expressions for the escape probabilities $P_f(E)$ and $P_m(E)$.

b. Show that, in the homogeneous limit where Δ_f and Δ_m go to 0 but Δ_f/Δ_m remains constant,

$$P_f(E) = \frac{\bar{\Sigma}_{t(m)}(E)}{\bar{\Sigma}_t(E)},$$
$$P_m(E) = \frac{\bar{\Sigma}_{t(f)}(E)}{\bar{\Sigma}_t(E)},$$

where $\bar{\Sigma}_{t(m)}(E)$ is the total moderator interaction cross section *averaged over the entire cell volume*, $\bar{\Sigma}_{t(f)}(E)$ is the analogous total fuel cross section, and $\bar{\Sigma}_t(E) = \bar{\Sigma}_{t(m)}(E) + \bar{\Sigma}_{t(f)}(E)$.

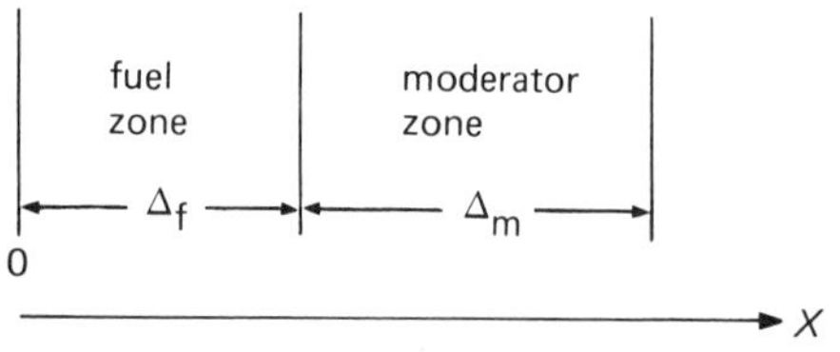

Figure 5.9

c. For this limiting condition $P_f + P_m = 1$. Explain physically why $P_f + P_m \neq 1$ when Δ_f and Δ_m are large.

8. Show that, if we make the Wigner rational approximation for $P_f^0(E)$ and for $\langle G_m(E)\rangle$, the Bell formula (5.6.34) for $P_f(E)$ is identical to the rational approximation for $P_f^0(E)$ with the mean cord length in the fuel rod augmented by an amount $\bar{R}_{sf}/(\bar{R}_{sm}\Sigma_{t(m)})$.

9. Show that, in the homogeneous limit, $V_m^* = V_m$.

10. Consider a fuel rod in which U^{28} (of concentration n^{28} atoms/cc) is the only neutron-absorbing material. Suppose this is embedded in an infinite amount of nonabsorbing moderator ($\Sigma_{t(m)}(E) = \Sigma_{s(m)}^p$).

a. Show that (5.7.16) for the homogeneous limit of $\bar{\sigma}_i^{28}$ becomes, under these conditions,

$$\bar{\sigma}_i^{28} = \int_{\Delta E_i} \frac{1}{1 + [\sigma_a^{28}(E)/\sigma_s^{NR}]} \frac{\sigma_a^{28}(E)}{E} \, dE,$$

where $\sigma_s^{NR} \equiv (V_f \Sigma_{s(f)}^p + V_m \Sigma_{s(m)}^p)/V_f n^{28}$, the narrow-resonance scattering cross section per atom of absorber.

b. Then show that (5.7.20) for the heterogeneous case becomes

$$\bar{\sigma}_i^{28} = \int_{\Delta E_i} \frac{1}{1 + [\sigma_a^{28}/(\sigma_{s(f)}^{NR} + \sigma_\Omega)]} \frac{\sigma_a^{28}}{E} \, dE,$$

where $\sigma_{s(f)}^{NR} \equiv \Sigma_{s(f)}^p/n^{28}$, the narrow-resonance scattering cross section of the fuel rod per absorber atom, and $\sigma_\Omega \equiv A/4n^{28}V_f$, where A is the surface area of the rod.

6 Fuel Depletion and Its Consequences

6.1 Introduction

To sustain a power level of one watt in a reactor requires that there be approximately 3×10^{10} fissions per second. Since only about 80 percent of the neutrons absorbed in fuel actually lead to fission, we conclude that the *destruction* rate of fissionable material in a reactor operating at a power level of one watt is approximately 3.75×10^{10} nuclei per second. Some simple arithmetic leads to the conclusion that a thousand-megawatt (electric) power plant operating at a thermal efficiency of 30 percent burns fissionable material at a rate of approximately 2.8 gm/min. Compared with, say, the fuel burned to get a jet transport into the air, this figure is astonishingly low. Nevertheless it implies that, in a year of such power operation, a nuclear reactor will burn about 1.5 metric tonnes of fissionable material.

The destruction of this amount of fuel and the consequent introduction of fission fragments and other products of neutron absorption will have a significant effect on the nuclear characteristics of the reactor.

From the point of view of reactor physics, the treatment of depletion and its consequences involves a straightforward extension of the methods we have developed for determining criticality conditions. Changes in the flux level and shape of a reactor operating under constant high-power conditions are so slow that there is no need to resort to time-dependent neutron equations. Instead a static calculation may be made for the reactor in its initial, hot, full-power, critical state, and the flux shape resulting from that calculation may be assumed to remain constant for an extended period of time (called a *depletion time step*), even though both depletion and the control-rod motion needed to maintain criticality are taking place in the real reactor during that time. Thus the flux shape assumed to persist throughout a depletion time step is correct (within the limits of group diffusion theory) at the beginning of the time step, but it deviates slightly as time proceeds from the values that would be computed if we were to solve the time-dependent group diffusion equations throughout the time interval. Changes in the concentrations of materials in the reactor due to depletion are calculated using this fixed flux shape throughout the time step in a manner to be described below. A new flux shape (and critical control-rod configuration) based on these changes is then determined, and this shape is assumed to persist during the next depletion time step.

Thus the conventional method of analyzing a reactor during its depletion lifetime utilizes a sequence of static criticality computations, interspersed with time-dependent depletion computations to determine the concentrations throughout the reactor of all isotopes affected by the depletion process. This tandem method of determining the spatial changes in flux shape and local isotopic content provides a surprisingly good approximation to the true space-time behavior during depletion. If certain modifications to the scheme are made, time steps of 1000 full-power hours or more can be taken without encountering serious errors due to the "tandem approximation."

We shall examine the phenomena associated with depletion in three stages. First we shall consider how the concentrations of nuclear materials are changed in the presence of a neutron flux. Next we shall examine how to translate these changes into the computation of energy-group parameters. Finally we shall discuss problems of reactor design and control arising from depletion and its consequences.

6.2 Changes in Isotopic Concentrations in the Presence of a Neutron Flux

Destruction Processes

Nuclear interactions in which neutrons are destroyed change the identity of the interacting nuclei. Thus, if we wish to compute the effect of neutron absorption on an isotope j having a concentration of $n^j(\mathbf{r}, t)$ nuclei per unit volume at time t, we simply find the total number of neutrons being absorbed in j per second in a volume dV. According to (2.2.5) and (2.2.8), this rate is

$$dV \frac{\partial n^j(\mathbf{r}, t)}{\partial t} = -n^j(\mathbf{r}, t)\, dV \int_0^\infty \sigma_a^j(E)\Phi(\mathbf{r}, E, t)\, dE, \tag{6.2.1}$$

where

$$\sigma_a^j(E) \equiv \sigma_t^j(E) - \sigma_s^j(E), \tag{6.2.2}$$

so that $\sigma_a^j(E)$ includes both fission and (n, 2n), (n, p), and (n, γ) reactions. The solution to (6.2.1) is

$$n^j(\mathbf{r}, t) = n^j(\mathbf{r}, t_0) \exp\left(- \int_{t_0}^t dt' \int_0^\infty \sigma_a^j(E)\Phi(\mathbf{r}, E, t')\, dE \right). \tag{6.2.3}$$

Strictly speaking, because of local temperature changes leading to Doppler broadening, $\sigma_a^j(E)$ in this expression can be time-dependent. However, even for such fertile materials as U^{28}, which may be present in high concentrations, time changes in $\sigma_a^j(E)$ at constant reactor-power level will be, at most, a few tenths of a percent. They may be ignored as far as depletion computations are concerned, and $\sigma_a^j(E)$ may be taken out of the time integral in (6.2.3). As a result the time integral of the flux $\int_{t_0}^t \Phi(\mathbf{r}, E, t')\, dt'$ appears as an isolated parameter in expressions such as (6.2.3); and it is therefore a common measure of depletion due to neutrons in the range $dE\, dV$. For this reason it is customary to use the energy-flux-time integral $\int_{t_0}^t dt' \int_0^\infty dE\, \Phi(\mathbf{r}, E, t')$ as a measure of the amount of exposure to which a given material in a given location in the reactor has been subject. The units of this double integral are neutrons/cm^2. However one usually speaks of an exposure of so many "nvt" (average neutron density × average neutron speed × time of exposure). Accordingly a sample of U^{25} having an average one-group absorption cross section of 600 barns and irradiated to 10^{21} nvt will be depleted by an amount $1 - \exp(-0.6)$ or 45 percent. The terms "flux time" and "fluence" are also used to refer to the magnitude of $\int_{t_0}^{t} dt' \int_0^\infty dE\, \Phi(\mathbf{r}, E, t')$.

Notice that there is a certain ambiguity in using nvt as a measure of exposure. The energy dependence of $\Phi(\mathbf{r}, E, t)$ will, in general, depend on location and change with time. Thus equality of $(\text{nvt})_k \equiv \int_{t_0}^{t} dt' \int_0^\infty dE\, \Phi(\mathbf{r}_k, E, t')$ $(k = 1, 2)$ for two different locations $\mathbf{r}_1$ and $\mathbf{r}_2$ and corresponding exposure times t_1 and t_2 does not imply equality of $\int_{t_0}^{t} dt' \int_0^\infty dE\, \sigma_a^j(E)\Phi(\mathbf{r}_k, E, t')$ $(k = 1, 2)$. In fact equality doesn't hold even if $\sigma_a^j(E)$ is a $1/v$ cross section. Nevertheless, since a given material in a reactor is likely to be exposed to the same flux-time history independent of its location, using nvt as a measure of that history is satisfactory for most purposes.

There is another measure of exposure very commonly applied to reactor fuels. This one is based on the correspondence between fuel exposure and total fission energy released as a result of that exposure. The unit in question is the *megawatt-day per tonne* (MWD/T). As the name suggests, fuel in a reactor operating at a specific power level of one megawatt per tonne of fuel initially present will be depleted by one MWD/T in one day.

The unit of depletion MWD/T suffers from the same lack of precision as does nvt. In addition there are different conventions regarding the words "tonne" and "fuel." When the unit was first introduced, some persons meant a tonne to be 2000 pounds. Today there is general agreement that it means 1000 kg (hence the spelling "tonne," rather than "ton"). The convention of what is meant by "fuel" is even more confusing. The unit MWD/T was initially applied to natural-uranium reactors where the fuel was in the metallic form. For reactors of this type the fact that two fuel rods have been depleted the same number of megawatt-days per tonne implies that they have roughly the same U^{25} and Pu^{49} content and have suffered roughly the same radiation damage. For reactors containing several slightly or fully enriched fuel materials in the form of oxides or alloys or of dispersions of particles in some sort of binding material, this correspondence is lost. Moreover the question of whether the oxygen, alloy, or binding material is to be considered as part of the "fuel" is an arbitrary matter. At present the convention is that, in the unit MWD/T, the fuel is composed of the *heavy* elements U, Pu, Th, etc. Thus, in fuel elements of uranium oxide or carbide or combinations of uraniun, plutonium, and stainless steel, only the tonnes of uranium and plutonium are counted in determining the exposure in MWD/T.

Creation Processes

Except for the case of fission reactions and some rare (n, 2n) and (n, p) reactions, destruction of a nucleus of isotope j by neutron interaction leads to creation of isotope $j+1$ along with the release of a γ ray. This isotope will itself absorb neutrons and may, in addition, be radioactive. Thus, if $\sigma_\gamma^j(E)$ is the (n, γ) cross section for isotope j and $\sigma_a^{j+1}(E)$ and λ^{j+1} are the absorption cross section and radioactive decay constant of isotope $j+1$, we have for the rate of change of concentration of isotope $j+1$:

$$\frac{\partial}{\partial t} n^{j+1}(\mathbf{r}, t) = n^j(\mathbf{r}, t) \int_0^\infty \sigma_\gamma^j(E)\Phi(\mathbf{r}, E, t)\, dE - n^{j+1}(\mathbf{r}, t)\left[\int_0^\infty \sigma_a^{j+1}(E)\Phi(\mathbf{r}, E, t)\, dE + \lambda^{j+1}\right]. \tag{6.2.4}$$

If $\Phi(\mathbf{r}, E, t)$ is known, solving this equation simultaneously with (6.2.1) will yield the time dependence of $n^{j+1}(\mathbf{r}, t)$. The creation rate of isotope $j+2$ of the original element can then be found by using (6.2.4), with j replaced by $j+1$. In addition the creation rate of the isotope of mass $j+1$ for a *different* element, the one to which $j+1$ decays, is given by $\lambda^{j+1}n^{j+1}(\mathbf{r}, t)$. (It is necessary to generalize notation so that atomic number as well as mass number is identified in order to write an equation describing this latter process.)

Thus neutron capture and radioactive decay lead to the creation of a great many isotopes not originally present in the reactor. The calculation of the concentration of these isotopes at any location in the reactor and at any time requires that we know $\Phi(\mathbf{r}, E, t)$ and the appropriate cross section and that we solve a number of coupled time-dependent equations.

Fortunately, from a criticality viewpoint, only a few chains of isotopes resulting from neutron irradiation in a reactor need to be considered. There is, for example, generally a negligible effect on the critical condition of isotopes created by neutron capture in the moderator, coolant, or structural material. This is because the materials used for these constituents are purposely chosen to have small absorption cross sections. As a consequence the amount of daughter material created is small. For example, in a PWR for which zirconium is the fuel clad and structural material, a typical value of the one-group flux $\int_0^\infty \Phi(\mathbf{r}, E, t)\, dE$ is 3×10^{13} neutrons/cm^2/sec. Thus, with a one-group absorption cross section for zirconium of approximately 0.18 b, the fractional amount of that material destroyed in one year (approximately 3×10^7 sec) of constant full-power operation is, according to (6.2.3), $1 - \exp(-0.18 \times 10^{-24} \times 9 \times 10^{20}) = 0.00016$. Since *all* the zirconium in the reactor absorbs only a small fraction of the neutrons, this change in absorption rate due to zirconium depletion produces a completely negligible effect on criticality. In fact the total amount of zirconium initially present is not known to that degree of accuracy. Accordingly, in determining the criticality history of a reactor containing zirconium structural material, we simply ignore the depletion of the zirconium.

In this regard, however, it should be kept in mind that the products of neutron capture may be important because of considerations other than that of criticality. In a boron–stainless-steel control rod, for example, the helium created by neutron capture in B^{10} has a negligible effect on criticality. Being a gas, however, it may change significantly the metallurgical properties of the rod. In addition many products of neutron capture, while of no importance to criticality, may present serious problems from a radiation-protection viewpoint. Tritium formed from neutron capture in a D_2O-moderated reactor is an example. (Even in light-water-moderated reactors, tritium formed from ternary fission and neutron capture in B^{10} is sufficiently plentiful to require special precautions against its release.) The isotopes Na^{24} (15.0-hr half-life; 2.75-MeV γ ray released) and Co^{60} (5.27-year half-life; 1.33-MeV γ ray released) formed by the Na^{23} (n, γ) Na^{24} reaction in sodium coolant and by the Co^{59} (n, γ) Co^{60} reaction in some stainless steels are other examples of isotopes which are negligibly important from a criticality viewpoint but which cause severe radiation-protection problems.

Of course not all the nuclear reactions in which neutrons are absorbed produce a negligible effect on the criticality of the system. Reactions with fissile or fertile materials effect criticality in a major way. The extent to which chains of reactions emanating from such reactions must be followed depends on the concentration of isotopes initially present in the material and on the amount of exposure to which the material is subject. Thus it is more important to compute the concentrations of higher isotopes of plutonium (Pu^{41} and Pu^{42}) in a material consisting initially of plutonium mixed with natural uranium, than it is in a material consisting initially of slightly enriched uranium containing no plutonium.

Rather than presenting all the coupled equations, similar to (6.2.1) and (6.2.4), relating the concentrations of important isotopes associated with a given exposure history, we shall merely show the chains themselves for the most common reactor fuels. Accordingly, with half-lives and cross sections at 0.025 eV as indicated, the parts of these chains that effect the criticality of a reactor are displayed in Figure 6.1 for the U^{28} chain and Figure 6.2 for the Th^{02} chain.

Only the reactions important to criticality and breeding for reactors in use or under consideration today are indicated in these two chains. For example the α-decay rates of the various heavy isotopes are not displayed. The shortest-lived of them (Pu^{41}) has a half-life of 12.9 years (and the half-lives of the others run from thousands to millions of

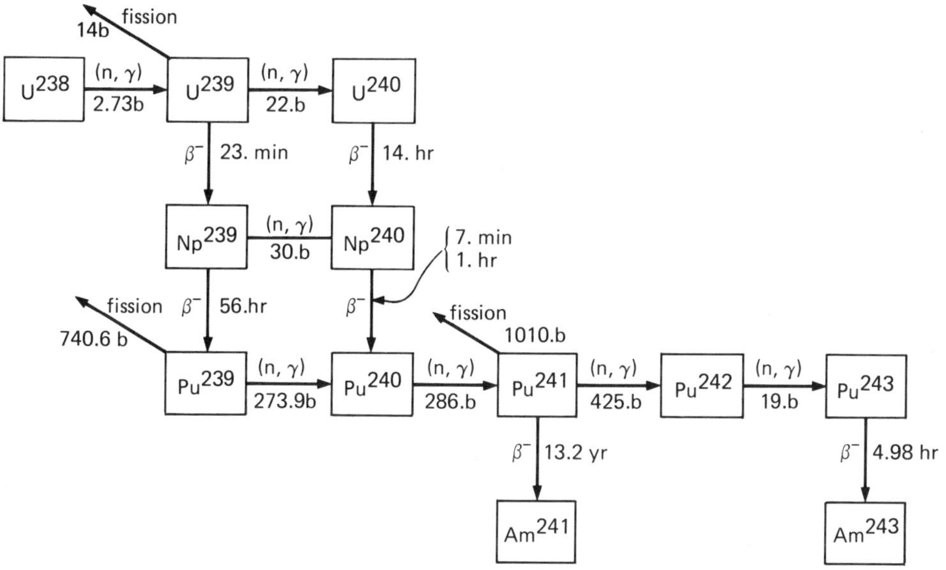

Figure 6.1 The chain of isotopes created by neutron irradiation of U^{238}. (The mass-number form of notation is used here to emphasize mass relationships in the chain.)

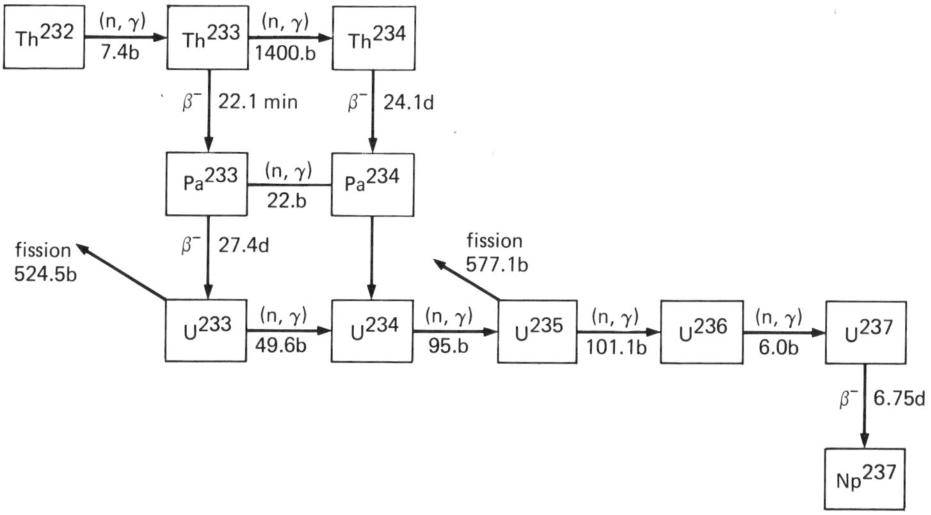

Figure 6.2 The chain of isotopes created by neutron irradiation of Th^{232}.

years). Thus, although α emission causes a severe handling problem for plutonium, it has only a negligible effect on criticality.

There are, however, some reactions not shown on the decay-chain diagrams that are important for criticality. Thus the fact that U^{28} and Th^{02} (as well as most of the other members of the chains) will fission if struck by high-energy neutrons is not indicated. Also, since the (n, γ) cross sections shown are at 0.025 eV, the importance of resonance absorption is not indicated. These reactions create no heavy isotopes not already shown on the chains. However they do alter the absorption and fission rates implied by the 0.025-eV cross sections shown and must be included in calculations of how the isotope concentrations in the chains change with time.

A Mathematical Description of the Depletion Process

It is a straightforward procedure to derive equations describing the rate of growth and decay of the isotopes shown in the chains. Such equations have the form of (6.2.4) with an extra production term due to radioactive decay included. However, for a given fuel composition in a given reactor, it is usually not necessary to consider all the members of the chain. To illustrate this point and to indicate the physical reasoning applied to determine what isotopes of a chain should be considered in criticality calculations, we shall develop in detail the equations specifying the changes in composition during core lifetime in a fuel element composed of highly enriched uranium.

The initial concentrations of uranium isotopes in such a fuel element will depend on whether the fuel material has come from a diffusion plant (in which case it can contain only U^{24}, U^{25}, and U^{28}, the isotopes that occur in nature) or from a reprocessing plant (in which case it can also contain U^{26}). In any event, highly enriched uranium contains approximately 93 percent U^{25}, so that isotopes formed from U^{24}, U^{26}, and U^{28} will be of only minor importance. Accordingly, right at the start, we shall limit the isotopes under consideration to U^{24}, U^{25}, U^{26}, U^{28}, U^{29}, Np^{39}, and Pu^{49} (and subsequently we shall show that most of these are unimportant).

If, then, we define the one-group flux and cross section for process α at point $\mathbf{r}$ and time t by

$$\begin{aligned} \Phi_1 &= \Phi_1(\mathbf{r}, t) \equiv \int_0^\infty \Phi(\mathbf{r}, E, t)\, dE, \\ \sigma_\alpha^j &= \sigma_\alpha^j(\mathbf{r}, t) \equiv \frac{\int_0^\infty \sigma_\alpha^j(E)\Phi(\mathbf{r}, E, t)\, dE}{\int_0^\infty \Phi(\mathbf{r}, E, t)\, dE}, \end{aligned} \tag{6.2.5}$$

and if we adopt the convention introduced in Section 1.6 of using for the supercript j a two-symbol identification, the first symbol being the last digit of the atomic number of the nucleus and the second being the last digit of the mass number, the concentrations of the isotopes of interest change according to the equations

$$\begin{aligned}
\frac{\partial n^{24}}{\partial t} &= -\sigma_{a}^{24}\Phi_1 n^{24}, \\
\frac{\partial n^{25}}{\partial t} &= \sigma_{\gamma}^{24}\Phi_1 n^{24} - \sigma_{a}^{25}\Phi_1 n^{25}, \\
\frac{\partial n^{26}}{\partial t} &= \sigma_{\gamma}^{25}\Phi_1 n^{25} - \sigma_{a}^{26}\Phi_1 n^{26}, \\
\frac{\partial n^{28}}{\partial t} &= -\sigma_{\alpha}^{28}\Phi_1 n^{28}, \\
\frac{\partial n^{29}}{\partial t} &= \sigma_{\gamma}^{28}\Phi_1 n^{28} - (\sigma_{a}^{29}\Phi_1 + \lambda^{29})n^{29}, \\
\frac{\partial n^{39}}{\partial t} &= \lambda^{29}n^{29} - (\sigma_{a}^{39}\Phi_1 + \lambda^{39})n^{39}, \\
\frac{\partial n^{49}}{\partial t} &= \lambda^{39}n^{39} - \sigma_{a}^{49}\Phi_1 n^{49}.
\end{aligned} \tag{6.2.6}$$

It is important in these equations to distinguish the σ_a^j, defined in (6.2.2), which include fission, from the σ_γ^j, which are cross sections for the (n, γ) reactions only. (For example σ_a^{25} is about seven times greater than σ_γ^{25}.) It should also be kept in mind that the σ_a^j in (6.2.6) are one-group cross sections defined by (6.2.5); they are only approximately equal to the 0.025-eV values displayed on the chain diagrams. In fact, for isotopes that absorb strongly in the resonance range, they may be several times larger than the numbers shown on the chain diagrams. Nevertheless we can get a rough idea of the magnitudes of the one-group numbers σ_a^j by using the 0.025-eV values of the corresponding energy-dependent cross sections. Accordingly Table 6.1 lists approximate numerical values for the parameters in the depletion equations (6.2.6) based on 0.025-eV values of the σ^j (except for U^{28}, for which σ_γ^{28} has arbitrarily been set equal to 10 b to account for reso-

Table 6.1 Parameters in the Depletion Chain for Highly Enriched Uranium

Isotope	$\sigma_a^j\Phi_1(\text{sec}^{-1})$	$\sigma_\gamma^j\Phi_1(\text{sec}^{-1})$	$\lambda^j(\text{sec}^{-1})$
24	4.8×10^{-9}	4.8×10^{-9}	
25	3.4×10^{-8}	5.1×10^{-9}	
26	3.0×10^{-10}	3.0×10^{-10}	
28	5.0×10^{-10}	5.0×10^{-10}	
29	1.7×10^{-9}	1.1×10^{-9}	5.0×10^{-4}
39	1.5×10^{-9}	1.5×10^{-9}	3.4×10^{-6}
49	5.1×10^{-8}	1.4×10^{-8}	

nance absorption). In compiling Table 6.1 Φ_1 has been taken to be 5×10^{13} neutrons/cm^2/sec and the λ^j have been found from the corresponding half-lives T^j by applying the definitional equation

$$\exp(-\lambda^j T^j) = \frac{1}{2} \quad \text{or} \quad \lambda^j = \frac{\ln 2}{T^j} = \frac{.693}{T^j}. \tag{6.2.7}$$

It is clear immediately from Table 6.1 that in (6.2.6) we may neglect $\sigma_a^{29}\Phi_1$ in comparison with λ^{29} and $\sigma_a^{39}\Phi_1$ in comparison with λ^{39}. It is also clear that the time scale for significant depletion effects is hundreds of hours (4 months $\approx$ 3000 hr $\approx 10^7$ sec). For example the table shows that in 300 hours at a constant flux of 5×10^{13} neutrons/cm/sec, the amount of U^{25} initially present is reduced by the factor $\exp(-0.034) \approx 0.966$.

Further examination of (6.2.6) and Table 6.1 will lead us to neglect the effects of U^{24} and the entire U^{28} chain. We first consider the situation with respect to U^{24}.

For the highly enriched (93 percent U^{25}) uranium we are considering, the greatest U^{24} content is in material coming directly from the diffusion plant and thus containing no U^{26}. In such material the U^{24} content is about 4.7 percent. From the first equation of (6.2.6) we see that the amount of this material converted to U^{25} in a time t is $n^{24}(0)$ $(1 - \exp(-\sigma_a^{24}\Phi_1 t))$. Now, during a time t, a fraction $1 - \exp(-\sigma_a^{25}\Phi_1 t)$ of the U^{25} initially present is destroyed, and for $t = 3 \times 10^7$ sec ($\approx$9000 hr) at $\Phi_1 = 5 \times 10^{13}$, the values shown in the table give 0.64 for this fraction. Since a highly enriched reactor with 64 percent of its initial U^{25} loading destroyed is unlikely to remain critical, the time 3×10^7 sec is a reasonable upper limit for the criticality lifetime of the fuel. Again, the figures in the table show that a total of 14 percent of the U^{24} initially present has been converted to U^{25} during that time. Since the U^{24} is at most 4.7 percent of the U^{25} to begin with, this total conversion leads to an increase of 0.65 percent in the U^{25} content throughout life. The effect is of marginal importance and can justifiably be ignored in rough calculations.

The justification for neglecting the entire U^{28} chain is much stronger. Uranium from a diffusion plant enriched to 93 percent contains about 2.3 percent U^{28}. The table shows that in 3×10^7 sec a fraction $1 - \exp(-0.015) \approx 0.015$ of this is destroyed. If all of this is assumed to be converted to Pu^{49}, the amount of Pu^{49} thereby created amounts to only a fraction $0.015 \times 0.023 = 0.000345$ of the initial fuel present. Even in detailed calculations there is little reason to account for this small amount.

We conclude that, for highly enriched fuel, at most the first three of the equations in (6.2.6) need be considered, and for many cases it is legitimate to neglect the effect of U^{24} entirely.

For slightly enriched reactors or reactors initially fueled with plutonium, the last four equations in (6.2.6) become quite important and, in fact, must be augmented by equations describing the concentrations of the higher isotopes of plutonium appearing in Figure 6.1. We have already seen from Table 6.1 that the double β decay of U^{29} to Pu^{49} is far more probable than the formation of U^{20} and Np^{30}. Thus the equations required in addition to (6.2.6) to describe the fission chains in a slightly enriched reactor or one initially fueled with plutonium are

$$\frac{\partial n^{40}}{\partial t} = \sigma_\gamma^{49}\Phi_1 n^{49} - \sigma_a^{40}\Phi_1 n^{40},$$
$$\frac{\partial n^{41}}{\partial t} = \sigma_\gamma^{40}\Phi_1 n^{40} - (\sigma_a^{41}\Phi_1 + \lambda^{41})n^{41}, \qquad (6.2.8)$$
$$\frac{\partial n^{42}}{\partial t} = \sigma_\gamma^{41}\Phi_1 n^{41} - \sigma_a^{42}\Phi_1 n^{42}.$$

Solution of the Depletion Equations

If the time dependence of Φ_1 and the σ_a^j is ignored, the solution of (6.2.6) and (6.2.8) is straightforward. The Laplace-transform technique is well suited to the problem, and the resultant solutions for the n^j are sums and differences of exponentials. Generally, however, many algebraically complicated terms are involved. Some simplification occurs because the (24–25–26) chain is completely decoupled from the (28–42) chain, so that the problem of solving ten coupled differential equations is reduced to that of solving two separate, smaller sets of equations, one containing three unknowns and the other containing seven. In highly enriched reactors the difficulties disappear because, as we have seen, the (28–42) chain and the effect of U^{24} depletion may generally be ignored, so that, in practice, only n^{25} and n^{26} are determined. To solve for these two concentrations is a trivial problem.

For more complicated fuels, the usual situation encountered in power reactors, this great simplification is not legitimate, and the complicated sum of exponentials has to be accepted if an exact, closed-form solution is desired. Unfortunately, in addition to its algebraic complexity, such a solution frequently involves differences of exponentials that are nearly equal, so that the exponentials themselves must be determined to a large number of significant figures if errors are to be avoided.

Because of these difficulties and because the assumption of a constant value of Φ_1 during a time step is an approximation in the first place, finite-difference methods for solving (6.2.6) and (6.2.8) become attractive. They have the great advantage of being easily extensible to more accurate mathematical models of reactor behavior in which changes in Φ_1 during a time step are accounted for.

One way to obtain a finite-difference solution of the depletion equations is simply to integrate over the duration of a time step. As an example we shall consider the equation for n^{24}. Integrating that equation over a time interval $t_{n+1} - t_n \equiv \Delta t$ gives

$$n^{24}(t_{n+1}) - n^{24}(t_n) = -\int_{t_n}^{t_{n+1}} \sigma_a^{24}(\tau)\Phi_1(\tau)n^{24}(\tau)\,d\tau. \tag{6.2.9}$$

If we approximate the integral on the right-hand side of this equation by

$$\int_{t_n}^{t_{n+1}} \sigma_a^{24}(\tau)\Phi_1(\tau)n^{24}(\tau)\,d\tau \approx \tfrac{1}{2}[\sigma_a^{24}(t_{n+1})\Phi_1(t_{n+1})n^{24}(t_{n+1}) + \sigma_a^{24}(t_n)\Phi_1(t_n)n^{24}(t_n)]\Delta t, \tag{6.2.10}$$

we get

$$n^{24}(t_{n+1}) \approx \frac{1 - \frac{1}{2}\sigma_a^{24}(t_n)\Phi_1(t_n)\Delta t}{1 + \frac{1}{2}\sigma_a^{24}(t_{n+1})\Phi_1(t_{n+1})\Delta t} n^{24}(t_n), \tag{6.2.11}$$

where it is assumed that we already know $n^{24}(t_n)$, $\sigma_a^{24}(t_n)$, and $\Phi_1(t_n)$ and that we can estimate $\Phi_1(t_{n+1})$ and $\sigma_a^{24}(t_{n+1})$ by some auxiliary calculations.

By extending this technique to the other equations in (6.2.6) and (6.2.8), we can convert the whole set of differential equations into algebraic ones. Moreover these algebraic equations are very simple to solve. Knowing $n^{24}(t_{n+1})$ from (6.2.11) permits us to solve immediately for $n^{25}(t_{n+1})$ when we difference the second equation in (6.2.6); knowledge of $n^{25}(t_{n+1})$, in turn, permits us to solve for $n^{26}(t_{n+1})$; and so on.

Not only is this finite-difference solution simple to determine, it is also very accurate. For example, for the case of $\sigma_a^{24}\Phi_1$ being independent of time, the difference between (6.2.11) and the exact solution $n^{24}(t_{n+1}) = n^{24}(t_n)\exp(-\sigma_a^{24}\Phi_1 t)$ can be shown (by expanding both solutions) to be less than $(\sigma_a^{24}\Phi_1\Delta t)^3 n^{24}(t_n)/12$. Thus, even with a time step four months ($\sigma_a^{24}\Phi_1\Delta t \approx 0.048$), after which $n^{24}(t_{n+1})/n^{24}(t_n) \approx 0.952$, the error in the finite-difference result for this ratio as computed by (6.2.11) is less than 0.0000092.

There is, however, one severe difficulty involved in the direct application of difference equations of the type (6.2.9) and (6.2.10) to the solution of (6.2.6) and (6.2.8). The problem is that a finite-difference approximation of this type will be inaccurate unless *all* the $\sigma_a^j\Phi_1\Delta t$, as well as the $\lambda^j\Delta t$, are small (say, less than or equal to 0.05). But $\lambda^{29}\Delta t \leq 0.05$ implies $\Delta t \leq 100$ sec, and time steps of that size make it prohibitively costly to obtain solutions to (6.2.6) for the time durations of interest (thousands of hours). It is not hard to avoid this problem by using a finite-difference technique different from that on which (6.2.11) is based. However the difficulty can be avoided even more easily by applying an argument that is appealing on physical grounds and that can be justified mathematically. To see how this argument goes we first rewrite the last four equations in (6.2.6), neglecting $\sigma_a^{29}\Phi_1$ in comparison with λ^{29} and $\sigma_a^{39}\Phi_1$ in comparison with λ^{39}. (We have seen that the figures in Table 6.1 justify this simplification.) Then, with σ_a^{28} set equal to σ_γ^{28} (another excellent approximation), we obtain

$$\begin{aligned}\frac{\partial n^{28}}{\partial t} &= -\sigma_\gamma^{28}\Phi_1 n^{28},\\ \frac{\partial n^{29}}{\partial t} &= \sigma_\gamma^{28}\Phi_1 n^{28} - \lambda^{29}n^{29},\\ \frac{\partial n^{39}}{\partial t} &= \lambda^{29}n^{29} - \lambda^{39}n^{39},\\ \frac{\partial n^{49}}{\partial t} &= \lambda^{39}n^{39} - \sigma_a^{49}\Phi_1 n^{49}.\end{aligned} \tag{6.2.12}$$

The physical argument for simplifying these equations is based on the fact that the decay constants of U^{29} and Np^{39} are so large compared to the time constant $\sigma_\gamma^{28}\Phi_1$ for the creation of U^{29} that we can think of the U^{29} nuclei as becoming Pu^{49} nuclei instantaneously. In other words we can replace the creation rate of Pu^{49} by the creation rate of U^{29} and condense the equations in (6.2.12) to

$$\begin{aligned}\frac{\partial n^{28}}{\partial t} &= -\sigma_\gamma^{28}\Phi_1 n^{28},\\ \frac{\partial n^{49}}{\partial t} &= \sigma_\gamma\Phi_1 n^{28} - \sigma_a^{49}\Phi_1 n^{49}.\end{aligned} \tag{6.2.13}$$

Note that the *amount* of U^{28} converted to Pu^{49} is not affected by this simplification; rather the *time* at which that conversion is completed is shortened.

To justify this approximation mathematically we first note that the solutions to the first two equations in (6.2.12) are

$$\begin{aligned}n^{28}(t) &= n^{28}(0)\exp(-\sigma_a^{28}\Phi_1 t),\\ n^{29}(t) &= \frac{\sigma_\gamma^{28}\Phi_1 n^{28}(0)}{\lambda^{29} - \sigma_a^{28}\Phi_1}[\exp(-\sigma_a^{28}\Phi_1 t) - \exp(-\lambda^{29}t)].\end{aligned} \tag{6.2.14}$$

Now $\sigma_a^{28}\Phi_1 \ll \lambda^{29}$ and $\exp(-\lambda^{29}t)$ becomes quite small for t greater than 2×10^4 sec (6 hours). Thus

$$n^{29}(t) \approx \frac{\sigma_\gamma^{28}\Phi_1 n^{28}(0)}{\lambda^{29}}\exp(-\sigma_a^{28}\Phi_1 t) = \frac{\sigma_\gamma^{28}\Phi_1}{\lambda^{29}}n^{28}(t) \quad \text{for } t > 6 \text{ hr}. \tag{6.2.15}$$

Thus the third equation in (6.2.12) becomes

$$\frac{\partial n^{39}}{\partial t} = \sigma_\gamma^{28}\Phi_1 n^{28}(t) - \lambda^{39}n^{39} \quad \text{for } t > 6 \text{ hr}. \tag{6.2.16}$$

By solving this equation along with the first equation in (6.2.12), we obtain the form (6.2.14) with superscripts 29 replaced by 39. Thus, for $\lambda^{39}t \gg 1$, we obtain, in analogy with (6.2.15),

$$n^{39}(t) \approx \frac{\sigma_{\gamma}^{28}\Phi_1 n^{28}(t)}{\lambda^{39}}. \tag{6.2.17}$$

Substituting this result into the last equation in (6.2.12) leads to (6.2.13), the result we are trying to justify.

The second stage of this argument will be valid only for times greater than 10^6 sec (300 hr). This might seem to be a significant amount of time. However the figures in the table indicate that during that time only a fraction $1 - \exp(-0.0005) \approx 0.0005$ of the U^{28} initially present and $1 - \exp(-0.051) \approx 0.052$ of any Pu^{49} initially present will be destroyed. Thus the total effect on the predicted concentration of plutonium isotopes will be small.

We conclude that the solutions of (6.2.13) and (6.2.12) generally differ by a negligible amount. The rest of the equations in the extended plutonium chain involve time constants having magnitudes close to those in the first two columns of Table 6.1. Hence time steps on the order of 10^8 seconds may now be taken in applying finite-difference techniques of the type exemplified by equations (6.2.9) to (6.2.11), and such techniques again become a very economical method for determining the $n^j(t)$.

6.3 Fission-Product Poisoning

The fragments into which a nucleus splits when it fissions are radioactive and give rise to chains of isotopes as they decay to stable nuclei. The members of these chains have cross sections for neutron absorption at 0.025 eV which vary from a few tenths of a barn to over two million barns. Moreover, when neutrons are absorbed in these nuclei, other neutron-absorbing isotopes are created. Thus a detailed description of how $\Sigma_a(\mathbf{r}, E)$ changes with time because of the creation of fission products requires that the concentrations of hundreds of nuclei be determined as a function of time. And these creation rates depend on both the magnitude and the energy dependence of $\Phi(\mathbf{r}, E, t)$ as well as on the time (i.e., they are not just a function of $\int_{t_0}^{t} dt' \int_0^{\infty} dE\, \Phi(\mathbf{r}, E, t')$).

The creation rate of one of these isotopes, j, is given by

$$\frac{\partial n^j(\mathbf{r}, t)}{\partial t} = \gamma^j \int_0^{\infty} \Sigma_f(\mathbf{r}, E, t)\Phi(\mathbf{r}, E, t)\, dE + \lambda^k n^k(\mathbf{r}, t) - \left(\lambda^j + \int_0^{\infty} \sigma_a^j(E)\Phi(\mathbf{r}, E, t)\, dE\right) n^j(\mathbf{r}, t). \tag{6.3.1}$$

In this equation γ^j is an average *fission yield* of isotope j. To be precise, γ^j is the average number of nuclei of isotope j created per fission in the material present at location $\mathbf{r}$ at time t. It is an average not only over a large number of fissions but also over the particular nuclei fissioning near point $\mathbf{r}$ and over the energies of the neutrons causing fission. (We could break the term in γ^j down into a sum over fissionable isotopes and energy groups each having its own yield, but for notational simplicity we shall not.) Note that, since almost all fissions yield two fragments, the sum of the γ^j over all fragments is very close to 2.0.

The term $\lambda^k n^k(\mathbf{r}, t)$ in (6.3.1) represents a creation rate of isotope j due to decay of some other isotope k. For many important isotopes the direct yield from fission may be quite small, so that this mode of formation is dominant.

There are computer programs capable of solving coupled equations of the type (6.3.1) needed to determine the concentrations of all fissions products of significance. Results obtained using such programs show that, for a given fuel material in a given reactor, it is sufficient to treat precisely only a few of the fission-fragment chains (those having large γ^j and σ_a^j). The effect of the rest of the chains on neutron absorption can be represented quite accurately by defining a few fictitious isotopes having artificial neutron-absorption cross sections and radioactive-decay constants. These fictitious isotopes are defined as obeying (6.3.1) (generally with the λ^k set equal to zero). The fictitious parameters γ^j, λ^j, and $\sigma_a^j(E)$ are found by matching the energy-group approximation for the sum

$$\sum_{j'} n^{j'}(\mathbf{r}, t)\, \sigma_a^{j'}(E)$$

(j' being used to denote fictitious isotopes) to the corresponding sum obtained by treating all the actual physical fission fragments explicitly (except those few with large γ^j and σ_a^j).

The simplest approximation of this type is to assume that there is only one artificial isotope, that it does not decay, and that, when it absorbs a neutron in accord with its fictitious absorption cross section σ_a^f, the product of that absorption process—and of all subsequent absorptions in the chain—also has the fictitious cross section σ_a^f.

This last assumption implies that, in effect, when a neutron is absorbed in the fictitious isotope, the isotope is not destroyed. Thus (6.3.1) for this case may be written

$$\frac{\partial n^{ff}(\mathbf{r}, t)}{\partial t} = \gamma^{ff} \int_0^\infty \Sigma_f(\mathbf{r}, E, t)\Phi(\mathbf{r}, E, t)\, dE, \tag{6.3.2}$$

where the superscript ff stands for "fictitious fission fragment."

If σ_a^{ff} is the one-group microscopic absorption cross section for the fictitious fission fragment, the corresponding macroscopic cross section at a time t in the life of the fuel is

$$\Sigma^{ff}(\mathbf{r}, t) \equiv n^{ff}(\mathbf{r}, t)\sigma_a^{ff} = \gamma^{ff}\sigma_a^{ff} \int_0^t dt' \int_0^\infty dE\, \Sigma_f(\mathbf{r}, E, t')\Phi(\mathbf{r}, E, t'). \tag{6.3.3}$$

Since $\int_0^t dt' \int_0^\infty dE\, \Sigma_f(\mathbf{r}, E, t')\Phi(\mathbf{r}, E, t')$ is the number of fissions per unit volume that have occurred at **r** in time t, and since we are assuming, in effect, that a fictitious nucleus of this type, once created, is never destroyed, the quantity $\gamma^{ff}\sigma_a^{ff}$ is the approximate number of "barns per fission" introduced into the reactor due to the fission fragments for which γ^j and σ_a^j are not unduly large. Values of $\gamma^{ff}\sigma_a^{ff}$ for thermal reactors generally lie in the range 40–50 barns per fission. Thus, if the average fission density in the fuel is 100 watts/cc, the value of $\Sigma^{ff}(\mathbf{r}, t)$ at the end of four months (10^7 sec) is in the range 0.0012–0.0015 cm^{-1}. A typical value for the one-group macroscopic fission cross section in a thermal reactor is 0.1 cm^{-1}. Thus the fraction of neutrons absorbed in the less important fission products after four months of continuous full-power operation is still small compared to the fraction absorbed in the fuel. Of course, as reactor depletion goes on, $\Sigma^{ff}(\mathbf{r}, t)$ continues to increase and eventually accounts for a significant fraction of the neutrons absorbed in the reactor. Unless these fission fragments are removed from the fuel (as in a molten-salt reactor), they will eventually make it impossible to maintain criticality.

The representation of this parasitic absorption in fission products (called *poisoning*) by a single fictitious isotope is adequate when the amount of depletion is small. To predict the criticality lifetime of a long-lived reactor, however, requires that a number of fictitious isotopes be introduced. We shall not go into the details of this process.

The Sm^{149} Fission-Product Chain

There are several fission-fragment chains that involve large yields γ^j and large absorption cross sections. The poisoning effect of these isotopes must be treated explicitly.

The two most important chains of this class are the ones containing Sm^{149} and Xe^{135}. We shall examine these two chains in detail.

The isotope Sm^{149} has a 0.025-eV cross section of 40,800 barns and an infinite-dilution resonance integral of 3400 barns. It appears in a reactor as the result of the chain of events shown in Figure 6.3. (The 1.13 percent yield is that appropriate to U^{25} in thermal reactors.)

For practical purposes the 2-hour half-life of neodymium-149 is so much faster than the 54-hour half-life of promethium-149 that we may assume that the latter isotope appears directly from fission. Thus, if we use $P(\mathbf{r}, t)$ and $S(\mathbf{r}, t)$ to represent the concentra-

Figure 6.3 The Sm^{149} fission-product chain.

tion of Pm^{149} and Sm^{149} at point **r** and time t, we have from (6.3.1)

$$\frac{\partial P(\mathbf{r}, t)}{\partial t} = \gamma^{\mathrm{Nd}} \int_0^\infty \Sigma_{\mathrm{f}}(\mathbf{r}, E, t)\Phi(\mathbf{r}, E, t)\, dE - \lambda^{\mathrm{Pm}} P(\mathbf{r}, t),$$
$$\frac{\partial S(\mathbf{r}, t)}{\partial t} = \lambda^{\mathrm{Pm}} P(\mathbf{r}, t) - S(\mathbf{r}, t) \int_0^\infty \sigma_{\mathrm{a}}^{\mathrm{Sm}}(E)\Phi(\mathbf{r}, E, t)\, dE. \tag{6.3.4}$$

For the one-group parameters $\Sigma_{\mathrm{f}1}(\mathbf{r})$, $\Phi_1(\mathbf{r})$, and $\sigma_{\mathrm{a}1}^{\mathrm{Sm}}(\mathbf{r})$ constant in time, the solutions of these equations are (with **r** dependence suppressed)

$$P(t) = \frac{\gamma^{\mathrm{Nd}}\Sigma_{\mathrm{f}1}\Phi_1}{\lambda^{\mathrm{Pm}}}(1 - \exp(-\lambda^{\mathrm{Pm}}t)) + P(0)\exp(-\lambda^{\mathrm{Pm}}t) \tag{6.3.5}$$

and

$$S(t) = S(0)\exp(-\sigma_{\mathrm{a}1}^{\mathrm{Sm}}\Phi_1 t) + \frac{\gamma^{\mathrm{Nd}}\Sigma_{\mathrm{f}1}}{\sigma_{\mathrm{a}1}^{\mathrm{Sm}}}[1 - \exp(\sigma_{\mathrm{a}1}^{\mathrm{Sm}}\Phi_1 t)]$$
$$- \frac{\gamma^{\mathrm{Nd}}\Sigma_{\mathrm{f}1}\Phi_1 - \lambda^{\mathrm{Pm}}P(0)}{\lambda^{\mathrm{Pm}} - \sigma_{\mathrm{a}1}^{\mathrm{Sm}}\Phi_1}[\exp(-\sigma_{\mathrm{a}1}^{\mathrm{Sm}}\Phi_1 t) - \exp(-\lambda^{\mathrm{Pm}}t)], \tag{6.3.6}$$

where $P(0)$ and $S(0)$ are the concentrations of Pm^{149} and Sm^{149} at time zero.

Because of its large absorption cross section, the presence of Sm^{149} causes the k_{eff} of the reactor to decrease. To get some feeling for the importance of this effect and the time scale associated with it, we shall consider two simple situations involving the growth or decay of Sm^{149}.

At the Beginning of Life. If, in a reactor, the concentrations of Pm^{149} and Sm^{149} are both initially zero and the reactor is turned on and operated at a constant flux level, the concentrations $P(t)$ and $S(t)$ grow in accord with (6.3.5) and (6.3.6), with $P(0) = S(0) = 0$. If we take Φ_1 to be 5×10^{13} neutrons/cm^2/sec and $\sigma_{\mathrm{a}1}^{\mathrm{Sm}}$ to be 40,800 b, then $\sigma_{\mathrm{a}1}^{\mathrm{Sm}}\Phi_1$ becomes 2.04×10^{-6} sec^{-1}. Thus, with the 54-hour half-life of Pm^{149} giving $\lambda^{\mathrm{Pm}} = 3.56 \times 10^{-6}$ sec^{-1}, we see that the exponentials in (6.3.5) and (6.3.6) will die out after several million seconds (10^6 sec $\approx$ 300 hr), and the following equilibrium concentrations of Pm^{149} and Sm^{149} will result:

$$P(\infty) = \frac{\gamma^{\mathrm{Nd}}\Sigma_{\mathrm{f}1}\Phi_1}{\lambda^{\mathrm{Pm}}}, \tag{6.3.7}$$

$$S(\infty) = \frac{\gamma^{\mathrm{Nd}}\Sigma_{\mathrm{f}1}}{\sigma_{\mathrm{a}1}^{\mathrm{Sm}}} \tag{6.3.8}$$

(Note that these steady-state values can be obtained immediately from (6.3.4) by setting the time derivatives to zero.)

Equation (6.3.8) shows that $S(\infty)\sigma_{a1}^{Sm}$, the asymptotic one-group absorption cross section for Sm^{149}, is γ^{Nd} (1.13 percent) times the one-group fission cross section. Thus γ^{Nd} gives a measure of the competition for neutrons between the fission process and absorption in Sm^{149} once an equilibrium concentration has been attained. Moreover we see that this equilibrium concentration is independent of the flux level Φ_1. Thus, after a reactor has been operating at high power for a month or so, an equilibrium of Sm^{149} is attained and remains in competition for neutrons for the rest of the lifetime of the core.

For a low-power reactor the same potential competition exists. However it takes so much longer for the Sm^{149} concentration to reach equilibrium that, in many cases, its effect throughout lifetime is negligible. This fact can be seen from (6.3.6), which shows that, if $\sigma_{a1}^{Sm}\Phi_1 \ll \lambda^{Pm}$ and $S(0) = 0$,

$$S(t) = \frac{\gamma^{Nd}\Sigma_{f1}}{\sigma_{a1}} [1 - \exp(-\sigma_{a1}^{Sm}\Phi_1 t)]. \tag{6.3.9}$$

If Φ_1 is, say, 10^8 neutrons/cm^2/sec, $\exp(-\sigma_{a1}^{Sm}\Phi_1 t)$ will be close to unity until t exceeds 10^9 seconds (about 13 years), at which time it will be approximately 0.996.

After Shutdown from Equilibrium High-Power Operation. If equilibrium conditions have been attained and the reactor has then been shut down, the time behavior of Pm^{149} and Sm^{149} is given by (6.3.5) and (6.3.6) with $\Phi_1 = 0$ and $P(0)$ and $S(0)$ replaced by the equilibrium values (6.3.7) and (6.3.8). The result is

$$\begin{aligned} P(t) &= \frac{\gamma^{Nd}\Sigma_{f1}\Phi_1(0)}{\lambda^{Pm}} \exp(-\lambda^{Pm}t), \\ S(t) &= \frac{\gamma^{Nd}\Sigma_{f1}}{\sigma_{a1}^{Sm}} + \frac{\gamma^{Nd}\Sigma_{f1}\Phi_1(0)}{\lambda^{Pm}} [1 - \exp(-\lambda^{Pm}t)], \end{aligned} \tag{6.3.10}$$

where t is now the time after shutdown and $\Phi_1(0)$ is the flux *prior* to shutdown (i.e., the flux associated with the initial equilibrium concentrations of Pm^{149} and Sm^{149}).

Physically these equations state that the equilibrium samarium present at shutdown, $\gamma^{Nd}\Sigma_{f1}/\sigma_{a1}^{Sm}$, is augmented by an amount $\gamma^{Nd}\Sigma_{f1}\phi_1(0)/\lambda^{Pm}$ as the equilibrium promethium decays into samarium. This final concentration *does* depend on the value of Φ_1 prior to shutdown, and the increase in concentration will be significant for

$$\frac{\lambda^{Pm}}{\Phi_1} = \frac{3.5 \times 10^{-6}}{\Phi_1} < \sigma_{a1}^{Sm} = 4 \times 10^{-20}.$$

Thus, when Φ_1 is slightly less than 10^{14}, the concentration of Sm^{149} present after shutdown is about twice its equilibrium, full-power concentration. After the reactor is turned on again, the extra samarium will burn out and the equilibrium conditions (6.3.7) and

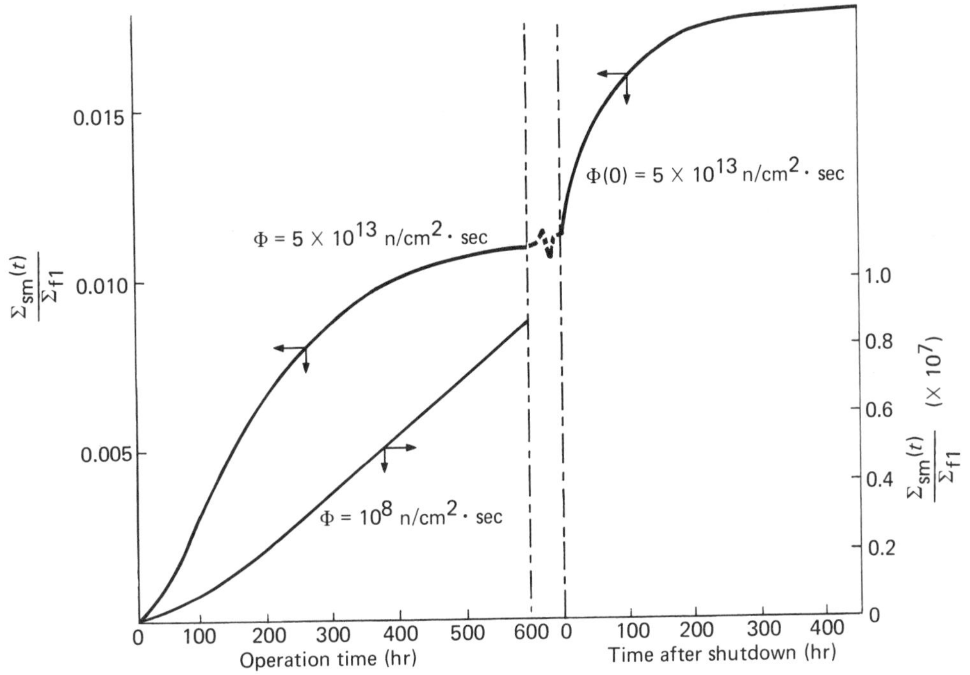

Figure 6.4 Behavior of Sm^{149} in several transient situations.

(6.3.8) will return. Thus, provided the reactor contains enough excess fuel to go critical despite the increased absorption in Sm^{149} after shutdown, that extra absorption produces no lasting effects.

Figure 6.4 shows the behavior of Sm^{149} under several transient conditions.

The Xe^{135} Fission-Product Chain

From the viewpoint of criticality and control, the isotope Xe^{135} is the most important of all the fission products. It has a large absorption resonance that peaks at $E = 0.082$ eV and results in an aborption cross section of approximately 2.7×10^6 barns at 0.025 eV. Xenon-135 is formed from the decay of iodine-135 (6.7-hour half-life) and is itself radioactive (9.2-hour half-life). It is part of the fission chain shown in Figure 6.5, where all the decay constants λ^j are in sec^{-1} and the fission fractions γ^{Te} and γ^{Xe} are those appropriate to thermal fission of U^{235}. None of the absorption cross sections in the chain except that of Xe^{135} are large enough to be of any significance.

The decay of Te^{135} is so fast, and that of Cs^{135} is so slow (2.6-million-year half-life), that we may assume for our purposes that I^{135} is formed directly from fission with a yield $\gamma^I = 0.064$ and that the chain ends with the destruction, by β decay or neutron

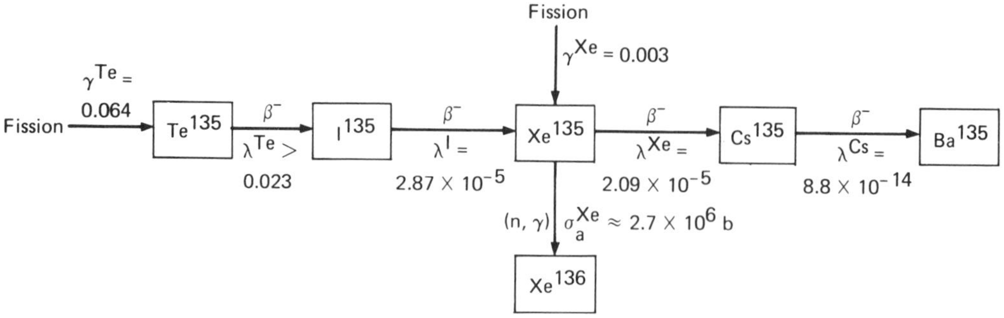

Figure 6.5 The Xe^{135} fission-product chain.

absorption, of Xe^{135}. Thus, if $I(t)$ and $X(t)$ represent the concentrations of I^{135} and Xe^{135} at some location $\mathbf{r}$ in the reactor, the equations specifying the time dependence of these concentrations may be written (with the $\mathbf{r}$ dependence suppressed),

$$\begin{aligned}\frac{\partial I(t)}{\partial t} &= \gamma^{\mathrm{I}}\Sigma_{\mathrm{f}1}\Phi_1(t) - \lambda^{\mathrm{I}}I(t),\\ \frac{\partial X(t)}{\partial t} &= \gamma^{\mathrm{Xe}}\Sigma_{\mathrm{f}1}\Phi_1(t) + \lambda^{\mathrm{I}}I(t) - [\sigma_{\mathrm{a}1}^{\mathrm{Xe}}\Phi_1(t) + \lambda^{\mathrm{Xe}}]\,X(t),\end{aligned} \tag{6.3.11}$$

where $\Sigma_{\mathrm{f}1}$, $\Phi_1(t)$, and $\sigma_{\mathrm{a}1}^{\mathrm{Xe}}$ are the one-group, macroscopic fission cross section, flux, and microscopic absorption cross section for Xe^{135}. As with the samaraium chain, we shall think of the yields γ^{I} and γ^{Xe} as representing averages over the fissionable isotopes present at location $\mathbf{r}$ and over the energy spectrum of the neutrons causing fission at that point. Strictly speaking $\Sigma_{\mathrm{f}1}$ and $\sigma_{\mathrm{a}1}^{\mathrm{Xe}}$ are also time-dependent. However, in the time scale (tens of hours) of xenon transients, this time dependence may be neglected. (Note, however, that changes of $\sigma_{\mathrm{a}1}^{\mathrm{Xe}}$ with time due to changes in temperature may have a short-term effect on the criticality of the reactor. We shall deal with such matters when we consider reactor kinetics.)

For $\Phi_1(t)$ constant the solution of (6.3.11) is

$$\begin{aligned}I(t) &= \frac{\gamma^{\mathrm{I}}\Sigma_{\mathrm{f}1}\Phi_1}{\lambda^{\mathrm{I}}}[1 - \exp(-\lambda^{\mathrm{I}}t)] + I(0)\exp(-\lambda^{\mathrm{I}}t),\\ X(t) &= X(0)\exp(-(\sigma_{\mathrm{a}1}^{\mathrm{Xe}}\Phi_1 + \lambda^{\mathrm{Xe}})t) + \frac{\gamma\Sigma_{\mathrm{f}1}\Phi_1}{\sigma_{\mathrm{a}1}^{\mathrm{Xe}}\Phi_1 + \lambda^{\mathrm{Xe}}}[1 - \exp(-(\sigma_{\mathrm{a}1}^{\mathrm{Xe}}\Phi_1 + \lambda^{\mathrm{Xe}})t)]\\ &\quad - \frac{\gamma^{\mathrm{I}}\Sigma_{\mathrm{f}1}\Phi_1 - \lambda^{\mathrm{I}}I(0)}{\lambda^{\mathrm{I}} - \lambda^{\mathrm{Xe}} - \sigma_{\mathrm{a}1}^{\mathrm{Xe}}\Phi_1}[\exp(-(\sigma_{\mathrm{a}1}^{\mathrm{Xe}}\Phi_1 + \lambda^{\mathrm{Xe}})t) - \exp(-\lambda^{\mathrm{I}}t)],\end{aligned} \tag{6.3.12}$$

where $\gamma \equiv \gamma^{\mathrm{I}} + \gamma^{\mathrm{Xe}}$.

Again, we examine what these equations predict about the approach to equilibrium at constant flux and the change in xenon absorption after shutdown from equilibrium operating conditions.

***X(t)* following Start-Up.** If a reactor, initially containing no Xe^{135} or I^{135}, is brought to a flux level Φ_1 at time $t = 0$ and operated at Φ_1 for a long enough time, (6.3.12) shows that, asymptotically, the concentrations of I^{135} and Xe^{135} become

$$I(\infty) = \frac{\gamma^{\mathrm{I}}\Sigma_{f1}\Phi_1}{\lambda^{\mathrm{I}}},$$
$$X(\infty) = \frac{\gamma\Sigma_{f1}\Phi_1}{\sigma_{a1}^{Xe}\Phi_1 + \lambda^{Xe}}. \tag{6.3.13}$$

Equation (6.3.12) also shows that the speed with which this equilibrium state is reached depends on $\lambda^{\mathrm{I}} = 2.87 \times 10^{-5}\ \mathrm{sec}^{-1}$ and on $(\sigma_{a1}^{Xe}\Phi_1 + \lambda^{Xe}) = (2.7 \times 10^{-18}\Phi_1 + 2.09 \times 10^{-5})\ \mathrm{sec}^{-1}$ and thus is about 10^5 sec ($\approx$30 hr) even if Φ_1 is very small. The situation is thus different from the Sm^{149} case, for which the speed of approach to equilibrium is crucially dependent on Φ_1. Physically the reason for this difference is that Xe^{135} decays radioactively and will thus come into an equilibrium condition, in which the rate of creation of Xe^{135} equals its rate of disappearance, in a time characteristic of both its own half-life and that of I^{135}. If there are neutrons present to add to the destruction rate of Xe^{135} due to radioactive decay, $X(t)$ will approach equilibrium faster. However the 6.7-hour half-life of I^{135} will limit the speed at which equilibrium conditions are reached. Thus 30 hours after start-up is a good estimate of the time to reach equilibrium conditions for any value of Φ_1.

Another important consequence of the fact that Xe^{135} is radioactive—and another characteristic of xenon behavior that differs from samarium behavior—is that $X(\infty)$ is flux-dependent. The "equilibrium xenon poison" $\sigma_{a1}^{Xe}X(\infty)$ associated with operation at a constant flux level Φ_1 is

$$\sigma_{a1}^{Xe}X(\infty) = \frac{1}{1 + \lambda^{Xe}/(\sigma_{a1}^{Xe}\Phi_1)}\gamma\Sigma_{f1}. \tag{6.3.14}$$

For Φ_1 equal to, say, 10 neutrons/cm^2/sec, this macroscopic xenon cross section is only about $9 \times 10^{-7}\ \Sigma_{f1}$. Thus, in a reactor at very low power, Xe^{135} offers a negligible competition for neutrons. However, for $\sigma_{a1}^{Xe}\Phi_1 \approx \lambda^{Xe}$, $\sigma_{a1}^{Xe}X(\infty)$ approaches $\gamma\Sigma_{f1}$ ($=0.067\ \Sigma_{f1}$). For example, for $\Phi_1 = 5 \times 10^{13}$ neutrons/cm^2/sec, $\sigma_{a1}^{Xe}X(\infty) = 0.058\ \Sigma_{f1}$. The rate of absorption of neutrons by Xe^{135} in a thermal reactor operating at such a flux level is quite comparable to their rate of absorption in the moderator or structural material. As a consequence the presence of "equilibrium xenon" in a high-power thermal reactor has a significant effect on the critical condition of the reactor. Unless an excess of fuel is provided to overcome the poisoning effects of this isotope, it will not be possible to maintain criticality for more than a few hours of full-power operation.

One of the advantages of fast reactors is that they do not have this problem. The energy-dependent microscopic xenon cross section $\sigma_a^{135}(E)$ is large only in the thermal energy region. Hence the one-group number σ_{a1}^{Xe} will be a million or so times smaller if it is obtained by averaging over a fast-reactor spectrum.

$X(t)$ **after Shutdown from Equilibrium Operating Conditions.** If we insert (6.3.13) for $I(0)$ and $X(0)$ in (6.3.12) we find that, when a reactor is shut down to $\Phi_1 = 0$ after operating under equilibrium conditions during which the one-group flux at point $\mathbf{r}$ has been $\Phi_1(0)$, the I^{135} and Xe^{135} concentrations at $\mathbf{r}$ behave according to

$$I(t) = \frac{\gamma^{I}\Sigma_{f1}\Phi_1(0)}{\lambda^{I}} \exp(-\lambda^{I}t),$$

$$X(t) = \frac{\gamma\Sigma_{f1}\Phi_1(0)}{\sigma_{a1}^{Xe}\Phi_1(0) + \lambda^{Xe}} \exp(-\lambda^{Xe}t) + \frac{\gamma^{I}\Sigma_{f1}\Phi_1(0)}{\lambda^{I} - \lambda^{Xe}} [\exp(-\lambda^{Xe}t) - \exp(-\lambda^{I}t)], \tag{6.3.15}$$

where, again, time is measured from the instant of shutdown, so that $X(0)$ of (6.3.15) is the equilibrium, preshutdown concentration $X(\infty)$ of (6.3.13).

Equation (6.3.15) shows that $I(t)$ and $X(t)$ both approach zero asymptotically. However the time derivative of $X(t)$ evaluated at $t = 0$ is

$$\left.\frac{\partial X(t)}{\partial t}\right|_{t=0} = \Sigma_{f1}\Phi_1(0)\left(\frac{\sigma_{a1}^{Xe}\gamma^{I}\Phi_1(0) - \gamma^{Xe}\lambda^{Xe}}{\sigma_{a1}^{Xe}\Phi_1(0) + \lambda^{Xe}}\right). \tag{6.3.16}$$

Thus, if the equilibrium, preshutdown flux $\Phi_1(0)$ exceeds $\gamma^{Xe}\lambda^{Xe}/\gamma^{I}\sigma_{a1}^{Xe}$ ($\approx 3 \times 10^{11}$ neutrons/cm^2/sec), the derivative $\partial X(t)/\partial t|_{t=0}$ will be positive, and $X(t)$ will at first grow. In fact, if the initial equilibrium concentration is high, the net amount of xenon present can increase by a factor of two or more before the reactor runs out of I^{135} and the net concentration of xenon begins to decrease. The turnover point depends on the equilibrium conditions at location $\mathbf{r}$ but generally occurs about ten hours after shutdown. It may, however, be necessary to wait 40 or 50 hours after shutdown for the decaying concentration $X(t)$ to return to its initial equilibrium value.

Figure 6.6 shows the behavior of Xe^{135} under various transient conditons.

6.4 Accounting for Depletion Effects in Mathematical Models of Reactor Behavior

Fuel burnup and the build-up of nuclei in the reactor resulting from neutron absorption are in principle easily accounted for in the group diffusion-theory model that forms the basis of most nuclear-reactor design computations. One simply determines multigroup

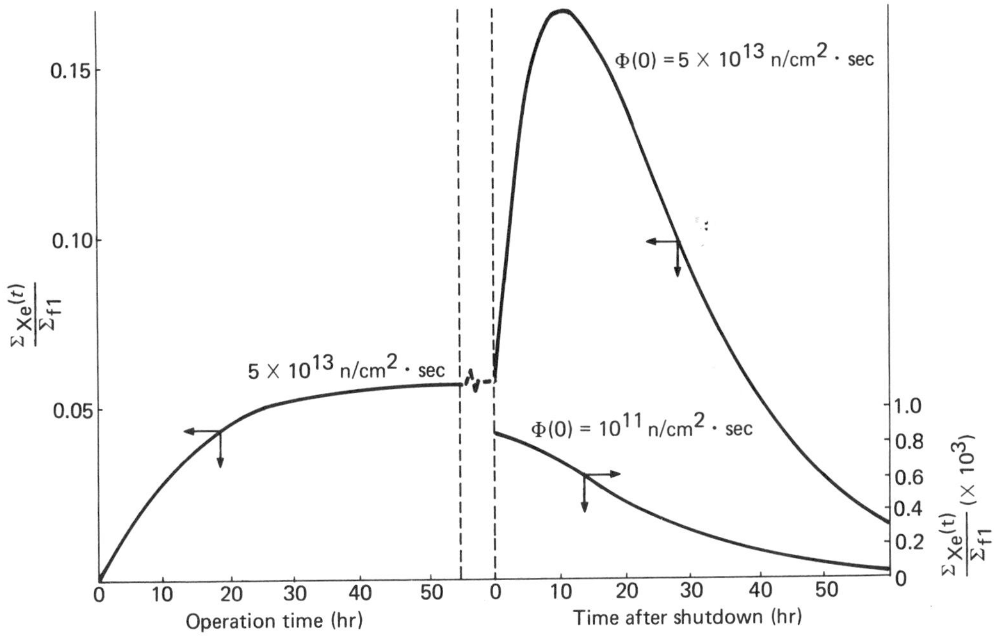

Figure 6.6 Behavior of Xe^{135} in several transient situations.

cross sections throughout the reactor at the beginning of a depletion time step from equations like (5.3.6) (or (5.5.8) if the material is not isotopically homogeneous). These may be used directly in the multigroup diffusion equations of (4.13.7); or they may be used to compute asymptotic spectra, via (5.3.7), for each composition in the reactor so that the spectrum-averaged constants of (5.3.8) and (5.3.9) may be found for use in few-group diffusion-theory calculations. They, by solving these group diffusion equations, we can determine the spatial dependence of the flux for each group throughout the reactor. These flux shapes are assumed to remain constant throughout each time step, and changes in the concentrations of the important nuclei are found by solving equations such as (6.2.6) or (6.2.8), with the one-group reaction-time constants $\sigma_\alpha^j \Phi_1$ replaced by the corresponding few-group or multigroup expressions $\sum_g \sigma_{\alpha g}^j \Phi_g$. With the new material concentrations determined in this manner for the end of the depletion time step, new energy-group parameters are found, and the tandem depletion process is continued.

In practice a great deal of ingenuity is necessary in making this procedure economically feasible. Problems associated with searching for critical conditions, with keeping track of the number densities of all the time-dependent isotopes throughout the reactor, and with using large time intervals for the depletion calculations, all require that special strategies be developed if the running time for a depletion calculation is not to become unacceptably long. We shall discuss these difficulties and their resolution in a qualitative way only. There are as many particular strategies for overcoming the difficulties as there are reactor-design groups throughout the world, and examination of details, even to a superficial extent, is not practical.

Criticality Considerations

The flux shape found at the beginning of each depletion time step must correspond to the reactor in its full-power operating condition. Thus the control rods must be in their critical positions and the moderator and coolant densities and temperatures throughout the reactor must be consistent with local power levels and heat-transfer conditions. Moreover, since equilibrium xenon and samarium concentrations will appear at a very early stage of a depletion time step, they must be found at the beginning of the time step, and their influence on flux shape and control-rod position must be accounted for.

Finding a flux shape consistent with the temperature-density profile throughout the reactor requires that some steady-state heat-transfer calculations be performed. These will ultimately relate the values of the energy-group parameters to the fluxes throughout the core. Because of this and because the equilibrium xenon at a given location also depends on the local flux, as shown in (6.3.14), the constants in the group diffusion equations will depend indirectly on those fluxes. Thus the criticality problem of a reactor at power is nonlinear.

Fortunately the dependence of the group diffusion-theory parameters on local fluxes is small. Hence we can find a solution to the nonlinear equations by an iterative procedure involving only linear equations at each stage.

For example we can at first neglect the spatial temperature-density and xenon effects and search for that control-rod configuration which makes the reactor critical ($k_{\text{eff}} = \lambda = 1$) with temperatures, densities, and equilibrium xenon concentrations all based on the (known) *average* power level in the reactor. Then, using the power *shape* resulting from the first iteration, we can compute position-dependent temperatures, densities, and xenon concentrations throughout the core and, hence, determine new position-dependent group parameters. Doing another search to find again the critical control-rod positions provides a second approximation to the critical flux shape. This iterative procedure is then continued to convergence.

In practice the search strategy is a bit more involved than what has just been described. The cost of solving the static group diffusion equations for a large, geometrically complex reactor is quite high. Thus many tricks are used to minimize the number of such calculations that must be performed to determine a totally consistent solution. In general the strategy is to converge first those variables to which the flux shape is most sensitive, but to limit the degree of convergence during the intermediate steps so that time is not spent determining changes in flux shapes smaller than those that will result when the next variable in the iterative process is considered.

For example, if, in a boiling-water reactor, criticality is to be obtained by fully inserting or fully withdrawing a number of lightly absorbing control rods (not a procedure used in practice), we might start to converge first on a consistent flux shape, coolant density, and xenon distribution without regard for the fact that the k_{eff} associated with this intermediate stage will not be unity. However, before complete convergence at this stage is obtained, a change in the number of fully inserted control rods expected to bring the reactor closer to criticality would be introduced. The coolant densities, xenon concentrations, and flux shapes consistent with this new control-rod configuration would then be found—this time a little more accurately—and the iterative process would be continued to convergence. By concentrating first on that variable (coolant density) to which the flux shape is most sensitive and by not converging all the way until a control-rod configuration closer to the critical one has been introduced, the total number of static criticality calculations needed to obtain a totally consistent solution for the temperature, density, xenon concentration, and control-rod position is reduced.

Not all reactors require such elaborate searches to determine the critical flux shape at the beginning of each time step. In fast reactors, in particular, the flux shapes are quite insensitive to the coolant density and temperature profiles, and the capture rate in Xe^{135} is negligibly small. Thus, for this type of reactor, only the critical control-rod configuration need be found at the beginning of each time step.

Bookkeeping Strategies

Even if a simplified description of the fission-fragment concentrations is employed, the problem of keeping track of all the material concentrations that change with time throughout the core during depletion is formidable. If the diffusion equations are solved by a finite-difference procedure that associates an average flux with each spatial mesh cube, there can easily be a quarter of a million different concentrations changing during a depletion time step. It is necessary, not only to determine changes in all these number densities, but also to calculate new values of the group diffusion-theory parameters that reflect these changes. None of this is difficult to do mathematically. However the sheer bulk of information that has to be manipulated can lead to excessively large computing costs unless some simplifying assumptions are adopted. We shall consider two of these, the first aimed at easing the problem of keeping track of nuclear concentrations and the second at simplifying the determination of group constants.

Block Depletion. To ease the bookkeeping problem with respect to isotopes created or destroyed in each mesh cube during a depletion time step, we simply use larger mesh cubes for depletion than we do for flux computations. Thus we partition the core into "depletion blocks," each block containing a number of mesh cubes. The blocks are chosen so that, within them, the group parameters at start of life are constant (possibly as the result of some homogenization procedures) and the flux shapes are reasonably flat. Having found the flux at all mesh points, we then determine the *average* flux shape within each block and deplete the homogeneous block material using this average flux. As a result all the isotopic concentrations in a given block change uniformly, and we need remember only "depletion-block" concentrations and need only determine new group constants for *blocks* rather than for individual mesh cubes within a block. Since, even for two-dimensional problems, a depletion block will often contain a hundred or more flux mesh cubes, the resultant saving in storage and computing time is quite sizeable.

This idea can be further extended to the area of the heat-transfer computations, where we can define "thermal hydraulic regions" containing many flux mesh cubes (and usually a fair number of depletion blocks) and thus replace position-dependent power shapes and temperature-density profiles by average values. Again the data-handling problem is reduced.

Fitted Constants. Even with the use of "depletion blocks" each made up of many flux mesh cubes, the determination of new few-group diffusion constants at the end of each depletion time step is a lengthy process. In principle multigroup asymptotic spectra have to be computed for each depletion block, and averages then have to be performed to get the few-group numbers. Symmetry considerations cut down on some of the computational work (i.e., the criticality searches and depletion calculations are performed for only a quarter or an eighth of the core). But even after advantage has been taken of any such simplifying geometrical structure, there is still a need to find new few-group constants for all the depletion blocks at the end of each depletion time step, even though, at the beginning of life, many of these blocks had the same composition.

One way around this difficulty is to make use of effective "fitted" few-group microscopic constants. The idea is based on the fact that, if a spectrum appropriate to a given material composition is known, it is possible to define effective microscopic group constants which, when multiplied by the actual physical concentrations of materials making up the composition, will reproduce the macroscopic group constants. Thus, if the fluxes $\{\Phi_n: n = 1, 2, \ldots, N\}$ form a multigroup spectrum for a certain composition, the total microscopic reaction cross section for isotope j in that composition for group g of a few-group scheme is

$$\sigma_{tg}^{j} \equiv \frac{\sum_{n \subset g} \sigma_{tn}^{j} \Phi_n}{\sum_{n \subset g} \Phi_n}, \tag{6.4.1}$$

σ_{tg}^j being the (known) *multigroup* microscopic cross section for isotope j and the sum being over all multigroups n contained in g.

The corresponding total macroscopic reaction cross section for group g of the few-group scheme is then

$$\Sigma_{tg} = \sum_j n^j \sigma_{tg}^j. \qquad (6.4.2)$$

Since we can determine analogous fitted microscopic constants for all other processes (absorption, fission, scattering, transport) and for all other groups in the few-group scheme, we can, by equations analogous to (6.4.2), express all the few-group parameters directly in terms of the number densities n^j of the materials making up a given composition.

To do this we must, of course, have available all the fitted, few-group microscopic cross sections $\sigma_{\alpha g}^j$ (α = t, s, f, tr, . . . , etc.) defined by expressions analogous to (6.4.1). Thus it would at first appear that using fitted constants will not ease the calculational problem of having to find the spectrum $\{\Phi_n\}$ for each depletion block as a function of lifetime. That is, (6.4.2) may very well provide a simple way of finding Σ_{tg}, but we need to know the σ_{tg}^j, and hence $\{\Phi_n\}$, before (6.4.2) can be applied.

The resolution of this difficulty is based on the fact that the multigroup spectrum at a given location corresponding to a particular set of number densities n^j can be assumed to be close to the asymptotic spectrum associated with those n^j. (This assumption is the basis of the few-group approximation.) Moreover the energy shape of this asymptotic spectrum within a given few-group energy range ΔE_g does not change significantly throughout lifetime. This is because the scattering and slowing-down properties of a given composition are due primarily to nuclei of the moderator, coolant, structural, and fertile materials present; and, of these, only the fertile materials change at all significantly during depletion. Hence the asymptotic material spectrum varies only a moderate amount throughout lifetime, that variation being due primarily to changes in the absorption rate caused by the depletion or build-up of isotopes associated with the fertile-fissile material chains or with burnable poisons.

It follows that, if the changes in n^j due to depletion can be estimated beforehand and if the asymptotic spectra and fitted few-group microscopic cross sections associated with a few sets of these n^j can be found by using the standard multigroup procedure of (5.3.7) and (6.4.1), fairly accurate values of the fitted cross sections for intermediate sets of number densities can be determined by interpolation.

The crucial problem, then, is to estimate the relative changes in n^j for a given composition *before* the actual depletion is carried out. To do this we make use of a "point-depletion" calculation for a given initial composition by assuming that, throughout the composition, we may replace $D_g \nabla^2 \Phi_g$ for each group of a few-group scheme by $-D_g B_m^2 \Phi_g$, where B_m^2 is the materials buckling for the composition. With this assumption, the few-group diffusion equations become algebraic. Determining the few-group fluxes Φ_g is then simple, and these fluxes can be used throughout a time step to compute changes in the n^j in accord with the few-group generalization of (6.2.6) and (6.2.8). The new values of n^j estimated in this fashion at the end of the time step can be used to generate another multigroup spectrum from which few-group constants and fitted microscopic cross sections for the new set of n^j (along with a new value of B_m^2) can be determined. The point depletion can then be continued. The final output of this overall procedure will be a set of fitted few-group microscopic cross sections $\sigma_{\alpha g}^j$ as a function of depletion history.

We thus have a procedure for determining few-group parameters appropriate to any mesh cube or depletion block during any stage of depletion. We simply perform point depletions beforehand for each depletable composition making up the reactor and prepare tables of the effective microscopic group parameters for each composition as a function of the degree of depletion (for example as a function of the total amount of energy that has been generated in the material up to that time). Then, to find the few-group parameters for any mesh cube or depletion block, we use equations analogous to (6.4.2), the n^j being those for the mesh cube of interest and the $\sigma_{\alpha g}^j$ being those appropriate to the amount of depletion experienced by the material within that mesh cube (these $\sigma_{\alpha g}^j$ being found by interpolation of the values previously determined for the tables).

Notice that the *point* depletions are performed only to compute the $\sigma_{\alpha g}^j$ for the tables. The n^j used to construct the few-group parameters for the *reactor* come from the depletion computations for the reactor itself.

This method of finding few-group parameters during a depletion calculation circumvents the problem of determining asymptotic spectra for each depletion block at the end of each time step. Instead spectra throughout lifetime are obtained only for each composition. The fraction by which the required number of spectrum calculations is decreased is roughly equal to the number of different depletable compositions initially present in the core divided by the number of depletion blocks. With the number of initial compositions rarely exceeding four and the number of depletion blocks often exceeding a thousand, the savings is substantial.

In practice it is found that many of the $\sigma_{\alpha g}^j$ can be taken as constant during depletion, and the number of tables that have to be stored is thus reduced. For a boiling-water reactor, however, it may be necessary to construct two-dimensional tables in which some of the $\sigma_{\alpha g}^j$ are functions of both the amount of depletion and the moderator density.

Clearly a careful study is required to determine an optimum table set structure for a given reactor type.

Depletion Strategies

There are several strategies associated with the depletion process that reduce computing costs by permitting longer depletion steps to be taken.

One of the simplest of these is to "renormalize" the flux *level* at several subintervals throughout a depletion time step. The renormalization consists of (1) constructing new few-group cross sections using the number densities and tables of fitted cross sections at the end of a given subinterval and (2) changing the flux *levels* so that the total power $\int_{\text{core}} \sum_g \Sigma_{fg}(\mathbf{r})\Phi_g(\mathbf{r})\, dV$ in fissions per second corresponds to the desired value when the updated fission cross sections $\Sigma_{fg}(\mathbf{r})$, are used. Thus changes in spectrum and flux levels due to fuel depletion are more properly accounted for.

The manipulations carried out in the renormalization are exactly those of a regular depletion time step *except* that the expensive flux-shape calculations, involving the simultaneous control-rod, temperature-density, and equilibrium-xenon-concentration searches, are omitted. Because of this omission renormalization is not an expensive procedure.

For thermal reactors it has been found empirically that, if only the *thermal* group flux level is raised or lowered to obtain the desired overall power level, the renormalized fluxes have a group-to-group ratio that is a better approximation to what would be found if a full criticality calculation were performed. Renormalization, combined with tricks of this nature, permits the use of time steps of over a thousand full-power hours for thermal reactors.

6.5 Physical Phenomena Associated with Depletion

The operation of a reactor at a high power level produces a great many secondary effects. Some of these are due directly to fuel depletion and the associated creation of fission products and conversion of fertile material. Others are due more to the large magnitude of the neutron flux in a reactor operating at high power. In designing a reactor, one must first predict quantitatively the physical consequences of all these phenomena and then optimize the design so that favorable effects are enhanced and unfavorable ones minimized. The proper determination of an optimum design requires a very elaborate and intelligent application of mathematical procedures such as those we have developed along with the exercise of sound engineering judgment.

It would be an extremely long undertaking even to summarize all the considerations that enter into the design of a power reactor. Accordingly we shall restrict our attention to matters relating to reactor physics and, at that, consider only qualitatively a few of the most important problems raised by these. It should be remembered that, in a balanced reactor design, problems of comparable complexity exist in the areas of metallurgy (fuel manufacture; radiation damage), chemistry (corrosion; mass transfer; reprocessing—particularly on-line reprocessing in a molten-salt reactor), heat transfer (across gaps between cracked oxide fuel and its cladding and from clad to coolant under conditions near the burnout point; also during accidental—even catastrophic—power excursions), fluid flow (particularly two-phase flow under boiling conditions), mechanical and structural engineering (control-rod drives; remote control of assembly and operation), and electrical, control engineering (power for pumps; control-rod drives systems; safety, control, and warning systems to determine reactor power and radiation levels throughout the plant).

It is convenient to summarize those aspects of power-reactor design involving reactor-physics considerations into two categories, the first related to breeding and conversion and the second to control. We shall consider these two categories in turn.

Conversion and Breeding Considerations

Since only one of the 2.1 to 2.3 neutrons emitted in fission per neutron absorbed in fissile material (the "eta factor" defined by (3.5.6)) is needed to continue the chain reaction, it is possible, in principle, to design a reactor so that the amount of fissile material created by neutron absorption in fertile material (U^{28} or Th^{02}) exceeds that destroyed. However, in practice, because there are inevitable neutron losses due to absorption in coolant, moderator, structural, and reflector material, the design of a breeder reactor is extremely difficult. For thermal reactors it appears (marginally) possible only if the Th–U^{23} cycle is used, and in fast reactors it appears attractive only if the U^{28}–Pu cycle is used. With the present generation of light-water-moderated PWR's and BWR's, breeding is not possible, and these reactors are currently regarded as interim designs to be superseded by fast breeders once they are developed. The reasons for this viewpoint are ultimately economic: a fast-breeder reactor is predicted, in the long run (when uranium ore becomes more expensive), to produce power at lower cost.

Converters. Even though present thermal power reactors, containing U^{25} as the initial fissile material (and possibly Pu^{49} in the future), cannot be made to breed, they are converters and create large amounts of plutonium from the U^{28} present. In fact, at the beginning of life such reactors tend to become supercritical as they deplete. The reason is not that they are making more fissionable material than they are burning, but rather that they are making a significant amount of plutonium, and the fission cross section of Pu^{49} is greater than that of U^{25}. Every attempt is made in designing the present generation of power reactors to take economic advantage of this conversion capability. Accordingly zirconium (with its small absorption cross section) is used as structural and cladding material, and fuel-loading patterns are chosen that (1) flatten the power distribution (so that the maximum power permitted by heat-transfer considerations can be extracted from each fuel element) and (2) minimize the amount of external control poison present during power operation (so as not to waste neutrons that otherwise could be absorbed in U^{28}).

Very elaborate calculations using the models we have developed are required to optimize the fuel-loading pattern in order to achieve and maintain the desired power shape as fuel depletes. The power shape can be altered by varying the amount of moderator in different regions of the core, by introducing neutron-absorbing material in particular regions, or by varying the enrichment of fuel in particular regions. Since the first method leads to mechanical design and heat-transfer difficulties and the second method destroys neutrons which could otherwise be absorbed in U^{28} to make plutonium, only the last scheme (spatial variation of enrichment) has been generally used. Even with the number of possibilities limited in this manner, creating an optimum design is a complex problem. In early reactors there was a tendency to use concentric zones of fuels with different enrichments (in the range 1.5 to 2.5 percent) in order to flatten the power. However, for reasons of economy in the manufacture of fuel elements, only a few zones could be used, and problems arose of keeping the peak-to-average power ratio close to unity within a given zone and of maintaining an even balance of power between different zones. Accordingly it is common in current designs to use a "scatter loading" in which fuel subassemblies (of approximate dimensions 8″ × 8″ × 12′) of different initial enrichment (or different degrees of depletion of the same initial enrichment) are loaded in a checkerboard pattern. With this loading the peak-to-average power ratio within a given subassembly is fairly low and the assembly-to-assembly power stays more constant during depletion. Fuel assemblies are cycled through such a design as they deplete, being introduced in one of the locations requiring the highest enrichment and subsequently transferred to locations appropriate to lower enrichments. Determining what loading pattern to adopt, what enrichments are the best, and what fuel-cycle patterns are optimum requires considerable ingenuity.

Breeders. There are many measurements of the breeding potential of a given reactor design. The simplest of these is the *instantaneous breeding ratio* (B.R.). This is just the total rate at which fissile nuclei are being created in a reactor at a given instant (by decay of Np^{39} or Pa^{13}, absorption in U^{24} or Pu^{40}, etc.) divided by the total rate at which the fissile nuclei then present are being destroyed. (In a converter the same quantity is called the *conversion ratio*.) An associated parameter is the *instantaneous breeding gain* defined as B.R. − 1. Note that the breeding ratio will, in general, vary as the reactor depletes. A more meaningful measure of the breeding capability of a given design is the so-called doubling time, defined as the time (in years) for the initial amount of fissile material in the core to double. If there is some cycle of replenishing and reprocessing of fuel during this period, the computation of doubling time becomes most complicated, and, until an equilibrium cycle is attained, its value will also be dependent on how long the reactor has been in operation.

No matter what measure is adopted, the Pu–U^{28} fast breeder has the highest breeding potential of any reactor. This is primarily because the $\eta(E)$ for plutonium at high energies increases from about 2.35 to 3.0 in the range 10^5 to 10^6 eV and is a number higher than that for U^{25} or U^{23} over the entire range. Moreover the fact that the neutrons in a fast breeder have such a high average energy also enhances breeding potential since both fission products and structural and coolant materials have relatively small capture cross sections for high-energy neutrons.

The currently favored design for fast-breeder reactors consists of a core constructed from approximately quarter-inch-diameter, stainless-steel-clad fuel elements, the fuel being $U^{25}O_2$ mixed with natural or depleted uranium for the first core, with PuO_2 eventually replacing the $U^{25}O_2$. The coolant is liquid sodium, and the blanket that surrounds the core is composed of rods of natural (or depleted) uranium oxide, clad with stainless steel and also cooled with sodium.

The chief technical concern with this type of design from a reactor-physics viewpoint relates to its inherent stability. Both theoretical and experimental analyses predict that the Doppler broadening of the absorption resonances during a fuel-temperature excursion will exceed that of the fission resonances, so that the net effect from Doppler broadening on the k_{eff} of the reactor will be to decrease it. However it is also predicted that reduction in the sodium density in some parts of the core (such as would occur if a coolant channel were blocked and the sodium therein boiled) will *increase* k_{eff} (because the spectrum will harden, causing the average value of η to rise). Care must thus be taken that the Doppler broadening always provides an overriding stabilizing effect, so that, if for some accidental reason reactor temperatures suddenly start to rise, the reactor will have a natural tendency to become subcritical. Unfortunately design modifications aimed at enhancing this type of stability seem always to decrease the predicted breeding gain.

The chief difficulty in analyzing fast reactors is generally stated to be that of obtaining reliable cross-sectional information in the high and intermediate energy ranges. Measurements of high resolution are most difficult in this range. In fact the lower part of it is known as the "unresolved-resonance range" since it is not possible to analyze the data obtained in cross-sectional measurements with enough precision to deduce the Breit-Wigner resonance parameters. Nevertheless comparisons between critical experiments and theoretical calculations based on presently available data and present mathematical modeis of the reactor yield values of k_{eff} usually within a few tenths of a percent of what is measured.

Control Problems

The second design aspect of a power reactor involving reactor-physics considerations relates to control problems.

In a power reactor the value of k_{eff} changes as the core heats up, as the fuel depletes, and as fertile material is converted to fissile. It is necessary to provide an external control system for the reactor so that it can be kept critical during all these changes and—more important, at least from the viewpoint of safety—so that it can always be made *sub*critical. In addition control rods are often used to obtain desired power distributions in the reactor; and they are also used as a safety-protection system to shut down the reactor suddenly if some unforeseen accident situation arises. We will discuss each of these aspects in turn.

Maintaining Criticality. A control system capable of maintaining criticality as well as shutdown under all reactor conditions (including a maximum xenon transient) must have the capability of altering k_{eff} by as much as 20 or 30 percent. This represents a severe demand on a complement of control rods, particularly in a thermal reactor where the thermal diffusion length is small and the region of local influence of an isolated rod is only a few inches. Moreover, if, as happens, it is required for safety-license purposes that one be able to shut down the reactor with any given control rod stuck in its fully withdrawn position, the problem is aggravated.

The original solution of this problem for thermal power reactors was to use large cross-shaped rods (usually containing B^{10} in stainless steel). Early BWR and PWR designs generally had one such rod per fuel cluster, each rod being driven by a separate motor. BWR's still use such a control method. However, in PWR's there has been a switch to many-pronged rods having seven to fifteen "fingers" attached to a single shaft. These take the place of the former cross (see Figure 6.7).

One method to cut down on the number or strength of control rods is to control by varying the concentration of a soluble absorber dissolved in the coolant. This scheme has

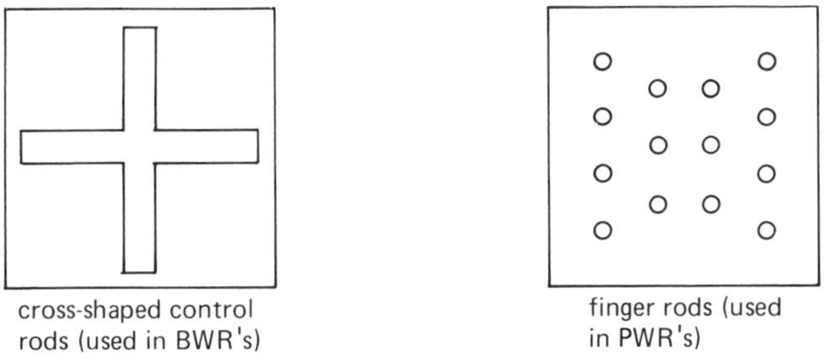

Figure 6.7 Control elements for a fuel cluster.

been adopted for PWR's, in which boric acid is dissolved in the water. The soluble boron is used to compensate for the change in k_{eff} as the operator goes from cold to hot condition and to maintain criticality as the reactor depletes. It is never used in a BWR because of the danger that it will plate out on boiling surfaces. While such deposition is not of concern in PWR's, it is still important to keep the concentration bounded since the soluble poison tends to make the core unstable with respect to increases in coolant temperature. (As coolant containing soluble poison of fixed concentration gets hot and expands, the total soluble-poison content of the core decreases and k_{eff} tends to increase.) The soluble-poison method is not applied to fast reactors since the control problems are not as severe, the region of influence of a given rod being greater.

Another method used to decrease the need for movable rods is to introduce some "burnable poison" into the fuel assemblies. The idea is to match the decrease in k_{eff} due to fuel depletion by an increase due to poison burnout. Lumping the poison (usually B^{10}) so that it is self-shielded in its initial concentrated condition permits a better match to be made between these two compensating processes. Burnable poisons, fixed in position, do not have the destabilizing disadvantage of soluble poisons (nor the chemistry problems). In fact, if they are lumped, they tend to make k_{eff} go down as the temperature increases. (A lower-density moderator implies a longer thermal diffusion length, so that more neutrons are permitted to reach the lump.) Their use does require, however, the ability to calculate, in advance, exactly how such poisons will behave throughout reactor lifetime. If they burn out too quickly, there may not be enough control in the control-rod complement to shut down the reactor. If they burn out too slowly, they will reduce the lifetime of the core.

One final advantage of burnable poisons: they can be positioned so that they reduce local power peaks due to water holes.

Shaping Reactor Power. Control rods are also used in a reactor to achieve a desired power shape. Generally such rods are not as highly absorbing as those designed to shut down the reactor. Moreover they are frequently only "part-length rods" so that, if they are driven from the top of the core (as is the usual practice in PWR's), it is possible to poison the central or lower portions of the reactor without, of necessity, simultaneously poisoning the top portion.

Calculating the proper distribution of control rods necessary to effect a desired power shape requires a full three-dimensional analysis of the reactor in which the control rods are either represented explicitly or are replaced by equivalent homogenized group absorption cross sections. One aspect of using part-length rods requires extremely careful analysis: one must always know whether k_{eff} will increase or decrease when the rod is moved in a given direction. There could be situations where moving the rod in *either* direction will increase k_{eff}. Clearly, if there is need to shut down the reactor quickly, such a rod should not be moved at all.

Control of Transients. As we have seen, power reactors are designed to be inherently stable against power increases. If the power does go up, k_{eff} will decrease and the reactor will start to shut down; if the power goes down, the opposite behavior will occur. Thus, in ordinary full-power operation, the control rods in a reactor remain in a stationary position. In fact the reactor is usually designed so that the full-length rods are either fully inserted or fully withdrawn. If an accident situation arises, all fully withdrawn rods are then "scrammed" into the core to achieve as much fast "shutdown" as possible.

Occasions do arise in normal operation when rod motion is required. On a long time scale, depletion changes k_{eff}, and control-rod readjustment is necessary to compensate for this change. On a shorter time scale, for PWR's, changes in power level (accompanied by changes in temperature or density distribution), or changes in the total xenon content of the core, require compensating rod motion if criticality is to be maintained. (BWR's use a recirculation pump to maintain a critical coolant density during power changes.)

Changes in local xenon concentrations lead to another potential problem in the control of large, high-powered, nonboiling, thermal reactors (PWR's; D_2O- or graphite-moderated reactors). This is the problem of "xenon tilt," in which the concentration of Xe^{135} shifts from one side of the reactor to the other, thereby causing the neutron flux shape, and hence the power shape, to tilt in a similar fashion.

To see qualitatively how the phenomenon can occur, consider a slab reactor in which, because of a good reflector and a preferential loading of fissile materials in the outer zones of the core, the flux level across the slab is flattened (see Figure 6.8). Suppose that an equilibrium xenon level has been attained throughout the reactor and that, subsequently (perhaps because of a control-rod mismatch), the flux on the left-hand side of the reactor is depressed slightly. The total power output being maintained constant, the flux on the right-hand side will rise a compensating amount. As a consequence the xenon

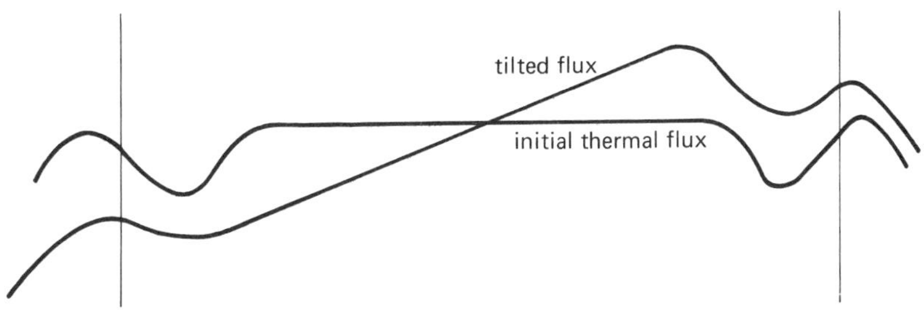

Figure 6.8 Flux shape during a xenon-tilt transient.

on the right-hand side will burn out faster than that on the left-hand side. This will decrease the amount of poison on the right-hand side and cause the flux to tilt more. The power will then continue to rise on the right-hand side until the extra I^{135} created by this higher fissioning rate begins to decay to Xe^{135} in substantial amounts. (The half-life of I^{135} is 6.7 hours, while that of Xe^{135} is 9.2 hours.) When this happens, the xenon concentration on the right-hand side will again begin to increase.

Meanwhile the xenon concentration on the left-hand side, which, because of the decrease in burnout associated with the lower flux level on that side, initially increased and further pushed the flux to the right-hand side, will start to decrease because the amount of I^{135} available for decay into Xe^{135} is reduced and because the Xe^{135} is itself radioactive.

Thus the Xe^{135} concentration will now grow on the right-hand side and diminish on the left, and the flux level will now be depressed on the right-hand side and increased on the left.

There is thus a chance that the xenon concentration and power will oscillate spatially in such a reactor. Moreover, since the total amount of xenon in the reactor is not altered much during the oscillation, the effect on k_{eff} is small, and it is necessary to monitor local flux levels in order to detect whether an oscillation is actually taking place.

It has been determined both theoretically and experimentally that the oscillation will occur only in large reactors at high power. It will not occur in fast reactors because, in them, Xe^{135} is of such little importance; it will not occur in boiling-water reactors because they are so inherently stable. (A local increase in power leads immediately to extra boiling in that area, and this so increases neutron leakage *out* of that area that the power drops to its initial value.)

When xenon oscillations do occur, the period of oscillation is about 24 hours and the amount of the tilt can be as great as 3:1. In a power reactor this cannot be tolerated and must either be avoided by design or, if it starts to occur, immediately damped out by control-rod motion. Part-length control rods, because of their ability to shape the flux distribution, can be used to good advantage for this purpose.

References

Breen, R. S., Marlowe, O. J., and Pfeifer, C. J., 1965. *Harmony: System for Nuclear Reactor Depletion Computation*, WAPD-TM-478, Bettis Atomic Power Laboratory.

Cadwell, W. R., 1967. *PDQ-7 Reference Manual*, WAPD-TM-678, Bettis Atomic Power Laboratory.

Fowler, T. B., Vondy, D. R., and Cunningham, G. W., 1969. *Nuclear Reactor Core Analysis Code: CITATION*, ORNL-TM-2406, Oak Ridge National Laboratory.

Poncelet, C. G., 1965. *Burnup Physics of Heterogeneous Reactor Lattices*, WCAP-6069, Westinghouse Electric Corporation.

Poncelet, C. G., 1966. *LASER-A Depletion Program for Lattice Calculation Based on MUFT and THERMOS*, WCAP-6073, Westinghouse Electric Corporation.

Radkowsky, A., ed., 1964. *Naval Reactor Physics Handbook*, Volume 1 (Washington, D. C.: U. S. Government Printing Office), pp. 751–852.

Problems

1. A reactor loaded initially with 100 kg of pure U^{25} in carbide form (UC_2) depletes for a total nvt of 2×10^{21} neutrons/cm^2. Assuming the one-group microscopic fission cross section for U^{25} is 500 b, find the average fuel burnup in MWD/T.

2. For an initial mixture of U^{25} and U^{28} the eta factor, defined by (3.5.6), near the beginning of an irradiation at a low flux level is given by

$$\eta(t) \approx \frac{\eta^{25}\Sigma_a^{25}(t) + \eta^{49}\Sigma_a^{49}(t)}{\Sigma_a^{28}(0) + \Sigma_a^{25}(t) + \Sigma_a^{49}(t)},$$

where η^{25} and η^{49} are the parameters for the pure isotopes and the Σ_a^j are one-group macroscopic cross sections; also it has been assumed that Pu^{49} is produced instantaneously by the absorption of a neutron in U^{28} and the effects on η of the depletion of U^{28} as well as the formation of fission products and higher plutonium isotopes have been neglected.

Find that enrichment of U^{25} such that, at $t = 0$, $d\eta/dt = 0$. (Use $\eta^{25} = 2.08$; $\eta^{49} = 2.12$; $\sigma_a^{49} = 1015$ b; $\sigma_a^{25} = 680$ b; $\sigma_a^{28} = 7$ b.)

3. At what U^{25} enrichment will a mixture of U^{25} and U^{28} initially breed (i.e., be such that at $t = 0$ the concentration of fissile material $n^{25} + n^{49}$ increases with time). (Use the one-group cross sections given for Problem 2.)

4. Show that, for $\sigma_a^{24}\,\Phi_1$ independent of time, and $\sigma_a^{24}\,\Phi_1\Delta t < 1$, the error in (6.2.11) is less than $(\sigma_a^{24}\Phi_1\Delta t)^3/12$.

5. Two alternate approximations analogous to (6.2.10) are

$$\int_{t_n}^{t_{n+1}} \sigma_a^{24}(\tau)\Phi_1(\tau)\,d\tau \approx \sigma_a^{24}(t_n)\Phi_1(t_n)n^{24}(t_n)\Delta t,$$

$$\int_{t_n}^{t_{n+1}} \sigma_a^{24}(\tau)\Phi_1(\tau)\,d\tau \approx \sigma_a^{24}(t_{n+1})\Phi_1(t_{n+1})n^{4}(t_{n+1})\Delta t.$$

a. For $\sigma_a^{24}\Phi_1$ independent of time and $\sigma_a^{24}\Phi_1\Delta t < 1$, find the lead error term (in the form of a constant times $(\Delta t)^m$) when these two approximations are used together with (6.2.9) to estimate $n^{24}(t_{n+1})$

b. Which of these two approximations plus (6.2.11) gives the smallest error if $\sigma_a^{24}\Phi_1\Delta t$ is very large (say, $\sigma_a^{24}\Phi_1\Delta t = 100$)?

6. Suppose that a sample of uranium having an initial number density ratio $n^{25}(0)/n^{28}(0) = 0.03$ is irradiated in a constant flux to a flux time nvt $= 10^{22}$ neutrons/cm^2. Let $\eta(t)$ be defined as in Problem 2 except that $\Sigma_a^{28}(t)$ is time-dependent and may include $\Sigma^{ff}(\mathbf{r}, t)$ given by (6.3.3). What (to two-place accuracy) is the ratio of the value of η at nvt $= 10^{22}$ with Σ^{ff} included in the denominator to the value with Σ^{ff} excluded. (In addition to the parameters given for Problem 2, use $\sigma_f^{25} = 580$ b; $\sigma_f^{49} = 740$ b; $\gamma^{ff}\sigma_a^{ff} = 50$ b.)

7. We have seen that the equilibrium concentration of Sm^{149} in a power reactor is independent of the flux level and that, if the reactor is shut down from an equilibrium condition, the Sm^{149}

concentration increases. Determine whether the Sm^{149} concentration can ever become lower than its equilibrium concentration once that value has been reached.

8. Suppose that an equilibrium Xe^{135} has built up in a reactor at some location where the steady, operating, one-group flux $\Phi_1(0)$ exceeds $\gamma^{Xe}\lambda^{Xe}/\gamma^{I}\sigma_{aI}^{Xe}$ but that the half-life of I^{135} is greater, rather than less, than that of Xe^{135}.

a. Will the xenon concentration still peak before dying away if $\Phi_1(0)$ is decreased to zero?

b. If so, determine, for the same initial equilibrium I^{135} concentration, whether the peak will be greater or less than the case for which I^{135} has a half-life less than that of Xe^{135}.

c. Answer part a under the assumption that I^{135} is a stable nucleus.

9. Suppose that resonance absorption in a reactor fueled by natural uranium is such that the one-group absorption cross section for U^{28} is 8 b (instead of the 2.7 b appropriate to purely thermal absorption). If the corresponding absorption cross section for U^{25} is 680 b, what is the instantaneous breeding ratio at the beginning of life?

This number will turn out to exceed unity. Why then can one not make a breeder with natural-uranium fuel?

10. A quantitative notion of the xenon stability characteristics of a reactor can be obtained by considering a simple "two-point" reactor consisting of two homogeneous multiplying slabs, each of thickness Δ, separated by a purely scattering medium.

If we let Φ_1 and Φ_2 be the average one-group fluxes in the two multiplying slabs and represent the net currents across the interfaces between the multiplying slabs and the central scattering region by

$$\text{net current from slab 1 to central zone} = -D\frac{\Phi_2 - \Phi_1}{L},$$

$$\text{net current from slab 2 to central zone} = -D\frac{\Phi_1 - \Phi_2}{L},$$

where D is an effective diffusion constant and L is a characteristic thickness of the scattering material, integration of the one-group diffusion equation (4.10.11) over each multiplying slab yields, for the critical condition,

$$J_1 - D\frac{\Phi_2 - \Phi_1}{L} + (\Sigma_1 + X_1\sigma^{Xe} - \nu\Sigma_{f1})\Phi_1\Delta = 0,$$

$$J_2 - D\frac{\Phi_1 - \Phi_2}{L} + (\Sigma_2 + X_2\sigma^{Xe} + \nu\Sigma_{f2})\Phi_2\Delta = 0,$$

where J_1 and J_2 are the net currents both directed outwards across the external surfaces of the two multiplying slabs and X_1 and X_2 are (flux-weighted) average xenon concentrations in the two slabs.

a. Show that, if, except for xenon concentrations, the properties of both slabs are the same and $J_1/\Phi_1 = J_2/\Phi_2 =$ constant, then

$$Y = -\frac{2D}{\bar{X}\sigma_x L\Delta}\frac{Z}{1 - Z^2},$$

where

$$Y \equiv \frac{X_1 - X_2}{2\bar{X}}, \qquad \bar{X} \equiv \frac{1}{2}(X_1 + X_2),$$

$$Z \equiv \frac{\Phi_1 - \Phi_2}{2\bar{\Phi}}, \qquad \Phi \equiv \frac{1}{2}(\bar{\Phi}_1 + \Phi_2).$$

b. Next, by making the linear approximation of neglecting Z^2 (valid where Z is small) and assuming $\bar{X} = \gamma\Sigma_f\Phi/(\lambda^{Xe} + \sigma^{Xe}\bar{\Phi})$ is a constant in time, show that

$$\frac{d^2Y}{dt^2} + \left[\lambda^{I} + \lambda^{Xe} + \sigma^{Xe}\bar{\Phi} - \frac{\Sigma_f L\Delta}{2D}\frac{\sigma^{Xe}\bar{\Phi}}{\lambda^{Xe} + \sigma^{Xe}\bar{\Phi}}(\gamma^{I}\sigma^{Xe}\bar{\Phi} - \gamma^{Xe}\lambda^{Xe})\right]\frac{dY}{dt}$$
$$+ \lambda^{I}\left[\lambda^{Xe} + \sigma^{Xe}\bar{\Phi}\left(1 + \frac{\Sigma_f L\Delta}{2D}\frac{\lambda^{Xe}\gamma}{\lambda^{Xe} + \sigma^{Xe}\bar{\Phi}}\right)\right] Y = 0.$$

c. Finally, from this result, show that the linear analysis predicts that Y (and hence $(\Phi_1 - \Phi_2)/2\overline{\Phi}$) will oscillate in a divergent manner if

$$\frac{\sigma^{\mathrm{Xe}}\overline{\Phi}}{\lambda^{\mathrm{Xe}} + \sigma^{\mathrm{Xe}}\overline{\Phi}} \frac{\gamma^{\mathrm{I}}\sigma^{\mathrm{Xe}}\overline{\Phi} - \gamma^{\mathrm{Xe}}\lambda^{\mathrm{Xe}}}{\lambda^{\mathrm{I}} + \lambda^{\mathrm{Xe}} + \sigma^{\mathrm{Xe}}\overline{\Phi}} > \frac{2D}{\Sigma_{\mathrm{f}} L \Delta}$$

and that, if it is just on the verge of instability, it will oscillate with a period

$$T = \frac{2\pi}{\sqrt{\left[\lambda^{\mathrm{I}}\left(\lambda^{\mathrm{Xe}} + \sigma^{\mathrm{Xe}}\overline{\Phi} + \lambda^{\mathrm{Xe}}\gamma \dfrac{\lambda^{\mathrm{I}} + \lambda^{\mathrm{Xe}} + \sigma^{\mathrm{Xe}}\overline{\Phi}}{\gamma^{\mathrm{I}}\sigma^{\mathrm{Xe}}\overline{\Phi} - \gamma^{\mathrm{Xe}}\lambda^{\mathrm{Xe}}}\right)\right]}}.$$

What will this period be in hours for $\overline{\Phi} = 10^{14}$ neutrons/cm^2/sec?

7 Reactor Kinetics

7.1 Introduction

The critical condition in a reactor represents a delicate balance between two dynamic processes: the creation rate and the destruction rate of neutrons. In a reactor operating at 4000 megawatts of thermal power, for example, there are $\sim 3 \times 10^{20}$ neutrons created per second and $\sim 3 \times 10^{20}$ destroyed. The average lifetime of these neutrons lies in the range 10^{-7} to 10^{-4} sec depending on the type of reactor. It follows that, if the production rate were to exceed the destruction rate by one part in a thousand for one second with the average neutron lifetime equal to, say, 5×10^{-5} sec, the power level at the end of that time would increase by a factor of approximately $(1.001)^{1/(5\times 10^{-5})} = 4.8 \times 10^8$. At that instant the reactor, if it were still in one piece—which it wouldn't be—would be producing around a quarter of a ton of neutrons per second (and absorbing all but one part in a thousand of them).

Balancing the neutron creation and destruction rates in a core is thus a delicate matter, and the study of how that balance is reestablished when it is disturbed is quite important.

Reactor kinetics is the area of reactor physics concerned with predicting what happens to the neutron flux density $\Phi(\mathbf{r}, E, t)$ when the balance condition associated with the critical state is disturbed. More generally, it concerns the space-time-energy behavior of the neutron population in an assembly of material—critical or not—into which neutrons are introduced.

In its most fundamental form reactor kinetics deals with time-dependent probability distributions. The individual nuclear events that take place in a reactor are not predictable in a deterministic fashion. Thus, if a neutron of known energy is absorbed in a U^{25} nucleus, it is not possible to predict beforehand whether there will be a fission or whether a U^{26} nucleus will result. All that can be said is that there are certain *probabilities* that one or the other of these two events will occur. As a result, if, on the average, only a few neutrons are present in a reactor, there will be fluctuations in the number and energy of the neutrons present in any small volume and in the reactor as a whole.

The study of the fluctuation phenomena induced under such conditions is one of the branches of reactor kinetics. It is, in fact, possible to use the statistical properties of the neutron population to infer information about the reactor—for example an approximate value of k_{eff}. Such studies are generally referred to as *noise analysis*.

We shall not deal here with the theory that underlies these fluctuation phenomena; we shall instead develop the analysis applicable to reactors with statistically large numbers of neutrons present. Such reactors are said to behave "deterministically," by which is ment that the value of the flux density at some future time can, in principle, be predicted with certainty if its present value and the (possibly time-dependent) characteristics of the reactor are known.

Thus we shall set as our goal the determination of the time-dependent flux density $\Phi(\mathbf{r}, E, t)$ throughout a reactor whose properties may themselves be varying with time.

7.2 The Basic Equation of Deterministic Reactor Kinetics

The basic equation of deterministic reactor kinetics may be obtained by simply extending the basic flux equation (4.4.2) to the time-dependent case. However, since we shall be concerned exclusively with the diffusion-theory approximation, we shall immediately apply Fick's Law and base our analysis on the continuous-energy diffusion equation (4.6.1). We must then incorporate into this equation the facts that the neutron density may be changing with time and that some of the neutrons produced as the result of fission events may be delayed.

In Section 1.4 it was shown that delayed neutrons come from certain fission fragments which, after a β decay, happen to be unstable with respect to the number of neutrons in their nuclei. Once the β decay occurs, a neutron appears essentially instantaneously. To determine the rate of neutron emission from this source it is necessary to keep account of the concentration of all the "delayed precursors" which, when they emit a β particle, become neutron-unstable. If the concentration of the ith one of these precursors at location $\mathbf{r}$ in the reactor is $c_i(\mathbf{r}, t)$ (precursor nuclei/cc) and its time constant for β decay is λ_i (sec^{-1}), we have:

Total number of delayed neutrons emitted per cc per sec at point $\mathbf{r}$ and time t

$$= \sum_{i=1}^{I} \lambda_i c_i(\mathbf{r}, t), \tag{7.2.1}$$

where the sum is over all the different fission fragments that are delayed-neutron emitters. There are about twenty of these, but most of them have such a small yield that it is possible to fit the data by assuming $I = 6$.

The delayed emitters are created by fission. Thus, if β_i^j is the fraction of the neutrons from a fission of isotope j (U^{25}, Pu^{49}, U^{28}, etc.) that eventually appear from the decay of precursor i, we have:

Number of delayed precursors of the ith kind created per cc per sec at $\mathbf{r}$ and t

$$= \sum_j \beta_i^j \int_0^\infty \nu^j \Sigma_f^j(\mathbf{r}, E, t)\Phi(\mathbf{r}, E, t). \tag{7.2.2}$$

It follows that the rate of change of $c_i(\mathbf{r}, t)$ for fuel that is stationary in space is given by

$$\frac{\partial c_i(\mathbf{r}, t)}{\partial t} = \sum_j \beta_i^j \int_0^\infty \nu^j \Sigma_f^j(\mathbf{r}, E, t)\Phi(\mathbf{r}, E, t) - \lambda_i c_i(\mathbf{r}, t). \tag{7.2.3}$$

(If the fuel is moving, as in a homogeneous reactor, another term accounting for this fact has to be added to (7.2.3). We shall not deal with this case in our analysis.)

Delayed neutrons are emitted with energies that are, on the average, lower than those of prompt neutrons. Designating the spectrum of emission energies for the ith precursor by $\chi_i(E)$ we have from (7.2.1):

Rate at which neutrons appear in $dV\,dE$ because of the decay of delayed precursors

$$= \sum_i \chi_i(E)\lambda_i c_i(\mathbf{r}, E)dV\,dE. \tag{7.2.4}$$

With χ_p^j defined as the corresponding neutron spectrum of prompt neutrons emitted by isotope j, we also have:

Rate at which prompt neutrons appear in $dV\,dE$

$$= \left[\sum_j \chi_p^j(1-\beta^j)\int_0^\infty \nu^j\Sigma_f^j(\mathbf{r}, E', t)\Phi(\mathbf{r}, E', t)\,dE'\right] dV\,dE, \tag{7.2.5}$$

where β^j is the total yield of delayed precursors from isotope j:

$$\beta^j \equiv \sum_{i=1}^{I} \beta_i^j. \tag{7.2.6}$$

Table 7.1 summarizes delayed-neutron data for several isotopes. Note that the fractions β_i^j differ significantly for the different fissionable isotopes. They also depend on the energy spectrum of the neutrons causing the fissions. Note also that the sets of λ_i values differ for the different isotopes. This is because we are representing the twenty or so physically real fission fragments that emit delayed neutrons by six "equivalent" groups for each fissionable isotope. The physically real precursors would presumably have decay constants independent of their fissionable ancestors. "Equivalent" precursors, which have the nature of a best fit to the data, need not. In applications of (7.2.3) and (7.2.4) it has been found sufficiently accurate to use only one set of λ_i, that corresponding to the principal fissioning isotope.

We also assume, for the sake of completeness, that there is an extraneous neutron source (usually referred to as the "external" source and taken to include, for example, neutrons from the interactions of radium γ rays with beryllium or of deuterons with deuterium in an accelerator) which isotropically emits $q(\mathbf{r}, E, t)dV\,dE$ neutrons per second into $dV\,dE$.

We can now extend (4.6.1) to the time-dependent case by expressing mathematically the fact that the net rate of increase of neutrons in $dV\,dE$ is the difference between their

Table 7.1 Delayed-Neutron Data

Group	Decay Constant λ_i(sec^{-1})	Yield (delayed neutrons/fission)
U^{25}(thermal fission; $\beta^{25} = 0.0065$)		
1	0.0124	0.00022
2	0.0305	0.00142
3	0.111	0.00127
4	0.301	0.00257
5	1.14	0.00075
6	3.01	0.00027
Pu^{49}(fast fission; $\beta^{49} = 0.0020$)		
1	0.0129	0.000076
2	0.0311	0.000560
3	0.134	0.000432
4	0.331	0.000656
5	1.26	0.000206
6	3.21	0.000070
U^{23} (thermal fission; $\beta^{23} = 0.0027$)		
1	0.0126	0.00023
2	0.0337	0.00081
3	0.139	0.00068
4	0.305	0.00075
5	1.13	0.00014
6	2.50	0.00009

Source: Keepin (1965).

rate of appearance and their rate of disappearance:

$$\frac{d}{dt}\left[\frac{1}{v}\Phi(\mathbf{r}, E, t)\,dV\,dE\right] = \left[\sum_j \chi_p^j(E)(1 - \beta^j)\int_0^\infty \nu^j\Sigma_f^j(\mathbf{r}, E', t)\Phi(\mathbf{r}, E', t)\,dE'\right] dV\,dE$$
$$+ \sum_i \chi_i(E)\lambda_i c_i(\mathbf{r}, t)\,dV\,dE + q(\mathbf{r}, E, t)\,dV\,dE + \nabla\cdot D(\mathbf{r}, E, t)\nabla\Phi(\mathbf{r}, E, t)\,dV\,dE$$
$$- \Sigma_t(\mathbf{r}, E, t)\Phi(\mathbf{r}, E, t)\,dV\,dE + \left[\int_0^\infty \Sigma_s(\mathbf{r}, E' \to E, t)\Phi(\mathbf{r}, E', t)\,dE'\right] dV\,dE. \qquad (7.2.7)$$

To simplify notation we introduce two integral operators A and F^j defined through their operation on any function $f(\mathbf{r}, E, t)$ by

$$Af \equiv \Sigma_t(\mathbf{r}, E, t)f(\mathbf{r}, E, t) - \int_0^\infty \Sigma_s(\mathbf{r}, E' \to E, t)f(\mathbf{r}, E', t)\,dE',$$
$$F^j f \equiv \int_0^\infty \nu^j\Sigma_f^j(\mathbf{r}, E', t)f(\mathbf{r}, E', t)\,dE'. \qquad (7.2.8)$$

Then, with all functional dependence suppressed, (7.2.7) and (7.2.3) become

$$\begin{aligned} &\nabla \cdot D\nabla\Phi - A\Phi + \sum_j \chi_p^j(1 - \beta^j)F^j\Phi + \sum_{i=1}^{I} \chi_i\lambda_i c_i + q = \frac{1}{v}\frac{\partial \Phi}{\partial t}, \\ &\sum_j \beta_i^j F^j\Phi - \lambda_i c_i = \frac{\partial c_i}{\partial t} \qquad (i = 1, 2, \ldots, I). \end{aligned} \tag{7.2.9}$$

These are the time-dependent continuous-energy diffusion equations for a reactor in which the fuel is stationary. We shall use them as the basis of our development of reactor kinetics.

7.3 The Point Kinetics Equations

Since mean free paths are fairly long and since the lifetimes of neutrons in a reactor are quite short, the effects of local perturbations on $\Phi(\mathbf{r}, E, t)$ will quickly spread throughout a reactor. The immediate consequence of perturbing a reactor locally (for example by changing a control-rod position slightly) is thus a readjustment in the shape of the flux. In many cases this readjustment is slight and is completed in a few milliseconds; after that the readjusted shape rises or falls as a whole depending on whether the initial perturbation increased or decreased k_{eff}. For reactors in which transients proceed in this manner, merely being able to predict the change in the *level* (i.e., *average* value) of the flux is sufficient to permit a very accurate prediction of the consequences of perturbation. Thus, instead of having to face the very difficult problem of solving (7.2.9) in full detail, we shall find a simple set of equations that specify how the overall *magnitude* of the flux changes with time.

A measure of the *level* of the flux is the integral of $\Phi(\mathbf{r}, E, t)$ over all energy and over the reactor volume. However, to conform to convention, and because it is conceptually simpler, we shall deal with the integral of the number density $\Phi(\mathbf{r}, E, t)/v(E)$. Moreover, for mathematical reasons that will become clear later, it will be convenient to weight this number density with a function $W(\mathbf{r}, E)$. For the present we shall specify the weight function no further than to say it is defined over the same spatial and energy domain as $\Phi(\mathbf{r}, E, t)$ and is *independent* of time. With this amount of generality left floating, we then introduce a quantity $T(t)$, sometimes called the *amplitude function*, defined by

$$T(t) \equiv \int_{\text{reactor}} dV \int_0^\infty dE\, W(\mathbf{r}, E)\frac{1}{v(E)}\Phi(\mathbf{r}, E, t). \tag{7.3.1}$$

Clearly $T(t)$ is a weighted integral of the total number of neutrons present in the reactor at any time. (If $W(\mathbf{r}, E) = 1$, $T(t)$ is *exactly* the total number of neutrons present.)

To derive equations for the amplitude function we simply multiply the first equation in (7.2.9) by $W(\mathbf{r}, E)$ and the second by $\chi_i(E)W(\mathbf{r}, E)$ and integrate them both over energy and volume. It is convenient, however, to make a clearer distinction between the *shape* of the flux and its amplitude. Accordingly we define a "shape function" $S(\mathbf{r}, E, t)$ by

$$S(\mathbf{r}, E, t) \equiv \frac{\Phi(\mathbf{r}, E, t)}{T(t)}, \tag{7.3.2}$$

and substitute the product $S(\mathbf{r}, E, t)T(t)$, rather than $\Phi(\mathbf{r}, E, t)$, into (7.2.9). Also, in the first equation in (7.2.9), to conform to convention, we add and subtract the term $\sum_j \sum_{i=1}^{I} \chi_i \beta_i^j F^j \Phi$. The result of this substituting, weighting, and integrating procedure is then

$$\begin{aligned}
&\int dV \int dE\, W\left[\nabla\cdot D\nabla S - AS + \sum_j \left\{\chi_p^j(1-\beta^j) + \sum_i \chi_i \beta_i^j\right\} F^j S\right] T(t) \\
&- \left[\int dV \int dE\, W \sum_{i,j} \chi_i \beta_i^j F^j S\right] T(t) + \sum_{i=1}^{I} \lambda_i \left[\int dV \int dE\, W \chi_i c_i\right] + \int dV \int dE\, Wq \\
&= \frac{\partial}{\partial t}\left[\int dV \int dE\, W \frac{1}{v} S T(t)\right],
\end{aligned} \tag{7.3.3}$$

$$\begin{aligned}
&\left[\int dV \int dE\, W\chi_i \sum_j \beta_i^j F^j S\right] T(t) - \lambda_i \left[\int dV \int dE\, W\chi_i c_i\right] \\
&= \frac{\partial}{\partial t}\left[\int dV \int dE\, W\chi_i c_i\right] \qquad (i = 1, 2, \ldots, I),
\end{aligned}$$

where the integrals are over the total volume of the reactor and the total range of neutron energies ($0 \le E < \infty$).

Now, because of (7.3.1) and (7.3.2),

$$T(t) = \int dV \int dE\, W \frac{1}{v} \Phi = \left[\int dV \int dE\, W \frac{1}{v} S\right] T(t). \tag{7.3.4}$$

As a result, even though $S(\mathbf{r}, E, t)$ is in general time-dependent,

$$\int dV \int dE\, W(\mathbf{r}, E) \frac{1}{v(E)} S(\mathbf{r}, E, t) = 1 \quad \text{for all } t. \tag{7.3.5}$$

That is, the normalization of the (time-dependent) shape function is such that the integral (7.3.5) is a constant, independent of time.

It follows that we may rearrange the right-hand sides of the equations in (7.3.3) as

$$\frac{\partial}{\partial t}\left[\int dV \int dE\, W\frac{1}{v}ST(t)\right] = \left[\int dV \int dE\, W\frac{1}{v}S\right]\frac{dT(t)}{dt},$$
$$\frac{\partial}{\partial t}\left[\int dV \int dE\, W\chi_i c_i\right] = \left[\int dV \int dE\, W\frac{1}{v}S\right]\frac{d}{dt}\left[\frac{\int dV \int dE\, W\chi_i c_i}{\int dV \int dE\, W(1/v)S}\right]. \tag{7.3.6}$$

Then, if we divide (7.3.3) by

$$\int dV \int dE\, W\sum_j\left[\chi_p^j(1-\beta^j) + \sum_i \chi_i\beta_i^j\right]F^jS$$

and define

$$\rho(t) \equiv \frac{\int dV \int dE\, W[\nabla\cdot D\nabla S - AS + \sum_j\{\chi_p^j(1-\beta^j) + \sum_i\chi_i\beta_i^j\}F^jS]}{\int dV \int dE\, W\sum_j\{\chi_p^j(1-\beta^j) + \sum_i\chi_i\beta_i^j\}F^jS}, \tag{7.3.7}$$

$$\beta_i(t) \equiv \frac{\int dV \int dE\, W\sum_j \chi_i\beta_i^jF^jS}{\int dV \int dE\, W\sum_j\{\chi_p^j(1-\beta^j) + \sum_i\chi_i\beta_i^j\}F^jS}, \qquad \beta(t) \equiv \sum_{i=1}^{I}\beta_i(t), \tag{7.3.8}$$

$$\Lambda(t) \equiv \frac{\int dV \int dE\, W(1/v)S}{\int dV \int dE\, W\sum_j\{\chi_p^j(1-\beta^j) + \sum_i\chi_i\beta_i^j\}F^jS}, \tag{7.3.9}$$

$$Q(t) \equiv \frac{\int dV \int dE\, Wq}{\int dV \int dE\, W(1/v)S}, \tag{7.3.10}$$

and

$$C_i(t) \equiv \frac{\int dV \int dE\, W\chi_i c_i(\mathbf{r}, t)}{\int dV \int dE\, W(1/v)S}, \tag{7.3.11}$$

we obtain

$$(\rho - \beta)T(t) + \Lambda\sum_{i=1}^{I}\lambda_iC_i(t) + \Lambda\, Q(t) = \Lambda\frac{dT(t)}{dt},$$
$$\beta_iT(t) - \lambda_i\Lambda C_i(t) = \Lambda\frac{dC_i(t)}{dt} \qquad (i = 1, 2, \ldots, I), \tag{7.3.12}$$

or

$$\frac{dT(t)}{dt} = \frac{\rho - \beta}{\Lambda}T(t) + \sum_{i=1}^{I}\lambda_iC_i(t) + Q(t),$$
$$\frac{dC_i(t)}{dt} = \frac{\beta_i}{\Lambda}T(t) - \lambda_iC_i(t) \qquad (i = 1, 2, \ldots, I). \tag{7.3.13}$$

Equations (7.3.13) are called the *point kinetics equations*. They form the basis of almost all the transient design analysis performed for reactors in operation today. Their structure is very simple. Nevertheless we have derived them in a formal fashion from (7.2.9) without making any approximations. In fact it is possible to derive this same form directly from the time-dependent transport equation.

This formal exactness implies that, if we use the values of the "*kinetics parameters*" ρ, β_i, Λ, Q, and C_i as defined by (7.3.7)–(7.3.11), the solution of (7.3.13) for $T(t)$ will be exactly the same as if we had solved (7.2.9) for $\Phi(\mathbf{r}, E, t)$ and then applied the definition (7.3.1). Such correspondence assures us that there *exist* parameters ρ, β_i, Λ, and Q for which the solutions of (7.3.13) will be exact. However, since we must know $\Phi(\mathbf{r}, E, t)$ to get the kinetics parameters, the practical utility of the kinetics equations depends on our ability to obtain a reasonably accurate value of $S(\mathbf{r}, E, t)$ *without* actually solving (7.2.9). The great success of the form (7.3.13) is due to the fact that, for many transients of practical interest, the flux *shape* changes very little and, provided we select an appropriate value of the weight function (in accord with a procedure to be described later), use of the *initial* flux shape throughout the entire transient to determine ρ, β_i, Λ, and Q yields excellent results for $T(t)$.

7.4 The Kinetics Parameters

The exact physical interpretations of the parameters C_i, Q, Λ, β_i, and ρ will depend on the choice of a weight function $W(\mathbf{r}, E)$. Moreover, even after that selection is made, if the quantities ρ, β_i, and Λ are all multiplied by any arbitrary constant, the solution for $T(t)$ will be unaltered. This is because equations (7.3.13) for $T(t)$ and the C_i depend only on the *ratios* ρ/Λ and β_i/Λ. In fact it would avoid a lot of confusion in the field if one of the three parameters ρ, β, or Λ were eliminated and we talked about reactor kinetics only in terms of the two remaining ratios. (Note, for instance, that, for an assembly containing no fissionable material, ρ/Λ and β_i/Λ are defined whereas ρ, β_i, and Λ, individually, are not.) The individual parameters, however, are so much a tradition that making the change at this date is probably a lost cause. Accordingly we shall discuss the physical interpretation of the parameters as defined in the conventional manner, keeping in mind, however, that they have no intrinsic physical significance in the sense that values of them can be inferred from experiment alone.

Interpretation of $Q(t)$ and the $C_i(t)$

In view of the normalization of $S(\mathbf{r}, E, t)$ and $W(\mathbf{r}, E)$ implied by (7.3.5), the denominators in the definitions (7.3.10) and (7.3.11) for $Q(t)$ and the $C_i(t)$ are unity. This fact permits us to interpret $Q(t)$ as a weighted integral of the total number of neutrons being introduced into the reactor per second from extraneous sources at time t, and the $C_i(t)$ as weighted integrals of the total number of delayed-neutron precursors present in the reactor at time t. Note in particular that, if $W(\mathbf{r}, E)$ is taken as unity,

$$\begin{aligned} Q(t) &= \int dV \int dE\, q(\mathbf{r}, E, t), \\ C_i(t) &= \int c_i(\mathbf{r}, t)dV, \end{aligned} \tag{7.4.1}$$

where we have used the fact that the $\chi_i(E)$ are normalized so that $\int_0^\infty \chi_i(E)dE = 1$. Thus, for $W(\mathbf{r}, E) = 1$, $Q(t)$ is the total number of source neutrons introduced per second at time t, and the $C_i(t)$ are the total number of precursors present in the reactor at that time.

Interpretation of the Reactivity $\rho(t)$

The parameter $\rho(t)$, defined by (7.3.7), is called the *reactivity* of the reactor. The present definition is an alternative to the one introduced in Chapter 3. To provide a physical interpretation we note that, since $T(t)$ depends only on time, the shape function $S(\mathbf{r}, E, t)$ in the expressions for $\rho(t)$ can be replaced by $S(\mathbf{r}, E, t)T(t)$ (i.e., $\Phi(\mathbf{r}, E, t)$) in the numerator and denominator. Also we recognize that the expression $\chi_p^j(1 - \beta^j) + \sum_i \chi_i \beta_i^j$ is the *total* fission spectrum χ^j defined in Section 1.4 for isotope j. Thus (7.3.7) may be written

$$\rho(t) \equiv \frac{\int dV \int dE\, W(\mathbf{r}, E)[\nabla \cdot D(\mathbf{r}, E, t)\nabla\Phi(\mathbf{r}, E, t) - A\Phi(\mathbf{r}, E, t) + \sum_j \chi^j(E)F^j\Phi(\mathbf{r}, E, t)]}{\int dV \int dE\, W(\mathbf{r}, E) \sum_j \chi^j(E)F^j\Phi(\mathbf{r}, E, t)}. \tag{7.4.2}$$

The term $\sum_j \chi^j(E)F^j\Phi(\mathbf{r}, E, t)$ is the rate at which events that are occurring at time t will ultimately (i.e., after delayed precursors decay) produce neutrons in $dV\, dE$. We shall call it the *instantaneous production rate due to fission*. It follows that the denominator of (7.4.2) is a weighted integral over the whole reactor of this instantaneous production rate. If $W(\mathbf{r}, E)$ were unity, it would be *exactly* the instantaneous production rate throughout the whole reactor of neutrons due to fission. Even then, the term is a bit artificial in that only $1 - \beta^j$ of the neutrons "instantaneously produced" at time t from fissions in isotope j appear physically at that time.

The term $-\nabla \cdot D\nabla\Phi + A\Phi$ in the numerator of (7.4.2) is the destruction rate of neutrons in $dV\, dE$ at time t. (This destruction rate truly *is* "instantaneous.") Thus the numerator of $\rho(t)$ is a weighted integral of the net "instantaneous" production rate of neutrons in the reactor.

Reactivity is, then, the ratio of this net weighted rate to the weighted production rate due to fission. It can be thought of loosely as the fractional increase per second (or per "generation") in the number of neutrons produced in the core. However, because of the weighting function and because delayed neutrons generated at time t don't appear until their precursors decay, this way of thinking about ρ is not precise.

The reactivity of a reactor is, in general, time-dependent, primarily because D, A, and the F^j depend on time, but also because of the time dependence of the shape part of $\Phi(\mathbf{r}, E, t)$. Since, in a critical reactor, D, A, and F^j are independent of time and the flux shape is the critical flux shape, (4.6.1) and (7.2.8) show us that the numerator of $\rho(t)$, and hence $\rho(t)$ itself, is zero for a critical reactor. If the production rate exceeds the destruction rate, $\rho(t) > 0$ and the reactor is said to be *supercritical*; if the production rate is less than the destruction rate, $\rho(t) < 0$ and the reactor is said to be *subcritical.* Also, if $\rho = \beta$, the reactor is *prompt-critical*, and, if $\rho > \beta$, it is *super-prompt-critical.* We shall investigate the implications of being in any of these states shortly.

Reactivity, being a ratio of rates, is dimensionless. It is thus often expressed as a pure number ("0.001 in reactivity") or in percent ("0.1 percent in reactivity"). It is also expressed in multiples of β, and, when this is done, the units of "dollars" and "cents" are used. Thus $\rho = \beta$ implies a reactivity of one dollar or 100 cents. Thus, if $\beta = 0.007$, a reactivity of 0.0035 is 0.35 percent, half a dollar, or fifty cents. Since we have already seen that ratios of ρ, β_i, and Λ are more meaningful than the individual quantities themselves, the dollar is the aesthetically more satisfying unit.

Interpretation of the Effective Delayed-Neutron Fraction β_i

The parameter β_i is called the *effective delayed-neutron fraction* for the ith precursor group. The sum β is, then, the "total effective delayed-neutron fraction." As with $\rho(t)$, we can rewrite (7.3.8) in the form

$$\beta_i(t) \equiv \frac{\int dV \int dE\, W(\mathbf{r}, E) \sum_j \chi_i \beta_i^j F^j \Phi(\mathbf{r}, E, t)}{\int dV \int dE\, W(\mathbf{r}, E) \sum_j \chi^j F^j \Phi(\mathbf{r}, E, t)} \qquad (i = 1, 2, \ldots, I). \tag{7.4.3}$$

In view of the earlier discussion, we can immediately interpret $\beta_i(t)$ as the "instantaneous" weighted rate of production of *delayed* neutrons belonging to group i throughout the reactor divided by the "instantaneous" weighted rate of *all* neutron production due to fission. If $W(\mathbf{r}, E)$ is taken as unity and there is only one fissionable isotope in the reactor (i.e., one term in the sum over j), $\beta_i(t)$ becomes the β_i of that isotope. If several fissioning isotopes (U^{25}, U^{28}, Pu^{49}, etc.) are present, and again if $W(\mathbf{r}, E) = 1$, the $\beta_i(t)$ become averages of the different β_i^j weighted by the amount of fissioning due to each isotope. If $W(\mathbf{r}, E) \neq 1$, the fact that the spectrum with which delayed neutrons are emitted is lower than the total fission spectrum causes a further deviation of the $\beta_i(t)$ from physical values, the amount and sign of this deviation being dependent on $W(\mathbf{r}, E)$. For the usual choices of $W(\mathbf{r}, E)$, however, the $\beta_i(t)$ are within 20 or 30 percent of the range 0.0001 to 0.0026 of the physical β_i^j (see Table 7.1).

Since the time dependence of $F^j S(\mathbf{r}, E, t)$ is slight in most cases and since this quantity appears in both the numerator and denominator of (7.3.8), the $\beta_i(t)$ actually vary only very slightly with time. In practice this dependence is quite justifiably neglected.

Interpretation of the Prompt-Neutron Lifetime $\Lambda(t)$

The final kinetics parameter, $\Lambda(t)$, is called the *prompt-neutron lifetime.* As with ρ and β_i, its mathematical definition can be rewritten as

$$\Lambda \equiv \frac{\int dV \int dE\, W(\mathbf{r}, E)[1/v(E)]\Phi(\mathbf{r}, E, t)}{\int dV \int dE\, W(\mathbf{r}, E) \sum_j \chi^j(E) F^j \Phi(\mathbf{r}, E, t)}. \tag{7.4.4}$$

An approximate physical interpretation of Λ can be seen by first assuming that the reactor is critical and that $W(\mathbf{r}, E)$ is unity. Then (7.4.4) tells us that

$$\int dV \int dE \sum_j \nu^j \Sigma_{f0}^j(\mathbf{r}, E)\Phi_0(\mathbf{r}, E) = \frac{\int dV \int dE[1/v(E)]\Phi_0(\mathbf{r}, E)}{\Lambda}, \tag{7.4.5}$$

i.e., the instantaneous production rate of neutrons in the reactor is equal to the number present divided by Λ. It follows by analogy with the radioactive-decay equation that Λ can be interpreted as the time an average neutron survives *after* it appears as either a prompt neutron or one emitted from a delayed-neutron precursor.

For the non-steady-state case (but with $W(\mathbf{r}, E)$ still taken as unity), this interpretation is only approximate. However, for the reactor close to critical ($\rho \approx 0$) and the flux shape $S(\mathbf{r}, E, t)$ close to the critical flux shape, Λ is still quite close to the average time a neutron lives. For $W(\mathbf{r}, E) \neq 1$, we shall think of Λ as the average time some *weighted* neutron population lives.

Values of Λ fall in the range 10^{-3} sec (for very thermal reactors) to 10^{-7} sec (for very fast reactors). As with the β_i, the time dependence of $\Lambda(t)$ is generally neglected. There are, however, cases involving very fast transients—either negative or positive—for which the shape function changes significantly, and failing to account for changes in $\Lambda(t)$ will result in serious errors.

Before examining the methods used to compute the kinetics parameters, we shall turn to the problem of solving the kinetics equations themselves. Thus, for the time being, we shall assume we know how to find ρ, Λ, and the β_i for a given transient in a given reactor. After examining the nature of the solutions to the equations in which these parameters appear and discussing how ρ/Λ and β/Λ may be inferred from experiment, we shall return to the problem of their analytical determination.

7.5 The Determination and Use of Solutions to the Point Kinetics Equations

With the time dependence of Λ and the β_i of negligible significance, the reactivity $\rho(t)$ and the effective source $Q(t)$ become the driving terms of the point kinetics equations.

One might then expect to be able to attack these equations analytically. However the fact that $\rho(t)$ multiplies $T(t)$ makes this impractical except when $\rho(t)$ changes in a step fashion and then remains constant. Finding an analytical solution for any other behavior of $\rho(t)$—even a ramp shape with $Q(t) = 0$—is most difficult. Numerical methods are best applied to these cases.

We shall thus restrict the examination of analytical methods for solving (7.3.13) to the case in which ρ is constant throughout the period of interest. The solution we shall obtain, although only rarely useful for practical applications, will provide considerable insight into how the neutron population in a reactor behaves under non-steady-state conditions.

Analytical Solutions

For ρ/Λ and β_i/Λ independent of time, solution of the point kinetics equations (7.3.13) is straightforward. However, with six delayed groups and $Q(t)$ nonzero, it is necessary to deal with a seventh-order algebraic equation to obtain this solution, and insight into the dependence of $T(t)$ on the reactor parameters is lost. Accordingly we shall carry out the algebraic details of the analytical solution with $Q(t)$ set equal to zero (no external source) and with the assumption that there is only one delayed-neutron group. Equation (7.3.13) then reduces to

$$\frac{d}{dt}\begin{bmatrix} T(t) \\ C(t) \end{bmatrix} = \begin{bmatrix} \dfrac{\rho - \beta}{\Lambda} & \lambda \\ \dfrac{\beta}{\Lambda} & -\lambda \end{bmatrix} \begin{bmatrix} T(t) \\ C(t) \end{bmatrix}, \tag{7.5.1}$$

where we have written the equations in matrix form and dropped subscripts i, there being only one precursor group.

The Laplace-transform method provides a standard way of finding a solution to these two coupled linear equations. We shall, however, apply the equally standard approach of assuming particular solutions such that $T(t) \approx C(t) \approx \exp(\omega t)$ and then taking linear combinations of these particular solutions to find the general solution obeying the initial conditions of the problem.

If the assumed exponential behavior is correct, it follows that $dT(t)/dt = \omega T(t)$ and $dC(t)/dt = \omega C(t)$. These relationships can be valid only if

$$\begin{bmatrix} \dfrac{\rho - \beta}{\Lambda} - \omega & \lambda \\ \dfrac{\beta}{\Lambda} & -(\omega + \lambda) \end{bmatrix} \begin{bmatrix} T(t) \\ C(t) \end{bmatrix} = 0, \tag{7.5.2}$$

a relationship which, in turn, implies that either $T = C = 0$ or the determinant of the matrix vanishes. Rejecting the first possibility as uninteresting and setting the determinant to zero then gives

$$\omega^2 - \left(\frac{\rho - \beta}{\Lambda} - \lambda\right)\omega - \frac{\lambda\rho}{\Lambda} = 0, \tag{7.5.3}$$

the solution of which is

$$\omega = \frac{1}{2}\left(\frac{\rho - \beta}{\Lambda} - \lambda\right) \pm \sqrt{\left[\frac{1}{4}\left(\frac{\rho - \beta}{\Lambda} + \lambda\right)^2 + \frac{\beta\lambda}{\Lambda}\right]}. \tag{7.5.4a}$$

We have arranged the terms under the square root in this expression to make it obvious that the ω are real for all values of ρ. An alternate algebraic arrangement of this term leads to

$$\omega = \frac{1}{2}\left(\frac{\rho - \beta}{\Lambda} - \lambda\right) \pm \sqrt{\left[\frac{1}{4}\left(\frac{\rho - \beta}{\Lambda} - \lambda\right)^2 + \frac{\lambda\rho}{\Lambda}\right]}, \tag{7.5.4b}$$

from which it is apparent that: for $\rho > 0$, one root is positive and the other is negative; for $\rho = 0$, one root is zero and the other is $-(\beta/\Lambda) - \lambda$; and, for $\rho < 0$, both roots are negative.

Equation (7.5.2) implies that, for each of the two roots ω_i of (7.5.4b), $T(t)$ and $C(t)$ have a fixed ratio. Thus we have

$$\left.\frac{C(t)}{T(t)}\right|_i = -\frac{[(\rho - \beta)/\Lambda] - \omega_i}{\lambda} = \frac{\beta}{\Lambda(\omega_i + \lambda)}. \tag{7.5.5}$$

It follows that the general solution of (7.5.1) may be written as

$$\begin{bmatrix} T(t) \\ C(t) \end{bmatrix} = a_1 \begin{bmatrix} 1 \\ \dfrac{\beta}{\Lambda(\omega_1 + \lambda)} \end{bmatrix} \exp(\omega_1 t) + a_2 \begin{bmatrix} 1 \\ \dfrac{\beta}{\Lambda(\omega_2 + \lambda)} \end{bmatrix} \exp(\omega_2 t), \tag{7.5.6}$$

where a_1 and a_2 are constants to be determined by the initial conditions.

The complete solutions of (7.5.1) can thus be written entirely in terms of ρ, β, Λ, λ, $T(0)$, $C(0)$, and t. However the resultant formula is complicated, and it is difficult to understand how its character depends on ρ without performing many numerical calculations. Accordingly, since our principal interest in deriving an analytical solution is to gain physical insight, we shall make an approximation that will greatly simplify the algebraic structure of the solution (7.5.6).

The approximation is based on the fact that, except for ρ close to β, one root of (7.5.4b) is very large in magnitude and the other is very small. Knowing this, we can determine an approximate solution of (7.5.3) by noting that for the root ω_1 of large magnitude, $\omega_1^2 \gg \lambda\rho/\Lambda$, and $\lambda\rho/\Lambda$ in (7.5.3) can be neglected. Similarly, for the root ω_2 of small magnitude, $\omega_2^2 \ll \lambda\rho/\Lambda$, and ω_2^2 in (7.5.3) can be neglected. Thus, for $\omega_1^2 \gg \omega_2^2$,

$$\begin{aligned} \omega_1 &\approx -\frac{\beta - \rho}{\Lambda}, \\ \omega_2 &\approx \frac{\lambda\rho}{\beta - \rho}, \end{aligned} \tag{7.5.7}$$

where we have also neglected λ in comparison with $(\rho - \beta)/\Lambda$. For values $\beta = 0.0075$, $\Lambda = 6 \times 10^{-5}$ sec, and $\lambda = 0.08\ \text{sec}^{-1}$ (typical of a light-water-moderated power reactor), the ratios $|\omega_1|/|\omega_2|$ and $|\rho - \beta|/\Lambda\lambda$ both exceed 250 for $|\rho - \beta| > 0.5\beta$ (i.e., for $\rho > 1.5\beta$ or $\rho < 0.5\beta$). The approximation (7.5.7) should thus be valid in this range. For a fast reactor ($\Lambda \approx 10^{-7}$ sec) it should be even better.

Using the approximate roots (7.5.7), we find that the general solution (7.5.6) becomes

$$\begin{bmatrix} T(t) \\ C(t) \end{bmatrix} \approx a_1 \begin{bmatrix} 1 \\ -\dfrac{\beta}{\beta - \rho} \end{bmatrix} \exp\left(-\frac{\beta - \rho}{\Lambda} t\right) + a_2 \begin{bmatrix} 1 \\ \dfrac{\beta - \rho}{\Lambda\lambda} \end{bmatrix} \exp\left(\frac{\lambda\rho}{\beta - \rho} t\right), \tag{7.5.8}$$

where we have once more neglected λ in comparison with ω_1.

We see immediately from this result that, if $\rho = 0$, the first term dies away quickly* and steady-state values of $T(t)$ and $C(t)$ are given by

$$\begin{bmatrix} T_0 \\ C_0 \end{bmatrix} = a_2 \begin{bmatrix} 1 \\ \dfrac{\beta}{\Lambda\lambda} \end{bmatrix} = T_0 \begin{bmatrix} 1 \\ \dfrac{\beta}{\Lambda\lambda} \end{bmatrix}. \tag{7.5.9}$$

Since C_0 is roughly the total number of precursors in the reactor and T_0 is roughly the total number of neutrons, it is evident that, in the steady state, the precursor population far outweighs the neutron population (by 1600 : 1 for the typical figures for β, Λ, and λ just quoted).

Let us now consider what the solution (7.5.8) predicts will happen when a reactor, initially in a critical condition, is perturbed by a step change in reactivity.

Values of a_1 and a_2 that give the correct initial condition (7.5.9) for this transient are (from (7.5.8) with $t = 0$)

$$\begin{aligned} a_1 &= \frac{-\rho T_0}{\beta - \rho + [\Lambda\lambda\beta/(\beta - \rho)]} \approx -\frac{\rho T_0}{\beta - \rho}, \\ a_2 &= \frac{\beta + [\Lambda\lambda\beta/(\beta - \rho)]}{\beta - \rho + [\Lambda\lambda\beta/(\beta - \rho)]} T_0 \approx \frac{\beta T_0}{\beta - \rho}. \end{aligned} \tag{7.5.10}$$

* With more than one group of delayed neutrons represented, attainment of the steady-state condition would be much slower.

Thus (7.5.8) becomes

$$\begin{bmatrix} T(t) \\ C(t) \end{bmatrix} \approx T_0 \left\{ \begin{bmatrix} \dfrac{-\rho}{\beta - \rho} \\ \dfrac{\rho\beta}{(\beta - \rho)^2} \end{bmatrix} \exp\left(-\frac{\beta - \rho}{\Lambda} t \right) + \begin{bmatrix} \dfrac{\beta}{\beta - \rho} \\ \dfrac{\beta}{\Lambda\lambda} \end{bmatrix} \exp\left(\frac{\lambda\rho}{\beta - \rho} t \right) \right\}. \tag{7.5.11}$$

We shall now examine the characteristics of this solution for three practical situations: $\rho = -0.05$; $\rho = 0.0015$; and $\rho = 0.0115$.

Shutdown of a Reactor. A step change from $\rho = 0$ to $\rho = -0.05$ represents a sudden shutdown or "scram" of the reactor. With the typical values of β, Λ, and λ we have been using, (7.5.11) for this case becomes

$$\begin{bmatrix} T(t) \\ C(t) \end{bmatrix} = T_0 \left\{ \begin{bmatrix} 0.87 \\ 0.0113 \end{bmatrix} \exp(-958.t) + \begin{bmatrix} 0.13 \\ 1563. \end{bmatrix} \exp(-0.068t) \right\}. \tag{7.5.12}$$

A plot of the behavior implied by this equation is shown in Figure 7.1. We see that, because of the large negative exponential, 87 percent of the neutron population $T(t)$ dies off in less than 0.01 sec. However only a very small fraction of the precursors decay in that time. Thereafter the rest of the neutron and precursor populations die away with about a 15-second "decay period" ($\equiv 1/\omega_2$). Physically the sudden decay of the first term in (7.5.12) is due to the readjustment of the neutron population to the subcritical condition of the reactor. If the fractional increase in the number of neutrons born per second is ~ -0.05, approximately 95 neutrons appear in the second generation for 100 appearing in the first. Thus, since a prompt generation dies in 6×10^{-5} sec, the initial neutron population is decreased by roughly $(0.95)^{10^5/6}$ in one second. The fact that a neutron population does persist, equal to ~ 13 percent of that initially present, is due to the neu-

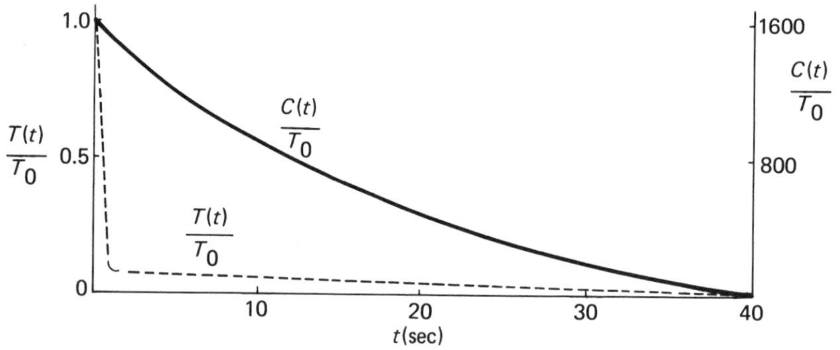

Figure 7.1 Neutron and precursor population following a change in reactivity of -0.05 from an initially critical state.

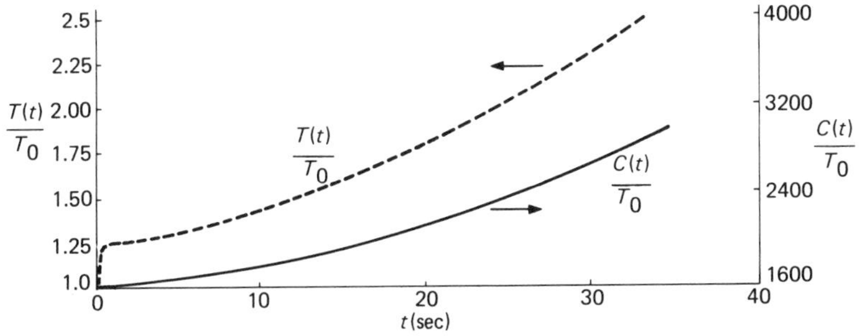

Figure 7.2 Neutron and precursor population following a change in reactivity of +0.0015 from an initially critical state.

trons emitted by the very large number of delayed-neutron precursors initially present and the fact that these decay as $\exp(-0.08t)$.

One might expect by this argument that the second term in (7.5.12) should decay as $\exp(-0.08t)$ rather than $\exp(-0.068t)$. It does not because those delayed neutrons emitted by precursors produce *some* prompt neutrons as they die. Thus the overall decay rate of the neutron population is retarded slightly. Notice in this connection that, as ρ becomes more and more negative, ω_2 as given by (7.5.7) approaches $-\lambda$.

A Supercritical Transient. If the initially critical reactor here being analyzed is subject to a step increase in reactivity of 0.0015, (7.5.11) becomes

$$\begin{bmatrix} T(t) \\ C(t) \end{bmatrix} = T_0 \left\{ \begin{bmatrix} -0.25 \\ 0.3125 \end{bmatrix} \exp(-100.t) + \begin{bmatrix} 1.25 \\ 1562.5 \end{bmatrix} \exp(0.02t) \right\}. \tag{7.5.13}$$

A plot of this result appears as Figure 7.2.

Again we see that $T(t)$ behaves in essentially a step fashion during the first few hundredths of a second, whereas $C(t)$ changes negligibly during that time. After the first term dies out, both populations continue to rise on an *asymptotic period* defined as $1/\omega_2$ ($=50$ seconds here).

The initial step in $T(t)$ is due again to the readjustment of prompt neutrons to the new multiplying conditions. This step is called the *prompt jump*. Physically it represents the start of a nuclear runaway. However, since the excess multiplication is only ~ 0.0015 and since 0.0075 neutrons per cycle are held up in precursors, the runaway cannot continue. Note that, according to (7.5.11), the magnitude of the prompt jump (which is due mathematically to the *dying out* of a negative term) increases as the step increase in ρ gets closer in magnitude to β. (Recall, however, that (7.5.11) is not valid for ρ very close to β.)

The prompt jump occurs so quickly that, in many cases, it can be assumed to be instantaneous. Physically such an assumption implies that $T(t)$ is in instant equilibrium with the $C_i(t)$. Mathematically this situation can be expressed by neglecting $dT(t)/dt$ in (7.5.1) or (7.3.13) in comparison to $[(\rho - \beta)/\Lambda]T$. Such neglect simplifies the point kinetics equations. Thus (7.5.1) becomes, for $t > 0$,

$$\begin{aligned} 0 &= \frac{\rho - \beta}{\Lambda} T(t) + \lambda C(t), \\ \frac{dC(t)}{dt} &= \frac{\beta}{\Lambda} T(t) - \lambda C(t), \end{aligned} \tag{7.5.14}$$

which is equivalent to the first-order differential equation

$$\frac{dC(t)}{dt} = \frac{\rho\lambda}{\beta - \rho} C(t). \tag{7.5.15}$$

The solution for $C(t)$ and $T(t)$ for $t > 0$ is, then,

$$\begin{aligned} C(t) &= C_0 \exp\left(\frac{\rho\lambda}{\beta - \rho} t\right), \\ T(t) &= \frac{\lambda\Lambda}{\beta - \rho} C(t) = \frac{\lambda\Lambda}{\beta - \rho} C_0 \exp\left(\frac{\rho\lambda}{\beta - \rho} t\right). \end{aligned} \tag{7.5.16}$$

If we recognize that C_0 is related to T_0 by $C_0 = \beta T_0/\Lambda\lambda$, where T_0 is the value of $T(t)$ at $t = 0$ (i.e., *before* the step change in ρ), we see that (7.5.16) is identical with the approximate solution to the second-order equations (7.5.1) after the exponential $\exp((\rho - \beta)t/\Lambda)$ has died out.

The procedure of neglecting dT/dt as in (7.5.14) is called the *prompt-jump approximation.*

A Super-Prompt-Critical Transient. If the step increase in ρ exceeds β, the reactor is said to be super-prompt-critical. The consequences are disastrous. For example, if $\rho = 0.0115$, (7.5.11) becomes

$$\begin{bmatrix} T(t) \\ C(t) \end{bmatrix} = T_0 \left\{ \begin{bmatrix} 2.9 \\ 5.4 \end{bmatrix} \exp(66.7t) + \begin{bmatrix} -1.9 \\ 1563. \end{bmatrix} \exp(-0.23t) \right\}.$$

In this case the reactor runs away. The first term rises by a factor of ~770 in a tenth of a second (on a 1.5-millisecond period); and, unless some inherent feedback mechanism in the reactor (such as the Doppler effect) annuls the reactivity step, the reactor will be destroyed.

Physically this excursion occurs because the reactivity exceeds β. When this happens, even with the complete neglect of the 0.0075 fraction of neutrons tied up each cycle in precursors, there is still a fractional increase of approximately $0.0115 - 0.0075 = 0.004$

in the neutron population occurring roughly each 6×10^{-5} sec. The 1.5-millisecond runaway results.

Clearly the super-prompt-critical condition is to be avoided in power reactors. The control systems for such reactors must restrict the value of reactivity so that it never gets more than one or two tenths of a percent supercritical ($\rho < 0.002$). At the same time ρ must be zero if the reactor is to be critical at all. Thus the allowable reactivity band in which a reactor can be operated is extremely small.

In fast reactors the situation is even more delicate since β^j for Pu^{49} is less than that for U^{25} (0.0021 versus 0.0065) and Λ may be a hundred times smaller (so that the periods associated with a runaway are a hundred times shorter).

It is because of this narrow working margin that reactors are designed in such a way that the immediate consequence of any feedback effect due to an increase in power level is a decrease in the criticality of the system, that is, a reduction of the reactivity.

The Inhour Formula

If six groups of delayed neutrons are accounted for, the solutions of the point kinetics equations for $T(t)$ and the $C_i(t)$ are all sums of seven exponentials $\sum_{i=0}^{6} \alpha_i \exp(\omega_i t)$. The ω_i are solutions of a seventh-order algebraic equation and cannot (as was possible in (7.5.4)) be found from a formula; they must be determined numerically. Simple explicit expressions for $T(t)$ and the $C_i(t)$ analogous to (7.5.11) and depending only on ρ, Λ, and the β_i can no longer be obtained, and acquiring insight into the physical nature of any given transient is more difficult.

Fortunately the response of a reactor to step changes in reactivity is qualitatively similar to the response we investigated in accounting for only one group of delayed neutrons. The prompt-jump and asymptotic-period behaviors are still observed, and the reactor again runs away for $\rho > \beta$. About the only qualitative difference is that, because it is necessary to wait for *six* exponentials (rather than one) to die out before an asymptotic period is established, it takes somewhat longer for the reactor to reach that state.

We shall restrict our quantitative examination of this more general case to a study of the nature of the seven ω_i involved in the exponential behavior.

If, with $Q(t) = 0$, we seek those values of ω which, for a constant value of ρ, make $T(t) \approx C_i(t) \approx \exp(\omega t)$ a particular solution of (7.3.13), we find that ω must be such that

$$\begin{aligned} \omega T &= \frac{\rho - \beta}{\Lambda} T + \sum_{i=1}^{I} \lambda_i C_i, \\ \omega C_i &= \frac{\beta_i}{\Lambda} T - \lambda_i C_i \qquad (i = 1, 2, \ldots, I). \end{aligned} \tag{7.5.17}$$

Using the last I equations to eliminate the C_i from the first and doing some rearranging yields

$$\rho = \omega\Lambda + \sum_{i=1}^{I} \frac{\beta_i \omega}{\omega + \lambda_i}. \qquad (7.5.18)$$

This result is known as the *inhour formula.* The name comes from the fact that, in the early days of reactor technology, values of ω were quoted in "inverse hours." The value of ρ such that $\omega = 1\ \text{hr}^{-1}$ is one inhour.

If, for $I = 6$, (7.5.18) is multiplied by $\prod_{i=1}^{6} (\omega + \lambda_i)$, a polynomial equation of the seventh order in ω results. Thus, for every value of ρ (Λ and the β_i being assumed fixed for a given reactor), there are seven roots ω_i to (7.5.18). To get some feeling for the nature of these roots we plot in Figure 7.3 the function

$$f(\omega) = \omega\Lambda + \sum_{i=1}^{6} \frac{\beta_i \omega}{\omega + \lambda_i}.$$

(Note that the scale is logarithmic; $-\lambda_6 = -3.0\ \text{sec}^{-1}$, whereas $-\beta/\Lambda$ might be $-125\ \text{sec}^{-1}$.)

To find the ω_i corresponding to a given value of ρ we draw a line parallel to the ω axis at a distance above or below it such that $f(\omega) = \rho$. The seven values ω_i corresponding to the points at which this line crosses the curve $f(\omega)$ versus ω (for instance the crosses and circles on Figure 7.3) are values such that $f(\omega_i) = \rho$. These ω_i are thus the desired roots.

We see that the roots are always real and that five of them always lie in the range $-\lambda_6 < \omega < -\lambda_1$. At most one of the roots is positive, and then only when $\rho > 0$. Equation (7.5.18) shows that, when $|\omega| \gg \lambda_i$, $\rho \approx \omega\Lambda + \beta$ or

$$\omega \approx \frac{\rho - \beta}{\Lambda}. \qquad (7.5.19)$$

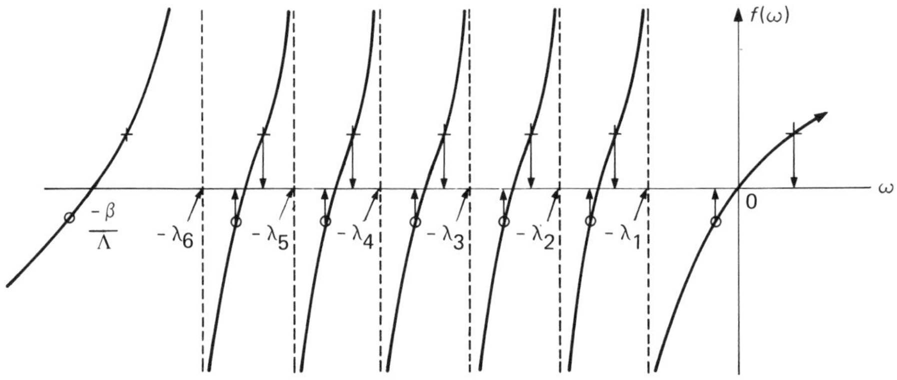

Figure 7.3 A plot of the inhour formula.

Thus, since $\beta/\Lambda \gg \lambda_i$, there is always one large negative root for $\rho \leq 0$. The corresponding most positive (least negative) root for this case is seen from Figure 7.3 to lie between $-\lambda_1$ and zero. For positive values of reactivity the condition $\omega \gg \lambda_i$ will occur for $\rho \approx 1.5\beta$, and the most negative root will move over toward the negative side of $\omega = -\lambda_6$ for this super-prompt-critical case. Note that the estimate of the large roots (7.5.19) for this case of six precursor groups is the same as that in (7.5.7) for one. We thus expect that the qualitative behavior derived for the one-precursor-group case ought to be similar to what we would find if all precursor groups were considered.

Measurement of the Kinetics Parameters

The fact that there is a most positive (least negative) root—call it ω_0—of the inhour equation implies that, if we wait long enough, $T(t)$ will behave as $\exp(\omega_0 t)$. This exponential rate can be measured since it it possible to show that the flux shape associated with a sustained constant value of ρ becomes independent of time ($S(\mathbf{r}, E, t) \rightarrow S(\mathbf{r}, E)$). Thus the flux $\Phi(\mathbf{r}, E, t) = S(\mathbf{r}, E)T(t)$ behaves in time as $T(t)$ at all points $\mathbf{r}$ and at all energies E. It follows that a counter located anywhere in the reactor and sensitive to neutrons of any energy will produce a count rate proportional to $T(t)$. The slope of the logarithm of this count rate plotted versus time is then ω_0, the asymptotic time constant; $1/\omega_0$ is the asymptotic period. If we know Λ and the β_i, we can then determine ρ by substituting the measured value of this asymptotic time constant into the inhour formula (7.5.18).

In practice this experiment is almost always performed for long positive periods (small positive values of ω_0) corresponding to positive values of ρ well below prompt-critical. (Putting a reactor on a short period is not a good safety practice, and the values of ω_0 for $\rho < 0$ soon get very close to $-\lambda_1$, where, as can be seen from Figure 7.3, the accuracy with which ω_0 must be known to infer an accurate value of ρ is very great.) If ω_0 *is* small, the term $\omega_0\Lambda$ in the inhour formula is generally negligible in comparison with the sum. Thus (7.5.18) becomes

$$\frac{\rho}{\beta} \approx \sum_{i=1}^{I} \frac{\beta_i}{\beta} \frac{\omega_0}{\omega_0 + \lambda_i}, \qquad \Lambda \ll \sum_{i=1}^{I} \frac{\beta_i}{\omega_0 + \lambda_i}. \tag{7.5.20}$$

In general the β_i/β in this expression have to be computed before ρ/β can be inferred from ω_0. However, if there is only one fissionable isotope j in the reactor, β_i/β will be very close to the physical β_i^j/β^j for that isotope. (The two will be exactly equal if the delayed-neutron spectra $\chi_i(E)$ are the same for all i; see (7.3.8).) For this case the measurement of a long, positive, asymptotic period gives us the reactivity in dollars directly, without the need to perform any calculations. (Of course, if we want ρ in percent, it is necessary to compute β.)

For a reactor that is subcritical in the range $-0.05 \leq \rho < 0$, it is often possible to measure ρ/Λ by the "pulsed-source technique." As the name suggests, this is a method in which pulses of neutrons are introduced into a subcritical reactor from an external source. We shall not go into the analysis of this experiment. It is possible to show, though, that, provided the pulses are intermittent and have a frequency such that the time between pulses is much less than the decay period of any of the delayed-neutron precursors, those precursors act like a constant low-level background of neutrons. This background can be subtracted off, leaving a signal that, throughout much of its decay, dies off as $\exp(\omega_6 t)$, ω_6 being the large negative root of the inhour formula that is given quite accurately by (7.5.19).

If we can isolate this fast decay period from other fast decays due to readjustment of the shape function $S(\mathbf{r}, E, t)$, this method gives us $(\rho - \beta)/\Lambda$ directly. If we then perform a similar experiment with $\rho = 0$, we get $-\beta/\Lambda|_{\text{critical}}$.

Unfortunately, because the shape function associated with the critical state may not be the same as the shape function of the shutdown reactor, which decays as $\exp(\omega_6 t)$, $\beta/\Lambda|_{\text{critical}}$ may differ from the β/Λ in the expression for ω_6. If, however, calculations show that the two are very nearly the same, or if calculations are used to correct the two, the pulsed-source method gives us a very direct measurement of ρ/Λ. Moreover, since it is necessary to know β/Λ to infer ρ/Λ from ω_6, the experiment also gives us ρ/β, the reactivity of the subcritical reactor in dollars.

The pulsed-source method requires special apparatus and a neutron background that is low in comparison with the signal induced by the pulse. Both these requirements make it difficult to perform pulsed experiments on an operating power reactor. Pressure vessels and shields make it difficult to introduce a pulsing beam, and the (γ, n) activity due to fission-product γ rays is very high in a reactor that has recently operated at high power.

A method called the *rod-drop experiment* avoids these difficulties. The idea in this method is to start with the reactor critical and simply scram the control rods. As we saw in analyzing the one-precursor-group calculation (7.5.11), the value of $T(t)$ changes from T_0 to $\beta T_0/(\beta - \rho)$ very quickly after the insertion of a large negative step of reactivity. Thus ρ/β can be inferred from the magnitude of this negative prompt jump. It is not possible to infer $\beta/\Lambda|_{\text{critical}}$ by this technique, however, and it must therefore be calculated if we wish to determine ρ/Λ or β/Λ.

The problems involved in avoiding errors due to fast changes in the shape function superimposed on the decay of $T(t)$ can be extremely serious with the rod-drop technique. Very careful analysis is required to avoid difficulties of this nature when the method is used.

Numerical Solution of the Point Kinetics Equations

Even for the case of step changes, it is necessary to solve a seventh-order algebraic equation to get the ω_i needed for the analytical solution of the point kinetics equations. Thus a numerical procedure is necessary to find the "analytical" solution.

Once this is accepted, it is reasonable to inquire if there is not some *overall* numerical procedure that would be just as accurate, but shorter-running, than the program for finding the roots numerically and then computing the exponentials. It turns out that there are a number of such methods, some of them an order of magnitude or more faster then the direct approach.

Thus, even for the step case, numerical methods are preferred if a numerical answer is what is wanted. For the case of $\rho(t)$ a continuously changing function, numerical methods are almost mandatory. We shall not discuss such methods other than to note that the "central-difference" approach introduced in connection with the depletion equations (see (6.2.11)) is one scheme that can be used, and that there are others that are significantly faster. An essential numerical difficulty arises, however, when difference methods are applied to the kinetics equations. It is related to the fact that the time constants ω_i associated with the behavior of $T(t)$ and the $C_i(t)$ vary over a range of several decades. (The difference between ω_6 and ω_0 can easily be 10^4 sec^{-1}.) Equations of this nature are said to be "stiff." Finding economical ways to obtain numerical solutions turns out not to be a trivial problem. The prompt-jump approximation ((7.5.14) in the case of one precursor group) greatly helps in countering this stiffness problem since it suppresses the root of large magnitude. It is, however, valid only when $\rho \leq 0.5\beta$. Many equally fast procedures that are not limited by this restriction have been devised and are discussed in the literature.

7.6 Calculation of the Kinetics Parameters

The Perturbation Formula for Reactivity

Let us return now to investigate methods for computing values of the kinetics parameters ρ, Λ, and the β_i. To make this computation practical it is necessary to find some way of approximating the shape function $S(\mathbf{r}, E, t)$ without actually having to solve the space-time equations (7.2.9). The fact that we are still free to choose the weight function $W(\mathbf{r}, E)$ will be helpful for this purpose.

Since in practice Λ and the β_i are taken to be independent of time, the main problem is to select $S(\mathbf{r}, E, t)$ and $W(\mathbf{r}, E)$ so that $\rho(t)$ is obtained as accurately as possible for all arbitrary, time-dependent perturbations of the system.

On the grounds that most perturbations disturb the *shape* of the flux in only a minor way, the standard procedure for approximating $S(\mathbf{r}, E, t)$ is to replace it by some time-*independent* shape $S_0(\mathbf{r}, E)$, determined by a *static* solution of the diffusion equation. Generally $S_0(\mathbf{r}, E)$ is taken to be the initial unperturbed flux shape. However we are free to choose any other fixed shape (for example the *final* shape) if the nature of the problem suggests that the true shape will be closer to that choice throughout most of the transient.

Having approximated $S(\mathbf{r}, E, t)$ by $S_0(\mathbf{r}, E)$, we may now select the weight function $W(\mathbf{r}, E)$ so that the error in determining $\rho(t)$ by using $S_0(\mathbf{r}, E)$ is minimized. To accomplish this we start by defining a (hopefully small) error $\delta S(\mathbf{r}, E, t)$ in $S(\mathbf{r}, E, t)$ by

$$\delta S(\mathbf{r}, E, t) \equiv S(\mathbf{r}, E, t) - S_0(\mathbf{r}, E). \tag{7.6.1}$$

We similarly express the time-dependent reactor parameters $D(\mathbf{r}, E, t)$ and those that appear in the operators A and and F^j as

$$\begin{aligned} D(\mathbf{r}, E, t) &= D_0(\mathbf{r}, E) + \delta D(\mathbf{r}, E, t), \\ A &= A_0 + \delta A, \\ F^j &= F_0^j + \delta F^j, \end{aligned} \tag{7.6.2}$$

where D_0, A_0, and F_0^j are the steady-state quantities that lead to $S_0(\mathbf{r}, E)$.

As a result of these definitions we have

$$-\nabla \cdot D_0 \nabla S_0 + A_0 S_0 - \sum_j \chi^j F_0^j S_0 = 0. \tag{7.6.3}$$

Then the expression (7.3.7), or (7.4.2), for $\rho(t)$ becomes

$$\rho(t) = \frac{\begin{array}{c}\int dV \int dE\, W(\mathbf{r}, E)[\nabla \cdot D_0 \nabla S_0 - A_0 S_0 + \chi F_0 S_0 + \nabla \cdot (\delta D) \nabla S_0 - (\delta A) S_0 + \delta(\chi F) S_0 \\ + \nabla \cdot D_0 \nabla(\delta S) - A_0(\delta S) + \chi F_0(\delta S) + \nabla \cdot (\delta D) \nabla(\delta S) - (\delta A)(\delta S) + \delta(\chi F_0)(\delta S)]\end{array}}{\int dV \int dE\, W(\mathbf{r}, E)[\chi F_0 S_0 + \delta(\chi F) S_0 + \chi F(\delta S) + \delta(\chi F)(\delta S)]}, \tag{7.6.4}$$

where, to simplify notation, we have assumed that only one of the isotopes is fissionable and have hence dropped the superscripts j, and we have also omitted the functional dependencies of the functions in the brackets.

Equation (7.6.3) permits us to get rid of the first three terms in the numerator of (7.6.4), and the last three terms, along with the last term in the denominator, may generally be neglected since they contain products of small changes. Those remaining terms that involve perturbations in the reactor properties (δD, δA, $\delta(\chi F)$) give us no trouble since we know what these perturbations are and since they are multiplied into S_0, which we also know. However the terms in the numerator containing δS present a problem since we do not know δS and since a term like $A_0(\delta S)$ may be roughly the same size as $(\delta A)S_0$ and therefore cannot legitimately be neglected. (The term $\chi F(\delta S)$ in the denominator does not cause any difficulty since it can be neglected in comparison with $\chi F_0 S_0$.)

We shall now show that, by a particular selection of the weight function $W(\mathbf{r}, E)$, the unwanted terms can be caused to vanish.

We first introduce the symbol Δ for the unwanted terms:

$$\Delta \equiv \int dV \int dE\, W(\mathbf{r}, E)[\nabla \cdot D_0 \nabla(\delta S) - A_0(\delta S) + \chi F_0(\delta S)]. \tag{7.6.5}$$

We want to select $W(\mathbf{r}, E)$ so that Δ will not depend in a first-order manner on δS.

To proceed we note that, in view of (7.6.3), (7.6.5) can be rewritten in terms of $S = S_0 + \delta S$. Doing so, and expressing the operators A_0 and F_0 explicitly, we obtain

$$\begin{aligned}\Delta = \int dV \int dE\, W(\mathbf{r}, E)\Big\{&\nabla \cdot D_0(\mathbf{r}, E)\nabla S(\mathbf{r}, E, t) - \Sigma_{t0}(\mathbf{r}, E) S(\mathbf{r}, E, t) \\ &+ \int dE'[\Sigma_{s0}(\mathbf{r}, E' \to E) + \chi(E)\nu\Sigma_{f0}(\mathbf{r}, E')]S(\mathbf{r}, E', t)\Big\}.\end{aligned} \tag{7.6.6}$$

Now, since

$$\begin{aligned}&\int_0^\infty dE\, W(\mathbf{r}, E) \int_0^\infty dE'\, \Sigma_{s0}(\mathbf{r}, E' \to E) S(\mathbf{r}, E', t) \\ &= \int_0^\infty dE \int_0^\infty dE'\, W(\mathbf{r}, E)\Sigma_{s0}(\mathbf{r}, E' \to E) S(\mathbf{r}, E', t) \\ &= \int_0^\infty dE' \int_0^\infty dE\, W(\mathbf{r}, E')\Sigma_{s0}(\mathbf{r}, E \to E') S(\mathbf{r}, E, t) \\ &= \int_0^\infty dE\, S(\mathbf{r}, E, t) \int_0^\infty dE'\, \Sigma_{s0}(\mathbf{r}, E \to E') W(\mathbf{r}, E'),\end{aligned} \tag{7.6.7}$$

we can immediately interchange W and S in all terms in (7.6.6) except the one involving D_0.

If we *could* equate

$$\int dV \int_0^\infty dE\, W(\mathbf{r}, E)\nabla \cdot D_0(\mathbf{r}, E)\nabla S(\mathbf{r}, E, t)$$

to

$$\int dV \int_0^\infty dE\, S(\mathbf{r}, E, t)\nabla \cdot D_0(\mathbf{r}, E)\nabla W(\mathbf{r}, E),$$

then (7.6.6) could be rewritten so that the braces would contain

$$\begin{aligned}\Big\{&\nabla \cdot D_0(\mathbf{r}, E)\nabla W(\mathbf{r}, E) - \Sigma_{t0}(\mathbf{r}, E) W(\mathbf{r}, E) \\ &+ \int_0^\infty dE'\,[\Sigma_{s0}(\mathbf{r}, E \to E') + \chi(E')\nu\Sigma_{f0}(\mathbf{r}, E)]W(\mathbf{r}, E')\Big\},\end{aligned}$$

and by choosing $W(\mathbf{r}, E)$ so that this term vanished, we could make Δ vanish.

It turns out we cannot quite do this. Nevertheless this line of approach is a fruitful one if we follow it to its ultimate conclusion. Accordingly we choose for $W(\mathbf{r}, E)$, the function $\Phi_0^*(\mathbf{r}, E)$ satisfying

$$\nabla \cdot D_0(\mathbf{r}, E)\nabla\Phi_0^*(\mathbf{r}, E) - \Sigma_{t0}(\mathbf{r}, E)\Phi_0^*(\mathbf{r}, E)$$
$$+ \int_0^\infty dE' \,[\Sigma_{s0}(\mathbf{r}, E \to E') + \chi(E')\nu\Sigma_{f0}(\mathbf{r}, E)]\Phi_0^*(\mathbf{r}, E') = 0, \qquad (7.6.8)$$

with the boundary conditions that $\Phi_0^*(\mathbf{r}, E) = 0$ for $\mathbf{r}$ on the external surface of the reactor and that $\Phi_0^*(\mathbf{r}, E)$ and $\mathbf{n} \cdot D_0(\mathbf{r}, E)\nabla\Phi_0^*(\mathbf{r}, E)$ are continuous across any internal surface separating different material compositions ($\mathbf{n}$ being the normal to such a surface). Thus the boundary conditions imposed on $\Phi_0^*(\mathbf{r}, E)$ are the same as those imposed on $S_0(\mathbf{r}, E)$.

To proceed we rewrite (7.6.4) in the condensed form (7.4.2), replacing $W(\mathbf{r}, E)$ by $\Phi_0^*(\mathbf{r}, E)$:

$$\rho(t) = \frac{\int dV \int dE\, \Phi_0^*(\mathbf{r}, E)[\nabla \cdot D(\mathbf{r}, E, t)\nabla S(\mathbf{r}, E, t) - AS(\mathbf{r}, E, t) + \chi(E)FS(\mathbf{r}, E, t)]}{\int dV \int dE\, \Phi_0^*(\mathbf{r}, E)\chi(E)FS(\mathbf{r}, E, t)}. \qquad (7.6.9)$$

We then multiply (7.6.8) by $S(\mathbf{r}, E, t)$, integrate over energy and reactor volume, and subtract the result (which still equals zero) from the numerator of (7.6.9). These manipulations lead to

$$\rho(t) = \frac{\begin{aligned}&\int dV \int dE\, \{\Phi_0^*(\mathbf{r}, E)\nabla \cdot D(\mathbf{r}, E, t)\nabla S(\mathbf{r}, E, t) - S(\mathbf{r}, E, t)\nabla \cdot D_0(\mathbf{r}, E)\nabla\Phi_0^*(\mathbf{r}, E)\\ &\quad - [\Sigma_t(\mathbf{r}, E, t) - \Sigma_{t0}(\mathbf{r}, E)]\Phi_0^*(\mathbf{r}, E)S(\mathbf{r}, E, t) + \Phi_0^*(\mathbf{r}, E)\int dE'\,[\Sigma_s(\mathbf{r}, E' \to E, t)\\ &\quad + \chi(E)\nu\Sigma_f(\mathbf{r}, E', t)]S(\mathbf{r}, E', t) - S(\mathbf{r}, E, t)\int dE'\,[\Sigma_{s0}(\mathbf{r}, E \to E')\\ &\quad + \chi(E')\nu\Sigma_{f0}(\mathbf{r}, E)]\Phi_0^*(\mathbf{r}, E')]\}\end{aligned}}{\int dV \int dE\, \Phi_0^*(\mathbf{r}, E)\chi(E)FS(\mathbf{r}, E, t)}, \qquad (7.6.10)$$

which, by rearranging the double integral $\int dE \int dE'\, S(\mathbf{r}, E, t)[\Sigma_{s0}(\mathbf{r}, E \to E') + \chi(E')\nu\Sigma_{f0}(\mathbf{r}, E)]\Phi_0^*(\mathbf{r}, E')$ as in (7.6.7), immediately reduces to

$$\rho(t) = \frac{\begin{aligned}&\int dV \int dE\, \{\Phi_0^*(\mathbf{r}, E)\nabla \cdot D(\mathbf{r}, E, t)\nabla S(\mathbf{r}, E, t) - S(\mathbf{r}, E, t)\nabla \cdot D_0(\mathbf{r}, E)\nabla\Phi_0^*(\mathbf{r}, E)\\ &\quad - \delta\Sigma_t(\mathbf{r}, E, t)\Phi_0^*(\mathbf{r}, E)S(\mathbf{r}, E, t) + \Phi_0^*(\mathbf{r}, E)\int dE'\,[\delta\Sigma_s(\mathbf{r}, E' \to E, t)\\ &\quad + \chi(E)\delta\nu\Sigma_f(\mathbf{r}, E', t)]S(\mathbf{r}, E', t)\}\end{aligned}}{\int dV \int dE\, \Phi_0^*(\mathbf{r}, E)\chi(E)FS(\mathbf{r}, E, t)}, \qquad (7.6.11)$$

where $\delta\Sigma_t$, $\delta\Sigma_s$, and $\delta(\nu\Sigma_f)$ are defined analogously to δS.

To simplify the divergence terms in this result we make use of a general relationship, valid for any differentiable scalar function $f(\mathbf{r})$ and any differentiable vector function $\mathbf{g}(\mathbf{r})$, namely,

$$\nabla\cdot(f\mathbf{g}) = f\nabla\cdot\mathbf{g} + (\nabla f)\cdot\mathbf{g}, \tag{7.6.12}$$

which, if $\mathbf{g}$ happens to be representable in the form $c(\mathbf{r})\nabla h(\mathbf{r})$ with $c(\mathbf{r})$ and $h(\mathbf{r})$ any two functions such that $\mathbf{g}$ is differentiable, yields

$$\nabla\cdot[fc\nabla h] = f\nabla\cdot(c\nabla h) + (\nabla f)\cdot c(\nabla h). \tag{7.6.13}$$

Applying this result to the first two terms in the numerator of (7.6.11), we obtain

$$\begin{aligned}
&\int dV \int dE\,[\Phi_0^*(\mathbf{r}, E)\nabla\cdot D(\mathbf{r}, E, t)\nabla S(\mathbf{r}, E, t) - S(\mathbf{r}, E, t)\nabla\cdot D_0(\mathbf{r}, E)\nabla\Phi_0^*(\mathbf{r}, E)]\\
&= \int dV \int dE\,\{\nabla\cdot[\Phi_0^*(\mathbf{r}, E)D(\mathbf{r}, E, t)\nabla S(\mathbf{r}, E, t)] - \nabla\Phi_0^*(\mathbf{r}, E)\cdot D(\mathbf{r}, E, t)\nabla S(\mathbf{r}, E, t)\\
&\quad - \nabla\cdot[S(\mathbf{r}, E, t)D_0(\mathbf{r}, E)\nabla\Phi_0^*(\mathbf{r}, E)] + \nabla S(\mathbf{r}, E, t)\cdot D_0(\mathbf{r}, E)\nabla\Phi_0^*(\mathbf{r}, E)\\
&= \int dV \int dE\,\{\nabla\cdot[\Phi_0^*(\mathbf{r}, E)D(\mathbf{r}, E, t)\nabla S(\mathbf{r}, E, t) - S(\mathbf{r}, E, t)D_0(\mathbf{r}, E)\nabla\Phi_0^*(\mathbf{r}, E)]\\
&\quad - \nabla\Phi_0^*(\mathbf{r}, E)\cdot\delta D(\mathbf{r}, E, t)\nabla S(\mathbf{r}, E, t)\}.
\end{aligned} \tag{7.6.14}$$

The divergence theorem applied to the first term on the right-hand side yields

$$\int dV \int dE\,\nabla\cdot[\Phi_0^* D\nabla S - SD_0\nabla\Phi_0^*] = \int_{G_0} dG \int dE[\Phi_0^* D\nabla S - SD_0\nabla\Phi_0^*]\cdot\mathbf{n}_i, \tag{7.6.15}$$

where G_0 is the external surface of the reactor and $\mathbf{n}_i$ is the outward normal perpendicular to that surface. Notice that, unless $[\Phi_0^* D\nabla S - SD_0\nabla\Phi_0^*]\cdot\mathbf{n}_i$ is continuous across all internal surfaces between different material compositions, there will be additional *internal* surface terms in (7.6.15), as required by the divergence theorem. In our case there are no such internal discontinuities; the time-dependent shape function $S(\mathbf{r}, E, t)$ and the net current $\mathbf{n}\cdot D(\mathbf{r}, E, t)\nabla S(\mathbf{r}, E, t)$ across internal surfaces separating material discontinuities must be continuous since these are the physical boundary conditions we have imposed on $\Phi = ST$ and $\mathbf{n}\cdot D\nabla\Phi$; the functions Φ_0^* and $\mathbf{n}\cdot D_0\nabla\Phi_0^*$ have the required properties because these are the conditions we imposed on (7.6.8) defining Φ_0^*.

Now, since Φ_0^* and S both vanish on the outer surface G_0, the right-hand side of (7.6.15) vanishes, and, hence, so does the divergence term in (7.6.14).

Collecting all these results and dropping functional dependencies, we obtain for the reactivity $\rho(t)$ of (7.6.11) the expression

$$\rho(t) = \frac{\int dV \int dE\,[-\nabla\Phi_0^*\cdot(\delta D)\nabla S - \Phi_0^*(\delta A)S + \Phi_0^*\delta(\chi F)S]}{\int dV \int dE\,\Phi_0^*\chi FS}, \tag{7.6.16}$$

where the operators δA and $\delta(\chi F)$ are defined by (7.2.8) with Σ_t, $\Sigma_s(E' \to E)$, and $\nu\Sigma_f$ replaced by $\delta\Sigma_t$, $\delta\Sigma_s(E' \to E)$, and $\delta(\nu\Sigma_f)$.

This result is less general than (7.3.7), (7.4.2), or (7.6.4) in that a specific weight function, the solution $\Phi_0^*(\mathbf{r}, E)$ of (7.6.8), is employed. However it is still formally exact in that use of the rigorous value of the shape function $S(\mathbf{r}, E, t)$ to compute $\rho(t)$ by (7.6.16) and subsequent solution of the point kinetics equations (7.3.13) will yield an exact value for $T(t) = \int dV \int dE\, \Phi_0^*(\mathbf{r}, E)[1/v(E)]\Phi(\mathbf{r}, E, t)$.

The great advantage of the form (7.6.16) (and hence of using $\Phi_0^*(\mathbf{r}, E)$ as the weight function) is that, if we now replace $S(\mathbf{r}, E, t)$ by $S_0(\mathbf{r}, E)$, the error in $\rho(t)$ consists of the terms $-\nabla\Phi_0^* \cdot (\delta D)\nabla(\delta S) - \Phi_0^*(\delta A)(\delta S) + \Phi_0^*\delta(\chi F)\delta S$ in the numerator and $\Phi_0^*\chi F\delta S$ in the denominator. In both cases the neglected error terms should be smaller than the retained terms roughly in the ratio $\delta S/S_0$. When such a replacement is made and $S_0(\mathbf{r}, E)$ is taken to be the unperturbed flux shape, the resulting equation is called the *perturbation-theory expression* for the reactivity:

$$\rho(t) \approx \frac{\int dV \int dE\,[-\nabla\Phi_0^* \cdot (\delta D)\nabla S_0 - \Phi_0^*(\delta A)S_0 + \Phi_0^*\delta(\chi F)S_0]}{\int dV \int dE\, \Phi_0^*\chi F S_0}. \tag{7.6.17}$$

Thus, by introducing Φ_0^* for the weight function, we have found an approximate method of computing reactivity which does not depend on knowing δS, but in which, nevertheless, we can be certain that the terms neglected are of "lower order" (i.e., contain a higher multiple of a perturbed quantity) than those retained.

The Physical Interpretation of $\Phi_0^*(\mathbf{r}, E)$

The function $\Phi_0^*(\mathbf{r}, E)$ is called the *adjoint flux* of the critical reactor. Equation (7.6.8), which defines it, is very similar to (7.6.3), differing only in that the E and E' variables are reversed in the operators A_0 and F_0. Since (7.6.8), like (7.6.3), is a homogeneous equation, it will have a nontrivial solution for a given reactor only for certain values of the parameters appearing in D_0, A_0, and F_0. It turns out, however, that, if the reactor is critical, that is, if (7.6.3) has a nontrivial solution, then (7.6.8) will also have a nontrivial solution. We shall prove this fact in Chapter 8.

The physical significance of $\Phi_0^*(\mathbf{r}, E)$ is not obvious. The fact of having reversed E and E' in A_0 and F_0 makes it impossible to think of Φ_0^* as a flux of neutrons—even an artificial one in which the neutrons scatter or fission "up." That is, if $\Phi_0(E')$ is the flux density at E', $\Sigma_{s0}(E' \to E)\Phi_0(E')dE'\, dE\, dV$ is the rate at which neutrons scatter from dE' to dE in dV; but, even if we call $\Phi_0^*(E')$ a "flux density" at E', $\Sigma_{s0}(E \to E')\Phi_0^*(E')dE\, dE'\, dV$ is *not* the rate at which neutrons in dV scatter from dE to dE'.

It turns out that $\Phi_0^*(\mathbf{r}, E)$ *can* be interpreted physically as a quantity called the *neutron importance*. We shall deal quantitatively with the neutron-importance function in Chapter

8. For present purposes it will be sufficient to consider the following conceptual experiment:

Suppose a reactor is exactly critical but at zero power (i.e., it contains, at the start, no neutrons). If we add, at some point $\mathbf{r}$ within a volume dV, a sample $\eta \equiv N(\mathbf{r}, E)dV\,dE$, containing a definite number of neutrons—say 10^6—having energies in the range dE about E and an isotropic distribution of directions of travel, the power level of the reactor will start to rise. Since the reactor is just critical, the neutron population will eventually assume a steady-state distribution in energy and position such that the total number of neutrons created per second equals the total number lost, and such that the total population in the reactor remains constant. Now we know that the total number of neutrons in this asymptotic distribution will depend on the energy E and position $\mathbf{r}$ at which the sample is introduced. To determine that number, however, we would have to solve (7.2.9) with $\Phi(\mathbf{r}, E, t)$ initially zero and with a pulse $[\int_0^\infty dt\, q(\mathbf{r}, E, t)]dV\,dE = \eta$ introduced at time zero. We can see intuitively, though, that, if the original η neutrons are introduced in the reflector with a high energy, many of them will simply leak out of the reactor or slow down and be absorbed in it, and hence only a few of them will diffuse into the core and start the chain reactions that will eventually lead to the asymptotic distribution. Thus the total asymptotic neutron population per original neutron introduced will be relatively low. On the other hand, if the η neutrons have a low energy and are introduced in the fuel material, a large fraction of them will immediately cause a fission, and the asymptotic population per neutron introduced will be relatively high.

Thus the asymptotic increase in the total neutron population of a critical reactor per neutron added at time zero depends on where and with what energy the neutron is introduced. We call this increase the *importance* of a neutron "introduced isotropically" at location $\mathbf{r}$ and energy E. Importance is thus a function of $\mathbf{r}$ and E. From our operational definition it is seen to be dimensionless. In particular, unlike the flux $\Phi(\mathbf{r}, E)$, it is *not* a density. Note also that the importance of a sample $[1/v(E)]\Phi(\mathbf{r}, E)dV\,dE$ of the *actual* neutron population in a small volume element dV and a small energy increment dE is *not*, in general, $\Phi_0^*(\mathbf{r}, E)[1/v(E)]\Phi(\mathbf{r}, E)dV\,dE$. The neutrons belonging to such a sample would not usually have an isotropic distribution of directions of travel, as is required by the definition of $\Phi_0^*(\mathbf{r}, E)$.

It is possible to derive a differential equation for the importance of a neutron (along with external and internal boundary conditions that the importance function must obey). As will be demonstrated in Chapter 8, the equation turns out to be identical with the equation (7.6.8) for the adjoint flux $\Phi_0^*(\mathbf{r}, E)$, and the boundary conditions we imposed on $\Phi_0^*(\mathbf{r}, E)$ are exactly those that must be imposed on the importance function. Thus the adjoint flux for neutrons in a critical reactor can be interpreted as the importance function for those neutrons.

A More General Interpretation of the Kinetics Parameters

This physical interpretation of the weight function $\Phi_0^*(\mathbf{r}, E)$ permits us to interpret the kinetics parameters more precisely.

To do so we note that, since any given neutron of energy E at location $\mathbf{r}$ has an importance, all processes that introduce or destroy neutrons in $dV\,dE$ may be thought of as introducing or destroying "importance." Thus, since $q(\mathbf{r}, E, t)dV\,dE$ is the number of neutrons per second introduced isotropically into $dV\,dE$ and $W(\mathbf{r}, E) = \Phi_0^*(\mathbf{r}, E)$ is the importance that those neutrons would have if they were introduced into the critical reactor, $Q(t)$ of (7.3.10) is the total rate at which the source at time t would introduce importance into the critical reactor.

The reactivity (7.4.2), the effective delayed-neutron fraction (7.4.3), and the prompt-neutron lifetime (7.4.4) may be similarly interpreted. The denominators of these parameters represent the "instantaneous" rate at which importance is added to the critical reactor by the fission process. Thus $\rho(t)$ is the net instantaneous rate at which importance is added divided by the total rate due to fission neutrons. The effective delayed-neutron fraction β_i is the ratio of the instantaneous rate of addition of importance by delayed neutrons to the instantaneous rate of addition by all fission neutrons. Finally Λ is the total importance that the isotropically distributed portion of the neutrons present in the core at time t would have in the critical reactor divided by the instantaneous rate of production of importance by fission neutrons in the critical reactor at that same time t; it is thus a kind of lifetime for the importance of neutrons generated by fission.

These interpretations are rather abstract in that they refer to "instantaneous fission rates" (in which delayed neutrons are regarded as appearing instantaneously), to the isotropic component of the neutrons present, and to importance generation in the "critical reactor" (i.e., the one specified by D_0, A_0, and F_0). It would be more pleasing aesthetically to introduce time-dependent importance functions appropriate to the actual state of the reactor at time t, rather than to some fixed, critical condition. However, if $W(\mathbf{r}, E)$ is made time-dependent, additional terms appear in the point kinetics equations and they become more difficult to solve.

Alternate Schemes for Determining $\rho(t)$

Equation (7.6.16) provides us with a tractable procedure for determining $\rho(t)$ once we are given the perturbations δD, δA, and $\delta(\chi F)$. We simply solve (7.6.8) for $\Phi_0^*(\mathbf{r}, E)$ and (7.6.3) for $S_0(\mathbf{r}, E)$ and perform the indicated integrations. (In practice, of course, we would deal with the multigroup or few-group finite-difference approximations to these equations.) Thus the procedure requires that we perform two criticality-type computations, one for S_0 and one for Φ_0^*.

If we are interested only in computing $\rho(t)$ for a single, fixed, perturbed situation, there is a convenient alternative to this procedure that helps us avoid the problem of determining Φ_0^*. This alternate scheme is based on the fact that there always exists a fictitious eigenvalue λ that will provide a stationary solution for the equation

$$-\nabla\cdot D\nabla S + AS - \frac{1}{\lambda}\chi FS = 0. \tag{7.6.18}$$

Thus, if the initial, critical condition ($D = D_0$; $A = A_0$; $\chi F = \chi F_0$; $\lambda = \lambda_0 = 1$) of the reactor is perturbed, and, at the same time, a fictitious perturbation is introduced changing λ from one to some other value such that the stationary equation (7.6.18) is still valid, we can regard the final (fictitious) condition of the reactor as a critical one. Hence the reactivity ρ in this final condition will be zero. But this zero reactivity will be composed of two equal and opposite parts, one due to the perturbation (δD; δA; $\delta(\chi F)$) and the other to the fictitious perturbation $\delta(1/\lambda)$ ($\equiv(1/\lambda) - (1/\lambda_0) = (1/\lambda) - 1$). To see this note that

$$\begin{aligned}\delta\left(\frac{1}{\lambda}\chi F\right)S &= \left(\frac{1}{\lambda}\chi F - \chi F_0\right)S\\ &= \left\{\left[\delta\left(\frac{1}{\lambda}\right) + 1\right]\chi F - \chi F_0\right\}S\\ &= \left[\chi F\delta\left(\frac{1}{\lambda}\right) + \delta(\chi F)\right]S,\end{aligned} \tag{7.6.19}$$

so that, replacing χF in (7.6.16) by $(1/\lambda)\chi F$, we have

$$\rho = 0 = \frac{\int dV \int dE[-\nabla\Phi_0^*\cdot(\delta D)\nabla S - \Phi_0^*(\delta A)S + \Phi_0^*\delta(\chi F)S]}{\int dV \int dE\, \Phi_0^*\chi FS} + \delta\left(\frac{1}{\lambda}\right) \tag{7.6.20}$$

or

$$1 - \frac{1}{\lambda} = \frac{\int dV \int dE[-\nabla\Phi_0^*\cdot(\delta D)\nabla S - \Phi_0^*(\delta A)S + \Phi_0^*\delta(\chi F)S]}{\int dV \int dE\, \Phi_0^*\chi FS}. \tag{7.6.21}$$

If we now assume that the flux shape $S(\mathbf{r}, E)$ (the solution of (7.6.17) resulting from the simultaneous perturbations δD, δA, $\delta(\chi F)$, *and* $\delta(1/\lambda)$) is a good approximation to the shape that would result if we perturbed D, A, and χF alone, the right-hand side of (7.6.21) will be the reactivity we seek, namely, that associated with the perturbation (δD; δA; $\delta(\chi F)$).

Thus, within the limits of the perturbation-theory approximation, the reactivity associated with a given perturbation is given either by (7.6.16), where $\Phi_0^*(\mathbf{r}, E)$ is found by solving the adjoint equation (7.6.8) for the unperturbed core and $S(\mathbf{r}, E, t)$ is approximated by the unperturbed flux shape $S_0(\mathbf{r}, E)$, or by $1 - (1/\lambda)$, where λ is the eigenvalue k_{eff} of the perturbed core.

The fact that the reactivity is approximately $(k_{\text{eff}} - 1)/k_{\text{eff}}$ is in keeping with its interpretation as the fractional increase in the number of neutrons produced per generation. In fact some people like to define reactivity formally as $1 - (1/\lambda)$ and to regard the definition (7.3.7) as an approximation. (We did this earlier with (3.3.24).) Since λ is precisely defined by (7.6.18) for any reactor geometry and for any values of D, A, and F, this procedure is perfectly unambiguous. However the fact that $\rho(t)$ as defined by (7.3.7) depends on the instantaneous flux shape $S(\mathbf{r}, E, t)$, as well as on D, A, and F, makes it more general. Only when the shape function $S(\mathbf{r}, E, t)$ is the solution of (7.6.18) corresponding to the fictitiously critical condition will the two definitions yield precisely the same number. For most practical cases, however, they will be very close, and $1 - (1/\lambda)$ has the great advantage of being easier to calculate.

When the reactivity corresponding to a great variety of different small perturbations δD, δA, and $\delta(\chi F)$ is desired, the perturbation expression is preferable to the determination of $\delta(1/\lambda)$. This is because the application of (7.6.17) requires only that we know Φ_0^* and S_0, which may be obtained from two criticality-type calculations, whereas the determination of reactivities for a number of different perturbations from the changes $\delta(1/\lambda)$ requires that we do a separate criticality calcuation for each perturbation.

The perturbation expression (7.6.17) also provides us with the means for making a quantitative comparison between the effects on a reactor of changing various parameters: we simply look at the reactivity changes these produce. Thus we can compare a change in the average temperature of the fuel to the number of inches a particular control rod moves or establish an equivalence between a certain amount of fuel depletion and the removal of a certain amount of soluble boron from the core.

It is extremely important to remember, if comparisons of this sort are made, that they are valid only when the flux shapes associated with *all* the conditions intercompared may legitimately be approximated by $S_0(\mathbf{r}, E)$. We can see from (7.6.17) that, if $\rho(\delta A_1)$ is the reactivity associated with perturbing one of the cross sections in A a certain amount, and $\rho(\delta A_2)$ is the reactivity due to perturbing A a different amount, then

$$\rho(\delta A_1) + \rho(\delta A_2) = \rho(\delta A_1 + \delta A_2) \tag{7.6.22}$$

if perturbation theory is valid here.

Since, however, the true shape functions associated with δA_1 and δA_2 will, in general, differ from each other and from that associated with the perturbation $\delta A_1 + \delta A_2$, the *exact* reactivities defined by (7.3.7) or (7.6.20) are *not* additive. The larger the perturbation, the more in error (7.6.22) is likely to be. Failure to recognize that large reactivity changes are not additive can lead to serious errors in interpreting the results of reactor calculations or measurements.

7.7 Application of Reactor Kinetics to Design

We have developed the subject of reactor kinetics using the continuous-energy diffusion-theory model of reactor behavior. The algebraic formulas are a bit less cumbersome when this formalism is used.

In practice, of course, the energy-group approximation must be made if we wish to determine actual numerical results for a physical problem. The point equations (7.3.13) will have the same form, but the formulas for the kinetics parameters will be in terms of sums over energy groups rather than integrals over energy. Thus (7.6.17), the perturbation expression for reactivity, will become

$$\rho(t) = \frac{\begin{array}{l}\int dV \sum_{g=1}^{G} \{-\nabla\Phi_g^*(\mathbf{r})\cdot\delta D_g(\mathbf{r}, t)\nabla S_{0g}(\mathbf{r}) - \Phi_{0g}^*(\mathbf{r})[\![\delta\Sigma_{tg}(\mathbf{r}, t)S_{0g}(\mathbf{r}) \\ \quad - \sum_{g'=1}^{G} [\delta\Sigma_{tgg'}(\mathbf{r}, t) + \chi_g\delta\nu\Sigma_{fg'}(\mathbf{r}, t)]S_{0g'}(\mathbf{r})]\!]\}\end{array}}{\int dV \sum_{g=1}^{G} \Phi_{0g}^*(\mathbf{r}) \sum_{g'=1}^{G} \chi_g\nu\Sigma_{fg'}(\mathbf{r})S_{0g'}(\mathbf{r})}, \tag{7.7.1}$$

where S_{0g} and Φ_{0g}^* are the energy-group expressions for the flux and adjoint flux of the critical core. Moreover, in practice, the volume integrals in this expression are evaluated by summing over spatial mesh cubes.

Similar expressions may be written for Λ and the β_i. Many computer programs will, on option, solve the energy-group adjoint equations as well as the regular flux equations and then determine the integrals needed to compute Λ, ρ, and the β_i.

In this last step the integrals are usually performed over "thermal hydraulic" regions, i.e., regions where the average temperature and density are assumed to be spatially constant. In such regions, if the D_g and Σ_g change because of changes in density or temperature, they will do so uniformly, and the total value of $\rho(t)$ at any time can be expressed as the sum over all such regions of the changes in the D_g and Σ_g for each region weighted by the appropriate flux-adjoint volume integral for that region. Thus, if $\rho_j(t)$ represents the contribution to $\rho(t)$ from changes in the group parameters of region j, and if the changes $\delta D_{gj}(t)$, $\delta\Sigma_{tg}(t)$, etc., are spatially constant throughout region j, we have

$$\rho_j(t) = \frac{\begin{array}{l}\sum_{g=1}^{G}\{\delta D_{gj}(t)\int_{V_j} dV[-\nabla\Phi_{0g}^*(\mathbf{r})\cdot\nabla S_{0g}(\mathbf{r})] - \delta\Sigma_{tgj}(t)\int_{V_j} dV\,\Phi_{0g}^*(\mathbf{r})S_{0g}(\mathbf{r}) \\ \quad + \sum_{g'=1}^{G}[\delta\Sigma_{gg'j}(t)\int_{V_j} dV\,\Phi_{0g}^*(\mathbf{r})S_{g'}(\mathbf{r}) + \chi_g\delta\nu\Sigma_{fg'j}(t)\int_{V_j} dV\,\Phi_{0g}^*(\mathbf{r})S_{g'}(\mathbf{r})]\}\end{array}}{\sum_j \int_{V_j} dV \sum_{g=1}^{G}\Phi_{0g}^*(\mathbf{r})\sum_{g'=1}^{G}\chi_g\nu\Sigma_{f0g'j}S_{0g'}(\mathbf{r})}, \tag{7.7.2}$$

$$\rho(t) = \sum_j \rho_j(t).$$

As we specialize to a less general mathematical model, the notational complexity gets more and more forbidding. Thus we encounter symbols such as $\Sigma_{f0g'j}$, the unperturbed fission cross section for group g' in thermal hydraulic region j. Equation (7.7.2) is really numerically simpler than (7.6.16), but it doesn't look that way.

The ρ_j change with time if the average temperature or density changes in region j. Because of this, it is possible to define "temperature and density coefficients of reactivity" in region j that reflect the changes in D_{gj}, Σ_{tgj}, etc., due to changes in the temperature or density of that region. Thus we speak of a "temperature coefficient of reactivity" $\partial\rho_j/\partial\theta_j$ for region j (where θ_j is the average temperature of the region). These coefficients can be computed by replacing terms such as $\delta D_{gj}(t)$ in (7.7.2) by $\partial D_{gj}/\partial\theta_j$, assumed to be independent of temperature and hence of time. Thus we can compute, for a given reactor at a given stage of depletion, $\partial\rho_j/\partial\theta_j$ for all thermal hydraulic regions j and analogous coefficients for changes in density δd_j. The total reactivity for any distribution of temperature or density changes then becomes

$$\rho(t) = \sum_j \left[\left(\frac{\partial\rho_j}{\partial\theta_j}\right)\delta\theta_j(t) + \left(\frac{\partial\rho_j}{\partial d_j}\right)\delta d_j(t) \right]. \tag{7.7.3}$$

Overall reactor transient behavior can thus be predicted by a tandem process:

1. The initial change in reactivity that starts the transient is found for each region from (7.7.2), and the change for the whole reactor is then determined from (7.7.3).
2. The kinetics equations (7.3.13) are solved for a single time step to find $T(t_1)$ at the end of that time step, and from it the fluxes $S_g(\mathbf{r})T(t_1)$ are computed.
3. The power level associated with these fluxes is used to predict temperature and density changes at the end of the next time step, and from these changes the reactivity at the end of that second time step is predicted according to (7.7.3).
4. This predicted reactivity and that at the beginning of the second time step are used in the kinetics equations to predict the flux behavior during the second time step.
5. The fluxes at the end of the second time step are then used to determine the power level and, from it, the temperature and density changes at the end of the third time step; and the tandem computation of power levels and changes in temperature and density is continued.

If the temperature or density changes are the same throughout the reactor, (7.7.3) becomes

$$\rho(t) = \left(\sum_j \frac{\partial\rho_j}{\partial\theta_j}\right)\delta\theta(t) + \left(\sum_j \frac{\partial\rho_j}{\partial d_j}\right)\delta d(t), \tag{7.7.4}$$

and we define

$$\frac{\partial\rho}{\partial\theta} \equiv \sum_j \frac{\partial\rho_j}{\partial\theta_j} \quad \text{and} \quad \frac{\partial\rho}{\partial d} \equiv \sum_j \frac{\partial\rho_j}{\partial d_j}$$

as the temperature and density coefficients of reactivity for the reactor. To insure stability it is always desirable for the temperature coefficient to be negative and for the density coefficient to be positive.

This combination of the point kinetics equations, driven by a total reactivity that is, in turn, determined by changes in reactor temperature and density multiplied into appropriate "reactivity coefficients," is the standard mathematical model used to predict the transient characteristics of most reactors. The essential assumption on which the whole approach has been based is that the flux shape $S(\mathbf{r}, E, t)$ during a transient can be approximated by some steady-state value $S_0(\mathbf{r}, E)$. If this assumption is valid, the point kinetics approach, applied correctly, will predict transient behavior quite accurately. If it is not, we must then face the problem of solving the true space-time equations (7.2.9) in more detail, accounting for changes in flux shape induced by the transient. A finite-difference solution in both space and time is the most direct method for obtaining such a solution. It is, however, quite expensive. We shall see in Chapter 11 that there are alternate methods for obtaining detailed solutions of the group diffusion equations. Extending such schemes to the time-dependent case will provide the most efficient way to predict the flux behavior when it changes locally during a transient.

References

Ash, M., 1965. *Nuclear Reactor Kinetics* (New York: McGraw-Hill).

Hetrick, D. L., 1971, *Dynamics of Nuclear Reactors* (Chicago: The University of Chicago Press).

Hetrick, D. L., ed., 1972. *Dynamics of Nuclear Systems* (Tucson: The University of Arizona Press).

Keepin, G. R., 1965. *Physics of Nuclear Kinetics* (Reading, Mass.: Addison-Wesley).

Stacey, W. M., Jr., 1969. *Space-Time Nuclear Reactor Kinetics* (New York: Academic Press).

Problems

1. To what do the equations (7.3.13) reduce if there is no fissionable material present in the assembly? Use this form to determine $T(t)$ following the removal of a source Q at $t = 0$ from a homogeneous sphere of nonfissionable material. Under the assumption that

$$S(\mathbf{r}, t) = \frac{\sin B_r r}{B_r r}\psi(E)$$

and taking $W(B_r, E) = (\sin B_r r)/B_r r$, express the result in terms of integrals involving $v(E)$, $D(E)$, and $\Sigma_a(E)$.

2. Suppose that in deriving the point kinetics equations (7.3.13) we abandon the definition

(7.3.1) of $T(t)$ and, instead, first normalize $S(\mathbf{r}, E, t)$ so that, at all times,

$$\int dV \int dE\, S(\mathbf{r}, E, t) = \int dV \int dE\, \Phi(\mathbf{r}, E, 0).$$

The formal definition of $T(t)$ then becomes

$$T(t) \equiv \frac{\int dV \int dE\, \Phi(\mathbf{r}, E, t)}{\int dV \int dE\, \Phi(\mathbf{r}, E, 0)},$$

and (7.3.2) again defines $S(\mathbf{r}, E, t)$ precisely, with (7.3.5) now being replaced by the new normalization.

Show that, if this is done and if the definitions (7.3.7)–(7.3.11) are not altered, the point kinetics equations take the form

$$\frac{dT(t)}{dt} = \left(\frac{\rho - \beta}{\Lambda} - \lambda_s(t)\right) T + \sum_i \lambda_i C_i(t) + Q(t),$$

$$\frac{dC_i(t)}{dt} = \frac{\beta_i}{\Lambda} T(t) - (\lambda_i + \lambda_s(t)) C_i(t) \qquad (i = 1, 2, \ldots, I),$$

where

$$\lambda_s(t) \equiv \frac{\int dV \int dE\, [W(\mathbf{r}, E)/v(E)][\partial S(\mathbf{r}, E, t)/\partial t]}{\int dV \int dE\, [W(\mathbf{r}, E)/v(E)]\, S(\mathbf{r}, E, t)}.$$

3. Suppose that in deriving the point kinetics equations (7.3.13) we abandon the definition (7.3.1) but retain (7.3.2), leaving the precise normalization of $S(\mathbf{r}, E, t)$ (analogous to (7.3.5)) to be specified as may be convenient for the problem at hand. Suppose, further, that we let the weight function be time-dependent but also let its normalization be specified as desired.

Show that, if the definitions (7.3.7)–(7.3.11) are retained, the point kinetics equations take the form

$$\frac{dT(t)}{dt} = \left(\frac{\rho - \beta}{\Lambda} - \lambda_s(t)\right) T(t) + \sum_i \lambda_i C_i(t) + Q(t),$$

$$\frac{dC_i(t)}{dt} = \frac{\beta_i}{\Lambda} T(t) - (\lambda_i + \lambda_s(t) + \lambda_w(t) - \lambda_{w_i}(t)) C_i(t) \qquad (i = 1, 2, \ldots, I),$$

where

$$\lambda_s(t) \equiv \frac{\int dV \int dE\, [W(\mathbf{r}, E, t)/v(E)][\partial S(\mathbf{r}, E, t)/\partial t]}{\int dV \int dE\, [W(\mathbf{r}, E, t)/v(E)]\, S(\mathbf{r}, E, t)},$$

$$\lambda_w(t) \equiv \frac{\int dV \int dE\, [\partial W(\mathbf{r}, E, t)/\partial t][1/v(E)]\, S(\mathbf{r}, E, t)}{\int dV \int dE\, [W(\mathbf{r}, E, t)/v(E)]\, S(\mathbf{r}, E, t)},$$

$$\lambda_{w_i}(t) \equiv \frac{\int dV \int dE\, [\partial W(\mathbf{r}, E, t)/\partial t]\, \chi_i(E)\, c_i(\mathbf{r}, E, t)}{\int dV \int dE\, W(\mathbf{r}, E, t)\, \chi_i(E)\, c_i(\mathbf{r}, E, t)} \qquad (i = 1, 2, \ldots, I).$$

Note that, if we choose to normalize $S(\mathbf{r}, E, t)$ and $W(\mathbf{r}, E, t)$ so that the denominator of $\lambda_s(t)$ is a constant at all times and the numerator vanishes at all times (i.e., so that $\lambda_s(t) = 0$), $\lambda_w(t)$ will also vanish. Unfortunately the (time-dependent) $\lambda_{w_i}(t)$ will still remain.

4. Consider a one-group model of a slab reactor of thickness L uniformly and homogeneously loaded so that, at time $t = 0$, the reactor is critical with the group parameters D_0, $\nu\Sigma_{f0}$, and Σ_{a0}. Suppose that, for $t > 0$, the neutron-production cross section becomes

$$\nu\Sigma_f(t) = \nu\Sigma_{f0} + \Gamma(x)\alpha t,$$

where

$$\Gamma(x) = \begin{cases} 0 \text{ for } 0 \le x < L/2, \\ 1 \text{ for } L/2 \le x \le L, \end{cases}$$

while D_0 and Σ_{a0} retain their initial valves. Suppose it is also known that, for $t \ge 0$, the flux shape $\Phi(x, t)$ can be represented by

$$\Phi(x, t) = a_1(t) \sin \frac{\pi x}{L} + a_2(t) \sin \frac{2\pi x}{L},$$

where $a_1(0) = 1$ and $a_2(0) = 0$. Determine algebraic expressions, independent of D_0 and Σ_{a0}, for $S(x, t)$, $\rho(t)$, $\beta(t)$, and $\Lambda(t)$, assuming that $W(x) = \Phi(x, 0)$ and that there is only one fissionable isotope present.

5. Consider a reactor that is operating in a steady-state critical condition and suppose that a constant source of neutrons $q(\mathbf{r}, E)$, is suddenly turned on, no other changes in the system being made. Neglecting any changes in flux shape, derive an expression for $T(t)$ when $q(\mathbf{r}, E)$ is present. (Use one group of delayed neutrons.) Assuming $Q/T_0 = 1$, plot $T(t)/T_0$ versus t on the interval $0 \leq t \leq 100$ sec for $\beta = 0.0075$, $\Lambda = 6 \times 10^{-5}$ sec, and $\lambda = 0.08$ sec^{-1}.

Will the exact value of the reactivity given by (7.3.7) remain equal to zero after the source has been turned on?

6. Suppose that the transient depicted on Figure 7.2 runs for 20 sec as indicated and that ρ is then set to zero. Find the algebraic expression analogous to (7.5.13) for $t \geq 20$ sec and replot Figure 7.2 for this altered situation.

7. For super-prompt-critical transients it is usually a satisfactory approximation to assume the delayed-neutron source $\Sigma_i \lambda_i C_i(T)$ in (7.3.13) remains constant during the period of interest. Making this assumption, and with $Q = 0$, show that, if a reactor is initially in a steady-state critical condition with $T = T_0$ and ρ becomes $\beta + \alpha t$ for $t > 0$, the amplitude function behaves approximately as

$$\frac{T(t)}{T_0} \approx \exp\left(\frac{1}{2}\frac{\alpha}{\Lambda}t^2\right)\left[1 + \int_0^t \frac{\beta}{\Lambda} \exp\left(-\frac{1}{2}\frac{\alpha}{\Lambda}\tau^2\right) d\tau\right].$$

For $\beta = 0.004$, $\Lambda = 10^{-7}$ sec, and $\alpha = 1\$/\text{sec}$, find $T(t)/T_0$ at $t = 0.01$ sec and at $t = 0.1$ sec.

8. For $\beta = 0.0075$, $\Lambda = 10^{-4}$ sec, and the ratio of the physical β_i^j to the effective β_i being the same for all i, find the number of cents of reactivity equivalent to one inhour when the reactor is thermal and the fuel is U^{25}.

Answer the same question for $\beta = 0.0023$, $\Lambda = 10^{-7}$ sec, and the reactor a fast one, fueled with Pu^{49}.

9. Another scheme for inferring the reactivity of a subcritical reactor is called the "source-jerk" technique. It consists of a sudden removal of an external source $q(\mathbf{r}, E)$ from a subcritical reactor, the source having been present long enough so that, initially, the precursor concentrations are in equilibrium with the neutron flux. The immediate change in a counter reading following the source removal can (after a number of theoretical corrections) be used to infer the reactivity of the assembly.

Let the source jerk commence at t_0, and let t_1 be a short time later when the fast transient resulting from the source removal has died out but when the precursor concentrations may still be assumed to have their initial values $C_i(t_0)$. Assuming that ρ, Λ, and the β_i are unchanged by the source jerk and that $T(t_0)$ and $T(t_1)$ are proportional to the detector readings $D(t_0)$ and $D(t_1)$ at those times, derive a formula for the reactivity in dollars in terms of $D(t_0)$ and $D(t_1)$.

10. Derive finite-difference solutions for the point kinetics equations by integrating them over a time step. Assume that $Q = 0$ and that $\rho =$ constant, and treat only one group of precursors ($I = 1$). Approximate the integrals $\int_0^\Delta T(\tau)\, d\tau$ and $\int_0^\Delta C(\tau)\, d\tau$ by:

a. $(\Delta/2)\,[T(\Delta) + T(0)]$ and $(\Delta/2)\,[C(\Delta) + C(0)]$;

b. $\Delta T(\Delta)$ and $\Delta C(\Delta)$.

For $\beta = 0.0075$, $\Lambda = 6 \times 10^{-5}$ sec, and $\lambda = 0.08$ sec^{-1}, compute $T(\Delta)$ and $C(\Delta)$ by both schemes for a reactor, initially critical, for which, at $t = 0$, ρ becomes:

c. $\rho = -0.05$;
d. $\rho = 0.0115$.

For part c use the single time step $\Delta = 0.5$ sec; for part d use $\Delta = 0.001$ sec. Compare these results with those found using the more exact formulas (7.5.12) and (7.5.16).

11. Consider a slab reactor extending from $x = 0$ to $x = L$ and suppose it is initially critical with homogeneous one-group parameters D_0, $\nu\Sigma_{f0}$, and Σ_0 such that $-[D_0(\pi/L)^2 + \Sigma_0] + \nu\Sigma_{f0} = 0$. With $\nu\Sigma_{f0} = 0.2$ cm^{-1}, suppose this quantity is increased by an amount 0.0023 cm^{-1} in the range $0 \leq x \leq L/2$. With the weight function used to determine the point kinetics parameters taken as unity, calculate the asymptotic period under the assumption that the shape function is:
a. $S_1(r) = \sin(\pi x/L)$;
b. $S_1(r) = \sin(\pi x/L) + \frac{1}{2}\sin(2\pi x/L)$.
Use one group of delayed neutrons with $\beta = 0.007$, $\lambda = 0.08$ sec^{-1}, and $v = 3 \times 10^5$ cm/sec.

12. Consider the flux equation

$$-\nabla\cdot D(\mathbf{r}, E)\nabla\Phi(\mathbf{r}, E) + \Sigma_t(\mathbf{r}, E)\Phi(\mathbf{r}, E) - \int_0^\infty dE'\,\Sigma_{s0}(\mathbf{r}, E' \to E)\Phi(\mathbf{r}, E')$$
$$= \frac{1}{\lambda}\int_0^\infty dE'\,\chi(E)\nu\Sigma_f(\mathbf{r}, E')\Phi(\mathbf{r}, E')$$

and the corresponding importance equation

$$-\nabla\cdot D(\mathbf{r}, E)\nabla\Phi^*(\mathbf{r}, E) + \Sigma_t(\mathbf{r}, E)\Phi^*(\mathbf{r}, E) - \int_0^\infty dE'\,\Sigma_{s0}(\mathbf{r}, E \to E')\Phi^*(\mathbf{r}, E)$$
$$= \frac{1}{\lambda^*}\int_0^\infty dE'\,\chi(E')\nu\Sigma_f(\mathbf{r}, E)\Phi^*(\mathbf{r}, E').$$

Prove that, for solutions $\Phi(\mathbf{r}, E)$ and $\Phi^*(\mathbf{r}, E)$ that are positive for all $\mathbf{r}$ within the reactor, $\lambda = \lambda^*$.

Hint: Under this condition,

$$\int dV \int_0^\infty dE \int_0^\infty dE'\,\Phi^*(\mathbf{r}, E)\chi(E)\nu\Sigma_f(\mathbf{r}, E')\Phi(\mathbf{r}, E') > 0.$$

13. Consider a homogeneous slab reactor extending from $x = -L/2$ to $x = +L/2$ such that the one-group flux shape is $\Phi_0 \cos(\pi x/L)$. Suppose that, if a uniform homogeneous poison is removed from the region $0 \leq x \leq L/2$ in such a way that $\delta\Sigma$ in that region is -0.001 cm^{-1}, the flux shape becomes $\Phi_0[\cos(\pi x/L) + \frac{1}{2}\sin(2\pi x/L)]$. With $\nu\Sigma_f = 0.1$ cm^{-1}, compute the reactivity according to (7.6.16) and then according to the perturbation formula (7.6.17). In both cases use $\Phi_0^* = \cos(\pi x/L)$.

Show by finding the actual values that, if $\delta\Sigma = -0.001$ cm^{-1} throughout the core, the change in reactivity does not equal the sum of the exact reactivities resulting from removal of the poison from the two halves individually.

14. Suppose that a homogeneous slab reactor of thickness L is being analyzed by a two-group model and that the only effect of a uniform temperature change is a uniform change in the group diffusion constant D_1. Assuming that $\nu\Sigma_{f1}$ and χ_2 are zero, use (7.7.2) to find the temperature coefficient $\partial\rho/\partial\theta$ in terms only of the two-group parameters and the thickness L.

8 The Neutron Transport Equation

8.1 Introduction

By using collision theory (Sections 5.5–5.7) and assuming Fick's Law (4.5.1), we have been able to develop the physical theory underlying most of the reactor-design methods in use today. There are a number of simplifying physical and numerical approximations that we have not yet dealt with, and there are a great many "prescriptions," peculiar to individual design groups, that we shall not attempt to deal with (for example schemes for finding effective homogeneous cross sections or artificial cross sections to represent fission-product effects). However the basic theory that supports reactor-design procedures is now largely complete.

The few gaps that remain in this basic theory are nevertheless very important. For example, in stating the fundamental relationship (Fick's Law) between the net current density $\mathbf{J}(\mathbf{r}, E)$ and the scalar flux density $\Phi(\mathbf{r}, E)$, we have provided only an approximate formula (4.5.3) for the diffusion constant $D(\mathbf{r}, E)$. Moreover we have not derived Fick's Law, but have, rather, merely stated it and described qualitatively the conditions under which it is valid. In addition, except for the development of collision theory—itself an approximate scheme—we have not yet found out how to determine $\Phi(\mathbf{r}, E)$ under conditions where Fick's Law is not valid.

To correct these deficiencies it will be necessary to derive the neutron transport equation and to develop practical schemes for solving it. This will be done in the present chapter.

We shall start by reviewing the definitions of the various neutron densities, fluxes, and currents used to describe characteristics of the neutron population in a reactor and then deriving the transport equation for the directional neutron flux density $\Psi(\mathbf{r}, \mathbf{\Omega}, E)$ (defined below). After discussing a few of the mathematical properties of the transport equation, we shall introduce a generalization of neutron importance and show that it obeys an equation adjoint to the transport equation. Finally we shall outline briefly a few of the methods used to solve the transport equation for cases of interest to reactor designers.

8.2 Derivation of the Static Transport Equation

Basic Definitions of the Quantities of Interest

Recall that the fundamental quantity of interest in static reactor theory is the *neutron density in phase space*, $N(\mathbf{r}, \mathbf{\Omega}, E)$. It is defined in such a way that, in a small volume element dV surrounding the point $\mathbf{r}$ and in a small element of solid angle $d\Omega$ "surrounding" the direction $\mathbf{\Omega}$ and in a small energy band dE containing the energy E, the number of neutrons is $N(\mathbf{r}, \mathbf{\Omega}, E)dV\, d\Omega\, dE$.

We speak loosely of $N(\mathbf{r}, \mathbf{\Omega}, E)$ being the number of neutrons "at" $\mathbf{r}$ having direction

$\mathbf{\Omega}$ and energy E. But this is only a way of talking, and it should be remembered that N is a *density*.

Since we shall now be concerned specifically with the directions in which neutrons in a reactor are traveling, it may be helpful to spell out in detail how the unit vector $\mathbf{\Omega}$ and the element of solid angle $d\mathbf{\Omega}$ may be defined. It will be sufficient to specify $\mathbf{\Omega}$ and $d\mathbf{\Omega}$ relative to a Cartesian reference system for $\mathbf{r}$. Figure 8.1 shows how this is done.

In the figure point $\mathbf{r}$ is specified relative to the fixed Cartesian system XYZ by coordinates (x, y, z) (as just stated, other coordinate systems, spherical, cylindrical, etc., could also be used). To specify the direction of $\mathbf{\Omega}$ we introduce another Cartesian system $X'Y'Z'$ with its origin at $\mathbf{r}$ and its axes *for all locations of* $\mathbf{r}$ parallel to XYZ. The direction of $\mathbf{\Omega}$ is then specified by the angles θ and φ as shown. Since $\mathbf{\Omega}$ is to be a unit vector, its length $\rho = 1$. In this coordinate system the directions of all parallel vectors $\mathbf{\Omega}$ are specified by the same values of θ and φ no matter where $\mathbf{r}$ is located.

The element of solid angle $d\mathbf{\Omega}$ surrounding $\mathbf{\Omega}$ is defined as the differential area surrounding the point at which $\mathbf{\Omega}$ pierces the sphere of radius ρ with origin at $\mathbf{r}$, divided by the total area of that sphere:

$$d\mathbf{\Omega} \equiv \frac{(\rho\, d\theta)(\rho \sin\theta\, d\varphi)}{4\pi\rho^2} = \frac{1}{4\pi}\sin\theta\, d\theta\, d\varphi. \tag{8.2.1}$$

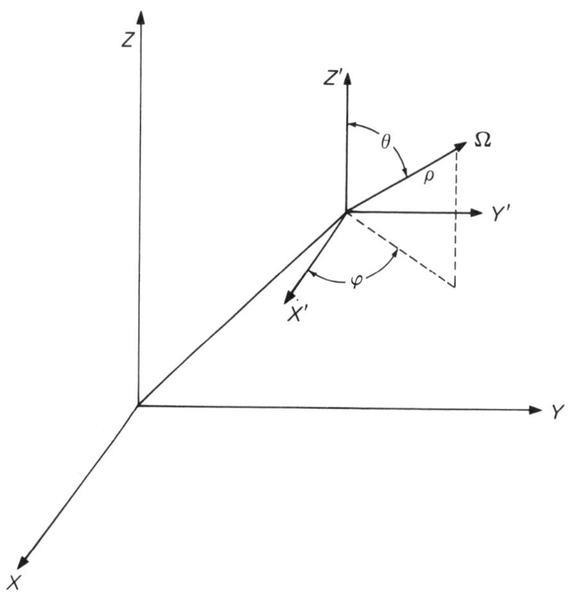

Figure 8.1 A specification of the unit vector $\mathbf{\Omega}$ relative to a system of Cartesian coordinates.

It follows from this definition of $d\Omega$ that

$$\int_{\text{all }\boldsymbol{\Omega}} d\Omega \equiv \int_0^{2\pi} d\varphi \int_0^{\pi} d\theta \frac{1}{4\pi} \sin\theta = 1. \tag{8.2.2}$$

(Some papers in the literature use the normalization $\int d\Omega = 4\pi$ so that $d\Omega$ conforms to the mathematical definition of an element of solid angle. However this procedure usually leads to a number of extra 4π's in equations and thus seems undesirable.)

In the expressions (2.2.3), (2.2.6), etc., specifying the interaction rates between neutrons traveling in a beam and nuclei in the medium through which the beam is traveling, the neutron density in phase space is always multiplied by $v(E)$, the speed of neutrons of kinetic energy E relative to the laboratory system. Accordingly it is convenient to introduce a special expression for the product $v(E)N(\mathbf{r}, \boldsymbol{\Omega}, E)$ and call it the *directional flux density*:

$$\Psi(\mathbf{r}, \boldsymbol{\Omega}, E) \equiv v(E)N(\mathbf{r}, \boldsymbol{\Omega}, E). \tag{8.2.3}$$

A closely related quantity which we shall have very little occasion to use (but which appears in some literature) is the *vector flux density*, defined as $\boldsymbol{\Omega}\, v(E)N(\mathbf{r}, \boldsymbol{\Omega}, E)$. Frequently Ψ and $\boldsymbol{\Omega}\Psi$ are referred to simply as the directional flux and the vector flux. It should be kept in mind, however, that both are densities in space, solid angle, and energy.

By constructing a small right cylinder of base dS and height $v\,dt$ parallel to $\boldsymbol{\Omega}$, we have already shown in (2.2.1) that $\Psi(\mathbf{r}, \boldsymbol{\Omega}, E)dS\, d\Omega\, dE\, dt$ is the number of neutrons having energies in the range E to $E + dE$ that cross dS in time dt.

Two very important quantities derived from $\Psi(\mathbf{r}, \boldsymbol{\Omega}, E)$ are the scalar flux (density) $\Phi(\mathbf{r}, E)$ defined by (2.2.7),

$$\Phi(\mathbf{r}, E) \equiv \int \Psi(\mathbf{r}, \boldsymbol{\Omega}, E)\, d\Omega = v(E) \int N(\mathbf{r}, \boldsymbol{\Omega}, E)\, d\Omega, \tag{8.2.4}$$

and the net current (density) $\mathbf{J}(\mathbf{r}, E)$ defined by (4.2.8),

$$\mathbf{J}(\mathbf{r}, E) \equiv \int \boldsymbol{\Omega}\Psi(\mathbf{r}, \boldsymbol{\Omega}, E)\, d\Omega. \tag{8.2.5}$$

It is clear from (8.2.4) that $\Phi(\mathbf{r}, E)dV\, dE$ is the total number of neutrons in dE and dV (without regard to direction of travel) multiplied by their speed. The physical significance of $\mathbf{J}$ is as follows:

$\mathbf{J}(\mathbf{r}, E)\, dE =$ The maximum over all orientations of a unit surface at $\mathbf{r}$ of the net number of neutrons with energies between E and $E + dE$ crossing that unit surface per second, the direction of $\mathbf{J}$ being the direction of this maximum net flow. (8.2.6)

(Actually in Chapter 4 we *defined* **J** by (8.2.6)—the same as (4.2.1)—and then showed that (8.2.5)—the same as (4.2.8)—followed. The reverse procedure, that is, going from (8.2.5) to (8.2.6), is equally satisfactory.)

It is important to keep clearly in mind the physical distinction between the directional flux density Ψ and the net current density **J**. The former concerns a current due to a beam of neutrons that are *all* traveling in essentially the same direction. The latter concerns a *net* flow of neutrons that may all be traveling in different directions. (Recall, in fact, that *none* of the neutrons making up $\mathbf{J}(\mathbf{r}, E)$ need actually be traveling in a direction $\boldsymbol{\Omega}$ parallel to **J**.)

The Balance Condition

The microscopic cross section $\sigma_\alpha^j(E)$ for nuclear reaction α between a neutron of kinetic energy E relative to the laboratory system and an isotope j specifies the relative *probability* that interaction α will take place. Thus, if a neutron with energy E is released in a medium contaning $n^j(\mathbf{r})$ atoms of isotope j per unit volume and if we examine what happens to that neutron in time dt and then repeat the experiment a great number of times, the expected fraction of the experiments in which the neutron will undergo interaction α in time dt is $v(E)n^j\sigma_\alpha^j(E)dt$. Recall further that expressing σ_α^j as a function of E implies that an average over the thermal motions of the nuclei of isotope j has been made (see Section 5.2).

It follows that, if there are $N(\mathbf{r}, \boldsymbol{\Omega}, E)dV\, d\Omega\, dE$ neutrons in an element $dV\, d\Omega\, dE$ of phase space, the expected number of interactions of type α in time dt will be

$$N(\mathbf{r}, \boldsymbol{\Omega}, E)v(E)n^j(\mathbf{r})\sigma_\alpha^j(E)dV\, d\Omega\, dE\, dt = \Psi(\mathbf{r}, \boldsymbol{\Omega}, E)n^j(\mathbf{r})\sigma_\alpha^j(E)dV\, d\Omega\, dE\, dt.$$

We discussed in Chapter 2 the fact that the essential step in deriving a deterministic rather than a probabilistic description of the behavior of neutrons in a reactor is to assume that the *actual* reaction rate is the expected one. We were therefore led to state in (2.2.3) that, when the neutron population is large:

$$\text{Number of interactions per second of type } \alpha \text{ in } dV\, d\Omega\, dE$$
$$= \sum_j \Psi(\mathbf{r}, \boldsymbol{\Omega}, E)n^j(\mathbf{r})\sigma_\alpha^j(E)\, dV\, d\Omega\, dE \equiv \Sigma_\alpha(\mathbf{r}, E)\Psi(\mathbf{r}, \boldsymbol{\Omega}, E)\, dV\, d\Omega\, dE. \tag{8.2.7}$$

In our previous derivation of a fundamental balance condition between neutron production and destruction rates (Section 4.4), we considered neutrons in $dV\, dE$ without regard to their direction of travel. The result was a relationship (4.4.2) involving **J** and Φ, and it was necessary to postulate a second, approximate relationship (Fick's Law) in order to eliminate **J** and obtain an equation (the energy-dependent diffusion equation) in $\Phi(\mathbf{r}, E)$ alone. A more rigorous procedure is to derive an equation obeyed by the directional flux density $\Psi(\mathbf{r}, \boldsymbol{\Omega}, E)$ and *then* to find Φ and **J** by applying the definitions (8.2.4) and (8.2.5). To do this we consider a portion of phase space, $dV\, d\Omega\, dE =$

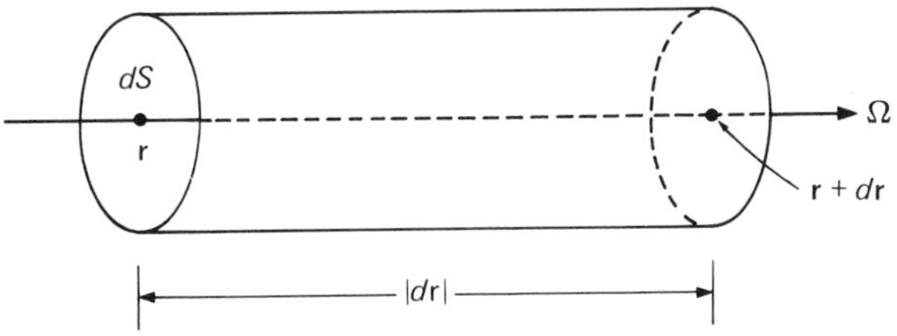

Figure 8.2 A cylinder for use in the derivation of a fundamental neutron-balance condition.

$dS|d\mathbf{r}|d\Omega\, dE$, as shown in Figure 8.2 and write a balance condition for neutrons appearing in and disappearing from this hypervolume.

We fix attention on a beam of neutrons with energies in the range E to $E + dE$ and traveling in directions within the solid angle $d\Omega$ surrounding $\mathbf{\Omega}$. For this beam:

Rate at which neutrons in $dE\, d\Omega$ are removed (by absorption and scattering) from the cylinder

$$= \Sigma_t(\mathbf{r}, E)\Psi(\mathbf{r}, \mathbf{\Omega}, E)\, dE\, d\Omega\, dV, \tag{8.2.8}$$

where $\Sigma_t(\mathbf{r}, E)$ is the total macroscopic interaction cross section. Now, since the rate at which neutrons in $dE\, d\Omega$ enter the cylinder is $\Psi(\mathbf{r}, \mathbf{\Omega}, E)dE\, d\Omega\, dS$ and the rate at which they leave the cylinder is $\Psi(\mathbf{r} + d\mathbf{r}, \mathbf{\Omega}, E)dE\, d\Omega\, dS$, we have:

Rate at which the number of neutrons in the cylinder decreases because of leakage

$$\begin{aligned} &= [\Psi(\mathbf{r} + d\mathbf{r}, \mathbf{\Omega}, E) - \Psi(\mathbf{r}, \mathbf{\Omega}, E)]\, dE\, d\Omega\, dS \\ &= [\Psi(\mathbf{r}, \mathbf{\Omega}, E) + \mathbf{\Omega}\cdot\nabla\Psi(\mathbf{r}, \mathbf{\Omega}, E)|d\mathbf{r}| + \cdots - \Psi(\mathbf{r}, \mathbf{\Omega}, E)]dE\, d\Omega\, dS \\ &= \mathbf{\Omega}\cdot\nabla\Psi(\mathbf{r}, \mathbf{\Omega}, E)dE\, d\Omega\, dV + \text{terms in } dE\, d\Omega\, dV\, |d\mathbf{r}|, \end{aligned} \tag{8.2.9}$$

where we have expanded $\Psi(\mathbf{r} + d\mathbf{r}, \mathbf{\Omega}, E)$ in a Taylor series about $\mathbf{r}$, $\mathbf{\Omega}\cdot\nabla\Psi$ being the spatial derivative of Ψ in direction $\mathbf{\Omega}$. As dV (and hence $|d\mathbf{r}|$) $\to 0$, the terms of order $dE\, d\Omega\, dV\, |d\mathbf{r}|$ and higher will vanish in comparison with the first term of (8.2.9).

The expressions (8.2.8) and (8.2.9) give the rates at which neutrons are lost from the beam. Note that the "leakage term" $\mathbf{\Omega}\cdot\nabla\Psi$ involves only neutrons traveling in the direction $\mathbf{\Omega}$ and crossing the ends of the cylinder. Neutrons originally in $dE\, d\Omega$ that escape through the sides of the cylinder must first have experienced a scattering collision and hence no longer belong to the beam $dE\, d\Omega$. (Their loss is accounted for by the scattering part of $\Sigma_t\Psi dE\, d\Omega\, dV$.)

Neutrons are added to $dE\, d\Omega\, dV$ by fission, by scattering from other energies and directions, and by external sources. Thus:

Rate at which fission neutrons due to fission of nuclei of isotope j appear in the cylinder

$$= \left[\int d\Omega' \int dE' \nu^j \Sigma_f^j(\mathbf{r}, E')\Psi(\mathbf{r}, \mathbf{\Omega}', E')\right] dV, \tag{8.2.10}$$

where $\nu^j\Sigma_f^j(\mathbf{r}, E')$ is the macroscopic cross section for neutron production due to fission at energy E' for nuclei of isotope j.

Rate at which fission neutrons in the range $dE\,d\Omega$ appear in dV

$$= \sum_j f^j(E)\,dE\,d\Omega\left[\int d\Omega' \int dE'\,\nu^j\Sigma_f^j(\mathbf{r}, E')\Psi(\mathbf{r}, \mathbf{\Omega}', E')\right] dV, \tag{8.2.11}$$

where $f^j(E)$ is the fission spectrum for isotope j and $f^j(E)dE\,d\Omega$ is the fraction of neutrons from fissions of isotope j that appear in $dE\,d\Omega$. (It is assumed that fission neutrons are emitted isotropically. Hence the probability that they will pass through any given portion of the surface of a unit sphere with origin at the point of fission is just that portion of the surface area divided by the whole surface area, that is, $d\Omega$.)

A term similar to (8.2.11) can be used to represent neutrons added to $dE\,d\Omega\,dV$ by (n, 2n) reactions. The only complication is that, for (n, 2n) reactions, it is not valid to assume that the spectrum of emitted neutrons is independent of the energy E' of the neutron causing the event. We shall not represent this effect explicitly in our equations since it is generally quite small and since no special mathematical procedures are required to account for (n, 2n) reactions.

Returning to the list of processes that add neutrons to $dE\,d\Omega\,dV$ we have:

Rate at which neutrons enter $dE\,d\Omega$ in the cylinder due to scattering

$$= \int dE' \int d\Omega'[\Sigma_s(\mathbf{r}, \mathbf{\Omega}' \to \mathbf{\Omega}, E' \to E)\,d\Omega\,dE]\Psi(\mathbf{r}, \mathbf{\Omega}', E')\,dV, \tag{8.2.12}$$

where $\Sigma_s(\mathbf{r}, \mathbf{\Omega}' \to \mathbf{\Omega}, E' \to E)d\Omega\,dE$ is the differential scattering cross section from energy E' and direction $\mathbf{\Omega}'$ into the interval $dE\,d\Omega$. Finally:

Rate at which neutrons enter $dE\,d\Omega$ in the cylinder because of nonfission sources

$$= Q(\mathbf{r}, \mathbf{\Omega}, E)\,dV\,d\Omega\,dE, \tag{8.2.13}$$

where $Q(\mathbf{r}, \mathbf{\Omega}, E)$ is the "extraneous" (i.e., nonfission) source density in phase space.

By definition a steady-state condition is one in which the rate of neutron loss in $dE\,d\Omega\,dV$ equals the rate of gain. Thus equating the sum of (8.2.9) and (8.2.8) to the sum of (8.2.11), (8.2.12), and (8.2.13) leads to the fundamental balance condition

$$\mathbf{\Omega}\cdot\nabla\Psi(\mathbf{r}, \mathbf{\Omega}, E) + \Sigma_t(\mathbf{r}, E)\Psi(\mathbf{r}, \mathbf{\Omega}, E)$$

$$= \int dE' \int d\Omega'\left[\sum_j f^j(E)\nu^j\Sigma_f^j(\mathbf{r}, E') + \Sigma_s(\mathbf{r}, \mathbf{\Omega}' \to \mathbf{\Omega}, E' \to E)\right]\Psi(\mathbf{r}, \mathbf{\Omega}', E') + Q(\mathbf{r}, \mathbf{\Omega}, E). \tag{8.2.14}$$

Equation (8.2.14) is the steady-state Boltzmann transport equation for the directional flux density $\Psi(\mathbf{r}, \mathbf{\Omega}, E)$. If we could solve it (and its time-dependent generalization) efficiently for the material and geometrical configurations of interest, the accuracy of the physics of reactor design would be limited only by the accuracy of our knowledge of cross sections.

Continuity, Boundary, and Solution Conditions

The physical nature of Ψ suggests that we seek solutions for the transport equations subject to the following conditions:

$$\begin{aligned} &\Psi(\mathbf{r}, \mathbf{\Omega}, E) \text{ real and } \geq 0 \text{ for all } \mathbf{r}, \mathbf{\Omega}, E; \\ &\Psi(\mathbf{r}, \mathbf{\Omega}, E) \text{ continuous in } \mathbf{r} \text{ in the direction } \mathbf{\Omega}. \end{aligned} \tag{8.2.15}$$

Note that Ψ need not be continuous in $\mathbf{\Omega}$ or E and need only be continuous in $\mathbf{r}$ in the direction $\mathbf{\Omega}$. Physically this means a beam of neutrons cannot disappear abruptly as it crosses some surface. (It may be attenuated very quickly, but it cannot vanish in a discontinuous fashion.) On the other hand, in spatial directions transverse to the direction of travel of a beam, Ψ can drop to zero discontinuously. (It will do so if, for example, we move crosswise from the inside to the outside of a well-defined beam.)

Mathematically this continuity condition implies that $\mathbf{\Omega}\cdot\nabla\Psi(= \nabla\cdot\mathbf{\Omega}\Psi)$, the spatial derivative of Ψ in the direction $\mathbf{\Omega}$, exists throughout the region of interest, so that the transport equation (8.2.14) is meaningful. Note, however, that expressing the operator $\mathbf{\Omega}\cdot\nabla$ as $\Omega_x(\partial/\partial x) + \Omega_y(\partial/\partial y) + \Omega_z(\partial/\partial z)$ may not be proper since the individual partial derivatives may not be defined at all points.

The usual boundary conditions imposed on $\Psi(\mathbf{r}, \mathbf{\Omega}, E)$ on the outer surface of the reactor (see (4.6.3)) are based on the assumption that the reactor may be treated as a convex region imbedded in a source-free vacuum (so that no neutrons enter the reactor from the vacuum). Thus we require that, if $\mathbf{n}$ is a unit normal vector pointing outwards from any portion of the surface S_0 of the reactor,

$$\Psi(\mathbf{r}, \mathbf{\Omega}, E) = 0 \quad \text{for } \mathbf{r} \text{ on } S_0 \text{ and } \mathbf{n}\cdot\mathbf{\Omega} < 0. \tag{8.2.16}$$

8.3 Neutron Importance

The adjoint flux $\Phi_0^*(\mathbf{r}, E)$ was introduced in Chapter 7 as a function obeying an integro-differential equation (7.6.8) that was rather similar to the diffusion equation (7.6.3) for the scalar flux density $\Phi_0(\mathbf{r}, E)$. We showed that using Φ_0^* as a weighting function in the expression (7.4.2) leads to an approximation for the reactivity that is accurate to first order when the exact flux shape during a reactor transient is not known. We then defined the "importance" of a sample of η neutrons all having energies in the range dE and being introduced into a volume element dV of a critical reactor with an isotropic distribution of initial directions of travel. We shall now generalize the definition of neutron importance to the case of a beam of neutrons traveling in a specific direction. It will then be possible to derive an equation which this generalized importance must obey and to show that this equation is mathematically adjoint to the source-free transport equation for the critical reactor.

To proceed we define the neutron importance $\Psi^*(\mathbf{r}, \mathbf{\Omega}, E)$ for a critical reactor by considering a group of η neutrons all having energies in the range dE and directions of travel in the small cone specified by the solid angle $d\Omega$ and all being introduced into the reactor in a small volume element dV. These η neutrons will scatter, leak, be absorbed, and fission in amounts dependent on their initial location, energy, and direction of travel. Because of fissioning, subsequent generations of neutrons will be born and diffuse throughout the medium. Since the reactor is exactly critical, the net result after many generations (including time for delayed neutrons to come into equilibrium) will be a small increase in the overall neutron population of the reactor. Specifically, if the population level is initially N_0, it will eventually increase by an amount ΔN and, the reactor being critical, it will remain in a new steady-state chain-reacting condition at a population level $N_0 + \Delta N$. Thus the neutrons introduced into the phase-space element $dV\,d\Omega\,dE$ "at" $\mathbf{r}$, $\mathbf{\Omega}$, and E lead asymptotically to an increase of ΔN neutrons in the overall neutron population of the reactor.

The size of ΔN will depend on where in phase space the η neutrons are introduced. If they are introduced at the center of the reactor, with energy in the thermal range, it is likely that a large number of them will cause fissions, so that the final ΔN will be relatively large (and only slightly dependent on the initial direction of the neutrons in the sample). If they are introduced in the reflector, particularly if their initial direction of travel is away from the center of the reactor, ΔN is likely to be small. Thus the asymptotic increase in overall population level resulting from introducing the η neutrons into the critical reactor will be a function of the initial location $\mathbf{r}$, the direction of travel $\mathbf{\Omega}$, and the energy E of the neutrons introduced.

We define the neutron-importance function $\Psi^*(\mathbf{r}, \mathbf{\Omega}, E)$ for neutrons introduced into a critical reactor in the interval $dV\,d\Omega\,dE$ as $\Delta N/\eta$. It is thus the increase in asymptotic population level caused by one neutron introduced at $\mathbf{r}$ with energy E and direction of travel $\mathbf{\Omega}$. Note that, unlike $\Psi(\mathbf{r}, \mathbf{\Omega}, E)$, the neutron-importance function is not a density; it is, rather, a property of a single neutron—or, more precisely, of the "point" $(\mathbf{r}, \mathbf{\Omega}, E)$ in phase space where that neutron is introduced.

As we did with $\Psi(\mathbf{r}, \mathbf{\Omega}, E)$, we here neglect statistical fluctuations and so regard the importance $\Psi^*(\mathbf{r}, \mathbf{\Omega}, E)$ as a deterministic quantity, the average expected $\Delta N/\eta$ for a great many samples. This is why we speak of a "sample" of neutrons introduced (rather than a single neutron).

To find the less general importance function $\Phi_0^*(\mathbf{r}, E)$ for a neutron of energy E introduced "isotropically" into the reactor at point $\mathbf{r}$, we simply multiply $\Psi^*(\mathbf{r}, \mathbf{\Omega}, E)$ by the probability that a neutron which is part of an isotropic distribution will have a direction $\mathbf{\Omega}$ in a cone $d\Omega$ and then integrate over all directions. Because of (8.2.1) and (8.2.2), this probability is $d\Omega$ itself. Hence we obtain

$$\Phi_0^*(\mathbf{r}, E) = \int \Psi^*(\mathbf{r}, \mathbf{\Omega}, E)\, d\Omega. \tag{8.3.1}$$

Note that the isotropic importance $\Phi_0^*(\mathbf{r}, E)$ bears the same relation to $\Psi^*(\mathbf{r}, \mathbf{\Omega}, E)$ as the scalar flux $\Phi(\mathbf{r}, E)$ does to the directional flux $\Psi(\mathbf{r}, \mathbf{\Omega}, E)$ (see (8.2.4)).

Derivation of the Equation for Neutron Importance

In order to derive an equation for the neutron importance we make use of the physical circumstance, implicit in the definition of $\Psi^*(\mathbf{r}, \mathbf{\Omega}, E)$, that neutron importance is always conserved. Thus the importance $\eta\Psi^*(\mathbf{r}, \mathbf{\Omega}, E)$ of the original sample must equal the sum of the importances at some later stage of all the progeny of the original η neutrons. Specifically, if we compute the importance of all the neutrons still left in the original beam after these uncollided neutrons have moved a distance $|d\mathbf{r}|$ in the $\mathbf{\Omega}$ direction and add to it the importances of all the neutrons that have interacted in traveling the distance $|d\mathbf{r}|$, we must get the original importance $\eta\Psi^*(\mathbf{r}, \mathbf{\Omega}, E)$. Physically this is because the ΔN associated with a given sample of neutrons must equal the ΔN associated with the progeny of that sample at some later point in history.

To treat the problem mathematically we start with η neutrons moving a short distance $|d\mathbf{r}|$ in the $\mathbf{\Omega}$ direction between the point $\mathbf{r}$ and the point $\mathbf{r} + d\mathbf{r}$ (see Figure 8.2). The initial importance of these neutrons is $\eta\Psi^*(\mathbf{r}, \mathbf{\Omega}, E)$.

The number of these neutrons that reach the point $\mathbf{r} + d\mathbf{r}$ without interacting is, by (2.3.5), $\eta \exp(-\Sigma_t(\mathbf{r}, E)|d\mathbf{r}|)$ and their importance is $\eta\Psi^*(\mathbf{r} + d\mathbf{r}, \mathbf{\Omega}, E)\exp(-\Sigma_t(\mathbf{r}, E)|d\mathbf{r}|)$.

The number of neutrons interacting as the beam traverses the distance $|d\mathbf{r}|$ is $\eta[1 - \exp(-\Sigma_t(\mathbf{r}, E)|d\mathbf{r}|)]$. Of these a fraction

$$\frac{\Sigma_s(\mathbf{r}, E \to E', \mathbf{\Omega} \to \mathbf{\Omega}')\, dE'\, d\Omega'}{\Sigma_t(\mathbf{r}, E)}$$

scatter into the energy range dE' and into the solid angle $d\Omega'$. These scattered neutrons have an importance $\Psi^*(\mathbf{r}, \mathbf{\Omega}', E')$. The total importance of scattered neutrons is then

$$\eta[1 - \exp(-\Sigma_t(\mathbf{r}, E)|d\mathbf{r}|)] \int dE' \int d\Omega' \frac{\Sigma_s(\mathbf{r}, E \to E', \mathbf{\Omega} \to \mathbf{\Omega}')}{\Sigma_t(\mathbf{r}, E)} \Psi^*(\mathbf{r}, \mathbf{\Omega}', E').$$

Similarly the number of the neutrons from the original beam that interact in traveling the distance $|d\mathbf{r}|$ to cause a fission with fissionable isotope j is

$$\eta[1 - \exp(-\Sigma_t(\mathbf{r}, E)|d\mathbf{r}|)]\frac{\Sigma_f^j(\mathbf{r}, E)}{\Sigma_t(\mathbf{r}, E)}.$$

These fissions lead to the eventual production (including delayed neutrons) of

$$\eta[1 - \exp(-\Sigma_t(\mathbf{r}, E)|d\mathbf{r}|)]\frac{\nu^j\Sigma_f^j(\mathbf{r}, E)}{\Sigma_t(\mathbf{r}, E)}f^j(E')dE'\,d\Omega'$$

neutrons in the energy interval dE' and the solid-angle interval $d\Omega'$, where $f^j(E')$ is the total fission spectrum for neutrons emitted when isotope j fissions, and the fission process is assumed to be isotropic so that the fraction of neutrons emitted with directions in the cone specified by $d\Omega'$ is $d\Omega'$ itself. The total importance of all neutrons of this kind is then

$$\eta[1 - \exp(-\Sigma_t(\mathbf{r}, E)|d\mathbf{r}|)]\int dE'\int d\Omega'\sum_j\frac{\nu^j\Sigma_f^j(\mathbf{r}, E)}{\Sigma_t(\mathbf{r}, E)}f^j(E')\Psi^*(\mathbf{r}, \mathbf{\Omega}', E').$$

Equating the importance of the original η neutrons to the importance of all these progeny we obtain

$$\eta\Psi^*(\mathbf{r}, \mathbf{\Omega}, E) = \eta\Psi^*(\mathbf{r} + d\mathbf{r}, \mathbf{\Omega}, E)\exp(-\Sigma_t(\mathbf{r}, E)|d\mathbf{r}|) + \eta[1 - \exp(-\Sigma_t(\mathbf{r}, E)|d\mathbf{r}|)]$$
$$\times\left\{\int dE'\int d\Omega'\left[\frac{\Sigma_s(\mathbf{r}, E\to E', \mathbf{\Omega}\to\mathbf{\Omega}')}{\Sigma_t(\mathbf{r}, E)} + \sum_j\frac{\nu^j\Sigma_f^j(\mathbf{r}, E)}{\Sigma_t(\mathbf{r}, E)}f^j(E')\right]\Psi^*(\mathbf{r}, \mathbf{\Omega}', E')\right\}. \tag{8.3.2}$$

In order to simplify this equation we note that, for small $|d\mathbf{r}|$,

$$\exp(-\Sigma_t(\mathbf{r}, E)|d\mathbf{r}|) \approx 1 - \Sigma_t(\mathbf{r}, E)|d\mathbf{r}| \tag{8.3.3}$$

and

$$\Psi^*(\mathbf{r} + d\mathbf{r}, \mathbf{\Omega}, E) \approx \Psi^*(\mathbf{r}, \mathbf{\Omega}, E) + \mathbf{\Omega}\cdot\nabla\Psi^*(\mathbf{r}, \mathbf{\Omega}, E)|d\mathbf{r}|, \tag{8.3.4}$$

where $\mathbf{\Omega}\cdot\nabla\Psi^*$ is the partial derivative of Ψ^* with respect to position in the direction $\mathbf{\Omega}$. Thus the importance conservation equation becomes

$$\Psi^*(\mathbf{r}, \mathbf{\Omega}, E) = [\Psi^*(\mathbf{r}, \mathbf{\Omega}, E) + \mathbf{\Omega}\cdot\nabla\Psi^*(\mathbf{r}, \mathbf{\Omega}, E)|d\mathbf{r}|][1 - \Sigma_t(\mathbf{r}, E)|d\mathbf{r}|]$$
$$+ |d\mathbf{r}|\int dE'\int d\Omega'\left[\Sigma_s(\mathbf{r}, E\to E', \mathbf{\Omega}\to\mathbf{\Omega}') + \sum_j\nu^j\Sigma_f^j(\mathbf{r}, E)f^j(E')\right]\Psi^*(\mathbf{r}, \mathbf{\Omega}', E') \tag{8.3.5}$$

or, neglecting $|d\mathbf{r}|^2$ terms,

$$\mathbf{\Omega}\cdot\nabla\Psi^*(\mathbf{r}, \mathbf{\Omega}, E) - \Sigma_t(\mathbf{r}, E)\Psi^*(\mathbf{r}, \mathbf{\Omega}, E)$$
$$+ \int dE'\int d\Omega'\left[\sum_j\nu^j\Sigma_f^j(\mathbf{r}, E)f^j(E') + \Sigma_s(\mathbf{r}, E\to E', \mathbf{\Omega}\to\mathbf{\Omega}')\right]\Psi^*(\mathbf{r}, \mathbf{\Omega}', E') = 0. \tag{8.3.6}$$

The physical definition of neutron importance implies that we should seek solutions $\Psi^*(\mathbf{r}, \mathbf{\Omega}, E)$ to (8.3.6) that are real and nonnegative (adding a neutron to a critical system cannot *decrease* the power level) and continuous in $\mathbf{r}$ in the direction $\mathbf{\Omega}$. (The asymptotic change in power level due to addition at $\mathbf{r}$ of a neutron of energy E traveling in direction $\mathbf{\Omega}$ can differ only infinitesimally from the corresponding change in power level due to addition of a neutron of the same energy and traveling in the same direction at a point $\mathbf{r} + \mathbf{\Omega}|d\mathbf{r}|$, an infinitesimal distance $|d\mathbf{r}|$ further along the path that the first neutron will travel.)

The physical definition of importance further implies that, on a convex surface of a reactor-reflector system imbedded in a vacuum, the importance of neutrons crossing the surface into the vacuum will be zero. (These neutrons will never return and hence can cause no asymptotic ΔN.) Thus, if $\mathbf{n}$ is an outwardly directed normal on the surface S_0 of the reactor, we have

$$\Psi^*(\mathbf{r}, \mathbf{\Omega}, E) = 0 \quad \text{for } \mathbf{r} \text{ on } S_0 \text{ and } \mathbf{n}\cdot\mathbf{\Omega} > 0. \tag{8.3.7}$$

Note that this boundary condition is complementary to the condition on $\Psi(\mathbf{r}, \mathbf{\Omega}, E)$ on the outer surface. For the latter we have required in (8.2.16) that no neutrons enter the core *from* the vacuum.

Operator Notation and the Equation Adjoint to the Transport Equation

There is a great similarity between the equation (8.3.6) specifying the behavior of the neutron importance and the transport equation (8.2.14) specifying the behavior of the directional flux for a critical reactor ($Q = 0$). The mathematical nature of this relationship and some of the consequences that follow from it can be discussed more conveniently if we adopt a compact linear-operator notation. Accordingly we define two linear operators, L, a net-loss operator, and M, a fission-production operator, by the following equations:

$$\begin{aligned} LG(\mathbf{r}, \mathbf{\Omega}, E) &\equiv [\mathbf{\Omega}\cdot\nabla + \Sigma_t(\mathbf{r}, E)]G(\mathbf{r}, \mathbf{\Omega}, E) \\ &\quad - \int_{\mathbf{\Omega}'} d\Omega' \int_0^\infty dE' \Sigma_s(\mathbf{r}, \mathbf{\Omega}' \to \mathbf{\Omega}, E' \to E) G(\mathbf{r}, \mathbf{\Omega}', E'), \\ MG(\mathbf{r}, \mathbf{\Omega}, E) &\equiv \int_{\mathbf{\Omega}'} d\Omega' \int_0^\infty dE' \sum_j f^j(E) \nu^j \Sigma_f^j(\mathbf{r}, E') G(\mathbf{r}, \mathbf{\Omega}', E'), \end{aligned} \tag{8.3.8}$$

where $G(\mathbf{r}, \mathbf{\Omega}, E)$ is any square-integrable function, piecewise analytic in the component of $\mathbf{r}$ in the direction $\mathbf{\Omega}$ and obeying the boundary conditions of $\Psi(\mathbf{r}, \mathbf{\Omega}, E)$ on the outer surface of the reactor. Equation (8.2.14) then becomes

$$L\Psi = M\Psi + Q. \tag{8.3.9}$$

Similarly we define linear operators L^* and M^* by their operation on any square-integrable function $G^*(\mathbf{r}, \mathbf{\Omega}, E)$ analytic in the component of $\mathbf{r}$ in the direction $\mathbf{\Omega}$ and obeying the external boundary conditions imposed on the importance function $\Psi^*(\mathbf{r}, \mathbf{\Omega}, E)$:

$$\begin{aligned} L^*G^*(\mathbf{r}, \mathbf{\Omega}, E) &\equiv [-\mathbf{\Omega}\cdot\nabla + \Sigma_t(\mathbf{r}, E)]G^*(\mathbf{r}, \mathbf{\Omega}, E) \\ &\quad - \int_{\mathbf{\Omega}'} d\Omega' \int_0^\infty dE'\, \Sigma_s(\mathbf{r}, \mathbf{\Omega} \to \mathbf{\Omega}', E \to E')G^*(\mathbf{r}, \mathbf{\Omega}', E'), \\ M^*G^*(\mathbf{r}, \mathbf{\Omega}, E) &\equiv \int_{\mathbf{\Omega}'} d\Omega' \int_0^\infty dE' \sum_j f^j(E')\nu^j\Sigma_f^j(\mathbf{r}, E)G^*(\mathbf{r}, \mathbf{\Omega}', E'). \end{aligned} \tag{8.3.10}$$

Then the neutron-importance equation can be written

$$L^*\Psi^* = M^*\Psi^*. \tag{8.3.11}$$

The operators L and L^*, M and M^* are related mathematically in a very important way. To show this in compact form we introduce the Dirac notation for an inner product: if $G^*(\mathbf{r}, \mathbf{\Omega}, E)$ and $G(\mathbf{r}, \mathbf{\Omega}, E)$ are any two square-integrable functions, piecewise analytic in the component of $\mathbf{r}$ in the $\mathbf{\Omega}$ direction and obeying respectively the outer boundary conditions but not necessarily the continuity conditions imposed on Ψ^* and Ψ, we define an *inner product* (a number—possibly complex) by

$$\langle G^*|G\rangle \equiv \int dV \int d\Omega \int dE\, G^*(\mathbf{r}, \mathbf{\Omega}, E)G(\mathbf{r}, \mathbf{\Omega}, E). \tag{8.3.12}$$

Since the function $LG(\mathbf{r}, \mathbf{\Omega}, E)$ defined by (8.3.8) is in the same space as $G(\mathbf{r}, \mathbf{\Omega}, E)$, we have at once that

$$\begin{aligned} \langle G^*|LG\rangle \equiv \langle G^*|L|G\rangle &= \int dV \int d\Omega \int dE\, G^*(\mathbf{r}, \mathbf{\Omega}, E)LG(\mathbf{r}, \mathbf{\Omega}, E) \\ &= \int dV \int d\Omega \int dE\, G^*(\mathbf{r}, \mathbf{\Omega}, E)\Big\{[\mathbf{\Omega}\cdot\nabla + \Sigma_t(\mathbf{r}, E)]G(\mathbf{r}, \mathbf{\Omega}, E) \\ &\quad - \int d\Omega' \int dE'\, \Sigma_s(\mathbf{r}, E' \to E, \mathbf{\Omega}' \to \mathbf{\Omega})G(\mathbf{r}, \mathbf{\Omega}', E')\Big\}. \end{aligned} \tag{8.3.13}$$

In order to make the mathematical relationship between L^* and L as simple as possible we now restrict attention to spaces of functions less general than those to which G and G^* belong, but more general than those to which Ψ and Ψ^* belong. Specifically we consider the space of all functions that obey the continuity and boundary conditions—but not necessarily the positivity and reality conditions—imposed on the directional flux $\Psi(\mathbf{r}, \mathbf{\Omega}, E)$. Let $F(\mathbf{r}, \mathbf{\Omega}, E)$ be any function in this space. We also consider a "dual space" of all functions obeying the continuity and boundary conditions imposed on $\Psi^*(\mathbf{r}, \mathbf{\Omega}, E)$. Let $F^*(\mathbf{r}, \mathbf{\Omega}, E)$ be any function in this space. Integration by parts then yields for the inner product involving the operator $\nabla\cdot\mathbf{\Omega}$

$$\begin{aligned}\langle F^*|\nabla\cdot\mathbf{\Omega}|F\rangle &\equiv \int dV \int d\Omega \int dE\, F^*\nabla\cdot\mathbf{\Omega}F \\ &= \int dV \int d\Omega \int dE\, [\nabla\cdot(\mathbf{\Omega}F^*F) - F\nabla\cdot(\mathbf{\Omega}F^*)] \\ &= \int d\Omega \int dE \int_{S_0} dS\, \mathbf{n}\cdot\mathbf{\Omega}F^*F - \int dV \int d\Omega \int dE\, F\nabla\cdot\mathbf{\Omega}F^* \\ &\equiv \langle F|-\nabla\cdot\mathbf{\Omega}|F^*\rangle, \end{aligned} \tag{8.3.14a}$$

where application of the divergence theorem to $\int \nabla\cdot(\mathbf{\Omega}F^*F)$ results in no terms at internal surfaces since F^*F is continuous in the direction $\mathbf{\Omega}$, and where the integral over the outer surface vanishes since, there, $F^* = 0$ for $\mathbf{n}\cdot\mathbf{\Omega} > 0$ and $F = 0$ for $\mathbf{n}\cdot\mathbf{\Omega} < 0$.

Also, since we can change the order of integration and the symbols over which we are integrating,

$$\begin{aligned}&\int dV \int d\Omega \int dE\, F^*(\mathbf{r}, \mathbf{\Omega}, E) \int d\Omega' \int dE'\, \Sigma_s(\mathbf{r}, E' \to E, \mathbf{\Omega}' \to \mathbf{\Omega})F(\mathbf{r}, \mathbf{\Omega}', E') \\ &= \int dV \int d\Omega \int dE \int d\Omega' \int dE'\, F^*(\mathbf{r}, \mathbf{\Omega}, E)\Sigma_s(\mathbf{r}, E' \to E, \mathbf{\Omega}' \to \mathbf{\Omega})F(\mathbf{r}, \mathbf{\Omega}', E') \\ &= \int dV \int d\Omega' \int dE' \int d\Omega \int dE\, F^*(\mathbf{r}, \mathbf{\Omega}', E')\Sigma_s(\mathbf{r}, E \to E', \mathbf{\Omega} \to \mathbf{\Omega}')F(\mathbf{r}, \mathbf{\Omega}, E) \\ &= \int dV \int d\Omega \int dE\, F(\mathbf{r}, \mathbf{\Omega}, E) \int d\Omega' \int dE'\, \Sigma_s(\mathbf{r}, E \to E', \mathbf{\Omega} \to \mathbf{\Omega}')F^*(\mathbf{r}, \mathbf{\Omega}', E'). \end{aligned} \tag{8.3.14b}$$

Thus (8.3.13) becomes

$$\begin{aligned}\langle F^*|L|F\rangle = \int dV \int d\Omega \int dE\, F(\mathbf{r}, \mathbf{\Omega}, E)\Big\{[-\mathbf{\Omega}\cdot\nabla + \Sigma_t(\mathbf{r}, E)]F^*(\mathbf{r}, \mathbf{\Omega}, E) \\ - \int d\Omega' \int dE'\, \Sigma_s(\mathbf{r}, E \to E', \mathbf{\Omega} \to \mathbf{\Omega}')F^*(\mathbf{r}, \mathbf{\Omega}', E')\Big\} = \langle F|L^*|F^*\rangle. \end{aligned} \tag{8.3.15}$$

It can similarly be shown that

$$\langle F^*|M|F\rangle = \langle F|M^*|F^*\rangle. \tag{8.3.16}$$

Operators related to each other by equations like (8.3.15) and (8.3.16) (where F and F^* are *any* functions in the given spaces that are defined by the continuity and boundary conditions on Ψ and Ψ^* respectively), are said to be *adjoint* to each other, and the equation (8.3.11) for the neutron importance is said to be the equation *adjoint* to the transport equation (8.3.9) (with $Q = 0$). Thus the neutron importance is frequently called the *adjoint flux*.

Equation (7.6.8), which we introduced in Chapter 7 as the adjoint to the energy-dependent diffusion equation (7.6.3), can now be seen to conform to the mathematical definition of an adjoint equation provided we restrict the space of functions to those obeying the continuity and external boundary conditions imposed on the scalar flux and its adjoint. (With the space of functions so restricted, each operator acting on Φ_0^* in (7.6.8) obeys a relationship similar to (8.3.15) with respect to the corresponding operator in (7.6.3).)

Eigenfunctions Associated with the Transport Equation and Its Adjoint

For $Q = 0$, the transport equation (8.3.9) becomes homogeneous and will have non-trivial (i.e., nonzero) solutions only under certain conditions. As we have seen before, this mathematical behavior reflects the physical fact that a reactor will be critical only when the fuel loading is such that the neutron-production rate exactly equals the neutron-loss rate. We have also seen that it is always possible to force criticality in an artificial manner by dividing the neutron-production rate due to fission ($M\Psi$ of (8.3.8) in the present case) by a positive real number λ, the k_{eff} of the system. Thus we expect that there is always a positive real eigenvalue λ such that a physically acceptable solution Ψ exists for the equation

$$L\Psi = \frac{1}{\lambda} M\Psi. \tag{8.3.17}$$

In Chapter 3 we encountered the analogous (but very much simpler) equation (3.3.11) when the multigroup model was applied to an infinite homogeneous medium. It was shown there that other solutions, corresponding to different λ values, could be found. As is to be expected, solutions other than the physically acceptable one also exist for (8.3.17). In fact there are an infinite number of other eigenvalues for which solutions to (8.3.17) obeying the continuity and external boundary conditions (but *not* the positivity conditions) exist. These eigenvalues may be real or complex, discrete or continuous; we shall represent them (as if they were discrete) by the symbol λ_n and their corresponding eigenfunctions by $\Psi_n^{(\lambda)}(\mathbf{r}, \mathbf{\Omega}, E)$. Equation (8.3.17) then becomes

$$L\Psi_n^{(\lambda)} = \frac{1}{\lambda_n} M\Psi_n^{(\lambda)} \tag{8.3.18}$$

or, if we assume the existence of an operator L^{-1}, the inverse to the operator L (such that $L^{-1}L =$ the identity operator),

$$L^{-1}M\Psi_n^{(\lambda)} = \lambda_n \Psi_n^{(\lambda)}, \tag{8.3.19}$$

the latter form being a more common way to write an eigenvalue equation.

Little is known about the nature of the eigenvalues λ_n. In particular it is not known whether the eigenfunctions belonging to these eigenvalues form a complete set (in the sense that *any* function obeying the boundary and continuity conditions of Ψ can be expanded in terms of them). It would be an advantage to be able to prove completeness, since formal proofs of various properties of solutions to the transport equations could then be obtained using eigenfunction expansions. It *is* known that the eigenfunctions belonging to the eigenvalues λ_n are not complete unless the eigenvalue $\lambda = 0$ is included. Moreover there are an infinite number of linearly independent eigenfunctions $\Psi_n^{(0)}$ corresponding to that eigenvalue. They are solutions of

$$M\Psi_n^{(0)} = 0, \tag{8.3.20}$$

where the subscript n for this case only refers to independent eigenfunctions of the particular eigenvalue zero.

Despite the lack of positive evidence, it is current practice to assume that, with the $\Psi_n^{(0)}$ included, the set of eigenfunctions are complete. Thus it is assumed that any arbitrary function $F(\mathbf{r}, \mathbf{\Omega}, E)$ obeying the continuity and boundary conditions imposed on the directional flux can be expanded as

$$F(\mathbf{r}, \mathbf{\Omega}, E) = \sum_{n=0}^{\infty} a_n \Psi_n^{(\lambda)}(\mathbf{r}, \mathbf{\Omega}, E) + \sum_{n=0}^{\infty} a_n^0 \Psi_n^{(0)}(\mathbf{r}, \mathbf{\Omega}, E), \tag{8.3.21}$$

where the parts of the expansion involving eigenfunctions belonging to nonzero and zero eigenvalues have been written separately.

There are completely analogous sets of eigenvalues and eigenfunctions corresponding to the adjoint equation (8.3.11). Thus we have

$$L^*\Psi_m^{(\lambda)*} = \frac{1}{\lambda_m^*} M^*\Psi_m^{(\lambda)*} \tag{8.3.22}$$

or, if we assume that the inverse $(L^*)^{-1}$ exists,

$$(L^*)^{-1} M^*\Psi_m^{(\lambda)*} = \lambda_m^* \Psi_m^{(\lambda)*}. \tag{8.3.23}$$

Moreover, for the nonzero eigenvalues, an orthogonality condition between the $\Psi_n^{(\lambda)}$ and $\Psi_n^{(\lambda)*}$ can be found and used to determine the a_n in an expansion such as (8.3.21). Thus, for λ_n, $\lambda_m^* \neq 0$, we multiply (8.3.18) by $\Psi_m^{(\lambda)*}$ and (8.3.22) by $\Psi_n^{(\lambda)}$ and integrate over V, $\mathbf{\Omega}$, and E to obtain

$$\begin{aligned} \langle \Psi_m^{(\lambda)*} | L | \Psi_n^{(\lambda)} \rangle &= \frac{1}{\lambda_n} \langle \Psi_m^{(\lambda)*} | M | \Psi_n^{(\lambda)} \rangle, \\ \langle \Psi_n^{(\lambda)} | L^* | \Psi_m^{(\lambda)*} \rangle &= \frac{1}{\lambda_m^*} \langle \Psi_n^{(\lambda)} | M^* | \Psi_m^{(\lambda)*} \rangle. \end{aligned} \tag{8.3.24}$$

But, by (8.3.15) and (8.3.16),

$$\langle \Psi_m^{(\lambda)*}|L|\Psi_n^{(\lambda)}\rangle = \langle \Psi_n^{(\lambda)}|L^*|\Psi_m^{(\lambda)*}\rangle,$$
$$\langle \Psi_m^{(\lambda)*}|M|\Psi_n^{(\lambda)}\rangle = \langle \Psi_n^{(\lambda)}|M^*|\Psi_m^{(\lambda)*}\rangle. \tag{8.3.25}$$

Therefore

$$0 = \left(\frac{1}{\lambda_n} - \frac{1}{\lambda_m^*}\right)\langle \Psi_m^{(\lambda)*}|M|\Psi_n^{(\lambda)}\rangle. \tag{8.3.26}$$

Thus, if $\lambda_n \neq \lambda_m^*$,

$$\langle \Psi_m^{(\lambda)*}|M|\Psi_n^{(\lambda)}\rangle = 0 \tag{8.3.27}$$

and we have the desired orthogonality condition.

Further, if the nonzero λ_n and λ_m^* are distinct and there exists one eigenfunction, call it $\Psi_n^{(\lambda)*}$, for each $\Psi_n^{(\lambda)}$ such that $\langle \Psi_n^{(\lambda)*}|M|\Psi_n^{(\lambda)}\rangle \neq 0$, then $\lambda_n = \lambda_n^*$. Thus, if these conditions are met, the regular and adjoint equations will have the same set of eigenvalues, and the eigenfunctions can thus be set into unique correspondence. For the "fundamental" (everywhere-positive) eigenfunctions $\Psi_0^{(\lambda)}$ and $\Psi_0^{(\lambda)*}$, assumed to exist on physical grounds, the nature of M is such that $\langle \Psi_0^{(\lambda)*}|M|\Psi_0^{(\lambda)}\rangle$ must be positive. Thus we know that $\lambda_0 = \lambda_0^*$. Moreover, if, in addition, λ_0 is distinct, we know from (8.3.27) and (8.3.16) that

$$\langle \Psi_m^{(\lambda)*}|M|\Psi_0^{(\lambda)}\rangle = 0 = \langle \Psi_n^{(\lambda)}|M^*|\Psi_0^{(\lambda)*}\rangle \qquad m, n \neq 0.$$

Again the natures of M and M^* are such that $\Psi_m^{(\lambda)*}$ and $\Psi_n^{(\lambda)}$ must be somewhere zero or negative. These results are generalizations of the mathematical properties of the multigroup flux vectors in an infinite homogeneous medium discussed in Section 3.3. Unfortunately they are far less rigorous mathematically since we have had to depend on physical arguments to establish them.

If the λ_n are distinct and the condition $\langle \Psi_n^{(\lambda)*}|M|\Psi_n^{(\lambda)}\rangle \neq 0$ holds, the expansion coefficients a_n in (8.3.21) can be found immediately by operating on (8.3.21) with $\Psi_m^{(\lambda)*}M$ and integrating over the phase space of the reactor. We obtain

$$\langle \Psi_m^{(\lambda)*}|M|F\rangle = \sum_{n=0}^{\infty} a_n \langle \Psi_m^{(\lambda)*}|M|\Psi_n^{(\lambda)}\rangle + \sum_{n=0}^{\infty} a_n^0 \langle \Psi_m^{(\lambda)*}|M|\Psi_n^{(0)}\rangle \tag{8.3.28}$$

which, because of (8.3.27) and (8.3.20), yields

$$a_m = \frac{\langle \Psi_m^{(\lambda)*}|M|F\rangle}{\langle \Psi_m^{(\lambda)*}|M|\Psi_m^{(\lambda)}\rangle}\,. \tag{8.3.29}$$

Unfortunately we cannot obtain the a_n^0 in any such simple fashion. As a result the "λ modes" $\Psi_n^{(\lambda)}$ and $\Psi_n^{(0)}$ are not very useful unless it is known that none of the $\Psi_n^{(0)}$ are needed (for example if we are trying to expand MF rather than F).

It is far more common to use eigenfunction expansions to determine solutions of equations than to express arbitrary functions. Suppose, for example, that we wish to determine the fission-neutron source term $M\Psi$ due to an arbitrary external source Q. The direct procedure is to solve (8.3.9) for Ψ and then compute $M\Psi$ by (8.3.8). An alternate way is to expand Ψ as

$$\Psi(\mathbf{r}, \mathbf{\Omega}, E) = \sum_{n=0}^{\infty} a_n \Psi_n^{(\lambda)}(\mathbf{r}, \mathbf{\Omega}, E) + \sum_{n=0}^{\infty} a_n^0 \Psi_n^{(0)}(\mathbf{r}, \mathbf{\Omega}, E). \tag{8.3.30}$$

Since we are interested in $M\Psi$, we need only find the a_n. (Equation (8.3.20) shows that none of the terms in the second sum will contribute to $M\Psi$.)

To determine the a_n we substitute (8.3.30) into (8.3.9), obtaining

$$\sum_{n=0}^{\infty} a_n L\Psi_n^{(\lambda)} + \sum_{n=0}^{\infty} a_n^0 L\Psi_n^{(0)} = \sum_{n=0}^{\infty} a_n M\Psi_n^{(\lambda)} + Q. \tag{8.3.31}$$

Making use of the defining equation (8.3.18), we find that

$$\sum_{n=0}^{\infty} a_n\left(\frac{1}{\lambda_n} - 1\right) M\Psi_n^{(\lambda)} + \sum_{n=0}^{\infty} a_n^0 L\Psi_n^{(0)} = Q. \tag{8.3.32}$$

Then, operating on this result with $\Psi_m^{(\lambda)*}$ and integrating leads to

$$a_m\left(\frac{1}{\lambda_m} - 1\right)\langle\Psi_m^{(\lambda)*}|M|\Psi_m^{(\lambda)}\rangle + \sum_{n=0}^{\infty} a_n^0 \langle\Psi_m^{(\lambda)*}|L|\Psi_n^{(0)}\rangle = \langle\Psi_m^{(\lambda)*}|Q\rangle. \tag{8.3.33}$$

But, from (8.3.25), (8.3.22), and (8.3.20), we have, for all n,

$$\langle\Psi_m^{(\lambda)*}|L|\Psi_n^{(0)}\rangle = \langle\Psi_n^{(0)}|L^*|\Psi_m^{(\lambda)*}\rangle = \frac{1}{\lambda_m^*}\langle\Psi_n^{(0)}|M^*|\Psi_m^{(\lambda)*}\rangle = \frac{1}{\lambda_m^*}\langle\Psi_m^{(\lambda)*}|M|\Psi_n^{(0)}\rangle = 0. \tag{8.3.34}$$

(Note that by this double use of the relationship between an operator and its adjoint we avoid having to deal directly with $\lambda^{-1}M\Psi_n^{(0)}$ with $\lambda = 0$ and $M\Psi_n^{(0)} = 0$.)

From (8.3.33) and (8.3.34) we then obtain

$$a_m = \frac{\lambda_m\langle\Psi_m^{(\lambda)*}|Q\rangle}{(1 - \lambda_m)\langle\Psi_m^{(\lambda)*}|M|\Psi_m^{(\lambda)}\rangle}, \tag{8.3.35}$$

so that, applying (8.3.30) and (8.3.20), we may express the fission-production rate in a reactor with a source as

$$M\Psi(\mathbf{r}, \boldsymbol{\Omega}, E) = \sum_{m=0}^{\infty} \frac{\lambda_m \langle \Psi_m^{(\lambda)*} | Q \rangle}{(1 - \lambda_m)\langle \Psi_m^{(\lambda)*} | M | \Psi_m^{(\lambda)} \rangle} M\Psi_m^{(\lambda)}(\mathbf{r}, \boldsymbol{\Omega}, E). \tag{8.3.36}$$

Thus the problem of solving the inhomogeneous equation (8.3.9) for Ψ has been replaced by the problem of solving a number of homogeneous equations of the type (8.3.18) and (8.3.22). This exchange may provide a practical advantage if all but a few of the a_m are negligibly small and if, for example, we are concerned with analyzing a large number of situations involving external sources $Q(\mathbf{r}, \boldsymbol{\Omega}, E)$ that are at different locations and that release neutrons having different angular and energy distributions (so that (8.3.9) will have to be solved for each case, whereas the same eigenfunctions $\Psi_m^{(\lambda)}$ and $\Psi_m^{(\lambda)*}$ can be used for all cases). Note, in this connection, that if the reactor is almost critical, so that $\lambda_0 \equiv k_{\text{eff}} \approx 1$, the $m = 0$ term in (8.3.36) greatly exceeds all other terms in magnitude. Thus, for a reactor close to critical, $M\Psi$ approaches the fundamental-λ-mode fission source $M\Psi_0^{(\lambda)}$, no matter what the nature of $Q(\mathbf{r}, \boldsymbol{\Omega}, E)$. Note also that, for k_{eff} slightly greater than unity, $M\Psi$ becomes negative. This unphysical result tells us that no steady-state condition can be achieved in a supercritical reactor containing an external source.

The λ modes are not the only set of eigenfunctions that can be generated by the operators L and M. Several other sets, as a class called ω *modes* or *period eigenfunctions*, are often used for the analysis of transient situations. The most mathematically simple of these classes of ω modes may be derived by assuming that, with L and M constant in time and with delayed neutrons assumed to appear instantaneously (i.e., on the same time scale as prompt neutrons), the reactor (Q being zero) is on an asymptotic period so that the neutron density in phase space $N(\mathbf{r}, \boldsymbol{\Omega}, E, t)$ $(= \Psi(\mathbf{r}, \boldsymbol{\Omega}, E, t)/v(E))$ is behaving in time as $N(\mathbf{r}, \boldsymbol{\Omega}, E)\exp(\omega t)$. Then, since the rate of rise of N equals the difference between the fission-production rate and the net-loss rate,

$$\frac{\partial}{\partial t}\left[\frac{\Psi(\mathbf{r}, \boldsymbol{\Omega}, E, t)}{v(E)}\right] = M\Psi - L\Psi = \frac{\omega}{v}\Psi. \tag{8.3.37}$$

Multiplication by $v(E)$ yields the eigenvalue equation

$$v(M - L)\Psi_n^{(\omega)} = \omega_n \Psi_n^{(\omega)}, \tag{8.3.38}$$

where, on physical grounds, we expect for a reactor of finite size (provided it is not in an extremely subcritical condition) that there will be one isolated, least negative eigenvalue ω_0 and a corresponding everywhere-positive eigenfunction $\Psi_0^{(\omega)}$ such that at asymptotic times the directional flux $\Psi(\mathbf{r}, \boldsymbol{\Omega}, E, t)$ will behave as $\Psi_0^{(\omega)}\exp(\omega_0 t)$. Since the physical picture on which this particular class of ω modes is based assumes that delayed neutrons appear as soon as their precursors are formed, we would expect ω_0 to be large in magnitude unless the reactor is critical (in which case $\omega_0 = 0$).

If we define a set of adjoint ω modes by

$$v(M^* - L^*)\Psi_m^{(\omega)*} = \omega_m^* \Psi_m^{(\omega)*}, \tag{8.3.39}$$

it is easy to show that

$$(\omega_n - \omega_m^*)\left\langle \Psi_m^{(\omega)*} \left| \frac{1}{v} \right| \Psi_n^{(\omega)} \right\rangle = 0. \tag{8.3.40}$$

Thus, if the ω_n and ω_m^* are distinct and for each $\Psi_n^{(\omega)}$ there is a corresponding adjoint mode—call it $\Psi_n^{(\omega)*}$—such that $\langle \Psi_n^{(\omega)*} | 1/v | \Psi_n^{(\omega)} \rangle \neq 0$, then $\omega_n = \omega_n^*$.

No anomalous ω modes, analogous to the $\Psi_n^{(0)}$, are known. If $\omega_0 = 0$ happens to be an eigenvalue of (8.3.38), it will be the only zero eigenvalue.

A Transport-Theory Expression for Reactivity

The adjoint flux was first introduced in Chapter 7 to improve the accuracy of the perturbation expression for reactivity obtained from the diffusion-theory model. Use of operator notation and the mathematical definition of an adjoint operator simplifies the extension of this result to the transport-theory model.

As in (7.4.2) we define the reactivity $\rho(t)$ as a weighted integral of the net neutron-production rate divided by a weighted integral of the fission-production rate and use the fundamental-λ-mode importance function $\Psi_0^{(\lambda)*}$ of the unperturbed reactor as the weight function. With M and L the (now time-dependent) fission-production and net-loss operators and $\Psi(\mathbf{r}, \mathbf{\Omega}, E, t)$ the instantaneous directional flux, we have

$$\rho(t) \equiv \frac{\langle \Psi_0^* | M - L | \Psi(\mathbf{r}, \mathbf{\Omega}, E, t\rangle}{\langle \Psi_0^* | M | \Psi(\mathbf{r}, \mathbf{\Omega}, E, t)\rangle}. \tag{8.3.41}$$

Thus, physically, the reactivity is the instantaneous net rate at which the neutrons in the actual reactor would produce importance in the unperturbed reactor divided by the instantaneous rate at which the fission neutrons (both the prompt neutrons being produced at t and the delayed neutrons that will eventually come from precursors being produced at t) would produce importance in the unperturbed reactor. (Since $\lambda_0 = 1$ and $\omega_0 = 0$, $\Psi_0^{(\lambda)*} = \Psi_0^{(\omega)*}$, and we have simplified notation accordingly.)

Because of (8.3.15) and (8.3.16) and the fact that $\Psi(\mathbf{r}, \mathbf{\Omega}, E, t)$ and Ψ_0^* are in the proper function space, we find that

$$\rho(t) = \frac{\langle \Psi(\mathbf{r}, \mathbf{\Omega}, E, t) | M^* - L^* | \Psi_0^* \rangle}{\langle \Psi(\mathbf{r}, \mathbf{\Omega}, E, t) | M^* | \Psi_0^* \rangle}. \tag{8.3.42}$$

If we now express the perturbed operators L, M, L^*, and M^* as

$$L = L_0 + \delta L, \quad M = M_0 + \delta M,$$
$$L^* = L_0^* + \delta L^*, \quad M^* = M_0^* + \delta M^*, \tag{8.3.43}$$

we have, since $(M_0^* - L_0^*)\Psi_0^* = 0$,

$$\rho(t) = \frac{\langle \Psi(\mathbf{r}, \mathbf{\Omega}, E, t) | \delta M^* - \delta L^* | \Psi_0^* \rangle}{\langle \Psi(\mathbf{r}, \mathbf{\Omega}, E, t) | M^* | \Psi_0^* \rangle} = \frac{\langle \Psi_0^* | \delta M - \delta L | \Psi(\mathbf{r}, \mathbf{\Omega}, E, t) \rangle}{\langle \Psi_0^* | M | \Psi(\mathbf{r}, \mathbf{\Omega}, E, t) \rangle}. \tag{8.3.44}$$

If, in analogy with (7.3.1), we now define

$$T(t) \equiv \left\langle \Psi_0^* \left| \frac{1}{v(E)} \right| \Psi(\mathbf{r}, \mathbf{\Omega}, E, t) \right\rangle$$

and if we also define a shape function $S(\mathbf{r}, \mathbf{\Omega}, E, t)$ by

$$\Psi(\mathbf{r}, \mathbf{\Omega}, E, t) = S(\mathbf{r}, \mathbf{\Omega}, E, t) T(t), \tag{8.3.45}$$

we obtain

$$\rho(t) = \frac{\langle \Psi_0^* | \delta M - \delta L | S(\mathbf{r}, \mathbf{\Omega}, E, t) \rangle}{\langle \Psi_0^* | M | S(\mathbf{r}, \mathbf{\Omega}, E, t) \rangle}, \tag{8.3.46}$$

where $S(\mathbf{r}, \mathbf{\Omega}, E, t)$ doesn't grow or decay in time but, at most, changes shape.

If we now replace $S(\mathbf{r}, \mathbf{\Omega}, E, t)$ by $S = S_0 + \delta S$, where S_0 is some approximation to S obtained without solving the time-dependent transport equation, (8.3.44) becomes

$$\rho(t) = \frac{\langle \Psi_0^* | \delta M - \delta L | S_0 + \delta S \rangle}{\langle \Psi_0^* | M | S_0 + \delta S \rangle} = \frac{\langle \Psi_0^* | \delta M - \delta L | S_0 \rangle + \langle \Psi_0^* | \delta M - \delta L | \delta S \rangle}{\langle \Psi_0^* | M | S_0 \rangle + \langle \Psi_0^* | M | \delta S \rangle}. \tag{8.3.47}$$

Then, since the second terms in both the numerator and denominator of (8.3.47) are expected to be small in comparison with the corresponding first terms, we may write a "perturbation formula" for $\rho(t)$:

$$\rho(t) = \frac{\langle \Psi_0^* | \delta M - \delta L | S_0 \rangle}{\langle \Psi_0^* | M | S_0 \rangle}. \tag{8.3.48}$$

Note that, had we not been able to use the fact that $(M_0^* - L_0^*)\Psi_0^* = 0$, we would have had to neglect a term

$$\frac{\langle \Psi_0^* | M_0 - L_0 | \delta S \rangle}{\langle \Psi_0^* | M | S \rangle}$$

to obtain (8.3.48). This term could be of the same order as (8.3.48) itself.

Usually the simplest quantity to take for $S_0(\mathbf{r}, \mathbf{\Omega}, E, t)$ is the initial, unperturbed flux

$\Psi_0(\mathbf{r}, \mathbf{\Omega}, E)$, the solution of $L_0\Psi_0 = M_0\Psi_0$. We have seen that the approximation associated with this choice is called *first-order perturbation theory.* A better approximation is to take S_0 to be $\Psi_0^{(\lambda)}(t)$, the fundamental solution of the static equation

$$L\Psi_0^{(\lambda)}(t) = \frac{1}{\lambda} M\Psi_0^{(\lambda)}(t), \tag{8.3.49}$$

where L and M are the *instantaneous* (i.e., time-dependent) operators of the system. Since some of the time-dependent changes in the system are reflected in the shape of $\Psi_0^{(\lambda)}(t)$, it is expected to provide a better approximation to the shape of $\Psi(\mathbf{r}, \mathbf{\Omega}, E, t)$ than will the unperturbed flux Ψ_0.

Because of (8.3.49), use of $\Psi_0^{(\lambda)}(t)$ in (8.3.41) yields a result identical with (7.6.21), which was derived for a constant perturbation in Chapter 7:

$$\rho(t) = \frac{\langle\Psi_0^*|M - (1/\lambda)M|\Psi_0^{(\lambda)}(t)\rangle}{\langle\Psi_0^*|M|\Psi_0^{(\lambda)}(t)\rangle} = 1 - \frac{1}{\lambda}, \tag{8.3.50}$$

where, since (8.3.49) is to be solved at successive instants of time, the eigenvalue λ is, in effect, time-dependent. This procedure for finding a time-dependent value of the reactivity is called the *adiabatic approximation.*

8.4 Methods for Solving the Transport Equation

Using eigenfunctions associated with the transport equation to analyze reactor problems is, of course, only a formal procedure. Although, as we have seen, certain general properties of the solutions can be deduced, finding numerical values of the directional flux density $\Psi(\mathbf{r}, \mathbf{\Omega}, E)$ requires the prior determination of the $\Psi_n^{(\lambda)}(\mathbf{r}, \mathbf{\Omega}, E)$ or $\Psi_n^{(\omega)}(\mathbf{r}, \mathbf{\Omega}, E)$, and this latter problem is often more difficult than the former one.

In the present section we shall sketch four approximate methods for attacking the transport equation: the spherical-harmonics method; a Fourier-transform approach; the discrete-ordinates technique, and the Monte Carlo method. The first of these schemes forms the theoretical basis for diffusion theory; the second justifies the few-group approximation and provides a practical set of equations for determining asymptotic spectra. The latter two schemes are the ones most relied upon today to analyze reactor problems in which an explicit accounting of transport-theory effects is important.

The discussion of these four methods will be extremely elementary. The references at the end of the chapter provide a much more thorough coverage. Moreover a number of methods for solving the transport equation will not be described at all. For example we shall not discuss the Weiner-Hopf technique or Case's singular-eigenfunction method. Such analytical schemes provide very useful numerical standards against which to check more approximate procedures. However the mathematics involved is beyond the scope of the present book, and application to many practical reactor problems (for example the determination of the neutron-capture rate in a nonuniformly depleted square cluster of cylindrical fuel elements) has not been found to be a tractable problem.

The Spherical-Harmonics Method

Spherical harmonics are a complete set of functions in the angular variables μ ($\equiv \cos\theta$) and φ that determine the direction of the unit vector $\mathbf{\Omega}$ (see Figure 8.1); they may be defined mathematically by

$$Y_n^m(\mathbf{\Omega}) = Y_n^m(\mu, \varphi) \equiv \left\{\frac{(2n+1)(n-m)!}{(n+m)!}\right\}^{1/2} P_n^m(\mu)\exp(im\varphi), \tag{8.4.1}$$

where n is a positive integer or zero, $-n \leq m \leq n$, and the $P_n^m(\mu)$ are the *associated Legendre polynomials* defined by

$$\begin{aligned} P_n^m(\mu) &\equiv (1-\mu^2)^{m/2}\frac{d^m}{d\mu^m}P_n(\mu), \\ P_n^{-m}(\mu) &\equiv (-1)^m\frac{(n-m)!}{(n+m)!}P_n^m(\mu). \end{aligned} \tag{8.4.2}$$

Here, in turn, the $P_n(\mu)$ are *Legendre polynomials* defined so that

$$P_0(\mu) = 1, \qquad P_1(\mu) = \mu, \tag{8.4.3}$$

and $P_n(\mu)$ for $n > 1$ may be found from the recursion relationship

$$(2n+1)\mu P_n(\mu) = (n+1)P_{n+1}(\mu) + nP_{n-1}(\mu). \tag{8.4.4}$$

The $Y_n^m(\mathbf{\Omega})$ obey the orthogonality relationship

$$\int_{\mathbf{\Omega}} \bar{Y}_r^s(\mathbf{\Omega})Y_n^m(\mathbf{\Omega})d\Omega = \int_{-1}^{1}\frac{d\mu}{2}\int_0^{2\pi}\frac{d\varphi}{2\pi}\bar{Y}_r^s(\mu,\varphi)Y_n^m(\mu,\varphi) = \delta_{rn}\delta_{sm}, \tag{8.4.5}$$

where $\bar{Y}_r^s$ is the complex conjugate of Y_r^s and δ_{rn} is the Kronecker delta.

Since the Y_n^m form a complete set, we may expand $\Psi(\mathbf{r}, \mathbf{\Omega}, E)$ as

$$\Psi(\mathbf{r}, \mathbf{\Omega}, E) = \sum_{n=0}^{\infty}\sum_{m=-n}^{n}\Psi_n^m(\mathbf{r}, E)Y_n^m(\mathbf{\Omega}), \tag{8.4.6}$$

where, applying (8.4.5), we can formally define the expansion coefficients by

$$\Psi_n^m(\mathbf{r}, E) = \int d\Omega\,\bar{Y}_n^m(\mathbf{\Omega})\Psi(\mathbf{r}, \mathbf{\Omega}, E) \tag{8.4.7}$$

and find them (approximately) by substituting (8.4.6), terminated at $n = N$, into the transport equation (8.2.14), multiplying by $\bar{Y}_r^s(\mathbf{\Omega})$ for all r and s consistent with $r \leq N$, and solving the resultant coupled equations.

The expansion (8.4.6) is suggested by the physics of the problem. Within material media typical values of $\Sigma_t(\mathbf{r}, E)$ are in the range 0.1 to 1.0 cm^{-1}. Thus mean free paths $1/\Sigma_t(\mathbf{r}, E)$ range from 1 to 10 cm, and, as a result, $\Psi(\mathbf{r}, \mathbf{\Omega}, E)$ is fairly isotropic. (Figure 4.4 shows some polar plots of what $N(\mathbf{r}, \mathbf{\Omega}, E) = \Psi(\mathbf{r}, \mathbf{\Omega}, E)/v(E)$ might look like within a typical reactor.) Because the angular dependence of $\Psi(\mathbf{r}, \mathbf{\Omega}, E)$ is often close to isotropic, it is often possible to represent $\Psi(\mathbf{r}, \mathbf{\Omega}, E)$ accurately with only a few spherical harmonics.

Spherical-harmonics expansions have the attractive characteristic that they are invariant with respect to axis orientation. Thus, while a choice of orientation may greatly facilitate finding a solution, it will not alter the numerical values of the approximation to Ψ. (Most discrete-ordinates approximations do not exhibit this characteristic.) In addition there is an appealing property associated with the expansion of any function $F(\mu)$ to order N in the Legendre polynomials (to which the spherical harmonics reduce in one dimension). That expansion is

$$F(\mu) \approx F_N(\mu) \equiv \sum_{n=0}^{N} f_n(2n + 1)P_n(\mu), \tag{8.4.8}$$

where $P_n(\mu)$ is the Legendre polynomial of order n and the numbers f_n are expansion coefficients which may be found by making use of the following orthogonality relationship for Legendre polynomials (obtained from (8.4.5) with $m = 0$):

$$\int_{-1}^{1} \frac{d\mu}{2}(2n + 1)P_r(\mu)P_n(\mu) = \delta_{rn}. \tag{8.4.9}$$

Thus

$$f_r \equiv \int_{-1}^{1} \frac{d\mu}{2} P_r(\mu)F(\mu) \qquad (r = 0, 1, 2, \ldots). \tag{8.4.10}$$

(Note that, in accord with custom, we have here expanded in $(2n + 1)P_n(\mu)$ and weighted it with $P_r(\mu)$. Application of (8.4.1) and (8.4.5) to an expansion such as (8.4.6) with $m = 0$ would be equivalent to expanding in $(2n + 1)^{1/2}\, P_n(\mu)$ and weighting with $(2r + 1)^{1/2}\, P_r(\mu)$. Equation (8.4.9) would be the same, but the f_r would differ.)

The appealing property we refer to in the expansion (8.4.8) is that the quantity

$$\int_{-1}^{1} [F(\mu) - F_N(\mu)]^2\, d\mu$$

is less than it would be if any polynomials up to order N other than Legendre polynomials were used. Thus Legendre polynomials are the "best" expansion functions for $F(\mu)$ in a least-squares sense.

In order to illustrate the spherical-harmonics approximation with a minimum of algebraic complication we shall consider in detail only the one-dimensional (slab-geometry) case in which $\Psi(\mathbf{r}, \mathbf{\Omega}, E)$ depends only on z and μ (μ being $\mathbf{\Omega}\cdot\mathbf{k}$, where $\mathbf{k}$ is the unit vector specifying the direction of the Z axis). For this simplified situation the transport equation (8.2.14) becomes

$$\mu \frac{\partial}{\partial z}\Psi(z, \mu, E) + \Sigma_t(z, E)\Psi(z, \mu, E) = \int_0^\infty dE' \int_{-1}^{1} \frac{d\mu'}{2} \sum_j f^j(E)\nu^j\Sigma_f^j(z, E')\Psi(z, \mu', E')$$
$$+ \int_0^\infty dE' \int_0^{2\pi} \frac{d\varphi'}{2\pi} \int_{-1}^{1} \frac{d\mu'}{2} \Sigma_s(z, E' \to E, \mu_0)\Psi(z, \mu', E'), \tag{8.4.11}$$

where Q has been dropped to avoid algebraic detail and the assumption that the differential scattering cross section depends only on $\mu_0 = \mathbf{\Omega}'\cdot\mathbf{\Omega}$ has been made. (Neutron-polarization effects invalidate this assumption, but they have been found to be small.)

We now expand

$$\Psi(z, \mu, E) = \sum_{n=0}^{\infty} (2n + 1)\Psi_n(z, E)P_n(\mu),$$
$$\Sigma_s(z, E' \to E, \mu_0) = \sum_{n=0}^{\infty} (2n + 1)\Sigma_{sn}(z, E' \to E)P_n(\mu_0), \tag{8.4.12}$$

where

$$\Sigma_{sn}(z, E' \to E) \equiv \int_{-1}^{1} \frac{d\mu_0}{2} \Sigma_s(z, E' \to E, \mu_0)P_n(\mu_0),$$

and by substituting into (8.4.11) obtain

$$\sum_{n=0}^{\infty}\left\{(2n + 1)\mu P_n(\mu)\frac{\partial}{\partial z}\Psi_n(z, E) + \Sigma_t(z, E)(2n + 1)P_n(\mu)\Psi_n(z, E)\right\}$$
$$= \int_0^\infty dE' \int_{-1}^{1} \frac{d\mu'}{2} \sum_j f^j(E)\nu^j\Sigma_f^j(z, E') \sum_{n=0}^{\infty} (2n + 1)P_n(\mu')\Psi_n(z, E')$$
$$+ \int_0^\infty dE' \int_0^{2\pi} \frac{d\varphi'}{2\pi} \int_{-1}^{1} \frac{d\mu'}{2}\left[\sum_{n=0}^{\infty} (2n + 1)\Sigma_{sn}(z, E' \to E)P_n(\mu_0)\right]$$
$$\times\left[\sum_{m=0}^{\infty} (2m + 1)P_m(\mu')\Psi_m(E')\right]. \tag{8.4.13}$$

To express the $P_n(\mu_0)$ in terms of μ', μ, φ', and φ we make use of the law of angular

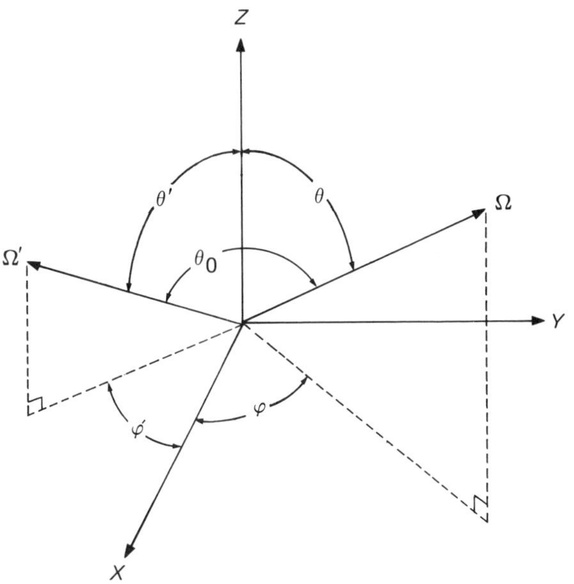

Figure 8.3 The angles involved in the addition law for Legendre polynomials.

addition of Legendre polynomials:

$$P_n(\mu_0) = P_n(\mu')P_n(\mu) + 2\sum_{m=1}^{n} \frac{(n-m)!}{(n+m)!} P_n^m(\mu')P_n^m(\mu) \cos m(\varphi' - \varphi), \tag{8.4.14}$$

where $\mu_0 = \cos\theta_0$, $\mu = \cos\theta$, and $\mu' = \cos\theta'$, the angles θ_0, θ, and θ' (as well as φ and φ') being shown in Figure 8.3 and the associated Legendre functions $P_n^m(\mu)$ being defined by (8.4.2).

Substituting this result into (8.4.13), making use of (8.4.4) to eliminate the terms $\mu P_n(\mu)$, and applying the orthogonality relationship (8.4.9) yields

$$\begin{aligned}
&\sum_{n=0}^{\infty}\left\{[(n+1)P_{n+1}(\mu) + nP_{n-1}(\mu)]\frac{\partial}{\partial z}\Psi_n(z,E) + (2n+1)\Sigma_t(z,E)P_n(\mu)\Psi_n(z,E)\right\}\\
&= \int_0^{\infty} dE' \sum_j f^j(E)\nu^j\Sigma_f^j(z,E')\Psi_0(z,E')\\
&\quad + \int_0^{\infty} dE' \int_{-1}^{1} \frac{d\mu'}{2}\left[\sum_n (2n+1)\Sigma_{sn}(z,E'\to E)P_n(\mu')P_n(\mu)\right]\left[\sum_m (2m+1)P_m(\mu')\Psi_m(E')\right]\\
&= \sum_{n=0}^{\infty} nP_n(\mu)\frac{\partial}{\partial z}\Psi_{n-1}(z,E) + (n+1)P_n(\mu)\frac{\partial}{\partial z}\Psi_{n+1}(z,E) + (2n+1)\Sigma_t(z,E)P_n(\mu)\Psi_n(z,E)\\
&= \int_0^{\infty} dE' \sum_j f^j(E)\nu^j\Sigma_f^j(z,E')\Psi_0(z,E') + \sum_{n=0}^{\infty}\int_0^{\infty} dE'(2n+1)P_n(\mu)\Sigma_{sn}(E'\to E)\Psi_n(E'),
\end{aligned} \tag{8.4.15}$$

where the sums involving $\partial\Psi_n/\partial z$ have been rewritten using

$$\sum_{n'=0}^{\infty}(n'+1)P_{n'+1}(\mu)\frac{\partial}{\partial z}\Psi_{n'}(z,E)=\sum_{n=1}^{\infty}nP_n(\mu)\frac{\partial}{\partial z}\Psi_{n-1}(z,E)=\sum_{n=0}^{\infty}nP_n(\mu)\frac{\partial}{\partial z}\Psi_{n-1}(z,E),$$

$$\sum_{n'=0}^{\infty}n'P_{n'-1}(\mu)\frac{\partial}{\partial z}\Psi_{n'}(z,E)=\sum_{n=-1}^{\infty}(n+1)P_n(\mu)\frac{\partial}{\partial z}\Psi_{n+1}(z,E)$$
$$=\sum_{n=0}^{\infty}(n+1)P_n(\mu)\frac{\partial}{\partial z}\Psi_{n+1}(z,E).$$

It then follows from the orthogonality of the $P_n(\mu)$ that

$$n\frac{\partial}{\partial z}\Psi_{n-1}(z,E)+(n+1)\frac{\partial}{\partial z}\Psi_{n+1}(z,E)+(2n+1)\Sigma_t(z,E)\Psi_n(z,E)$$
$$=\delta_{0n}\int_0^{\infty}dE'\sum_j f^j(E)\nu^j\Sigma_f^j(z,E')\Psi_0(z,E')+(2n+1)\int_0^{\infty}dE'\Sigma_{sn}(z,E'\to E)\Psi_n(z,E'). \tag{8.4.16}$$

This infinite set of coupled equations in z and E is entirely equivalent to (8.4.11) in z, E, and μ. The "P_N approximation" results when $\partial\Psi_{N+1}/\partial z$ is neglected and the first N equations are solved for $\Psi_0, \Psi_1, \ldots, \Psi_N$. Note that such solutions are only approximations to the actual expansion coefficients $\int_{-1}^{1}\frac{1}{2}d\mu P_n(\mu)\Psi(z,\mu,E)$ of $\Psi(z,\mu,E)$. If they were exact components, reactor theory would be exceptionally simple since Ψ_0 alone would be sufficient to give the power at all points in a reactor and a P_0 approximation would therefore suffice.

Boundary Conditions. At internal interfaces, since $\Psi(z,\mu,E)$ is to be continuous in the direction defined by μ, we want the $\Psi_n(z,E)$ to be continuous in z. At a surface next to a vacuum two classical sets of boundary conditions are applied to the $\Psi_n(z,E)$, one due to Marshak and the other to Mark.

Marshak boundary conditions direct that, for N odd and for a slab reactor extending from $z=0$ to $z=L$, we set

$$\left.\begin{aligned}\int_{-1}^{0}P_n(\mu)\Psi(L,\mu,E)\,d\mu&=0\\ \int_{0}^{1}P_n(\mu)\Psi(0,\mu,E)\,d\mu&=0\end{aligned}\right\}\qquad(n=1,3,5,\ldots,N),$$

where $\Psi(L,\mu,E)$ and $\Psi(0,\mu,E)$ are approximated by the Nth-order expansions of $\Psi(z,\mu,E)$ at $z=L$ and $z=0$, the coefficients $\Psi_n(z,E)$ in these expansions being unknowns in the P_N equations evaluated at $z=L$ and $z=0$. For N even (not a commonly used approximation) it is necessary to add some other conditions.

Marshak conditions are conveniently justified on the basis that the condition for $n=1$ is the P_N expression for zero returning net current.

Mark boundary conditions are obtained by setting $\Sigma_a > 0$ and $\Sigma_s = 0$ beyond the physical boundaries of the reactor. This procedure is equivalent to setting $\Psi(\mu_n) = 0$ for μ_n a particular choice of incoming ordinates in an equivalent discrete-ordinates scheme.

For cylinders and spheres Marshak boundary conditions can be extended naturally; Mark conditions cannot.

In the early literature the statement is frequently found that Marshak conditions are the most accurate for low-order expansions ($N = 1, 3$) but that Mark are better for higher-order expansions. The statement is probably based on the fact that Mark conditions are physically more appealing. Whatever the reason, the statement is wrong. Numerical studies show that (up to a P_{19} approximation) applying the Marshak conditions leads to consistently more accurate results.

The P_N equations obtained from (8.4.16) by neglecting $\partial\Psi_{N+1}/\partial z$ are a set of $N + 1$ coupled integrodifferential equations in position and energy. Finding the $N + 1$ unknown expansion coefficients $\Psi_n(z, E)$ ($n = 0, 1, 2, \ldots, N$) is thus still a formidable job. However it is one that can be and has been accomplished by approximating the spatial derivatives by finite-difference expressions and by applying a multigroup procedure to deal with the energy part of the problem. Thus 100-energy-group computer programs capable of solving up to, say, a P_7 approximation by spatial finite-difference techniques are available.

Unfortunately the extension of such schemes to two and three dimensions, for which all the $\Psi_n^m(\mathbf{r}, E)$ have to be determined, leads to such cumbersome equations that, except for the P_1 approximation, which we shall see to be the basis of diffusion theory, alternate methods for attacking such problems are preferred.

Fourier-Transform Techniques

The spherical-harmonics method is essentially an analytical procedure for determining the dependence on direction of the directional flux. One might expect to be able to attack the spatial part of the problem in an analogous way, that is by expanding the **r** dependence of $\Psi(\mathbf{r}, \mathbf{\Omega}, E)$ in a complete, predetermined set of functions of position such as a Fourier series. This procedure, however, has not been applied successfully except for the analysis of reactors that are extremely simple geometrically. The reason is that, for the analysis of most real reactors, a great many spatial expansion functions must be employed to provide an accurate approximation to the spatial shape of $\Psi(\mathbf{r}, \mathbf{\Omega}, E)$ (whereas a P_1 approximation, involving only four expansion functions $\Psi_n^m(\mathbf{r}, E)$, often adequately describes the dependence on $\mathbf{\Omega}$). Under such circumstances a finite-difference approximation has been found a more efficient method for obtaining a result of any desired accuracy.

There is an alternate analytical approach to determining the spatial dependence of $\Psi(\mathbf{r}, \mathbf{\Omega}, E)$ which, while it also appears to be less efficient than a finite-difference approach, is of considerable theoretical interest. Moreover it provides a theoretical justification for expecting (or, in some cases, for *not* expecting) that the spherical-harmonics components $\Psi_n^m(\mathbf{r}, E)$ of $\Psi(\mathbf{r}, \mathbf{\Omega}, E)$, given by (8.4.7), can be written as separable functions of space and energy asymptotically far inside a homogeneous material composition. (Such separability is the basis of the few-group approximation).

We shall refer to this scheme as the *Fourier-transform approach.* When it is applied only to determine the spatial behavior of $\Psi(\mathbf{r}, \mathbf{\Omega}, E)$ far inside a homogeneous medium, the scheme is frequently referred to as *asymptotic transport theory*. Some extension is required to describe the detailed solution for $\Psi(\mathbf{r}, \mathbf{\Omega}, E)$ near interfaces. One systematic scheme for accomplishing such an extension involving the use of singular integral equations is called *Case's method.* A description of Case's method is beyond the scope of this book. In fact our whole treatment of the Fourier-transform approach will be extremely qualitative, the prime motivation being to derive equations that can be solved for the asymptotic spectra associated with the spherical-harmonics components of $\Psi(\mathbf{r}, \mathbf{\Omega}, E)$.

The basic assumption of the Fourier-transform approach is that, within a given homogeneous material composition, a "particular" solution for $\Psi(\mathbf{r}, \mathbf{\Omega}, E)$ may be written

$$\Psi(\mathbf{r}, \mathbf{\Omega}, E) = F(\mathbf{\Omega}, E) \exp(-i\mathbf{B}\cdot\mathbf{r}). \tag{8.4.17}$$

Substitution into the transport equation (8.2.14) (with $Q = 0$, the Σ's independent of $\mathbf{r}$, and, for simplicity, only one fissionable isotope j considered) then yields an integral equation for $F(\mathbf{\Omega}, E)$:

$$-i\mathbf{\Omega}\cdot\mathbf{B}F(\mathbf{\Omega}, E) + \Sigma_t(E)F(\mathbf{\Omega}, E) = \int d\Omega' \int dE'[f(E)\nu\Sigma_f(E') + \Sigma_s(E' \to E, \mathbf{\Omega}'\cdot\mathbf{\Omega})]F(\mathbf{\Omega}', E'). \tag{8.4.18}$$

Equation (8.4.18) is homogeneous in $F(\mathbf{\Omega}, E)$, and we expect it to have a nontrivial solution only for particular values of the vector $\mathbf{B}$. Moreover, on physical grounds, since $F(\mathbf{\Omega}, E)$ is associated with an infinite homogeneous medium, we expect any eigenfunction $F(\mathbf{\Omega}, E)$ associated with a particular value of $\mathbf{B}$ to be essentially independent of the *direction* of $\mathbf{B}$, in the sense that solutions corresponding to an arbitrary direction for $\mathbf{B}$ can be found by transforming those found for a fixed direction. Thus we may select the direction of $\mathbf{B}$ as parallel to the Z axis and, designating μ as the cosine of the angle between $\mathbf{\Omega}$ and that axis, simplify (8.4.18) to

$$-i\mu BF(\mathbf{\Omega}, E) + \Sigma_t(E)F(\mathbf{\Omega}, E) = \int d\Omega' \int dE'[f(E)\nu\Sigma_f(E') + \Sigma_s(E' \to E, \mu_0)]F(\mathbf{\Omega}', E'), \tag{8.4.19}$$

where B is now a number (usually complex) that yields a nontrivial solution for $F(\mathbf{\Omega}, E)$. From symmetry considerations we expect that, if $+B$ yields a nontrivial solution, so will $-B$. We thus speak of solutions of (8.4.19) corresponding to values of B^2, where B^2 may be positive or negative depending on whether B is real or pure imaginary. For most materials—certain nonmultiplying materials provide an exception—there exist nontrivial solutions $F_{\pm B_m}(\mathbf{\Omega}, E)$ corresponding to a most positive (least negative) isolated value of B^2. We call this most positive eigenvalue of (8.4.19) the *materials buckling* and designate it as B_m^2. It is the same quantity (now specified more accurately as an eigenvalue of the asymptotic transport equation (8.4.19)) that we encountered in (4.10.3) in trying to determine the energy shape that the scalar flux assumes far in the interior of a given material. However, in the former case, the (so far) vaguely defined diffusion constant $D(E)$ of the material appeared in the eigenvalue equation (4.10.3). Equation (8.4.19) suffers from no such limitation. It is, however, a much more difficult equation to solve.

In order to arrive at (8.4.19) we chose the simplifying orientation $\mathbf{B} = \mathbf{k}B$, where $\mathbf{k}$ is a unit vector in the Z direction. As a result the only particular solutions (8.4.17) associated with B_m^2 are $F_{B_m}(\mathbf{\Omega}, E)\exp(-iB_m z)$ and $F_{-B_m}(\mathbf{\Omega}, E)\exp(iB_m z)$. Thus, if $B_m^2 > 0$, the solution associated with B_m^2 is expected to involve $\cos B_m z$ and $\sin B_m z$. If $B_m^2 < 0$, we let $L_m^2 \equiv -1/B_m^2$, so that the associated solutions will be combinations of $\exp(-z/L_m)$ and $\exp(z/L_m)$. Solutions of (8.4.19) corresponding to more negative values of $B^2 \equiv -1/L^2$ $(L < L_m)$ will involve more quickly decaying or rising exponential shapes. Moreover the scalar fluxes $\int d\Omega\, F_{\pm B_m}(\mathbf{\Omega}, E)$ associated with these "higher-order modes" turn out to be negative for some part of the energy range. Thus shapes associated with these more negative values of B^2 will die out in the interior of a homogeneous material medium. (If we choose this coordinate system so that $z = 0$ at the center plane of the medium and assume the medium extends from $-Z$ to $+Z$, a term in $\exp(+Z/L)$ will have to have a small magnitude near $z = 0$; otherwise it will be huge at $z = +Z$ and thus make the overall scalar flux unphysically negative for some part of the energy range. Similarly a term in $\exp(-z/L)$ will have to be small near $z = 0$; otherwise it will be huge at $z = -Z$.) As a result the only terms $F_{\pm B}(\mathbf{r}, E)\exp(\pm iBz)$ left in the interior of a large homogeneous material region will be those associated with the materials buckling B_m^2.

For the direction of $\mathbf{B}$ general, the simplification (8.4.19) is not possible, and we must therefore deal with (8.4.18). We expect that the same eigenvalues B^2 that yield a solution for (8.4.19) will yield one for (8.4.18). The corresponding functions $F_{\mathbf{B}}(\mathbf{\Omega}, E)$ will, however, depend specifically on the components B_x, B_y, and B_z of $\mathbf{B}$. They thus provide particular solutions of the form $F_{\mathbf{B}}(\mathbf{\Omega}, E)\exp(\pm iB_x x \pm iB_y y \pm iB_z z)$, where $B_x^2 + B_y^2 + B_z^2 = B^2$ and the notation $F_{\mathbf{B}}(\mathbf{\Omega}, E)$ is meant to indicate an explicit dependence on B_x, B_y, and B_z.

Again we expect that particular solutions corresponding to the materials buckling will be the only ones present far within a material medium. Thus, if we expose a subcritical amount of material to an external source of neutrons (for example by placing it next to an operating reactor), we expect that, within the material *near* the interface, the overall flux $\Psi(\mathbf{r}, \mathbf{\Omega}, E)$ will be made up of a number of solutions to (8.4.18). In general it will be necessary to include particular solutions corresponding to all the different values of B^2 (both discrete values and those that are part of a continuum), each particular solution behaving spatially in accord with its particular B_x, B_y, and B_z. However, far in the interior of the material, we expect the only $F_{\mathbf{B}}(\mathbf{\Omega}, E)$ surviving will be those associated with the particular solutions $\exp(\pm iB_{mx}x)\exp(\pm iB_{my}y)\exp(\pm iB_{mz}z)$, where $B_{mx}^2 + B_{my}^2 + B_{mz}^2 = B_m^2$, the most positive (least negative) value of B^2 for which a solution to (8.4.18) exists.

Spectra Associated with Fourier Components: P_L and B_L Approximations

If both discrete and continuous values of B^2 are admitted, it is thought that the particular solutions $F_{\mathbf{B}}(\mathbf{\Omega}, E)\exp(\pm i\mathbf{B}\cdot\mathbf{r})$ (written here as though the $\mathbf{B}$ were discrete) form a complete set, so that any arbitrary $\Psi(\mathbf{r}, \mathbf{\Omega}, E)$ can be expressed in terms of them. In principle, then, the solution for $\Psi(\mathbf{r}, \mathbf{\Omega}, E)$ throughout a reactor made up of a number of homogeneous material regions can be determined by selecting a combination of particular solutions for each different homogeneous material region and requiring that that combination obey the boundary and continuity conditions imposed on $\Psi(\mathbf{r}, \mathbf{\Omega}, E)$. (The particular solutions making up the combination in any given region are solutions of (8.4.18) for the Σ_t's, $\nu\Sigma_f$'s, etc., of that region.) Unfortunately it appears to be impractical to carry out this procedure for most cases of interest. It is simply too difficult to determine the $F_{\mathbf{B}}(\mathbf{\Omega}, E)$—particularly those associated with the continuum of eigenvalues.

The Fourier-transform approach is, nevertheless, very useful in that it leads us to expect that, far within most homogeneous materials, the spatial shape of the directional flux will be a combination of particular solutions $F_{\mathbf{B}_m}(\mathbf{\Omega}, E)\exp(\pm i\mathbf{B}_m\cdot\mathbf{r})$, where B_m^2 is the materials buckling of the medium. With only one B_m^2 involved, we can solve (8.4.18) for $F_{\mathbf{B}_m}(\mathbf{\Omega}, E)$ by a multigroup spherical-harmonics scheme and thereby obtain much useful information about the spatially asymptotic neutron population. To be specific, we can expand $F_{\mathbf{B}_m}(\mathbf{\Omega}, E)$ as

$$F_{\mathbf{B}_m}(\mathbf{\Omega}, E) = \sum_{n=0}^{\infty} \sum_{m=-n}^{n} F_n^m(E) Y_n^m(\mathbf{\Omega}) \tag{8.4.20}$$

and, by substituting this into (8.4.18), multiplying the result by $Y_r^p(\mathbf{\Omega})$, integrating over $\mathbf{\Omega}$, and applying the orthogonality condition (8.4.5), obtain coupled integral equations in the $F_r^p(E)$ which can be solved by the multigroup method.

If this procedure is carried out, it is found that the energy dependence of the $F_r^p(E)$ for a fixed value of r is the same for all p $(-r \leq p \leq r)$ and dependent only on B_m^2 rather than on the components of $\mathbf{B}_m$. Thus the ratio $F_r^p(E)/F_r^{p'}(E)$ for $p \neq p'$ is not dependent on energy (although it will usually depend on the numbers B_x, B_y, and B_z). It follows that, if we are only searching for the energy shapes of the spherical-harmonics components of $F_{\mathbf{B}_m}(\mathbf{\Omega}, E)$, we may choose the direction of $\mathbf{B}_m$ to suit our convenience and search for only one of the $2r + 1$ functions $F_r^p(E)$. Accordingly we make $\mathbf{B}_m$ parallel to the Z axis and try to determine only the $F_r^0(E)$ $(r = 0, 1, 2, \ldots)$. The first choice permits use of the simpler equation (8.4.19) and the second, according to (8.4.20) and (8.4.1), implies that we are looking for the components of $F_{\mathbf{B}_m}(\mathbf{\Omega}, E)$ that are independent of the angle φ, and hence that we are searching for the solutions $F_{\mathbf{B}_m}(\mu, E)$ of (8.4.19) that are independent of φ. Equation (8.4.19) thus becomes

$$-i\mu BF(\mu, E) + \Sigma_t(E)F(\mu, E) = \int d\Omega' \int dE'[f(E)\nu\Sigma_f(E') + \Sigma_s(E' \to E, \mu_0)]F(\mu', E'). \tag{8.4.21}$$

Since (8.4.21) is just a special case of the one-dimensional transport equation (8.4.11) with $\Psi(z, \mu, E) \to F(\mu, E) \exp(-iBz)$, we may immediately use the P_L solution found in connection with that approximation. Thus, with $F(\mu, E)$ expanded as

$$F(\mu, E) = \sum_{l=0}^{\infty} (2l + 1)F_l(E)P_l(\mu), \tag{8.4.22}$$

we get from (8.4.16), for each l,

$$\begin{aligned} &-iB[lF_{l-1}(E) + (l + 1)F_{l+1}(E)] + (2l + 1)\Sigma_t(E)F_l(E) \\ &= \delta_{0l}\int_0^{\infty} dE' f(E)\nu\Sigma_f(E')F_0(E') + (2l + 1)\int_0^{\infty} dE'\,\Sigma_{sl}(E' \to E)F_l(E'). \end{aligned} \tag{8.4.23}$$

Neglecting $F_{L+1}(E)$ in (8.4.23) with $l = L$ and neglecting all $l > L$ then gives a "P_L approximation" of the energy spectra belonging to the spherical-harmonics components of the directional flux asymptotically far inside a homogeneous medium. The $L + 1$ coupled, homogeneous, integral equations that result are solved by multigroup methods, with the number B^2 treated as an eigenvalue. (The solution corresponding to the most positive B^2 is the one here sought.)

There is another way to solve (8.4.21) for the components $F_l(E)$. Unlike the P_L approximation it cannot be applied to the general spatially dependent problem (8.4.13), but it *is* applicable to (8.4.21) and leads to results whose validity depends only on neglecting all $\Sigma_{sl}(E' \to E)$ for $l > L$ rather than neglecting $F_{L+1}(E)$.

The essence of the method is to avoid the use of (8.4.4), which is the equation that introduces the F_{l-1} and F_{l+1} terms on the left-hand side of (8.4.23). To avoid this coupling we simply divide (8.4.21) by $\Sigma_t(E) - iB\mu$ before making use of the expansion (8.4.22). The result is

$$\sum_{n=0}^{\infty}(2n+1)P_n(\mu)F_n(E) = \int_0^{\infty} dE' \frac{f(E)\nu\Sigma_f(E')}{\Sigma_t(E) - iB\mu} F_0(E') + \sum_{n=0}^{\infty}\int_0^{\infty} dE'(2n+1)\Sigma_{sn}(E' \to E)\frac{P_n(\mu)}{\Sigma_t(E) - iB\mu} F_n(E'), \tag{8.4.24}$$

where the manipulations on $\Sigma_s'(E' \to E, \mu)$ that led to the right-hand side of (8.4.15) have been repeated. Multiplication by $P_l(\mu)$ and integration over μ then yields

$$F_l(E) = \int_0^{\infty} dE'\left[f(E)\nu\Sigma_f(E')F_0(E')\int_{-1}^{1}\frac{d\mu}{2}\frac{P_l(\mu)}{\Sigma_t(E) - iB\mu} + \sum_{n=0}^{\infty}\Sigma_{sn}(E' \to E)F_n(E')\int_{-1}^{1}\frac{d\mu}{2}\frac{(2n+1)P_l(\mu)P_n(\mu)}{\Sigma_t(E) - iB\mu}\right] \quad (l = 0, 1, 2, \ldots). \tag{8.4.25}$$

We obtain a "B_L approximation" for the $F_l(E)$ by neglecting all the components $\Sigma_{sn}(E' \to E)$ of the differential scattering cross section for $n > L$ and solving the resulting $L + 1$ equations for the first $L + 1$ components $F_0(E), F_1(E), \ldots, F_L(E)$. The B_L approximation has the property of "finality," which means that, if the $\Sigma_{sn}(E' \to E)$ for $n > L$ may be assumed to vanish, and the components $F_l(E)$ $(l = 0, 1, \ldots, L)$ have been computed, computation of components for $l > L$ doesn't change the value of the F_l already found for $l \leq L$. Thus, insofar as neglect of the $\Sigma_{sn}(E' \to E)$ for $n > L$ is legitimate, the $F_l(E)$ are exact. The P_L approximation does not have this finality property. Even if the $\Sigma_{sn}(E' \to E)$ vanish for $n > L$, determining values of the higher components by including more equations of the form (8.4.23) alters the components found by a lower-order computation. For example $F_0(E)$ as determined by a P_3 approximation is different from $F_0(E)$ as determined by a P_5 approximation.

Application to a Bare, Homogeneous Slab Reactor

It has already been pointed out that application of the Fourier-transform and spherical-harmonics methods appears to be impractical for reactors having any geometrical complexity. When the geometry is simple, however, such schemes can provide much detailed information about the nature of the directional flux $\Psi(\mathbf{r}, \mathbf{\Omega}, E)$ throughout most of the reactor. As an illustration we shall consider one of the simplest of all examples: the bare, homogeneous, critical, slab reactor. Moreover we shall limit our analysis to finding the asymptotic flux in the interior of the slab and shall restrict the analysis of the angular part of the solution to the P_1 approximation. Then, if we take the Z axis to be perpendicular to the faces of the slab and choose $z = 0$ at the center plane so that the slab extends from $-Z$ to $+Z$ and is infinite in the X and Y directions, we know from symmetry and the requirement that $\Psi(\mathbf{r}, \mathbf{\Omega}, E)$ be nonnegative that the only acceptable, asymptotic, particular solutions $F(\mathbf{\Omega}, E)\exp(-i\mathbf{B}\cdot\mathbf{r})$ are such that $B_x = B_y = 0$ and $F(\mathbf{\Omega}, E) = F(\mu, E)$. Equation (8.4.22) (restricted to a P_1 expansion) then shows that the general solution asymptotically far inside the slab is to be constructed of the particular solutions $F_{0(B)}(E)\exp(-iBz) + 3\mu F_{1(B)}(E)\exp(-iBz)$ and $F_{0(-B)}(E)\exp(+iBz) + 3\mu F_{1(-B)}(E)\exp(+iBz)$, where B^2 must be the materials buckling and the P_0 and P_1 spectrum functions $F_{0(\pm B)}$ and $F_{1(\pm B)}$ are solutions of (8.4.23) for $l = 0$ and 1, $F_2(E)$ being neglected. That is, $F_{0(\pm B)}(E)$ and $F_{1(\pm B)}(E)$ obey

$$
\begin{aligned}
-iBF_{1(B)}(E) + \Sigma_t(E)F_{0(B)}(E) &= \int_0^\infty dE'[f(E)\nu\Sigma_f(E') + \Sigma_{s0}(E' \to E)]F_{0(B)}(E'), \\
-iBF_{0(B)}(E) + 3\Sigma_t(E)F_{1(B)}(E) &= 3\int_0^\infty dE'\,\Sigma_{s1}(E' \to E)F_{1(B)}(E').
\end{aligned}
\tag{8.4.26}
$$

Making use of the first of these equations to eliminate $F_{1(B)}(E)$ from the second and multiplying the result by iB yields

$$
\begin{aligned}
&B^2F_{0(B)}(E) + 3\Sigma_t(E)\left\{\Sigma_t(E)F_{0(B)}(E) - \int_0^\infty dE'[f(E)\nu\Sigma_f(E') + \Sigma_{s0}(E' \to E)]F_{0(B)}(E')\right\} \\
&= 3\int_0^\infty dE'\,\Sigma_{s1}(E' \to E)\left\{\Sigma_t(E')F_{0(B)}(E) - \int_0^\infty dE''[f(E')\nu\Sigma_f(E'') + \Sigma_{s0}(E'' \to E')]F_{0(B)}(E'')\right\}.
\end{aligned}
\tag{8.4.27}
$$

It follows that the eigenvalues of (8.4.26) are all particular values of B^2. Thus, if $F_{0(+B)}(E)$ is a solution of (8.4.26), $F_{0(-B)}(E)$ will also be a solution. Moreover, if $F_{0(\pm B)}(E)$ are solutions, their complex conjugates $\bar{F}_{0(\pm B)}(E)$ will also be solutions. We want to combine these particular solutions so that $\Psi(z, \mu, E)$ far inside the slab will be real, nonnegative, and such that the corresponding scalar flux is symmetric about $z = 0$. If B^2 is the materials buckling and is positive, we assume it is possible to form a combination having these desired characteristics whether the solution $F_{0(B)}(E)$, first found when (8.4.27) is solved, is real or complex, even or odd in B. However, in forming the desired combination, we will find it convenient to start with a function $F_0^B(E)$ that is real, nonnegative, and even in B. Accordingly, if the original $F_{0(B)}(E)$ does not have these characteristics, we can find constants a_i such that the two combinations $F_0^B(E)$ and $F_1^B(E)$, given by

$$
F_l^B(E) = a_1F_{l(B)}(E) + a_2F_{l(-B)}(E) + a_3\bar{F}_{l(B)}(E) + a_4\bar{F}_{l(-B)}(E) \qquad (l = 0, 1), \tag{8.4.28}
$$

are solutions of (8.4.26) and such that $F_0^B(E)$ is real and even in B. Then the first equation in (8.4.26) shows that, since $F_0^B(E)$ is real and even in B, $iBF_1^B(E)$ must also be real and even in B. Thus $F_1^B(E)$ must be purely imaginary and odd in B, that is,

$$F_1^B(E) = -F_1^{-B}(E). \tag{8.4.29}$$

It is convenient to exhibit these properties of $F_1^B(E)$ explicitly by defining

$$F_1^B(E) \equiv iBG_1^B(E), \tag{8.4.30}$$

where $G_1^B(E)$ is real and even in B. The particular solutions from which we wish to form the asymptotic solution inside the slab are, then, $[F_0(E) + 3iB\mu G_1(E)]\exp(-iBz)$ and $[F_0(E) - 3iB\mu G_1(E)]\exp(+iBz)$, where superscripts B have been dropped from F_0 and G_1 since they are both even in B.

The scalar flux $\Phi(z, E)$ in the slab, given by $\int_{-1}^{1} \Psi(z, \mu, E)d\mu/2$, must be a linear combination of $F_0(E)\exp(-iBz)$ and $F_0(E)\exp(+iBz)$. Since we want $\Phi(z, E)$ to be symmetric about $z = 0$, the proper combination is $F_0(E)[\exp(-iBz) + \exp(+iBz)]$. Simply adding the two particular solutions for $\Psi(z, \mu, E)$ will thus provide the final desired result. This is

$$\Psi(z, \mu, E) = 2F_0(E)\cos B_m z + 6\mu B_m G_1(E)\sin B_m z, \tag{8.4.31}$$

where we now indicate explicitly that the appropriate value of B^2 is the materials buckling. The corresponding scalar flux density is

$$\Phi(z, E) \equiv \int_{-1}^{1} \Psi(z, \mu, E)\frac{d\mu}{2} = 2F_0(E)\cos B_m z \tag{8.4.32}$$

and the corresponding net current density is

$$\begin{aligned}\mathbf{J}(z, E) &\equiv \int \mathbf{\Omega}\Psi(z, \mu, E)\,d\Omega \\ &= \int_{-1}^{1}\frac{d\mu}{2}\int_0^{2\pi} d\varphi[\mathbf{i}(1-\mu^2)^{1/2}\cos\varphi + \mathbf{j}(1-\mu^2)^{1/2}\sin\varphi + \mathbf{k}\mu] \\ &\quad\times[2F_0(E)\cos B_m z + 6\mu B_m G_1(E)\sin B_m z] \\ &= 2\mathbf{k}B_m G_1(E)\sin B_m z.\end{aligned} \tag{8.4.33}$$

Recall that we have assumed B_m^2 to be positive in selecting the form of (8.4.31). If the materials buckling *is* positive, it can be shown that $G_1(E)$ is nonnegative and hence that the net leakage is in the $+Z$ direction for $0 < z \leq \pi/2B_m$ and in the $-Z$ direction for $-\pi/2B_m \leq z < 0$. Thus, since the thickness of the critical slab cannot exceed π/B_m if $\Phi(z, E)$ is to be nonnegative, there is net leakage out of the slab. For $B_m^2 < 0$, it is easy to show that there must be net leakage *into* the slab to sustain the asymptotic neutron population and, therefore, that a bare slab reactor with $B_m^2 < 0$ cannot be critical.

If $B_m^2 > 0$, (8.4.32) shows that the scalar flux will extrapolate to zero at the surfaces

$z = \pm\pi/2B_m$. These surfaces occur at a distance $\delta \equiv (\pi/2B_m) - Z$ beyond the actual physical surfaces $z = \pm Z$ of the reactor. This gives us a much better definition of the extrapolation distance than the one we introduced as (4.6.12) while discussing the diffusion-theory boundary conditions for the slab reactor. There we found an extrapolation distance $\delta(E) = 0.67\lambda_{tr}(E)$ which was a function of energy. We determined $\delta(E)$ by extrapolating the shape of $\Psi_0(Z, E)$ to zero at the *edge* of the core. Our present view differs in two respects. First of all we are not restricted to the diffusion-theory approximation. Equation (8.4.32) is valid for any order P_L approximation, and we can thus determine B_m^2, and hence the surfaces at which $\Phi(z, E)$ will extrapolate to zero, to any desired degree of accuracy. Second, as with the few-group solution used in Chapter 4 to define reflector savings, the flux shape we are extrapolating is one present in the *interior* portions of the slab, *not* at the physical surface. (Near the physical surface we expect that solutions of (8.4.21) corresponding to other values of B^2 will be needed if we are to construct an overall solution $\Psi(z, \mu, E)$ which is zero for μ directed inward. Thus $\Phi(z, E) = \int d\Omega\, \Psi(z, \mu, E)$ near the surface will no longer have the simple cosine shape.)

Just as a high-order solution of (8.4.23) will yield an accurate value of B_m^2, so high-order solution of the space-dependent spherical-harmonics equations (8.4.16) (along with Marshak boundary conditions) will yield—at much greater cost—a quite good prediction of the actual physical boundaries of the slab. It is thus possible to determine δ quite accurately. When such computations are run, it is found that δ is a small number. Hence the surfaces on which the extrapolated scalar flux falls to zero will be very close to the actual physical surfaces of the homogeneous material making up the reactor. It follows that determining B^2 and the corresponding spherical-harmonics components of $F(\mathbf{\Omega}, E)$ and employing an *estimated* δ will give us an excellent value for the amount of any particular homogeneous reactor material (with $B_m^2 > 0$) needed to construct a critical reactor. It will also give us the spectra of the spherical-harmonics components of $F(\mathbf{\Omega}, E)$ expected in that material several mean free paths from the boundary. Finally these components will be the same (asymptotically) whether that material constitutes the whole or only part of a reactor. This latter fact, as we shall see, is of great help in determining few-group parameters.

Discrete-Ordinates Methods

In order to comprehend more clearly why the spherical-harmonics method becomes intractable for complex geometrical situations and, at the same time, to provide a transformation of the transport equation which will serve as a starting point for discussing discrete-ordinates methods, it will be convenient to expand the directional flux density $\Psi(\mathbf{r}, \mathbf{\Omega}', E')$ appearing in the fission and scattering-in integrals in terms of spherical harmonics. Accordingly we expand $\Psi(\mathbf{r}, \mathbf{\Omega}', E')$ as in (8.4.6) and $\Sigma_s(\mathbf{r}, E' \to E, \mu_0)$ as in

(8.4.12) and obtain from (8.2.14)

$$\mathbf{\Omega}\cdot\nabla\Psi(\mathbf{r}, \mathbf{\Omega}, E) + \Sigma_t(\mathbf{r}, E)\Psi(\mathbf{r}, \mathbf{\Omega}, E) = \int d\Omega' \int dE' \Bigg[\sum_j f^j(E)\nu^j\Sigma_f^j(\mathbf{r}, E') + \sum_{p=0}^{\infty} (2p+1)\Sigma_{sp}(\mathbf{r}, E' \to E)P_p(\mu_0)\Bigg] \sum_{n=0}^{\infty} \sum_{m=-n}^{n} \Psi_n^m(\mathbf{r}, E')Y_n^m(\mathbf{\Omega}'). \tag{8.4.34}$$

Just as in the analogous one-dimensional case given by (8.4.13)–(8.4.15), if we expand the $P_p(\mu_0)$ as in (8.4.14) and apply the orthogonality condition (8.4.7), this result simplifies to

$$\Omega_x \frac{\partial}{\partial x}\Psi(\mathbf{r}, \mathbf{\Omega}, E) + \Omega_y \frac{\partial}{\partial y}\Psi(\mathbf{r}, \mathbf{\Omega}, E) + \Omega_z \frac{\partial}{\partial z}\Psi(\mathbf{r}, \mathbf{\Omega}, E) + \Sigma_t(\mathbf{r}, E)\Psi(\mathbf{r}, \mathbf{\Omega}, E) = \int_0^{\infty} dE' \Bigg[\sum_j f^j(E)\nu^j\Sigma_f^j(\mathbf{\Omega}, E')\Psi_0^0(\mathbf{r}, E) + \sum_{n=0}^{\infty} \sum_{m=-n}^{n} \Sigma_{sn}(\mathbf{r}, E' \to E)\Psi_n^m(\mathbf{r}, E')Y_n^m(\mathbf{\Omega})\Bigg], \tag{8.4.35}$$

where we have expressed $\mathbf{\Omega}\cdot\nabla$ in terms of its components in the X, Y, and Z directions. (Note that, because of the difference between our normalization of the $Y_n^m(\mathbf{\Omega})$ (8.4.5) and of the $P_n(\mu)$ (8.4.9), the spherical-harmonics coefficients $\Psi_n^m(\mathbf{r}, E')$, if applied to the one-dimensional case, would be related to the $\Psi_n(z, E')$ of (8.4.15) by $\Psi_n^0(z, E') = (2n+1)^{1/2}\Psi_n(z, E')$.)

If we now expand $\Psi(\mathbf{r}, \mathbf{\Omega}, E)$ on the left-hand side of (8.4.35) in terms of spherical harmonics, express the $\Omega_x Y_n^m(\mathbf{\Omega})$, $\Omega_y Y_n^m(\mathbf{\Omega})$, and $\Omega_z Y_n^m(\mathbf{\Omega})$ in terms of the higher- and lower-order harmonics $Y_{n\pm 1}^{m\pm 1}$—the extension of (8.4.4) to the more general case—and apply the orthogonality condition (8.4.5), an infinite set of coupled integrodifferential equations, entirely equivalent to the transport equation, results:

$$\begin{aligned}
&\left[\frac{(n+m+1)(n-m+1)}{(2n+1)(n+3)}\right]^{1/2} \frac{\partial}{\partial z}\Psi_{n+1}^m(\mathbf{r}, E) + \left[\frac{(n-m)(n+m)}{(2n+1)(2n-1)}\right]^{1/2} \frac{\partial}{\partial z}\Psi_{n-1}^m(\mathbf{r}, E) \\
&+ \frac{1}{2}\left(\frac{\partial}{\partial x} - i\frac{\partial}{\partial y}\right)\Bigg\{\left[\frac{(n+m)(n+m-1)}{(2n+1)(2n-1)}\right]^{1/2} \Psi_{n-1}^{m-1}(\mathbf{r}, E) \\
&- \left[\frac{(n-m+2)(n-m+1)}{(2n+3)(2n+1)}\right]^{1/2} \Psi_{n+1}^{m-1}(\mathbf{r}, E)\Bigg\} \\
&+ \frac{1}{2}\left(\frac{\partial}{\partial x} + i\frac{\partial}{\partial y}\right)\Bigg\{\left[\frac{(n+m+1)(n+m+2)}{(2n+3)(2n+1)}\right]^{1/2} \Psi_{n+1}^{m+1}(\mathbf{r}, E) \\
&- \left[\frac{(n-m)(n-m-1)}{(2n+1)(2n-1)}\right]^{1/2} \Psi_{n-1}^{m+1}(\mathbf{r}, E)\Bigg\} \\
&+ \Sigma_t(\mathbf{r}, E)\Psi_n^m(\mathbf{r}, E) = \delta_{n0}\delta_{m0}\int dE' \sum_j f^j(E)\nu^j\Sigma_f^j(\mathbf{r}, E')\Psi_0^0(\mathbf{r}, E') \\
&+ \int dE'\,\Sigma_{sn}(\mathbf{r}, E' \to E)\Psi_n^m(\mathbf{r}, E') \qquad (n = 0, 1, 2, \ldots;\ m = -n, -n+1, \ldots, +n).
\end{aligned} \tag{8.4.36}$$

If we terminate this set by neglecting derivatives of Ψ_{N+1}^{m} and Ψ_{N+1}^{m+1}, the Nth-order spherical-harmonics approximation to the transport equation results.

It can be seen that, in general, each equation of the type (8.4.36) couples together seven of the expansion functions. It is this extensive coupling within each equation (which must still be differenced in space and energy in order to obtain a solution) that makes it impractical to apply the spherical-harmonics method to cases of even moderate geometrical complexity.

The discrete-ordinates approximation circumvents this difficulty by attempting only to find $\Psi(\mathbf{r}, \mathbf{\Omega}, E)$ for a discrete number of directions $\mathbf{\Omega}_d$ $(d = 1, 2, \ldots, D)$, where $\mathbf{\Omega}_d = \mathbf{i}\Omega_{xd} + \mathbf{j}\Omega_{yd} + \mathbf{k}\Omega_{zd}$. Thus the unknown function $\Psi(\mathbf{r}, \mathbf{\Omega}, E)$ is replaced by the D functions $\Psi(\mathbf{r}, \mathbf{\Omega}_d, E)$, and it is assumed that the left-hand side of (8.4.35) may be replaced by

$$\Omega_{xd}\frac{\partial}{\partial x}\Psi(\mathbf{r}, \mathbf{\Omega}_d, E) + \Omega_{yd}\frac{\partial}{\partial y}\Psi(\mathbf{r}, \mathbf{\Omega}_d, E) + \Omega_{zd}\frac{\partial}{\partial z}\Psi(\mathbf{r}, \mathbf{\Omega}_d, E) + \Sigma_t(\mathbf{r}, E)\Psi(\mathbf{r}, \mathbf{\Omega}_d, E).$$

On the right-hand side of (8.4.35) the coefficients $\Psi_n^m(\mathbf{r}, E')$, defined by (8.4.7) as $\int d\Omega\, \bar{Y}_n^m(\mathbf{\Omega})\Psi(\mathbf{r}, \mathbf{\Omega}, E')$, are replaced by sums $\sum_{d=1}^{D} w_d \bar{Y}_n^m(\mathbf{\Omega}_d)\Psi(\mathbf{r}, \mathbf{\Omega}_d, E')$, where a set of weighting constants w_d $(d = 1, 2, \ldots, D)$ has been chosen so that the sums match the integrals "as closely as possible." The final result of these manipulations is to replace (8.4.35) by the D equations

$$\begin{aligned}
&\Omega_{xd}\frac{\partial}{\partial x}\Psi(\mathbf{r}, \mathbf{\Omega}_d, E) + \Omega_{yd}\frac{\partial}{\partial y}\Psi(\mathbf{r}, \mathbf{\Omega}_d, E) + \Omega_{zd}\frac{\partial}{\partial z}\Psi(\mathbf{r}, \mathbf{\Omega}_d, E) + \Sigma_t(\mathbf{r}, E)\Psi(\mathbf{r}, \mathbf{\Omega}_d, E)\\
&= \int_0^\infty dE'\Bigg\{\sum_j f^j(E)\nu^j\Sigma_f^j(\mathbf{r}, E')\sum_{d'=1}^{D} w_{d'}\Psi(\mathbf{r}, \mathbf{\Omega}_{d'}, E') \qquad (8.4.37)\\
&+ \sum_{n=0}^{\infty}\sum_{m=-n}^{n}\Sigma_{sn}(\mathbf{r}, E' \to E)\left[\sum_{d'=1}^{D} w_{d'}\bar{Y}_n^m(\mathbf{\Omega}_{d'})\Psi(\mathbf{r}, \mathbf{\Omega}_{d'}, E')\right]Y_n^m(\mathbf{\Omega}_d)\Bigg\} \qquad (d = 1, 2, \ldots, D).
\end{aligned}$$

These discrete-ordinates equations, with a particular choice of w_d, are often called S_n *equations*. They couple the directional flux in a given direction $\mathbf{\Omega}_d$ to those in all the other chosen directions. The mathematical form of the equations is, however, far better suited than (8.4.36) to solution by iterative procedures. Values of $\Psi(\mathbf{r}, \mathbf{\Omega}_d, E)$ for $\mathbf{r}$ on the outer surface and $\mathbf{\Omega}_d$ directed inward are set to zero at all times, and, on the pth iterate of the calculation, the source terms on the right-hand side of (8.4.37) are computed from the $\Psi(\mathbf{r}, \mathbf{\Omega}_{d'}, E')$ found during the $(p - 1)$th iterate. The left-hand side of any one of the coupled equations involves only the function $\Psi(\mathbf{r}, \mathbf{\Omega}_{d'}, E)$ associated with *one* of the discrete ordinates. As a result, within a given energy group, the directional flux densities for the different $\mathbf{\Omega}_d$ can be found one at a time (by replacing the spatial derivatives with finite-difference expressions). The various $\Psi(\mathbf{r}, \mathbf{\Omega}_d, E)$ are coupled together only when the fission and scattering sources are found for the next iteration.

In order to demonstrate this point more clearly we rewrite (8.4.37) for the special case of two dimensions ($\partial\Psi/\partial z = 0$), isotropic scattering ($\Sigma_{sn}(\mathbf{r}, E' \to E) = 0$ for $n > 0$), and only four discrete ordinates, one for each quadrant. If we select the directions of the four ordinates as ($\Omega_{x1} = 1/\sqrt{3}$; $\Omega_{y1} = 1/\sqrt{3}$), ($\Omega_{x2} = -1/\sqrt{3}$; $\Omega_{y2} = 1/\sqrt{3}$), ($\Omega_{x3} = -1/\sqrt{3}$; $\Omega_{y3} = -1/\sqrt{3}$), and ($\Omega_{x4} = 1/\sqrt{3}$; $\Omega_{y4} = -1/\sqrt{3}$) and take the weighting constants w_d all equal to 1/4, (8.4.37) becomes

$$\Omega_{xd}\frac{\partial}{\partial x}\Psi(x, y, \mathbf{\Omega}_d, E) + \Omega_{yd}\frac{\partial}{\partial y}\Psi(x, y, \mathbf{\Omega}_d, E) + \Sigma_t(x, y, E)\Psi(x, y, \mathbf{\Omega}_d, E)$$
$$= \int_0^\infty dE'\left[\sum_j f^j(E)\nu^j\Sigma_f^j(x, y, E') + \Sigma_{s0}(x, y, E' \to E)\sum_{d'=1}^{4} \tfrac{1}{4}\Psi(x, y, \mathbf{\Omega}_{d'}, E')\right]$$
$$(d = 1, 2, 3, 4). \qquad (8.4.38)$$

The solution of these equations (accomplished by splitting the energy into groups and approximating the derivatives by finite-difference expressions) is a fairly lengthy, but quite tractable, procedure.

Discrete-ordinates methods are extremely useful for the analysis of reactor problems for which diffusion theory is an inadequate approximation. They do, however, suffer from one defect that can be serious in certain types of problems. The defect referred to, called the *ray effect*, arises because neutrons that can move only in certain discrete directions will, of necessity, have certain portions of the reactor that they will not be able to reach by direct flight. An extreme example of this difficulty occurs if neutrons are born isotropically at a point in a pure absorber. Under these circumstances the directional fluxes $\Psi(x, y, \mathbf{\Omega}_d, E)$ will be finite only for points (x, y) lying on rays emanating from the point source in the directions $\mathbf{\Omega}_d$. The scalar flux $\Phi(\mathbf{r}, E) = \sum_{d=1}^{D} w_d\Psi(\mathbf{r}, \mathbf{\Omega}_d, E)$ will thus be zero over large, extended regions of space. Even for situations where the scattering is *not* zero, so that all regions of space are accessible, the scalar flux tends to be anomolously low in regions far from the trajectories of the uncollided source neutrons.

The situation persists for extended sources. As an example consider the square isotropic source shown in Figure 8.4. For the low-order, four-ordinate approximation represented by (8.4.38), only points lying in the areas (*acfe*) and (*bhgd*) "see" unscattered neutrons emanating from the source region (*abcd*). In particular the triangular area (*gif*) will be inaccessible to such neutrons. Because of these ray effects, a plot of the scalar flux will exhibit an unphysical scalloped character far from the source.

In most cases ray effects can be avoided by simply employing more ordinates. There are, however, situations, in which a surprisingly large number of ordinates must be used to avoid the effect. There do exist mathematical refinements of the theory through which the whole problem can be avoided, but these are somewhat costly in terms of computational effort.

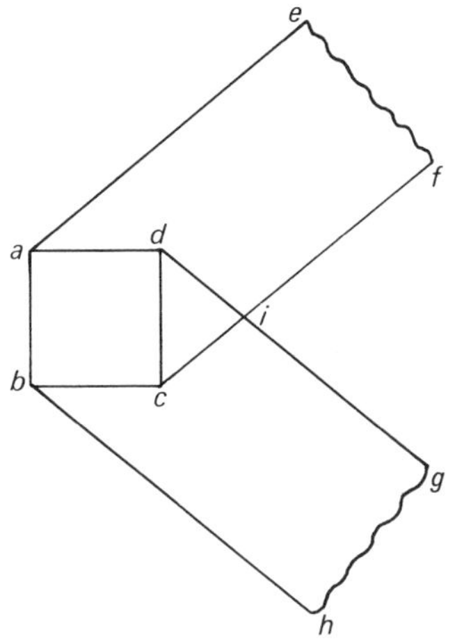

Figure 8.4 Ray effects from a square source.

This brief description of discrete-ordinates methods has been intended only to provide a qualitative notion of the subject. We have omitted any discussion of systematic procedures for determining either the $\boldsymbol{\Omega}_d$ or the weighting constants w_d of (8.4.37). Moreover we have touched on neither the numerical problems that arise in forming the spatial difference equations from (8.4.37) nor the iterative strategies applied to solve them. The extended literature in these areas is beyond the scope of this book. It exists, however, and should be consulted before any attempt to apply discrete-ordinates techniques to a particular problem is made.

The Monte Carlo Method: General Characteristics

As the name suggests, the Monte Carlo method (in its most basic form) is a statistical procedure wherein the expected characteristics of a neutron population in an assembly are estimated by drawing samples from a large number of "case histories" of neutrons whose individual "lives" are simulated by a computer. The method is best suited for the analysis of geometrically complex assemblies for which analytical approaches are virtually impossible and numerical schemes are extremely time-consuming. On the other hand, because of its inherently statistical nature, it is poorly suited for predicting accurate values of such quantities as $\Psi(\mathbf{r}, \boldsymbol{\Omega}, E)$ over extended ranges of the variables $\mathbf{r}$, $\boldsymbol{\Omega}$, and E.

The steps required to carry out the "analogue" Monte Carlo method are conceptually simple. However the details of how these procedures are implemented and of how, in order to increase computational efficiency, direct analogue models are replaced by fictitious models (or even models for which there is no direct physical analogue) can become quite involved. We shall not treat the subject in any such depth, nor shall we even touch upon the sophisticated measure theory that provides the underlying mathematical support for the Monte Carlo method. Even with these omissions, however, it will be possible to form a fairly accurate qualitative picture of the general approach.

As just stated, expected characteristics of the neutron population in an assembly are predicted in the analogue method by averaging over individual neutron "case histories." The procedure is to determine the individual case histories and then to perform the appropriate averages.

For our purposes an "event" is something that happens to a neutron. (It appears at a certain location with a certain energy and direction of travel, interacts at a particular location, loses a certain amount of energy, causes fission, crosses a certain boundary, is absorbed, etc.) A "case history" is, then, a sequence of events beginning and ending according to some preselected criteria. For example, if we are interested in the rate at which neutrons introduced uniformly and isotropically with energy E in an infinite cylinder of some nonfissioning material will escape into a vacuum surrounding that material, all case histories should begin with the appearance of one of these source neutrons and end either with that neutron being absorbed or with it crossing the outer boundary of the cylinder. It seems intuitively clear that the escape probability we seek can be found by determining the fraction of all case histories that are terminated by the neutron's crossing the boundary of the cylinder.

Determining the Outcome of Particular Events

The particular events that make up an individual case history are determined from the laws of probability that govern the event under consideration through application of a table of "random numbers." Random numbers consist of a sequence of positive real numbers ρ $(0 \leq \rho \leq 1)$ such that the nth number in the sequence is "equally likely" to be any number in the range. (We shall not refine further the intuitive notion of "equally likely." Such a refinement, as well as the methods of generating and testing the "randomness" of random numbers, are nontrivial mathematical problems.) We use the numbers from the random-number table to decide what particular event, out of all the possible events, actually happens next as the case history develops. For example, if we know that a neutron having energy E has interacted at point $\mathbf{r}$ and want to decide what particular interaction to select to continue the case history, we split the interval 0–1 into fractions $\Sigma_c(\mathbf{r}, E)/\Sigma_t(\mathbf{r}, E)$, $\Sigma_s^j(\mathbf{r}, E)/\Sigma_t(\mathbf{r}, E)$, $\Sigma_f^j(\mathbf{r}, E)/\Sigma_t(\mathbf{r}, E)$, etc., in accord with the known nuclear data and number densities $n^j(\mathbf{r})$. Then, if these fractions are, for example, $\Sigma_c/\Sigma_t = .2$, $\Sigma_s/\Sigma_t = .3$, and $\Sigma_f/\Sigma_t = .5$, there being only one isotope present at $\mathbf{r}$, we pick the next, unused random number from the table and, if that number is between 0 and .2, take the capture reaction as the next event. If the random number is between .2 and .5, we take the scattering interaction as the next event; and, if it is between .5 and 1, we take the fission interaction.

Now, if the interaction decided by the value of the random number is scattering and if in the case being analyzed all scattering is isotropic in the center-of-mass system, we know that the energy of the neutron after the interaction can take on any value in the range αE–E with equal probability (see (2.5.13), (2.5.17), and (2.5.19)). Thus, if ρ is the next random number on the list, we take the energy of the emergent neutron to be $[E - \rho(E - \alpha E)]$.

Not all of the variables to be selected by the random-number process are distributed with uniform probability throughout their range. For example, if a neutron is traveling with a given energy in an infinite homogeneous medium, the probability of its undergoing its first interaction at a location near its point of origin is much larger than that of its not interacting until it is far from that point. Thus we cannot determine the location of the interaction point directly by the selection of a random number. Instead we must find some other statistical property of the neutron traveling in the infinite medium that *is* distributed randomly and try to infer a probable interaction point from the randomly selected value of *that* property.

To illustrate we consider the example just cited. The probability that the neutron will travel a distance x in an infinite homogeneous medium without interacting is $\exp(-\Sigma_t x)$, where Σ_t is the total macroscopic interaction cross section for neutrons of energy E (see (2.3.5)). The probability it will travel a distance $x + \Delta x$ without interacting is $\exp(-\Sigma_t(x + \Delta x))$. Thus the probability that it *will* interact in the interval Δx is $\exp(-\Sigma_t x)[1 - \exp(-\Sigma_t \Delta x)]$, which for small Δx becomes $\Sigma_t \Delta x \exp(-\Sigma_t \Delta x)$. The function $f(x) \equiv \Sigma_t \exp(-\Sigma_t x)$ is called the *probability distribution function* for neutron interaction. Note that

$$\int_0^\infty f(x)dx = \int_0^\infty \Sigma_t \exp(-\Sigma_t x)\, dx = 1. \tag{8.4.39}$$

If we sample from a large number N_0 of independent observations of where a neutron, born at $x = 0$, first interacts, we shall find that there are $N_0 \Sigma_t \exp(-\Sigma_t x)\, \Delta x$ interactions, on the average, between x and $x + \Delta x$.

Another probability distribution function of great importance is $F(x)$, the *cumulative distribution function*, defined in general by

$$F(x) \equiv \int_0^x f(x')dx', \tag{8.4.40}$$

so that, in the particular case at hand,

$$F(x) = 1 - \exp(-\Sigma_t x). \tag{8.4.41}$$

As x varies over its accessible range, $F(x)$ increases monotonically from 0 to 1. Thus, corresponding to every observed interaction location x_i between $x = 0$ and $x = \infty$, there is a number $F_i(\equiv F(x_i))$ between 0 and 1. Sampling the distances x_i will then lead to a probability distribution of the numbers F_i. Conversely, sampling the F_i (if that is experimentally possible) will lead to the probability distribution $f(x)$ of the x_i. This last observation permits us to use random numbers to predict $f(x)$, for we shall now show that, in general, assuming the x_i are distributed according to $f(x)$ implies that the F_i are distributed randomly.

To prove this assertion we define $P(F)dF$ as the probability that a given value of F will be between F and $F + dF$. Then, since there is a one-to-one correspondence between values of F and values of x, we have, in general,

$$P(F)dF = f(x)dx. \tag{8.4.42}$$

But, from the definition (8.4.40),

$$dF = \frac{dF}{dx}dx = f(x)\,dx. \tag{8.4.43}$$

Thus $P(F) = 1$; that is, the probability distribution of the numbers F is uniform. Accordingly we can use a random-number table to select a value of F for a given event and determine the corresponding value of x from (8.4.41). The result is

$$x = -\frac{1}{\Sigma_t}\ln(1 - F). \tag{8.4.44}$$

This formula permits us, for any given neutron history, to pick the point x at which the neutron, born at $x = 0$ in an infinite homogeneous medium, next interacts. Note that the random distribution of the values of the cumulative distribution function $F(x)$ is a general result depending only on the definitions of $F(x)$ and $f(x)$. To take advantage of the relationship for any particular situation requires that an expression such as (8.4.44) explicitly relating x to F be found.

Not all of the choices made in the course of following a case history are based directly on physical probability laws. For example, if we wish to decide at what radial location in an infinite cylindrical rod a particular neutron belonging to a uniformly distributed source first appears, we can use a geometrical method: If the radius of the cylinder is R, we select two random numbers ρ_1 and ρ_2 and use them to specify the coordinates of a point (x, y) relative to an axis system with origin on the center line of the cylinder (and the XY plane perpendicular to the axis of the cylinder). The relationship between (x, y) and (ρ_1, ρ_2) is taken to be

$$\begin{aligned} x &= 2(\rho_1 - 0.5)R, \\ y &= 2(\rho_2 - 0.5)R, \end{aligned} \tag{8.4.45}$$

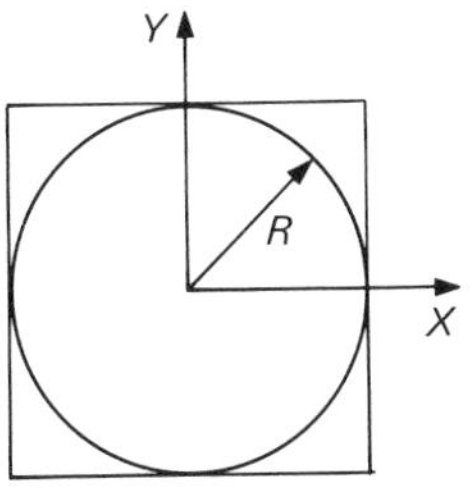

Figure 8.5 A rejection scheme for finding points randomly distributed in a circle.

so that, as seen in Figure 8.5, with $0 \le (\rho_1, \rho_2) \le 1$, the point (x, y) lies randomly in the square $(-R \le x \le R;\ -R \le y \le R)$. To find points randomly distributed in the circle we simply reject all points outside the circle. Thus, if

$$x^2 + y^2 \le R^2, \tag{8.4.46}$$

we retain the point (x, y). If not, we reject it and select another pair of random numbers (ρ_1, ρ_2) until we find a pair such that the coordinates of the corresponding point (x, y) obey (8.4.46). It is clear that the retained pairs of coordinates (x, y) represent just what we want: points randomly located in the circle of radius R. Methods of this type are called *rejection techniques*. They will clearly be most useful when the fraction of samples that have to be rejected is small.

In practice determining the particular events that make up a case history may not be as simple as it has been in the examples given, since, for example, many probability functions are far more complicated than the one illustrated by (8.4.41). Thus it may often not be possible to write the analogue of (8.4.44) as a simple formula. So called *table look-up procedures* can then be used, in which tables of ρ versus x are constructed and the x corresponding to a particular random number ρ is found by interpolation. It may also be a very time-consuming job to determine the various $\Sigma_\alpha(\mathbf{r}, E)$ (α = t, s, f, c, etc.) needed at the different stages of a given history (for example if materials having Doppler-broadened resonance cross sections are involved). Table look-up procedures (as well as many more involved schemes) are also useful for this problem. Finally, when the problem being analyzed involves a complicated geometrical arrangement, it requires considerable ingenuity just to keep track of where in space the neutron is at any given point in its history.

These difficulties have largely been overcome, and computer programs capable of processing tens of thousands of case histories under extremely complicated geometrical conditions operate with acceptable running times. It is even possible to carry out full-core criticality calculations by the Monte Carlo method. The scheme thus provides a practical theoretical standard against which to check more economical procedures.

Statistical Analysis

Constructing a set of case histories constitutes only the first half of a Monte Carlo analysis. To obtain the desired information from such a set a statistical analysis must be performed. We shall not present any of the details of that analysis but shall merely introduce the definitions of a few frequently used terms and state the *central-limit theorem*, which provides a way of estimating the expected accuracy of any given Monte Carlo analysis.

To make matters quantitative we associate a real number ξ, called a *random variable*, with every possible outcome of a case history. Suppose, for example, that we are trying to find the fraction of those neutrons having a certain known initial distribution that are absorbed in U^{28} in a certain region of the core before they are absorbed either in other materials or in other regions or slow down past a certain energy. We first create case histories for neutrons selected from the given initial distribution and follow these histories until the neutrons involved either are absorbed or slow down past the energy cut point. The outcome of each case history is, then, either that the neutron is absorbed in the U^{28} in the energy range and region of interest (a "success"), or it is not (a "failure"). To describe this situation quantitatively we let the random variable ξ_i be one if the outcome of the case history is a success and zero if the outcome is a failure. (Note that a random variable has the nature of a *function*. It should not be confused with a random *number*.)

Suppose again that we are interested in the average energy of neutrons escaping from some portion of a shield. We select from all case histories a subset every member of which terminates with the escape of a neutron from the portion of interest, and we let the random variable ξ_i be the energy of the escaping neutron for the ith member of this subset.

In both examples we are interested in the "expected" or "mean" value $\langle\xi\rangle$ of the random variable ξ, and we estimate that expected value as the average $\bar{\xi}_N$ over the N case histories in the sample. That is, we determine

$$\bar{\xi}_N = \frac{1}{N}\sum_{i=1}^{N} \xi_i. \tag{8.4.47}$$

We expect $\bar{\xi}_N$ to approach $\langle\xi\rangle$ as the number of case histories increases, and we would like to have some estimate of how close $\bar{\xi}_N$ is to $\langle\xi\rangle$ for a given value of N. The central-limit theorem provides such an estimate. It states that, as N becomes very large, the probability that $|\bar{\xi}_N - \langle\xi\rangle|$ will be less than a number ε is given by

$$P\{|\bar{\xi}_N - \langle\xi\rangle| < \varepsilon\} \xrightarrow[N\to\infty]{} \left(\frac{2}{\pi}\right)^{1/2} \int_0^{(\varepsilon\sqrt{N})/\sigma} \exp\left(-\frac{t^2}{2}\right) dt, \tag{8.4.48}$$

where

$$\sigma^2 \equiv \langle \xi^2 \rangle - [\langle \xi \rangle]^2, \tag{8.4.49}$$

$\langle \xi^2 \rangle$ being the expected value of

$$\xi^2 = \lim_{N \to \infty} \frac{1}{N} \sum_{i=1}^{N} \xi_i^2.$$

Since $\int_0^\infty \exp(-t^2/2)dt = \sqrt{(\pi/2)}$, we see that (provided $\langle \xi \rangle$ and $\langle \xi^2 \rangle$ exist) the probability that $|\bar{\xi}_N - \langle \xi \rangle|$ will be less than ε becomes unity for fixed ε as $N \to \infty$. Moreover, if we have, for a particular value of N, an ε such that $P\{|\bar{\xi}_N - \langle \xi \rangle| < \varepsilon\}$ is, say, 0.99 and we wish to cut this value of ε in half, we must process *four* times as many cases so that $\varepsilon\sqrt{N}$, and thus the value of the right-hand side of (8.4.48), will remain constant.

The quantity σ^2 is called the *variance* of the random variable ξ, and σ is called the *standard deviation*. Equation (8.4.48) shows that it is desirable to have the variance of ξ as small as possible if $\bar{\xi}_N$ is to be close to $\langle \xi \rangle$.

Since use of a finite number of case histories does not yield exact values of $\langle \xi^2 \rangle$ and $\langle \xi \rangle$, the variance can only be estimated in carrying out any particular statistical analysis. Thus application of (8.4.48) is itself an approximation. This uncertainty is not, however, thought to be a serious limitation to the formula.

We speak of the right-hand side of (8.4.48) as the *confidence level* of the error estimate. Thus Monte Carlo results are said to produce an estimate of $\langle \xi \rangle$ with a particular "confidence" (usually expressed in percent) that the error is $\pm\varepsilon$.

The Monte Carlo method is by no means restricted exclusively to solving problems that have a direct analogue in the physical world. Indeed the method can be applied directly to the solution of a partial-differential equation. Moreover, even when analogue models can be constructed, it may not always be economical to use them. Attempting to determine how many neutrons escape through a shield by tracking a large number of individual neutrons until they either escape or are captured is not, for example, an efficient way to solve this problem. If for this case we let the random variable ξ_i be one for a case history ending in escape and zero otherwise, $\langle \xi \rangle$, the average value of the ξ_i, will be an extremely small number and $\langle \xi^2 \rangle$ will differ significantly from $\langle \xi \rangle^2$. Specifically, since each ξ_i is either one or zero, $\sum_i \xi_i = \sum_i \xi_i^2$, and we expect that $\langle \xi \rangle$ and $\langle \xi^2 \rangle$ will be equal. Thus, if $\langle \xi \rangle$ is, say, 10^{-8}, σ will be $\sim 10^{-4}$, and, in order for the fractional error $(\bar{\xi}_N - \langle \xi \rangle)/\langle \xi \rangle$ to be, say, 10 percent with 90 percent confidence, ε must be 10^{-9} and N must be such that

$$\left(\frac{2}{\pi}\right)^{1/2} \int_0^{10^{-5}\sqrt{N}} \exp\left(\frac{-t^2}{2}\right) dt = 0.9.$$

The required value of N is 2.7×10^{10} case histories. Hence, even if a history could be followed in 0.01 seconds on a computer, 7.5×10^4 hours of running time would be required to obtain the desired result. Clearly it is not practical to solve this problem by a straightforward analogue Monte Carlo technique.

A number of schemes have been devised for circumventing difficulties of this nature. One class of such schemes is based on interfering with the random case histories in such a way that the variance is reduced. For example, in the shield problem, if at the very beginning of a case history a neutron is randomly selected that happens to be headed toward the center of the core rather than toward the reflector, the probability that that neutron will be absorbed is artificially increased to a high value. Thus $\Sigma_a(E)$ may be increased from its physical value, which might be, say, 0.01, to, say, 0.90. The "weight" of those few particles that still avoid capture is then increased in such a way that the statistical game remains "unbiased." Case histories that are likely to terminate in neutron absorption will end much more quickly under these circumstances, and the time on the computer to create a large number of such histories will thereby be reduced.

This procedure is called *Russian roulette*. A complementary technique called *splitting* is used to enhance the number of "successful" histories. Thus, in the shielding example, if a particle, during the course of its history, is headed in a "favorable" direction (toward the boundary of the reactor), it is replaced by n particles each having a weight $1/n$, and independent case histories of these n particles are followed thereafter.

Both of these examples may be viewed as special cases of a technique called *importance sampling*, the essential idea of which is to distort the basic probabilities that determine the individual case histories in such a way that the occurrence of "favorable" events is enhanced. Great care must be taken to weight the "particles" involved in these nonanalogue histories so that the statistical properties of the case histories are not distorted. In principle importance sampling can reduce the variance to zero, so that following one case history will provide an accurate answer to the problem posed. However it is necessary to know a kind of adjoint function similar to $\Psi^*(\mathbf{r}, \mathbf{\Omega}, E)$ in order to do this. Thus, in practice, while (with a considerable increase in coding effort) importance sampling permits the analysis of many problems that would otherwise be intractable, it still requires that a significant number of case histories be processed.

Other schemes for improving on the efficiency of analogue Monte Carlo methods include *superposition methods*, in which a simplified part of the problem is solved analytically and corrections to this solution are determined by Monte Carlo procedures, and *reciprocity methods*, which involve a Monte Carlo solution of a type of adjoint equation.

A Generalized Reciprocity Relationship

This last class of schemes can be shown to result from an inherent property of adjoint operators. To illustrate we shall consider the problem of determining the value of the directional flux $\Psi(\mathbf{r}, \mathbf{\Omega}, E)$ at a particular point $(\mathbf{r}_1, \mathbf{\Omega}_1, E_1)$ in phase space within a subcritical reactor containing an extended external source $Q(\mathbf{r}, \mathbf{\Omega}, E)$. Conventional Monte Carlo procedures cannot provide a solution to this problem since the probability that during a given case history a neutron will pass exactly through the point $\mathbf{r}$ traveling in direction $\mathbf{\Omega}$ and having energy E, so that it will contribute to the statistical determination of $\Psi(\mathbf{r}, \mathbf{\Omega}, E)$, is zero (more precisely "a set of measure zero").

The function whose value we seek at $(\mathbf{r}_1, \mathbf{\Omega}_1, E_1)$ is the solution $\Psi(\mathbf{r}, \mathbf{\Omega}, E)$ of (8.3.9):

$$(L - M)\Psi(\mathbf{r}, \mathbf{\Omega}, E) = Q(\mathbf{r}, \mathbf{\Omega}, E).$$

To apply the reciprocity method we consider the adjoint equation

$$(L^* - M^*)\Psi^*(\mathbf{r}, \mathbf{\Omega}, E) = \delta(\mathbf{r} - \mathbf{r}_1, \mathbf{\Omega} - \mathbf{\Omega}_1, E - E_1), \tag{8.4.50}$$

where L^* and M^* are the adjoint operators defined by (8.3.10) and the right-hand side is the Dirac delta function, defined by

$$\int dV \int d\Omega \int dE\, F(\mathbf{r}, \mathbf{\Omega}, E)\delta(\mathbf{r} - \mathbf{r}_1, \mathbf{\Omega} - \mathbf{\Omega}_1, E - E_1) = F(\mathbf{r}_1, \mathbf{\Omega}_1, E_1) \text{ for all } F. \tag{8.4.51}$$

Note that (8.4.50) is not restricted to a critical condition. The physical interpretation of $\Psi^*(\mathbf{r}, \mathbf{\Omega}, E)$ is thus not the same as that of the neutron-importance function we defined earlier. We shall see how to interpret this more general importance function shortly.

Returning to the determination of $\Psi(\mathbf{r}, \mathbf{\Omega}, E)$, we multiply (8.3.9) by $\Psi^*(\mathbf{r}, \mathbf{\Omega}, E)$ and (8.4.50) by $\Psi(\mathbf{r}, \mathbf{\Omega}, E)$ and integrate over all of phase space. In terms of the Dirac notation (8.3.12), the result is

$$\begin{aligned} \langle \Psi^* | L - M | \Psi \rangle &= \langle \Psi^* | Q \rangle, \\ \langle \Psi | L^* - M^* | \Psi^* \rangle &= \langle \Psi | \delta(\mathbf{r} - \mathbf{r}_1, \mathbf{\Omega} - \mathbf{\Omega}_1, E - E_1) \rangle. \end{aligned} \tag{8.4.52}$$

By (8.3.15) and (8.3.16), however, the left-hand sides of these two equations are equal, and (8.4.51) shows that the right-hand side of the second equation is $\Psi(\mathbf{r}_1, \mathbf{\Omega}_1, E_1)$. Thus

$$\Psi(\mathbf{r}_1, \mathbf{\Omega}_1, E_1) = \langle \Psi^* | Q \rangle. \tag{8.4.53}$$

Since $Q(\mathbf{r}, \mathbf{\Omega}, E)$ is an *extended* source, the integral on the right-hand side of (8.4.53) is over an extended portion of phase space. Hence, if the values of $\Psi^*(\mathbf{r}, \mathbf{\Omega}, E)$ needed to determine $\langle \Psi^* | Q \rangle$ are estimated by a Monte Carlo procedure, many case histories will contribute, and the statistical accuracy of $\langle \Psi^* | Q \rangle$, and hence of $\Psi(\mathbf{r}_1, \mathbf{\Omega}_1, E_1)$, will be high. Monte Carlo reciprocity methods can thus be used to attack an adjoint problem to determine indirectly some of the properties of the regular problem.

Equation (8.4.53), on which the example just given was based, is a special case of a general relationship between the directional flux in a subcritical reactor and the solution to the companion adjoint equation for that reactor:

$$L^*\Psi^* = M^*\Psi^* + Q^*, \tag{8.4.54}$$

where $Q^*(\mathbf{r}, \boldsymbol{\Omega}, E)$ is an arbitrary function which we can choose to serve some desired purpose (as we did in (8.4.50)). We call $\Psi^*(\mathbf{r}, \boldsymbol{\Omega}, E)$ the *adjoint flux* due to the *adjoint source* Q^*.

Proceeding as we did to obtain (8.4.53) yields the general relationship

$$\langle\Psi|Q^*\rangle = \langle\Psi^*|Q\rangle. \tag{8.4.55}$$

One immediate consequence of this result is the *generalized reciprocity relationship*. If we take

$$\begin{aligned} Q^* &= \delta(\mathbf{r} - \mathbf{r}_2, \boldsymbol{\Omega} - \boldsymbol{\Omega}_2, E - E_2), \\ Q &= \delta(\mathbf{r} - \mathbf{r}_1, \boldsymbol{\Omega} - \boldsymbol{\Omega}_1, E - E_1), \end{aligned} \tag{8.4.56}$$

we get at once that

$$\Psi(\mathbf{r}_2, \boldsymbol{\Omega}_2, E_2) = \Psi^*(\mathbf{r}_1, \boldsymbol{\Omega}_1, E_1). \tag{8.4.57}$$

In words: The directional flux at $(\mathbf{r}_2, \boldsymbol{\Omega}_2, E_2)$ due to a delta-function source of neutrons at $(\mathbf{r}_1, \boldsymbol{\Omega}_1, E_1)$ equals the adjoint flux at $(\mathbf{r}_1, \boldsymbol{\Omega}_1, E_1)$ due to a delta-function adjoint source at $(\mathbf{r}_2, \boldsymbol{\Omega}_2, E_2)$. We used a very restricted version of this general principle (5.6.5) in determining the relationship between the escape probabilities associated with collision-theory methods.

Equation (8.4.55) also permits us to generalize the notion of neutron importance so that it will apply to a subcritical reactor. To do this we select Q^* to be a macroscopic cross section $\Sigma_\alpha(\mathbf{r}, \boldsymbol{\Omega}, E)$, so that $\langle\Psi|Q^*\rangle = \langle\Psi|\Sigma_\alpha\rangle$ is an interaction rate. For example we might select $\Sigma_\alpha(\mathbf{r}, \boldsymbol{\Omega}, E)$ to be $\Sigma_c(\mathbf{r}, E)$, the capture cross section of a neutron counter located at $\mathbf{r}$ and capturing neutrons of a given energy with a probability that is independent of their direction of travel. Then, if Q is taken as the delta function $\delta(\mathbf{r} - \mathbf{r}_1, \boldsymbol{\Omega} - \boldsymbol{\Omega}_1, E - E_1)$, we get from (8.4.56) that

$$\Psi^*(\mathbf{r}_1, \boldsymbol{\Omega}_1, E_1) = \langle\Psi|\Sigma_\alpha\rangle. \tag{8.4.58}$$

In words: The adjoint flux, interpreted as a generalized importance with respect to the interaction α, is the rate of that interaction which results when a delta-function source of neutrons is located at $(\mathbf{r}_1, \boldsymbol{\Omega}_1, E_1)$. Note that, to get the importance with respect to interaction α at some *other* point in phase space $(\mathbf{r}_2, \boldsymbol{\Omega}_2, E_2)$, we must move the source to

$\mathbf{r}_2$ and change its direction to $\mathbf{\Omega}_2$ and its energy to E_2. Note also that, in general, $\Psi^*(\mathbf{r}, \mathbf{\Omega}, E)$ depends on the nature of the interaction α. Counters located in different locations or sensitive to neutrons of different energies will result in different functions $\Psi^*(\mathbf{r}, \mathbf{\Omega}, E)$.

References

Bell, G. I., and Glasstone, S., 1970. *Nuclear Reactor Theory* (New York: Van Nostrand Reinhold).

Case, K. M., and Zweifel, P. F., 1967. *Linear Transport Theory* (Reading, Mass.: Addison-Wesley).

Greenspan, H., Kelber, C. N., and Okrent, D., 1968. *Computing Methods in Reactor Physics* (New York: Gordon and Breach).

Ferziger, J. H., and Zweifel, P. F., 1966. *The Theory of Neutron Slowing Down in Nuclear Reactors* (New York: Wiley).

Meghreblian, R. V., and Holmes, D. K., 1960. *Reactor Analysis* (New York: McGraw-Hill), Chapter 7.

New Developments in Reactor Mathematics and Applications, 1971. CONF-710302, Volume 2, USAEC Division of Technical Information.

Numerical Reactor Calculations, 1972. Proceedings of a seminar held in Vienna, Austria, 17–21 January 1972, Vienna IAEA, ST1/Pub/307, Sessions I–IV, IX, X.

Ussachoff, L. N., 1955. "Equations for Importance of Neutrons, Reactor Kinetics and Perturbation Theory," in *Proceedings of the International Conference on the Peaceful Uses of Atomic Energy*, Volume 5, pp. 503–570.

Problems

1. Show that the mathematical definition (8.2.5) of $\mathbf{J}(\mathbf{r}, E)$ implies the physical interpretation (8.2.6).

2. Derive the time-dependent neutron transport equation and the associated delayed-neutron precursor equations.

3. In the absence of feedback effects the neutron transport equation is linear since, in a reactor, neutron-neutron interactions are improbable compared to the interactions between neutrons and the materials constituting the reactor. Assuming that the cross section for neutron-neutron interactions equals that for neutron-proton interactions and that the one-group flux in a reactor rarely exceeds 10^{16} neutrons/cm^2/sec, provide a rough numerical validation of this neglect of nonlinear terms.

4. An infinite slab of purely absorbing material ($\Sigma_s = \Sigma_f = 0$) having a (spatially constant) macroscopic absorption cross section $\Sigma_a(E)$ and a thickness $2t$ is imbedded in an infinite homogeneous medium of reactor material as shown in Figure 8.6. Suppose that the directional flux

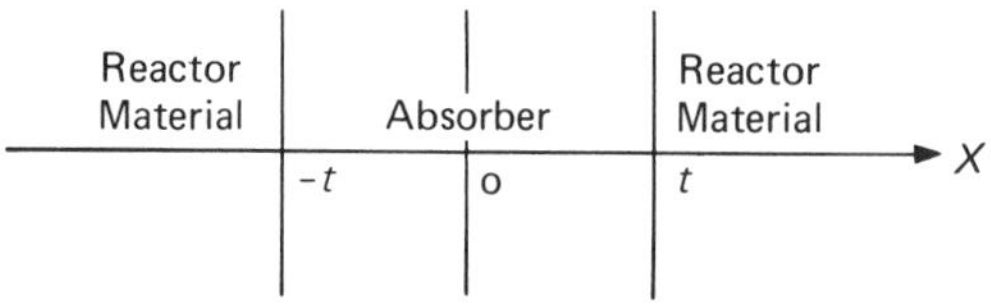

Figure 8.6

$\Psi(\mathbf{r}, \boldsymbol{\Omega}, E)$ at the surfaces of the slab are (for all y and z)

$$\Psi(t, \mu, E) = \Phi(t, E) + \mu j(t, E) \quad \text{for } -1 \leq \mu < 0,$$
$$\Psi(-t, \mu, E) = \Phi(-t, E) + \mu j(-t, E) \quad \text{for } 0 \leq \mu < 1,$$

where $\mu \equiv \boldsymbol{\Omega} \cdot \mathbf{i}$ ($\mathbf{i}$ being a unit vector in the $+X$ direction) and $\Phi(r, E), j(t, E), \Phi(-t, E)$, and $j(-t, E)$ are given functions, independent of μ.

a. Using the transport equation, find the directional flux at $x = t$ for μ in the range $0 < \mu < 1$.

b. Find the net current $\mathbf{J}(x, E)$ at $x = t$. (The result may be left in the form of an integral over μ.)

5. One very common method for expressing the angular dependence of the directional flux density is to write
$\Psi(\mathbf{r}, \boldsymbol{\Omega}, E) \approx \Psi_a(\mathbf{r}, \boldsymbol{\Omega}, E) \equiv F_0(\mathbf{r}, E) + \boldsymbol{\Omega} \cdot \mathbf{F}_1(\mathbf{r}, E)$, where $F_0(\mathbf{r}, E)$ is a scalar function of x, y, z, and E and $\mathbf{F}_1(\mathbf{r}, E)$ is a vector each component of which is a function of x, y, z, and E. Show that it is consistent with this approximation to interpret $F_0(\mathbf{r}, E)$ as the scalar flux density $\Phi(\mathbf{r}, E)$ and $\mathbf{F}_1(\mathbf{r}, E)$ as $3\mathbf{J}(\mathbf{r}, E)$, where $\mathbf{J}(\mathbf{r}, E)$ is the net current density.

6. Suppose $\Psi(\mathbf{r}, \boldsymbol{\Omega}, E)$ represents the directional flux throughout a critical reactor. Derive an expression for the total importance of all the neutrons in that reactor. How does this differ mathematically and physically from the total importance of the neutrons making up the scalar flux density $\Phi(\mathbf{r}, E)$?

7. Prove (8.3.16).

8. Prove (8.3.40).

9. With the eigenfunctions $\Psi_n^{(\omega)}$ defined by (8.3.38), find the operator equation for a set of adjoint fuctions W_m^* and corresponding eigenvalues ω_m^* such that $(\omega_n - \omega_m^*)\langle W_m^* | \Psi_n^{(\omega)} \rangle = 0$.

10. Show that a solution for (8.3.9) in terms of the eigenfunctions (8.3.18) and (8.3.22) is

$$\Psi(\mathbf{r}, \boldsymbol{\Omega}, E) = L^{-1}Q + \sum_{n=0}^{\infty} \frac{\lambda_n^2}{1 - \lambda_n} \frac{\langle \Psi_n^{(\lambda)*} | L^{-1} | Q \rangle}{\langle \Psi_n^{(\lambda)*} | M | \Psi_n^{(\lambda)} \rangle} \Psi_n^{(\lambda)}(\mathbf{r}, \boldsymbol{\Omega}, E).$$

11. Show that a solution for (8.3.9) in terms of the eigenfunctions (8.3.39) and (8.3.40) is

$$\Psi(\mathbf{r}, \boldsymbol{\Omega}, E) = \sum_{n=0}^{\infty} \left(-\frac{1}{\omega_n}\right) \frac{\langle \Psi_n^{(\omega)*} | Q \rangle}{\langle \Psi_n^{(\omega)*} | v^{-1} | \Psi_n^{(\omega)} \rangle} \Psi_n^{(\omega)}(\mathbf{r}, \boldsymbol{\Omega}, E).$$

12. In slab geometry it is sometimes useful to express $\Psi(z, \mu, E)$ in what is called a *double-P_n expansion*:

$$\Psi(z, \mu, E) = \begin{cases} \sum_{n=0}^{\infty} 2(2n + 1)\, \Psi_n^+(z, E) P_n(2\mu - 1) & \text{for } 0 < \mu \leq 1, \\ \sum_{n=0}^{\infty} 2(2n + 1) \Psi_n^-(z, E) P_n(2\mu + 1) & \text{for } -1 \leq \mu < 0, \end{cases}$$

where the coefficients $\Psi_n^{\pm}(z, E)$ specify the directional flux for neutrons having a component of velocity in the $+Z$ direction and the independent coefficients $\Psi_n^-(z, E)$ specify the flux of neutrons with velocity components in the $-Z$ direction.

Show that the scalar flux density and net current density are given by

$$\Phi(z, E) = \Psi_0^+(z, E) + \Psi_0^-(z, E),$$
$$\mathbf{J}(z, E) = \mathbf{k}\{\tfrac{1}{2}[\Psi_0^+(z, E) - \Psi_0^-(z, E)] + \tfrac{1}{2}[\Psi_1^+(z, E) + \Psi_1^-(z, E)]\}.$$

13. Consider an infinite-slab reactor of thickness L extending from $z = 0$ to $z = L$. If a P_1 approximation is made for the directional flux $\Psi(z, \mu, E)$, find the ratio of $\Psi_1(L, E)$ to $\Psi_0(L, E)$ according to:

a. Marshak boundary conditions;
b. Mark boundary conditions.

14. The transport equation for a one-group model of a homogeneous slab reactor is

$$\mu \frac{\partial \Psi(z, \mu)}{\partial z} + \Sigma_t \Psi(z, \mu) = \int_{-1}^{1} \frac{d\mu'}{2} \left[\nu\Sigma_f + \int_0^{2\pi} \frac{d\varphi'}{2\pi} \Sigma_s(\mu_0)\right] \Psi(z, \mu').$$

Suppose that the scattering is isotropic in the laboratory system and that we are trying to find particular solutions for $\Psi(z, \mu)$ within the slab by assuming that $\Psi(z, \mu) \approx \exp(iBz) f_B(\mu)$, where $i = \sqrt{-1}$, B is a number, and $f_B(\mu)$ is a function of μ which may be different for different values of B.

Suppose further that we wish to find the functions $f_B(\mu)$ by the P_3 approximation. Show that B must be one of the roots of the equation

$$9B^4 + [55\Sigma_t(\Sigma_t - \nu\Sigma_f - \Sigma_{s0}) + 35\Sigma_t^2]B^2 + 105(\Sigma_t - \nu\Sigma_f - \Sigma_{s0})\Sigma_t^3 = 0.$$

15. Write the equations representing the P_3 approximation to the transport equation in slab geometry for an assemblage of slabs which contain no fissionable material but in which an extraneous source $Q(z, \mu, E)$ is present.

16. Assuming that scattering is isotropic in the laboratory system, write the equations whose solution will yield:

a. the B_0 approximation for the Legendre component $F_0(E)$ of the asymptotic spectrum $F(\mu, E)$ given by (8.4.22);
b. the B_0 approximation for the components $F_0(E)$ and $F_1(E)$.

How do these results illustrate the finality property of the B_L approximation?

17. In one-dimensional slab geometry discrete-ordinates equations may be found by integrating (8.4.37) over the azimuthal angle φ (Figure 8.1) and letting the unknowns be the directional fluxes $\Psi(z, \mu_d, E)$ in the conical shells specified by the cosines μ_d ($d = 1, 2, \ldots, D$). The result is

$$\mu_d \frac{\partial}{\partial z} \Psi(z, \mu_d, E) + \Sigma_t(z, E)\, \Psi(z, \mu_d, E) = \int_0^\infty dE' f(E)\, \nu\Sigma_f(z, E') \sum_{d'=1}^{D} w_{d'} \Psi(z, \mu_{d'}, E')$$
$$+ \sum_{n=0}^{\infty} \int_0^\infty dE' \Sigma_{sn}(z, E' \to E)\, Y_n^0(\mu_d) \sum_{d'=1}^{\infty} w_{d'} \bar{Y}_n^0(\mu_{d'}) \Psi(z, \mu_{d'}, E').$$

a. For a two-ordinate approximation ($D = 2$), express the P_1 components $\Psi_0(z, E)$ and $\Psi_1(z, E)$ of the directional flux (8.4.12) in terms of $\Psi(z, \mu_1, E)$ and $\Psi(z, \mu_2, E)$.

b. By taking $w_1 = w_2 = \frac{1}{2}$ and assuming μ_1 and μ_2 to be the roots of $P_2(\mu) = 0$ (where $P_2(\mu)$ is the Legendre polynomial of order two), show that the equations for $\Psi_0(z, E)$ and $\Psi_1(z, E)$ derived from the equations for $\Psi(z, \mu_1, E)$ and $\Psi(z, \mu_2, E)$ are identical with the P_1 equations (i.e., (8.4.16) with $n = 0, 1$ and $\partial\Psi_2/\partial z = 0$).

Note: This exact correspondence between discrete-ordinates and P_n equations can be shown to be true for all orders provided that, for a given order N, the $\Sigma_{sn}(z, E' \to E)$ for $n > N$ are set to zero. There is no such simple correspondence in two and three dimensions.

18. Assuming that a particular solution for the transport equation in a given material may be

written in the form $F(\mathbf{r}, E) \exp(-i\mathbf{B}\cdot\mathbf{r})$ implies that the spherical-harmonics components $\Psi_r^{p*}(\mathbf{r}, E)$ in (8.4.36) may be written $F_r^p(E) \exp(-i(B_x x + B_y y + B_z z))$.

Starting with this assumption and (8.4.36), show that for a P_1 approximation the spherical-harmonics components of $F(\mathbf{r}, E)$ are related by

$$F_1^0(E) = \frac{iB_z}{\sqrt{3}} S^{-1} F_0^0(E),$$

$$F_1^{-1}(E) = \frac{-iB_x + B_y}{\sqrt{6}} S^{-1} F_0^0(E),$$

$$F_1^1(E) = \frac{iB_x + B_y}{\sqrt{6}} S^{-1} F_0^0(E),$$

where S^{-1} is the inverse of the linear operators defined by

$$SF(E) = \Sigma_t(E)\, F(E) - \int_0^\infty dE' \Sigma_{s1}(E' \to E)\, F(E').$$

Note that the result validates, for the P_1 approximation, the assertion that the energy shape of the $F_r^p(E)$ is the same for all values of p corresponding to a fixed value of $\mathbf{r}$.

19. Suppose that, in the X direction, material 1, having a total interaction cross section Σ_1 at some energy E, extends from $x = 0$ to $x = X$, and material 2, with total interaction cross section Σ_2 at E, extends from X to ∞. Show that neutrons having energy E, starting at $x = 0$ and traveling in the X direction, will interact at a distribution of locations x_i given by

$$x_i = x_1 - \frac{\Sigma_1}{\Sigma_2} x_1 - \frac{1}{\Sigma_2} \ln(1 - \rho_i) \quad \text{for } x_i \geq x_1,$$

$$x_i = -\frac{1}{\Sigma_1} \ln(1 - \rho_i) \quad \text{for } 0 \leq x_i \leq x_1,$$

where ρ_i is a random number.

20. Suppose that a large number of case histories have been run for neutrons born by fission in a fuel element and we have determined that the fraction of those case histories for which the fission neutron leaves the element without undergoing any interaction is 0.90. If we wish to have a 95 percent confidence that this fraction is within 1 percent of the true probability that a neutron will leave the element without interacting, approximately how many case histories should we have included in the sample?

21. Suppose that two identical counters characterized by interaction cross sections $\Sigma_{c1}(\mathbf{r}, E)$ and $\Sigma_{c2}(\mathbf{r}, E)$ are placed at different locations in or about a subcritical reactor. Show that the generalized importance functions $\Psi_1^*(\mathbf{r}, \boldsymbol{\Omega}, E)$ and $\Psi_2^*(\mathbf{r}, \boldsymbol{\Omega}, E)$ defined by (8.4.58) for these two counters are such that $\Psi_1^*(\mathbf{r}, \boldsymbol{\Omega}, E)/\Psi_2^*(\mathbf{r}, \boldsymbol{\Omega}, E)$ approaches a constant as the reactor approaches a critical condition. Explain how this result can be used to show that the importance for a critical reactor as defined physically in Section 8.3 is a special case of (8.4.58).

Hint: Use the results of Problem 7.

22. If there is no fissionable material present in an assembly, we expect on physical grounds that, for a point source of high-energy neutrons at $(\mathbf{r}_1, \boldsymbol{\Omega}_1, E_1)$, the directional flux density at all phase points $(\mathbf{r}_2, \boldsymbol{\Omega}_2, E_2)$ will be zero if $E_2 > E_1$.

Equation (8.4.57) states that under these conditions $\Psi^*(\mathbf{r}_1, \boldsymbol{\Omega}_1, E_1)$ will also be zero. Explain mathematically in terms of (8.4.54) and physically in terms of the meaning of generalized importance implied by (8.4.58) why this result is expected.

23. Consider a slab of homogeneous, purely absorbing material ($\Sigma_a(\mathbf{r}, E) = \Sigma_a(E)$; $\Sigma_s(\mathbf{r}, E) = 0$; $\Sigma_f(\mathbf{r}, E) = 0$) infinite in the X and Y dimensions and extending from $z = 0$ to $z = L$ in the Z dimension. Suppose the slab is in a vacuum but has within it at point z_1, a plane source $Q \equiv S(\mu, E)\, \delta(z - z_1)$, where $\delta(z - z_1)$ is the Dirac delta function.

a. Find $\Psi(z, \mu, E)$ within the slab for all μ and E.

b. Find the generalized importance function $\Psi^*(z, \mu, E)$ for neutrons within the slab for all μ and E when that importance is defined with respect to the spatially flat absorption cross section $\Sigma_a(E)$ of the slab.

c. Show that the results of a and b yield $\langle \Psi | \Sigma_a \rangle = \langle \Psi^* | Q \rangle$.

9 Group Diffusion Theory Derived from the Transport Equation

9.1 Introduction

The derivation of diffusion theory presented in Chapter 4 was based on an assumed relationship between the net current density $\mathbf{J}(\mathbf{r}, E)$ and the scalar flux density $\Phi(\mathbf{r}, E)$. That relationship, called Fick's Law, states that $\mathbf{J}(\mathbf{r}, E)$ is proportional to the gradient of the flux:

$$\mathbf{J}(\mathbf{r}, E) = -D(\mathbf{r}, E)\nabla\Phi(\mathbf{r}, E). \tag{9.1.1}$$

In Chapter 4 we merely presented Fick's Law, argued that it was physically plausible, and gave some approximate formulas (4.5.3) expressing the diffusion constant $D(\mathbf{r}, E)$ in terms of previously defined parameters.

In the present chapter we shall derive diffusion theory—first the continuous-energy form; then the multigroup and few-group forms—directly from the P_1 approximation to the transport equation. It will develop that additional assumptions must be made before (9.1.1) can be obtained by this procedure, and, to make the nature of these complications more clear, we shall first derive the continuous-energy form of the diffusion equation for the one-dimensional case. Extension to three dimensions will then be a straightforward matter. In making this extension, however, we shall take a slight detour and derive the three-dimensional P_1 equations directly from the transport equation by the weighted-residual method rather than as a special case of the spherical-harmonics equation (8.4.36).

The derivation of energy-group diffusion equations will then be carried out in two slightly different ways that will lead to different forms of equations. The first of these is expected to be the more accurate approximation; however it involves extra terms of the from $\partial[\tilde{D}_{gg'}(\partial\Phi_{g'}/\partial x)]/\partial x$ connecting groups g' and g. The second is the conventional form (4.13.7); we shall see, however, that, for many situations, this form should only be used for the few-group model. The two forms become identical if there is no anisotropic slowing down, that is, if the P_1 component $\Sigma_{s1}(\mathbf{r}, E' \to E)$ of $\Sigma_s(\mathbf{r}, E' \to E, \mu_0)$ vanishes. Accordingly, in the final part of the chapter we shall discuss the mathematical form of $\Sigma_{s1}(\mathbf{r}, E' \to E)$ and, more generally, of all the P_n components $\Sigma_{sn}(\mathbf{r}, E' \to E)$ of $\Sigma_s(\mathbf{r}, E' \to E, \mu_0)$. We shall see that it is possible to replace integrals such as $\int_0^\infty \Sigma_{sn}(z, E' \to E)\, \Psi_n(z, E')dE'$ in the P_n equations (8.4.16) by infinite series involving derivatives with respect to energy and in this way to convert the integral equation (8.4.16) into a set of differential equations. Such a procedure is called *continuous-slowing-down theory*. It provides a generalization of the age approximation discussed in Chapter 4 and leads to a set of approximate multigroup diffusion equations containing no terms of the type $\partial[\tilde{D}_{gg'}(\partial\Phi_{g'}/\partial x)]/\partial x$.

9.2 Derivation from the One-Dimensional P_1 Approximation

In one-dimensional slab geometry the P_1 equations, obtained from (8.4.16), are

$$\frac{\partial \Psi_1(z, E)}{\partial z} + \Sigma_t(z, E)\Psi_0(z, E) = \int_0^\infty dE'\left[\sum_j f^j(E)\nu^j\Sigma_f^j(E') + \Sigma_{s0}(z, E' \to E)\right]\Psi_0(z, E'),$$
$$\frac{\partial \Psi_0(z, E)}{\partial z} + 3\Sigma_t(z, E)\Psi_1(z, E) = 3\int_0^\infty dE'\,\Sigma_{s1}(z, E' \to E)\Psi_1(E'). \qquad (9.2.1)$$

The P_1 approximation to the directional flux $\Psi(z, \mu, E)$ is then $\Psi_0(z, E) + 3\mu\Psi_1(z, E)$, where Ψ_0 and Ψ_1 are the solutions of (9.2.1).

It follows from the definitions (8.2.4) and (8.2.5) that the P_1 approximations to the scalar flux density $\Phi(z, E)$ and the net current density $\mathbf{J}(z, E)$ are, for this case,

$$\Phi(z, E) = \int_{-1}^{1} \frac{d\mu}{2}\int_0^{2\pi}\frac{d\varphi}{2\pi}[\Psi_0(z, E) + 3\mu\Psi_1(z, E)] = \Psi_0(z, E),$$
$$\mathbf{J}(z, E) = \int_{-1}^{1} \frac{d\mu}{2}\int_0^{2\pi}\frac{d\varphi}{2\pi}\boldsymbol{\Omega}[\Psi_0(z, E) + 3\mu\Psi_1(z, E)] = \mathbf{k}\Psi_1(z, E). \qquad (9.2.2)$$

The second equation in (9.2.1) can then be written

$$\frac{\partial \Phi(z, E)}{\partial z} = -3\left[\Sigma_t(z, E)J(z, E) - \int_0^\infty dE'\Sigma_{s1}(z, E' \to E)J(z, E')\right]$$
$$\equiv -3A_1 J(z, E), \qquad (9.2.3)$$

where $J(z, E)$ is the magnitude of $\mathbf{J}(z, E)$ in the $+Z$ direction and the operator A_1 is defined by the second equation.

It would appear from this result that, to obtain Fick's Law, $J = -D(\partial\Phi/\partial z)$, we must associate D with the operator

$$\tilde{D} \equiv \tfrac{1}{3}A_1^{-1}, \qquad (9.2.4)$$

where the operator A_1^{-1} is such that $A_1^{-1}A_1$ is the identity operator. Using this definition, we obtain from (9.2.3)

$$J(z, E) = -\tilde{D}\frac{\partial \Phi(z, E)}{\partial z}, \qquad (9.2.5)$$

a result which, when introduced into the term $\partial\Psi_1/\partial z$ $(= \partial J/\partial z)$ in the first equation in (9.2.1), yields

$$\frac{\partial}{\partial z}\left[\tilde{D}\frac{\partial}{\partial z}\Phi(z, E)\right] + \Sigma_t(z, E)\Phi(z, E)$$
$$= \int_0^\infty dE'\left[\sum_j f^j(E)\nu^j\Sigma_f^j(z, E') + \Sigma_{s0}(z, E' \to E)\right]\Phi(z, E'). \qquad (9.2.6)$$

Equation (9.2.6) differs from the continuous-energy diffusion equation (4.6.1) for the one-dimensional case only in the leakage term $\partial[\tilde{D}(\partial\Phi/\partial z)]/\partial z$, which here involves the operator $\tilde{D}$ rather than the function $D(z, E)$. Mathematically, however, the difference is major, and, although we shall see that the operator $\tilde{D}$ can be readily determined in the energy-group approximation, we would much rather replace $\tilde{D}$ by a function of z and E. We can achieve this objective in a formal way by rewriting (9.2.3) as

$$\frac{\partial}{\partial z}\Phi(z, E) = -3\left[\Sigma_t(z, E) - \frac{\int_0^\infty dE'\,\Sigma_{s1}(z, E' \to E)J(z, E')}{J(z, E)}\right]J(z, E) \tag{9.2.7}$$

and defining the functional $\Sigma_{tr}(z, E)$ by

$$\Sigma_{tr}(z, E) \equiv \Sigma_t(z, E) - \frac{\int_0^\infty dE'\,\Sigma_{s1}(z, E' \to E)J(z, E')}{J(z, E)} \tag{9.2.8}$$

and the diffusion constant $D(z, E)$ by

$$D(z, E) \equiv \frac{1}{3\Sigma_{tr}(z, E)} = \frac{1}{3\left[\Sigma_t(z, E) - \frac{\int_0^\infty dE'\,\Sigma_{s1}(z, E' \to E)J(z, E')}{J(z, E)}\right]}. \tag{9.2.9}$$

Then from (9.2.7) and the first equation in (9.2.1) we obtain

$$-\frac{\partial}{\partial z}\left[D(z, E)\frac{\partial}{\partial z}\Phi(z, E)\right] + \Sigma_t(z, E)\Phi(z, E) = \int_0^\infty dE'\left[\sum_j f^j(E)\nu^j\Sigma_f^j(z, E') + \Sigma_{s0}(z, E' \to E)\right]\Phi(z, E'). \tag{9.2.10}$$

This procedure does yield the continuous-energy diffusion equation in its usual form. It is, however, entirely formal since $J(z, E)$ must be known to find $D(z, E)$. To implement (9.2.10) it is thus necessary first to make an estimate of $J(z, E)$ by some auxiliary method. We might use an iterative procedure, approximating $\Sigma_{s1}(z, E' \to E)$ in (9.2.8) by $\Sigma_{s1}(z, E')\delta(E' - E)$ for the first iteration ($\delta(E' - E)$ being the Dirac delta function defined by (8.4.51)), finding $J(z, E)$ for the ith iteration from (9.2.7), and using that value of $J(z, E)$ in (9.2.8) to determine $\Sigma_{tr}(z, E)$ for the $(i + 1)$th iteration. However a far more common way to estimate $J(z, E)$ in (9.2.9) is to approximate it by an infinite-medium spectrum. We shall discuss this procedure further when we develop the energy-group approximation.

Since the Legendre components $\Psi_n(z, E)$ are required to be continuous across internal interfaces, we impose on the solution of (9.2.6) the conditions that $\Phi(z, E)$ and $\tilde{D}(\partial\Phi(z, E)/\partial z)$ be continuous in z and on the solution of (9.2.10) that $\Phi(z, E)$ and

$D(z, E)(\partial\Phi(z, E)/\partial z)$ be continuous in z. We shall see how to implement the condition involving the operator $\tilde{D}$ when we derive the energy-group approximation.

On the outer surfaces Marshak boundary conditions can be imposed. However, as discussed in Chapter 4 and again in Section 8.4, it is simpler and legitimate for a great many cases just to set $\Phi(z, E)$ equal to zero on the outer surfaces.

9.3 The Continuous-Energy Diffusion Equation in Three Dimensions

The difficulties involved in defining a diffusion constant that will yield solutions $\Phi(z, E)$ to one-dimensional diffusion equations such as (9.2.3) or (9.2.6) that are identical with those obtained by solving the corresponding P_1 equation (9.2.1) are due entirely to $\Sigma_{s1}(z, E' \to E)$, the P_1 component of the differential scattering cross section $\Sigma_s(z, E' \to E)$. If this term vanishes (isotropic scattering in the laboratory system) or if energy changes due to anisotropic scattering can be neglected (i.e., if $\Sigma_{s1}(z, E' \to E) = \Sigma_{s1}(z, E')\delta(E' - E)$), the functional $D(z, E)$ and the operator $\tilde{D}$ become identical and, as can be seen from (9.2.3) and (9.2.9), equal to

$$D(z, E) = \frac{1}{3\Sigma_t(z, E)} \quad \text{for } \Sigma_{s1}(z, E' \to E) = 0 \tag{9.3.1}$$

and

$$D(z, E) = \frac{1}{3[\Sigma_t(z, E) - \Sigma_{s1}(z, E)]} \quad \text{for } \Sigma_{s1}(z, E' \to E) = \Sigma_{s1}(z, E')\delta(E' - E). \tag{9.3.2}$$

(Note that the latter result is essentially the simple formula (4.5.3) quoted in Chapter 4.)

The difficulties involved in defining an energy-dependent diffusion constant thus stem from the fact that this constant accounts, not only for the gross flow of neutrons of energy E, but also for the flow of neutrons appearing in the range E to $E + dE$ because of anisotropic scattering at other energies. For the general three-dimensional situation these difficulties are augmented by the fact that it is necessary to define different values of the diffusion constant for different coordinate directions. To explore the nature of these added difficulties it will be necessary to start with the P_1 equations in three dimensions. We could obtain these equations directly from the general result (8.4.36). However this is a convenient place to introduce another useful tool, the *weighted-residual method*, a straightforward but extremely powerful procedure for finding approximate solutions to equations such as the transport equations when some information concerning the nature of these solutions is already available.

The Weighted-Residual Method

In the weighted-residual method one assumes for the solution to the equation at hand a general mathematical form containing a number of undertermined parameters or functions. The particular form selected reflects any knowledge of the nature of the solution obtained by past experience. For example, in one dimension, on the grounds that the directional flux is not strongly dependent on μ, one might approximate $\Psi(z, E, \mu)$ by a trial function

$$\Psi_t(z, E, \mu) = A(z, E) + B(z, E)\mu + C(z, E)\mu^2, \tag{9.3.3}$$

where A, B, and C are functions to be determined.

Generally the form chosen is not sufficiently general that an exact representation of $\Psi(z, E, \mu)$ (i.e., one obeying the transport equation for all z, E, and μ) is possible. We can, however, force the solution to be correct in a weighted-integral sense and in this way determine the unknown parameters or functions of the general expression. Such a procedure is justified on the grounds that, if H is, for example, the operator representing the transport equation, the solution Ψ of $H\Psi = 0$ can be regarded either as a function which obeys the equation $H\Psi = 0$ or equally well as one which obeys

$$\int dV \int d\Omega \int dE\, W(\mathbf{r}, \mathbf{\Omega}, E) H\Psi(\mathbf{r}, \mathbf{\Omega}, E) = 0 \quad \text{for all } W(\mathbf{r}, \mathbf{\Omega}, E). \tag{9.3.4}$$

For a trial function $\Psi_t(\mathbf{r}, \mathbf{\Omega}, E)$ of limited generality, (9.3.4) cannot be made valid for all $W(\mathbf{r}, \mathbf{\Omega}, E)$. Attempting to do so will only lead to the conclusion that $\Psi_t(\mathbf{r}, \mathbf{\Omega}, E)$ must be zero. However it is possible to obtain a nontrivial result by requiring that (9.3.4) be obeyed for a limited number of weight functions. If, then, for a particular trial function Ψ_t containing N undetermined parameters or functions (for example A, B, and C in the example just cited) we form the *residual*, $H\Psi - H\Psi_t$, we can determine appropriate values for these parameters by requiring that

$$\int dV \int d\Omega \int dE\, W_i(\mathbf{r}, \mathbf{\Omega}, E)[H\Psi - H\Psi_t] = 0 \qquad (i = 1, 2, \ldots, N), \tag{9.3.5}$$

for as many linearly independent weight functions W_i as there are undetermined parameters in Ψ_t. Since $H\Psi = 0$, the first term in the integral vanishes, and the procedure is thus equivalent to substituting Ψ_t into the transport equation and then requiring weighted integrals of the resulting expression to vanish. The Ψ_t so determined is an approximate solution to $H\Psi = 0$ in a weighted-integral sense.

The weight functions $W_i(\mathbf{r}, \mathbf{\Omega}, E)$ are always restricted as little as possible. Thus, in (9.3.3), since the z and E dependence of $\Psi_t(z, E, \mu)$ is still unspecified, the z and E dependence of the $W_i(z, \mu, E)$ can also be completely general. We can, however, choose only three different expressions for the μ-dependent part of the W_i. Thus, for the specific case, we have

$$W_i(z, \mu, E) = V_i(z, E)U_i(\mu) \qquad (i = 1, 2, 3), \tag{9.3.6}$$

where the $V_i(z, E)$ are *any* functions of z and E and the $U_i(\mu)$ are three particular, but arbitrarily chosen, functions of μ.

In view of (9.3.3) the choice $U_i = \mu^{i-1}$ $(i = 1, 2, 3)$ would be a natural one. Note that selecting unity as one of the weight functions forces the trial function to maintain overall neutron balance.

Whatever the choice of the $U_i(\mu)$, (9.3.4) for this example becomes

$$\int dz \int dE\, V_i(z, E)\left[\int_{-1}^{1} \frac{d\mu}{2} U_i(\mu) H\Psi_t(z, \mu, E)\right] = 0 \qquad (i = 1, 2, 3), \tag{9.3.7}$$

with Ψ_t given by (9.3.3). Then, since the $V_i(z, E)$ are completely arbitrary functions of z and E, this result in turn reduces to

$$\int_{-1}^{1} \frac{d\mu}{2} U_i(\mu) H\Psi_t(z, \mu, E) = 0 \qquad (i = 1, 2, 3). \tag{9.3.8}$$

When the indicated integration is carried out, three integrodifferential equations involving $A(z, E)$, $B(z, E)$, and $C(z, E)$, but independent of μ, result. Their solution permits us to reconstruct $\Psi_t(z, \mu, E)$.

In general, then, the extent to which a trial function is forced, in an integral sense, to obey the governing equation depends on the form selected for that trial function. If the form selected still allows for the general behavior with respect to certain variables, the weight functions may be chosen to be independent of those variables. The integrations over those variables in (9.3.5) should then be omitted. Thus, if we assume some specific mathematical form for the spatial part of $\Psi(\mathbf{r}, \mathbf{\Omega}, E)$ but leave the angular and energy parts unspecified, the weight functions should be functions of $\mathbf{r}$ only, and (9.3.5) should be replaced by

$$\int dV\, W_i(\mathbf{r}) H\Psi_t = 0 \qquad (i = 1, 2, \ldots, N). \tag{9.3.9}$$

The equations that result from applying (9.3.9) will then be integral equations in the variables $\mathbf{\Omega}$ and E.

It follows that, the more specific the trial function, the simpler the equations that will result from applying the method. For example, if the only undetermined parts of the trial function are a set of constants, application of (9.3.5) will yield a set of algebraic equations that will determine these constants.

The weighted-residual method is clearly very general. Both trial and weight functions (within the limits just discussed) may be chosen arbitrarily. The freedom that thus results can be used to incorporate our previous experience with the problem at hand into the form of Ψ_t, so that, for a given amount of computational effort, a more accurate result can be obtained.

Of course this freedom can lead to disastrous results if it is not employed intelligently. A poor choice of trial function can yield solutions that are seriously in error. A successful application of the method thus requires far more intuitive judgment than is demanded by, say, the spherical-harmonics technique.

The P_1 Equations in Three Dimensions

To obtain the P_1 equations in three dimensions by the weighted-residual method we take advantage of our experience in one dimension which, according to (8.4.12) and (9.2.2), tells us that

$$\begin{aligned}\Psi(z, \mu, E) &\approx \Phi(z, E) + 3\mu J(z, E)\\ &= \Phi(z, E) + 3\Omega_z J_z(z, E).\end{aligned} \tag{9.3.10}$$

In three dimensions it seems plausible to assume an analogous behavior for the angular part of $\Psi(\mathbf{r}, \mathbf{\Omega}, E)$ in each dimension and thus to write

$$\Psi_t(\mathbf{r}, \mathbf{\Omega}, E) = \Phi(\mathbf{r}, E) + 3\mathbf{\Omega}\cdot\mathbf{J}(\mathbf{r}, E). \tag{9.3.11}$$

Note, in addition, that the definitions (8.2.4) and (8.2.5) show that $\Phi(\mathbf{r}, E)$ and $\mathbf{J}(\mathbf{r}, E)$ are the scalar flux and net current densities consistent with approximating the directional flux by $\Psi_t(\mathbf{r}, \mathbf{\Omega}, E)$.

Since we obtained the P_1 equations in one dimension by, in effect, weighting first by $P_0(\mu)$ $(=1)$ and then by $P_1(\mu)$ $(= \mu = \Omega_z)$, it seems plausible in three dimensions to weight by 1, Ω_x, Ω_y, and Ω_z, that is, by 1 and $\mathbf{\Omega}$. Thus, to obtain the weighted-residual equations for Φ and $\mathbf{J}$ resulting from these choices of trial and weight functions, we substitute (9.3.11) for $\Psi(\mathbf{r}, \mathbf{\Omega}, E)$ into the transport equation (8.2.14) and require the result to be valid when weighted by unity and integrated over all directions and also when weighted by $\mathbf{\Omega}$ and integrated over all directions. The result is

$$\begin{aligned}&\int d\Omega[\mathbf{\Omega}\cdot\nabla(\Phi + 3\mathbf{\Omega}\cdot\mathbf{J}) + \Sigma_t(\Phi + 3\mathbf{\Omega}\cdot\mathbf{J})]\\ &= \int_{\mathbf{\Omega}} d\Omega \int_0^\infty dE' \int_{\mathbf{\Omega}'} d\Omega'\left[f(E)\nu\Sigma_f(E') + \sum_n (2n+1)\Sigma_{sn}(E' \to E)P_n(\mu_0)\right][\Phi(E') + 3\mathbf{\Omega}'\cdot\mathbf{J}(E')]\end{aligned} \tag{9.3.12}$$

and

$$\int d\Omega\, \mathbf{\Omega}[\mathbf{\Omega}\cdot\nabla(\Phi + 3\mathbf{\Omega}\cdot\mathbf{J}) + \Sigma_t(\Phi + 3\mathbf{\Omega}\cdot\mathbf{J})]$$

$$= \int_{\mathbf{\Omega}} d\Omega \int_0^{\infty} dE' \int_{\mathbf{\Omega}'} d\Omega' \mathbf{\Omega}\left[f(E)\nu\Sigma_f(E') + \sum_n (2n+1)\Sigma_{sn}(E' \to E)P_n(\mu_0)\right][\Phi(E') + 3\mathbf{\Omega}'\cdot\mathbf{J}(E')], \tag{9.3.13}$$

where we have simplified to one fissionable isotope, set $Q = 0$, and expanded the term $\Sigma_s(\mathbf{r}, E' \to E)$ in Legendre polynomials as in (8.4.12). Writing out (9.3.12) in terms of the components of $\mathbf{\Omega}$, ∇, and $\mathbf{J}$ yields

$$\int d\Omega \left\{\left(\Omega_x \frac{\partial}{\partial x} + \Omega_y \frac{\partial}{\partial y} + \Omega_z \frac{\partial}{\partial z}\right)[\Phi + 3(\Omega_x J_x + \Omega_y J_y + \Omega_z J_z)]\right.$$

$$\left. + \Sigma_t[\Phi + 3(\Omega_x J_x + \Omega_y J_y + \Omega_z J_z)]\right\} = \int_{\mathbf{\Omega}} d\Omega \int_0^{\infty} dE' \int_{\mathbf{\Omega}'} d\Omega' \Big[f(E)\nu\Sigma_f(E') + \Sigma_{s0}(E' \to E)$$

$$+ 3\Sigma_{s1}(E' \to E)(\Omega_x\Omega_x' + \Omega_y\Omega_y' + \Omega_z\Omega_z') + \sum_{n=2}^{\infty}(2n+1)\Sigma_{sn}(E' \to E)P_n(\mu_0)\Big]$$

$$\times [\Phi(E') + 3(\Omega_x' J_x(E') + \Omega_y' J_y(E') + \Omega_z' J_z(E')], \tag{9.3.14}$$

and (9.3.13) may be similarly expanded with multiplication by $\mathbf{\Omega}$ included in the integrands.

Terms involving $\Sigma_{sn}(E' \to E)$ for $n \geq 2$ have been written separately since they all vanish (in both (9.3.13) and (9.3.14)) when the integration over $d\Omega$ is performed. This happens because the expansion (8.4.14) of $P_n(\mu_0)$ contains terms in $P_n(\mu)$ and $P_n^m(\mu)$ which, for $n \geq 2$, vanish when the integration over μ is performed.

To carry out the remaining integrations we write

$$\Omega_x = \sin\theta\cos\varphi, \qquad \Omega_y = \sin\theta\sin\varphi, \qquad \Omega_z = \cos\theta, \qquad d\Omega = \frac{1}{4\pi}\sin\theta\, d\theta\, d\varphi, \tag{9.3.15}$$

and note that

$$\begin{aligned}
&\int d\Omega = \frac{1}{4\pi}\int_0^{2\pi} d\varphi \int_0^{\pi} \sin\theta\, d\theta = 1,\\
&\int \Omega_x^2\, d\Omega = \int \Omega_y^2\, d\Omega = \int \Omega_z^2\, d\Omega = \tfrac{1}{3},\\
&\int \Omega_x\Omega_y\, d\Omega = \int \Omega_y\Omega_z\, d\Omega = \int \Omega_z\Omega_x\, d\Omega = 0,\\
&\int \Omega_x\, d\Omega = \int \Omega_y\, d\Omega = \int \Omega_z\, d\Omega = 0,\\
&\int \Omega_x^3 d\Omega = \int \Omega_y^3\, d\Omega = \int \Omega_z^3\, d\Omega = 0.
\end{aligned} \tag{9.3.16}$$

Application of these results to (9.3.14) and to (9.3.13) written out in a form similar to (9.3.14) yields the three-dimensional P_1 equations

$$\nabla\cdot\mathbf{J}(\mathbf{r}, E) + \Sigma_t(\mathbf{r}, E)\Phi(\mathbf{r}, E) = \int_0^\infty dE'[f(E)\nu\Sigma_f(\mathbf{r}, E') + \Sigma_{s0}(\mathbf{r}, E' \to E)]\Phi(\mathbf{r}, E') \quad (9.3.17)$$

and

$$\tfrac{1}{3}\nabla\Phi(\mathbf{r}, E) + \Sigma_t(\mathbf{r}, E)\mathbf{J}(\mathbf{r}, E) = \int_0^\infty dE'\,\Sigma_{s1}(\mathbf{r}, E' \to E)\mathbf{J}(\mathbf{r}, E'), \quad (9.3.18)$$

where the latter result, being a relationship between vectors, is equivalent to three scalar equations.

It should be noted that the justification for stating that (9.3.17) and (9.3.18) are the P_1 equations in three dimensions is that they can also be derived from the general spherical-harmonics equations (8.4.36). A different choice of trial function or weight functions—for example weighting by Ω^2 and $\mathbf{\Omega}$ rather than by 1 and $\mathbf{\Omega}$—would have produced quite different results.

Directionally Dependent Diffusion Constants

In order to convert (9.3.17) and (9.3.18) into a single scalar equation for $\Phi(\mathbf{r}, E)$ we would like to be able to derive from the second equation a relationship of the form

$$\mathbf{J}(\mathbf{r}, E) = -D(\mathbf{r}, E)\nabla\Phi(\mathbf{r}, E). \quad (9.3.19)$$

However, just as in the one-dimensional case, we can do this rigorously only if $\Sigma_{s1}(\mathbf{r}, E' \to E)$ is either zero or a delta function in energy (in which case formulas for $D(\mathbf{r}, E)$ completely analogous to (9.3.1) and (9.3.2) are obtained). When energy changes due to anisotropic scattering in the laboratory system are significant (as, for example, with hydrogen) the diffusion constant in (9.3.19) must be replaced by an operator or a functional depending on the function $\mathbf{J}(\mathbf{r}, E)$. Moreover, in the three-dimensional case there is an additional complication: a relationship of the form (9.3.19) implies that the direction of the vector $\mathbf{J}$ for each energy E is the same as the direction of $\nabla\Phi$. For this to be true, the direction of the right-hand side of (9.3.18) (which involves the limit of an infinite sum of vectors whose directions are those of the net currents at energies E' different from E) must be the same as the direction of $\mathbf{J}$. In general this will not be the case. Hence $\mathbf{J}$ and $\nabla\Phi$ are not parallel, and (9.3.19) is improper.

To correct this situation we simply write the vector equation (9.3.18) as three separate scalar equations, one for each component of $\nabla\Phi$ and the corresponding component of $\mathbf{J}$, and repeat the one-dimensional analysis of Section 9.2 for each coordinate direction.

Thus, in complete analogy with (9.2.3) and (9.2.4), we define an operator A_1 by

$$\frac{\partial}{\partial u}\Phi(\mathbf{r}, E) = -3\left[\Sigma_t(\mathbf{r}, E)J_u(\mathbf{r}, E) - \int_0^\infty dE'\, \Sigma_{s1}(\mathbf{r}, E' \to E)J_u(\mathbf{r}, E')\right]$$
$$\equiv -3A_1 J_u(\mathbf{r}, E) \qquad (u = x, y, z) \tag{9.3.20}$$

and an analogous diffusion-constant operator $\tilde{D}$ by

$$\tilde{D} \equiv \tfrac{1}{3}A_1^{-1}, \tag{9.3.21}$$

so that

$$J_u(\mathbf{r}, E) = -\tilde{D}\frac{\partial}{\partial u}\Phi(\mathbf{r}, E) \qquad (u = x, y, z) \tag{9.3.22}$$

or

$$J(\mathbf{r}, E) = -\tilde{D}\nabla\Phi(\mathbf{r}, E). \tag{9.3.23}$$

Equation (9.3.17) can then be written

$$-\nabla\cdot\tilde{D}\nabla\Phi - \Sigma_t(\mathbf{r}, E)\Phi(\mathbf{r}, E) = \int_0^\infty dE'[f(E)\nu\Sigma_f(\mathbf{r}, E') + \Sigma_{s0}(\mathbf{r}, E' \to E)]\Phi(\mathbf{r}, E'). \tag{9.3.24}$$

If, in analogy with (9.2.8) and (9.2.9), we want to deal with functionals rather than operators, matters become more complicated. We define *three* functions $\Sigma_{tr,u}(\mathbf{r}, E)$ by

$$\Sigma_{tr,u}(\mathbf{r}, E) \equiv \Sigma_t(\mathbf{r}, E) - \frac{\int_0^\infty dE'\, \Sigma_{s1}(\mathbf{r}, E' \to E)J_u(\mathbf{r}, E')}{J_u(\mathbf{r}, E)} \tag{9.3.25}$$

and *three* components of a directional diffusion constant

$$D_u(\mathbf{r}, E) \equiv \frac{1}{3\Sigma_{tr,u}(\mathbf{r}, E)}. \tag{9.3.26}$$

Since, in general, J_x, J_y, and J_z will not be the same functions of energy, the three components $D_u(\mathbf{r}, E)$ $(u = x, y, z)$ will usually be different. To account for this fact (9.3.17) is now written

$$-\left[\frac{\partial}{\partial x}D_x\frac{\partial}{\partial x} + \frac{\partial}{\partial y}D_y\frac{\partial}{\partial y} + \frac{\partial}{\partial z}D_z\frac{\partial}{\partial z} - \Sigma_t(\mathbf{r}, E)\right]\Phi(\mathbf{r}, E)$$
$$= \int_0^\infty dE'[f(E)\nu\Sigma_f(\mathbf{r}, E') + \Sigma_{s0}(\mathbf{r}, E' \to E)]\Phi(\mathbf{r}, E'). \tag{9.3.27}$$

Both (9.3.23) and (9.3.26) are formally equivalent to the three-dimensional P_1 equations (9.3.17) and (9.3.18). However they are considerably more complicated than the usual continuous-energy diffusion equation (4.6.1). We shall see in the next section that most of this complexity disappears when we make the multigroup and few-group approximations.

9.4 Reduction to Group Diffusion-Theory Models

In order to derive the energy-group counterparts of the $\tilde{D}$ of (9.3.21) and the $D_u(\mathbf{r}, E)$ of (9.3.26) it will be necessary to return to the P_1 equations (9.3.17) and (9.3.18) and convert these to the energy-group form. We shall do this in a formal way and then derive formally exact group diffusion-theory equations. The simplifications that result when the multigroup and few-group approximations are made will be discussed in the subsequent two sections.

Formal Procedure

To reduce the continuous-energy P_1 equations to energy-group form in a formal way we partition the energy range from E_0 ($\approx$ 15 MeV) to $E_G(=0)$ into subintervals

$$\Delta E_g \equiv E_{g-1} - E_g \qquad (g = 1, 2, \ldots, G) \tag{9.4.1}$$

and define group fluxes $\Phi_g(\mathbf{r})$ and the components of group currents $J_{gu}(\mathbf{r})$ ($u = x, y, z$) by

$$\begin{aligned} \Phi_g(\mathbf{r}) &= \int_{E_g}^{E_{g-1}} \Phi(\mathbf{r}, E)\, dE \qquad (g = 1, 2, \ldots, G), \\ J_{gu}(\mathbf{r}) &= \int_{E_g}^{E_{g-1}} J_u(\mathbf{r}, E)\, dE \qquad (g = 1, 2, \ldots, G;\ u = x, y, z). \end{aligned} \tag{9.4.2}$$

Note that, as in Chapter 4, the Φ_g and J_{gu} are *total* fluxes and currents, not densities in energy like $\Phi(\mathbf{r}, E)$ and $J_u(\mathbf{r}, E)$. Next, group parameters are defined by

$$\begin{aligned} \Sigma_{\alpha g}(\mathbf{r}) &\equiv \frac{\int_{E_g}^{E_{g-1}} \Sigma_\alpha(\mathbf{r}, E)\Phi(\mathbf{r}, E) dE}{\Phi_g(\mathbf{r})} \qquad (\alpha = \mathrm{t}, \mathrm{f}), \\ \chi_g &\equiv \int_{E_g}^{E_{g-1}} f(E)\, dE, \\ \Sigma_{gg'}(\mathbf{r}) &\equiv \frac{\int_{E_g}^{E_{g-1}} dE \int_{E_{g'}}^{E_{g'-1}} dE'\, \Sigma_{s0}(\mathbf{r}, E' \to E)\Phi(\mathbf{r}, E')}{\Phi_{g'}(\mathbf{r})}, \\ A_{1gg',u} &\equiv \frac{\int_{E_g}^{E_{g-1}} dE \int_{E_{g'}}^{E_{g'-1}} dE' [\Sigma_t(\mathbf{r}, E)\delta(E' - E) - \Sigma_{s1}(\mathbf{r}, E' \to E)] J_u(\mathbf{r}, E')}{J_{g'u}(\mathbf{r})}. \end{aligned} \tag{9.4.3}$$

Notice that in this case, because the $J_u(\mathbf{r}, E)$ may be different functions of energy in different groups g', the parameters $A_{1gg',u}$ depend on u. In the limit of many groups, however, this directional dependence will vanish and the $A_{1gg'}$ (the same now for all u) will become more like the operator A_1 of (9.3.20).

Integrating (9.3.17) and (9.3.18) over ΔE_g in the usual way yields the P_1 energy-group equations:

$$\frac{\partial}{\partial x} J_{gx}(\mathbf{r}) + \frac{\partial}{\partial y} J_{gy}(\mathbf{r}) + \frac{\partial}{\partial z} J_{gz}(\mathbf{r}) + \Sigma_{tg}(\mathbf{r})\Phi_g(\mathbf{r}) = \sum_{g'=1}^{G} [\chi_g \nu \Sigma_{fg'}(\mathbf{r}) + \Sigma_{gg'}(\mathbf{r})]\Phi_{g'}(\mathbf{r}) \quad (g = 1, 2, \ldots, G) \tag{9.4.4}$$

and

$$\frac{1}{3}\frac{\partial}{\partial u}\Phi_g(\mathbf{r}) = -\sum_{g'=1}^{G} A_{1gg',u}(\mathbf{r}) J_{g'u}(\mathbf{r}) \qquad (g = 1, 2, \ldots, G;\ u = x, y, z). \tag{9.4.5}$$

To obtain the energy-group analogue of the operator $\tilde{D}$ we write (9.4.5) in matrix form:*

$$\frac{1}{3}\frac{\partial}{\partial u}[\Phi(\mathbf{r})] = -[A_{1,u}(\mathbf{r})][J_u(\mathbf{r})], \tag{9.4.6}$$

where

$$\begin{aligned} [\Phi(\mathbf{r})] &= \mathrm{Col}\{\Phi_1(\mathbf{r}), \Phi_2(\mathbf{r}), \ldots, \Phi_G(\mathbf{r})\}, \\ [J_u(\mathbf{r})] &= \mathrm{Col}\{J_{1u}(\mathbf{r}), J_{2u}(\mathbf{r}), \ldots, J_{Gu}(\mathbf{r})\}, \\ [A_{1,u}(\mathbf{r})] &= \{A_{1gg',u}(\mathbf{r})\}, \text{ a } G \times G \text{ matrix.} \end{aligned} \tag{9.4.7}$$

We then define $G \times G$ diffusion-constant matrices $[\tilde{D}_u(\mathbf{r})]$ by

$$[\tilde{D}_u(\mathbf{r})] = \tfrac{1}{3}[A_{1,u}(\mathbf{r})]^{-1} \qquad (u = x, y, z). \tag{9.4.8}$$

Thus, in the general energy-group formalism, the analogue of the operator $\tilde{D}$ is a set of well-defined $G \times G$ matrices having elements that we shall designate as $\tilde{D}_{gg',u}$. It can be shown (essentially because $\Sigma_t(\mathbf{r}, E)$ exceeds $\Sigma_s(\mathbf{r}, E)$ for at least some part of the energy range) that the matrices $[A_{1,u}]^{-1}$ and hence the matrices $[\tilde{D}_u]$ always exist—in fact all the elements of the $[\tilde{D}_u]$ can be proved to be positive.

Operating on (9.4.6) with $-[A_{1,u}(\mathbf{r})]^{-1}$ and writing out the resultant vector equation in scalar form then yields

$$J_{gu}(\mathbf{r}) = -\sum_{g'=1}^{G} \tilde{D}_{gg',u}(\mathbf{r}) \frac{\partial}{\partial u}\Phi_{g'}(u) \qquad (g = 1, 2, \ldots, G;\ u = x, y, z), \tag{9.4.9}$$

and substituting this result into (9.4.4) leads to a set of energy-group diffusion equations

* Since we have been using boldface letters to represent vectors in the vector-analysis sense, we shall henceforth use square brackets to represent matrices—both column matrices (column "vectors") like $[\Phi]$ and square matrices like $[A_{1,u}]$. Braces will be used to indicate the elements of a column matrix or a typical element of a square matrix.

analogous to (9.3.24):

$$-\sum_{g'=1}^{G}\left[\frac{\partial}{\partial x}\tilde{D}_{gg',x}(\mathbf{r})\frac{\partial}{\partial x}+\frac{\partial}{\partial y}\tilde{D}_{gg',y}(\mathbf{r})\frac{\partial}{\partial y}+\frac{\partial}{\partial z}\tilde{D}_{gg',z}(\mathbf{r})\frac{\partial}{\partial z}\right]\Phi_{g'}(\mathbf{r})+\Sigma_{tg}(\mathbf{r})\Phi_{g}(\mathbf{r})$$
$$=\sum_{g'=1}^{G}[\chi_g\nu\Sigma_{fg'}(\mathbf{r})+\Sigma_{gg'}(\mathbf{r})]\Phi_{g'}(\mathbf{r}) \qquad (g=1,2,\ldots,G). \tag{9.4.10}$$

To get the energy-group analogue of (9.3.27) we rewrite (9.4.5) in the rather artificial form

$$\frac{1}{3}\frac{\partial}{\partial u}\Phi_g(\mathbf{r}) = -\frac{\sum_{g'=1}^{G}A_{1gg',u}(\mathbf{r})J_{g'u}(\mathbf{r})}{J_{gu}(\mathbf{r})}J_{gu}(\mathbf{r}) \qquad (g=1,2,\ldots,G;\ u=x,y,z), \tag{9.4.11}$$

and then define

$$\Sigma_{\mathrm{tr},g,u}(\mathbf{r}) \equiv \frac{\sum_{g'=1}^{G}A_{1gg',u}(\mathbf{r})J_{g'u}(\mathbf{r})}{J_{gu}(\mathbf{r})}, \tag{9.4.12}$$

$$D_{gu}(\mathbf{r}) \equiv \frac{1}{3\Sigma_{\mathrm{tr},g,u}(\mathbf{r})}. \tag{9.4.13}$$

The last three equations yield

$$J_{gu}(\mathbf{r}) = -D_{gu}(\mathbf{r})\frac{\partial}{\partial u}\Phi_g(\mathbf{r}) \qquad (g=1,2,\ldots,G;\ u=x,y,z). \tag{9.4.14}$$

Substitution into (9.4.4) then results in

$$\left[-\frac{\partial}{\partial x}D_{gx}(\mathbf{r})\frac{\partial}{\partial x}-\frac{\partial}{\partial y}D_{gy}(\mathbf{r})\frac{\partial}{\partial y}-\frac{\partial}{\partial z}D_{gz}(\mathbf{r})\frac{\partial}{\partial z}+\Sigma_{tg}(\mathbf{r})\right]\Phi_g(\mathbf{r})$$
$$=\sum_{g'=1}^{G}[\chi_g\nu\Sigma_{fg'}(\mathbf{r})+\Sigma_{gg'}(\mathbf{r})]\Phi_{g'}(\mathbf{r}) \qquad (g=1,2,\ldots,G). \tag{9.4.15}$$

Both (9.4.10) and (9.4.15) are formally equivalent to the continuous-energy P_1 equations (9.3.17) and (9.3.18) in the sense that, if group parameters based on $\Phi(\mathbf{r}, E)$ and $\mathbf{J}(\mathbf{r}, E)$ could be determined, solution of either (9.4.10) or (9.4.15) would yield exactly $\int_{\Delta E_g}\Phi(\mathbf{r}, E)\,dE$.

Neither equation, however, has the conventional group diffusion-theory form (4.13.7). In (9.4.15) there are different values of $D_{gu}(\mathbf{r})$ associated with each coordinate direction, and the $\tilde{D}_{gg',u}(\mathbf{r})$ of (9.4.10), in addition to being directionally dependent, result in a contribution of neutrons from other groups g' to the leakage term in group g. Fortunately, as we shall now see, some of these complications disappear when we make the multigroup and few-group approximations.

The Multigroup Approximation

In Chapter 4 it was pointed out that the essential characteristic of a multigroup approximation is the determination of group constants such as the $\Sigma_{\alpha g}(\mathbf{r})$ and $\Sigma_{gg'}(\mathbf{r})$ of (9.4.3) by writing $\Phi(\mathbf{r}, E)$ in the form

$$\Phi(\mathbf{r}, E) = \Phi_g(\mathbf{r})F_{0g}(E) \tag{9.4.16}$$

for E in the interval ΔE_g and all $\mathbf{r}$. The $F_{0g}(E)$ are preselected functions of energy, normalized so that $\int_{\Delta E_g} F_{0g}(E)dE = 1$ and possibly differing from group to group, but the same for all $\mathbf{r}$ throughout the reactor. Thus the $F_{0g}(E)$ may be the constants $1/\Delta E_g$, or pieces of a fission spectrum or of a $1/E$ slowing-down spectrum, or pieces of a "hyperfine" multigroup flux involving thousands of groups. In all cases, however, the same $F_{0g}(E)$ are used for a wide range of material compositions.

To extend the multigroup approximation for present purposes we merely assume that, in addition to approximating $\Phi(\mathbf{r}, E)$ in (9.4.3) by (9.4.16), we may approximate the $J_u(\mathbf{r}, E)$ by

$$J_u(\mathbf{r}, E) \equiv J_{gu}(\mathbf{r})F_{1g}(E) \qquad (u = x, y, z) \tag{9.4.17}$$

for E in ΔE_g and all $\mathbf{r}$, where, as with (9.4.16), the $F_{1g}(E)$ are preselected shapes, the same for all $\mathbf{r}$ and all u and normalized so that $\int_{\Delta E_g} F_{1g}(E)dE = 1$.

Equation (9.4.3) shows immediately that, because we have taken the energy shapes $F_{1g}(E)$ to be independent of direction, the matrix elements $A_{1gg',u}$ are the same for $u = x, y$, or z. It follows from (9.4.8) that the matrix $[\tilde{D}_u(\mathbf{r})]$ must be independent of u. Thus (9.4.10) reduces to

$$-\sum_{g'=1}^{G} \nabla\cdot\tilde{D}_{gg'}(\mathbf{r})\nabla\Phi_{g'}(\mathbf{r}) + \Sigma_{tg}(\mathbf{r})\Phi_g(\mathbf{r}) = \sum_{g'=1}^{G} [\chi_g\nu\Sigma_{fg'}(\mathbf{r}) + \Sigma_{gg'}(\mathbf{r})]\Phi_{g'}(\mathbf{r}). \tag{9.4.18}$$

No such simplification is justified for (9.4.15). Even though the $A_{1gg'}(\mathbf{r})$ in (9.4.12) are independent of u, the ratios $J_{g'u}(\mathbf{r})/J_{gu}(\mathbf{r})$ will still generally be different for different coordinate directions. Thus the $\Sigma_{tr,g,u}(\mathbf{r})$ and hence the $D_{gu}(\mathbf{r})$ will still have a directional dependence. Moreover the $J_{g'u}(\mathbf{r})$ will not be known a priori. Hence, if (9.4.15) is to be applied, it is necessary either to determine the $\Sigma_{tr,g,u}(\mathbf{r})$ iteratively as the calculation proceeds or to make some arbitrary assumption about the behavior of the $J_{g'u}(\mathbf{r})$ needed for (9.4.12). Neither possibility is very attractive, and we conclude, at this stage, that, if $\Sigma_{s1}(\mathbf{r}, E' \to E)$ is significant, one should either deal directly with the P_1 equations (9.4.4) and (9.4.5) or use the form (9.4.18).

Interface Conditions

The alternate forms (9.4.15) and (9.4.18) of the multigroup diffusion equations require the application of different continuity conditions across internal interfaces separating materials of different composition. Across such interfaces $\Sigma_t(\mathbf{r}, E)$ and $\Sigma_{s1}(\mathbf{r}, E' \to E)$ change discontinuously, as, therefore, do the $\tilde{D}_{gg'}$ and the D_{gu}. A plausible counterpart in the P_1 approximation to the requirement that the directional flux $\Psi(\mathbf{r}, \boldsymbol{\Omega}, E)$ be continuous in $\mathbf{r}$ in the direction $\boldsymbol{\Omega}$ is that the Legendre components $\Phi(\mathbf{r}, E)$ and $\mathbf{J}(\mathbf{r}, E)$ (i.e., the integrals $\int \Psi\, d\Omega$ and $\int \Psi\boldsymbol{\Omega}\, d\Omega$) be continuous in $\mathbf{r}$. However, for reasons to be discussed shortly, we apply instead the following somewhat weaker conditions:

$\Phi(\mathbf{r}, E)$ and $\mathbf{n}\cdot\mathbf{J}(\mathbf{r}, E)$ must be continuous for points $\mathbf{r}$ on any surface across which there is a material discontinuity, $\mathbf{n}$ being a normal to that surface. (9.4.19)

When the energy-group approximation is made, it is thus plausible to require that $\Phi_g(\mathbf{r})$ and $\mathbf{n}\cdot\mathbf{J}_g(\mathbf{r})$ be continuous across such surfaces. In view of (9.4.9) and (9.4.14) the multigroup approximation then yields

$$\begin{aligned} &\Phi_g(\mathbf{r})|_{S^-} = \Phi_g(\mathbf{r})|_{S^+}, \\ &\sum_{g'=1}^{G} \tilde{D}_{gg'}(\mathbf{r})\frac{\partial}{\partial n}\Phi_{g'}(\mathbf{r})\bigg|_{S^-} = \sum_{g'=1}^{G} \tilde{D}_{gg'}(\mathbf{r})\frac{\partial}{\partial n}\Phi_{g'}(\mathbf{r})\bigg|_{S^+}, \\ &D_{gn}(\mathbf{r})\frac{\partial}{\partial n}\Phi_g(\mathbf{r})\bigg|_{S^-} = D_{gn}(\mathbf{r})\frac{\partial}{\partial n}\Phi_g(\mathbf{r})\bigg|_{S^+}, \end{aligned} \tag{9.4.20}$$

where subscripts S^- and S^+ indicate that quantities are to be evaluated on the $(-)$ and $(+)$ sides of a surface, $\partial\Phi_g(\mathbf{r})/\partial n$ stands for the partial derivative with respect to the direction normal to the surface, and D_{gn} is the value of the group-g diffusion constant (9.4.13) for the direction $\mathbf{n}$. (As has already been noted, the multigroup values of the $\tilde{D}_{gg'}(\mathbf{r})$ are the same for all component directions.)

Note that the conditions also apply to the degenerate case of a surface across which there is *no* discontinuity. Generally there is no need to apply the conditions explicitly at such surfaces. Nevertheless they are being imposed implicitly on the solution of the multigroup equations by the way in which we construct the general solution within a homogeneous medium.

To understand the difficulty that arises if we try to make the vector $\mathbf{J}(\mathbf{r}, E)$ continuous across the surface of a material discontinuity, consider the two-dimensional situation depicted by Figure 9.1 and simplify to the case where $\Sigma_{s1}(\mathbf{r}, E' \to E) = 0$ for both materials. Then, just as in (9.3.1), the continuous-energy diffusion constant $D(\mathbf{r}, E)$ becomes the algebraic function $1/3\Sigma_t(\mathbf{r}, E)$. If we define $D^{(1)}(E)$ and $D^{(2)}(E)$ to be this diffusion constant in materials (1) and (2), we have

$$\left.\begin{aligned} \mathbf{J}(x, y, E) &= -D^{(1)}(E)\nabla\Phi(x, y, E) \\ J_y(x, y, E) &= -D^{(1)}(E)\frac{\partial}{\partial y}\Phi(x, y, E) \end{aligned}\right\}(x, y) \text{ in Material (1),} \tag{9.4.21}$$

$$\left.\begin{aligned} \mathbf{J}(x, y, E) &= -D^{(2)}(E)\nabla\Phi(x, y, E) \\ J_y(x, y, E) &= -D^{(2)}(E)\frac{\partial}{\partial y}\Phi(x, y, E) \end{aligned}\right\}(x, y) \text{ in Material (2).} \tag{9.4.22}$$

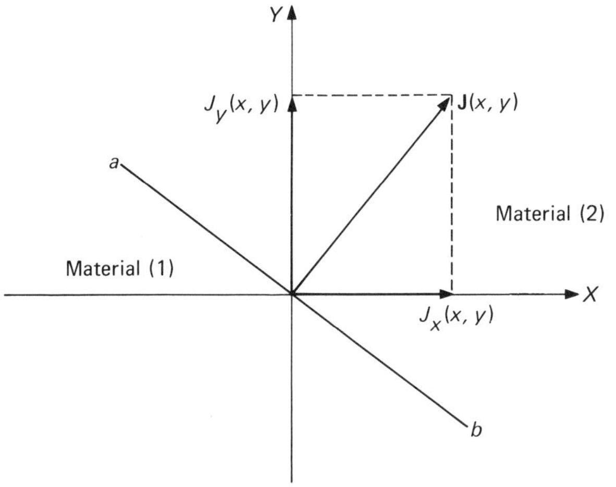

Figure 9.1 Current-continuity conditions across a surface separating two materials.

Now, if $x = 0^+$, $y = 0$, is a point just to the right of the Y axis and $x = 0^-$, $y = 0$, is a point just to the left, requiring $\mathbf{J}(x, y, E)$ to be continuous in x and y at (0, 0) implies that

$$\begin{aligned} -D^{(1)}(E)\nabla\Phi(0^-, 0, E) &= -D^{(2)}(E)\nabla\Phi(0^+, 0, E), \\ -D^{(1)}(E)\frac{\partial}{\partial x}\Phi(0^-, 0, E) &= -D^{(2)}(E)\frac{\partial}{\partial x}\Phi(0^+, 0, E), \\ -D^{(1)}(E)\frac{\partial}{\partial y}\Phi(0^-, 0, E) &= -D^{(2)}(E)\frac{\partial}{\partial y}\Phi(0^+, 0, E). \end{aligned} \tag{9.4.23}$$

The last of these relationships is inconsistent with the requirement that $\Phi(x, y, E)$ be continuous across the Y axis, for, if $\Phi(0^-, y, E) = \Phi(0^+, y, E)$ for all y, $\partial\Phi(0^-, y, E)/\partial y$ must equal $\partial\Phi(0^+, y, E)/\partial y$ for all y and in particular for $y = 0$. But, for finite values of $\partial\Phi(0^\pm, 0, E)/\partial y$, the last equation in (9.4.23) permits this only if $D^{(1)}(E) = D^{(2)}(E)$ or if $\partial\Phi(0^-, 0, E)/\partial y = \partial\Phi(0^+, 0, E)/\partial y = 0$ (and, by extension, if $\partial\Phi(0^-, y, E)/\partial y = \partial\Phi(0^+, y, E)/\partial y = 0$ for all y). The first of these possibilities contradicts the initial hypothesis and the second is clearly absurd. Thus, as the continuity condition across interfaces separating different compositions, we require only that the second equation in (9.4.23) be valid. This leads to (9.4.20) in energy-group theory for the more general situation where $\Sigma_{s1}(\mathbf{r}, E' \to E) \neq 0$. The fact that the first equation in (9.4.23) is invalid is no cause for concern physically since it refers to only *one point* on the surface (*ab*) in Figure 9.1. From point *a* to $(0^-, 0)$ all three equations in (9.4.23) are valid since $D^{(2)}(E)$ is replaced by $D^{(1)}(E)$ in that range. Similarly all three equations in (9.4.23) are valid in the range $(0^+, 0)$ to b. Thus the net flow of neutrons is continuous across (ab) except at the point (0, 0). Since it is not physically meaningful to speak of a net current through a *point*, there is no reason to be concerned that unphysical consequences will result from the apparent ànomaly at point (0, 0).

An analogous but more serious difficulty arises at corner points where there is a material discontinuity in both the X and Y directions. At such a point even the somewhat restricted interface conditions (9.4.19) cause difficulty in that, if they are applied, both $\partial\Phi/\partial x$ and $\partial\Phi/\partial y$ become infinite (although Φ itself remains finite). Again the singularity is of no physical concern. However it turns out to affect the rate at which finite-difference solutions of the group diffusion equations converge to the correct answer as the spatial-mesh spacings approach zero.

The Few-Group Approximation

Few-group diffusion theory results when the spectrum functions $F_{0g}(E)$ and $F_{1g}(E)$, associated with $\Phi(\mathbf{r}, E)$ and the $J_u(\mathbf{r}, E)$ respectively, are determined from the particular material composition present at a given point $\mathbf{r}$. In our earlier discussion of the theory in Chapter 4 we noted that the spectrum function generally assumed for the scalar flux at point $\mathbf{r}$ is the one that would be present asymptotically far inside a region composed entirely of the material at that $\mathbf{r}$. We determined that spectrum by solving (4.10.3) by the multigroup method, having found this equation by assuming that, far inside a homogeneous material consisting of composition k, the leakage term $D^k(E)\nabla^2\Phi(\mathbf{r}, E)$ becomes $-D^k(E)\,(B_m^k)^2 R^k(\mathbf{r})\psi^k(E)$, where $(B_m^k)^2$ is the materials buckling for material k, $\psi^k(E)$ is the spectrum we seek, and $R^k(\mathbf{r})$ is the separable spatial part of the scalar flux which has been factored out in obtaining (4.10.3). Extending this asymptotic assumption to the present case leads to the suggestion that the appropriate expressions for $F_{0g}(E)$ and $F_{1g}(E)$ are the P_0 and P_1 components of the asymptotic spectrum associated with the materials buckling for the material at point $\mathbf{r}$. These spectra may be found by solving in multigroup fashion the asymptotic P_1 equations (8.4.23) or the B_1 equations (8.4.25) for the everywhere-positive functions $F_0(E)$ and $iBF_1(E)$ associated with the most positive eigenvalue B_m^2.

We saw in Section 8.4 that the energy shape of the P_1 component of an asymptotic spectrum is the same for all orientations of the axis system. Thus (9.4.17) with $F_{1g}(E)$ independent of u is valid asymptotically far inside a material region even if ΔE_g is a large energy range. The last equation in (9.4.3) thus becomes, for material k and $u = x, y$, or z,

$$A_{1gg'}^k = \frac{\int_{E_g}^{E_{g-1}} dE \int_{E_{g'}}^{E_{g'-1}} dE'[\Sigma_t^k(E)\delta(E' - E) - \Sigma_{s1}^k(E' \to E)]F_1^k(E')}{\int_{E_{g'}}^{E_{g'-1}} dE'\, F_1^k(E')}, \tag{9.4.24}$$

where $F_1^k(E')$ is the P_1 component of the asymptotic spectrum appropriate to material k

and we have written the denominator explicitly since the $F_1^k(E')$ will not in general be normalized in such a way that $\int_{E_{g'}}^{E_{g'-1}} dE'\ F_1^k(E')$ is unity.

Equation (9.4.24) is written as if $F_1^k(E)$ were a continuous function of energy. Actually it is the result of a multigroup solution of (8.4.23) or (8.4.25). Thus $F_1^k(E')$ is approximated by a set of multigroup fluxes F_{1n}^k ($n = 1, 2, \ldots, N$), where N is the total (large) number of groups employed in the multigroup model. In terms of these fluxes and the corresponding multigroup cross sections, (9.4.24) becomes

$$A_{1gg'}^k = \frac{\sum_{n\subset g} \Delta E_n \sum_{n'\subset g'}[\Sigma_{\text{t}n'}^k \delta_{n'n} - \Sigma_{\text{s}1;nn'}^k]F_{1n'}^k}{\sum_{n'\subset g'} F_{1n'}^k}, \tag{9.4.25}$$

where the sums are over all narrow groups n (or n') contained in the wide group g (or g').

Corresponding expressions for the other group parameters in (9.4.3), written for material k in terms of the multigroup approximation to $F_0^k(E)$, are

$$\begin{aligned} \Sigma_{\alpha g}^k &= \frac{\sum_{n\subset g} \Sigma_{\alpha n}^k F_{0n}^k}{\sum_{n\subset g} F_{0n}^k} \qquad (\alpha = \text{f, t}), \\ \chi_g &= \sum_{n\subset g} \chi_n, \\ \Sigma_{gg'}^k &= \frac{\sum_{n\subset g} \Delta E_n \sum_{n'\subset g'} \Sigma_{\text{s}0;nn'}^k F_{0n'}^k}{\sum_{n'\subset g'} F_{0n'}^k}. \end{aligned} \tag{9.4.26}$$

These parameters along with the components $\tilde{D}_{gg'}^k$ of the few-group (direction-independent) diffusion-constant matrix obtained from (9.4.25), (9.4.7), and (9.4.8) lead to few-group equations identical in form with (9.4.18). The first and second continuity conditions in (9.4.20) are thus appropriate. (We note that this form of the few-group equations still involves coupling between groups g and g' through the leakage terms.)

To obtain the few-group analogue of the group-g transport cross section (9.4.12) for material k in order to derive the conventional few-group equations, we assume that in (9.4.12) $J_{g'u}(\mathbf{r})$ and $J_{gu}(\mathbf{r})$ can be approximated by unnormalized pieces of the P_1 spectrum function $F_1^k(E)$. Since this function is independent of direction, the resulting few-group transport cross section will also be independent of u. Thus, replacing the $A_{1gg',u}$ by their definitions (9.4.3), we find (in integral form) that, for a wide group g,

$$\Sigma_{\text{tr},g}^k = \frac{\sum_{g'=1}^{G} \int_{E_g}^{E_{g-1}} dE \int_{E_{g'}}^{E_{g'-1}} dE'\ [\Sigma_{\text{t}}^k(E)\delta(E'-E) - \Sigma_{\text{s}1}^k(E'\to E)]F_1^k(E')}{\int_{E_g}^{E_{g-1}} dE'\, F_1^k(E')}, \tag{9.4.27}$$

or, recognizing that $F_1(E')$ will be a multigroup result,

$$\Sigma_{\text{tr},g}^k = \frac{\sum_{n\subset g} \Delta E_n \sum_{n'=1}^{N} [\Sigma_{\text{t}n}^k \delta_{nn'} - \Sigma_{\text{s}1;nn'}^k]F_{1n'}^k}{\sum_{n\subset g} F_{1n}^k}, \tag{9.4.28}$$

where the second sum in the numerator is over all the narrow groups n' used in the multigroup model.

Since $\Sigma_{tr,g}^k$ is independent of direction, the diffusion constant D_g^k for group g, defined by (9.4.13), will also be independent of direction, and (9.4.15) reduces to the conventional form

$$-\nabla \cdot D_g \nabla \Phi_g + \Sigma_{tg}(\mathbf{r})\Phi_g(\mathbf{r}) = \sum_{g'=1}^{G} [\chi_g \nu \Sigma_{fg'}(\mathbf{r}) + \Sigma_{gg'}(\mathbf{r})]\Phi_{g'}(\mathbf{r}) \qquad (g = 1, 2, \ldots, G), \qquad (9.4.29)$$

where the parameters other than $D_g(\mathbf{r})$ are given by (9.4.26).

In this result D_g^k, Σ_{tg}^k, etc., have been replaced with the more general notation $D_g(\mathbf{r})$, $\Sigma_{tg}(\mathbf{r})$, etc., to allow for the possibility that the few-group parameters may vary slightly with position in region k (for example because of nonuniform depletion effects). The notation $D_g(\mathbf{r})$ can account for minor variations of this nature as well as for the discontinuous step changes between regions of different composition. The first and third continuity conditions (9.4.20) are the appropriate ones for this equation.

Few-group equations of the type (9.4.29), which when cast in matrix form involve a diagonal matrix $[D]$, are easier to deal with numerically than those of the form (9.4.18) involving the full matrix $[\tilde{D}]$. However the approximations made in deriving (9.4.29) seem somewhat less mild than those made to obtain (9.4.18). Both forms use the parameters Σ_{tg}, $\chi_g \nu \Sigma_{fg}$, and $\Sigma_{gg'}$, determined by (9.4.26) from the multigroup approximation to $F_0^k(E)$, the P_0 component of the asymptotic spectrum associated with material k. But determining the $A_{1gg'}$ from (9.4.24), or, more precisely, from its multigroup counterpart (9.4.25), and using them to find the $D_{gg'}$ requires only that we approximate the energy shape of the currents $J_u(\mathbf{r}, E')$ within material k for the limited energy range $\Delta E_{g'}$, whereas determining $\Sigma_{tr,g}^k$ from (9.4.27) or (9.4.28) requires that we estimate the energy shape of $J_u(\mathbf{r}, E')$ over the entire energy range. Approximating the energy shape of $J_u(\mathbf{r}, E')$ by $F_1^k(E')$ over a limited energy range is more accurate than so approximating it over the entire energy range. The error is not as significant as it might appear, however, since the $\Sigma_{s1;nn'}$ tend to be small and are zero for elastic scattering in heavy elements if group n' is far from group n. Thus the values of the $F_{1n'}^k$ in (9.4.28) for narrow groups n' far outside the energy range of wide group g are not too important. For the same reason, if the full matrix $[\tilde{D}]$ is used for a few-group scheme, the $A_{1gg'}$ of (9.4.25)—and hence the $\tilde{D}_{gg'}$—will be small for $g' \neq g$. Thus the full matrix $[\tilde{D}]$ and the diagonal matrix $[D]$ will not be as different as might at first be thought. For these reasons, and because the equations are simpler to deal with numerically, the form (9.4.29) of the few-group diffusion equations is the one almost universally used.

Conclusions

We have seen that the energy-group P_1 equations (9.4.4) and (9.4.5) may be reduced to diffusion-theory form in two ways depending on how one defines the group diffusion constants. In the formally exact case, moreover, we have found that the diffusion constants can depend on direction. Before summarizing and discussing the several possible group diffusion-theory models we have constructed, we will find it useful to rewrite the equations defining the models in compact matrix form. The matrix form will be particularly convenient for the discussion of approximation procedures in Chapter 10.

Accordingly, in terms of the general group parameters (9.4.3), we define the column vectors

$$[\nu^j\Sigma_f^j] \equiv \mathrm{Col}\{\nu_1^j\Sigma_{f1}^j(\mathbf{r}), \nu_2^j\Sigma_{f2}^j(\mathbf{r}), \ldots, \nu_G^j\Sigma_{fG}^j(\mathbf{r})\},$$
$$[\chi^j] \equiv \mathrm{Col}\{\chi_1^j, \chi_2^j, \ldots, \chi_G^j\}, \tag{9.4.30}$$

and the $G \times G$ matrix $[A_0]$ having as its (gg') element

$$A_{0gg'} \equiv \Sigma_{tg}(\mathbf{r})\delta_{gg'} - \Sigma_{gg'}(\mathbf{r}). \tag{9.4.31}$$

We further define the $G \times G$ matrix $[M]$ by

$$[M] \equiv \sum_j [\chi^j][\nu^j\Sigma_f^j]^{\mathrm{T}}, \tag{9.4.32}$$

where $[\nu^j\Sigma_f^j]^{\mathrm{T}}$ is a G-element row vector, the transpose of the column vector $[\nu^j\Sigma_f^j]$. Note that in defining $[M]$ we have returned to the more general case of fissioning by several isotopes. The individual $G \times G$ matrices $[\chi^j][\nu^j\Sigma_f^j]^{\mathrm{T}}$ of which $[M]$ is composed are called *dyads*, column vectors multiplied into row vectors. They must not be confused with *scalar products* such as $[\nu^j\Sigma_f^j]^{\mathrm{T}}[\chi^j]$, which, being rows multiplied into columns, are numbers.

We shall use the notation (9.4.30)–(9.4.32) for all energy-group models whether the group parameters are exact, as with (9.4.3), or approximate, as with the few-group constants (9.4.26). With this understanding and using the column vectors $[\Phi]$ and $[J_u]$ and the matrix $[\tilde{D}]$ defined by (9.4.7) and (9.4.8), we may rewrite (9.4.18) as

$$-\nabla\cdot[\tilde{D}]\nabla[\Phi] + [A_0][\Phi] = [M][\Phi]. \tag{9.4.33}$$

This is an unconventional form of group diffusion equations in that the diffusion-constant matrix $[\tilde{D}]$ has off-diagonal terms. We have seen that, in both the multigroup and few-group approximations, a single value of $[\tilde{D}]$ applies for all coordinate directions. (If this were not so, the term $\nabla\cdot[\tilde{D}]\nabla[\Phi]$ would have to be written

$$\sum_u \frac{\partial}{\partial u}\left([\tilde{D}]\frac{\partial}{\partial u}[\Phi]\right).)$$

To write the conventional form (9.4.29) of the group diffusion equations in matrix form we define the diagonal matrix $[D]$ by

$$[D] = \text{Diag}\{D_1(\mathbf{r}), D_2(\mathbf{r}), \ldots, D_G(\mathbf{r})\}, \tag{9.4.34}$$

where $D_g(\mathbf{r}) = 1/3\Sigma_{tr,g}(\mathbf{r})$ and $\Sigma_{tr,g}(\mathbf{r})$ is given, for the few-group case, by (9.4.28). Equation (9.4.29) may then be written

$$-\nabla\cdot[D]\nabla[\Phi] + [A_0][\Phi] = [M][\Phi]. \tag{9.4.35}$$

We have seen that this conventional form of the energy-group equations applies, in general, only to a few-group model. For a multigroup model there will be different values of $[D]$ for different coordinate directions unless there is negligible anisotropic scattering or unless the energy loss associated with anisotropic scattering is very small. Moreover, when the $[D_u]$ are directionally dependent, their values ought to be found iteratively as the calculation proceeds.

Table 9.1 lists the equations defining the model and the elements of the corresponding diffusion-constant matrices.

It should be remembered that, if $\Sigma_{s1}(\mathbf{r}, E' \to E)$ vanishes or can be well approximated by $\Sigma_{s1}(\mathbf{r})\delta(E' \to E)$, $[D] = [\tilde{D}]$ in both the few-group and multigroup models and the directional dependence of $[D]$ vanishes in the multigroup model. Moreover, in the continuous-energy case, the operator $\tilde{D}$ given by (9.3.21) and the three functionals $D_u(\mathbf{r}, E)$ given by (9.3.26) all become equal to the algebraic function $1/3\{\Sigma_t(\mathbf{r}, E) - \Sigma_{s1}(\mathbf{r}, E)\}$. The complication to the group diffusion-theory model due to energy changes associated with anisotropic scattering are thus major.

One further point concerning direction-dependent diffusion constants should be made: the directional dependence we have encountered in these last two sections arises essentially from the fact that the spectrum of $\mathbf{J}(\mathbf{r}, E)$ may be different in different directions. So-called *anisotropic djffusion constants* are also encountered when one tries to force diffusion theory to predict what is essentially a transport-theory phenomenon. Thus, even

Table 9.1 Equations Defining Models and the Corresponding Elements of Diffusion-Constant Matrices

Model	Nature of Diffusion Matrix	Directionally Dependent?	Equations Defining Element of $[D]$ or $[\tilde{D}]$*	Energy-Group Equations**
Multigroup	Full	No	3, 7, 8	18, 33
Multigroup	Diagonal	Yes	3, 12, 13	15
Few-group	Full	No	25, 7, 8	18, 33
Few-group	Diagonal	No	28, 13	29, 35

*Notation: 3 means (9.4.3), etc.
**18 means (9.4.18), etc.

in a one-group model with isotropic scattering, if $\mathbf{J}$ and Φ have been found by a transport calculation, trying to find a value of D such that $\mathbf{J} = -D\nabla\Phi$ will usually require that D have directional components. A similar need to employ direction-dependent D's arises if we try to find homogeneous diffusion-theory parameters that will reproduce, in an integral sense, the reaction rates associated with a heterogeneous geometrical region (for example a fuel subassembly). We shall encounter such situations when we discuss homogenization procedures in Chapter 10.

9.5 Approximations Made in Determining Few-Group Parameters

The few-group procedure of approximating the P_0 and P_1 components of the spectrum within a wide group g at location $\mathbf{r}$ by the asymptotic functions $F_0(E)$ and $F_1(E)$ appropriate to the material at $\mathbf{r}$ is justified only under certain circumstances. As was discussed in Chapter 4, the situation in which the procedure is most likely to be valid occurs when the reactor is composed of a number of large, isotopically homogeneous regions. In addition we saw in Chapter 5 that collision-theory methods can be used to determine homogenized multigroup cross sections for a heterogeneous cell composed of a fuel element and its associated moderator (see Section 5.7). A review of this procedure will show that it is based on assuming there is no net leakage from the cell and on finding spatially averaged values of the multigroup fluxes within the fuel rod and within the moderator. These are then used to determine the homogenized multigroup constants $\bar{\Sigma}_{tn}, \overline{\nu\Sigma}^{j}_{fn}$, and $\bar{\Sigma}_{nn'}$ according to (5.5.4) or (5.5.8). One can insert these homogenized constants into the multigroup counterpart of (8.4.23) and, since the equivalent homogenized buckling is zero (there being no net leakage from the cell), obtain a multigroup equation for the components $\bar{F}_{0n}$ of the homogenized multigroup flux:

$$\bar{\Sigma}_{tn}\bar{F}_{0n} = \sum_{n'} [\chi_n \overline{\nu\Sigma}_{fn'} + \bar{\Sigma}_{nn'}]\bar{F}_{0n'}. \tag{9.5.1}$$

Equation (9.4.26) can then be used to reduce to few-group parameters. Thus, for lattices for which k_∞ is unity, equivalent homogenized few-group interaction cross sections can readily be found. Unfortunately this procedure fails to yield values for the few-group diffusion constants. The collision-theory method developed in Chapter 5 provides no information about spatially averaged values of the net current density. Thus it does not permit us to determine current-weighted values of $\Sigma_t(\mathbf{r}, E)$ and $\Sigma_{s1}(\mathbf{r}, E' \to E)$ for the cell (and hence, by applying equations like (9.4.25) and (9.4.8), or (9.4.28) and (9.4.13), to determine diffusion constants).

An entirely satisfactory resolution of this problem is not a simple task. (We shall discuss the matter further in Chapter 10 when we consider equivalent homogenized diffusion-theory parameters in a general way.) Fortunately the problem is thought not to be serious since the fuel-element cells of fast reactors tend to be nuclearly homogeneous and the cell-to-cell neutron-leakage rates of large thermal reactors tend to be small. Accordingly the present design practice is to determine homogenized values for the $A_{1nn'}$ of (9.4.3) for a fuel-element cell either by treating the cell as a homogeneous medium in finding the multigroup cross sections that make up $\bar{A}_{1nn'}$ or, in effect, by spatially flux-weighting $\Sigma_t(\mathbf{r}, E)$ and $\Sigma_{s1}(\mathbf{r}, E' \to E)$ in (9.4.3) by $\Phi(\mathbf{r}, E)$ rather than by $\mathbf{J}(\mathbf{r}, E)$. Thus $\bar{A}_{1nn'}$ is treated like $\bar{\Sigma}_{tn}$, $\overline{\nu\Sigma^j_{fn}}$, and $\bar{\Sigma}_{nn'}$ in (5.5.4) or (5.5.8). In the remainder of the present section we shall assume that, if few-group parameters are being obtained for a material composition containing fuel-element cells, a set of multigroup cross sections for the homogenized cell material is already available. This assumption having been made, we shall first examine more closely the difference between a P_1 and a B_1 approximation for finding $F_0(E)$ and $F_1(E)$ and then discuss some of the additional approximations made when the procedures for finding few-group parameters are applied in practice.

Comparison between the P_1 and B_1 Equations for $F_0(E)$ and $F_1(E)$

The decision whether to use the P_1 or B_1 approximation to find the spectrum components $F_0(E)$ and $F_1(E)$ is made easier if we write the B_1 equations in a form similar to that of the P_1 equations.

The P_1 equations (which, for simplicity, we shall use in their continuous-energy form) are obtained by writing (8.4.23) for $l = 0$ and $l = 1$ and neglecting $-2iBF_2(E)$ in the second equation. The result is

$$\begin{aligned} -iB_m F_1(E) + \Sigma_t(E)F_0(E) &= \int_0^\infty dE'[f(E)\nu\Sigma_f(E') + \Sigma_{s0}(E' \to E)]F_0(E'), \\ -iB_m F_0(E) + 3\Sigma_t(E)F_1(E) &= 3\int_0^\infty dE'\,\Sigma_{s1}(E' \to E)F_1(E'). \end{aligned} \tag{9.5.2}$$

The analogous B_1 equations are obtained by writing (8.4.25) for $l = 0$ and $l = 1$, neglecting $\Sigma_{sn}(E' \to E)$ in both equations for $n > 1$, and using the formulas

$$\int_{-1}^{1} \frac{1}{2}\frac{d\mu}{1 - i\alpha\mu} = \frac{\tan^{-1}\alpha}{\alpha}, \qquad \int_{-1}^{1} \frac{1}{2}\frac{\mu d\mu}{1 - i\alpha\mu} = i\frac{1 - \beta}{\alpha}, \qquad \int_{-1}^{1} \frac{1}{2}\frac{\mu^2 d\mu}{1 - i\alpha\mu} = \frac{1 - \beta}{\alpha^2},$$

$$\text{where } \alpha(E) \equiv \frac{B_m}{\Sigma_t(E)}, \qquad \beta(E) \equiv \frac{\tan^{-1}\alpha(E)}{\alpha(E)}. \tag{9.5.3}$$

The result is

$$\begin{aligned} F_0(E) &= \frac{1}{\Sigma_t}\int_0^\infty dE'\left\{[f(E)\nu\Sigma_f(E') + \Sigma_{s0}(E' \to E)]F_0(E')\beta + 3\Sigma_{s1}(E' \to E)i\frac{1 - \beta}{\alpha}F_1(E')\right\}, \\ F_1(E) &= \frac{1}{\Sigma_t}\int_0^\infty dE'\left\{[f(E)\nu\Sigma_f(E') + \Sigma_{s0}(E' \to E)]F_0(E')i\frac{1 - \beta}{\alpha}\right. \\ &\quad \left. + 3\Sigma_{s1}(E' \to E)F_1(E')\frac{1 - \beta}{\alpha^2}\right\}. \end{aligned} \tag{9.5.4}$$

To cast this result into a form similar to (9.5.2) we multiply the second equation in (9.5.4) by $i\alpha\beta/(1-\beta)$ and add it to the first. The result is

$$F_0 + \frac{i\alpha\beta}{1-\beta}F_1 = \frac{3i}{\Sigma_t\alpha}\int_0^\infty \Sigma_{s1}(E' \to E)F_1(E'), \tag{9.5.5}$$

from which, using $\alpha\Sigma_t = B_m$ and defining

$$\gamma(E) \equiv \frac{\beta\alpha^2}{3(1-\beta)} = \frac{\alpha\tan^{-1}\alpha}{3\{1-[(\tan^{-1}\alpha)/\alpha]\}} \approx 1 + \frac{4}{15}\alpha^2 = 1 + \frac{4}{15}\left(\frac{B_m}{\Sigma_t}\right)^2, \tag{9.5.6}$$

we obtain

$$-iB_mF_0(E) + 3\gamma(E)\Sigma_t(E)F_1(E) = 3\int_0^\infty dE'\Sigma_{s1}(E' \to E)F_1(E'). \tag{9.5.7}$$

Similarly, if we multiply the second equation in (9.5.4) by $i\alpha$, subtract it from the first, and multiply by $\Sigma_t(E)$, we obtain

$$-iB_mF_1(E) + \Sigma_t(E)F_0(E) = \int_0^\infty dE'[f(E)\nu\Sigma_f(E') + \Sigma_{s0}(E' \to E)]F_0(E'). \tag{9.5.8}$$

In this form the B_1 equations (9.5.7) and (9.5.8) differ from the P_1 equations (9.5.2) only by the presence of the function $\gamma(E)$ in (9.5.7), and (9.5.6) shows that $\gamma \to 1$ as $B_m^2 \to 0$. Thus the B_1 and P_1 approximations to the spectrum functions $F_0(E)$ and $F_1(E)$ become indistinguishable when the materials buckling is small (as it generally is in large thermal power reactors). For this reason, and in order to be consistent with the diffusion-theory (P_1) model, in which the group parameters are generated from $F_0(E)$ and $F_1(E)$, the P_1 equations (9.5.2) (in multigroup form) are the ones usually employed to determine the spectrum functions for power-reactor analyses.

For those rare cases in which a bare, homogeneous reactor is being analyzed, the B_1 equations are to be preferred. They provide more accurate values of $F_0(E)$ and $F_1(E)$—in fact *exact* values if $\Sigma_{sn}(E' \to E) = 0$ for $n > 1$.

It follows that the materials buckling found when a multigroup solution of (9.5.7) and (9.5.8) is carried out (treating the B_m^2 dependence of $\gamma(E)$ in (9.5.7) by an iterative procedure) should be quite accurate. The discussion in Section 8.4 shows that, if an estimate of the extrapolation distance δ is available, the value of B_m^2 immediately yields a prediction of the critical size of the bare, homogeneous reactor. Moreover it is not too difficult to show that the value of B_m^2 that results from the solution of the P_1 equations is smaller than that resulting from solution of the B_1 equations. Thus the critical size as predicted by the P_1 approximation (and hence by diffusion theory) will always exceed that predicted by transport theory.

Approximate Determination of $F_0(E)$ and $F_1(E)$ in the Slowing-Down Region

Using an iterative procedure, we can solve (9.5.2) or (9.5.7) and (9.5.8) for the materials buckling B_m^2 and the corresponding P_0 and P_1 spectra $F_0(E)$ and $F_1(E)$ over the entire energy range. In practice this is rarely done since it has been found that certain additional approximations greatly simplify the problem of determining $F_0(E)$ and $F_1(E)$ without introducing errors exceeding those already inherent in the basic few-group procedure (namely those arising from the assumption that $F_0(E)$ and $F_1(E)$ may be used as estimates of the P_0 and P_1 spectra throughout an *entire* material composition).

The chief approximation of this nature is to split the spectrum equations into separate equations over the fast and thermal energy ranges and to treat these as inhomogeneous equations with known sources. Thus it is customary to define an energy cut point E_c separating the "slowing-down region" from the "thermal region." It is assumed that no neutrons appear at any energy above E_c as the result of having gained energy in a scattering collision. Since there is, actually, a finite probability that a neutron may gain a considerable amount of energy in a thermal scattering collision, this assumption is an approximation. However E_c can easily be chosen to make any errors resulting from this approximation negligibly small. Taking E_c to be 1 eV, for example, leads to a negligible probability that a neutron with initial energy below E_c will emerge from a scattering collision with final energy greater than E_c. In fact taking E_c to be 0.625 eV, a traditional value for light-water-moderated thermal reactors, appears to be quite acceptable.

In the slowing-down range ($E > E_c$), $\Sigma_{s0}(E' \to E)$ and $\Sigma_{s1}(E' \to E)$ both vanish for $E' < E$. Thus (9.5.7) and (9.5.8) may be written, for $E > E_c$,

$$\begin{aligned} -iB_m F_1 + \Sigma_t F_0 &= \int_E^\infty dE'\,\Sigma_{s0}(E' \to E)F_0(E') + \sum_j f^j(E)\int_0^\infty dE'\,\nu^j\Sigma_f^j(E')F_0(E'), \\ -iB_m F_0 + 3\gamma\Sigma_t F_1 &= 3\int_E^\infty dE'\,\Sigma_{s1}(E' \to E)F_1(E'), \end{aligned} \tag{9.5.9}$$

where we use the B_1 form since any result based on it can be applied immediately to the P_1 case by setting $\gamma = 1$; also, since it matters for what we are about to do, we have returned to a notation which indicates that there may be a number of fissionable isotopes present.

To solve (9.5.9) (cast in multigroup form) in an approximate manner we *guess* at B_m and replace the last term of the first equation by $\bar{f}(E)$, some "average" fission spectrum appropriate to the composition in question. This latter approximation is justified on the grounds that (1) the fission spectra for the different fissionable isotopes are fairly similar, so that replacing each of the $f^j(E)$ by some average $\bar{f}(E)$ is a good approximation, and (2) $\sum_j \int_0^\infty dE'\,\nu^j\Sigma_f^j(E')F_0(E')$, being just a number, can be taken as unity no matter what

$F_0(E')$ turns out to be. Thus we can immediately find F_0 and F_1 in the slowing-down range to within a multiplicative constant. For determining few-group parameters, this is all that is needed.

Having to guess at a value for B_m is not as severe a limitation on accuracy as it might at first seem. The spectra $F_0(E)$ and $F_1(E)$ are not too sensitive to the guess, and a little experience soon makes the choice easy. In fact, when the k_∞ of the material is less than unity (so that B_m should be pure imaginary), it is customary, in order to avoid possible numerical instabilities in solving (9.5.9), to assume a totally artificial *real* value for B_m.

For a material composition in which no fissionable isotopes are present—for example reflector material—it is customary still to take $\bar{f}(E)$ as an inhomogeneous source in the first equation in (9.5.9). The justification for this is that such material usually comprises either the reflector or a small internal region within the reactor, near the fuel. In either case it is argued that the neutrons leaking into the material from the reactor and making their first collision there are largely the high-energy neutrons coming directly from fission and hence having an energy spectrum approximately proportional to $\bar{f}(E)$.

Approximate Determination of $F_0(E)$ and $F_1(E)$ in the Thermal Energy Range

In the thermal range below E_c both "up" and "down" scattering in energy must be dealt with, and the source of thermal neutrons comes from slowing down from above E_c rather than from fission ($f(E) = 0$ for $E \leq E_c$). Accordingly the appropriate simplification of (9.5.8) and (9.5.7) is, for $E < E_c$,

$$\begin{aligned} -iB_m F_1 + \Sigma_t F_0 &= \int_0^{E_c} dE'\,\Sigma_{s0}(E' \to E) F_0(E') + \int_{E_c}^{\infty} dE'\,\Sigma_{s0}(E' \to E) F_0(E'), \\ -iB_m F_0 + 3\gamma\Sigma_t F_1 &= 3\int_0^{E_c} dE'\Sigma_{s1}(E' \to E) F_1(E') + 3\int_{E_c}^{\infty} dE'\Sigma_{s1}(E' \to E) F_1(E'), \end{aligned} \tag{9.5.10}$$

where, if we are working in a P_1 rather than a B_1 approximation, γ is unity.

The last terms in these equations represent sources of neutrons slowing down into the thermal energy range from higher energies. The values of $F_0(E')$ and $F_1(E')$ in these integrals can be obtained from the approximate solution of (9.5.9). With the "slowing-down" sources determined in this manner and a value of B_m again chosen in accord with past experience, a multigroup solution of (9.5.10) for the thermal energy range yields a flux and current spectrum from which few-group thermal constants can be determined.

In practice this procedure is usually simplified by making a number of assumptions which, in order of decreasing accuracy, are as follows:

1. In the scattering-in integrals for the energy range $E_c < E < \infty$ we set

$$F_0(E') = \frac{1}{E'} \quad \text{and} \quad F_1(E') = \frac{iB_m F_0(E')}{3[\gamma\Sigma_t(E_c) - \Sigma_{s1}(E_c)]}.$$

(Most of the neutrons first appear in the thermal range after having been scattered from energies close to E_c. Even for hydrogen moderation about 90 percent of the neutrons scattered below E_c come from the energy range E_c to $10E_c$. Thus we assume that $F_0(E')$ for $E' > E_c$ has the $1/E$ energy shape characteristic of neutrons slowing down through elastic scattering in a weakly absorbing medium. The estimated ratio of $F_1(E')$ to $F_0(E')$ comes from the second equation in (9.5.10) evaluated at E just above E_c (at E_c^+) under the assumption that $\Sigma_{s1}(E_c^+ \to E) = \Sigma_{s1}(E_c)\delta(E_c^+ - E)$.)

2. The anisotropic source (last term of the second equation in (9.5.10)) is neglected. (Except when hydrogen is the moderator, $\Sigma_{s1}(E' \to E)$ tends to be much smaller than $\Sigma_{s0}(E' \to E)$. Thus the angular distribution of the source of neutrons scattered into the thermal energy range is largely isotropic.)

3. $\Sigma_{s1}(E' \to E)$ in the range $0 < E \leq E_c$, is taken as $\Sigma_{s1}\delta(E' - E)$. (Because of the effect that molecular binding has on neutron scattering cross sections, it is difficult to make any general statements supporting the validity of this assumption. However, possibly because $\Sigma_{s1}(E')$ is usually much smaller than $\Sigma_{s0}(E')$, the approximation has been found empirically to be satisfactory.)

When all three of these simplifying assumptions are made, (9.5.10) reduces to

$$\left[\frac{B_m^2}{3[\gamma\Sigma_t(E) - \Sigma_{s1}(E)]} + \Sigma_t(E)\right]F_0(E)$$
$$= \int_0^{E_c} dE'\,\Sigma_{s0}(E' \to E)F_0(E') + \int_{E_c}^{\infty} \Sigma_{s0}(E' \to E)\frac{dE'}{E'} \quad \text{for } E \leq E_c. \tag{9.5.11}$$

It should be noted that the multigroup values of $\Sigma_{nn'}$ and $A_{1nn'}$ given by (9.4.3) along with (9.4.16) and (9.4.17) are required if the multigroup forms of (9.5.10) or (9.5.11) are to be solved. As was mentioned in Chapter 3, this requires that scattering kernels reflecting temperature-dependent crystalline and chemical-binding effects of the moderating materials be available. It is also necessary to account for thermal motion of the nuclei comprising the material when multigroup values of absorption cross sections are computed. (Recall that, in writing $\Sigma_t(E)$, $\Sigma_{s0}(E' \to E)$, etc., we have assumed that an average over the thermal motion of the nuclear constituents of the medium has already been performed.) Thus the preparation of multigroup cross sections required before the multigroup counterparts of (9.5.10) or (9.5.11) can be solved is a formidable job.

Since it is common practice to use only a single wide group to represent the entire thermal energy range, the magnitude of the effort involved in manipulating the nuclear data and determining the multigroup thermal spectra may seem excessive. There are, in fact, prescriptions for finding thermal group constants that avoid all these difficulties by assuming $F_0(E)$ to be a Maxwellian distribution like (3.4.48) with either a temperature corresponding to that of the principal moderating material or an "effective" temperature. This effective temperature is chosen so that the average neutron kinetic energy implied by it equals that implied by $F_0(E)$ as determined by solving (9.5.10) or (9.5.11) for one particular composition in the range of interest. (A procedure involving use of a Maxwellian distribution was discussed in Section 3.4.) Prescriptions of this nature have been used in the analysis of highly thermal systems such as graphite- or D_2O-moderated natural-uranium reactors. However, for heavily loaded systems such as HTGR's, PWR's, or BWR's, they are not recommended; here the more cumbersome procedure of determining diffusion-theory parameters for the thermal group by first finding multigroup thermal spectra based on physical scattering kernels and then performing averages over these spectra seems preferable.

9.6 Continuous-Slowing-Down Theory

It has been noted several times that the essential source of all the difficulties involved in defining a diffusion constant that will make diffusion theory entirely equivalent to the P_1 approximation is the anisotropic, differential scattering cross section $\Sigma_{s1}(\mathbf{r}, E' \to E)$. It turns out that $\Sigma_{s1}(\mathbf{r}, E' \to E)$ and all the other Legendre components $\Sigma_{sl}(\mathbf{r}, E' \to E)$ of $\Sigma_s(\mathbf{r}, E' \to E, \mu_0)$ introduce an additional practical difficulty. They appear in the spherical-harmonics equations (8.4.36) and give rise to the *scattering-in integrals* $\int_0^\infty \Sigma_{sl}(\mathbf{r}, E' \to E)\Psi_l^m(\mathbf{r}, E')dE'$; hence they show up in all approximations based on the spherical-harmonics expansion: the P_l equations (8.4.16) in slab geometry; the P_l and B_l approximations (8.4.23) and (8.4.25) for the Legendre components of the asymptotic spectrum; and the diffusion-theory approximations (9.3.20) and (9.3.26).

The practical difficulty referred to is that of integrating over all energies E' in order to determine the scattering-in integrals at E. In multigroup terms this becomes the difficulty of evaluating the sums $\sum_{n'=0}^{N} \Sigma_{sl;nn'}(\mathbf{r})\Psi_{l;n'}^m(\mathbf{r})$. A numerical operation of this type is itself trivial. However, if hydrogen is present, there are $\sim\frac{1}{2}N^2$ values of the $\Sigma_{sl;nn'}(\mathbf{r})$ to be handled, and, even in space-independent spectrum calculations, storage and data-handling problems can become severe. (Note that the fission integrals

$$\delta_{n0}\delta_{m0}\int dE'\sum_{j=1}^{J} f^j(E)\nu^j\Sigma_f^j(\mathbf{r}, E')\Psi_0^0(\mathbf{r}, E')$$

in (8.4.36) present a much less serious problem; in an N-group approximation, at most $2JN$ numbers are needed. In fact, in an asymptotic spectrum calculation, if we make the usual approximations discussed in Section 9.5, only the nonzero group values of the average spectrum function $\bar{f}(E)$ need be stored.)

When multigroup methods were first applied to reactor problems, computing-machine limitations were such that the direct evaluation of the scattering-in integrals was almost prohibitively costly. *Continuous-slowing-down* theory was devised to overcome the difficulty. Computer capability is such that it is no longer a practical necessity to use the scheme. Nevertheless, since it can be made quite accurate, it is still retained in many computer programs. In fact for hyperfine group calculations ($N = 2000$, for example) its use results in significant savings in computer costs.

Since continuous-slowing-down theory is employed chiefly for asymptotic spectrum calculations, we shall develop the formalism for that case. Thus attention will be focused on scattering-in integrals for each isotope j of the form

$$\eta_l^j(E) \equiv \int_0^\infty \Sigma_{sl}^j(E' \to E)F_l(E')\,dE'. \tag{9.6.1}$$

Moreover, to reduce the number of subscripts, we shall carry out the development in the continuous-energy formalism. Extension of the results to the space-dependent situation and the multigroup formalism is straightforward.

The essential idea of continuous-slowing-down theory is to find differential equations for the $\eta_l^j(E)$ that involve only values *at* E of the flux components $F_l(E)$ (and not integrals of $F_l(E')$ over E'). We shall see that this can be done exactly for hydrogen in the slowing-down energy range. For heavier elements the theory will be restricted to elastic scattering (not necessarily isotropic) in the slowing-down energy range; it will be necessary to expand the $\Sigma_{sl}^j(E' \to E)F_l(E')$ in a Taylor series about E and then to drop the higher-order terms. Further refinements are necessary if the theory is to be applied to the thermal energy range or to inelastic-scattering terms.

We shall proceed by first developing a mathematical expression for the double-differential elastic-scattering cross section $\Sigma_s^j(E' \to E, \mu_0)dE\,d\mu_0$ that accounts explicitly for the fact that the cosine μ_0 of the scattering angle and the energy ratio E/E' are rigidly related when the scattering is elastic. From this result will follow general expressions and then differential equations for the $\eta_l^j(E)$—first for hydrogen, then for heavier elements.

An Expression for $\Sigma_s^j(E' \to E, \mu_0)$ When Scattering Is Elastic

We seek an analytical expression for the differential cross section for elastic scattering from E' to E and through an angle whose cosine is μ_0. To reduce notational complexity we shall first develop this expression for one isotope of mass M in terms of the mass ratio $A = M/m$, where m is the mass of the neutron and, for the moment, the superscript j identifying the isotope will be dropped. Since cross-sectional results are frequently quoted for the center-of-mass (CM) system, and since some of the relationships we shall need are in that system, we first recall some elementary relations and definitions connecting quantities measured in the CM system to those in the lab system. Thus, for elastic (but not necessarily isotropic) scattering, we have already derived (2.5.14) from the conservation of energy and momentum for an isotope of mass ratio A:

$$\frac{E}{E'} = \frac{A^2 + 2A\mu_c + 1}{(A + 1)^2}, \tag{9.6.2}$$

where E' and E are the initial and final energies of the neutron in the lab system and μ_c is the cosine of the scattering angle in the CM system. Equation (2.5.16) then gives μ_0, the cosine of the scattering angle in the lab system:

$$\mu_0 = \frac{A\mu_c + 1}{\sqrt{(A^2 + 2A\mu_c + 1)}}. \tag{9.6.3}$$

It will be convenient to switch from energies E and E' to lethargies u and u', so we note that the lethargy gain due to scattering is

$$U = u - u' = -\ln\frac{E}{E'} \tag{9.6.4}$$

or

$$\frac{E}{E'} = \exp(-(u - u')) = \exp(-U). \tag{9.6.5}$$

The corresponding transformed value of the differential scattering cross section $\Sigma_s(u' \to u, \mu_0)$ is given by

$$\Sigma_s(u' \to u, \mu_0)du\, d\mu_0 = \Sigma_s(E' \to E, \mu_0)dE\, d\mu_0. \tag{9.6.6}$$

It follows from (9.6.2) and (9.6.5) that

$$\mu_c = 1 - \frac{(A + 1)^2}{2A}[1 - \exp(-U)], \tag{9.6.7}$$

whence, from (9.6.3),

$$\mu_0 = \frac{A + 1}{2}\exp\left(-\frac{U}{2}\right) - \frac{A - 1}{2}\exp\left(\frac{U}{2}\right). \tag{9.6.8}$$

This result shows the rigid connection for elastic scattering between the lethargy gain U and μ_0, the cosine of the scattering angle in the lab system. We express this rigid connection by writing $\Sigma_s(u' \to u, \mu_0)$ as proportional to a delta function involving μ_0. Thus, for the lab system,

$$\Sigma_s(u' \to u, \mu_0) = \Sigma_s(u', \mu_0)C\delta\left(\mu_0 - \left[\frac{A + 1}{2}\exp\left(-\frac{U}{2}\right) - \frac{A - 1}{2}\exp\left(\frac{U}{2}\right)\right]\right), \tag{9.6.9}$$

where C, which at this stage could be a function of u', u, and μ_0, is to be determined by requiring that

$$\int du \int_{-1}^{1} \frac{d\mu_0}{2} \Sigma_s(u' \to u, \mu_0) = \int_{-1}^{1} \frac{d\mu_0}{2} \Sigma_s(u', \mu_0) = \Sigma_s(u'). \tag{9.6.10}$$

In this last equation $\Sigma_s(u')$ is the total scattering cross section at lethargy u', and the middle equation is really a definition of $\Sigma_s(u', \mu_0)$. The range of integration of u is over all values accessible as the result of the collision. Equation (9.6.2) shows that the maximum lethargy gain (maximum energy loss) obtainable is for $\mu_c = -1$ and, by (9.6.7), equals $-\ln \alpha$ where $\alpha \equiv (A-1)^2/(A+1)^2$. Thus the limits of the integration over u are from u' to $u' - \ln \alpha$. Inserting (9.6.9) into (9.6.10), defining the function

$$\mu(U) \equiv \frac{A+1}{2} \exp\left(-\frac{U}{2}\right) - \frac{A-1}{2} \exp\left(\frac{U}{2}\right), \tag{9.6.11}$$

and integrating over μ_0, we find

$$\int_{u'}^{u'-\ln\alpha} du \int_{-1}^{1} \frac{d\mu_0}{2} \Sigma_s(u', \mu_0) C\delta(\mu_0 - \mu(U)) = \int_{u'}^{u'-\ln\alpha} du \frac{C}{2} \Sigma_s(u', \mu(U)). \tag{9.6.12}$$

Now we first make a change of variable from u to U $(= u - u')$. Since u' is fixed, $du = dU$, and the limits of integration are from $U = 0$ to $U = -\ln \alpha$. We then make another change of variable from U to $\mu(U)$. Here $dU = (dU/d\mu)d\mu$ and the limits of integration on $\mu(U)$ are from

$$\frac{A+1}{2} - \frac{A-1}{2} = 1$$

to

$$\frac{A+1}{2} \exp(\tfrac{1}{2} \ln \alpha) - \frac{A-1}{2} \exp(-\tfrac{1}{2} \ln \alpha) = \frac{A-1}{2} - \frac{A+1}{2} = -1.$$

Thus, in view of (9.6.9), (9.6.10), and (9.6.12), we have

$$\int_{-1}^{1} \frac{d\mu_0}{2} \Sigma_s(u', \mu_0) = \int_{0}^{-\ln\alpha} dU \Sigma_s(u', \mu(U)) \frac{C}{2} = \int_{-1}^{1} \frac{d\mu}{2} \left[-C \frac{dU}{d\mu}\right] \Sigma_s(u', \mu). \tag{9.6.13}$$

Accordingly C must equal $-d\mu(U)/dU$, and the analytical expression (9.6.9) for the double-differential scattering cross section reflecting the fact that μ_0 and u are rigidly related becomes

$$\Sigma_s(u' \to u, \mu_0) = \Sigma_s(u', \mu_0)\left[-\frac{d\mu(U)}{dU}\right]\delta(\mu_0 - \mu(U)). \tag{9.6.14}$$

General Expressions for Scattering-in Integrals

From the definitions (9.6.1) for $\eta_l^j(E)$ and (8.4.12) for Σ_{sl}, with superscripts j again suppressed, we have

$$\eta_l(E) = \int_0^\infty dE' \, \Sigma_{sl}(E' \to E) F_l(E') = \int_0^\infty dE' \int_{-1}^{1} \frac{d\mu_0}{2} \Sigma_s(E' \to E, \mu_0) P_l(\mu_0) F_l(E'). \quad (9.6.15)$$

Switching from energy to lethargy through use of (9.6.6) and employing (9.6.14) leads to

$$\eta_l(u) = \int_{u+\ln\alpha}^{u} du' \int_{-1}^{1} \frac{d\mu_0}{2} P_l(\mu_0) F_l(u') \Sigma_s(u', \mu_0) \left[-\frac{d\mu(U)}{dU} \right] \delta(\mu_0 - \mu(U)), \quad (9.6.16)$$

where the lower limit of the u' integration is $u + \ln \alpha$, the lowest lethargy from which neutrons can reach lethargy u by collision with the isotope under consideration. Integration over μ_0 yields

$$\eta_l(u) = \frac{1}{2} \int_{u+\ln\alpha}^{u} du' P_l(\mu(U)) F_l(u') \Sigma_s(u', \mu(U)) \left[-\frac{d\mu(U)}{dU} \right]. \quad (9.6.17)$$

The functional form of $\Sigma_s(u', \mu(U))$ in this equation is determined by $\Sigma_s(u', \mu_0) = \int du \, \Sigma_s(u' \to u, \mu_0)$ (see (9.6.10)), where $\Sigma_s(u' \to u, \mu_0)$ is for the laboratory system and hence will show dependence on μ_0 even if the scattering in the CM system is isotropic. It is convenient to avoid this complication. Accordingly we define a function $\Sigma_s^c(u', \mu_c)$ in the CM system analogous to $\Sigma_s(u', \mu_0)$ in the lab system. This we can do because the physical definition of $\Sigma_s^c(u' \to u, \mu_c)$, the double-differential cross section for scattering from u' to u and through an angle in the CM system whose cosine is μ_c, is such that

$$\Sigma_s^c(u' \to u, \mu_c) d\mu_c = \Sigma_s(u' \to u, \mu_0) d\mu_0 \quad (9.6.18)$$

(see (2.4.21)). Thus

$$\Sigma_s^c(u', \mu_c) \, d\mu_c = \left[\int du \, \Sigma_s^c(u' \to u, \mu_c) \right] d\mu_c = \left[\int du \, \Sigma_s(u' \to u, \mu_0) \right] d\mu_0 \equiv \Sigma_s(u', \mu_0) \, d\mu_0, \quad (9.6.19)$$

and we can use

$$\Sigma_s(u', \mu(U)) = \Sigma_s^c(u', \mu_c(U)) \frac{d\mu_c(U)}{d\mu(U)}$$

in (9.6.17), with $\mu_c(U)$ given by (9.6.7). Thus (9.6.17) becomes

$$\eta_l(u) = -\frac{1}{2} \int_{u+\ln\alpha}^{u} du' \, P_l(\mu(U)) F_l(u') \Sigma_s^c(u', \mu_c(U)) \frac{d\mu_c(U)}{dU}. \quad (9.6.20)$$

An alternate expression is obtained by changing variables first from u' to U. This yields

$$\eta_l(u) = \frac{1}{2}\int_{-\ln\alpha}^{0} dU\, P_l(\mu(U))F_l(u - U)\Sigma_s^c(u - U, \mu_c(U))\frac{d\mu_c(U)}{dU}. \tag{9.6.21}$$

Next we switch variables from U to $\mu_c(U)$, using $dU = (dU/d\mu_c)d\mu_c$ and with the limits of integration now extending from -1 to $+1$ (see (9.6.7)). We find

$$\eta_l(u) = \int_{-1}^{1} \frac{d\mu_c}{2} P_l(\mu(\mu_c))F_l(u - U(\mu_c))\Sigma_s^c(u - U(\mu_c), \mu_c). \tag{9.6.22}$$

Equations (9.6.20) and (9.6.22) provide a means of converting the $\eta_l(u)$ from integral to derivative form.

The Scattering-in Integral for Hydrogen

We first consider the case of hydrogen. Here $A = 1$, $\alpha \equiv (A - 1)^2/(A + 1)^2 = 0$, $\mu(U) = \exp(-U/2)$, $\mu_c(U) = 2\exp(-U) - 1$, and $\Sigma_s^c(u', \mu_c)$ is not a function of the variable μ_c since hydrogen scattering is, to a high degree of approximation, isotropic in the CM system. Thus, from (9.6.10) and (9.6.19),

$$\begin{aligned}\Sigma_s(u') &= \int_{-1}^{1} \frac{d\mu_c}{2}\Sigma_s^c(u', \mu_c) = \Sigma_s^c(u', \mu_c)\int_{-1}^{1}\frac{d\mu_c}{2} \\ &= \Sigma_s^c(u', \mu_c) = \int_{-1}^{1}\frac{d\mu_c}{2}\Sigma_s(u').\end{aligned} \tag{9.6.23}$$

Equation (9.6.20) then yields

$$\begin{aligned}\eta_l^1(u) &= \int_{-\infty}^{u} -\frac{du'}{2} P_l\left[\exp\left(-\frac{U}{2}\right)\right]F_l(u')\Sigma_s^1(u')[-2\exp(-U)] \\ &= \int_{0}^{u} du'\, P_l\left[\exp\left(-\frac{U}{2}\right)\right]\exp(-U)F_l(u')\Sigma_s^1(u'),\end{aligned} \tag{9.6.24}$$

where superscripts 1 refer to hydrogen and the lower limit of integration has been shifted to $u' = 0$ since $F_l(u') = 0$ for all $u' < 0$. For $l = 0$ we thus get

$$\eta_0^1(u) = \int_0^u du' \exp(-(u - u'))F_0(u')\Sigma_s^1(u'), \tag{9.6.25}$$

so that

$$\frac{\partial\eta_0^1(u)}{\partial u} = F_0(u)\Sigma_s^1(u) - \eta_0^1(u). \tag{9.6.26}$$

Similarly for $l = 1$ we find

$$\eta_1^1(u) = \int_0^u du' \exp\left(-\frac{3}{2}(u - u')\right) F_1(u')\Sigma_s^1(u'), \tag{9.6.27}$$

so that

$$\frac{\partial \eta_1^1(u)}{\partial u} = F_1(u)\Sigma_s^1(u) - \frac{3}{2}\eta_1^1(u). \tag{9.6.28}$$

Continuation of this process yields further differential equations analogous to (9.6.26) and (9.6.28) except that, in equations for $l > 1$, terms involving lower $\eta_l^1(u)$ functions appear. Thus, for hydrogen, the scattering-in integrals η_l^1 are determined by solving differential equations (9.6.26), (9.6.28), and the analogous equations for higher l values. There is no approximation involved in determining the η_l^1 in this way.

Scattering-in Integrals for Isotopes Other than H^1

The computation of $\eta_l^j(u)$ for any isotope other than hydrogen is not nearly so simple, even if this isotope scatters isotropically in the CM system. Mathematically this is because $\ln \alpha$ in (9.6.20) is finite for this case and differentiation of $\eta_l^j(u)$ with respect to u fails to eliminate all integrals over u'. Thus the procedure will not convert the integral expression (9.6.1) into a differential equation.

In view of this we turn to an approximation based specifically on the fact that the maximum lethargy gain $-\ln \alpha$ is finite for $A > 1$ and becomes quite small for heavy elements. The idea is to expand $F_l(u - U(\mu_c))\Sigma_s^c(u - U(\mu_c), \mu_c)$ in (9.6.22) in a Taylor series about u. The result for a given element of mass ratio A^j is

$$\eta_l^j(u) = \int_{-1}^{1} \frac{d\mu_c}{2} P_l(\mu(\mu_c)) \sum_{n=0}^{\infty} \frac{1}{n!} \frac{\partial^n [F_l(u)\Sigma_s^{jc}(u, \mu_c)]}{\partial u^n} [-U(\mu_c)]^n. \tag{9.6.29}$$

Since the range of $U(\mu_c)$ is from 0 to $-2\ln((A^j - 1)/(A^j + 1))$, we expect to be able to approximate this series accurately with only a few terms when $A^j \gg 1$.

Equation (9.6.29) can be cast in a superficially simple form by defining a set of functions $G_l^{jn}(u)$ by

$$G_l^{jn}(u) \equiv \int_{-1}^{1} \frac{d\mu_c}{2} P_l(\mu(\mu_c)) \frac{1}{n!} \Sigma_s^{jc}(u, \mu_c)[-U(\mu_c)]^n. \tag{9.6.30}$$

(Note that the $G_l^{jn}(u)$ do not depend on the $F_l(u)$ and can thus be precomputed for each isotope j as a function of u.) Then (9.6.29) can be written

$$\eta_l^j(u) = \sum_{n=0}^{\infty} \frac{\partial^n}{\partial u^n} [F_l(u) G_l^{jn}(u)]. \tag{9.6.31}$$

Thus, to avoid computing the $\eta_l^j(E)$ by a direct integration over energy (9.6.1), an infinite-order differential equation must be solved. The method becomes tractable only because in practice—even for deuterium ($A = 2$)—retaining only one or two terms in the expansion appears to yield acceptably accurate results.

The most common of the approximations of this kind is age theory. Here it is assumed that for $l = 0$ only terms $n = 0$ and $n = 1$ need be retained in (9.6.31), while for $l \geq 1$ only the term $n = 0$ need be retained. Thus we get

$$\begin{aligned}\eta_0^h(u) &\equiv \sum_{j\neq 1} \eta_0^j(u) = F_0(u) \sum_{j\neq 1} G_0^{j0}(u) + \frac{\partial}{\partial u}\left[F_0(u) \sum_{j\neq 1} G_0^{j1}(u) \right],\\ \eta_l^h(u) &\equiv \sum_{j\neq 1} \eta_l^j(u) = F_l(u) \sum_{j\neq 1} G_l^{j0}(u) \quad \text{for } l \geq 1,\end{aligned} \tag{9.6.32}$$

where a superscript h is used to indicate that a sum over all elements excluding hydrogen has been made.

If we further assume that the scattering is isotropic in the CM system so that, in accord with (9.6.23), $\Sigma_s^{jc}(u, \mu_c)$ may be replaced by $\Sigma_s^j(u)$, (9.6.30) yields

$$\begin{aligned}G_0^{j0}(u) &= \Sigma_s^j(u),\\ G_1^{j0}(u) &= \Sigma_s^j(u) \int_{-1}^{1} \frac{d\mu_c}{2} \mu(\mu_c) \equiv \Sigma_s^j(u)\bar{\mu}^j = \Sigma_s^j(u) \int_{-1}^{1} \frac{d\mu_c}{2} \frac{A^j\mu_c + 1}{[(A^j)^2 + 2A^j\mu_c + 1]^{1/2}} = \frac{2\Sigma_s^j(u)}{3A_j},\\ G_0^{j1}(u) &= -\Sigma_s^j(u) \int_{-1}^{1} \frac{d\mu_c}{2} U(\mu_c) \equiv -\Sigma_s^j(u)\xi^j = -\Sigma(u)\left(1 + \frac{\alpha^j}{1-\alpha^j} \ln \alpha^j\right),\\ G_l^{j0}(u) &= \int_{-1}^{1} \frac{d\mu_0}{2} P_l(\mu_0)\Sigma_s^j(u, \mu_0) \equiv \Sigma_{sl}^j(u),\end{aligned} \tag{9.6.33}$$

where $\bar{\mu}^j$ and ξ^j are defined by the second and third equations in (9.6.33) and can be interpreted physically as the average cosine of the scattering angle in the lab system and the average loss in the logarithm of the energy due to isotropic scattering (see (2.5.24) and (2.5.26)). The last equation in (9.6.33) comes from introducing (9.6.19) into (9.6.30). The $\Sigma_{sl}^j(u)$ are thus the P_l components of the differential scattering cross section for isotope j in the lab system.

The first and last of these results are general for elastic scattering. The last parts of the middle two are valid only if the elastic scattering is isotropic in the CM system.

Applications of Continuous-Slowing-Down Theory

Spectrum Calculations. As a first illustration of continuous-slowing-down theory we consider an asymptotic P_1-spectrum calculation in the slowing-down region. When we guess at the value of the materials buckling and assume that the fission spectra of the various fissionable isotopes can be approximated by the average spectrum function $\bar{f}(E)$, the P_1 equations (9.5.7) and (9.5.8), cast into lethargy form and with the number

$\sum_j \int_0^\infty du' v^j \Sigma_f^j(u') F_0(u')$ arbitrarily set equal to one, become

$$
\begin{aligned}
-iB_m F_1(u) + \Sigma_t(u) F_0(u) &= \int_0^u du' \, \Sigma_{s0}(u' \to u) F_0(u') + \bar{f}(u), \\
-iB_m F_0(u) + 3\Sigma_t(u) F_1(u) &= 3 \int_0^u du' \, \Sigma_{s1}(u' \to u) F_1(u').
\end{aligned}
\tag{9.6.34}
$$

Assuming that inelastic scattering is isotropic in the laboratory system ($\Sigma_{s1}^{in}(u' \to u) = 0$) and replacing the elastic scattering-in integrals for hydrogen by $\eta_0^1(u)$ and $\eta_1^1(u)$, and those for isotopes other than H^1 by (9.6.31) with terms for $n > 1$ neglected, permits us to replace (9.6.34) by

$$
\begin{aligned}
&-iB_m F_1(u) + \Sigma_t(u) F_0(u) = \int_0^\infty du' \, \Sigma_{s0}^{in}(u' \to u) F_0(u') + \eta_0^1(u) + \sum_{j \neq 1} \eta_0^j(u) + \bar{f}(u), \\
&-iB_m F_0(u) + 3\Sigma_t(u) F_1(u) = 3\eta_1^1(u) + 3 \sum_{j \neq 1} \eta_1^j(u), \\
&\frac{d}{du} \eta_0^1(u) = F_0(u) \Sigma_s^1(u) - \eta_0^1(u), \\
&\frac{d}{du} \eta_1^1(u) = F_1(u) \Sigma_s^1(u) - \frac{3}{2} \eta_1^1(u), \\
&\eta_l^j(u) = F_l(u) G_l^{j0}(u) + \frac{d}{du} [F_l(u) G_l^{j1}(u)] \quad \text{for } l = 0, 1 \text{ and } j > 1,
\end{aligned}
\tag{9.6.35}
$$

where $\Sigma_{s0}^{in}(u' \to u)$ is the P_0 component of the differential cross section for inelastic scattering from u' into du.

Since the only sums it is necessary to deal with are $\sum_{j \neq 1} \eta_l^j(u)$, there is no need to treat the last equation for each isotope separately. Accordingly we define

$$
\begin{aligned}
\eta_l^h(u) &\equiv \sum_{j \neq 1} \eta_l^j(u), \\
G_l^{hn}(u) &\equiv \sum_{j \neq 1} G_l^{jn}(u),
\end{aligned}
\tag{9.6.36}
$$

and deal only with the sum of the last equation in (9.6.35) over all $j \neq 1$.

If we now cast (9.6.35) into multigroup form by integrating over group n, of lethargy width $\Delta u_n \equiv u_{n+1} - u_n$ ($u_0 = 0$ at $E = E_0 \approx 15$ MeV), and apply the usual definitions of multigroup quantities, we obtain

$$
\begin{aligned}
&-iB_m F_{1n} + \Sigma_{tn} F_{0n} = \sum_{n'=0}^{n} \Sigma_{nn'}^{in} F_{0n'} + \eta_{0n}^1 + \eta_{0n}^h + \chi_n, \\
&-iB_m F_{0n} + 3\Sigma_{tn} F_{1n} = 3(\eta_{1n}^1 + \eta_{1n}^h), \\
&\eta_0^1(u_{n+1}) - \eta_0^1(u_n) = \Sigma_{sn}^1 F_{0n} - \eta_{0n}^1, \\
&\eta_1^1(u_{n+1}) - \eta_1^1(u_n) = \Sigma_{sn}^1 F_{1n} - \tfrac{3}{2} \eta_{1n}^1, \\
&\eta_{ln}^h = F_{ln} G_{ln}^{h0} + F_l(u_{n+1}) G_l^{h1}(u_{n+1}) - F_l(u_n) G_l^{h1}(u_n) \quad \text{for } l = 0, 1.
\end{aligned}
\tag{9.6.37}
$$

To complete these equations it is necessary to assume some relationship between the group-integrated and end-point values of the η_l^1 and F_l. One simple choice is

$$\eta_{ln}^1 \approx \tfrac{1}{2}[\eta_l^1(u_{n+1}) + \eta_l^1(u_n)]\Delta u_n,$$
$$F_{ln} \approx \tfrac{1}{2}[F_l(u_n) + F_l(u_{n+1})]\Delta u_n. \tag{9.6.38}$$

A more sophisticated relationship is sometimes required to avoid numerical oscillations in the solution of the equations.

The only $N \times N$ array of numbers required for the solution of (9.6.37) is the inelastic-scattering matrix $\{\Sigma_{nn'}^{\text{in}}\}$, and most of its entries will generally be zero. Thus the data-storage and manipulation problems associated with the solution of (9.6.37) are greatly reduced over what would be required for a straightforward multigroup solution of (9.6.34). Note, however, that, since the matrix $\{\Sigma_{s1;nn'}\}$ is not available, the $A_{1gg'}$ of (9.4.25) and thus the few-group diffusion matrix $\{\tilde{D}_{gg'}\}$ cannot be obtained. In fact the conventional few-group diagonal matrix $\{D_g\}$ cannot be determined in the usual manner employing (9.4.28) and (9.4.13), but must instead be found from the fact that we want D_g to be such that $F_{1g} = iB_m D_g F_{0g}$. Thus, to get D_g, we solve (9.6.37) for the multigroup fluxes F_{0n} and F_{1n} and then compute

$$D_g = \frac{\sum_{n \subset g} F_{1n}}{iB_m \sum_{n \subset g} F_{0n}}. \tag{9.6.39}$$

The Age Approximation. As another example of continuous-slowing-down theory we examine the age approximation applied to the P_1 equations (9.2.1) in slab geometry. We shall assume that no hydrogen is present in the system.

Switching from energy to lethargy using the definitions (9.2.2), and replacing the elastic scattering-in integrals by the definitions (9.6.1) and (9.6.36), converts (9.2.1) into

$$\frac{\partial}{\partial z} J(z, u) + \Sigma_t(z, u)\Phi(z, u) = \int_0^\infty du' \left[\sum_j f^j(u)\nu^j \Sigma_f^j(z, u') + \Sigma_{s0}^{\text{in}}(z, u' \to u) \right] \Phi(z, u') + \eta_0^h(z, u),$$
$$\frac{\partial}{\partial z} \Phi(z, u) + 3\Sigma_t(z, u) J(z, u) = 3\eta_1^h(z, u). \tag{9.6.40}$$

The age approximation (9.6.32) and (9.6.33) then yields for the P_0 and P_1 components $\Phi(z, u)$ and $J(z, u)$ at each z

$$\eta_0^h(z, u) = \Sigma_s(z, u)\Phi(z, u) - \frac{\partial}{\partial u}[\xi \Sigma_s(z, u)\Phi(z, u)],$$
$$\eta_1^h(z, u) = \Sigma_{s1}(z, u) J(z, u). \tag{9.6.41}$$

Substituting the second equation in (9.6.41) into the second one in (9.6.40) gives

$$J(z, u) = -\frac{1}{3[\Sigma_t(z, u) - \Sigma_{s1}(z, u)]} \frac{\partial}{\partial z} \Phi(z, u) \equiv -D(z, u) \frac{\partial}{\partial z} \Phi(z, u). \tag{9.6.42}$$

Comparison with (9.3.2) thus shows that making the age-theory approximation (9.6.41) for $\eta_1^h(z, u)$ is equivalent to assuming that $\Sigma_{s1}(z, u' \to u) = \Sigma_{s1}(z, u)\delta(u' - u)$.

If (9.6.42) and the first equation in (9.6.41) are now substituted into the first equation in (9.6.40), the *age-diffusion equation* results:

$$-\frac{\partial}{\partial z}\left[D(z, u)\frac{\partial}{\partial z}\Phi(z, u)\right] + \Sigma_a(z, u)\Phi(z, u)$$
$$= \int_0^\infty du'\left[\sum_j f^j(u)\nu^j\Sigma_f^j(z, u') + \Sigma_{s0}^{in}(z, u' \to u)\right]\Phi(z, u') - \frac{\partial}{\partial u}[\xi\Sigma_s(z, u)\Phi(z, u)]. \qquad (9.6.43)$$

Applying this result to a homogeneous slab reactor and neglecting inelastic scattering then leads to (4.8.1), which, as we have seen in Section 4.8, is the basis of the Fermi age equation (4.8.9).

The fact that the Fermi age equation applies only to homogeneous systems (and also that it cannot account for hydrogen moderation and inelastic scattering) greatly limits its utility. As a result continuous-energy slowing-down theory is used today mostly for spectrum computations, as in (9.6.37), or in the "multigroup age-diffusion" approximation derived from (9.6.43).

References

Henry, A. F., 1967. "Few Group Approximations Based on a Variational Principle," *Nuclear Science and Engineering* **27**: 493–510.

Michelini, M., 1972. "Anisotropic Diffusion Coefficients in Generalized *XY* Geometry: Theory and Application to Diffusion Codes," *Nuclear Science and Engineering* **47**: 116–126.

Stacey, W. M., Jr., 1971. "The Effect of Anisotropic Scattering upon the Elastic Moderation of Fast Neutrons," *Nuclear Science and Engineering* **44**: 194–203.

Suich, J. E., and Honeck, H. C., 1967. *The HAMMER System*, DP-1064, Savannah River Laboratory.

Problems

1. The double-P_N approximation for $\Psi(z, \mu, E)$ is obtained by substituting the double-P_n expansion given in Problem 12, Chapter 8, into the one-dimensional transport equation (8.4.11), terminating the expansion at $n = N$, and expanding $\Sigma_s(z, E' \to E, \mu_0)$ in Legendre polynomials (8.4.12) up to $n = 2N + 1$. The resultant equation is first multiplied successively by $P_n(2\mu - 1)$ ($n = 0, 1, \ldots, N$) and integrated over the range $0 \le \mu \le 1$ and then multiplied successively by $P_n(2\mu + 1)$ ($n = 0, 1, \ldots, N$) and integrated over the range $-1 \le \mu < 0$. Coupled integro-differential equations in the double-P_n components $\Psi_n^+(z, E)$ and $\Psi_n^-(z, E)$ ($n = 0, 1, \ldots, N$) result.

a. Derive the energy-dependent double-P_0 equations analogous to (9.2.1).

b. Assuming that $\Sigma_{s1}(z, E' \to E)$ may be approximated by $\Sigma_{s1}(z, E')\,\delta(E' - E)$ and making use of the expressions for $\Phi(z, E)$ and $J(z, E)$ determined in Problem 12, Chapter 8, show that, according to the double-P_0 approximation,

$$J(z. E) = -\frac{1}{4\Sigma_t(z, E) - 3\Sigma_{s1}(z, E)} \frac{\partial \Phi(z, E)}{\partial z}.$$

2. Derive the three-dimensional P_1 equations (9.3.17) and (9.3.18) directly from the spherical-harmonics equations (8.4.36).

3. Assume that scattering is isotropic in the laboratory system and that we wish to find an approximate solution for the one-dimensional transport equation (8.4.11) having the form

$$\Psi_t(z, E, \mu) \approx A(z, E) + B(z, E)\mu,$$

where $A(z, E)$ and $B(z, E)$ are to be determined. Find by the weighted-residual method coupled integrodifferential equations for $A(z, E)$ and $B(z, E)$, using as weight functions:

a. $W_1 = 1, \qquad W_2 = \mu \text{ for } -1 \le \mu \le 1.$

b. $W_1 = \begin{cases} 1 \text{ for } -1 \le \mu \le 0, \\ 0 \text{ for } 0 < \mu \le 1, \end{cases} \qquad W_2 = \begin{cases} 0 \text{ for } -1 \le \mu < 0, \\ 1 \text{ for } 0 \le \mu \le 1. \end{cases}$

c. $W_1 = \delta\left(\mu - \frac{\pi}{4}\right), \qquad W_2 = \delta\left(\mu + \frac{\pi}{4}\right).$

4. Starting with the transport equation for slab geometry (8.4.11) and combining (8.4.31) with (8.4.30) to give a trial function

$$\Psi_t(z, \mu, E) = 2F_0(E) \cos B_m z - 6\,\mu i F_1(E) \sin B_m z$$

for a homogeneous reactor extending from $z = -\pi/(2B_m)$ to $z = +\pi/(2B_m)$, derive the B_1 equations ((8.4.25) with $l = 0, 1$) for $F_0(E)$ and $F_1(E)$. (Assume $\Sigma_{sn}(E' \to E) = 0$ for $n > 1$.)

5. Consider, for the two-dimensional case, the leakage term

$$-\sum_{g'=1}^{G} \left[\frac{\partial}{\partial x} \tilde{D}_{gg',x} \frac{\partial}{\partial x} + \frac{\partial}{\partial y} \tilde{D}_{gg',y} \frac{\partial}{\partial y}\right] \Phi_{g'}(x, y)$$

in (9.4.10). Suppose the $\tilde{D}_{gg',x}$ and $\tilde{D}_{gg',y}$ are known and we wish to solve (9.4.10) using a coordinate system $X'Y'$ formed by rotating the initial system an angle θ about the Z axis (so that the X' axis makes an angle θ with the X axis and the Y' axis makes an angle θ with the Y axis). Find the leakage term in terms of derivatives $\partial/\partial x'$ and $\partial/\partial y'$, taken in this new coordinate system. (This result shows that the directional components of the $\tilde{D}_{gg'}$ are elements not of a vector, but rather of a symmetric tensor of rank 2. For this reason the direction-dependent diffusion constant is often called a "tensor D.")

6. If **i**, **j**, and **k** are unit victors in the X, Y, and Z directions, the quantities **ii**, **ij**, **ji**, **jj**, **jk**, **kj**, **ik**, **ki**, and **kk** are called unit dyadics. A dot product between a dyadic and a vector yields a vector. Thus

$$(\mathbf{ii}) \cdot \mathbf{i} = \mathbf{i}(\mathbf{i} \cdot \mathbf{i}) = \mathbf{i},$$
$$(\mathbf{ji}) \cdot \mathbf{i} = \mathbf{j}(\mathbf{i} \cdot \mathbf{i}) = \mathbf{j},$$
$$(\mathbf{ij}) \cdot \mathbf{i} = \mathbf{i}(\mathbf{j} \cdot \mathbf{i}) = 0,$$
$$\mathbf{i} \cdot (\mathbf{ij}) = (\mathbf{i} \cdot \mathbf{i})\mathbf{j} = \mathbf{j}, \text{ etc.}$$

Show that, if we define a $G \times G$ matrix $[\tilde{\mathbf{D}}]$ as

$$[\tilde{\mathbf{D}}] = [\tilde{D}_x]\mathbf{ii} + [\tilde{D}_y]\mathbf{jj} + [\tilde{D}_z]\mathbf{kk},$$

then (9.4.33), generalized to the case of direction-dependent diffusion constants, becomes

$$-\nabla \cdot [\tilde{\mathbf{D}}] \cdot \nabla[\Phi] + [A_0][\Phi] = [M][\Phi].$$

7. For a three-energy-group picture in which a neutron cannot gain energy in a scattering collision, write explicit expressions for the diffusion-constant matrices $[\tilde{D}_u]$ ($u = x, y, z$) and $[D_u]$ under the assumption that the energy dependence of $\mathbf{J}(\mathbf{r}, E)$ in group g can be approximated as $\gamma_g(\mathbf{r})\alpha_g(E)$. Express matrix elements in terms of $\Sigma_t(\mathbf{r}, E)$, $\Sigma_{s0}(\mathbf{r}, E' \to E)$, and $\Sigma_{s1}(\mathbf{r}, E' \to E)$.

8. Suppose that, for a material having a given value of $\Sigma_t(E)$ and $\Sigma_{s1}(E' \to E)$, there exists a complete set of expansion functions $Z_n(E)$ and corresponding weight functions $Z_m^*(E)$ such that

$$\int_0^\infty Z_m^*(E) A_1 Z_n(E)\, dE = \delta_{nm} \qquad (n = 1, 2, \ldots),$$

where A_1 is the linear operator defined by (9.2.3).

a. Show that the operator $\tilde{D}$ of (9.2.4) may be expressed as

$$\tilde{D} = \tfrac{1}{3} \sum_{n=1}^{\infty} Z_n(E) \left[\int_0^\infty dE' Z_n^*(E') \right]_{\text{op}}$$

where the integral-operator notation $[\int_0^\infty dE'\ Z_n^*(E')]_{\text{op}}$ means that operating on a function $g(E)$ with this operator yields $\int_0^\infty dE'\ Z_n^*(E')g(E')$.

b. The operator $\tilde{D}$ is constructed so that $\tilde{D}A_1 = \frac{1}{3} I$, where I is the identity operator. Show that $A_1\tilde{D}$ is also $\frac{1}{3}I$. (*Hint*: If the operator H is such that $Hf = 0$ for all f, $H = 0$.)

9. Using the definition of an adjoint operator, write the equations mathematically adjoint to the few-group equations in the form (9.4.18). Then write the mathematical adjoints of the P_1 equations (9.3.17) and (9.3.18) and, by making the few-group assumptions

$$\Psi^*(\mathbf{r}, E) = \Phi_g^*(\mathbf{r}) F_{0g}^{k*}(E), \qquad J_u^*(\mathbf{r}, E) = J_{ug}^*(\mathbf{r}) F_{1g}^{k*}(E),$$

for E in E_g and $\mathbf{r}$ in region k, derive few-group adjoint equations directly. (Note that the two sets of few-group equations are not the same.)

10. Assuming that, for the purpose of determining one-group constants, $F_0(E)$ and $F_1(E)$ are the same whether computed by the P_1 or the B_1 apporximation, show that B_m^2 from a P_1 approximation must be less than B_m^2 from a B_1 approximation.

11. For elastic scattering involving an isotope of mass ratio A, derive a general expression for the Legendre components $\Sigma_{sl}(u' \to u)$ of the double-differential scattering cross section $\Sigma_s(u' \to u, \mu_0)$. Express the answer in terms of U, A, and $\Sigma_s^c(u', \mu_c(U))$.

From this result, and assuming scattering is isotropic in the center-of-mass system, find a formula for the P_1 component $\Sigma_{s1}(E' \to E)$ in terms of E', E, A, and $\Sigma_s^c(E')$.

12. With superscripts j omitted, we define the quantities

$$q_l(u) \equiv \sum_{n=0}^{\infty} \frac{\partial^n}{\partial u^n} G_l^{n+1}(u) F_l(u), \qquad \lambda_l(u) \equiv -\frac{G_l^2(u)}{G_l^1(u)}.$$

Show that

$$\frac{\partial q_l(u)}{\partial u} = G_l^0(u) F_l(u) - \eta_l(u)$$

and

$$q_l + \lambda_l \frac{\partial q_l}{\partial u} = -G_l^1 \Psi_l \left(1 - \frac{\partial \lambda_l}{\partial u}\right) - \sum_{n=2}^{\infty} \left[\frac{\partial^n}{\partial u^n} G_l^{n+1} F_l + \lambda_l \frac{\partial^n}{\partial u^n} G_l^n F_l\right].$$

Also show that, for isotropic scattering in the center-of-mass system, λ_l is not a function of u.

The second result for $l = 0$ and $l = 1$ with the sum over n neglected constitutes the "Greuling-Geortzel" approximation for q_0 and q_1. These two quantities can be shown to be the P_0 and P_1 components of the slowing-down density at lethargy u.

13. Cast (9.6.43) into an approximate multi-lethargy-group form. Be sure there are enough group equations to solve for all the unknown group fluxes.

14. Suppose that, in a four-group model, group two extends from 10 eV to 10^5 eV and group three extends from 1 eV to 10 eV. Assume that, in the range 1 eV–10^5 eV, both the P_0 and P_1 components of the spectra $F_0(E)$ and $F_1(E)$ are proportional to $1/E$. Find, for elastic scattering in hydrogen (assumed isotropic in the center-of-mass system), the ratios $A_{0,32}/A_{0,33}$ (9.4.31) and $A_{1,32}/A_{1,33}$ (9.4.24). (Neglect the absorption part of $\Sigma_t(E)$ for hydrogen.)

15. Consider a mixture of 10 percent (by volume) Pu^{49} and 90 percent graphite. If the densities are $\rho^{49} = 19.6$ g/cc and $\rho^{C} = 1.6$ g/cc, and if $\sigma_s^{49} = 10$ b and $\sigma_s^{C} = 5$ b, compute $A_{1gg'}$ for ΔE_g extending from 70 to 80 eV and for $\Delta E_{g'}$ extending from 100 to 110 eV, under the assumption that $F_1(E)$ is proportional to $1/E$.

10 The Generation of Equivalent Diffusion-Theory Parameters

10.1 Introduction

Most power reactors are geometrically quite heterogeneous because of the presence of fuel elements, control rods, local flux depressors, and burnable poisons. The theory developed so far (except for the discussion in Chapter 5 of the application of collision theory to resonance capture in fuel and lattices) has been geared, not to this situation, but rather to the case of reactors composed of large homogeneous regions in which the neutron flux is well described by a P_1 approximation spatially and in which (when the few-group method is employed) the spectra of the P_0 and P_1 components are fairly constant throughout a given material composition.

In principle local heterogeneities can be accounted for by using a high-order, multigroup, discrete-ordinates approximation or a Monte Carlo analysis. For three-dimensional situations, however, this procedure is at present almost prohibitively expensive.

The potential magnitude of the heterogeneity problem is illustrated (for a rather extreme case) by Figure 10.1, which shows the cross section of a fuel-element subassembly used in the highly enriched seed of a seed-blanket pressurized-water reactor. The subassembly is one of four making up a cluster surrounding a cross-shaped control rod. (The seed-blanket reactor itself consists of an approximately cylindrical ring of these highly enriched "seed" clusters embedded in a "blanket" of control-rod-free subassemblies fueled with *natural*-uranium elements.) The seed subassembly is about 3.5 inches square. It consists of highly enriched UO_2-ZrO_2 fuel wafers, compartmentalized and clad by zirconium. As indicated, there are three fuel zones, differing from each other in the ratio

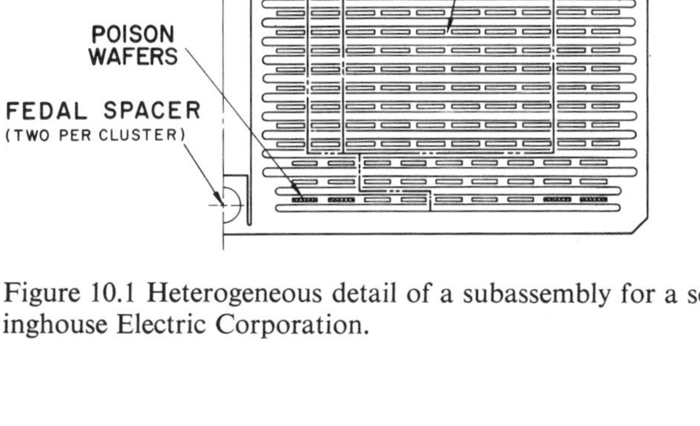

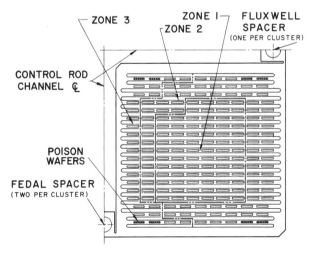

Figure 10.1 Heterogeneous detail of a subassembly for a seed-blanket PWR. Courtesy of Westinghouse Electric Corporation.

of enriched UO_2 to ZrO_2 in the wafer. At the four corners wafers of burnable poison (B^{10} in stainless steel) replace the fuel wafers. Light water is present in the coolant channels, the small zones that separate each cluster from its neighbors, and the control-rod channel.

The geometrical complexity of this seed subassembly exceeds considerably that of the subassemblies making up the slightly enriched PWR's being built today. (See, for example, Figure 1.11.) However it does illustrate—all in one assembly—a great many of the heterogeneity problems encountered in reactor design:

1. The fuel, the cross section of which has large narrow resonances in the epithermal range and is very large in the thermal range, is present in small lumps. Thus diffusion theory should not be valid.
2. Many of these lumps are not part of an extended, uniform lattice but are, rather, near a strong moderating region (the cross-shaped hole left when the control rod is removed). Thus the lattice-theory techniques developed in Chapter 5 are suspect.
3. Small lumps of strongly absorbing thermal poison are located in some cases where the gradient of the thermal flux is sizeable. Again a full transport-theory calculation is called for.
4. The control rod, when inserted, constitutes an extended region within which diffusion theory is not valid.
5. Even if the fuel wafers, associated clad, and moderator can be replaced by an equivalent homogeneous material, the fuel zones are sufficiently small that the few-group approximation of an asymptotic spectrum being present throughout much of any given fuel zone is suspect.

It is considerably encouraging to note that, despite this list, a reactor containing the subassemblies shown in Figure 10.1 was designed, built, and operated. Its actual performance was very close to design predictions.

The present chapter deals with methods that permit diffusion theory to be used for reactor analysis despite the heterogeneous nature of most reactors. By far the most common strategy for achieving this goal is to define sets of "effective" parameters which, when used in the group diffusion-theory equations, will reproduce correct interaction rates in some integral sense. (One might, for example, choose parameters that will make the *total* fission rate in a subassembly come out correct, although the rates in the individual fuel plates making up the assembly will be wrong.) Accordingly, although we shall begin by defining "ideal" equivalent constants in a formal way, most of the subsequent discussion will concern the definition and generation of parameters that are equivalent in a more approximate, integral sense. It will develop that, for two- and three-dimensional situations, even this more limited goal cannot be achieved exactly (i.e., the equivalent constants will not reproduce even integral reaction rates exactly). In the latter part of Chapter 11 we shall investigate practical schemes that account for heterogeneous effects in a more straightforward manner without the introduction of equivalent diffusion-theory parameters at an intermediate stage of the calculations.

10.2 Equivalent Group Diffusion-Theory Parameters

Although the problem is of great practical importance, the theoretical bases for most homogenizing procedures are surprisingly weak. Approximate ways of finding equivalent homogenized diffusion-theory constants can in some cases be described and defended in a systematic way. More often, however, they have the character of plausible prescriptions which have been found satisfactory for the cases tested but which cannot be heedlessly extended outside this range. Moreover they are frequently prescriptions which are needlessly elaborate but which no one has taken the time to simplify, either because they are thought adequate and cheap enough in computing time that the sizeable effort required to simplify them is not warranted or because they have been used at an early stage of a design study and hence must be used throughout (so that differences in calculations can be interpreted as physical effects rather than as outcomes of a change in the calculational model).

Part of the reason for this weak theoretical foundation lies in the nature of the problem itself. As we shall see, if we ask too much of the equivalent parameters, they may take on absurd values (equivalent homogenized D_g's and Σ_g's may become infinite, negative, or complex). In fact, in some cases, even if absurd values are accepted, it is not possible to find a set that will lead exactly to the reaction rates desired.

Despite the difficulties a firm theoretical foundation for the introduction of equivalent parameters seems desirable. Accordingly we shall begin by defining, as precisely as possible, just what we shall mean by ideal equivalent group diffusion parameters. This precise definition will permit us to find exact equivalent constants for a few extremely simple cases, and these results, while of limited practical use, will suggest approximate procedures for more realistic problems.

Definition of Ideal Equivalent Parameters

We begin by noting a theoretically comforting but practically useless result: If diffusion constants $D_u(\mathbf{r}, E)$ having different values for different component directions u are admitted, the diffusion equation (9.3.27) is general enough to yield a value of the scalar flux density $\Phi(\mathbf{r}, E)$ that is identical with what would be found by solving the transport equation for the directional flux density $\Psi(\mathbf{r}, \boldsymbol{\Omega}, E)$ and then integrating over all directions in accord with the definition (8.2.4).

To establish this result we first integrate the transport equation (8.2.14) over all directions $\mathbf{\Omega}$. Doing so, making use of the definitions (8.2.4) and (8.2.5) of $\Phi(\mathbf{r}, E)$ and $\mathbf{J}(\mathbf{r}, E)$, and introducing the operators M, defined by (8.3.8), and A, defined by

$$A = \Sigma_t(\mathbf{r}, E) - \left[\int_0^\infty dE' \, \Sigma_{s0}(\mathbf{r}, E' \to E)\right]_{op}, \tag{10.2.1}$$

where the subscript op indicates that the function operated on is to be included in the integral, yields (for the source-free case)

$$\nabla \cdot \mathbf{J}(\mathbf{r}, E) + A\Phi(\mathbf{r}, E) = \frac{1}{\lambda} M\Phi(\mathbf{r}, E), \tag{10.2.2}$$

where we have again represented explicitly the eigenvalue λ.

This exact result is identical with (4.4.2), which was obtained by physical arguments, and has the same form as the P_1 equation (9.3.17).

If components $D_u(\mathbf{r}, E)$ are then defined such that

$$J_u(\mathbf{r}, E) = -D_u(\mathbf{r}, E)\frac{\partial}{\partial u}\Phi(\mathbf{r}, E) \qquad (u = x, y, z), \tag{10.2.3}$$

the diffusion equation

$$-\sum_u \frac{\partial}{\partial u}\left[D_u(\mathbf{r}, E)\frac{\partial}{\partial u}\Phi(\mathbf{r}, E)\right] + A\Phi(\mathbf{r}, E) = \frac{1}{\lambda} M\Phi(\mathbf{r}, E) \tag{10.2.4}$$

is obtained.

Equation (10.2.4) has the same form as (9.3.27). Here, however, the components $D_u(\mathbf{r}, E)$ take care of both spectrum and transport effects. Of course one must already know Φ and $\mathbf{J}$ to find the D_u.* Hence the only value of (10.2.4) is that it demonstrates that the diffusion-equation form is *capable* of yielding exact results. Moreover the existence of a formally exact equation has one very nontrivial consequence. It focuses attention on the fact that most schemes for finding equivalent constants yield diffusion equations in a form that *cannot* provide completely accurate values of $\Phi(\mathbf{r}, E)$, even if $\Psi(\mathbf{r}, \mathbf{\Omega}, E)$ is known and used to determine the equivalent parameters. (We shall see why this is so in the following sections.)

Group Diffusion-Theory Parameters That Are Equivalent in an Integral Sense

The ideal equivalent diffusion-theory parameters defined in the previous section yield

*We shall not pursue the point, but it should be noted that, mathematically, the D_u are, not components of a vector, but rather the diagonal elements of a dyadic or tensor of rank 2. If the coordinate axes were rotated so that cross-derivative terms appeared in (10.2.4), nonzero components of the diffusion tensor corresponding to these cross terms would appear. (See Problems 5 and 6 in Chapter 9.)

values of $\Phi(\mathbf{r}, E)$ and corresponding reaction rates that are exact at every point $\mathbf{r}$ and every energy E. If we relax the requirement that there be point-by-point equivalence and instead try to find effective diffusion-theory parameters that will reproduce reaction rates integrated over some volume V_i or some energy range ΔE_g, it may be possible to define equivalent region- and group-averaged parameters that can be found without first going through the self-defeating process of solving the transport equation for the entire reactor. To investigate this possibility we shall first define the precise sense in which these parameters are to be equivalent and then go on to examine, for particular cases, whether such parameters exist and, if so, how to find them.

We want, then, to investigate whether there exist group diffusion matrices $[\bar{D}_u]$ (G-element diagonal matrices possibly differing for each of the coordinate directions $u = x, y$, or z), $[\bar{A}]$, and $[\bar{M}]$ (see (9.4.31) and (9.4.32)) that have the following properties:

1. They are spatially constant over a given volume V_i. (Depending on the circumstances V_i may be a homogeneous lump, a fuel-element cell, a subassembly, or a cluster.)
2. They are such that a solution of the group equations

$$\sum_u \frac{\partial}{\partial u}[\bar{D}_u]\frac{\partial}{\partial u}[\bar{\Phi}] - [\bar{A}][\bar{\Phi}] + \frac{1}{\bar{\lambda}}[\bar{M}][\bar{\Phi}] = 0 \tag{10.2.5}$$

yields an eigenvalue $\bar{\lambda}$ identical with that given by solution of the transport equation.
3. They are such that reaction rates in the whole of each V_i for each energy group are the same as those found using the solution of the transport equation.

To sharpen the conditions that $[\bar{D}_u]$, $[\bar{A}]$, and $[\bar{M}]$ must obey we assume that the directional flux $\Psi(\mathbf{r}, \mathbf{\Omega}, E)$ (and hence $\Phi(\mathbf{r}, E)$ and $\mathbf{J}(\mathbf{r}, E)$) is known. Then, converting the exact equation (10.2.2) to energy-group form in the usual formal way (see Section 9.4), we obtain

$$-\nabla\cdot[\mathbf{J}(\mathbf{r})] - [A(\mathbf{r})][\Phi(\mathbf{r})] + \frac{1}{\lambda}[M(\mathbf{r})][\Phi(\mathbf{r})] = 0. \tag{10.2.6}$$

If the three conditions are to be met, it is necessary that, for each volume element V_i,

$$-\int_{V_i}\frac{\partial}{\partial u}[J_u(\mathbf{r})]\,dV = \int_{V_i}\frac{\partial}{\partial u}\left\{[\bar{D}_u]_i\frac{\partial}{\partial u}[\bar{\Phi}(\mathbf{r})]\right\}dV = [\bar{D}_u]_i\int_{V_i}\frac{\partial^2}{\partial u^2}[\bar{\Phi}(\mathbf{r})]\,dV \qquad (u = x, y, z),$$

$$\int_{V_i}[A(\mathbf{r})][\Phi(\mathbf{r})]\,dV = [\bar{A}]_i\int_{V_i}[\bar{\Phi}(\mathbf{r})]\,dV, \tag{10.2.7}$$

$$\int_{V_i}[M(\mathbf{r})][\Phi(\mathbf{r})]\,dV = [\bar{M}]_i\int_{V_i}[\bar{\Phi}(\mathbf{r})]dV.$$

Under the assumption that $\Psi(\mathbf{r}, \mathbf{\Omega}, E)$ is known, $[\mathbf{J}(\mathbf{r})]$, $[\Phi(\mathbf{r})]$, and the integrals on the left-hand side of (10.2.7) can readily be found. However precise definitions of the elements of $[\bar{D}_u]_i$, $[\bar{A}]_i$, and $[\bar{M}]_i$ do not follow immediately for two reasons: First, we do not know $\bar{\Phi}(\mathbf{r})$ for the various cells. Second, equations (10.2.7) are between vectors and are thus not sufficient to determine all the elements of the matrices $[\bar{A}]_i$, $[\bar{D}_u]_i$, etc., even if $[\bar{\Phi}(\mathbf{r})]$ is known. Thus, for example, the second equation in (10.2.7), written out for region i and group g, is

$$\int_{V_i} A_{g1}(\mathbf{r})\Phi_1(\mathbf{r})\,dV + \int_{V_i} A_{g2}(\mathbf{r})\Phi_2(\mathbf{r})\,dV + \cdots + \int_{V_i} A_{gG}(\mathbf{r})\Phi_G(\mathbf{r})\,dV$$
$$= \bar{A}^i_{g1}\int_{V_i} \bar{\Phi}_1(\mathbf{r})\,dV + \bar{A}^i_{g2}\int_{V_i} \bar{\Phi}_2(\mathbf{r})\,dV + \cdots + \bar{A}^i_{gG}\int_{V_i} \bar{\Phi}_G(\mathbf{r})dV. \tag{10.2.8}$$

Even if the $\bar{\Phi}_g(\mathbf{r})$ were known, there would still be only G equations for the G^2 unknowns $A^i_{gg'}$ $(g, g' = 1, 2, \ldots, G)$ in cell i.

We get around this latter difficulty by matching the two sides of (10.2.8) term by term. Thus we require that

$$\int_{V_i} A_{gg'}(\mathbf{r})\Phi_{g'}(\mathbf{r})\,dV = \bar{A}^i_{gg'}\int_{V_i} \bar{\Phi}_{g'}(\mathbf{r})dV \qquad (g, g' = 1, 2, \ldots, G), \tag{10.2.9}$$

with a similar expression for the terms involving the $\bar{M}^i_{gg'}$.

We now note that (10.2.9) is still a sum of terms involving the different isotopes contained in region i. We equate these in detail to the extent that, in accord with condition 3, we wish individual reaction rates for the cell, obtained using the homogenized constants, to agree with the corresponding rates predicted by transport theory. For example, if we wanted the homogenized constants to yield the correct absorption rate in U^{25}, we would supplement (10.2.9) by the condition

$$\int_{V_i} \Sigma^{25}_{ag}(\mathbf{r})\Phi_g(\mathbf{r})dV = \bar{\Sigma}^{25(i)}_{ag}\int_{V_i} \bar{\Phi}_g(\mathbf{r})\,dV \qquad (g = 1, 2, \ldots, G). \tag{10.2.10}$$

Equations of the type (10.2.9) and (10.2.10) define the precise sense in which the parameters $\bar{A}^i_{gg'}$, etc., are to be equivalent. However they provide no indication of whether such parameters exist or, if they do, of how to find numerical values for them. The heart of the difficulty is the fact that the integrals $\int_{V_i} \bar{\Phi}_g(\mathbf{r})dV$ themselves depend on the equivalent constants. Thus the right-hand sides of (10.2.7), (10.2.9), etc., are *implicit* nonlinear functions of *all* the equivalent group parameters. Since the numerical values of the integrals are restricted to some extent by the fact that $[\bar{\Phi}(\mathbf{r})]$ is a solution of the group diffusion equation (10.2.5), there is no guarantee that there *exists* a set of equivalent constants—even admitting absurd values—that will permit (10.2.7) to be satisfied. In fact we shall see that the present state of development in this area is such that we must relax further the conditions (10.2.7) in order to find equivalent constants for most practical situations.

10.3 Equivalent Diffusion-Theory Constants for Slab Geometry

As so often happens, the slab-geometry case is much more tractable than any multidimensional situation. We shall examine it in some detail, both because it provides a practical method for dealing with bladed control rods and because it will give us some insight when we turn to two- and three-dimensional situations.

To find equivalent diffusion-theory parameters for the one-dimensional case we use the fact that parameters which preserve all the integrated reaction rates within all subregions V_i must necessarily preserve the net leakage rate out of each V_i. The first equation in (10.2.7) states this fact mathematically. It seems both plausible and desirable to extend this leakage-preservation requirement so that the integral of the net leakage between V_i and each one of its neighbors is preserved individually.

In slab geometry this extra restriction implies that, if we know from more accurate calculations the column vector $[J(x_i)]$ of group currents at the point x_i separating a zone of thickness Δ_{i-1} from its neighbor of thickness Δ_i, we want the equivalent diffusion-theory parameters for these zones to be such that solution of the group diffusion equations containing the effective constants will reproduce the $[J(x_i)]$. If we also require that the column vector of group fluxes $[\Phi(x_i)]$ at all the interfaces match those values known from more accurate calculations, a systematic procedure for finding the equivalent parameters emerges. It should be noted, however, that this latter requirement is by no means forced by physical considerations. Demanding that the equivalent group constants and associated fluxes reproduce the integrals of reaction rates over the zone Δ_i implies no *necessary* equality between the values of these fluxes at the *points* x_i and the rigorous flux values. On the other hand there is nothing *wrong* with requiring the $[\Phi(x_i)]$ to take on the rigorous values, provided equivalent parameters can be found that will accomplish this end. Moreover imposing this requirement insures that the artificial group fluxes corresponding to the equivalent group parameters will be continuous at the interfaces x_i. It has been found that equivalent parameters determined without specifically demanding that the correct values of the $[\Phi(x_i)]$ be preserved will reproduce the desired reaction rates only if discontinuities in the equivalent group fluxes at the x_i are permitted.

For these reasons we shall seek equivalent group diffusion-theory parameters that preserve both the $[J(x_i)]$ and the $[\Phi(x_i)]$. The basic idea will be to relate $[J(x_i)]$ and $[\Phi(x_i)]$ to $[J(x_{i+1})]$ and $[\Phi(x_{i+1})]$ in terms of the equivalent diffusion-theory parameters associated with the interval $\Delta_i(\equiv x_{i+1} - x_i)$ and then to try to infer the values of the equivalent parameters from this relationship. We shall first do this for the one-group case with the material inside Δ_i homogeneous and then show a useful application of the result (so-called *blackness theory*). Finally we shall extend the procedures to many groups and a heterogeneous arrangement of materials within Δ_i.

The One-Group Case for a Homogeneous Region

We consider first the one-group case with the material inside Δ_i homogeneous but not of a nature such that diffusion theory should be valid. If equivalent constants $\overline{\nu\Sigma}_f$, $\overline{\Sigma}_a$, and $\overline{D}$ can be found for this case, they will be such that within Δ_i the one-group P_1 equations (9.4.4) and (9.4.5) are

$$\begin{aligned} &\frac{d}{dx}J(x) + \overline{\Sigma}_a\Phi(x) = \overline{\nu\Sigma}_f\Phi(x), \\ &\frac{d}{dx}\Phi(x) + \frac{1}{\overline{D}}J(x) = 0. \end{aligned} \tag{10.3.1}$$

In matrix form these become

$$\frac{d}{dx}\begin{bmatrix}\Phi \\ J\end{bmatrix} = -\begin{bmatrix}0 & \dfrac{1}{\overline{D}} \\ \overline{\Sigma}_a - \overline{\nu\Sigma}_f & 0\end{bmatrix}\begin{bmatrix}\Phi \\ J\end{bmatrix}, \tag{10.3.2}$$

and, with the definitions

$$[u(x)] \equiv \begin{bmatrix}\Phi(x) \\ J(x)\end{bmatrix}, \qquad [N] \equiv \begin{bmatrix}0 & \dfrac{1}{\overline{D}} \\ \overline{\Sigma}_a - \overline{\nu\Sigma}_f & 0\end{bmatrix}, \tag{10.3.3}$$

(10.3.2) becomes

$$\frac{d}{dx}[u(x)] + [N][u(x)] = 0. \tag{10.3.4}$$

If we define the 2×2 matrix $\exp([N]x)$ by the infinite series

$$\exp([N]x) \equiv I + [N]x + \frac{1}{2}[N]^2x^2 + \frac{1}{3!}[N]^3x^3 + \cdots, \tag{10.3.5}$$

(10.3.4) may be rewritten

$$\exp(-[N]x)\frac{d}{dx}\{\exp([N]x)[u(x)]\} = 0. \tag{10.3.6}$$

Integration from x_i to x_{i+1} then yields

$$[u(x_i)] = \exp([N]\Delta)[u(x_{i+1})], \tag{10.3.7}$$

where we have dropped subscript i from Δ_i.

The fact that $\exp(-[N]\Delta)\exp([N]\Delta)$ is the identity matrix yields the inverse result

$$[u(x_{i+1})] = \exp(-[N]\Delta)[u(x_i)]. \tag{10.3.8}$$

Since $[N]$ depends essentially on two parameters, $\bar{D}$ and $(\bar{\Sigma}_a - \nu\bar{\Sigma}_f)$, and since (10.3.8) is equivalent to two algebraic equations, we can now find $\bar{D}$ and $(\bar{\Sigma}_a - \nu\bar{\Sigma}_f)$ provided we know the values of $[u(x_i)]$ corresponding to $[u(x_{i+1})]$ from some higher-order computation. Note, however, that values of the equivalent parameters found in this way will generally depend on the particular $[u(x_{i+1})]$ used to determine them. Unless this dependence is slight, the equivalent parameters will be of little practical value.

To determine the two nonzero components of $[N]$ we first expand the exponential. Defining

$$\kappa^2 = \frac{\bar{\Sigma}_a - \overline{\nu\Sigma}_f}{\bar{D}}, \tag{10.3.9}$$

we find that

$$\begin{aligned}
\exp([N]\Delta) &= I + [N]\Delta + \frac{1}{2}[N]^2\Delta^2 + \frac{1}{3!}[N]^3\Delta^3 + \frac{1}{4!}[N]^4\Delta^4 + \cdots \\
&= I + \begin{bmatrix} 0 & \bar{D}^{-1} \\ \bar{D}\kappa^2 & 0 \end{bmatrix}\Delta + \frac{1}{2}\begin{bmatrix} \kappa^2 & 0 \\ 0 & \kappa^2 \end{bmatrix}\Delta^2 + \frac{1}{3!}\begin{bmatrix} 0 & \bar{D}^{-1}\kappa^2 \\ \bar{D}\kappa^4 & 0 \end{bmatrix}\Delta^3 + \frac{1}{4!}\begin{bmatrix} \kappa^4 & 0 \\ 0 & \kappa^4 \end{bmatrix}\Delta^4 + \cdots \\
&= \begin{bmatrix} 1 + \frac{1}{2}\kappa^2\Delta^2 + \frac{1}{4!}\kappa^2\Delta^4 + \cdots & (\bar{D}\kappa)^{-1}\left(\kappa\Delta + \frac{1}{3!}\kappa^3\Delta^3 + \cdots\right) \\ \bar{D}\kappa\left(\kappa\Delta + \frac{1}{3!}\kappa^3\Delta^3 + \cdots\right) & 1 + \frac{1}{2}\kappa^2\Delta^2 + \frac{1}{4!}\kappa^4\Delta^4 + \cdots \end{bmatrix} \\
&= \begin{bmatrix} \cosh\kappa\Delta & (\bar{D}\kappa)^{-1}\sinh\kappa\Delta \\ \bar{D}\kappa\sinh\kappa\Delta & \cosh\kappa\Delta \end{bmatrix}.
\end{aligned} \tag{10.3.10}$$

Equation (10.3.7) thus becomes

$$\begin{bmatrix} \Phi(x_i) \\ J(x_i) \end{bmatrix} = \begin{bmatrix} \cosh\kappa\Delta & (\bar{D}\kappa)^{-1}\sinh\kappa\Delta \\ \bar{D}\kappa\sinh\kappa\Delta & \cosh\kappa\Delta \end{bmatrix}\begin{bmatrix} \Phi(x_{i+1}) \\ J(x_{i+1}) \end{bmatrix}. \tag{10.3.11a}$$

Since $\sinh\kappa\Delta = \sqrt{(\cosh^2\kappa\Delta - 1)}$ and we are assuming we know $[u(x_i)]$ and $[u(x_{i+1})]$ from some previous, highly accurate computations, (10.3.11a) can be solved for $\bar{D}$ and $\kappa\Delta$. Of the many ways to do this we shall employ one that is directly applicable to blackness theory and that can be extended fairly readily to the multigroup case.

Thus, noting that $\exp(-[N]\Delta)$ can be obtained from (10.3.10) by simply changing the sign of Δ, we express (10.3.8) as

$$\begin{bmatrix} \Phi(x_{i+1}) \\ J(x_{i+1}) \end{bmatrix} = \begin{bmatrix} \cosh\kappa\Delta & -(\bar{D}\kappa)^{-1}\sinh\kappa\Delta \\ -\bar{D}\kappa\sinh\kappa\Delta & \cosh\kappa\Delta \end{bmatrix}\begin{bmatrix} \Phi(x_i) \\ J(x_i) \end{bmatrix}. \tag{10.3.11b}$$

Then, by writing out the second of the algebraic equations implied by (10.3.11a,b), we can show that

$$\begin{aligned} J(x_i) + J(x_{i+1}) &= -\bar{D}\kappa \sinh \kappa\Delta\,(\Phi(x_i) - \Phi(x_{i+1})) + \cosh \kappa\Delta\,(J(x_i) + J(x_{i+1})), \\ J(x_i) - J(x_{i+1}) &= \bar{D}\kappa \sinh \kappa\Delta\,(\Phi(x_i) + \Phi(x_{i+1})) - \cosh \kappa\Delta\,(J(x_i) - J(x_{i+1})). \end{aligned} \tag{10.3.12}$$

If we now define *blackness coefficients* α and β by

$$\alpha \equiv \frac{J(x_i) - J(x_{i+1})}{\Phi(x_i) + \Phi(x_{i+1})}, \qquad \beta \equiv \frac{J(x_i) + J(x_{i+1})}{\Phi(x_i) - \Phi(x_{i+1})}, \tag{10.3.13}$$

(10.3.12) becomes

$$\begin{aligned} \beta &= -\bar{D}\kappa \sinh \kappa\Delta + \beta \cosh \kappa\Delta, \\ \alpha &= \bar{D}\kappa \sinh \kappa\Delta - \alpha \cosh \kappa\Delta, \end{aligned} \tag{10.3.14}$$

from which we obtain

$$\cosh \kappa\Delta = \frac{\beta + \alpha}{\beta - \alpha}. \tag{10.3.15}$$

Then multiplying the first equation in (10.3.14) by α and the second by β and adding yields

$$2\alpha\beta = (\beta - \alpha)\bar{D}\kappa \sinh \kappa\Delta. \tag{10.3.16}$$

Now, from (10.3.15),

$$\sinh \kappa\Delta = \sqrt{(\cosh^2\kappa\Delta - 1)} = \frac{2\sqrt{(\alpha\beta)}}{\beta - \alpha}, \tag{10.3.17}$$

so we can write (10.3.16) as

$$\bar{D}\kappa = \sqrt{(\alpha\beta)}. \tag{10.3.18}$$

It follows that, if the blackness coefficients α and β are known for the region Δ, κ may be found from (10.3.15), $\bar{D}$ from (10.3.18), and $\bar{\Sigma}_a - \overline{\nu\Sigma_f}$ from (10.3.9). The ratio $\overline{\nu\Sigma_f}/\bar{\Sigma}_a$, if it is desired, can be found from the ratio of the rate of fission production to that of absorption, as determined by some higher-order scheme.

Note that, since $J(x_i)$ and $J(x_{i+1})$ are net currents in the $+X$ direction, $J(x_i)$ and $-J(x_{i+1})$ are both currents *into* the slab. Thus the blackness coefficient α is a measure of the net absorption capability of the slab. If the rate of production of neutrons due to fission in the slab exceeds the rate of absorption, α will be negative, $\cosh \kappa\Delta$ (for $\beta + \alpha > 0$) will be less than one, κ will be pure imaginary, and κ^2 will be negative. Equation (10.3.18) shows that $\bar{D}$ will still be real. Thus $\bar{D}^2\kappa^2 = \bar{D}(\bar{\Sigma}_a - \overline{\nu\Sigma_f})$ will be negative, and $\overline{\nu\Sigma_f}$ will exceed $\bar{\Sigma}_a$ as we would expect.

Note also, however, that it is possible to contrive situations (arbitrarily selected values of $[u(x_i)]$ and corresponding "exact" values of $[u(x_{i+1})]$) for which α and β are such that the equivalent constants assume unphysical values (e.g., a negative or imaginary $\bar{D}$). Such equivalent constants are of little practical value.

There is no a priori reason to expect that the blackness coefficients for a given slab of materials will be the same for all $J(x_{i+1})$, $\Phi(x_{i+1})$, and the associated $J(x_i)$ and $\Phi(x_i)$ used to determine them. However, if they are, (10.3.15) and (10.3.18) show that the equivalent diffusion-theory parameters will also be independent of these input and output fluxes and currents. Thus the same matrix $[N]$ in (10.3.7) will yield the correct $[u(x_i)]$ for any input vector $[u(x_{i+1})]$. For this ideal situation (which we shall see actually occurs in certain practical cases) effective parameters can be determined directly from (10.3.7). We simply apply twice whatever "exact" method is being used to determine $[u(x_i)]$ when we are given $[u(x_{i+1})]$. Specifically we use this numerical standard to generate $[u^{(1)}(x_i)]$ from $[u^{(1)}(x_{i+1})]$ and then to generate $[u^{(2)}(x_i)]$ from $[u^{(2)}(x_{i+1})]$, where $[u^{(2)}(x_{i+1})]$ is linearly independent of $[u^{(1)}(x_{i+1})]$. The two input flux-current vectors may be written in the form of a 2×2 matrix

$$[v_{i+1}] \equiv \begin{bmatrix} \Phi^{(1)}(x_{i+1}) & \Phi^{(2)}(x_{i+1}) \\ J^{(1)}(x_{i+1}) & J^{(2)}(x_{i+1}) \end{bmatrix}, \tag{10.3.19}$$

and then, on the basis of (10.3.7), we have

$$[v_i] = \exp([N\Delta])[v_{i+1}]. \tag{10.3.20}$$

Since $[\Phi^{(1)}(x_{i+1})]$ and $[\Phi^{(2)}(x_{i+1})]$ are linearly independent, $[v_{i+1}]^{-1}$ exists and thus $[N]$ is given immediately as

$$[N] = \frac{1}{\Delta} \ln\left([v_i][v_{i+1}]^{-1}\right) \equiv \frac{1}{\Delta} \ln[w], \tag{10.3.21}$$

where the 2×2 matrix $\ln[w]$ may be defined by an infinite series.

In practice a straightforward way to determine $\ln[w]$ is to diagonalize $[v_i][v_{i+1}]^{-1}$. To outline this procedure we shall assume that the eigenvalues ω_i $(i = 1, 2)$ of $[w]$ are distinct. Then, if we designate the eigenvectors of $[w]$ corresponding to the ω_i as $\mathrm{Col}\{z_1^{(i)}, z_2^{(i)}\}$, we have

$$[w]\begin{bmatrix} z_1^{(i)} \\ z_2^{(i)} \end{bmatrix} = \omega_i \begin{bmatrix} z_1^{(i)} \\ z_2^{(i)} \end{bmatrix} \qquad (i = 1, 2), \tag{10.3.22}$$

so that, with the matrices $[Z]$ and $[\Omega]$ defined by

$$[Z] \equiv \begin{bmatrix} z_1^{(1)} & z_1^{(2)} \\ z_2^{(1)} & z_2^{(2)} \end{bmatrix}, \qquad [\Omega] \equiv \begin{bmatrix} \omega_i & 0 \\ 0 & \omega_2 \end{bmatrix}, \tag{10.3.23}$$

(10.3.22) may be expressed as

$$[w][Z] = [Z][\Omega]. \tag{10.3.24}$$

Since the ω_i are assumed distinct, the eigenvectors of $[w]$ are linearly independent. Thus $[Z]^{-1}$ exists, and we have

$$[Z]^{-1}[w][Z] = [\Omega]. \tag{10.3.25}$$

It follows that, since $\ln[w]$ may be expressed as a sum of polynomial fractions in $[w]$ and since

$$[Z]^{-1}[w]^m[Z] = [\Omega]^m = \begin{bmatrix} \omega_1^m & 0 \\ 0 & \omega_2^m \end{bmatrix} \qquad (m = 2, 3, \ldots), \tag{10.3.26}$$

where $[w]^m$ is the mth power of $[w]$,

$$\begin{aligned} [N] &= \frac{1}{\Delta}[Z][Z]^{-1}\{\ln[w]\}[Z][Z]^{-1} \\ &= \frac{1}{\Delta}[Z]\{\ln[\Omega]\}[Z]^{-1} \\ &= \frac{1}{\Delta}[Z]\begin{bmatrix} \ln\omega_1 & 0 \\ 0 & \ln\omega_2 \end{bmatrix}[Z]^{-1}. \end{aligned} \tag{10.3.27}$$

We have developed this procedure in order to determine equivalent constants that are the same for all the $[v_{i+1}]$ of (10.3.19). Thus we expect that the expression for $[N]$ found from (10.3.27) will have its diagonal elements equal to zero. Conversely, if we try to apply (10.3.27) to a more general situation for which $\bar{D}$ and $\bar{\Sigma}_a - \overline{\nu\Sigma_f}$ differ for different $[v_{i+1}]$, the resultant expression for $[N]$ will *not* have zero diagonal elements. Thus, if (10.3.27) predicts nonzero diagonal elements for $[N]$, we conclude that the equivalent one-group diffusion parameters for the region between x_i and x_{i+1} will depend on the reactor conditions external to that region and must be determined from (10.3.15) and (10.3.18) for the particular $[u(x_{i+1})]$ and related $[u(x_i)]$ at hand.

Blackness Theory

Blackness theory is a procedure for determining energy-dependent blackness coefficients for a given homogeneous slab of material under the following assumptions:

1. The directional flux $\Psi(x, \mu, E)$ has the P_1 form at the surfaces of the slab for neutrons directed into the slab.
2. There is no source of neutrons inside the slab due either to fission or to scattering from other energies.
3. The net current density $J(x, E)$ is continuous at $x = x_i$ and $x = x_{i+1}$.

The essential idea of the theory is to solve the transport equation (simplified because of assumption 2) within the slabs and to use the solution to determine $\Phi(x, E)$ and $J(x, E)$ at the surfaces. Values of the blackness coefficients at any energy can then be found from the definitions (10.3.13).

We shall carry out the procedure in detail for the case of a purely absorbing slab ($\Sigma_s(x, E) = 0$) since the result is of practical importance for slab control-rod calculations and since an analytical solution permitting insight into the nature of the resultant coefficients is possible for this simple case.

When assumption 2 is made and in addition all scattering within the slab is neglected, the transport equation (8.4.11) for the slab becomes the "one-speed" equation

$$\mu \frac{\partial}{\partial x} \Psi(x, \mu, E) + \Sigma_a(E)\Psi(x, \mu, E) = 0. \tag{10.3.28}$$

If we then define

$$\begin{aligned} &\Psi^+(\mu) \equiv \Psi(x_{i+1}, \mu, E), \\ &\Psi^-(\mu) \equiv \Psi(x_i, \mu, E), \\ &T \equiv \Sigma_a(E)\Delta, \text{ where } \Delta \equiv x_{i+1} - x_i, \end{aligned} \tag{10.3.29}$$

the solution of (10.3.28) shows that

$$\begin{aligned} \Psi^+(\mu) &= \Psi^-(\mu) \exp\left(-\frac{T}{\mu}\right) \quad \text{for } 0 < \mu \le 1, \\ \Psi^-(\mu) &= \Psi^+(\mu) \exp\left(\frac{T}{\mu}\right) \quad \text{for } -1 \le \mu < 0. \end{aligned} \tag{10.3.30}$$

Moreover assumption 1 that at the surfaces of the slab the directional fluxes entering the slab have the P_1 form leads to

$$\begin{aligned} \Psi^-(\mu) &= \Phi^- + 3\mu J^- \quad \text{for } 0 < \mu \le 1, \\ \Psi^+(\mu) &= \Phi^+ + 3\mu J^+ \quad \text{for } -1 \le \mu < 0. \end{aligned} \tag{10.3.31}$$

Since a full transport-theory solution is employed inside the absorber, the directional fluxes for directions *emergent* from the absorber will not have the simple P_1 form. We can nevertheless connect the solutions inside and outside the slab by applying assumption 3 that the P_1 *components* of the directional flux (i.e., the net currents J^- and J^+) are continuous at the interfaces. Since the P_1 components just outside the surfaces are J^- and J^+ while those just inside are

$$\int_{-1}^{1} \frac{d\mu}{2} \mu\Psi^-(\mu) \quad \text{and} \quad \int_{-1}^{1} \frac{d\mu}{2} \mu\Psi^+(\mu),$$

we find, using (10.3.30) and (10.3.31), that

$$\begin{aligned} J^- &= \int_{-1}^{1} \frac{d\mu}{2} \mu\Psi^-(\mu) = \int_{-1}^{0} \frac{d\mu}{2} \mu\left[\Psi^+(\mu)\exp\left(\frac{T}{\mu}\right)\right] + \int_0^1 \frac{d\mu}{2} \mu(\Phi^- + 3\mu J^-) \\ &= \int_{-1}^{0} \frac{d\mu}{2} \mu(\Phi^+ + 3\mu J^+)\exp\left(\frac{T}{\mu}\right) + \int_0^1 \frac{d\mu}{2} \mu(\Phi^- + 3\mu J^-) \end{aligned} \tag{10.3.32}$$

and

$$J^+ = \int_{=1}^{1} \frac{d\mu}{2} \mu\Psi^+(\mu) = \int_{-1}^{0} \frac{d\mu}{2} \mu(\Phi^+ + 3\mu J^+) + \int_0^1 \frac{d\mu}{2} \mu(\Phi^- + 3\mu J^-)\exp\left(-\frac{T}{\mu}\right). \tag{10.3.33}$$

The foregoing results can be expressed in terms of the *exponential integrals* defined by

$$E_{n+2}(T) \equiv \int_0^1 \mu^n \exp\left(-\frac{T}{\mu}\right) d\mu. \tag{10.3.34}$$

These integrals are nonnegative and have an upper limit

$$E_{n+2}(T) \le \int_0^1 \mu^n \, d\mu = \frac{1}{n+1},$$

reached when $T = 0$. They approach zero as $T \to \infty$. Use of them allows (10.3.32) and (10.3.33) to be rewritten as

$$\begin{aligned} \frac{J^-}{2} &= \frac{\Phi^-}{4} - \frac{1}{2}E_3(T)\Phi^+ + \frac{3}{2}E_4(T)J^+, \\ -\frac{J^+}{2} &= \frac{\Phi^+}{4} - \frac{1}{2}E_3(T)\Phi^- - \frac{3}{2}E_4(T)J^-. \end{aligned} \tag{10.3.35}$$

Adding and subtracting equations (10.3.35) and rearranging some of the terms yields the following analytical results for energy-dependent blackness coefficients analogous to the one-group expressions (10.3.13):

$$\begin{aligned} \alpha(T) &\equiv \frac{J^- - J^+}{\Phi^- + \Phi^+} = \frac{1 - 2E_3(T)}{2[1 + 3E_4(T)]}, \\ \beta(T) &\equiv \frac{J^- + J^+}{\Phi^- - \Phi^+} = \frac{1 + 2E_3(T)}{2[1 - 3E_4(T)]}. \end{aligned} \tag{10.3.36}$$

As $T(\equiv \Sigma_a(E)\Delta)$ goes from 0 to ∞, these blackness coefficients take on values in the ranges $0 \le \alpha(T) \le 0.5$ and $0.5 \le \beta(T) < \infty$.

Provided—and *only* provided—scattering is neglected, extension of blackness theory to the case where the absorption cross section is spatially dependent is trivial. Equation (10.3.30) continues to be a valid solution of (10.3.28) if we merely generalize the definition (10.3.29) of the *optical thickness* T to

$$T \equiv \int_{x_i}^{x_{i+1}} \Sigma_a(x, E)\, dx. \tag{10.3.37}$$

Extension to the case of nonzero scattering (still with spatially homogeneous material and with energy changes due to scattering neglected) is somewhat more complicated. The relevant form of the transport equation which must then be solved is

$$\mu \frac{\partial}{\partial x} \Psi(x, \mu, E) + \Sigma_t(E)\Psi(x, \mu, E) = \int_{-1}^{1} \frac{d\mu}{2} \int_0^\infty dE' \, \Sigma_s(E, \mu_0)\delta(E' - E)\Psi(x, \mu', E')$$
$$= \int_{-1}^{1} \frac{d\mu'}{2} \Sigma_s(E, \mu_0)\Psi(x, \mu', E). \tag{10.3.38}$$

An analytical solution of this equation is difficult, and a high-order spherical-harmonics or discrete-ordinates approximation is called for. However, since (10.3.38) is a one-dimensional, one-speed equation, spherical-harmonics or discrete-ordinates solutions will cost very little in computer time.

If $\Sigma_s(E)$ becomes much greater than $\Sigma_a(E)$ (and if assumption 2 is still imposed), the one-speed transport equation (10.3.38) inside the slab will be well approximated by the corresponding one-speed diffusion equation. Thus we expect for this case that $\bar{\Sigma}_a$ and $\bar{D}$ will approach the physical values $\Sigma_a(E)$ and $1/[3(\Sigma_t(E) - \Sigma_{s1}(E))]$. It follows that a plot of $\bar{\Sigma}_a\Delta$ versus $\Sigma_a(E)\Delta$ for $\Sigma_s(E) \gg \Sigma_a(E)$ will approach a straight line. Figure 10.2 shows such a plot along with the other limiting curve for $\Sigma_s(E) = 0$ obtained from (10.3.36), (10.3.9), (10.3.15), and (10.3.18). (It is not hard to show that these last three equations

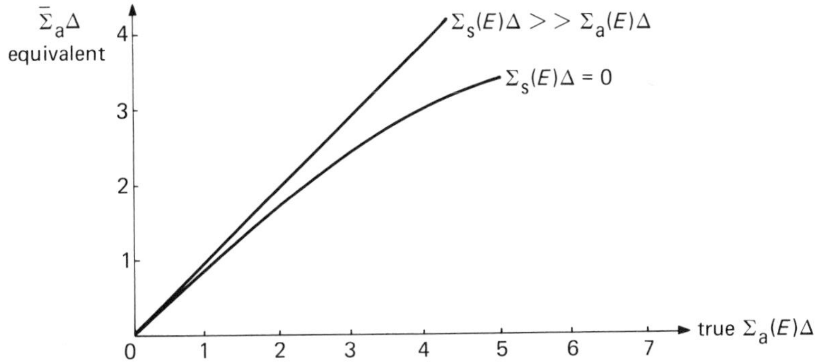

Figure 10.2 Limiting curves of effective versus true absorption cross sections for a slab of thickness Δ.

yield the formula

$$\bar{\Sigma}_a - \overline{\nu\Sigma}_f = \frac{\sqrt{(\alpha\beta)}}{\Delta} \ln \frac{\sqrt{\beta} + \sqrt{\alpha}}{\sqrt{\beta} - \sqrt{\alpha}}, \tag{10.3.39}$$

where in this case $\overline{\nu\Sigma}_f$ is zero.)

Values of $\bar{\Sigma}_a\Delta$ for the range of moderate scattering lie between the two limiting curves shown in Figure 10.2. We conclude in general that the equivalent optical thickness $\bar{\Sigma}_a\Delta$ is always less than the true optical thickness $\Sigma_a(E)\Delta$. The figure shows that the ratio of the two is about 0.7 for $\Sigma_a(E)\Delta = 1.0$ and falls to about 0.6 for $\Sigma_a(E)\Delta = 5.0$.

Because of assumption 2 that there is no fissioning and no neutron energy exchange when scattering occurs within the slab, we have been able to determine blackness coefficients and equivalent diffusion-theory parameters for any neutron energy of interest. To find corresponding parameters for an energy group we could simply average $\bar{\Sigma}_a(T)$ and $\bar{D}(T)$ over the energy range of the group, weighting them by some estimate of the flux spectra at the surfaces of the slab. However it is more rigorous to integrate (10.3.35) over the energy group, in that manner obtaining a relationship connecting the group currents and fluxes on the two surfaces of the absorbing slab. Doing so and making use of the definitions (10.3.36) yields, for energy group n,

$$\begin{aligned} \int_{\Delta E_n} [J^-(E) - J^+(E)]\,dE &= \int_{\Delta E_n} [\Phi^-(E) + \Phi^+(E)]\alpha(E)\,dE \equiv J_n^- - J_n^+, \\ \int_{\Delta E_n} [J^-(E) + J^+(E)]\,dE &= \int_{\Delta E_n} [\Phi^-(E) - \Phi^+(E)]\beta(E)\,dE \equiv J_n^- + J_n^+, \end{aligned} \tag{10.3.40}$$

where the dependence of α and β on energy is through the optical thickness T. We then define blackness coefficients for the group by

$$\begin{aligned} \alpha_n &\equiv \frac{\int_{\Delta E_n}[\Phi^-(E) + \Phi^+(E)]\alpha(E)dE}{\int_{\Delta E_n}[\Phi^-(E) + \Phi^+(E)]dE} = \frac{J_n^- - J_n^+}{\Phi_n^- + \Phi_n^+}, \\ \beta_n &\equiv \frac{\int_{\Delta E_n}[\Phi^-(E) - \Phi^+(E)]\beta(E)dE}{\int_{\Delta E_n}[\Phi^-(E) - \Phi^+(E)]dE} = \frac{J_n^- + J_n^+}{\Phi_n^- - \Phi_n^+}. \end{aligned} \tag{10.3.41}$$

Because of the assumption that there is no fissioning and that scattering leads to no energy exchange, the equivalent P_1 equations (9.4.4) for the slab for any energy group n reduce to

$$\begin{aligned} &\frac{d}{dx} J_n(x) + \bar{\Sigma}_{an}\Phi_n(x) = 0, \\ &\frac{d}{dx} \Phi_n(x) + \frac{1}{\bar{D}_n} J_n(x) = 0. \end{aligned} \tag{10.3.42}$$

Thus the one-speed analysis leading from (10.3.1) to (10.3.15) and (10.3.18) is immediately applicable to each individual group n. We conclude that effective values $\bar{D}_n$ and $\bar{\Sigma}_{an}$ for the slab can be found by using the appropriate α_n and β_n for group n in the one-speed expressions (10.3.15) and (10.3.18).

One problem remains. In order to compute α_n and β_n we must have an estimate of $\Phi^-(E)$ and $\Phi^+(E)$. A first-order approximation is to use the flux spectra appropriate to the materials in the two media adjacent to the slab (the spectra from which the energy-group constants for these media are obtained). A better approximation is to alter these spectra in a manner that reflects the presence of the absorber. Fairly good prescriptions have been devised for doing this. However we shall not pursue the matter further.

The most common application of blackness theory is to the determination of effective group diffusion parameters for slab control-rod elements. Even though the derivation of the effective parameters has been limited to a one-dimensional situation, these parameters have been found to provide an accurate prediction of the worth of control rods formed from slabs—cross-shaped rods for example. It should be noted, however, that the assumptions on which blackness theory is based imply that diffusion theory becomes immediately valid at locations just outside the absorbing slab. This is unlikely to be the case, and, as a result, the group flux shape determined by an overall diffusion-theory calculation in the immediate vicinity of a control slab is likely to be erroneous. Since the flux will be low at such locations, this situation is unlikely to cause any serious errors with regard to the prediction of local depletion effects. However the net currents $-D_g\nabla\Phi_g$ at such locations will also be in error, and this fact causes concern that the predicted rate at which neutrons are flowing into the control slab may be incorrect. The success of blackness theory as applied to control-rod calculations must then be regarded as due, to some extent, to a mutual cancelling of errors.

Finding Equivalent Parameters for a Multigroup Model

The energy-group blackness coefficients α_n and β_n defined by (10.3.41) permit the vector $\text{Col}\{\Phi_n^-, J_n^-\}$ for group n to be uniquely related to $\text{Col}\{\Phi_n^+, J_n^+\}$ in a manner that is totally independent of the surface fluxes and currents of neutrons belonging to other groups. This independence is due to the assumption that there is no fissioning within the slab and no energy exchange due to scattering. A neutron entering the slab retains its initial energy until it is captured or leaks out.

When fission takes place in the slab or when energy exchange due to neutron scattering is accounted for, blackness theory as we have developed it is no longer applicable. To find the group fluxes and currents transmitted by the slab a multigroup procedure for solving the transport equation must be employed. In this section we shall assume that this has been done. Thus we define the $2G$-element column vector of group fluxes and currents at position x by

$$[u(x)] \equiv \text{Col}\{\Phi_1(x), \Phi_2(x), \ldots, \Phi_G(x), J_1(x), J_2(x), \ldots, J_G(x)\} \tag{10.3.43}$$

and assume that we know the correct output vector $[u(x_i)]$ due to the transmission through the homogeneous slab of neutrons associated with the input vector $[u(x_{i+1})]$. We shall sketch methods for finding equivalent homogenized diffusion-theory parameters that will properly transmit the group fluxes and currents.

The procedure that can be used is a straightforward generalization of the one-group methods derived in the first part of this section. The P_1 equations (10.3.1) are first cast in multigroup form:

$$\frac{d}{dx}\begin{bmatrix}\Phi(x) \\ J(x)\end{bmatrix} + \begin{bmatrix}0 & D^{-1} \\ A_0 - M & 0\end{bmatrix}\begin{bmatrix}\Phi(x) \\ J(x)\end{bmatrix} = 0, \tag{10.3.44}$$

where the elements of the $G \times G$ matrices $[A_0]$ and $[M]$ have the form (9.4.31) and (9.4.32) and those of $[D]^{-1}$ have either the full matrix form as in (9.4.8) or the diagonal form (9.4.13).* The problem then becomes that of determining the matrix elements of $[A_0 - M]$ and $[D]^{-1}$. To solve it we proceed, as in the one-group case, to define the $2G \times 2G$ matrix

$$[N] \equiv \begin{bmatrix}0 & D^{-1} \\ A_0 - M & 0\end{bmatrix}. \tag{10.3.45}$$

Equations (10.3.4)–(10.3.7) are then seen to be valid for the G-group case, and we have immediately from (10.3.7)

$$[u(x_i)] = \exp([N]\Delta)[u(x_{i+1})], \tag{10.3.46}$$

where here $[u(x_i)]$ and $[u(x_{i+1})]$ are $2G$-element flux-current vectors and $\exp([N]\Delta)$ is a $2G \times 2G$ matrix.

If $[N]$ is known to be independent of the $[u(x_{i+1})]$, we may find it immediately from an equation analogous to (10.3.21). Thus we select $2G$ linearly independent vectors $[u^{(j)}(x_i)]$ $(j = 1, 2, \ldots, 2G)$ and form the $2G \times 2G$ matrix

$$[v_i] \equiv \{u^{(1)}(x_i), u^{(2)}(x_i), \ldots, u^{(2G)}(x_i)\}. \tag{10.3.47}$$

The analysis leading to (10.3.21) is immediately applicable, and we obtain

$$[N] = \frac{1}{\Delta}\ln\{[v_i][v_{i+1}]^{-1}\}. \tag{10.3.48}$$

*In this notation the matrix elements $A_0 - M$ and D^{-1} in (10.3.44) are themselves matrices, with the (k, l) element of $[A_0 - M]$ being $A_{0kl} - M_{kl}$ and the superscript -1 indicating an inverse matrix.

The diagonalization procedure of (10.3.22)–(10.3.27) then yields numbers for all the elements of $[N]$.

If $[N]$ is not independent of the inputs $[u(x_{i+1})]$, this method will fail in that the diagonal blocks of $[N]$ in (10.3.45) will not be found to equal zero. When this happens, it is necessary to extend the one-group procedures that led to (10.3.15) and (10.3.18).

To do so we start by expanding the $2G \times 2G$ matrix $\exp([N]\Delta)$ as in (10.3.10). It is very important to recognize, however, that $[D]^{-1}$ and $[A_0 - M]$ do not commute. With due care taken we obtain

$$\begin{bmatrix} \Phi(x_i) \\ J(x_i) \end{bmatrix} = \begin{bmatrix} \cosh \gamma\Delta & \sinh \gamma\Delta \, \gamma(A_0 - M)^{-1} \\ D\gamma \sinh \gamma\Delta & \cosh \gamma^*\Delta \end{bmatrix} \begin{bmatrix} \Phi(x_{i+1}) \\ J(x_{i+1}) \end{bmatrix}, \tag{10.3.49}$$

where

$$\gamma^2 \equiv D^{-1}(A_0 - M),$$
$$(\gamma^*)^2 \equiv (A_0 - M)D^{-1},$$

and where matrices of the form $\sqrt{(\alpha\beta)}$ are defined by $(\sqrt{(\alpha\beta)})^2 \equiv \alpha\beta$.

Multiplied out, (10.3.49) becomes two $G \times G$ matrix equations in the G-element column vectors $[\Phi(x_i)]$, $[J(x_i)]$, $[\Phi(x_{i+1})]$, and $[J(x_{i+1})]$. The G-group analogues of (10.3.12) and (10.3.13) are then written, and $G \times G$ generalizations of the blackness coefficients (10.3.13) are formed by employing some numerical standard to find the G values of $[\Phi(x_i)]$ and $[J(x_i)]$ arising from the G linearly independent input values of $[\Phi(x_{i+1})]$ and $[J(x_{i+1})]$. For example, if we define the $G \times G$ matrices $[\Phi(x)]$ and $[J(x)]$ by

$$[\Phi(x)] \equiv \{\Phi^{(1)}(x), \Phi^{(2)}(x), \ldots, \Phi^{(G)}(x)\},$$
$$[J(x)] \equiv \{J^{(1)}(x), J^{(2)}(x), \ldots, J^{(G)}(x)\}, \tag{10.3.50}$$

where the $\Phi^{(i)}(x)$ and $J^{(i)}(x)$ are now the G-element column vectors, the analogue of the scalar α defined by (10.3.13) becomes

$$[\alpha] \equiv [J(x_i) - J(x_{i+1})][\Phi(x_i) + \Phi(x_{i+1})]^{-1}. \tag{10.3.51}$$

The analogues of (10.3.14)–(10.3.18) for the G-group case can then be found by a matrix generalization of the one-group analysis.

Numerical values of the $G \times G$ matrices $[A_0 - M]$ and $[D]^{-1}$ emerge from this procedure, and, with some degree of arbitrariness (and nonzero off-diagonal values of the $[D]^{-1}$ matrix permitted), equivalent diffusion parameters can be obtained. In this case equivalent constants can be found even when they depend on the particular input matrices $[\Phi(x_{i+1})]$ and $[J(x_{i+1})]$ used for the determination. As in the one-group case, though, the utility of such constants for any given application will depend on how insensitive they are to changes in these input matrices.

Equivalent Constants for Multiregion Slabs

The extension from homogeneous to heterogeneous material configurations of the methods we have derived for finding equivalent diffusion-theory parameters is immediate. This is because the equivalent homogenized parameters are derived from accurate values of the fluxes and currents at the *edges* of the zones of interest. (See (10.3.48) or (10.3.51) for example.) Thus correct input and output values of $[\Phi]$ and $[J]$ will determine a set of equivalent parameters even if there is a complicated material structure within the zone under consideration.

When geometrical structure is present, there is a greater likelihood that equivalent constants will be sensitive to the values of the edge fluxes and currents used to determine them. There is, however, a fortunate, practical situation for which the equivalent constants are completely independent of the input (and corresponding output) group fluxes and currents. This situation arises when the material compositions making up the zone being homogenized are distributed symmetrically about the center line and when, in addition, the diffusion-theory model is valid throughout each such material composition (or—more importantly—when equivalent diffusion-theory parameters, independent of input fluxes and currents, can be found for each material composition). These conditions are approximately fulfilled when blackness theory is used to describe a control slab embedded in fuel material.

10.4 Equivalent Diffusion-Theory Constants for Two- and Three-Dimensional Situations

In evaluating any approximate theoretical method for analyzing reactors, we must ask two important questions:

1. Can the scheme take care of effects due to depletion?
2. Can it be extended to two and three dimensions?

What fragmentary evidence is available indicates that the procedures we have been outlining for slab geometry *can* take care of depletion effects. The G^2 elements of $[A_0 - M]$ determined by the methods just sketched do not specify equivalent group absorption cross sections directly. For example, in the simple two-group case with all neutrons born in group one and all fissioning in group two,

$$[A_0 - M] = \begin{bmatrix} \bar{\Sigma}_1 & -\overline{\nu\Sigma}_{f2} \\ -\bar{\Sigma}_{21} & \bar{\Sigma}_2 \end{bmatrix}, \tag{10.4.1}$$

and these four elements do not even specify $\bar{\Sigma}_{a1}$ and $\bar{\Sigma}_{a2}$ directly, let alone the equivalent absorption cross sections for isotope j. However the auxiliary cell calculations that yield $[A_0 - M]$ for the portion of the reactor being homogenized can be extended to give $\int_{V_i} \Sigma_{ag}^j(x)\Phi_g(x)dx$. Thus we can find numbers that will preserve the integrated reaction rates for isotopes j by defining

$$\overline{\Sigma}_{ag}^{j} \equiv \frac{\int_{V_i} \Sigma_{ag}^{j}(x)\Phi_g(x)\,dx}{\int_{V_i} \overline{\Phi}_g(x)\,dx}, \tag{10.4.2}$$

where $\Phi_g(x)$ is the group-g flux for the heterogeneous problem and $\overline{\Phi}_g(x)$ is that for the corresponding homogenized problem, both normalized to the same values at the edges of the portion of the slab reactor being homogenized. The integral $\int_{V_i} \overline{\Sigma}_{ag}\overline{\Phi}_g(x)dx$ is then the absorption rate of group-g neutrons in isotope j of region V_i.

The effective absorption cross sections conserve absorption rates only in an integral sense. To be consistent we should thus represent the depletion of an isotope j having concentration $n_{(i)}^{j}(t)$ within a homogenized region V_i as occurring in a uniform manner according to the equation

$$V_i \frac{d}{dt} n_{(i)}^{j}(t) = -\sum_{j=1}^{G} \overline{\Sigma}_{ag}^{j(i)} \int_{V_i} dV \overline{\Phi}_g(\mathbf{r}), \tag{10.4.3}$$

where $\overline{\Sigma}_{ag}^{j(i)}$ is the equivalent absorption cross section of isotope j for group-g neutrons in region V_i.

If depletion is carried out in this manner, a heterogeneous arrangement of fuel and absorber slabs will remain symmetric if it is symmetric initially. Thus any homogenization procedures valid initially will remain valid throughout the depletion history.

Of course representing as uniform the physically nonuniform depletion of a heterogeneous arrangement of materials may itself lead to a nonnegligible error. If so, a more refined approximation must be made. (For example it may be necessary to represent each naterial region making up V_i explicitly as a separate region having its own effective group parameters.)

Extension beyond slab geometry of the approach we have been investigating is a much more difficult matter. This is because the fluxes and currents on the surface of a two- or three-dimensional zone V_i are not generally constant (as they are over each of the two faces of a slab region). In fact, if V_i is very heterogeneous (containing a cross-shaped control rod for example), the surface fluxes and currents may be far from smooth. As a result we simply cannot find spatially constant group parameters that will reproduce exactly the *detailed* surface fluxes and currents as determined by higher-order computations. In fact it is not clear that equivalent group diffusion parameters can be found that will reproduce the correct surface fluxes and currents even in an integral sense. Thus, suppose we are trying to find equivalent constants for the two-dimensional region V_i shown in Figure 10.3 (where V_i might be a small lump of absorber or a geometrically complicated subassembly such as that shown in Figure 10.1). If $\overline{\Phi}_g(x, y)$ and $\overline{\mathbf{J}}_g(x, y)$

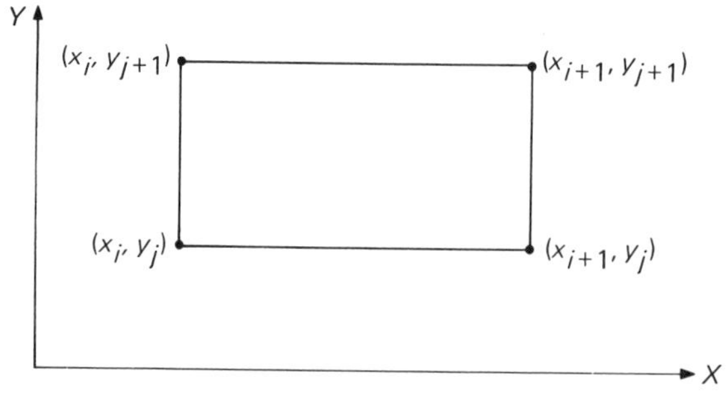

Figure 10.3 A two-dimensional region to be homogenized.

($\equiv \mathbf{i}\bar{J}_{xg}(x, y) + \mathbf{j}\bar{J}_{yg}(x, y)$) are the group-$g$ flux and current found when the group diffusion equations *based on the effective parameters* are solved, we might try to specify the effective parameters leading to $\bar{\Phi}_g(x, y)$ and $\bar{\mathbf{J}}_g(x, y)$ such that

$$\int_{x_i}^{x_{i+1}} \bar{\Phi}(x, y_i)dx, \int_{x_i}^{x_{i+1}} \bar{J}_y(x, y_i)\,dx, \int_{y_i}^{y_{i+1}} \bar{\Phi}(x_i, y)\,dy, \int_{y_i}^{y_{i+1}} \bar{J}_x(x_i, y)\,dy, \ldots, \int_{y_i}^{y_{i+1}} \bar{J}_x(x_{i+1}, y)dy$$

equal the corresponding integrals determined from a more exact analysis (a multigroup discrete-ordinates scheme for example). It is this problem which, we assert, may not have a solution.

We shall see that one potential resolution of this difficulty lies in abandoning the attempt to find diffusion-theory parameters that will yield the proper coupling between input and output currents and fluxes and instead trying to determine general "coupling matrices" having elements that are not constrained by the requirement that they be derivable from equations having the diffusion-theory form. The extensive effort we have already expended in solving equations cast in the group diffusion form provides, however, a strong motivation for further pursuing the problem of finding equivalent diffusion-theory parameters. This can be done if we relax somewhat the demands made of the equivalent parameters. Accordingly we shall investigate two methods for linking equivalent group diffusion parameters which are readily applied to two- and three-dimensional geometries but which do not yield constants such that the detailed reaction- and leakage-rate-conservation conditions (10.2.7) are fulfilled.

Equivalent Constants Found by Matching Particular Reaction Rates

The procedures we have developed for finding equivalent constants for slab geometry are based essentially on preserving the net neutron currents at the surfaces of the slab zone being homogenized. In the one-group case this requirement specifies $\bar{D}$ and $\bar{\Sigma}_a - \overline{\nu\Sigma_f}$ directly by (10.3.15) and (10.3.18), and there is generally enough information available from the "exact" computation for the output flux and current in terms of the input flux and current to specify in addition the ratio $\bar{\Sigma}_a/\overline{\nu\Sigma}_f$. A point to note is that, if $\bar{\Sigma}_a - \overline{\nu\Sigma}_f$ is correct but $\bar{\Sigma}_a$ and $\overline{\nu\Sigma}_f$ individually are such that the ratio of absorption to fission rates is incorrect, the criticality of the system will still be correctly predicted when the diffusion equation containing the equivalent constants is solved. Moreover, if we find correct equivalent constants for only one particular subregion of the reactor (for example a control slab), the net neutron absorption and transmission properties of that subregion will be rigorously correct even if other regions of the reactor are represented only approximately.

For two- and three-dimensional situations we shall alter the basic approach to finding equivalent parameters and, instead of attempting to preserve neutron currents, try directly to preserve individual reaction rates within the zone being homogenized. If effective constants can be found that do this for *all* reaction rates in all the zones into which the reactor is partitioned, criticality and currents between zones will again be predicted exactly when the equivalent parameters are used. However, if we find equivalent parameters that reproduce exact reaction rates for only a limited number of regions (for example control rods or fuel-clad-moderator cells), there is no guarantee that the neutron-transmission and neutron-leakage properties of these regions will be correctly represented.

Once this basic limitation is recognized, it seems palatable to make other simplifying approximations related to the leakage of neutrons into and out of the zone for which the equivalent reaction-rate cross sections are being found. In addition it seems acceptable to find parameters that force a match for only a few of the most important reactions. For example, in calculations of equivalent parameters in the slowing-down range for a fuel-rod cell (containing fuel material and its associated clad, moderator, and coolant), it is an almost universal practice to match the absorption rate in the fuel material only (and not worry about matching exactly the very small absorption rates in the clad, moderator, and coolant). In addition the equivalent constants are almost always found for conditions corresponding to no net leakage of neutrons across the outer boundary of the cell ("zero-current boundary conditions"). (Recall that this was an essential approximation made in Chapter 5 when we determined spatially homogenized group parameters for a fuel-rod cell in the resonance energy range.) No attempt to correct the parameters is made when the fuel-rod cell has a k_∞ different from unity, in which case, if it is part of a lattice, the net leakage into the cell is nonzero.

As a specific example of this type of calculation, suppose we wish to find, for the resonance range, an effective homogeneous absorption cross section for a cell consisting of a fuel rod and its associated clad, moderator, and coolant.

We first perform an "exact" computation and find for those energy groups covering the range in question the ratio of the number of neutrons absorbed in the fuel rod to the number appearing as the result of slowing down from higher energies (the "source"). (The methods developed in Chapter 5 or a Monte Carlo calculation might be used.) We then do an infinite-medium multigroup slowing-down computation with all the materials of the cell homogenized, correcting the absorption in each group in accord with the "exact" absorption-to-source ratio. The corrected spectrum and homogenized multigroup absorption cross sections that result can then be used to determine few-group cross sections for the resonance range.

Note that the equivalent group diffusion constant D_g for the cell must be determined by (9.6 39) from the ratio of the homogenized group current to the homogenized group flux. It is difficult to judge intuitively how accurately the group diffusion constants generated in this manner will measure the transmission of neutron currents across the cell. Moreover it is difficult to make this judgment empirically since the consequences of other approximations inherent in the overall scheme cannot be readily disentangled. Nevertheless the cell-homogenization procedure is thought to be adequate for fuel-element cells in the interior regions of a lattice, and it is used extensively for the analysis of slightly enriched, light-water-moderated reactors. The fact that absorption and slowing-down rates considerably exceed leakage rates in such reactors may partially account for its success.

Another example of determining equivalent group diffusion-theory parameters by matching reaction rates occurs when effective absorption cross sections are sought for burnable-poison flux depressors or for control rods. It is fairly common to find these constants homogenized for the absorber material alone, which is then treated as a separate, explicit region in subsequent diffusion-theory computations. In this way the approximate magnitudes of local flux perturbations caused by the absorber can be predicted. Under these circumstances, however, the search for the effective absorption cross sections requires spatial diffusion-theory calculations as well as the "exact" calculations.

For example, if it is desired to represent explicitly a lump of burnable poison embedded in a homogeneous moderator region, an "exact" calculation is first performed (often by Monte Carlo procedures) to determine the fraction of source neutrons absorbed in the lump. This computation is carried out for only a small portion of the reactor (for example the subassembly and quarter-control-rod water channel shown in Figure 10.1), zero-current boundary conditions being used on the surfaces. For each group the fraction of those neutrons appearing that are absorbed in the lump is determined. The appropriate equivalent diffusion-theory absorption cross section for the lump is then found by running a series of diffusion-theory calculations for the same geometry and zero-current boundary conditions. For each group the spatial source of neutrons for these computations is found from the "exact" calculation. Thermal absorption and diffusion constants are found for the surrounding material by the standard few-group method, and the diffusion constant for the lump may be estimated as $1/3\Sigma_{\text{tr},g}$, where $\Sigma_{\text{tr},g} = \Sigma_{sg} + \Sigma_{ag} - \Sigma_{s1g}$ is an average over the asymptotic group spectrum associated with the material outside the lump. A series of effective absorption cross sections Σ_{ag}^{eff} are tried for the lump until, by interpolation, a value for Σ_{ag}^{eff} is found such that the absorption rate in the lump for that group, as computed by the spatial diffusion-theory method, matches that predicted by the "exact" computation. These effective parameters for the poison lump are then assumed valid for all current boundary conditions on the surface of the subassembly.

Flux-Averaged Constants

One final method of determining equivalent homogenized constants must be discussed since it is a very common and intuitively appealing procedure. It applies to heterogeneous regions and consists simply of flux-weighting the position-dependent cross sections over the region to be homogenized.

We symbolize the sought-for homogenized cross sections by $\bar{\Sigma}_{\alpha g}$ (where α may be a, f, s, or tr) and represent the corresponding heterogeneous cross sections for group g by $\Sigma_{\alpha g}(\mathbf{r})$. In addition we symbolize the homogenized diffusion constant by $\bar{D}_g$. There may, however, be no counterpart to $\bar{D}_g$ in the heterogeneous case since diffusion theory may not be applicable there.

To find the $\bar{\Sigma}_{\alpha g}$ we first obtain, for each group g, a detailed scalar flux shape $\Phi_g(\mathbf{r})$ associated with the heterogeneous structure. This is usually done by employing some auxiliary calculation—usually a one-group, fixed-source computation. The homogenized cross sections $\bar{\Sigma}_{\alpha g}$ are then defined by

$$\bar{\Sigma}_{\alpha g} \equiv \frac{\int_{V_i} \Sigma_{\alpha g}(\mathbf{r})\Phi_g(\mathbf{r})\,dV}{\int_{V_i} \Phi_g(\mathbf{r})\,dV}. \tag{10.4.4}$$

To find $\bar{\Sigma}_{\alpha g}$ for an assembly consisting of a square cell of fuel elements containing a cross-shaped control rod, for example, we might first compute blackness coefficients for the blades and use them to determine equivalent group-g diffusion-theory parameters. Then, using group-g diffusion-theory parameters D_g and Σ_g for the fuel material found by means of an infinite-medium spectrum calculation, we might solve the diffusion equation

$$\nabla \cdot D_g(\mathbf{r})\nabla\Phi_g(\mathbf{r}) - \Sigma_g(\mathbf{r})\Phi_g(\mathbf{r}) + S_g(\mathbf{r}) = 0, \tag{10.4.5}$$

where $S_g(\mathbf{r})$ is a neutron source for group g that is assumed to be spatially flat in the fuel region and zero in the rod region. Inserting the group flux shape $\Phi_g(\mathbf{r})$ into (10.4.4) then yields a value for the equivalent homogenized cross section $\bar{\Sigma}_{\alpha g}$.

The exact sense in which the $\bar{\Sigma}_{\alpha g}$ are equivalent is very difficult to define. Since the "detailed flux" $\Phi_g(\mathbf{r})$ is found from a fixed-source calculation with the source taken as spatially flat in the moderator regions and zero elsewhere, the right-hand side of (10.4.4) is not the true ratio of the reaction rate α to the integrated flux in the assembly. Thus, if the effective constants are used when the assembly is part of a full core and an equivalent, full-core flux shape $\overline{\Phi}_g(\mathbf{r})$ is computed for group g, the integral $\int_{V_i} \overline{\Sigma}_{ag}\overline{\Phi}_g(\mathbf{r})dV$ will in general not be the correct reaction rate over the assembly V_i.

Moreover it is not at all clear how to define an effective homogenized diffusion constant by this procedure. A variety of prescriptions are used to circumvent the difficulty. However they are all hard to justify. Thus, when homogenized constants obtained in this way are used, the leakage rates from one cell to the next may not be very accurately predicted. This latter shortcoming is probably the most serious defect of flux-weighting methods.

Nevertheless, because of the relative ease of finding detailed flux shapes by a one-group, fixed-source computation, flux-weighting is a very commonly used technique for generating homogenized diffusion-theory constants. Many variants (usually referred to as refinements) on the procedure are used. However, since the justification for (10.4.4) is somewhat intuitive in the first place, it is rather difficult to predict which variants are likely to produce the best results.

A Numerical Example

Since any method for finding equivalent constants for two-dimensional situations must automatically apply to a one-dimensional case, we can gain some insight into the validity of the approximate flux-weighting procedure by comparing, for slab geometry, the flux-weighted parameters generated by (10.4.4) with the corresponding "exact" parameters generated by (10.3.15) and (10.3.18). A comparison of this kind is shown in Table 10.1 for a one-group model of a symmetric slab subassembly consisting of a central absorbing zone embedded in homogeneous fuel material.

The "exact" numbers were generated by selecting one-group equivalent diffusion-theory parameters for the control and fuel slabs separately and performing a three-region diffusion-theory computation to find the flux-current vector at the left-hand side of the slab in terms of the corresponding vector at the right-hand side. One-group blackness coefficients for the subassembly were then determined from (10.3.18), and (10.3.15) and

Table 10.1 Comparison of Exact and Flux-Weighted Equivalent Constants

Parameter	Exact homogenized value	Flux-weighted value	Renormalized flux-weighted value
$\overline{\nu\Sigma_f}(\text{cm}^{-1})$	0.04910	0.05734	0.04910
$\overline{\Sigma}_a(\text{cm}^{-1})$	0.05017	0.05875	0.05031
$\overline{D}(\text{cm})$	0.5535	0.8072	0.6912

(10.3.18) were used to find the "exact" equivalent values of $\bar{D}$ and $\bar{\Sigma}_a - \overline{\nu\Sigma_f}$. The first column of Table 10.1 shows the results, the individual values of $\overline{\nu\Sigma_f}$ and $\bar{\Sigma}_a$ having been found by requiring that the ratio of total fissions to total absorptions computed for the subassembly using the equivalent parameters and the corresponding flux shape match the correct result. The second column lists these same parameters as found by the conventional flux-weighting procedure (10.4.4).

The very sizeable discrepancies between corresponding entries in the first two columns do not have as serious consequences as might at first be feared. The group diffusion equations (10.2.5) are homogeneous. Hence multiplying all $\bar{D}$'s and $\bar{\Sigma}$'s in all regions and all groups by the same number (i.e., multiplying (10.2.5) by a constant) will have no effect on the solution. It follows that a more meaningful comparison between exact and flux-weighted constants can be obtained by renormalizing the entries in column 2 of the table. The result of doing so (in such a manner that the exact and flux-weighted values of $\overline{\nu\Sigma_f}$ match) appears as column 3. It can be seen that forcing equality of the $\overline{\nu\Sigma_f}$ values brings the corresponding $\bar{\Sigma}_a$ values into close agreement. However the renormalized value of the flux-averaged diffusion constant is still 20 percent higher than the exact result.*

This discrepancy appears to be irreconcilable and illustrates an unhappy characteristic of the flux-weighting technique, namely that it yields inaccurate values for equivalent diffusion constants. For a large reactor in which only a percent or so of the neutrons produced leak out of the core, changing the diffusion constant by 20 percent will not seriously affect criticality. However there may be a more serious effect on the gross power distribution.

In this connection it should be noted that the third column of Table 10.1 is the *best* that can be done in matching exact and flux-weighted equivalent constants. If there are other subassemblies making up the overall reactor (for example ones with the absorber slab replaced by moderator), the associated renormalization factor may differ from the one needed to make the $\overline{\nu\Sigma_f}$'s match in Table 10.1. But, if *any* renormalization of the equivalent $\bar{D}$'s and $\bar{\Sigma}$'s is carried out, *all* $\bar{D}$'s and $\bar{\Sigma}$'s *throughout the reactor* must be multiplied by the *same* factor. Thus, if flux-weighted $\bar{D}$'s and $\bar{\Sigma}$'s for one subassembly of a reactor are renormalized so that the $\overline{\nu\Sigma_f}$ for that subassembly is exact, similarly renormalized $\overline{\nu\Sigma_f}$'s for other subassemblies may *not* agree with the exact results. Thus these simple one-dimensional results suggest that there is some cause for concern, not only that the flux-weighted diffusion constants may be incorrect, but also that the ratio between the flux-weighted absorption and fission cross sections may differ from subassembly to subassembly.

*The occasionally suggested procedure of flux averaging $1/D(x)$ (i.e., $3\Sigma_{tr}(x)$) and taking the inverse to get $\bar{D}$ yields $\bar{D} = 0.8089$ cm (unnormalized) for this case and is thus equally unsatisfactory.

10.5 Summary

The attempt to define equivalent diffusion-theory parameters and to specify tractable procedures for finding them has been only partially successful.

For slab geometry a tractable method appears to be available. However there is little practical motivation to develop it in detail since real reactors are rarely one-dimensional and, for those infrequent occasions when an assembly built with slab geometry does have to be analyzed, there exist multigroup P_n or discrete-ordinates programs capable of dealing with the problem in an acceptably economical fashion.

In two and three dimensions the motivation for finding equivalent diffusion-theory parameters is strong, but the theoretical justification for the common procedures by which effective parameters are found is questionable and the prospects for developing a theoretically satsifying foundation are dim.

The heart of the difficulty is that the basic mathematical form of the group diffusion equations may not be general enough to *permit* matching known leakage rates and reaction rates through use of equivalent parameters that are spatially constant over the subregion for which they are defined (i.e., that obey (10.2.7) rather than (10.2.3)). If this conjecture is valid, the two- and three-dimensional situations are uniquely different from the one-dimensional case (for which (10.3.15) and (10.3.18) *always* yield exact equivalent values of $\bar{D}$ and $\bar{\Sigma}_a - \overline{\nu\Sigma_f}$), and the prescriptions outlined in the first two parts of Section 10.4 can never yield exact results.

We are thus left with an unsatisfactory theoretical situation. The current design prescriptions work fairly well (or, with rather minor adjustment based on comparison with more accurate calculations, can be made to do so). But we cannot (as we can for the one-dimensional case) guarantee beforehand that they *will* work. Clearly further study is needed in this area to establish a more quantitative evaluation of the inherent accuracy of the present design procedures. It may develop that attempting to determine equivalent diffusion-theory parameters for two- and three-dimensional geometrical situations is not the most economical long-range approach to the problem. At the end of the next chapter we shall examine an alternate scheme which may, in the long run, prove to be superior.

References

Becker, M., 1971. "Albedo Models for Nodal Reactor Simulation," *Transactions of the American Nuclear Society* **14**: 832.

Børresen, S., 1971. "A Simplified Coarse-Mesh, Three-Dimensional Diffusion Scheme for Calculating the Gross Power Distribution in a Boiling Water Reactor," *Nuclear Science and Engineering* **44**: 37–43.

Mueller, A., and Wagner, M. R., 1972. "Use of Collision Probability Methods for Multidimensional Neutron Flux Calculations," *Transactions of the American Nuclear Society* **15**: 280.

Numerical Reactor Calculations, 1972. Proceedings of a seminar held in Vienna, Austria, 17–21 January 1972, Vienna IAEA, STI/Pub/307, paper IAEA-SM-154/21.

Problems

1. The equivalent homogenized constants given by (10.3.15) and (10.3.18) were obtained by requiring that they correctly determine $\Phi(x_i)$ and $J(x_i)$ for given values of $\Phi(x_{i+1})$ and $J(x_{i+1})$. An alternate procedure would be to require that this condition be met only for $J(x)$ and to introduce the additional requirement that the homogenized constants preserve the correct *average flux*, so that if, for given values of $\Phi(x_{i+1})$ and $J(x_{i+1})$, $\Phi(x)$ is the true flux within Δ and $\overline{\Phi}(x)$ is the flux obtained using homogenized constants $\bar{D}$ and $\bar{\kappa}$ (10.3.9), we will have

$$\frac{1}{\Delta}\int_{\Delta} \Phi(x)\,dx = \frac{1}{\Delta}\int_{\Delta} \overline{\Phi}(x)\,dx.$$

Find the relationship between $\bar{D}$, $\bar{\kappa}$, and the given quantities $J(x_{i+1})$, $\Phi(x_{i+1})$, and $\int_{\Delta}\Phi(x)dx$ that results if this condition is imposed along with the requirement that $J(x_i)$ (but *not* necessarily $\Phi(x_i)$) be preserved by the equivalent constants.

This requirement is an attractive one since "flux-weighted" constants

$$\overline{\Sigma}_\alpha \equiv \frac{\int_{\Delta} \Sigma_\alpha(x)\Phi(x)\,dx}{\int_{\Delta} \Phi(x)\,dx}$$

will then be such that $\overline{\Sigma}_\alpha\int_{\Delta}\overline{\Phi}(x)dx = \int_{\Delta}\Sigma_\alpha(x)\Phi(x)dx$, so that the rate of reaction α is preserved by $\overline{\Sigma}_\alpha$ and the flux $\overline{\Phi}(x)$ that results from its use. Doing some simple problems on a computing machine will show, however, that values of $\bar{D}$ and $\bar{\kappa}$ found in this way will not lead to the correct criticality condition unless the fluxes at points x_i, x_{i+1}, etc., are permitted to be discontinuous.

2. Show that, if region (1), of thickness Δ_1 and properties specified by $[N_1]$ in (10.3.3), extends from x_1 to x_2 and region (2), of thickness Δ_2 and properties specified by $[N_2]$, extends from x_2 to x_3, then $[u(x_1)]$ in (10.3.3) is related to $[u(x_3)]$ by

$$[u(x_1)] = \exp([N_1]\Delta_1)\exp([N_2]\Delta_2)\,[u(x_3)].$$

One might conclude from this fact that equivalent diffusion-theory parameters $[\overline{N}]$ such that

$$[u(x_1)] = \exp([\overline{N}](\Delta_1 + \Delta_2))\,[u(x_3)]$$

might be determined by defining

$$[\overline{N}] \equiv \frac{[N_1]\Delta_1 + [N_2]\Delta_2}{\Delta_1 + \Delta_2}.$$

Why is this, in general, an erroneous conclusion? Under what conditions is it valid?

3. One very common way of approximating the matrix exponential $\exp([N]\Delta)$ is called the central difference, or semi-implicit, or Crank-Nicholson method. The approximation is

$$\exp([N]\Delta) \approx [I - \tfrac{1}{2}N\Delta]^{-1}[I + \tfrac{1}{2}N\Delta],$$

where $[I]$ is the identity matrix.

a. Show that this approximation is correct through order Δ^2. Compare it in this regard to $\exp([N]\Delta) \approx [I + N\Delta]$ and $\exp([N]\Delta) \approx [I - N\Delta]^{-1}$.

b. Derive finite-difference equations connecting $\Phi(x_{n-1})$, $\Phi(x_n)$, and $\Phi(x_{n+1})$ by assuming that

$$[u(x_n)] = [I - \tfrac{1}{2}N_n\Delta_n]^{-1}[I + \tfrac{1}{2}N_n\Delta_n]\,[u(x_{n+1})]$$

and

$$[u(x_{n-1})] = [I - \tfrac{1}{2}N_{n-1}\Delta_{n-1}]^{-1}\,[I + \tfrac{1}{2}N_{n-1}\Delta_{n-1}]\,[u(x_n)],$$

where $[N_n]$ is the transition matrix in the interval Δ_n between x_n and x_{n+1} and $[N_{n-1}]$ is the transition matrix in the interval Δ_{n-1} between x_{n-1} and x_n.

c. Find the conventional finite-difference equations (4.10.17) by an analogous procedure. (*Hint*: Use the other two approximations for exp $([N]\Delta)$ listed above and introduce $[u(x_n - \tfrac{1}{2}\Delta_n)]$ and $[u(x_n + \tfrac{1}{2}\Delta_{n+1})]$ temporarily.)

4. Derive (10.3.10) from (10.3.7) by making use of the eigenvectors of $[N]$ to diagonalize exp $([N]\Delta)$.

5. Derive new expressions for the blackness coefficients of a purely absorbing slab by replacing the assumption that $J(x, E)$ is continuous at x_i and x_{i+1} with the assumption that $\Phi(x, E)$ is continuous at those two surfaces.

6. Consider two slabs of different material and of thicknesses Δ_1 and Δ_2 and suppose that equivalent one-group diffusion-theory parameters, independent of input fluxes and current, are known for both of them. Show that, if the two slabs are joined to form a composite slab of thickness $\Delta_1 + \Delta_2$, the equivalent diffusion-theory parameters for this composite slab will *not* be independent of the input fluxes and currents. Then show that, if a third slab having material properties and a thickness identical with one of the first two is joined to form a *symmetric* arrangement of thickness $\Delta_1 + \Delta_2 + \Delta_1$, the equivalent diffusion-theory parameters for this symmetric arrangement of the *three* slabs *will* be independent of input fluxes and currents.

Hint: If a composite of slabs is to have equivalent parameters independent of $\Phi(x_{i+1})$ and $J(x_{i+1})$, the transmission matrix jointing the flux and current on the right-hand side of this composite to the flux and current on the left-hand side must have the structure of the matrix in (10.3.11).

7. Derive (10.3.49).

8. The most common procedure for finding flux-weighted constants for a heterogeneous fuel assembly containing control rods or lumps of burnable poison is first to determine the group fluxes $\Phi_g(\mathbf{r})$ throughout the assembly by doing a criticality or fixed-source computation using zero-net-current boundary conditions on the outer surfaces of the assembly. Thus, for a one-group diffusion-theory model in square XY geometry, one solves

$$-\nabla\cdot D(x, y)\nabla\Phi_c(x, y) + \Sigma_a(x, y)\Phi_c(x, y) = \frac{1}{\lambda}\nu\Sigma_f(x, y)\,\Phi_c(x, y)$$

subject to the condition that $\mathbf{n}\cdot\nabla\Phi_c(x, y) = 0$ on the outer surface S_0. The flux-averaged value of, for example, $\Sigma_a(x, y)$ is then

$$\Sigma_a \equiv \frac{\iint \Sigma_a(x, y)\Phi_c(x, y)\,dx\,dy}{\iint \Phi_c(x, y)\,dx\,dy}.$$

An alternate way to determine $\Phi(x, y)$ for this computation is to assume that the assembly is isolated in a vacuum but has a flat, isotropic source located on one of its outer surfaces. If Marshak boundary conditions are applied on all four surfaces of the assembly, a value of the resultant flux $\Phi_s(x, y)$ can be determined by solving the appropriate inhomogeneous equation. The value

$$\Sigma_a^{(s)} \equiv \frac{\iint \Sigma_a(x, y)\,\Phi_s(x, y)\,dx\,dy}{\iint \Phi_s(x, y)\,dx\,dy}$$

can then be determined.

Show that, if the assembly is symmetric with respect to a 90° rotation about the Z axis and with respect to reflection at the $x = 0$ and $y = 0$ planes, the value of $\Sigma_a^{(s)}$ will be unchanged if $\Phi_s(x, y)$ is taken to be the flux resulting from four flat, isotropic sources that have different magnitudes and are located on the four different surfaces of the assembly. (This result shows that values of $\Sigma_a^{(s)}$ found from a source that is constant over the whole periphery of the assembly may be used when the flux across the assembly has a net gradient.)

How do the conditions imposed for determining Φ_c and Φ_s differ from those experienced by the assembly when it is located off-center in the reactor?

11 Advanced Methods for Reactor Analysis

11.1 Introduction

It must be emphasized that, despite lamentations concerning the lack of a firm theoretical foundation, the methods currently applied to the determination of equivalent diffusion-theory parameters appear to be adequate for the design of the current generation of large power reactors. Moreover, even if the continual model-testing that goes on uncovers deficiencies in the present procedures, or if they prove to be inadequate for the analysis of more sophisticated designs, it is quite probable that more accurate methods for determining equivalent group diffusion-theory parameters can be devised. In either case the need to solve the group diffusion equations to some required degree of accuracy at a minimum cost will remain for some time to come.

The nature of the group diffusion equations containing equivalent parameters is frequently such that the conventional approach employing a spatial finite-difference solution (sketched in Chapter 4) may be either intractable or inefficient. This happens because the spatially constant equivalent group diffusion-theory parameters that result will be in some cases multigroup numbers pertaining to small homogeneous regions, such as a lump of burnable poison, and in others homogenized few-group numbers representing rather large heterogeneous clusters, such as the cross-shaped control rod and four surrounding fuel subassemblies shown in Figure 10.1. In the former case the equivalent group diffusion equations may involve a large number of regions having different group parameters, and their solution by finite-difference methods may be a practical impossibility. (Several million mesh points may be required just to represent the geometrical details.) In the latter case a relatively small number of homogenized regions will be present, and the conventional finite-difference method, while quite tractable, may be needlessly time-consuming. It follows that, if we cast complex reactor-analysis problems into equivalent group diffusion-theory form, we shall need more sophisticated methods for finding numerical solutions to such equations.

The present chapter is concerned primarily with the development of such methods. We shall first examine *synthesis schemes*, in which the scalar flux $\Phi_g(\mathbf{r}, E)$ for group g is represented as a linear combination of expansion functions, parts of which are preselected. (The P_1 expansion of (9.3.11) for $\Psi(\mathbf{r}, \mathbf{\Omega}, E)$ is an example of representing $\Psi(\mathbf{r}, \mathbf{\Omega}, E)$ as a linear combination of expansion functions whose angular parts have been preselected.) The synthesis approach will provide an alternate way of regarding the energy-group approximation, and we shall apply this viewpoint to develop so-called *overlapping-group models*, which are particularly useful when the material regions comprising a reactor are so small in extent that the neutron spectrum within them never approaches its asymptotic value. A far more important application of the synthesis idea will be to develop a tractable means for finding an approximate solution of the group diffusion equations when the geometry is so complex that hundreds of thousands—or even millions—of spatial mesh points would be required if conventional finite-difference methods were used.

Having seen how to attack problems for which conventional methods are extremely expensive, we shall then turn to the other extreme and seek more efficient methods for solving the group diffusion equations when the finite-difference approach is perfectly tractable, but needlessly time-consuming. Such cases arise when the reactor is composed of large homogeneous zones or when equivalent diffusion-theory parameters, spatially constant over large zones (such as a subassembly and associated control elements), can be found. The *finite-element method* will be applied to the resolution of problems of this class.

Finally we shall examine the *nodal* method of determining the group diffusion constants. The essential characteristic of this approach is to regard the *average* fluxes in subassembly-sized subregions or "nodes" of the reactor as the unknowns. Although space-averaged group reaction cross sections for the subregions are defined in this method, no attempt is made to find equivalent homogenized group diffusion constants. Instead the neutron-leakage rate across a surface common to two nodes is taken to be proportional to the difference of the nodal fluxes, the exact mathematical relationship between the current and the adjacent average nodal fluxes being expressed in terms of *coupling constants*. We shall see that certain extensions of this point of view permit us to incorporate into the nodal approach some of the best features of the synthesis method and the finite-element method.

11.2 Synthesis Methods Derived by the Weighted-Residual Method

We have already indicated that the essential characteristic of a synthesis approximation is to assume that the dependence of the function we seek (for example $\Phi_g(\mathbf{r}, E)$) on some of its arguments has a simple fixed form, specified beforehand on the basis of general knowledge of the physics of the problem. The remaining, still undetermined part of the unknown function is then found by requiring that the *overall* function obey its defining mathematical equation in some integral sense.

There are two standard procedures by which the remaining, unspecified part of the desired function can be found. These are the *weighted-residual method* and the *variational method*. The former is conceptually less mysterious, and we have encountered it before. Accordingly we shall make use of it for the first two examples of the synthesis approach. The latter is more powerful in that it automatically provides internal continuity conditions for the approximate solution and in addition suggests what weight functions should be used. The application we shall make of the variational procedure will take advantage of this greater generality.

The Overlapping-Energy-Group Method

We have seen that the basic assumption of few-group theory is that the space and energy dependence of the scalar flux $\Phi(\mathbf{r}, E)$ and the net current vector $\mathbf{J}(\mathbf{r}, E)$ are separable within each material composition. For a thermal reactor composed of several large material zones (for example several large zones of differently enriched fuel material and a reflector), this separability assumption is quite good throughout most of the core volume since the neutron spectrum reaches an equilibrium value (specified by the materials buckling B_m^2) a few mean free paths from the interfaces. In some reactors, however, the size of the differing material compositions is not great (the rather extreme case illustrated in Figure 10.1 provides an example). Moreover in fast reactors, because neutrons of high energy tend to have longer mean free paths, material compositions which are physically large (say seven inches thick) may be small from a neutron-coupling viewpoint. Thus the basic few-group assumption may be a questionable one for certain types of reactor.

The essential idea of the overlapping-group method is to assume that, within a given material composition, the scalar flux and net current vector can be expressed as a linear combination of several *trial spectra*, each having an independent spatial dependence. Thus, rather than assuming that within region i and energy group g the scalar flux $\Phi(\mathbf{r}, E)$ has the form

$$\Phi(\mathbf{r}, E) \approx R_i^g(\mathbf{r}) F_i^g(E), \tag{11.2.1}$$

we write for $\Phi(\mathbf{r}, E)$ the more general approximation

$$\Phi(\mathbf{r}, E) \approx \sum_{k=1}^{K} R_{ik}^g(\mathbf{r}) F_{ik}^g(E), \tag{11.2.2}$$

where the spectra $F_{ik}^g(E)$ are known energy shapes obtained from several asymptotic calculations using (9.5.7) and (9.5.8) and the space-dependent functions $R_{ik}^g(\mathbf{r})$ are to be found by a method we shall outline shortly. An analogous approximation is used for $\mathbf{J}(\mathbf{r}, E)$.

The expression (11.2.2) is clearly more general than (11.2.1). With it we can represent the flux at an interface between two different compositions as a combination of the asymptotic spectra appropriate to the individual compositions. (The flux $\Phi(\mathbf{r}, E)$ would then be continuous across the interface for all E in ΔE_g.) We can similarly express $\Phi(\mathbf{r}, E)$ throughout a given composition as a linear combination of its own asymptotic spectrum and the asymptotic spectra of all neighboring compositions.

Finding equations for the $R_{ik}^g(\mathbf{r})$ (and the corresponding coefficients in the expression for $\mathbf{J}(\mathbf{r}, E)$) is, in the general case, a very complicated algebraic problem. We can, however, illustrate many features of the general procedure by working out the details for a very simple example. Accordingly we shall consider the one-dimensional case with only two trial functions ($K = 2$), both defined over the *entire* energy range (i.e., only one energy group g) and both present throughout the entire reactor (i.e., only one region i). The goal will be to obtain an approximate solution to the one-dimensional energy-dependent P_1 equations (9.2.1):

$$\frac{\partial}{\partial z}\Psi_1(z, E) + \Sigma_t(z, E)\Psi_0(z, E) = \int_0^\infty dE'\left[\frac{f(E)}{\lambda}\nu\Sigma_f(z, E') + \Sigma_{s0}(z, E' \to E)\right]\Psi_0(z, E'),$$

$$\frac{\partial}{\partial z}\Psi_0(z, E) + 3\Sigma_t(z, E)\Psi_1(z, E) = 3\int_0^\infty dE'\,\Sigma_{s1}(z, E' \to E)\Psi_1(z, E'). \quad (9.2.1)$$

We seek this solution under the restriction that $\Psi_0(z, E)$ and $\Psi_1(z, E)$ have the approximate forms, for $0 \le E \le 15$ MeV and for all z,

$$\begin{aligned}\Psi_0(z, E) &\approx \Phi_1(z)U_1(E) + \Phi_2(z)U_2(E),\\ \Psi_1(z, E) &\approx J_1(z)V_1(E) + J_2(z)V_2(E),\end{aligned} \quad (11.2.3)$$

where U_1, U_2, V_1, and V_2 are known functions of energy defined over the range of importance for fission reactors (0–15 MeV). For simplicity we shall express them as continuous functions of energy, although in practice they would be multigroup solutions of equations such as (9.5.7) and (9.5.8) for two different material compositions.

To obtain equations for the "overlapping-group fluxes" $\Phi_1(z)$ and $\Phi_2(z)$ and the "overlapping-group currents" $J_1(z)$ and $J_2(z)$, we apply the weighted-residual method and require that (11.2.3) be a solution of the space-dependent P_1 equations (9.2.1) in an integral sense. Thus, since we need four equations to determine Φ_1, Φ_2, J_1, and J_2, we must substitute (11.2.3) into (9.2.1), multiply the resultant equations in succession by at least two linearly independent weight functions, and integrate the four equations that result over all energy.

To be somewhat more general we shall actually make use of four linearly independent weight functions designated as $U_1^*(E)$, $U_2^*(E)$, $V_1^*(E)$, and $V_2^*(E)$. We defer for the moment discussing the problem of selecting these weight functions.

Carrying out the weighted-residual process just outlined yields

$$\begin{aligned}&\int_0^\infty U_i^*(E)V_1(E)\,dE\frac{d}{dz}J_1(z) + \int_0^\infty U_i^*(E)V_2(E)\,dE\frac{d}{dz}J_2(z) + \int_0^\infty U_i^*(E)\Sigma_t(z, E)U_1(E)\,dE\,\Phi_1(z)\\ &+ \int_0^\infty U_i^*(E)\Sigma_t(z, E)U_2(E)\,dE\,\Phi_2(z) = \int_0^\infty dE\,U_i^*(E)\int_0^\infty dE'[\lambda^{-1}f(E)\nu\Sigma_f(z, E')\\ &\qquad + \Sigma_{s0}(z, E' \to E)][U_1(E')\Phi_1(z) + U_2(E')\Phi_2(z)],\\ &\int_0^\infty V_i^*(E)U_1(E)\,dE\frac{d}{dz}\Phi_1(z) + \int_0^\infty V_i^*(E)U_2(E)\,dE\frac{d}{dz}\Phi_2(z)\\ &+ 3\int_0^\infty V_i^*(E)\Sigma_t(z, E)[V_1(E')J_1(z) + V_2(E')J_2(z)]\,dE\\ &= 3\int_0^\infty dE\,V_i^*(E)\int_0^\infty dE'\,\Sigma_{s1}(z, E' \to E)[V_1(E')J_1(z) + V_2(E')J_2(z)] \quad (i = 1, 2).\end{aligned} \quad (11.2.4)$$

These four coupled differential equations must then be solved by standard means for Φ_1, Φ_2, J_1, and J_2.

To see the similarity between the structure of these equations and that of the ordinary equations for two nonoverlapping groups we rewrite (11.2.4) in matrix form as (11.2.5) and (11.2.6).

To simplify notation we then rewrite (11.2.5) and (11.2.6) as

$$[\eta]\frac{d}{dz}\begin{bmatrix} J_1(z) \\ J_2(z) \end{bmatrix} + [\mathscr{A}(z)]\begin{bmatrix} \Phi_1(z) \\ \Phi_2(z) \end{bmatrix} - \frac{1}{\lambda}[\mathscr{M}(z)]\begin{bmatrix} \Phi_1(z) \\ \Phi_2(z) \end{bmatrix} = 0 \tag{11.2.7}$$

and

$$[\eta^*]\frac{d}{dz}\begin{bmatrix} \Phi_1(z) \\ \Phi_2(z) \end{bmatrix} = -3[\Sigma_{tr}(z)]\begin{bmatrix} J_1(z) \\ J_2(z) \end{bmatrix}, \tag{11.2.8}$$

where the 2×2 matrices $[\eta]$, $[\mathscr{A}(z)]$, $[\mathscr{M}(z)]$, $[\eta^*]$, and $[\Sigma_{tr}(z)]$ are defined by comparison with (11.2.5) and (11.2.6). (Note that $[\eta]$ and $[\eta^*]$ do not depend on z.)

From (11.2.8), assuming $[\Sigma_{tr}(z)]^{-1}$ exists, we get

$$\begin{bmatrix} J_1(z) \\ J_2(z) \end{bmatrix} = -\frac{1}{3}[\Sigma_{tr}(z)]^{-1}[\eta^*]\frac{d}{dz}\begin{bmatrix} \Phi_1(z) \\ \Phi_2(z) \end{bmatrix}, \tag{11.2.9}$$

so that (11.2.7) yields

$$-\frac{1}{3}\frac{d}{dz}[\eta][\Sigma_{tr}(z)]^{-1}[\eta^*]\frac{d}{dz}\begin{bmatrix} \Phi_1(z) \\ \Phi_2(z) \end{bmatrix}[\mathscr{A}(z)]\begin{bmatrix} \Phi_1(z) \\ \Phi_2(z) \end{bmatrix} - \frac{1}{\lambda}[\mathscr{M}(z)]\begin{bmatrix} \Phi_1(z) \\ \Phi_2(z) \end{bmatrix} = 0 \tag{11.2.10}$$

or, with the 2×2 matrix $[\mathscr{D}(z)]$ defined by

$$[\mathscr{D}(z)] \equiv \frac{1}{3}[\eta][\Sigma_{tr}(z)]^{-1}[\eta^*], \tag{11.2.11}$$

$$\frac{d}{dz}[\mathscr{D}(z)]\frac{d}{dz}\begin{bmatrix} \Phi_1(z) \\ \Phi_2(z) \end{bmatrix} - [\mathscr{A}(z)]\begin{bmatrix} \Phi_1(z) \\ \Phi_2(z) \end{bmatrix} + \frac{1}{\lambda}[\mathscr{M}(z)]\begin{bmatrix} \Phi_1(z) \\ \Phi_2(z) \end{bmatrix} = 0. \tag{11.2.12}$$

The general structure of this equation is identical with that of the conventional energy-group equations (for example (9.4.33)). There are, however, differences. The most notable is that the matrices $[\mathscr{D}(z)]$ and $[\mathscr{A}(z)]$ are full. In addition the fission matrix $[\mathscr{M}(z)]$, while still a dyad

$$[\mathscr{M}(z)] = \begin{bmatrix} \int_0^\infty dE\, U_1^*(E)f(E) \\ \int_0^\infty dE\, U_2^*(E)f(E) \end{bmatrix}\left[\int_0^\infty dE'\, \nu\Sigma_f(z, E')U_1(E') \int_0^\infty dE'\, \nu\Sigma_f(z, E')U_2(E')\right],$$

$$
\begin{bmatrix}
\int_0^\infty U_1^*(E)V_1(E)\,dE & \int_0^\infty U_1^*(E)V_2(E)\,dE \\
\int_0^\infty U_2^*(E)V_1(E)\,dE & \int_0^\infty U_2^*(E)V_2(E)\,dE
\end{bmatrix}
\frac{d}{dz}
\begin{bmatrix} J_1(z) \\ J_2(z) \end{bmatrix}
$$

$$
+\begin{bmatrix}
\int_0^\infty U_1^*(E)\,dE \int_0^\infty dE'[\delta(E'-E)\Sigma_t(z,E') - \Sigma_{s0}(z,E'\to E)]U_1(E') & \int_0^\infty U_1^*(E)\,dE \int_0^\infty dE'[\delta(E'-E)\Sigma_t(z,E') - \Sigma_{s0}(z,E'\to E)]U_2(E') \\
\int_0^\infty U_2^*(E)\,dE \int_0^\infty dE'[\delta(E'-E)\Sigma_t(z,E') - \Sigma_{s0}(z,E'\to E)]U_1(E') & \int_0^\infty U_2^*(E)\,dE \int_0^\infty dE'[\delta(E'-E)\Sigma_t(z,E') - \Sigma_{s0}(z,E'\to E)]U_2(E')
\end{bmatrix}
\begin{bmatrix} \Phi_1(z) \\ \Phi_2(z) \end{bmatrix}
$$

$$
-\frac{1}{\lambda}\begin{bmatrix}
\int_0^\infty dE\,U_1^*(E)f(E)\int_0^\infty dE'\,\nu\Sigma_f(z,E')U_1(E') & \int_0^\infty dE\,U_1^*(E)f(E)\int_0^\infty dE'\,\nu\Sigma_f(z,E')U_2(E') \\
\int_0^\infty dE\,U_2^*(E)f(E)\int_0^\infty dE'\,\nu\Sigma_f(z,E')U_1(E') & \int_0^\infty dE\,U_2^*(E)f(E)\int_0^\infty dE'\,\nu\Sigma_f(z,E')U_2(E')
\end{bmatrix}
\begin{bmatrix} \Phi_1(z) \\ \Phi_2(z) \end{bmatrix} = 0. \qquad (11.2.5)
$$

$$
\begin{bmatrix}
\int_0^\infty dE\,V_1^*(E)U_1(E) & \int_0^\infty dE\,V_1^*(E)U_2(E) \\
\int_0^\infty dE\,V_2^*(E)U_1(E) & \int_0^\infty dE\,V_2^*(E)U_2(E)
\end{bmatrix}
\frac{d}{dz}
\begin{bmatrix} \Phi_1(z) \\ \Phi_2(z) \end{bmatrix}
$$

$$
= -3\begin{bmatrix}
\int_0^\infty V_1^*(E)\,dE \int_0^\infty dE'[\Sigma_t(z,E')\delta(E'-E) - \Sigma_{s1}(z,E'\to E)]V_1(E') & \int_0^\infty V_1^*(E)\,dE \int_0^\infty dE'[\Sigma_t(z,E')\delta(E'-E) - \Sigma_{s1}(z,E'\to E)]V_2(E') \\
\int_0^\infty V_2^*(E)\,dE \int_0^\infty dE'[\Sigma_t(z,E')\delta(E'-E) - \Sigma_{s1}(z,E'\to E)]V_1(E') & \int_0^\infty V_2^*(E)\,dE \int_0^\infty dE'[\Sigma_t(z,E')\delta(E'-E) - \Sigma_{s1}(z,E'\to E)]V_2(E')
\end{bmatrix}
\begin{bmatrix} J_1(z) \\ J_2(z) \end{bmatrix}. \qquad (11.2.6)
$$

introduces "group neutrons" into all "groups." Thus the overlapping-group equations (11.2.12) are more tightly coupled than the regular group equations. As a result trying to solve them in an iterative fashion, one group at a time, can be a very lengthy process which may not even converge. Instead they are best solved for both groups simultaneously—easy in one dimension but expensive in two and three dimensions. Moreover there is a general cause for concern about the utility of the overlapping-group method associated with the fact that there is no mathematical guarantee (as there *is* for the regular non-overlapping-group case) that the reconstructed solution (11.2.3) will be everywhere positive or that the corresponding eigenvalue will be real, isolated, and the most positive of all eigenvalues of the equation.

The overlapping-group method can also be applied to the determination of asymptotic spectra. Thus, rather than solving (9.5.7) and (9.5.8) for the asymptotic spectra associated with each of a large number of fairly similar material compositions, we find only a few bracketing spectra exactly and then determine spectra for intermediate compositions by the weighted-residual scheme. If only two trial spectra are used, for example, then the functions $\Phi_1(z)$, $\Phi_2(z)$, $J_1(z)$, and $J_2(z)$ in (11.2.3) become numbers, Φ_1 and Φ_2 being the proportions of $U_1(E)$ and $U_2(E)$ making up $\Psi_0(E)$ and J_1 and J_2 being the proportions of $U_1(E)$ and $U_2(E)$ making up $\Psi_1(E)$. These proportions can be found by setting

$$\frac{d}{dz}\begin{bmatrix} J_1 \\ J_2 \end{bmatrix} = iB_m \begin{bmatrix} J_1 \\ J_2 \end{bmatrix}$$

in (11.2.7) and

$$\frac{d}{dz}\begin{bmatrix} \Phi_1 \\ \Phi_2 \end{bmatrix} = iB_m \begin{bmatrix} \Phi_1 \\ \Phi_2 \end{bmatrix}$$

in (11.2.8) and solving the resulting four coupled algebraic equations with the elements of $[\mathscr{A}(z)]$, $[\mathscr{M}(z)]$, and $[\Sigma_{tr}(z)]$ computed for the particular material composition of interest.

For space-dependent problems the overlapping scheme has been applied most commonly to the thermal energy range only, the range above thermal being represented by the usual non-overlapping-group procedure. The reason for this is that, while the "normal" matrix $[A_0]$ of (9.4.33) and (9.4.35) has only zeros above the diagonal in the slowing-down range, it is (because of up-scattering) a full matrix in the thermal range. Thus the iteration difficulties associated with the off-diagonal elements of the overlapping-group matrix $[\mathscr{A}(z)]$ must be faced in the thermal range, whether $[A(z)]$ or $[\mathscr{A}(z)]$ is used, and it has been found that, for the thermal range, in the absence of Pu^{40} (with its large thermal capture resonance), use of two overlapping groups is equivalent in accuracy to use of about six nonoverlapping groups.

The trial functions $U_i(E)$ and $V_i(E)$ used in the overlapping-group method are usually suggested by the character of the problem at hand. The guiding principle is to pick functions that bracket the energy shapes expected throughout the reactor, and the most usual procedure is to use asymptotic spectra appropriate to the different compositions making up the reactor.

Selection of the weight functions $U_i^*(E)$ and $V_i^*(E)$ is a somewhat more involved process. A variational derivation of the overlapping-group equations suggests that one should use functions that are solutions to the equations adjoint to those from which the regular functions were derived. This is probably the best strategy if no special tricks are known. However it should be noted that this choice is not consistent with maintaining neutron balance (for which one of the $U_i^*(E)$ and one of the $V_i^*(E)$ should be taken equal to unity). When breeding-ratio determinations are important, use of equations that represent a neutron-balance condition may be important.

It is not difficult to generalize the overlapping-group model to more than one dimension. However generalization of the method so that different trial functions are used in different parts of the reactor gives rise to some severe mathematical problems concerning the appropriate continuity conditions on the $\Phi_i(z)$ and $J_i(z)$ to be applied at interfaces where one or more of the trial functions is replaced. Extreme care must be taken in carrying out a generalization of this kind.

Flux Synthesis and Point Synthesis

The overlapping-group method (as well, of course, as the few-group method itself) makes it possible to reduce to a manageable size the number of energy groups required for an accurate description of the criticality and power distribution throughout a reactor. There still remains the problem of the number of spatial mesh points required in a finite-difference solution of the diffusion equations if a detailed description of a geometrically complex reactor is to be provided. To attack this problem we shall examine so-called flux-synthesis methods, starting, for the sake of simplicity, with a few-group diffusion-theory model having a diagonal diffusion-constant matrix. No essentially new procedures are required to extend what we shall derive to overlapping-group models.

The simplest of the flux-synthesis methods is called *point synthesis* and consists of representing the flux $\Phi_g(\mathbf{r})$ in a given energy group g as a superposition of known fluxes

$$\Phi_g(\mathbf{r}) \approx \sum_{k=1}^{K} \Psi_k^g(\mathbf{r}) T_k^g \qquad (g = 1, 2, \ldots, G), \tag{11.2.13}$$

where the $\Psi_k^g(\mathbf{r})$ are predetermined "expansion functions" and the T_k^g are "mixing coefficients" to be found by requiring the form (11.2.13) to be a solution to the group diffusion equations in a weighted-residual sense.

Since the $\Psi_k^g(\mathbf{r})$ are themselves many-mesh-point, full-core solutions to the group diffusion equations, the point-synthesis method appears at first sight to be no improvement over finding the finite-difference solution for the group fluxes directly. The gain comes for situations where just a few expansion functions $\Psi_k^g(\mathbf{r})$ for each group can be used to synthesize many fluxes $\Phi_g(\mathbf{r})$. For example, if one wishes to find the eigenvalue λ for a large number of different concentrations of soluble poison dissolved in the reactor coolant, use of (11.2.13) with just two values of $\Psi_k^g(\mathbf{r})$ for each group (obtained by solving the group equations with the maximum and minimum concentrations of soluble poison present) should provide a good representation of the flux for all intermediate poison concentrations. Thus the (expensive) solution of only two detailed problems, which bracket, however, a range of operating conditions, permits us, by "blending" these two solutions, to find approximate solutions corresponding to many intermediate conditions.

Another situation where this "bracket and blend" procedure is useful arises in determining the depletion history of a core controlled during power operation by homogeneous poison. In order to determine the flux shape throughout lifetime we start by calculating detailed solutions appropriate to the beginning and the end of life. (The diffusion-theory parameters needed for the detailed solution at the end of life are found from a "one-shot" depletion calculation of the core done by assuming that the initial flux shape persists throughout the entire reactor lifetime.) We then assume that the flux for each group g at intermediate times can be approximated by the form (11.2.13) and determine the critical poison concentration, fuel depletion, and flux shape in this period by finding the mixing coefficients at a number of intermediate time steps between the beginning and end of life. Since the T_k^g will change at each time step and thus lead to a change in flux shape, this second depletion will result in a different (presumably more accurate) estimate of the fuel-concentration history throughout the reactor.

The "bracket and blend" procedure is also attractive for representing changes in flux shape due to transient xenon effects or during cold-to-hot transitions. There are thus a number of important design problems for which the finite-difference computation of a few detailed flux shapes associated with extreme conditions provides a means of determining many approximate shapes associated with intermediate conditions. It is for these problems that the approximation (11.2.13) is valuable.

In order to find equations for the T_k^g we simply substitute (11.2.13) into the group diffusion equations, multiply each group equation by a sequence of weight functions $W_j^g(\mathbf{r})$ ($j = 1, 2, \ldots, K$), and integrate over the volume of the reactor. The resulting $G \times K$ equations are just sufficient to determine the eigenvalue λ along with the $G \times K$ values of the T_k^g. We shall illustrate the procedure for the two-group, two-trial-function case and then generalize, by using matrix notation, to G groups with K expansion functions in each group. For notational simplicity we shall continue to write group equations in differential form, even though in practice the finite-difference form would be employed.

The two-group equations with $[D]$ diagonal (and with functional dependencies dropped) are then

$$\begin{aligned}
&\nabla \cdot D_1 \nabla \Phi_1 - \Sigma_1 \Phi_1 + \frac{1}{\lambda}(\nu\Sigma_{f1}\Phi_1 + \nu\Sigma_{f2}\Phi_2) = 0, \\
&\nabla \cdot D_2 \nabla \Phi_2 - \Sigma_2 \Phi_2 + \Sigma_{21}\Phi_1 = 0.
\end{aligned} \tag{11.2.14}$$

For two expansion functions, (11.2.13) is

$$\begin{aligned}
\Phi_1(\mathbf{r}) &\approx \Psi_1^1(\mathbf{r})T_1^1 + \Psi_2^1(\mathbf{r})T_2^1, \\
\Phi_2(\mathbf{r}) &\approx \Psi_1^2(\mathbf{r})T_1^2 + \Psi_2^2(\mathbf{r})T_2^2,
\end{aligned} \tag{11.2.15}$$

where the expansion functions $\Psi_1^1(\mathbf{r})$, $\Psi_2^1(\mathbf{r})$, $\Psi_1^2(\mathbf{r})$, and $\Psi_2^2(\mathbf{r})$ are found by solving equations of the form (11.2.14) with D's and Σ's that bracket the values corresponding to the cases at hand.

Substituting (11.2.15) into (11.2.14) leads to

$$\begin{aligned}
&\{\nabla \cdot D_1 \nabla \Psi_1^1 - \Sigma_1 \Psi_1^1\}T_1^1 + \{\nabla \cdot D_1 \nabla \Psi_2^1 - \Sigma_1 \Psi_2^1\}T_2^1 \\
&+ \frac{1}{\lambda}\{\nu\Sigma_{f1}\Psi_1^1 T_1^1 + \nu\Sigma_{f1}\Psi_2^1 T_2^1 + \nu\Sigma_{f2}\Psi_1^2 T_1^2 + \nu\Sigma_{f2}\Psi_2^2 T_2^2\} = 0, \\
&\{\nabla \cdot D_2 \nabla \Psi_1^2 - \Sigma_2 \Psi_1^2\}T_1^2 + \{\nabla \cdot D_2 \nabla \Psi_2^2 - \Sigma_2 \Psi_2^2\}T_2^2 + \Sigma_{21}\Psi_1^1 T_1^1 + \Sigma_{21}\Psi_2^1 T_2^1 = 0.
\end{aligned} \tag{11.2.16}$$

Since the Ψ_k^g in these equations are associated with D's and Σ's different from those in (11.2.14), the only solution for the four scalars T_k^g valid at all locations is the trivial solution $T_k^g = 0$ $(g, k = 1, 2)$. In accord with the weighted-residual procedure, however, we demand only that (11.2.16) be valid in a weighted-integral sense. Thus we introduce four arbitrary functions $W_1^1(\mathbf{r})$, $W_2^1(\mathbf{r})$, $W_1^2(\mathbf{r})$, and $W_2^2(\mathbf{r})$ (the choice of which we shall discuss later) and require that

$$\begin{aligned}
&\int dV\, W_i^1\{\nabla \cdot D_1 \nabla \Psi_1^1 - \Sigma_1 \Psi_2^1\}T_1^1 + \int dV\, W_i^1\{\nabla \cdot D_1 \nabla \Psi_2^1 - \Sigma_1 \Psi_2^1\}T_2^1 \\
&+ \frac{1}{\lambda}\left\{\int dV\, W_i^1 \nu\Sigma_{f1}\Psi_1^1 T_1^1 + \int dV\, W_i^1 \nu\Sigma_{f1}\Psi_2^1 T_2^1 + \int dV\, W_i^1 \nu\Sigma_{f2}\Psi_1^2 T_1^2\right. \\
&\left. + \int dV\, W_i^1 \nu\Sigma_{f2}\Psi_2^2 T_2^2\right\} = 0 \qquad (i = 1, 2), \\
&\int dV\, W_i^2\{\nabla \cdot D_2 \nabla \Psi_1^2 - \Sigma_2 \Psi_1^2\}T_1^2 + \int dV\, W_i^2\{\nabla \cdot D_2 \nabla \Psi_2^2 - \Sigma_2 \Psi_2^2\}T_2^2 \\
&+ \int dV\, W_i^2 \Sigma_{21}\Psi_1^1 T_1^1 + \int dV\, W_i^2 \Sigma_{21}\Psi_2^1 T_2^1 = 0 \qquad (i = 1, 2).
\end{aligned} \tag{11.2.17}$$

Since all the space-dependent quantities (the D's, Σ's, Ψ_k^g, and W_k^g) are known, we can perform the indicated integrations over the volume of the reactor and arrive at four homogeneous algebraic equations in the T_k^g. In matrix form, and using the Dirac inner-product notation of (8.3.12) for the volume integrals, we get

$$\begin{bmatrix} \langle W_1^1|\nabla\cdot D_1\nabla - \Sigma_1|\Psi_1^1\rangle & 0 & \langle W_1^1|\nabla\cdot D_1\nabla - \Sigma_1|\Psi_2^1\rangle & 0 \\ \langle W_1^2|\Sigma_{21}|\Psi_1^1\rangle & \langle W_1^2|\nabla\cdot D_2\nabla - \Sigma_2|\Psi_1^2\rangle & \langle W_1^2|\Sigma_{21}|\Psi_2^1\rangle & \langle W_1^2|\nabla\cdot D_2\nabla - \Sigma_2|\Psi_2^2\rangle \\ \langle W_2^1|\nabla\cdot D_1\nabla - \Sigma_1|\Psi_1^1\rangle & 0 & \langle W_2^1|\nabla\cdot D_1\nabla - \Sigma_1|\Psi_2^1\rangle & 0 \\ \langle W_2^2|\Sigma_{21}|\Psi_1^1\rangle & \langle W_2^2|\nabla\cdot D_2\nabla - \Sigma_2|\Psi_1^2\rangle & \langle W_2^2|\Sigma_{21}|\Psi_2^1\rangle & \langle W_2^2|\nabla\cdot D_2\nabla - \Sigma_2|\Psi_2^2\rangle \end{bmatrix} \begin{bmatrix} T_1^1 \\ T_1^2 \\ T_2^1 \\ T_2^2 \end{bmatrix}$$

$$+ \frac{1}{\lambda} \begin{bmatrix} \langle W_1^1|\nu\Sigma_{f1}|\Psi_1^1\rangle & \langle W_1^1|\nu\Sigma_{f2}|\Psi_1^2\rangle & \langle W_1^1|\nu\Sigma_{f1}|\Psi_2^1\rangle & \langle W_1^1|\nu\Sigma_{f2}|\Psi_2^2\rangle \\ 0 & 0 & 0 & 0 \\ \langle W_2^1|\nu\Sigma_{f1}|\Psi_1^1\rangle & \langle W_2^1|\nu\Sigma_{f2}|\Psi_1^2\rangle & \langle W_2^1|\nu\Sigma_{f1}|\Psi_2^1\rangle & \langle W_2^1|\nu\Sigma_{f2}|\Psi_2^2\rangle \\ 0 & 0 & 0 & 0 \end{bmatrix} \begin{bmatrix} T_1^1 \\ T_1^2 \\ T_2^1 \\ T_2^2 \end{bmatrix} = 0. \qquad (11.2.18)$$

There will be four solutions to these equations corresponding to four different values of the eigenvalue λ. On physical grounds we seek the one corresponding to the largest such value. Unfortunately the mathematics doesn't guarantee that this solution, when inserted into (11.2.15), will give the best match to $\Phi_1(\mathbf{r})$ and $\Phi_2(\mathbf{r})$ of all four solutions. Experience, however, indicates that, for "judicious" choices of weight and expansion functions, it will.

We are thus led to the central problem of synthesis methods, which makes reactor designers so uncomfortable about using them: How should the expansion and weight functions be chosen?

We have already indicated that the choice of expansion functions depends on the physics of the problem. We must use our knowledge of the physics to insure that (11.2.15) *can* represent, fairly accurately, the true flux shapes for the range of problems to which it is to be applied. The notion of "bracketing and blending," which we discussed earlier, is thus the key to the selection process. Experience indicates that, for the problems to which this method is usually applied, four or five expansion functions per group are sufficient. This is enough to get an accuracy of a few tenths of a percent in the eigenvalue and a maximum error of 5 or 10 percent in the flux shape. Adding more expansion functions merely increases the labor of the calculation without improving the accuracy.

Choice of the weight functions is a more difficult matter. As we shall see, deriving (11.2.18) by a variational principle rather than by the weighted-residual method suggests that the $W_k^g(\mathbf{r})$ should be such that linear combinations of them can provide a good approximation to the solution of the group equations adjoint to (11.2.14). This situation, in turn, suggests that we generate the weight functions by finding solutions to the adjoint equations corresponding to the *same* reactor conditions for which we found the expansion functions. In practice, however, one frequently avoids the extra expense of solving such adjoint equations and, instead, simply uses the $\Psi_k^g(\mathbf{r})$ themselves as weight functions. (This is called the *Galerkin method.*) Such a procedure is justified on the grounds that the shape of the adjoint flux for a given reactor configuration is not very much different from the shape of the regular flux. Moreover numerical tests indicate that there is usually a negligible difference between synthesized fluxes found by adjoint or Galerkin weighting.

There is a simpler form of the point-synthesis method called *collapsed-group synthesis.* In this scheme the group-to-group ratio of the expansion functions is kept fixed, and the kth expansion fluxes are all multiplied by a single mixing coefficient T_k. Thus (11.2.15) (written in matrix form) changes from

$$\begin{bmatrix}\Phi_1(\mathbf{r})\\ \Phi_2(\mathbf{r})\end{bmatrix} = \begin{bmatrix}\Psi_1^1(\mathbf{r}) & 0\\ 0 & \Psi_1^2(\mathbf{r})\end{bmatrix}\begin{bmatrix}T_1^1\\ T_1^2\end{bmatrix} + \begin{bmatrix}\Psi_2^1(\mathbf{r}) & 0\\ 0 & \Psi_2^2(\mathbf{r})\end{bmatrix}\begin{bmatrix}T_2^1\\ T_2^2\end{bmatrix} \tag{11.2.19}$$

to

$$\begin{bmatrix}\Phi_1(\mathbf{r})\\ \Phi_2(\mathbf{r})\end{bmatrix} = \begin{bmatrix}\Psi_1^1(\mathbf{r})\\ \Psi_1^2(\mathbf{r})\end{bmatrix} T_1 + \begin{bmatrix}\Psi_2^1(\mathbf{r})\\ \Psi_2^2(\mathbf{r})\end{bmatrix} T_2. \tag{11.2.20}$$

The simplification is thus equivalent to equating T_k^1 and T_k^2 in (11.2.15). To find the T_k one simply *adds* the two equations (11.2.17) for each separate value of k, the weight-function index, so that for each i there is only one equation (rather than G equations) in the unknowns T_1 and T_2.

When this procedure is used, it is important that each expansion vector $\mathrm{Col}\{\Psi_k^1, \Psi_k^2\}$ have the group-to-group ratio Ψ_k^1/Ψ_k^2 appropriate to the reactor conditions (D's and Σ's) used in its determination. Also, although one may still use the flux shapes ($W_k^1 = \Psi_k^1$ and $W_k^2 = \Psi_k^2$) as weight functions, the ratio W_k^1/W_k^2 should be approximately that of the *adjoint* flux vectors rather that that of the regular fluxes. It is usually sufficiently accurate to estimate this ratio in some simple, approximate manner that avoids the necessity of solving the equations adjoint to (11.2.14) (for example through the use of a guess based on past experience).

Let us now generalize the point-synthesis method to G energy groups and K trial and weight functions. A matrix notation greatly simplifies such a development. Thus we start with the group diffusion equations in the form (9.4.35):

$$\nabla\cdot[D]\nabla[\Phi] - [A_0][\Phi] + \frac{1}{\lambda}[M][\Phi] = 0, \tag{11.2.21}$$

where $[D]$, $[A_0]$, and $[M]$ are $G \times G$ matrices.

We simplify notation further by defining a net-loss operator $[L]$ by

$$[L] \equiv -\nabla\cdot[D]\nabla + [A_0], \tag{11.2.22}$$

so that (11.2.21) becomes

$$-[L][\Phi] + \frac{1}{\lambda}[M][\Phi] = 0. \tag{11.2.23}$$

Now we write the trial function for $[\Phi]$ as

$$[\Phi] \approx \sum_{k=1}^{K} [\Psi_k][T_k], \tag{11.2.24}$$

where each $[\Psi_k]$ is a $G \times G$ diagonal matrix of expansion shapes

$$[\Psi_k] = \text{Diag}\{\Psi_k^1(\mathbf{r}), \Psi_k^2(\mathbf{r}), \ldots, \Psi_k^G(\mathbf{r})\} \tag{11.2.25}$$

and each $[T_k]$ is a G-element column vector of unknown mixing coefficients

$$[T_k] = \text{Col}\{T_k^1, T_k^2, \ldots, T_k^G\}.$$

We now substitute (11.2.24) into (11.2.23), multiply successively by K linearly independent weight matrices $[W_j]$ $(j = 1, 2, \ldots, K)$, where

$$[W_j] = \text{Diag}\{W_j^1(\mathbf{r}), W_j^2(\mathbf{r}), \ldots, W_j^G(\mathbf{r})\},$$

and integrate over the reactor volume:

$$-\int [W_j][L]\sum_{k=1}^{K} [\Psi_k][T_k]dV + \frac{1}{\lambda}[W_j][M]\sum_{k=1}^{k} [\Psi_k][T_k]\,dV = 0. \tag{11.2.26}$$

This whole set of K matrix equations may be rewritten as a single "supermatrix" equation

$$\begin{bmatrix} \int [W_1][L][\Psi_1]\,dV & \int [W_1][L][\Psi_2]\,dV & \cdots & \int [W_1][L][\Psi_K]\,dV \\ \int [W_2][L][\Psi_1]\,dV & \int [W_2][L][\Psi_2]\,dV & \cdots & \int [W_2][L][\Psi_K]\,dV \\ \cdot & \cdot & \cdot & \cdot \\ \int [W_K][L][\Psi_1]\,dV & \int [W_K][L][\Psi_2]\,dV & \cdots & \int [W_K][L][\Psi_K]\,dV \end{bmatrix} \begin{bmatrix} T_1 \\ T_2 \\ \cdot \\ T_K \end{bmatrix}$$

$$= \frac{1}{\lambda}\begin{bmatrix} \int [W_1][M][\Psi_1]\,dV & \int [W_1][M][\Psi_2]\,dV & \cdots & \int [W_1][M][\Psi_K]\,dV \\ \int [W_2][M][\Psi_1]\,dV & \int [W_2][M][\Psi_2]\,dV & \cdots & \int [W_2][M][\Psi_K]\,dV \\ \cdot & \cdot & \cdot & \cdot \\ \int [W_K][M][\Psi_1]\,dV & \int [W_K][M][\Psi_2]\,dV & \cdots & \int [W_K][M][\Psi_K]\,dV \end{bmatrix} \begin{bmatrix} T_1 \\ T_2 \\ \cdot \\ T_K \end{bmatrix}. \tag{11.2.27}$$

Finally, if we define $K \times K$ matrices $[\mathscr{L}]$ and $[\mathscr{M}]$ by

$$\begin{aligned} [\mathscr{L}] &= \{\mathscr{L}_{jk}\} = \{\textstyle\int [W_j][L][\Psi_k]\,dV\}, \\ [\mathscr{M}] &= \{\mathscr{M}_{jk}\} = \{\textstyle\int [W_j][M][\Psi_k]\,dV\}, \end{aligned} \tag{11.2.28}$$

and a K-element column vector $[T]$ by

$$[T] = \text{Col}\{T_1, T_2, \ldots, T_K\}, \tag{11.2.29}$$

we may rewrite (11.2.27) in the compact form

$$[\mathscr{L}][T] = \frac{1}{\lambda}[\mathscr{M}][T]. \tag{11.2.30}$$

It is important to recognize that the elements of the $K \times K$ matrices $[\mathscr{L}]$ and $[\mathscr{M}]$ are themselves $G \times G$ matrices and that the elements of the K-element column vector $[T]$ are themselves G-element column vectors. Thus (11.2.30) is equivalent to $G \times K$ homogeneous algebraic equations in the mixing coefficients T_k^g. The size of (11.2.18) (which is just about the simplest point-synthesis model) makes it clear why an abbreviated notation is essential in writing matrix equations of this type.

It should be noted that the expansion functions and weight functions that have been discussed so far for the point-synthesis method are all fundamental-mode solutions (everywhere positive). They are linearly independent because they are solutions corresponding to different reactor conditions (D's and Σ's). In operator notation they are usually solutions to

$$\left.\begin{aligned} L_k\Psi_k^0 &= \frac{1}{\lambda_k^0} M_k\Psi_k^0 \\ L_k^*\Psi_k^{*0} &= \frac{1}{\lambda_k^0} M_k^*\Psi_k^{*0} \end{aligned}\right\} (k = 1, 2, \ldots, K), \tag{11.2.31}$$

where the L_k, M_k, L_k^*, and M_k^* are net-loss and fission-production operators and their adjoints corresponding to K different reactor conditions specified by values of D and Σ that bracket and blend with those of interest for the problem to be solved. The superscripts zero indicate fundamental-mode solutions.

This use of fundamental-mode solutions has two numerical consequences, one theoretically unsatisfying but practically trivial and the other theoretically trivial but practically unsatisfactory.

The first consequence is the fact that, since Ψ_k^0 and Ψ_j^{*0} correspond to different operators when $j \neq k$, there is no orthogonality relationship of the type

$$\langle\Psi_j^{*0}|M_k|\Psi_k^0\rangle = \delta_{jk}. \tag{11.2.32}$$

As a result the matrix elements (11.2.28) of $[\mathscr{L}]$ and $[\mathscr{M}]$ are all generally nonzero. This is unfortunate in that it necessitates more multiplications in the solution of (11.2.30). However it is unavoidable since, even if sets of functions are chosen such that (11.2.32) is valid for the operator $[M_k]$, the orthogonality will not hold for the case of interest since we are there concerned with an operator $[M]$ *different* from $[M_k]$. Thus only for a very limited class of problems, in which the reactor to be analyzed is unperturbed (the analysis of pulsed-source experiments is an instance), is there any practical advantage to using expansion and weight functions that obey an orthogonality condition.

The other numerical consequence of expanding in fundamental-mode solutions is of some practical concern. This is the fact that, in general, the matrix elements in (11.2.28) are all of nearly the same size. Under these circumstances it is possible to run into rounding-off errors in solving (11.2.30). This practical difficulty can be overcome by expanding in linear combinations of the original $[\Psi_k]$ and weighting by linear combinations of the $[W_j]$. For example use of $[\Psi_1]$, $[\Psi_1 - \Psi_2]$, $[\Psi_1 - \Psi_3]$,. . . , $[\Psi_1 - \Psi_K]$, and $[W_1]$, $[W_1 - W_2]$, $[W_1 - W_3]$, . . . , $[W_1 - W_K]$ in place of the $[\Psi_k]$ and $[W_j]$ is usually sufficient to avoid rounding-off errors. If not, an actual orthogonalization can be carried out. Such transformations do not affect the answer mathematically; hence, from that viewpoint, the procedure is unnecessary. It can, however, affect the numerical answer considerably.

The use of fundamental-mode solutions of the type (11.2.31) as expansion and weight functions is not a universal practice. Two other classes of expansion function are discussed in a number of papers and implemented in a few computer programs. The first class consists of the eigenfunctions $[\Psi_\lambda]$ or $[\Psi_\omega]$ of a single reactor equation (analogous to (8.3.18) and (8.3.38) with $[L]$ and $[M]$, the group diffusion-theory matrices, in place of L and M). The D's and Σ's in that single equation may be those corresponding to any state in the range of reactor conditions for which the approximate expansion is being used. Because of orthogonality relationships with the corresponding adjoint functions, the algebra of certain formal operations is sometimes simplified by the use of such expansion functions. However it is a very difficult numerical problem to find any λ or ω modes other than the fundamental (and, for symmetric cores, the first harmonic). Moreover, if approximate fluxes for a range of operating conditions are to be found by the synthesis procedure, the generation of expansion functions using an operator corresponding to only *one* condition in this range is intuitively less appealing than using operators bracketing the range of conditions. Using the modes of a single operator is, therefore, not recommended for most practical problems.

A third class of expansion functions has been suggested for performing point-synthesis calculations. This class consists of solutions to the Helmholtz equation

$$\nabla^2\Psi_k(\mathbf{r}) + B_k^2\Psi_k(\mathbf{r}) = 0 \qquad (11.2.33)$$

that vanish on the outer boundary of the reactor (usually defined as the outer boundary of the reflector). For example, for a reactor of dimensions $L_x \times L_y \times L_z$, these functions are

$$\Psi_{l,m,n}(x, y, z) = \sin\frac{l\pi}{L_x}\sin\frac{m\pi}{L_y}\sin\frac{n\pi}{L_z} \qquad (l, m, n = 1, 2, 3, \ldots),$$

so that

$$B^2_{l,m,n} = \left(\frac{l\pi}{L_x}\right)^2 + \left(\frac{m\pi}{L_y}\right)^2 + \left(\frac{n\pi}{L_z}\right)^2.$$

(Thus the index k in (11.2.33) stands for a triplet of numbers.) The same expansion functions are used for all energy groups.

The great advantage of the Helmholtz modes is that they are analytical and need not be found by solving auxiliary full-core problems (although finding the numerical value of fluxes at mesh points is relatively expensive). The great difficulty is that it may take hundreds of such modes to represent $\Phi_g(\mathbf{r})$; the solution of (11.2.30) thus becomes extremely time-consuming.

11.3 Variational Techniques as a Means of Deriving Approximation Methods

The variational techniques used to derive approximation methods for problems in reactor physics have had a flavor rather different from that of the standard applications. They have also required the use of functionals and function spaces somewhat more general than those encountered in quantum and classical mechanics. In particular, functionals take on stationary rather than maximum or minimum values, and the function spaces frequently contain discontinuous trial functions. Mathematical complications arise from these nonstandard aspects of the method; but these can be circumvented in a variety of ways, and we shall develop the method without emphasizing its finer mathematical aspects. Instead we shall build up a limited amount of the theory starting from the one-group diffusion equation and leading to several important approximation methods. The motivation and applications will thus accent the physical rather than the mathematical characteristics of the technique. The reader should therefore not lose sight of the fact that we are dealing with only a relatively small part of a fairly extensive field.

A Functional Made Minimum by the Solution to the One-Group Diffusion Equation

We start with the one-group diffusion-theory equation

$$-\nabla\cdot D\nabla\Phi_0(\mathbf{r}) + \Sigma(\mathbf{r})\Phi_0(\mathbf{r}) = \frac{1}{\lambda_0}\nu\Sigma_f(\mathbf{r})\Phi_0(\mathbf{r}), \qquad (11.3.1)$$

where D, Σ, and $\nu\Sigma_f$ would in general be functions (possibly discontinuous) of position; however to avoid mathematical complications we shall take D to be constant and

$\nu\Sigma_f(\mathbf{r}) > 0$ everywhere within the reactor (thus excluding the eigenvalue $\lambda_0 = 0$). The flux $\Phi_0(\mathbf{r})$ is nonnegative and vanishes on the external surface of the reactor; and both $\Phi_0(\mathbf{r})$ and the normal component of the net current, $-D\nabla\Phi_0(\mathbf{r})$, are continuous in the direction perpendicular to any internal surface.

It can be shown that there are an infinite number of other solutions to (11.3.1) for which the λ_n are all real and obey $\lambda_n < \lambda_0$ and that the corresponding fluxes $\Phi_n(\mathbf{r})$ are somewhere negative and somewhere positive. We shall assume that the $\Phi_n(\mathbf{r})$ (including $\Phi_0(\mathbf{r})$) form a complete set in the sense that any arbitrary function obeying the same conditions as those applied to the $\Phi_n(\mathbf{r})$ can be expanded in terms of them. We shall use this completeness property to show that a certain integral expression for the number $1/\lambda_0$ involving the function $\Phi_0(\mathbf{r})$ will be less when $\Phi_0(\mathbf{r})$ is used in it than it will be when any other function of $\mathbf{r}$ satisfying the conditions imposed on $\Phi_0(\mathbf{r})$ is used. Then we shall turn the argument around and use the integral expression for $1/\lambda_0$ to infer $\Phi_0(\mathbf{r})$ itself.

To get this expression for $1/\lambda_0$ we multiply (11.3.1) by $\Phi_0(\mathbf{r})$ and integrate over the entire volume of the reactor. This leads to

$$\frac{1}{\lambda_0} = \frac{\int \Phi_0[-\nabla\cdot D\nabla\Phi_0 + \Sigma\Phi_0]\,dV}{\int \Phi_0 \nu\Sigma_f \Phi_0\,dV} = \frac{\int [+D\nabla\Phi_0\cdot\nabla\Phi_0 + \Sigma\Phi_0^2]\,dV}{\int \Phi_0\nu\Sigma_f\Phi_0\,dV}, \tag{11.3.2}$$

where the second expression comes from applying Gauss's theorem and the relationship

$$\nabla\cdot(\Phi_0 D\nabla\Phi_0) = \nabla\Phi_0\cdot(D\nabla\Phi_0) + \Phi_0\nabla\cdot(D\nabla\Phi_0)$$

along with the internal continuity and external boundary conditions, and we have suppressed functional dependencies.

Now suppose that instead of Φ_0 in (11.3.2) we use some other function $\Phi_0 + \delta\Phi$ but require that $\delta\Phi$ obey the internal continuity and external boundary conditions imposed on Φ_0 and $D\nabla\Phi_0$. Substitution into the first expression (11.3.2) will yield an altered value of $1/\lambda_0$, namely

$$\frac{1}{\lambda_0} + \delta\left(\frac{1}{\lambda}\right) = \frac{\int (\Phi_0 + \delta\Phi)(-\nabla\cdot D\nabla + \Sigma)(\Phi_0 + \delta\Phi)\,dV}{\int (\Phi_0 + \delta\Phi)\nu\Sigma_f(\Phi_0 + \delta\Phi)\,dV}, \tag{11.3.3}$$

so that

$$\begin{aligned}\delta\left(\frac{1}{\lambda}\right) &= \frac{\int (\Phi_0 + \delta\Phi)(-\nabla\cdot D\nabla + \Sigma - \lambda_0^{-1}\nu\Sigma_f)(\Phi_0 + \delta\Phi)\,dV}{\int (\Phi_0 + \delta\Phi)\nu\Sigma_f(\Phi_0 + \delta\Phi)\,dV}\\ &= \frac{\int (\Phi_0 + \delta\Phi)(-\nabla\cdot D\nabla + \Sigma - \lambda_0^{-1}\nu\Sigma_f)\delta\Phi\,dV}{\int \nu\Sigma_f(\Phi_0 + \delta\Phi)^2\,dV}\\ &= \frac{\int \delta\Phi(-\nabla\cdot D\nabla + \Sigma - \lambda_0^{-1}\nu\Sigma_f)(\Phi_0 + \delta\Phi)\,dV}{\int \nu\Sigma_f(\Phi_0 + \delta\Phi)^2\,dV}\\ &= \frac{\int \delta\Phi(-\nabla\cdot D\nabla + \Sigma - \lambda_0^{-1}\nu\Sigma_f)\delta\Phi\,dV}{\int \nu\Sigma_f(\Phi_0 + \delta\Phi)\,dV},\end{aligned} \tag{11.3.4}$$

where the second and fourth expressions arise from applying (11.3.1) and the third expression comes from applying Gauss's theorem twice to the term

$$\int (\Phi_0 + \delta\Phi)(-\nabla \cdot D\nabla)\delta\Phi \, dV,$$

recognizing that, on the surface S of the reactor, $\delta\Phi = \Phi_0 = 0$, that is, from writing

$$\begin{aligned}\int (\Phi_0 + \delta\Phi)(-\nabla \cdot D\nabla)\delta\Phi \, dV &= \int_S -(\Phi_0 + \delta\Phi)\, D\nabla(\delta\Phi)\cdot d\mathbf{S} + \int \nabla(\Phi_0 + \delta\Phi)\cdot D\nabla(\delta\Phi)\, dV \\ &= \int_S \delta\Phi D\nabla(\Phi_0 + \delta\Phi)\cdot d\mathbf{S} - \int \delta\Phi\nabla\cdot D\nabla(\Phi_0 + \delta\Phi)\, dV\end{aligned} \tag{11.3.5}$$

and noting that the surface integrals vanish. (Note that (11.3.5) shows that the operator $\nabla \cdot D\nabla$ is self-adjoint with respect to the space of functions to which Φ_0 and $\delta\Phi$ belong.)

We now expand the perturbation $\delta\Phi$ in the eigenfunctions Φ_n of (11.3.1):

$$\delta\Phi = \sum_{n=0}^{\infty} a_n \Phi_n, \tag{11.3.6}$$

where the Φ_n obey

$$-\nabla \cdot D\nabla\Phi_n + \Sigma\Phi_n = \frac{1}{\lambda_n}\nu\Sigma_f\Phi_n, \text{ with } \lambda_0 > \lambda_1 \geq \lambda_2 \geq \cdots > 0. \tag{11.3.7}$$

Moreover (8.3.27) shows that

$$\int \Phi_m \nu\Sigma_f \Phi_n dV = 0 \quad \text{for } \lambda_m \neq \lambda_n. \tag{11.3.8}$$

Also, if $\lambda_m = \lambda_n$ ($m \neq n$), as can happen for $m, n \neq 0$, the corresponding Φ_m and Φ_n can be combined linearly into two other functions Φ'_m and Φ'_n such that $\int \Phi'_m \nu\Sigma_f \Phi'_n dV$ again equals zero.

Thus (11.3.4) becomes

$$\begin{aligned}\delta\left(\frac{1}{\lambda}\right) &= \frac{\int (\sum_m a_m\Phi_m)(-\nabla\cdot D\nabla + \Sigma_a - \lambda_0^{-1}\nu\Sigma_f)(\sum_n a_n\Phi_n)\, dV}{\int \nu\Sigma_f(\Phi_0 + \delta\Phi)^2\, dV} \\ &= \frac{\int \sum_n a_n^2(\lambda_n^{-1} - \lambda_0^{-1})\nu\Sigma_f\Phi_n^2\, dV}{\int \nu\Sigma_f(\Phi_0 + \delta\Phi)^2\, dV}.\end{aligned} \tag{11.3.9}$$

This result shows that $\delta(1/\lambda)$ is positive for any $\delta\Phi$ for which any of the coefficients a_n ($n > 0$) are finite (i.e., for any flux shape other than the fundamental, Φ_0).

We now exploit this minimum property in order to select the solution of (11.3.1) from a complete space of functions $\Phi(\mathbf{r})$. To do so we first define a functional $F(\Phi)$ for the class of all continuous real functions vanishing at the outer boundary of the reactor and having the property that the component of $D\nabla\Phi$ across any interface is continuous:

$$F(\Phi) = \frac{\int [D\nabla\Phi\cdot\nabla\Phi + \Sigma\Phi^2]}{\int \Phi^2 \nu\Sigma_f\, dV}. \tag{11.3.10}$$

There are several features of this functional that are typical of many functionals used as bases for variational techniques:

1. $F(\Phi)$ is a real number.
2. $F(\Phi)$ is defined precisely, but only for a certain limited class of "admissible" functions.
3. For the function Φ_0 that is the solution of (11.3.1), $F(\Phi_0)$ equals $1/\lambda_0$.
4. For all other functions belonging to the admissible class, $F(\Phi)$ is greater than $1/\lambda_0$. (This is shown by (11.3.9).)
5. The function that makes $F(\Phi)$ stationary with respect to first-order variation in Φ is the solution of (11.3.1).

The last property is the one we wish to exploit. It is demonstrated as follows: The definition (11.3.10) implies that

$$\begin{aligned} F(\Phi + \delta\Phi) - F(\Phi) &= \frac{\int [D\nabla(\Phi + \delta\Phi)\cdot\nabla(\Phi + \delta\Phi) + \Sigma(\Phi + \delta\Phi)^2]\, dV}{\int \nu\Sigma_f(\Phi + \delta\Phi)^2\, dV} \\ &\quad - \frac{\int [D\nabla\Phi\cdot\nabla\Phi + \Sigma\Phi^2]\, dV}{\int \nu\Sigma_f\Phi^2\, dV} \\ &= \left[F(\Phi) + \frac{2\int [D\nabla\Phi\cdot\nabla(\delta\Phi) + \Sigma\Phi\delta\Phi]\, dV}{\int \nu\Sigma_f\Phi^2\, dV}\right]\left[1 - \frac{2\int \nu\Sigma_f\Phi\delta\Phi\, dV}{\int \nu\Sigma_f\Phi^2\, dV}\right] \\ &\quad + 0((\delta\Phi)^2) - F(\Phi) \\ &= \frac{2\int \delta\Phi[-\nabla\cdot D\nabla\Phi + \Sigma\Phi - F(\Phi)\nu\Sigma_f\Phi]\, dV}{\int \nu\Sigma_f\Phi^2\, dV} + 0((\delta\Phi)^2). \end{aligned} \tag{11.3.11}$$

The last line of this equation, with terms of order $(\delta\Phi)^2$ and higher neglected, is called the *first-order variation* of the functional $F(\Phi)$. If this first-order variation is to vanish for completely arbitrary $\delta\Phi$, the expression $[-\nabla\cdot D\nabla\Phi + \Sigma\Phi - F(\Phi)\nu\Sigma_f\Phi]$ must vanish for all $\mathbf{r}$ in the range of integration. We see that, if we set $F(\Phi) = 1/\lambda_0$, the function that makes this happen is Φ_0, the solution of (11.3.1). We see further that, if $F(\Phi)$ is set equal to any of the eigenvalues $1/\lambda_n$ of (11.3.7), the corresponding higher harmonic Φ_n will also cause the first-order variation in $F(\Phi)$ to vanish. However none of the $\Phi_n(\mathbf{r})$ $(n > 0)$ is positive for all $\mathbf{r}$, and all of the λ_n are less than λ_0. Thus we conclude that the only *everywhere-positive* function that causes the first-order variation in $F(\Phi)$ to vanish also makes $F(\Phi)$ a minimum and that that function must be $\Phi_0(\mathbf{r})$, the solution of (11.3.1).

In principle one could find this flux by repeatedly evaluating (11.3.10) with different trial shapes until the everywhere-positive shape that minimizes $F(\Phi)$ is found. Such a procedure is, of course, not practical for the general case. Moreover the systematic way of minimizing $F(\Phi)$ from (11.3.11) merely leads back again to the differential equation (11.3.1). Thus, at first encounter, the variational approach seems to add nothing new to the standard method.

But this is not the case. There *is* a great practical value to the variational approach. It stems from the fact that the method can be used when the space of trial functions is limited and does not contain the correct solution as one of its elements. If out of such a limited space we find the function that minimizes $F(\Phi)$, we know it will yield the closest approximation to $1/\lambda_0$ obtainable using shapes from that limited space. Experience, moreover, shows that the flux shape that yields the minimum $F(\Phi)$ will be a good approximation to the correct solution to (11.3.1).

An Example

As a simple example of this technique we consider a slab reactor having a constant value of D and extending from $x = -l$ to $x = +l$. Suppose the spatial dependence of $\nu\Sigma_f(\mathbf{r})$ and $\Sigma(\mathbf{r})$ is such that we expect the correct solution to (11.3.1) to be well approximated by the linear combination

$$\Phi_t(x) = a\cos\frac{\pi x}{2l} + b\sin\frac{\pi x}{l}, \tag{11.3.12}$$

where a and b are constants to be determined.

With arbitrary ratios a/b permitted, Φ_t covers a limited subspace of the space of admissible functions for $F(\Phi)$.

Substitution of (11.3.12) into (11.3.10) yields

$$F(\Phi_t) = \frac{\begin{aligned} &a^2\int_{-l}^{l}\left[D\left(\frac{\pi}{2l}\right)^2\sin^2\frac{\pi x}{2l} + \Sigma(x)\cos^2\frac{\pi x}{2l}\right]dx - 2ab\int_{-l}^{l}\left[\frac{\pi^2}{2l^2}D\sin\frac{\pi x}{2l}\cos\frac{\pi x}{l}\right.\\ &\left.+\Sigma(x)\cos\frac{\pi x}{2l}\sin\frac{\pi x}{l}\right]dx + b^2\int_{-l}^{l}\left[D\left(\frac{\pi}{l}\right)^2\cos^2\frac{\pi x}{l} + \Sigma(x)\sin^2\frac{\pi x}{l}\right]dx\end{aligned}}{\int_{-l}^{l}\nu\Sigma_f(x)\left[a^2\cos^2\frac{\pi x}{2l} + 2ab\cos\frac{\pi x}{2l}\sin\frac{\pi x}{2l} + b^2\sin^2\frac{\pi x}{l}\right]dx}. \tag{11.3.13}$$

Now, the most general variation in Φ_t is $\delta a\cos(\pi x/2l) + \delta b\sin(\pi x/l)$, where δa and δb are arbitrary. We can thus find a and b by substituting this general variation of Φ_t and Φ_t itself into (11.3.11) and *requiring* the resulting expressions to be valid, first for $\delta b = 0$ and then for $\delta a = 0$. However we also know that we want a and b to be such that $F(\Phi)$ is a minimum, and to minimize $F(\Phi)$ over the subspace (11.3.12) we need merely set $\partial F(\Phi_t)/\partial a = \partial F(\Phi_t)/\partial b = 0$. This procedure leads to two homogeneous equations in a and b that will have a nontrivial solution only when $\nu\Sigma_f$ is multiplied by the number $F(\Phi_t) = 1/\lambda_t$. This quantity is the best estimate of $1/\lambda_0$ provided by trial functions of the form (11.3.12), and the corresponding function Φ_t obtained from the a/b ratio associated with $1/\lambda_t$ is then the approximate shape for $\Phi_0(x)$.

More insight into the nature of the approximation is obtained by noting that the same result can be found using the weighted-residual method. Thus, if (11.3.12) is substituted into (11.3.1) and the result is multiplied first by $\cos(\pi x/2l)$ and then by $\sin(\pi x/l)$ and integrated from $-l$ to $+l$, the equations for a and b that result will be identical with those found by setting the first-order variation of $F(\Phi_t)$ to zero. We conclude that the flux shape resulting from the variational procedure is an approximation to Φ_0 in a weighted-integral sense, a result that accounts to some extent for the fact that the flux shape that minimizes $1/\lambda$ is usually a good approximation to Φ_0.

Extension to the Multigroup Case and to Discontinuous Trial Functions

The variational technique based on the functional (11.3.10) is limited in two major respects. First, it is applicable only to the one-group case. Second, the space of trial functions for which $F(\Phi)$ is defined is quite restricted.

Getting around the first limitation will not be difficult. A functional analogous to (11.3.10) but depending on a column vector of group fluxes and a second, *independent* column of weight functions will be defined so that, if the weight functions are varied, the functional will be stationary about the solution $[\Phi_0] \equiv \mathrm{Col}\{\Phi_{10}, \Phi_{20}, \Phi_{30}, \ldots\}$ of the group diffusion equations. The functional so defined, however, will generally not assume a maximum or minimum value at its stationary point. Instead the stationary point will have more the nature of a point of inflection. That is, if a limited subspace of group flux vectors *not containing* $[\Phi_0]$ is examined and a vector out of that subspace is found that makes the functional stationary, the resultant value of the functional will *not* in general be the best value of $1/\lambda_0$ obtainable using vectors from the subspace. There will be other vectors in the limited subspace that will yield a more accurate value of $1/\lambda_0$; however no systematic way to find these vectors is known.

The second limitation (restricting the space of trial functions for which the functional is defined) is more difficult to circumvent in that opportunities to commit mathematical sins abound. To avoid such problems as much as possible we shall define a functional that is made stationary by a solution of the P_1 equations (9.4.4) and (9.4.5) in energy-group form (rather than by a solution of the group diffusion equations). To provide needed flexibility, however, the functional will be defined for a space of functions whose elements may be discontinuous.

To proceed we define G-element column vectors $[U]$, $[U^*]$, $[\mathbf{V}]$, and $[\mathbf{V}^*]$ by

$$[U] \equiv \begin{bmatrix} U^{(1)}(\mathbf{r}) \\ U^{(2)}(\mathbf{r}) \\ \vdots \\ U^{(G)}(\mathbf{r}) \end{bmatrix}, \quad [U^*] \equiv \begin{bmatrix} U^{(1)*}(\mathbf{r}) \\ U^{(2)*}(\mathbf{r}) \\ \vdots \\ U^{(G)*}(\mathbf{r}) \end{bmatrix}, \quad [\mathbf{V}] \equiv \begin{bmatrix} \mathbf{V}^{(1)}(\mathbf{r}) \\ \mathbf{V}^{(2)}(\mathbf{r}) \\ \vdots \\ \mathbf{V}^{(G)}(\mathbf{r}) \end{bmatrix}, \quad [\mathbf{V}^*] \equiv \begin{bmatrix} \mathbf{V}^{(1)*}(\mathbf{r}) \\ \mathbf{V}^{(2)*}(\mathbf{r}) \\ \vdots \\ \mathbf{V}^{(G)*}(\mathbf{r}) \end{bmatrix}. \quad (11.3.14)$$

The elements of $[U]$ and $[U^*]$ are real scalar functions of position, while those of $[\mathbf{V}]$ and $[\mathbf{V}^*]$ are real vector functions of position (in the sense that each element $\mathbf{V}^{(g)}(\mathbf{r})$ or $\mathbf{V}^{(g)*}(\mathbf{r})$ has three directional components). The vectors $[U]$ and $[U^*]$ are required to vanish on the exterior surface S_0 of the reactor.

In order to be explicit and precise about the definition of the functional to be introduced, we let S_{int} denote the interior surfaces across which *any* of the functions $[U]$, $[U^*]$, $[\mathbf{V}]$, or $[\mathbf{V}^*]$ are discontinuous and let R_i denote the interior regions over which they are *all* continuous. We also let $\mathbf{N}$ be a normal to any interior surface and denote by $(+)$ or $(-)$ evaluation on that side of the surface corresponding to the head $(+)$ or tail $(-)$ of the vector $\mathbf{N}$. Figure 11.1 illustrates this convention for two-dimensional rectangular geometry.

We then construct a functional $F_d([U], [U^*], [\mathbf{V}], [\mathbf{V}^*])$ according to the rule

$$\begin{aligned} F_d([U], [U^*], [\mathbf{V}], [\mathbf{V}^*]) = {} & \left\{ \int [U^{*\mathrm{T}}\ \mathbf{V}^{*\mathrm{T}}] \begin{bmatrix} \chi\nu\Sigma_f^{\mathrm{T}} & 0 \\ 0 & 0 \end{bmatrix} \begin{bmatrix} U \\ \mathbf{V} \end{bmatrix} d\mathbf{r} \right\}^{-1} \\ & \times \left\{ \sum_i \int_{R_i} [U^{*\mathrm{T}}\ \mathbf{V}^{*\mathrm{T}}] \begin{bmatrix} A & \nabla\cdot \\ -(\cdot\nabla) & -(\cdot D^{-1}) \end{bmatrix} \begin{bmatrix} U \\ \mathbf{V} \end{bmatrix} d\mathbf{r} \right. \\ & + \oint_{S_{\text{int}}} \frac{1}{2} \left([U^*(+) + U^*(-)]^{\mathrm{T}} [\mathbf{V}(+) - \mathbf{V}(-)] \right. \\ & \left. \left. - [\mathbf{V}^*(+) + \mathbf{V}^*(-)]^{\mathrm{T}} [U(+) - U(-)] \right) \cdot \mathbf{N}\, dS \right\}, \end{aligned} \tag{11.3.15}$$

where $[\chi\nu\Sigma_f^{\mathrm{T}}]$, $[A]$, and $[D^{-1}]$ are $G \times G$ energy-group matrices defined from the elements (9.4.13), (9.4.30), and (9.4.31). (For simplicity we shall treat $[D^{-1}]$ as a $G \times G$ diagonal matrix having elements that are the same for all coordinate directions.)

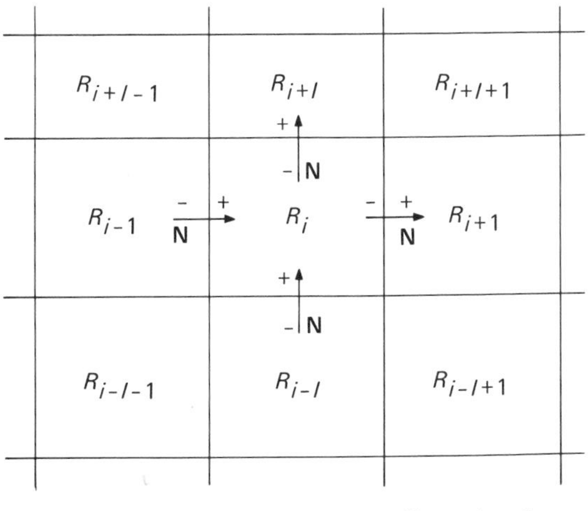

Figure 11.1 An illustration for two-dimensional rectangular geometry of the convention relating the $(+)$ and $(-)$ sides of the surfaces bounding a region R_i to the directions of normals to those surfaces going from R_i to neighboring regions.

The functional F_{d} is a real number unambiguously defined for any quartet of vectors $[U]$, $[U^*]$, $[\mathbf{V}]$, and $[\mathbf{V}^*]$ in the admissible domain. We now show that that quartet about which F_{d} is stationary with respect to first-order variations in its argument functions is a solution to the flux-current and adjoint-flux-current equations of group diffusion theory. To do so we take the first-order variation of F_{d} and set it equal to zero for completely arbitrary variations in the argument functions. Thus we write

$$
\begin{aligned}
\delta F_{\text{d}} &= F_{\text{d}}([U + \delta U], [U^* + \delta U^*], [\mathbf{V} + \delta \mathbf{V}], [\mathbf{V}^* + \delta \mathbf{V}^*]) - F_{\text{d}}([U], [U^*], [\mathbf{V}], [\mathbf{V}^*]) \\
&= \left\{ \int [U^*]^{\text{T}}[\chi\nu\Sigma_{\text{f}}^{\text{T}}][U] d\mathbf{r} + \int [\delta U^*]^{\text{T}}[\chi\nu\Sigma_{\text{f}}^{\text{T}}][U]\, d\mathbf{r} + \int [U^*]^{\text{T}}[\chi\nu\Sigma_{\text{f}}^{\text{T}}][\delta U]\, d\mathbf{r} \right\}^{-1} \\
&\times \left\{ \sum_i \int_{R_i} \left([U^{*\text{T}}\ \mathbf{V}^{*\text{T}}] \begin{bmatrix} A & \nabla\cdot \\ -(\cdot\nabla) & -(\cdot D^{-1}) \end{bmatrix} \begin{bmatrix} U \\ \mathbf{V} \end{bmatrix} + [\delta U^{*\text{T}}\ \delta \mathbf{V}^{*\text{T}}] \begin{bmatrix} A & \nabla\cdot \\ -(\cdot\nabla) & -(\cdot D^{-1}) \end{bmatrix} \begin{bmatrix} U \\ \mathbf{V} \end{bmatrix} \right.\right. \\
&+ \left. [U^{*\text{T}}\ \mathbf{V}^{*\text{T}}] \begin{bmatrix} A & \nabla\cdot \\ -(\cdot\nabla) & -(\cdot D^{-1}) \end{bmatrix} \begin{bmatrix} \delta U \\ \delta \mathbf{V} \end{bmatrix} \right) d\mathbf{r} + \oint_{S_{\text{int}}} \frac{1}{2}([U^*(+) + \delta U^*(+) + U^*(-) \\
&+ \delta U^*(-)]^{\text{T}}\, [\mathbf{V}(+) + \delta \mathbf{V}(+) - \mathbf{V}(-) - \delta \mathbf{V}(-)] - [\mathbf{V}^*(+) + \delta \mathbf{V}^*(+) + \mathbf{V}^*(-) \\
&+ \left. \delta \mathbf{V}^*(-)]^{\text{T}}[U(+) + \delta U(+) - U(-) - \delta U(-)]) \cdot \mathbf{N}\, dS \right\} - F_{\text{d}} + 0(\delta^2). \qquad (11.3.16)
\end{aligned}
$$

Expanded to first order in $[\delta U]$ and $[\delta U^*]$, the inverse on the right-hand side of (11.3.16) becomes

$$
\left\{ \int [U^*]^{\text{T}}[\chi\nu\Sigma_{\text{f}}^{\text{T}}][U]\, d\mathbf{r} \right\}^{-1} \left\{ 1 - \frac{\int [\delta U^*][\chi\nu\Sigma_{\text{f}}^{\text{T}}][U]\, d\mathbf{r} + \int [U^*]^{\text{T}}[\chi\nu\Sigma_{\text{f}}^{\text{T}}][\delta U]\, \delta\mathbf{r}}{\int [U^*]^{\text{T}}[\chi\nu\Sigma_{\text{f}}^{\text{T}}][U]\, d\mathbf{r}} \right\}.
$$

If the multiplication indicated in (11.3.16) is performed and the value of the number F_{d} is symbolized by $1/\lambda$, we obtain

$$
\begin{aligned}
\delta F_{\text{d}} &= \left\{ \int [U^*]^{\text{T}}[\chi\nu\Sigma_{\text{f}}^{\text{T}}][U]\, d\mathbf{r} \right\}^{-1} \left\{ \sum_i \int_{R_i} \left([\delta U^{*\text{T}}\ \delta \mathbf{V}^{*\text{T}}] \begin{bmatrix} A - (1/\lambda)\chi\nu\Sigma_{\text{f}}^{\text{T}} & \nabla\cdot \\ -(\cdot\nabla) & -(\cdot D^{-1}) \end{bmatrix} \begin{bmatrix} U \\ \mathbf{V} \end{bmatrix} \right.\right. \\
&+ \left. [U^{*\text{T}}\ \mathbf{V}^{*\text{T}}] \begin{bmatrix} A - (1/\lambda)\chi\nu\Sigma_{\text{f}}^{\text{T}} & \nabla\cdot \\ -(\cdot\nabla) & -(\cdot D^{-1}) \end{bmatrix} \begin{bmatrix} \delta U \\ \delta \mathbf{V} \end{bmatrix} \right) d\mathbf{r} + \oint_{S_{\text{int}}} \frac{1}{2}([\delta U^*(+) \\
&+ \delta U^*(-)]^{\text{T}}[\mathbf{V}(+) - \mathbf{V}(-)] + [U^*(+) + U^*(-)]^{\text{T}}[\delta \mathbf{V}(+) - \delta \mathbf{V}(-)] - [\delta \mathbf{V}^*(+) \\
&+ \left. \delta \mathbf{V}^*(-)]^{\text{T}}[U(+) - U(-)] - [\mathbf{V}^*(+) + \mathbf{V}^*(-)]^{\text{T}}[\delta U(+) - \delta U(-)]) \cdot \mathbf{N} dS \right\}.
\end{aligned}
$$

(11.3.17)

Writing out the second sum of volume integrals and applying the rule for differentiating a product,

$$
\nabla\cdot([\mathbf{f}(\mathbf{r})]^{\text{T}}[g(\mathbf{r})]) = [g(\mathbf{r})]^{\text{T}}\nabla\cdot[\mathbf{f}(\mathbf{r})] + [\mathbf{f}(\mathbf{r})]^{\text{T}}\cdot\nabla[g(\mathbf{r})],
$$

yields

$$\sum_i \int_{R_i} [U^{*T}\ \mathbf{V}^{*T}]\begin{bmatrix} A - (1/\lambda)\chi\nu\Sigma_f^T & \nabla\cdot \\ -\cdot\nabla & -(\cdot D^{-1}) \end{bmatrix}\begin{bmatrix} \delta U \\ \delta\mathbf{V} \end{bmatrix} d\mathbf{r}$$

$$= \sum_i \int_{R_i} \Big\{[U^*]^T\Big([A] - \frac{1}{\lambda}[\chi\nu\Sigma_f^T]\Big)[\delta U] - [\mathbf{V}^*]^T\cdot[D^{-1}][\delta\mathbf{V}] - [\delta\mathbf{V}]^T\cdot\nabla[U^*]$$

$$+ [\delta U]^T\nabla\cdot[\mathbf{V}^*] - \nabla\cdot([\mathbf{V}^*][\delta U]) + \nabla\cdot([U^*]^T[\delta\mathbf{V}])\Big\} d\mathbf{r}. \quad (11.3.18)$$

Next, maintaining the convention that the direction of **N** extends from (−) to (+) so that **N** is an outward-drawn normal for surfaces designated (−) and an inward-drawn normal for the adjacent (+) surfaces, we apply the divergence theorem to the last two terms of (11.3.18). The result is

$$\int_{R_i} \nabla\cdot([U^*]^T[\delta\mathbf{V}])d\mathbf{r} = \oint_{S_i} [U^*]^T[\delta\mathbf{V}]\cdot d\mathbf{S}$$

$$= \oint_{S_{i-}} [U^*(-)]^T[\delta\mathbf{V}(-)]\cdot\mathbf{N}\,dS - \oint_{S_{i+}} [U^*(+)]^T[\delta\mathbf{V}(+)]\cdot\mathbf{N}\,dS \quad (11.3.19)$$

and

$$\int_{R_i} \nabla\cdot([\mathbf{V}^*]^T[\delta U])\,d\mathbf{r} = \oint_{S_i} [\mathbf{V}^*]^T[\delta U]\cdot d\mathbf{S}$$

$$= \oint_{S_{i-}} [\mathbf{V}^*(-)]^T[\delta U(-)]\cdot\mathbf{N}\,dS - \int_{S_{i+}} [\mathbf{V}^*(+)]^T[\delta U(+)]\cdot\mathbf{N}\,dS, \quad (11.3.20)$$

where the surface integrals over S_{i-} are for those portions of the surface containing tails of the vectors **N** and those over S_{i+} are for portions containing heads (see Figure 11.1).

Note that the direction in which **N** is drawn between any two surfaces is arbitrary, but that, once it has been decided with respect to one of the regions R_i, the direction with respect to the surfaces of neighboring regions is fixed. Note also that, since all the components of the vectors $[U]$ and $[U^*]$ are required to vanish on the outer surface of the reactor, any portions of the integrals in (11.3.19) and (11.3.20) on such an outer surface will vanish.

Substituting (11.3.18)–(11.3.20) into (11.3.17), collecting terms, neglecting terms of order δ^2, and setting δF_d to zero yields

$$0 = \sum_i \int_{R_i} \Big\{[\delta U^{*T}\ \delta\mathbf{V}^{*T}]\begin{bmatrix} A - (1/\lambda)\chi\nu\Sigma_f^T & \nabla\cdot \\ -(\cdot\nabla) & -(\cdot D^{-1}) \end{bmatrix}\begin{bmatrix} U \\ \mathbf{V} \end{bmatrix}$$

(equation continued)

$$+\begin{bmatrix}\delta U^{\mathrm{T}} & \delta \mathbf{V}^{\mathrm{T}}\end{bmatrix}\begin{bmatrix} A^{\mathrm{T}} - (1/\lambda)\nu\Sigma_{\mathrm{f}}\chi^{\mathrm{T}} & \nabla\cdot \\ -(\cdot\nabla) & -(\cdot D^{-1}) \end{bmatrix}\begin{bmatrix} U^* \\ \mathbf{V}^* \end{bmatrix}\Bigg\} d\mathbf{r} + \frac{1}{2}\oint_{S_{\mathrm{int}}} \{[\delta U^*(+) + \delta U^*(-)]^{\mathrm{T}}$$
$$\times [\mathbf{V}(+) - \mathbf{V}(-)] - [\delta \mathbf{V}(+) + \delta \mathbf{V}(-)]^{\mathrm{T}}[U^*(+) - U^*(-)] - [\delta \mathbf{V}^*(+) + \delta \mathbf{V}^*(-)]^{\mathrm{T}}$$
$$\times [U(+) - U(-)] + [\delta U(+) + \delta U(-)]^{\mathrm{T}}[\mathbf{V}^*(+) - \mathbf{V}^*(-)]\}\cdot \mathbf{N}\, dS. \qquad (11.3.21)$$

Note that, in deriving this expression, we have made frequent use of the fact that an inner products such as $[\mathbf{V}^*(+)]^{\mathrm{T}}[\delta U]$ and $[U^*]^{\mathrm{T}}[A][\delta U]$ are identical with $[\delta U]^{\mathrm{T}}[\mathbf{V}^*(+)]$ and $[\delta U]^{\mathrm{T}}[A]^{\mathrm{T}}[U^*]$. Also note that the integral over S_0 is missing from (11.3.21) since the admissibility conditions on $[U]$ and $[U^*]$ require all vectors in their spaces to vanish on S_0.

Let us designate by $[\Phi]$, $[\Phi^*]$, $[\mathbf{J}]$, and $[\mathbf{J}^*]$ the quartet of column vectors about which F_{d} is stationary. Since the variations in $[U]$, $[U^*]$, $[\mathbf{V}]$, and $[\mathbf{V}^*]$ are entirely arbitrary, (11.3.21) shows that these stationary values must obey

$$\begin{aligned} &\nabla\cdot[\mathbf{J}] + [A][\Phi] - \frac{1}{\lambda}[\chi][\nu\Sigma_{\mathrm{f}}]^{\mathrm{T}}[\Phi] = 0, \qquad [\mathbf{J}] = -[D]\nabla[\Phi], \\ &\nabla\cdot[\mathbf{J}^*] + [A]^{\mathrm{T}}[\Phi^*] - \frac{1}{\lambda}[\nu\Sigma_{\mathrm{f}}][\chi]^{\mathrm{T}}[\Phi^*] = 0, \qquad [\mathbf{J}^*] = -[D]\nabla[\Phi^*], \end{aligned} \qquad (11.3.22)$$

with Φ, Φ^*, $\mathbf{N}\cdot\mathbf{J}$, and $\mathbf{N}\cdot\mathbf{J}^*$ all continuous across internal interfaces.

Thus, out of the set of all column vectors $[U]$, $[U^*]$, $[\mathbf{V}]$, and $[\mathbf{V}^*]$ having elements which may be discontinuous across any internal interface, the quartet that make F_{d} stationary are solutions to the energy-group equations for the neutron flux and current and their corresponding adjoint equations and obey the same continuity properties imposed on those solutions by physical considerations.

It should be noted that there are many other functionals besides F_{d} that are made stationary by the solutions of (11.3.22). They differ from F_{d} with respect to the spaces of functions for which they are defined, the treatment of the internal-surface terms, and the manner in which boundary conditions on the outer surface are accounted for. In addition it is possible to define functionals (extensions of (11.3.10) to two or more groups) that depend on two rather than four G-element vector functions, and functionals whose stationary value is, not $1/\lambda$, but some other eigenvalue (such as ω of (8.3.38)). Finally it is possible to define functionals that are made stationary by solutions of the transport equation, the continuous-energy diffusion equation, or the defining equations of just about any linear mathematical model of interest. Thus the functional F_{d} of (11.3.15), while extremely useful, is by no means unique.

One serious mathematical difficulty, extensively debated in the literature, appears when the functions $[U]$, $[U^*]$, $[\mathbf{V}]$, and $[\mathbf{V}^*]$ on which F_{d} depends are all permitted to be discontinuous across the same surface. The equations specifying which function out of the space will be selected by setting the variation of F_d to zero can be overdetermined under such circumstances. We shall avoid this difficulty in any application we make by simply limiting the space to trial functions such that, at most, two of the four functions on which F_d depends can be discontinuous across the same surface. Practical experience indicates that this limitation is not a serious one.

Another potential difficulty associated with applying F_d to find approximate solutions to the P_1 equations is the fact that, in general, there are many numerical values of F_d about which the functional is stationary. They correspond to eigenvalues λ other than the fundamental one. Again, practical experience indicates that this potential difficulty is rarely encountered. Probably because the space of trial functions is purposely chosen to contain elements that are in some sense close to the fundamental solution, the eigenvalue that results when the variation of F_d is set to zero is almost always a close approximation to λ_0, and the eigenfunction that results is thus an everywhere-positive approximation to the corresponding fundamental flux shape.

An Example: Conventional Finite-Difference Equations Derived Variationally

The great value of F_d in deriving approximation methods arises from the fact that it is defined for discontinuous trial functions. Thus, even if we limit the space of trial functions so that it contains *only* discontinuous functions, going through the variational procedure still yields an answer to the problem (one, which, in a weighted-residual sense, is the "best" obtainable from the limited class of trial functions employed). This situation is well illustrated when the functional F_d is used to obtain the conventional form of the finite-difference equations. To carry out this procedure we simply select a particular space of functions $[U]$, $[U^*]$, $[\mathbf{V}]$, and $[\mathbf{V}^*]$ and apply (11.3.21). To keep the algebraic details within bounds we shall consider only the slab-geometry case, for which $\mathbf{V}$ and $\mathbf{V}^*$ have but one component. Thus, with h_n representing the length of a mesh interval between points x_n and x_{n+1} (throughout which interval $[A]$, $[\nu\Sigma_f]$, and $[D]^{-1}$ are assumed to take on constant values $[A_n]$, $[\nu\Sigma_{fn}]$, and $[D_n]^{-1}$), we choose a space of trial functions and weight functions defined by

$$\left.\begin{aligned}[U(x)] &= [\Phi_n]\\ [U^*(x)] &= [\Phi_n^*]\end{aligned}\right\} \text{for } x_n - \frac{h_{n-1}}{2} < x < x_n + \frac{h_n}{2} \qquad (n = 1, 2, \ldots, N-1),$$

$$[U(x)] = [U^*(x)] = 0 \quad \text{for } 0 \le x < \frac{h_0}{2} \text{ and } x_N - \frac{h_N}{2} < x \le x_N, \tag{11.3.23}$$

$$\left.\begin{aligned}[V(x)] &= [J_n]\\ [V^*(x)] &= [J_n^*]\end{aligned}\right\} \text{for } x_n < x < x_{n+1} \qquad (n = 0, 1, 2, \ldots, N-1),$$

where $[\Phi_n]$, $[J_n]$, $[\Phi_n^*]$, and $[J_n^*]$ are G-element column vectors (G being the total number of energy groups). The elements of these vectors are taken as constants for a given value of the subscript n. Thus we are choosing the current and adjoint-current trial vectors to be spatially constant within the intervals h_n and discontinuous at interface points x_n, and we are choosing the flux and adjoint-flux trial vectors to be constant within the intervals $x_n - \frac{1}{2}h_{n-1}$ to $x_n + \frac{1}{2}h_n$ and discontinuous at the midpoints $x_n + \frac{1}{2}h_n$. This choice of trial functions is arbitrary. Many choices different from (11.3.23) could be made and would lead to other difference equations having different degrees of complexity and different orders of accuracy.

To cut down further on algebraic detail we shall find difference equations only for the vectors $[\Phi_n]$ and $[J_n]$. Thus we shall set the variations $[\delta U]$ and $[\delta V]$ to zero in (11.3.21) and shall find only those equations implied by the vanishing of δF_d for arbitrary variations $[\delta U^*]$ and $[\delta V^*]$. Substitution of (11.3.23) into (11.3.21) under these conditions yields

$$\begin{aligned} 0 &= \sum_{n=1}^{N-1} \int_{x_n - h_{n-1}/2}^{x_n + h_n/2} [\delta\Phi_n^*]^{\mathrm{T}}\left[A(x) - \frac{1}{\lambda}\chi\nu\Sigma_{\mathrm{f}}^{\mathrm{T}}(x)\right][\Phi_n]dx - \sum_{n=0}^{N-1}\int_{x_n}^{x_n+h_n} [\delta J_n^*]^{\mathrm{T}}[D^{-1}(x)][J_n]\,dx \\ &\quad + \sum_{n=1}^{N-1} [\delta\Phi_n^*]^{\mathrm{T}}[J_n - J_{n-1}] - \sum_{n=0}^{N-1}[\delta J_n^*]^{\mathrm{T}}[\Phi_{n+1} - \Phi_n] \\ &= \sum_{n=1}^{N-1}[\delta\Phi_n^*]^{\mathrm{T}}\left\{\frac{1}{2}h_{n-1}\left[A_{n-1} - \frac{1}{\lambda}\chi\nu\Sigma_{\mathrm{f},n-1}^{\mathrm{T}}\right][\Phi_n] + \frac{1}{2}h_n\left[A_n - \frac{1}{\lambda}\chi\nu\Sigma_{\mathrm{f}n}^{\mathrm{T}}\right][\Phi_n]\right. \\ &\quad \left. + [J_n - J_{n-1}]\right\} - \sum_{n=0}^{N-1}[\delta J_n^*]^{\mathrm{T}}\{h_n[D_n^{-1}][J_n] + [\Phi_{n+1} - \Phi_n]\}. \end{aligned} \tag{11.3.24}$$

Notice that, since the trial functions are spatially constant except at the points of discontinuity, the spatial derivatives in (11.3.21) vanish. The coupling of fluxes and currents across points of discontinuity is accomplished by the surface terms. If the weighted-residual method were being applied to this problem, such coupling would not be automatic but would, rather, have to be imposed arbitrarily.

Setting all variations in (11.3.24) except $[\delta\Phi_m^*]$ to zero yields

$$\left\{\frac{1}{2}h_{m-1}\left([A_{m-1}] - \frac{1}{\lambda}[\chi\nu\Sigma_{\mathrm{f},m-1}^{\mathrm{T}}]\right) + \frac{1}{2}h_m\left([A_m] - \frac{1}{\lambda}[\chi\nu\Sigma_{\mathrm{f}m}^{\mathrm{T}}]\right)\right\}[\Phi_m] + [J_m] - [J_{m-1}] = 0. \tag{11.3.25}$$

Similarly, setting all variations except $[\delta J_m^*]$ equal to zero yields

$$[J_m] = -[D_m]\frac{[\Phi_{m+1} - \Phi_m]}{h_m}, \tag{11.3.26}$$

where $[D_m] \equiv [D_m^{-1}]^{-1}$.

Using this last result to eliminate $[J_m]$ and $[J_{m-1}]$ leads to the difference equation

$$\left\{\frac{1}{2}h_{m-1}\left([A_{m-1}] - \frac{1}{\lambda}[\chi\nu\Sigma^{\mathrm{T}}_{\mathrm{f},m-1}]\right) + \frac{1}{2}h_m\left([A_m] - \frac{1}{\lambda}[\chi\nu\Sigma^{\mathrm{T}}_{\mathrm{f}m}]\right)\right\}[\Phi_m]$$

$$- \frac{1}{h_m}[D_m][\Phi_{m+1} - \Phi_m] + \frac{1}{h_{m-1}}[D_{m-1}][\Phi_m - \Phi_{m-1}] = 0. \tag{11.3.27}$$

Equation (11.3.27) is a generalization of the one-group result (4.10.16), derived in Chapter 4 by a direct integration procedure. We see that the present variational derivation leads us to interpret the $[\Phi_n]$ as spatially constant fluxes *throughout* intervals $x_n - \frac{1}{2}h_{n-1}$ to $x_n + \frac{1}{2}h_n$, rather than as fluxes *at* points x_n.

11.4 Synthesis Methods Derived from a Variational Principle

To illustrate the application of variational methods to the development of synthesis techniques we shall consider two very powerful schemes for determining approximate solutions to the group diffusion equations. The methods are known as "continuous" and "discontinuous" space-dependent synthesis.

Continuous Space-Dependent Synthesis

The point-synthesis method developed in Section 11.2 has two severe practical limitations in addition to the theoretical difficulties discussed earlier. The first is that, in the general case, a number of detailed, three-dimensional, finite-difference solutions for the group diffusion equations must be found if we are to obtain the basic expansion functions. Such solutions are very costly.

The second limitation is that the method works well only when the range of reactor conditions for which a given set of expansion functions is to be applied involves *homogeneous* changes in reactor properties (changes in homogeneous poison concentration, in overall reactor temperature, etc.). It is not possible, for example, to give an accurate representation of the detailed flux shapes corresponding to a *range* of control-rod positions using just two or three full-core expansion functions.

Both of these difficulties are largely circumvented by the use of the "space-dependent" synthesis method, the essential idea of which is to express the three-dimensional flux shape $\Phi_g(x, y, z)$ for group g as a linear combination of predetermined two-dimensional expansion functions $\Psi^g_k(x, y)$ multiplied by unknown one-dimensional mixing functions $T^g_k(z)$:

$$\Phi_g(x, y, z) \approx \sum_{k=1}^{K} \Psi^g_k(x, y) T^g_k(z). \tag{11.4.1}$$

The fact that the expansion functions $\Psi^g_k(x, y)$ are only two-dimensional results in their being relatively easy to calculate in detail (10 or 20 thousand mesh points will usually suffice to describe the most complicated geometrical layouts). Further, the fact that the mixing functions $T_k^g(z)$ depend on z permits the same $\Psi_k^g(x, y)$ to be used for many control-rod positions. For example, if we wish to determine the $\Phi_g(\mathbf{r})$ and the value of k_{eff} for a uniformly loaded, unreflected core controlled by a bank of control rods, we can take, for one of the $\Psi_k^g(x, y)$, a two-dimensional shape determined with all rods removed and, for another, a shape determined with all rods inserted. We then expect that for that range of z in which the rods are *not* present, the $T_k^g(z)$ for the rods-in function will be small and the $T_k^g(z)$ for the rods-out function will be large, while, for the range of z in which the rods *are* present, this amplitude behavior will be reversed. Thus we expect to be able to represent the $\Phi_g(\mathbf{r})$ for *any* location of the rod bank with just two expansion functions. If we used the *point*-synthesis method, it would take many three-dimensional expansion functions to achieve the same accuracy.

In order to apply the variational condition (11.3.21) directly, we must express the approximation (11.4.1) in matrix form and must in addition introduce trial functions for the three components of the neutron current. Accordingly we assume the following forms for the column vector $[U]$ and the three components of the column vector $[\mathbf{V}]$ to be used in (11.3.21):

$$\begin{aligned}
[U] &= \sum_{k=1}^{K} [\Psi_k(x, y)][T_k(z)] \\
[V_x] &= \sum_{k=1}^{K} [\xi_k(x, y)][X_k(z)], \\
[V_y] &= \sum_{k=1}^{K} [\eta_k(x, y)][Y_k(z)], \\
[V_z] &= \sum_{k=1}^{K} [\zeta_k(x, y)][Z_k(z)],
\end{aligned} \tag{11.4.2}$$

where the (precomputed) $G \times G$ diagonal matrices $[\Psi_k]$, $[\xi_k]$, $[\eta_k]$, and $[\zeta_k]$ (the "expansion functions") are continuous in x and y. Of the (unknown) G-element column vectors $[T_k]$, $[X_k]$, $[Y_k]$, and $[Z_k]$ (the "mixing functions") only $[T_k]$ and $[Z_k]$ need be continuous in z.

The two-dimensional flux expansion functions $\Psi_k^g(x, y)$ for group g are found by solving detailed two-dimensional problems having parameters and geometries suggested by intuition. The simplest procedure is to pick, for the two-dimensional problem, the geometry and material constituents appropriate to particular elevations (Z locations) in the full three-dimensional reactor. Provided there is fissionable material at every such elevation, this procedure is straightforward and can be made automatic in a computer program constructed to solve the overall problem. For reflector regions the $\Psi_k^g(x, y)$ are found by solving a fixed-source problem, the source of fast neutrons frequently being taken as the fission source for the adjacent core region.

The current expansion functions for $[V_x]$ and $[V_y]$ are usually found by applying Fick's Law, that is, by taking

$$[\xi_k(x, y)] = -[D_k(x, y)]\frac{\partial}{\partial x}[\Psi_k(x, y)],$$
$$[\eta_k(x, y)] = -[D_k(x, y)]\frac{\partial}{\partial y}[\Psi_k(x, y)], \tag{11.4.3}$$

where $[D_k(x, y)]$ is the diffusion-constant matrix used for the determination of $[\Psi_k]$. For the current expansion functions $[\zeta_k(x, y)]$ in the Z direction, the $[\Psi_k(x, y)]$ themselves are frequently used.

In practice all the expansion functions and the mixing functions will be obtained in finite-difference form. However the notational problems and the derivation of the equations for the mixing functions become even more complicated in the finite-difference approximation, and we shall postpone this additional complexity as long as possible. One consequence of taking the mixing functions as continuous rather than as step functions analogous to (11.3.23) will be that differential equations in z, rather than finite-difference equations, will result when the functional F_d is forced to be stationary with respect to first-order variations.

We define the weight functions analogous to (11.4.2) by

$$[U^*] = \sum_{k=1}^{K}[\Psi_k^*(x, y)][T_k^*(z)],$$
$$[V_x^*] = \sum_{k=1}^{K}[\xi_k^*(x, y)][X_k^*(z)],$$
$$[V_y^*] = \sum_{k=1}^{K}[\eta_k^*(x, y)][Y_k^*(z)], \tag{11.4.4}$$
$$[V_z^*] = \sum_{k=1}^{K}[\zeta_k^*(x, y)][Z_k^*(z)].$$

The fact that the functional F_d of (11.3.15) is made stationary when $[U^*]$ and $[\mathbf{V}^*]$ are the adjoint flux and current vectors suggests that we use for the expansion functions $[\Psi_k^*(x, y)]$ the solutions to the adjoints of the two-dimensional diffusion equations used to determine the $[\Psi_k(x, y)]$. However, because the solutions of two-dimensional diffusion-theory problems are moderately expensive to obtain, it is customary to take the weight functions $[\Psi_k^*(x, y)]$, $[\xi_k^*(x, y)]$, etc., to be the same as the corresponding trial functions $[\Psi_k(x, y)]$, $[\xi_k(x, y)]$, etc. Nevertheless we shall use independent weight functions in developing the formalism since the algebra involved is no more complicated than it is for the more restrictive Galerkin scheme.

For the sake of simplicity we shall derive only the equations that determine the $[T_k(z)]$, $[X_k(z)]$, $[Y_k(z)]$, and $[Z_k(z)]$. The equations for the corresponding starred quantities that determine the adjoint solutions can be derived in a completely analogous manner. Thus, to proceed, we set $\delta U = \delta \mathbf{V} = 0$ and write out (11.3.21) for the case at hand. Because of the choice of continuous functions, there is only one region R_i, and there are no surface terms. The result obtained is then

$$\begin{aligned} 0 = \iiint \Bigg\langle & \sum_{k=1}^{K} [\delta T_k^*(z)]^{\mathrm{T}} [\Psi_k^*(x, y)] \Bigg\{ \left[A - \frac{1}{\lambda} \chi \nu \Sigma_{\mathrm{f}}^{\mathrm{T}} \right] \sum_{l=1}^{K} [\Psi_l(x, y)][T_l(z)] \\ & + \sum_{l=1}^{K} \frac{d}{dx} [\xi_l(x, y)][X_l(z)] + \sum_{l=1}^{K} \frac{\partial}{\partial y} [\eta_l(x, y)][Y_l(z)] + \sum_{l=1}^{K} [\zeta_l(x, y)] \frac{\partial}{\partial z} [Z_l(z)] \Bigg\} \\ & - \sum_{k=1}^{K} [\delta X_k^*(z)]^{\mathrm{T}} [\xi_k^*(x, y)] \sum_{l=1}^{K} \frac{\partial}{\partial x} [\Psi_l(x, y)][T_l(z)] \\ & - \sum_{k=1}^{K} [\delta Y_k^*(z)]^{\mathrm{T}} [\eta_k^*(x, y)] \sum_{l=1}^{K} \frac{\partial}{\partial y} [\Psi_l(x, y)][T_l(z)] \\ & - \sum_{k=1}^{K} [\delta Z_k^*(z)]^{\mathrm{T}} [\zeta_k^*(x, y)] \sum_{l=1}^{K} [\Psi_l(x, y)] \frac{\partial}{\partial z} [T_l(z)] \\ & - \sum_{k=1}^{K} [\delta X_k^*(z)]^{\mathrm{T}} [\xi_k^*(x, y)][D^{-1}] \sum_{l=1}^{K} [\xi_l(x, y)][X_l(z)] \\ & - \sum_{k=1}^{K} [\delta Y_k^*]^{\mathrm{T}} [\eta_k^*(x, y)][D^{-1}] \sum_{l=1}^{K} [\eta_l(x, y)][Y_l(z)] \\ & - \sum_{k=1}^{K} [\delta Z_k^*(z)]^{\mathrm{T}} [\zeta_k^*(x, y)][D^{-1}] \sum_{l=1}^{K} [\zeta_l(x, y)][Z_l(z)] \Bigg\rangle \, dx\, dy\, dz. \end{aligned} \tag{11.4.5}$$

To simplify notation we define $G \times G$ matrices

$$\begin{aligned} [\mathscr{A}_{kl}(z)] &\equiv \iint [\Psi_k^*(x, y)][A(\mathbf{r})][\Psi_l(x, y)]\, dx\, dy, \\ [M_{kl}(z)] &\equiv \iint [\Psi_k^*(x, y)][\chi \nu \Sigma_{\mathrm{f}}^{\mathrm{T}}(\mathbf{r})][\Psi_l(x, y)]\, dx\, dy, \\ iB_x[\alpha_{kl}^{(x)}] &\equiv \iint [\Psi_k^*(x, y)] \frac{\partial}{\partial x} [\xi_l(x, y)]\, dx\, dy, \\ iB_y[\alpha_{kl}^{(y)}] &\equiv \iint [\Psi_k^*(x, y)] \frac{\partial}{\partial y} [\eta_l(x, y)]\, dx\, dy, \\ [\alpha_{kl}^{(z)}] &\equiv \iint [\Psi_k^*(x, y)][\zeta_l(x, y)]\, dx\, dy, \\ iB_x[\tilde{\alpha}_{kl}^{(x)}] &\equiv \iint [\xi_k^*(x, y)] \frac{\partial}{\partial x} [\Psi_l(x, y)]\, dx\, dy, \end{aligned} \tag{11.4.6}$$

$$iB_y[\tilde{\alpha}_{kl}^{(y)}] \equiv \iint [\eta_k^*(x, y)] \frac{\partial}{\partial y} [\Psi_l(x, y)]\, dx\, dy,$$

$$[\tilde{\alpha}_{kl}^{(z)}] \equiv \iint [\zeta_k^*(x, y)][\Psi_l(x, y)]\, dx\, dy,$$

$$[D_{x,kl}^{-1}(z)] \equiv \iint [\xi_k^*(x, y)][D^{-1}(\mathbf{r})][\xi_l(x, y)]\, dx\, dy,$$

$$[D_{y,kl}^{-1}(z)] \equiv \iint [\eta_k^*(x, y)][D^{-1}(\mathbf{r})][\eta_l(x, y)]\, dx\, dy,$$

$$[D_{z,kl}^{-1}(z)] \equiv \iint [\zeta_k^*(x, y)][D^{-1}(\mathbf{r})][\zeta_l(x, y)]\, dx\, dy,$$

where the transverse bucklings $B_x \equiv \pi/L_x$ and $B_y \equiv \pi/L_y$ (L_x and L_y being the X and Y dimensions of the reactor) have been arbitrarily and artificially introduced to induce a final matrix result very similar in form to the one-group, one-dimensional diffusion equations.

Since the two-dimensional expansion functions $[\Psi_k^*]$, $[\Psi_l]$, $[\xi_l]$, etc., are assumed to be known, all the matrices (11.4.6) (the elements of which are, as indicated, either numbers or functions of z) may be computed immediately.

The variations $[\delta T_k^*]$, $[\delta X_k^*]$, etc., being arbitrary and independent, (11.4.5) then implies that, for each k,

$$\begin{aligned}
&\sum_{l=1}^{K} \left\langle \left\{ [\mathscr{A}_{kl}(z)] - \frac{1}{\lambda}[M_{kl}(z)] \right\} [T_l(z)] + iB_x[\alpha_{kl}^{(x)}][X_l(z)] + iB_y[\alpha_{kl}^{(y)}][Y_l(z)] + [\alpha_{kl}^{(z)}]\frac{\partial}{\partial z}[Z_l(z)] \right\rangle = 0, \\
&\sum_{l=1}^{K} \{ iB_x[\tilde{\alpha}_{kl}^{(x)}][T_l(z)] + [D_{x,kl}^{-1}(z)][X_l(z)] \} = 0, \\
&\sum_{l=1}^{K} \{ iB_y[\tilde{\alpha}_{kl}^{(y)}][T_l(z)] + [D_{y,kl}^{-1}][Y_l(z)] \} = 0, \\
&\sum_{l=1}^{K} \left\{ [\tilde{\alpha}_{kl}^{(z)}]\frac{d}{dz}[T_l(z)] + [D_{z,kl}^{-1}(z)][Z_l(z)] \right\} = 0.
\end{aligned} \tag{11.4.7}$$

These coupled, first-order, one-dimensional, differential equations—there are $4K \times G$ of them—may now be solved for the critical eigenvalue λ and the mixing coefficients. Thus an intractable three-dimensional set of diffusion equations has been reduced by the approximation to a series of two-dimensional diffusion-theory equations (to get expansion and weight functions) plus a set of coupled one-dimensional equations for the mixing coefficients.

This latter set can be written in terms of $K \times K$ "supermatrices" $[A]$, $[M]$, $iB_x[\alpha^{(x)}]$, $iB_y[\alpha^{(y)}]$, etc., the klth elements of which are themselves the $G \times G$ matrices $[A_{kl}]$, $[M_{kl}]$, $iB_x[\alpha_{kl}^{(x)}]$, $iB_y[\alpha_{kl}^{(y)}]$, etc., defined by (11.4.6). Equations for the corresponding K-element "supervectors" $[T(z)]$, $[X(z)]$, $[Y(z)]$, and $[Z(z)]$, elements of which are themselves the

G-element vectors $[T_l(z)]$, $[X_l(z)]$, $[Y_l(z)]$, and $[Z_l(z)]$, can then be found by rewriting (11.4.7) as

$$\begin{aligned}
&\left\{[\mathscr{A}(z)] - \frac{1}{\lambda}[M(z)]\right\}[T(z)] + iB_x[\alpha^{(x)}][X(z)] + iB_y[\alpha^{(y)}][Y(z)] + [\alpha^{(z)}]\frac{d}{dz}[Z(z)] = 0,\\
&iB_x[\tilde{\alpha}^{(x)}][T(z)] + [D_x^{-1}(z)][X(z)] = 0,\\
&iB_y[\tilde{\alpha}^{(y)}][T(z)] + [D_y^{-1}(z)][Y(z)] = 0,\\
&[\tilde{\alpha}^{(z)}]\frac{d}{dz}[T(z)] + [D_z^{-1}(z)][Z(z)] = 0.
\end{aligned} \tag{11.4.8}$$

To simplify results further we assume that the inverses $[D_u(z)] \equiv [D_u^{-1}(z)]^{-1}$ $(u = x, y, z)$ exist, and we define

$$[\mathscr{D}_u(z)] \equiv [\alpha^{(u)}][D_u(z)][\tilde{\alpha}^{(u)}] \qquad (u = x, y, z). \tag{11.4.9}$$

Then the last three equation in (11.4.8) can be used to eliminate $[X(z)]$, $[Y(z)]$, and $[Z(z)]$ from the first equation, and we are left with a second-order equation for the supermatrix $[T(z)]$:

$$-\frac{d}{dz}[\mathscr{D}_z(z)]\frac{d}{dz}[T(z)] + \left\{[\mathscr{A}(z)] + B_x^2[\mathscr{D}_x(z)] + B_y^2[\mathscr{D}_y(z)] - \frac{1}{\lambda}[M(z)]\right\}[T(z)] = 0. \tag{11.4.10}$$

The similarity between the form of this matrix equation and the one-group, one-dimensional diffusion equation is evident. (Note also that (11.4.10) has the same form as (11.2.12), the equation that yields the overlapping-group fluxes for the one-dimensional case.) Moreover, since all the elements of the matrices $[T(z)]$ and $[Z(z)]$ are to be continuous in z and the elements of $[\alpha^{(z)}]$ and $[\tilde{\alpha}^{(z)}]$ are numbers, the elements of the column vector $[\alpha^{(z)}]$ $[Z(z)]$ must also be continuous in z. The last equation in (11.4.8) and the definition (11.4.9) then show that $[\mathscr{D}_z(z)]d[T(z)]/dz$ $(= -[\alpha^{(z)}][Z(z)])$ must be continuous in z—even though $[\mathscr{D}_z(z)]$ (because of step changes in material properties) may be discontinuous. Thus the internal continuity conditions imposed on $[T(z)]$ are also direct matrix generalizations of those imposed on the one-group flux.

It follows that a straightforward matrix generalization of the finite-difference procedures for solving the one-group, one-dimensional diffusion equation will yield a solution of (11.4.10). Actually, since the matrices $[\mathscr{D}_z(z)]$, $[\mathscr{A}(z)]$, etc., are $KG \times KG$ (where K is the number of expansion functions per energy group and G is the number of groups) and since the most straightforward solution procedure requires that some of these matrices be inverted, finding $[T(z)]$ by finite-difference methods is not a trivial problem. For $KG \lesssim 30$, however, it is quite tractable. As a result, if one is willing to make a substantial investment in programming effort and in time to gain experience with selecting suitable expansion functions, continuous space-synthesis methods permit the determination of group fluxes at literally millions of spatial mesh points in very acceptable running times. The accuracy of the method depends on the nature of the problem and the expansion functions. Numerical tests indicate that synthesized fluxes agree with three-dimensional finite-difference solutions to within a few percent, except in regions such as reflectors where the fluxes are very low; there, errors are found to be as great as 10 percent.

One of the major shortcomings of the synthesis method is the fact that no systematic a priori procedure for predicting the magnitude of errors is available. In particular there is no guarantee that increasing the number of expansion functions will improve the accuracy of the final result. This deficiency has resulted in a reluctance to use the method for problems in ranges where the scheme has not been validated by comparison with full three-dimensional, finite-difference computations.

Discontinuous Space-Dependent Synthesis

As has been mentioned, the solution of (11.4.10), while straightforward, is not a trivial problem for a large number of groups and expansion functions. The column vector $[T(z)]$ has $G \times K$ scalar elements in it, and the matrices $[\mathscr{D}(z)]$, $[\mathscr{A}(z)]$, and $[\mathscr{M}(z)]$ are full, having $(G \times K)^2$ scalar elements. Thus, for four energy groups and ten flux expansion matrices $[\Psi_k(x, y)]$, (11.4.10) is a coupled set of forty second-order differential equations.

There is a fairly obvious way to reduce the number of equations without seriously decreasing the accuracy of the scheme. One simply notes that, at a given elevation z, the most important expansion functions will be those characteristic of the radial planes close to z. The coefficients $[T_k(z)]$, $[X_k(z)]$, $[Y_k(z)]$, and $[Z_k(z)]$ of other expansion functions are expected to be small at such locations. At the z location corresponding to the midplane of the reactor, for example, we do not expect that the expansion functions appropriate to the top and bottom reflectors of the core will form a significant part of the sums (11.4.2) and (11.4.4). Accordingly we may decrease the number of terms in (11.4.10) by including at any particular elevation only those $[\Psi_k(x, y)]$ $[\xi_k(x, y)]$, $[\eta_k(x, y)]$, and $[\zeta_k(x, y)]$ that correspond to radial slices *near* that elevation. Thus, as we proceed up the Z axis, we alter the expansions (11.4.2) and (11.4.4) at certain elevations by dropping the $[\Psi_k(x, y)]$, $[\xi_k(x, y)]$, etc., corresponding to the *lowest* radial slice then being accounted for in the sums (11.4.2) and (11.4.4) and adding the $[\Psi_k(x, y)]$, $[\xi_k(x, y)]$, etc., corresponding to the *highest* radial slice not previously accounted for in those sums. For example, by retaining three expansion functions $[\Psi_k(x, y)]$ at any given elevation but switching to new functions at seven different axial elevations, we obtain an approximation (11.4.2) in which there are three expansion functions, but these are capable of representing the flux vector almost as well as an expansion with $K = 10$ functions present at all elevations. At the same time, for a four-group model, the number of scalar equations corresponding to (11.4.10) decreases from forty to twelve. The time to obtain a solution would be expected to decrease by about a factor of ten.

This type of synthesis is called a *spatially discontinuous flux synthesis.* The composite solution $[\Phi(\mathbf{r})]$ will now be discontinuous in the Z direction across the radial planes where one expansion function is dropped and another added. However, since both the dropped and the added functions account for only a few percent of the total sum at such locations, this discontinuity will be small. In fact it is usually not even discernible in a mesh-point plot of $[\Phi(\mathbf{r})]$ versus z at fixed xy.

For the case of axially discontinuous trial and weight functions the procedure of defining expressions analogous to (11.4.2)–(11.4.4) and of then determining the equations for the mixing functions that result when, with such expressions inserted, the functional F_{d} of (11.3.15) is required to be stationary, is in principle straightforward. The details of carrying out the procedure are, however, extremely complex. To reduce this algebraic detail we shall restrict attention to the case of cylindrical geometry and develop a synthesis procedure for finding $[\Phi(r, z)]$ in terms of predetermined radial functions blended together by z-dependent mixing functions.

The major problem is to define the trial and weight functions $[U(r, z)]$, $[\mathbf{V}(r, z)]$, $[U^*(r, z)]$, and $[\mathbf{V}^*(r, z)]$ to be used in the variational expression (11.3.21) in such a way that the mathematical difficulties which result when both $[U(r, z)]$ and $[\mathbf{V}(r, z)]$ are discontinuous at the same location are avoided. There is a method for doing this which is particularly attractive since it leads to finite-difference equations, rather than differential equations, for the mixing functions. The method consists of letting the mixing functions for the flux vector $[U(r, z)]$ and those for the Z component of the current vector $[V_z(r, z)]$ be overlapping step functions analogous to those we chose in (11.3.23) to derive the conventional finite-difference equation (11.3.27).

Accordingly, for the G-element column vector approximating the scalar flux, we write

$$[U(r, z)] = \sum_{k=1}^{K} \sum_{n=1}^{N-1} [\Psi_k^n(r)][T_k^n(z)], \tag{11.4.11}$$

where the $[\Psi_k^n(r)]$ are $G \times G$ diagonal matrices of r-dependent expansion functions. There are K linearly independent expansion matrices for every axial mesh interval $h_n = z_{n+1} - z_n$. The superscript n on the $[\Psi_k^n]$ indicates that each $[\Psi_k^n]$ can be replaced by another radial function at every mesh interface z_n. This general notation is convenient, although in practice the $[\Psi_k^n]$ will be replaced at only a few mesh interfaces, and, moreover, only the matrix corresponding to a single value of k will be replaced. (That is, at particular locations along the Z axis we shall replace one of the radial functions appropriate to a Z location now far away by one appropriate to a closer Z location.)

The $[T_k^n(z)]$ in (11.4.11) are G-element column vectors of the undetermined constants T_k^{gn}. They are regarded as functions of z in that they are restricted to nonzero values over an interval surrounding the mesh point z_n. That is,

$$T_k^{gn}(z) = \begin{cases} T_k^{gn} \text{ for } \left(z_n - \dfrac{h_{n-1}}{2}\right) \leq z \leq \left(z_n + \dfrac{h_n}{2}\right) & (k = 1, 2, \ldots, K;\ g = 1, 2, \ldots, G), \\ 0 \quad \text{for all other } z. \end{cases} \tag{11.4.12}$$

Thus, when we find the $[T_k^n]$ and reconstruct $[U(r, z)]$ according to (11.4.11), the result will have the shape of a series of steps in the Z direction. In this respect it will look like a conventional finite-difference solution in the Z direction.

Strictly speaking we should also represent the $[\Psi_k^n(\mathbf{r})]$ as step functions in $\mathbf{r}$ since they will usually be determined by finite-difference methods when they are precomputed. However nothing essential is lost and some algebraic complexity is avoided if we continue to treat the $[\Psi_k^n(\mathbf{r})]$ as continuous.

For the radial component $[V_r(r, z)]$ of the net current vector $[\mathbf{V}(r, z)]$ we use the trial function

$$[V_r(r, z)] = \sum_{k=1}^{K} \sum_{n=1}^{N-1} [\rho_k^n(r)][R_k^n(z)], \tag{11.4.13}$$

where the $[\rho_k^n(r)]$ are $G \times G$ diagonal matrices of continuous radial-current expansion functions (generally taken to be $-D_k^n(r)d[\Psi_k^n(r)]/dr$, where $D_k^n(r)$ is the diffusion constant used to determine $\Psi_k^n(\mathbf{r})$) and the $[R_k^n(z)]$ are defined by

$$[R_k^n(z)] = \begin{cases} [R_k^n] \text{ for } \left(z_n - \dfrac{h_{n-1}}{2}\right) \leq z \leq \left(z_n + \dfrac{h_n}{2}\right) & (k = 1, 2, \ldots, K), \\ 0 \quad \text{for all other } z, \end{cases} \tag{11.4.14}$$

where the $[R_k^n]$ are G-element column vectors of numbers to be found by solving the synthesis equations.

The trial function for the axial component $[V_z(r, z)]$ of the net current vector is expressed as

$$[V_z(r, z)] = \sum_{k=1}^{K} \sum_{n=1}^{N-1} [\zeta_k^n(r)][Z_k^n(z)], \tag{11.4.15}$$

where the $[\zeta_k^n(r)]$ are $G \times G$ diagonal matrices of axial-current functions (generally taken equal to the flux functions $[\Psi_k^n(r)]$) and the $[Z_k^n(z)]$ are defined by

$$[Z_k^n(z)] = \begin{cases} [Z_k^n] \text{ for } z_n \leq z \leq (z_n + h_n) & (k = 1, 2, \ldots, K), \\ 0 \quad \text{for all other } z. \end{cases} \tag{11.4.16}$$

Notice that, to avoid mathematical difficulties, we have defined the flux trial function $[U(r, z)]$ and the axial-current trial function $[V_z(r, z)]$ so that they are never both discontinuous across the same surface ($[U]$ is discontinuous at points $z_n + \frac{1}{2}h_n$, and $[V_z]$ is discontinuous at points z_n). No problems arise if the *radial* component of the current vector and the flux vectors are both discontinuous at the same *axial* planes. (Hence the similarity between (11.4.12) and (11.4.14).)

Notice also that the flux and radial-current vectors are taken to be zero about the end points z_0 and z_N ($[T_k^0] = [T_k^N] = [R_k^0] = [R_k^N] = 0$), whereas the *axial*-current vector is finite *within* the first and last mesh intervals h_0 and h_{N-1} ($[Z_k^0] \neq 0$; $[Z_k^{N-1}] \neq 0$).

The weight functions $[U^*(r, z)]$ and $[\mathbf{V}^*(r, z)]$ are defined in analogy with (11.4.11)–(11.4.16) by

$$
\begin{aligned}
[U^*(r, z)] &= \sum_{k=1}^{K} \sum_{n=1}^{N-1} [\Psi_k^{n*}(r)][T_k^{n*}(z)],\\
[T_k^{n*}(z)] &\equiv \begin{cases} [T_k^{n*}] \text{ for } \left(z_n - \dfrac{h_{n-1}}{2}\right) \le z_n \le \left(z_n + \dfrac{h_n}{2}\right) & (k = 1,2, \ldots, K),\\ 0 \quad \text{for all other } z, \end{cases}\\
[V_r^*(r, z)] &= \sum_{k=1}^{K} [\rho_k^{n*}(r)][R_k^{n*}(z)],\\
[R_k^{n*}(z)] &\equiv \begin{cases} [R_k^{n*}] \text{ for } \left(z_n - \dfrac{h_{n-1}}{2}\right) \le z \le \left(z_n + \dfrac{h_n}{2}\right) & (k = 1, 2, \ldots, K),\\ 0 \quad \text{for all other } z, \end{cases}\\
[V_z^*(r, z)] &= \sum_{k=1}^{K} \sum_{n=0}^{N-1} [\xi_k^{n*}(r)][Z_k^{n*}(z)],\\
[Z_k^{n*}(z)] &\equiv \begin{cases} [Z_k^{n*}] \text{ for } z_n \le z \le (z_n + h_n) & (k = 1, 2, \ldots, K),\\ 0 \quad \text{for all other } z. \end{cases}
\end{aligned}
\tag{11.4.17}
$$

Once the trial and weight functions have been selected, the application of the variational principle is a purely mechanical matter. One merely requires that (11.3.21) (specifying that the first-order variation of the functional F_d is to vanish) be valid for completely arbitrary $[\delta U^*]$, $[\delta \mathbf{V}^*]$, $[\delta U]$, and $[\delta \mathbf{V}]$. However carrying out this mechanical procedure properly requires extreme care. We shall sketch only how the regular (i.e., nonadjoint) equations are found. Thus, with $[\delta \mathbf{V}]$ and $[\delta U]$ taken as zero, we substitute (11.4.11)–(11.4.17) into (11.3.21). In so doing it is important to keep in mind the following facts:

1. There may be discontinuities in the trial and weight functions at any z_n and $z_n + \frac{1}{2}h_n$.
2. Any integral containing a product such as $[\delta U_k^{n*}]^{\mathrm{T}} \cdots [U_l^m]$ or $[\delta U_k^{n*}]^{\mathrm{T}} \cdots [R_l^m]$ will vanish for $n \neq m$.

3. Since the normals to surfaces of discontinuity are all parallel to the Z axis, the surface terms in (11.3.21) involving the $[V_r]$ part of $[\mathbf{V}]$ and the $[\delta V_r^*]$ part of $[\delta \mathbf{V}^*]$ will vanish even though $[V_r]$ and $[\delta V_r^*]$ are discontinuous across such surfaces.

4. Within the volume integrals where the trial functions are continuous, $\partial[V_z(r, z)]/\partial z = \partial[U(r, z)]/\partial z = 0$.

With these conditions imposed, (11.3.21) for the case at hand becomes

$$\begin{aligned}0 = \int 2\pi r\, dr \sum_{k=1}^{K} \Bigg(\sum_{n=1}^{N-1} [\delta T_k^{n*}]^{\mathrm{T}} \int_{z_n - \frac{1}{2}h_{n-1}}^{z_n + \frac{1}{2}h_n} [\Psi_k^{n*}(r)] \Bigg\{ \Bigg[A(r, z) - \frac{1}{\lambda}\chi\nu\Sigma_{\mathrm{f}}^{\mathrm{T}}(r, z) \Bigg] \sum_{l=1}^{K} [\Psi_l^n(r)][T_l^n] \\ + \sum_{l=1}^{K} \frac{d}{dr}[\rho_l^n(r)][R_l^n] \Bigg\} - \sum_{n=1}^{N-1} [\delta R_k^{n*}]^{\mathrm{T}} \int_{z_n - \frac{1}{2}h_{n-1}}^{z_n + \frac{1}{2}h_n} [\rho_k^{n*}(r)] \Bigg\{ \sum_{l=1}^{K} \frac{\partial}{\partial r}[\Psi_l^n(r)][T_l^n] \\ + [D^{-1}(r, z)][\rho_l^n(r)][R_l^n] \Bigg\} - \sum_{n=0}^{N-1} [\delta Z_k^{n*}]^{\mathrm{T}} \int_{z_n}^{z_n + h_n} [\zeta_k^{n*}(r)] \sum_{l=1}^{K} [D^{-1}(r, z)][\zeta_l^n(r)][Z_l^n] \Bigg) \\ + \int 2\pi r\, dr \sum_{k=1}^{K} \Bigg(\sum_{n=1}^{N-1} [\delta T_k^{n*}]^{\mathrm{T}} [\Psi_k^{n*}(r)] \sum_{l=1}^{K} \{[\zeta_l^n(r)][Z_l^n] - [\zeta_l^{n-1}(r)][Z_l^{n-1}]\} \\ - \sum_{n=0}^{N-1} [\delta Z_k^{n*}]^{\mathrm{T}} [\zeta_k^{n*}(r)] \sum_{l=1}^{K} \{[\Psi_l^{n+1}(r)][T_l^{n+1}] - [\Psi_l^n(r)][T_l^n]\} \Bigg).\end{aligned} \tag{11.4.18}$$

As in the case of spatially continuous flux synthesis, introducing special notation greatly simplifies the final form of the synthesis equations. Accordingly we define a set of $G \times G$ matrices as follows:

$$\begin{aligned}[\mathscr{A}_{kl}^n] &\equiv \int_0^R 2\pi r\, dr \int_{z_n - \frac{1}{2}h_{n-1}}^{z_n + \frac{1}{2}h_n} [\Psi_k^{n*}(r)][A(r, z)][\Psi_l^n(r)], \\ [\mathscr{M}_{kl}^n] &\equiv \int_0^R 2\pi r\, dr \int_{z_n - \frac{1}{2}h_{n-1}}^{z_n + \frac{1}{2}h_n} [\Psi_k^{n*}(r)][\chi\nu\Sigma_{\mathrm{f}}^{\mathrm{T}}(r, z)][\Psi_l^n(r)], \\ iB_r[\alpha_{kl}^n] &\equiv \frac{1}{2}(h_{n-1} + h_n) \int_0^R 2\pi r\, dr [\Psi_k^{n*}(r)] \frac{d}{dr}[\rho_l^n(r)], \\ iB_r[\tilde{\alpha}_{kl}^n] &\equiv \frac{1}{2}(h_{n-1} + h_n) \int_0^R 2\pi r\, dr [\rho_k^{n*}(r)] \frac{d}{dr}[\Psi_l^n(r)], \\ [D_{r,kl}^n]^{-1} &\equiv \int_0^R 2\pi r\, dr \int_{z_n - \frac{1}{2}h_{n-1}}^{z_n + \frac{1}{2}h_n} [\rho_k^{n*}(r)][D^{-1}(r, z)][\rho_l^n(r)], \\ [D_{z,kl}^n]^{-1} &\equiv \int_0^R 2\pi r\, dr \int_{z_n}^{z_n + h_n} [\zeta_k^{n*}(r)][D^{-1}(r, z)][\zeta_l^n(r)], \\ [\gamma_{kl}^{nm}] &\equiv \int_0^R 2\pi r\, dr [\Psi_k^{n*}(r)][\zeta_l^m(r)] \qquad (m = n, n-1), \\ [\tilde{\gamma}_{kl}^{nm}] &\equiv \int_0^R 2\pi r\, dr [\zeta_k^{n*}(r)][\Psi_l^m(r)] \qquad (m = n, n+1), \\ iB_r &\equiv \sqrt{(-1)}\left(\frac{2.403}{R}\right),\end{aligned} \tag{11.4.19}$$

where, again, the number B_r is introduced arbitrarily so that the final result will have a familiar form. (The fact should, therefore, not be taken to imply that material conditions are homogeneous in the radial direction.)

Using these definitions and requiring that (11.4.18) be valid for arbitrary variations of each element of the $[\delta T_k^{n*}]$, $[\delta R_k^{n*}]$, and $[\delta Z_k^{n*}]$ yields

$$\begin{gathered}\sum_{l=1}^{K}\left\{\left[\mathscr{A}_{kl}^{n}-\frac{1}{\lambda}\mathscr{M}_{kl}^{n}\right][T_l^n]+iB_r[\alpha_{kl}^n][R_l^n]+[\gamma_{kl}^{nn}][Z_l^n]-[\gamma_{kl}^{n,n-1}][Z_l^{n-1}]\right\}=0,\\ \sum_{l=1}^{K}\{iB_r[\tilde{\alpha}_{kl}^n][T_l^n]+[D_{r,kl}^n]^{-1}[R_l^n]\}=0\qquad(k=1,2,\ldots,K),\\ \sum_{l=1}^{K}\{[D_{z,kl}^n]^{-1}[Z_l^n]+[\tilde{\gamma}_{kl}^{n,n+1}][T_l^{n+1}]-[\tilde{\gamma}_{kl}^{nn}][T_l^n]\}=0.\end{gathered}\tag{11.4.20}$$

Defining $K\times K$ supermatrices $[\mathscr{A}^n]$, $[\mathscr{M}^n]$, $[\alpha^n]$, etc., having the $G\times G$ matrices $[\mathscr{A}_{kl}^n]$, $[\mathscr{M}_{kl}^n]$, $[\alpha_{kl}^n]$, etc., as their klth elements permits (11.4.20) to be rewritten as

$$\begin{gathered}\left[\mathscr{A}^{n}-\frac{1}{\lambda}\mathscr{M}^{n}\right][T^n]+iB_r[\alpha^n][R^n]+[\gamma^{nn}][Z^n]-[\gamma^{n,n-1}][Z^{n-1}]=0,\\ iB_r[\tilde{\alpha}^n][T^n]+[D_r^n]^{-1}[R^n]=0,\\ [D_z^n]^{-1}[Z^n]+[\tilde{\gamma}^{n,n+1}][T^{n+1}]-[\tilde{\gamma}^{nn}][T^n]=0,\end{gathered}\tag{11.4.21}$$

where $[T^n]$, $[Z^n]$, and $[R^n]$ are K-element column vectors the lth elements of which are the G-element column vectors $[T_l^n]$, $[Z_l^n]$, and $[R_l^n]$. Thus each of the three matrix equations in (11.4.21) is equivalent to $G\times K$ scalar equations.

If, finally, the last two equations in (11.4.21) are used to eliminate $[R^n]$ and $[Z^n]$ from the first equation, we obtain the basic difference equation of the spatially discontinuous synthesis approximation:

$$\begin{aligned}&-[\gamma^{nn}][D_z^n][\tilde{\gamma}^{n,n+1}][T^{n+1}]+\{[\gamma^{nn}][D_z^n][\tilde{\gamma}^{nn}]+[\gamma^{n,n-1}][D_z^{n-1}][\tilde{\gamma}^{n-1,n}]\}[T^n]\\ &-[\gamma^{n,n-1}][D_z^{n-1}][\tilde{\gamma}^{n-1,n-1}][T^{n-1}]+\left\{\left[\mathscr{A}^{n}-\frac{1}{\lambda}\mathscr{M}^{n}\right]+B_r^2[\alpha^n][D_r^n][\tilde{\alpha}^n]\right\}[T^n]=0.\end{aligned}\tag{11.4.22}$$

Notice the similarity in structure between this result and the one-dimensional finite-difference diffusion equation (11.3.27). Also note the similarity to the continuous synthesis result (11.4.10). If the trial and weight functions that determine the $[\gamma^{nm}]$ and $[\tilde{\gamma}^{nm}]$ were not discontinuous, $[\Psi_k^{n*}]$, $[\Psi_k^n]$, $[\rho_k^{n*}]$, etc., would all be independent of n, and so would $[\alpha^n]$, $[\tilde{\alpha}^n]$, $[\gamma^{nm}]$, and $[\tilde{\gamma}^{nm}]$. Diffusion-constant matrices $[D_u^n]$ analogous to (11.4.9) could then be defined, and (11.4.22) would be even closer in form to (11.4.10). (In fact it would be identical in form with the finite-difference approximation to (11.4.10).)

Thus the discontinuous synthesis method leads to difference equations only slightly more complicated than the corresponding continuous synthesis equations. The possibility of using fewer expansion functions with the discontinuous method (K may be three instead of ten) can thus lead to a real savings in computer running time.

Other Diffusion-Theory Synthesis Approximations

The basic procedure for constructing synthesis approximations should now be clear. Given the defining equations for the physical system and a variational functional that is defined for piecewise-continuous functions and made stationary by solutions to the defining equations and their adjoints, we select, on intuitive physical grounds, a space of trial functions (like (11.2.2), (11.2.24), (11.3.12), (11.3.23), (11.4.2), or (11.4.11)–(11.4.16)) such that the true solution of the problem will be closely approximated by some element in that space. In addition we select a space of weight functions such that the solution to the adjoint equation can be well approximated by some element in *that* space. Once the space of trial and weight functions has been chosen, finding the synthesis equations (the solutions of which will yield the desired synthesized flux and adjoint flux) is mechanical. The desired flux equations appear when the first-order variation of the functional is required to vanish for arbitrary variations in the weight functions; the desired adjoint-flux equations appear when it is required to vanish for arbitrary variations in the trial functions.

All the physics of the approximation is tied up with selecting the space of trial and weight functions, a characteristic which is both the strength and weakness of the method. Thus, if one has accurate prior knowledge concerning the nature of the true solution, a limited space of trial functions leading to very simple synthesis equations can be chosen with confidence that some element in that space will closely approximate the true solution. On the other hand, if one has very little knowledge of the nature of the true solution, *too* limited a space of trial functions may be chosen with the result that the synthesized solution will be seriously in error, or, if to avoid this potential difficulty one choses a very general space of trial functions, the synthesis equations may be as difficult to solve as the full set of finite-difference equations.

Attempts to find, for a given class of problems, trial functions providing maximum accuracy at minimum computational cost have led to the examination of a large number of trial-function spaces from which solutions to the group diffusion equations might be selected. We shall mention only two additional spaces, the first more general and the second less general than that defined by (11.4.2). The results of using vectors from these spaces in the variational expression (11.3.21) will thus be variants of the spatially continuous synthesis method. It will be sufficient in outlining them to describe alterations in the trial-function space only in terms of the flux expansion matrices $[\Psi_k(x, y)]$, it being understood that the current expansion matrices (11.4.3) should be altered in a similar fashion.

The first variant of the spatially continuous synthesis method to be sketched is obtained by permitting the $[\Psi_k(x, y)]$ to be discontinuous in the radial plane. For example the $[\Psi_k(x, y)]$ might be partitioned into four pieces, each covering only a quadrant of the core. Each piece would then have its own $[T_k(z)]$, which might be designated $[T_k(z)]_{(xy)}$. Equations coupling all these $[T_k(z)]$ together can be derived from the variational principle (11.3.21). This method is called *multichannel synthesis*. It permits a decrease in the number of $[\Psi_k(x, y)]$ needed and thus saves computing costs. However, since the $[\Psi_k(x, y)]$ are divided into pieces, each piece having its own mixing function, the number of unknowns in (11.4.10) is generally not decreased.

The discontinuities in flux shape in the XY plane that result when the multichannel method is applied can, for some situations, be much larger than those in the Z direction associated with the axially discontinuous synthesis method. Thus, unless there are very many channels (in which case the multichannel-synthesis scheme approaches the finite-difference method), the multichannel method as we have outlined it is intuitively less appealing than the axially discontinuous scheme.

A second variant of the continuous synthesis method is to reduce the number of unknowns in $[T_k(z)]$ by making the "collapsed-group approximation" in which the scalars $T_k^g(z)$ for a given k are assumed to be the same for all groups g. Thus the group-to-group flux ratios $\Psi_k^1(x, y): \Psi_k^2(x, y): \ldots : \Psi_k^G(x, y)$ are assumed to be fixed, so that, with $[\Psi_k(x, y)]$ taken as the column vector $\text{Col}\{\Psi_k^1(x, y), \Psi_k^2(x, y), \ldots, \Psi_k^G(x, y)\}$, the scalar flux is represented as a sum of K such column vectors. As discussed for the point-synthesis case, it is important, if the collapsed-group procedure is used, to normalize the weight functions so that they have a group-to-group ratio approximating that of the adjoint flux.

The collapsed-group method decreases by a factor G the number of unknowns in the supervector $[T(z)]$ of (11.4.10). (Each of the K, G-element column vectors $[T_k(z)]$ making up $[T(z)]$ is replaced by a scalar.) Thus, if it is used in conjunction with the discontinuous method, the equations analogous to (11.4.10), which might, by the discontinuous approximation, be reduced from having forty unknowns to having twelve, would be further reduced to equations having only three unknowns.

Unfortunately the collapsed-group scheme is not always acceptably accurate, and users frequently feel that the gain (reduction by a factor of G in the number of unknowns) is not worth the risk of a possible loss in accuracy.

11.5 The Finite-Element Method

To date no systematic procedure has been presented for insuring that the errors associated with synthesis approximations are acceptably small. As a result proponents of such methods must devise empirical tests (generally comparisons with limited-scale, three-dimensional, finite-difference solutions) to establish confidence in their schemes. Those who accept the investment in validation efforts do so because synthesis procedures provide the only practical means for determining solutions of the group diffusion equations when a great many significantly different material compositions are to be represented explicitly.

It would be worrisome if there were no other way of analyzing large, geometrically complex reactors. We saw, however, in Chapter 10 that there are methods (unfortunately rather difficult to justify) for defining equivalent homogenized diffusion-theory parameters. Thus the need to use millions of mesh points in order to represent geometrical detail disappears, and, in fact, if equivalent homogenized parameters can be found to represent entire fuel subassemblies, spatial mesh points 15 or 20 centimeters apart are sufficient to represent the geometrical properties of the resultant "homogenized reactor." (For example, if homogenized constants are known, the geometrical location of a cluster of four subassemblies of the sort shown in Figure 10.1 is specified by just the four extreme-corner mesh points.) Mesh intervals of this size are too large to be used as the intervals in the conventional finite-difference method of (4.11.13) or (11.3.27) for solving the group diffusion equations. Thus the numerical problem here is how to take advantage of the fact that very few (rather than very many) mesh points are needed to describe the geometry.

A class of approximation procedures called *finite-element methods* are particularly well suited for problems of this type. An essential characteristic of finite-element methods is the representation of the function to be determined by a sum of polynomials in its arguments, each polynomial in the sum being defined over only limited ranges of the arguments. For example a reactor may be partitioned into a number of rectangular parallelopiped blocks and the group flux $\Phi_g(x, y, z)$ within the kth such block may be approximated by

$$u_k^g(x, y, z) = \begin{cases} a_{1k}^g + a_{2k}^g x + a_{3k}^g y + a_{4k}^g z + a_{5k}^g x^2 + a_{6k}^g y^2 + a_{7k}^g z^2 + a_{8k}^g xy \\ \quad + a_{9k}^g xz + a_{10k}^g yz \quad \text{for } (x, y, z) \text{ in region } k, \\ 0 \quad \text{for } (x, y, z) \text{ not in region } k. \end{cases} \tag{11.5.1}$$

The overall group flux then has the form

$$\Phi_g(x, y, z) = \sum_k u_k^g(x, y, z). \tag{11.5.2}$$

If the material within the block is homogeneous, or if homogenized diffusion-theory parameters are available, one expects $\Phi_g(x, y, z)$ to be sufficiently smooth that the piecewise-polynomial expansion (11.5.2) should be able to represent it fairly accurately even if the block is, say, 20 cm on a side. Moreover the number of unknown coefficients in (11.5.1) is only 10, whereas, if a finite-difference approximation with, say, 4 cm mesh spacings were used to approximate $\Phi_g(x, y, z)$, the number of unknowns associated with the block would be 125. One thus expects that the finite-element determinations of $\Phi_g(x, y, z)$ ought to be cheaper. Moreover the finite-element procedure will yield a value of $\Phi_g(x, y, z)$ at *all* points (x, y, z) within the block—not just at the 125 mesh points.

We shall apply a variational method to determine the unknown coefficients of the polynomials used in the finite-element method. Since the only applications will be to obtain solutions for the group diffusion equations, we could use the functional F_d defined by (11.3.15) directly for this purpose. However it will be considerably simpler and quite adequate for the purpose at hand to derive from (11.3.15) a less general functional restricted to current functions $[\mathbf{V}]$ and $[\mathbf{V}^*]$ that obey Fick's Law and to flux functions $[U]$ and $[U^*]$ that are everywhere continuous. Once this functional is defined, we shall indicate how it can be applied to determine the equations characterizing three typical finite-element solutions for the group diffusion equations. These examples are intended primarily to give the flavor of the method. We shall not discuss analogous applications of the finite-element approach to the solution of transport-theory problems, spectrum problems, or kinetics problems, nor shall we cover the very extensive history of the use of the scheme in the field of stress analysis.

A Second Functional Made Stationary by the Solution of the Group Diffusion Equations

The functional $F_d(U, U^*, \mathbf{V}, \mathbf{V}^*)$ defined by (11.3.15) is made stationary by the solutions to the P_1 group equations and their adjoints. We could thus obtain equations to be obeyed by the polynomial coefficients a_{ik}^g of (11.5.1) by writing

$$[U(x, y, z)] = \sum_k [u_k(x, y, z)], \tag{11.5.3}$$

where the gth element of the column vector $[u_k]$ is the function u_k^g defined by (11.5.1). With analogous definitions for $[U^*]$, $[\mathbf{V}]$, and $[\mathbf{V}^*]$, setting to zero the first-order variation of F_d would yield equations relating the a_{ik}^g to corresponding coefficients associated with $[U^*]$, $[\mathbf{V}]$, and $[\mathbf{V}^*]$. The algebraic complexity of such a procedure is, however, formidable. We can avoid it by restricting the space of functions over which F_d is defined to one in which $[U]$ and $[U^*]$ are continuous and $[\mathbf{V}]$ and $[\mathbf{V}^*]$ are related to $[U]$ and $[U^*]$ by Fick's Law; that is, $[\mathbf{V}]$ and $[\mathbf{V}^*]$ are not allowed to be independent functions but are instead specified by

$$\begin{aligned} [\mathbf{V}(\mathbf{r})] &= -[D(\mathbf{r})]\nabla[U(\mathbf{r})], \\ [\mathbf{V}^*(\mathbf{r})] &= -[D(\mathbf{r})]\nabla[U^*(\mathbf{r})]. \end{aligned} \tag{11.5.4}$$

Since continuity of flux and Fick's Law are both physical characteristics of diffusion theory, this "limitation" of the function space is entirely acceptable. In fact, if we start with a more general space of discontinuous $[U]$'s and $[U^*]$'s (such as (11.5.1)) and completely independent $[\mathbf{V}]$'s and $[\mathbf{V}^*]$'s, the variational principle (11.3.21) will often select from the general space continuous fluxes and currents obeying (11.5.4). Much labor is saved by imposing these conditions, where possible, at the outset.

With $[U]$ and $[U^*]$ continuous and $[\mathbf{V}]$ and $[\mathbf{V}^*]$ given by (11.5.4), the variational principle (11.3.21) becomes

$$\begin{aligned}0 = \sum_i \int_{R_i} \Big\{ & [\delta U^*]^{\mathrm{T}}\Big[A - \frac{1}{\lambda}\chi\nu\Sigma_{\mathrm{f}}^{\mathrm{T}}\Big][U] - [\delta U^*]^{\mathrm{T}}\nabla\cdot[D]\nabla[U] + [\delta U]^{\mathrm{T}}\Big[A^{\mathrm{T}} - \frac{1}{\lambda}\nu\Sigma_{\mathrm{f}}\chi^{\mathrm{T}}\Big][U^*] \\ & - [\delta U]^{\mathrm{T}}\nabla\cdot[D]\nabla[U^*]\Big\} d\mathbf{r} + \int_{S_{\mathrm{int}}} \langle [\delta U^*]^{\mathrm{T}}\{-[D(+)]\nabla[U(+)] + [D(-)]\nabla[U(-)]\} \\ & + [\delta U]^{\mathrm{T}}\{-[D(+)]\nabla[U^*(+)] + [D(-)]\nabla[U^*(-)]\}\rangle\cdot\mathbf{N}\, dS,\end{aligned} \tag{11.5.5}$$

where the volume terms weighted by $[\mathbf{V}^*]^{\mathrm{T}}$ and $[\mathbf{V}]^{\mathrm{T}}$ have vanished because of (11.5.4) and the surface terms involving $[U(+)] - [U(-)]$, etc., have vanished because $[U]$ and $[U^*]$ belong to a space of continuous functions. Notice that the corresponding surface-current terms need *not* be continuous. That is, we are *not* restricting the spaces $[U]$ and $[U^*]$ in such a way that the normal components of $[D]\nabla[U]$ and $[D]\nabla[U^*]$ will be continuous. For certain choices of element functions the polynomial expressions for $[U]$ may turn out to be such that $[D]\nabla[U]\cdot\mathbf{N}$ *is* continuous; but for other choices this will not happen. The anomalous singularities at corner points of material interfaces in two- and three-dimensional geometries (discussed in Section 9.4) can be taken care of automatically by the variational principle if we allow the components of the net currents at such points to be discontinuous.

Equation (11.5.5) can be simplified if the terms $[\delta U^*]^{\mathrm{T}}\nabla\cdot[D]\nabla[U]$ and $[\delta U]^{\mathrm{T}}\nabla\cdot[D]\nabla[U^*]$ are integrated by parts. If we apply the convention that the direction of the vector $\mathbf{N}$ perpendicular to an internal surface is from $(-)$ to $(+)$ (see Figure 11.1), we obtain, for example,

$$\begin{aligned}\int_{R_i} [\delta U^*]^{\mathrm{T}}\nabla\cdot[D]\nabla[U]\, d\mathbf{r} = & \int_{S_i(-)} [\delta U^*]^{\mathrm{T}}[D(-)]\nabla[U(-)]\cdot\mathbf{N}\, dS \\ & - \int_{S_i(+)} [\delta U^*]^{\mathrm{T}}[D(+)]\nabla[U(+)]\cdot\mathbf{N}\, dS - \int_{R_i} \nabla[\delta U^*]^{\mathrm{T}}\cdot[D]\nabla[U]\, d\mathbf{r},\end{aligned} \tag{11.5.6}$$

where $S_i(-)$ is that portion of the surface of R_i for which the vectors $\mathbf{N}$ are pointing outward, and $S_i(+)$ is that portion for which they are pointed inward. (Note that, here, both $[D(-)]$ and $[D(+)]$ are within R_i.) Since $[U]$ and $[U^*]$ are required to vanish on the outer surface of the reactor and since the second integral in (11.5.5) is over *all* internal surfaces, substitution of (11.5.6) and the analogous expression for $[\delta U]^{\mathrm{T}}\nabla\cdot[D]\nabla[U^*]$ results in cancellation of all the surface integrals. We thus obtain

$$0 = \sum_i \int_{R_i} \left\{[\delta U^*]^{\mathrm{T}}\left[A - \frac{1}{\lambda}\chi\nu\Sigma_{\mathrm{f}}^{\mathrm{T}}\right][U] + \nabla[\delta U^*]\cdot[D]\nabla[U] + [\delta U]^{\mathrm{T}}\left[A^{\mathrm{T}} - \frac{1}{\lambda}\nu\Sigma_{\mathrm{f}}\chi^{\mathrm{T}}\right][U^*] + \nabla[\delta U]^{\mathrm{T}}\cdot[D]\nabla[U^*]\right\} d\mathbf{r}. \tag{11.5.7}$$

The variational principle we are seeking now states that, for $[U]$ and $[U^*]$ selected from the space of all continuous functions vanishing on the surface of the reactor, the function $[U]$ that causes the first-order variation (11.5.7) to vanish for an arbitrary variation $[U^*]$ is a solution of the group diffusion equations (11.2.21). Moreover, by retracing our steps from (11.5.7) to the completely equivalent expression (11.5.5), we see (by making $[\delta U^*]$ tend to zero within R_i and remain large and arbitrary on S_{int}) that the $[U]$ selected by the principle must obey

$$[D(+)]\nabla[U(+)]\cdot\mathbf{N} = [D(-)]\nabla[U(-)]\cdot\mathbf{N} \tag{11.5.8}$$

across all internal surfaces. In particular (11.5.8) must be valid across surfaces for which $[D(+)] \neq [D(-)]$. Arbitrary variation of $[\delta U]$ leads to analogous results for the adjoint equations.

With $[\delta U]$ set to zero, $[\delta U^*]$ regarded as an arbitrary, continuous weight function vanishing on the outer boundary of the reactor, and $\sum_i\int_{R_i}\ldots d\mathbf{r}$ interpreted as an integral over the entire reactor, (11.5.7) is known as the *weak form* of the group diffusion equations. It has the mathematical advantage over the conventional differential-equation form of avoiding the term $\nabla\cdot[D]\nabla[\Phi]$, which causes theoretical problems at singular points.

One-Dimensional Difference Equations Obtained by the Finite-Element Method

We return now to the problem of applying the finite-element method to determining an approximate solution of the group diffusion equations for a reactor composed of large homogenized material regions. As a very simple first example we consider piecewise-linear elements. Thus we define G-element column vectors

$$[u_k(x)] \equiv \begin{cases} \dfrac{x_{k+1} - x}{h_k}[\Phi_k] & \text{for } x_k \le x \le x_{k+1} \equiv x_k + h_k, \\ \dfrac{x - x_{k-1}}{h_{k-1}}[\Phi_k] & \text{for } x_{k-1} \le x \le x_k, \\ 0 & \text{for all other } x, \end{cases} \tag{11.5.9}$$

and

$$[U(x)] = \sum_{k=1}^{K-1}[u_k(x)], \tag{11.5.10}$$

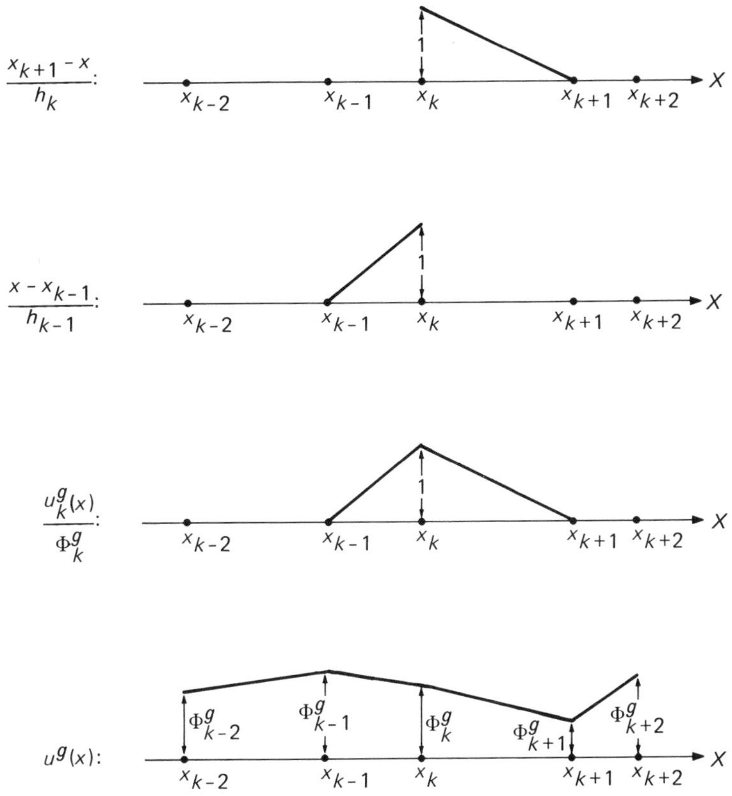

Figure 11.2 The linear finite-element trial function for energy group g.

where the $[\Phi_k]$ are G-element column vectors to be found by the variational procedure. Clearly the $[\Phi_k]$ are to be interpreted as the finite-element estimates of the group fluxes at mesh points x_k.

The elements of $[u_k(x)]$ are linear "tent functions" having values $[\Phi_k]$ at $x = x_k$ and going to zero at x_{k-1} and x_{k+1} (see Figure 11.2); $[\Phi_0]$ and $[\Phi_K]$ are taken to be zero. Thus, as must be the case if we are to use the principle (11.5.7), the trial function $[U(x)]$ vanishes at x_0 and x_K and is continuous in the interval between. Note that we could not legitimately insert the "step" trial function (11.3.23) into the expression (11.5.7). Note also that the net current vectors $[D]d[u(x)]/dx$ associated with the trial function (11.5.10) will in general be discontinuous at the x_K—even if the diffusion-constant matrix $[D]$ is spatially constant.

Notions of symmetry suggest that we select the trial function $[U^*]$ for the adjoint flux to be used in (11.5.7) in a fashion completely analogous to (11.5.10). Thus we define

$$[u_k^*(x)] \equiv \begin{cases} \dfrac{x_{k+1} - x}{h_k}[\Phi_k^*] \text{ for } x_k \le x \le x_{k+1}, \\ \dfrac{x - x_{k-1}}{h_{k-1}}[\Phi_k^*] \text{ for } x_{k-1} \le x \le x_k, \\ 0 \qquad\qquad \text{for all other } x, \end{cases} \tag{11.5.11}$$

$$[U^*(x)] = \sum_{k=1}^{K-1} [u_k^*(x)].$$

The regions R_i over which volume integrals are taken in the variational expression (11.5.7) are, according to our original definition, zones in which all the vectors $[U]$, $[U^*]$, $-[D]\nabla[U]$, and $-[D]\nabla[U^*]$ are continuous. Thus, in the present instance, the R_i coincide with the h_k. If we assume that the group parameters $[A]$, $[\nu\Sigma_f]$, and $[D]$ are constant over each h_k and take on the values $[A_k]$, $[\nu\Sigma_{fk}]$, and $[D_k]$, and if we vary $[U^*]$ by varying arbitrarily each element of one of the vectors $[\Phi_k^*]$, (11.5.7) yields

$$0 = [\delta\Phi_k^*]^{\mathrm{T}}\left\langle \int_{x_{k-1}}^{x_k} \left\{\left(\frac{x - x_{k-1}}{h_{k-1}}\right)\left[A_{k-1} - \frac{1}{\lambda}\chi\nu\Sigma_{\mathrm{f},k-1}^{\mathrm{T}}\right]\sum_{l=1}^{K-1}[u_l(x)] + \frac{1}{h_{k-1}}[D_{k-1}]\sum_{l=1}^{K}\frac{d}{dx}[u_l(x)]\right\} dx + \int_{x_k}^{x_{k+1}}\left\{\left(\frac{x_{k+1} - x}{h_k}\right)\left[A_k - \frac{1}{\lambda}\chi\nu\Sigma_{\mathrm{f}k}^{\mathrm{T}}\right]\sum_{l=1}^{K-1}[u_l(x)] - \frac{1}{h_k}[D_k]\sum_{l=1}^{K-1}\frac{d}{dx}[u_l(x)]\right\} dx\right\rangle \quad (k = 1, 2, \ldots, K-1), \tag{11.5.12}$$

whence, since each of the G elements of each $[\delta\Phi_k^*]$ can be varied independently, we conclude that each element of the $K - 1$ column vectors into which the $[\delta\Phi_k^*]^{\mathrm{T}}$ are multiplied (and hence those $K - 1$ column vectors themselves) must vanish.

Because of the tent-function character of the $[u_l(x)]$, only the functions for $l = k - 1$ and $l = k$ contribute to the integral over the range x_{k-1}–x_k in (11.5.12); similarly, only those for $l = k$ and $l = k + 1$ contribute to the integral from x_k to x_{k+1}. Substituting the definitions (11.5.9) of the $[u_l]$ into (11.5.12) and performing the indicated integrations then yields

$$-\frac{1}{h_k}[D_k][\Phi_{k+1} - \Phi_k] + \frac{1}{h_{k-1}}[D_{k-1}][\Phi_k - \Phi_{k-1}] + h_k\left[A_k - \frac{1}{\lambda}\chi\nu\Sigma_{\mathrm{f}k}^{\mathrm{T}}\right]\left\{\frac{1}{3}[\Phi_k] + \frac{1}{6}[\Phi_{k+1}]\right\} + h_{k-1}\left[A_{k-1} - \frac{1}{\lambda}\chi\nu\Sigma_{\mathrm{f},k-1}^{\mathrm{T}}\right]\left\{\frac{1}{3}[\Phi_k] + \frac{1}{6}[\Phi_{k-1}]\right\} = 0 \quad (k = 1, 2, \ldots, K-1). \tag{11.5.13}$$

Application of the finite-element technique has thus led to difference equations connecting the flux at mesh point k to the fluxes at the two nearest neighboring mesh points.

In this respect the technique is similar to the conventional finite-difference approximation (11.3.27). Empirical studies, however, indicate that, for a given mesh size, the finite-element technique is more accurate. The practical consequence is that, to attain a desired degree of accuracy, larger mesh intervals (and hence fewer unknowns $[\Phi_k]$) may be used with (11.5.13) than with (11.3.27). The coefficients of the $[\Phi_k]$ in (11.5.13) are, of course, somewhat more complicated. Nevertheless there is usually a net decrease in computing time if this "higher-order" difference form is used.

It should be noted that in the derivation of neither (11.3.27) nor (11.5.13) is it necessary mathematically to require that the group diffusion matrices $[A]$, $[\chi\nu\Sigma_f^T]$, and $[D]$ be spatially constant within the mesh intervals h_k. If they are not constant, the integrals in (11.5.12) must be performed numerically, and this will complicate computer programming and somewhat increase running time. For smooth changes in the core parameters within a mesh interval (for example changes due to depletion, xenon distribution, or temperature-density effects), however, accurate results may still be obtained with large mesh intervals.

Higher-Order Difference Equations

It is not surprising physically that the finite-element difference equations (11.5.13) are more accurate than the conventional set (11.3.27). The piecewise-linear form for the flux (11.5.10) is simply closer to reality than the step form (11.3.23). This observation suggests that we can effect a further improvement in accuracy by selecting element functions $[u_k(x)]$ having some curvature. Accordingly we shall sketch the application of the finite-element method when the element functions are cubic polynomials.

As in the linear case, it is conventional to define the polynomials so that the coefficients to be found by the variational procedure have a simple physical interpretation. To achieve this end we define G-element vectors $[u_k^0(x)]$, $[u_k^{1-}(x)]$, and $[u_k^{1+}(x)]$ ($k = 1, 2, \ldots, K-1$) by

$$[u_k^0(x)] \equiv \begin{cases} \left\{3\left(\dfrac{x-x_{k-1}}{h_{k-1}}\right)^2 - 2\left(\dfrac{x-x_{k-1}}{h_{k-1}}\right)^3\right\}[\Phi_k] & \text{for } x_{k-1} \le x \le x_k, \\ \left\{3\left(\dfrac{x_{k+1}-x}{h_k}\right)^2 - 2\left(\dfrac{x_{k+1}-x}{h_k}\right)^3\right\}[\Phi_k] & \text{for } x_k \le x \le x_{k+1}, \\ 0 & \text{for all other } x, \end{cases} \tag{11.5.14}$$

$$[u_k^{1-}(x)] \equiv \begin{cases} \left\{-\left(\dfrac{x-x_{k-1}}{h_{k-1}}\right)^2 + \left(\dfrac{x-x_{k-1}}{h_{k-1}}\right)^3\right\}h_{k-1}[\Phi_k^{1-}] & \text{for } x_{k-1} \le x \le x_k, \\ 0 & \text{for all other } x, \end{cases} \tag{11.5.15}$$

$$[u_k^{1+}(x)] \equiv \begin{cases} \left\{\left(\dfrac{x_{k+1}-x}{h_k}\right)^2 - \left(\dfrac{x_{k+1}-x}{h_k}\right)^3\right\}h_k[\Phi_k^{1+}] & \text{for } x_k \le x \le x_{k+1}, \\ 0 & \text{for all other } x, \end{cases} \tag{11.5.16}$$

where $[\Phi_k]$, $[\Phi_k^{1-}]$, and $[\Phi_k^{1+}]$ are G-element vectors of numbers that we shall shortly see can be interpreted as the group fluxes and their left-hand and right-hand derivatives at x_k.

In terms of these expansion functions we define a trial function $[U(x)]$ by

$$[U(x)] \equiv \sum_{k=1}^{K-1} [u_k^0(x) + u_k^{1-}(x) + u_k^{1+}(x)]. \tag{11.5.17}$$

Its definition shows that $[U(x)]$ is a G-element column vector of piecewise-cubic polynomials, each polynomial being nonzero over only a limited region of space.

Figure 11.3 shows the behavior of the polynomial "basis functions" of which $[U(x)]$ is composed. It can be seen that the first two functions are unity, but have zero derivatives, at x_k; they and and their derivatives vanish at x_{k-1} and x_{k+1}, respectively. The last two functions have zero value but unit slope at x_k; they and their derivatives also vanish at x_{k-1} and x_{k+1}, respectively. Such polynomials are called *cubic Hermite interpolating polynomials.* (The functions $(x_{k+1} - x)/h_k$ and $(x - x_{k-1})/h_{k-1}$ of Figure 11.2 are linear Hermite interpolating polynomials.)

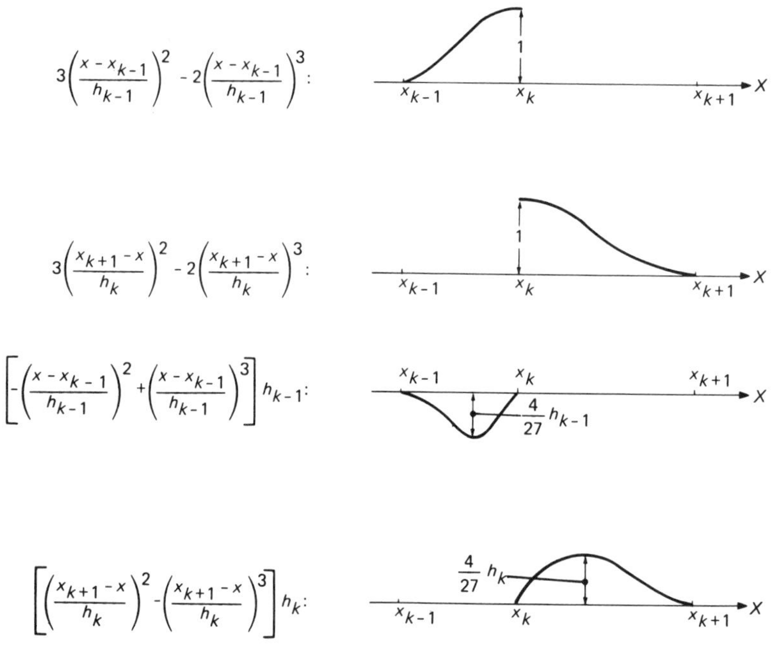

Figure 11.3 Cubic Hermite interpolating polynomials.

Figure 11.3 and the definitions (11.5.14)–(11.5.17) make it clear that

$$[U(x_k)] = [\Phi_k],$$
$$\frac{d}{dx}[U(x)]_{x_{k-}} = [\Phi_k^{1-}], \qquad (11.5.18)$$
$$\frac{d}{dx}[U(x)]_{x_{k+}} = [\Phi_k^{1+}].$$

Moreover the fact that $[u_k^{1-}(x_k)] = [u_k^{1+}(x_k)] = 0$ insures that $[U(x)]$ is continuous at $x = x_k$. Thus the variational principle (11.5.7) will be applicable.

Notice that we have not required a priori that there be continuity of the group currents at the x_k. That is, we have not restricted $[u_k^1(x)]$ in such a way that $[D_{k-1}][\Phi_k^{1-}] = [D_k][\Phi_k^{1+}]$. We could easily do so by replacing (11.5.15)–(11.5.17) by

$$v_k(x) \equiv \begin{cases} \left\{-\left(\dfrac{x - x_{k-1}}{h_{k-1}}\right)^2 + \left(\dfrac{x - x_{k-1}}{h_{k-1}}\right)^3\right\} h_k[D_{k-1}]^{-1}[J_k] & \text{for } x_{k-1} \le x \le x_k, \\ \left\{\left(\dfrac{x_{k+1} - x}{h_k}\right)^2 - \left(\dfrac{x_{k+1} - x}{h_k}\right)^3\right\} h_k[D_k]^{-1}[J_k] & \text{for } x_k \le x \le x_{k+1}, \\ 0 & \text{for all other } x, \end{cases} \qquad (11.5.19)$$

and

$$[U_J(x)] \equiv \sum_{k=1}^{K-1} [u_k^0(x) + v_k(x)]. \qquad (11.5.20)$$

This trial function is so constructed that both $[U_J(x)]$ and $[D(x)]d[U_J(x)]/dx$ are continuous at the points x_k. ($[J_k]$ is the group current vector at x_k.) Moreover $[U_J(x)]$ has one-third fewer unknowns than $[U(x)]$. Thus the difference equations that result from applying the variational method should be easier to solve. We have avoided this more attractive choice of trial function in order to illustrate that one always has the option of letting the variational principle account for continuity of current. In two- and three-dimensional situations this option is most important since (as we saw in Section 9.4) the derivatives of the flux at interior corner points are not defined. One very satisfactory way around this difficulty is to let the derivatives at such points be discontinuous, with the magnitude of the discontinuity determined automatically by the difference equations.

In keeping with the usual finite-element procedure, we define weight functions $[u_k^{0*}(x)]$, $[u_k^{1-*}(x)]$, and $[u_k^{1+*}(x)]$ by equations identical with (11.5.14)–(11.5.16) except that $[\Phi_k]$, $[\Phi_k^{1-}]$, and $[\Phi_k^{1+}]$ are replaced by $[\Phi_k^*]$, $[\Phi_k^{1-*}]$, and $[\Phi_k^{1+*}]$. An adjoint trial function $[U^*(x)]$ completely analogous to (11.5.17) is then constructed, and the variational expression (11.5.7) with the trial function (11.5.17) used for $[U(x)]$ is made stationary for completely arbitrary variations in $[U^*(x)]$, that is, for completely arbitrary variations in all elements of each of the $[\Phi_k^*]$, $[\Phi_k^{1-*}]$, and $[\Phi_k^{1+*}]$. The result is three sets of equations analogous to (11.5.12). If it is again assumed that the reactor parameters are constant within individual mesh intervals, the integrations in (11.5.7) can be performed analytically, and three coupled vector equations analogous to the single vector equation (11.5.13) will result. By judicious adding and subtracting, these equations can be cast in the form

$$\begin{aligned}
&-\frac{1}{h_k}[D_k][\Phi_{k+1}-\Phi_k]+\frac{1}{h_{k-1}}[D_{k-1}][\Phi_k-\Phi_{k-1}] \\
&+h_k\left[A_k-\frac{1}{\lambda}\chi\nu\Sigma_{fk}^{\mathrm{T}}\right]\left\{\frac{7}{20}[\Phi_k]+\frac{3}{20}[\Phi_{k+1}]-\frac{1}{30}h_k[\Phi_{k+1}^{1-}]+\frac{1}{20}h_k[\Phi_k^{1+}]\right\} \\
&+h_{k-1}\left[A_{k-1}-\frac{1}{\lambda}\chi\nu\Sigma_{f,k-1}^{\mathrm{T}}\right]\left\{\frac{7}{20}[\Phi_k]+\frac{3}{20}[\Phi_{k-1}]+\frac{1}{30}h_{k-1}[\Phi_{k-1}^{1+}]-\frac{1}{20}h_{k-1}[\Phi_k^{1-}]\right\} \\
&=0 \qquad (k=1,2,\ldots,K-1), \qquad (11.5.21)
\end{aligned}$$

$$\begin{aligned}
&h_k\left[A_k-\frac{1}{\lambda}\chi\nu\Sigma_{fk}^{\mathrm{T}}\right]\left\{-\frac{3}{140}[\Phi_{k+1}-\Phi_k]+\frac{h_k}{420}[\Phi_{k+1}^{1-}+\Phi_k^{1+}]\right\} \\
&+\frac{1}{h_k}[D_k]\left\{-\frac{1}{5}[\Phi_{k+1}-\Phi_k]+\frac{h_k}{10}[\Phi_{k+1}^{1-}+\Phi_k^{1+}]\right\}=0 \qquad (k=1,2,\ldots,K-1), \qquad (11.5.22)
\end{aligned}$$

and

$$\begin{aligned}
&h_k\left[A_k-\frac{1}{\lambda}\chi\nu\Sigma_{fk}^{\mathrm{T}}\right]\left\{-\frac{1}{12}[\Phi_{k+1}+\Phi_k]+\frac{h_k}{60}[\Phi_{k+1}^{1-}-\Phi_k^{1+}]\right\}+\frac{1}{6}[D_k][\Phi_{k+1}^{1-}-\Phi_k^{1+}]=0 \\
&(k=1,2,\ldots,K-1). \qquad (11.5.23)
\end{aligned}$$

Note the great similarity between these results and the corresponding linear finite-element equation (11.5.13). The derivative terms in $[\Phi_{k\mp 1}^{1\pm}]$ appear as corrections in (11.5.21). They vanish as the h_k tend to zero (in which case it is easy to show that both difference forms (11.5.21) and (11.5.13) reduce to the group equations in differential form).

The $3(K-1)$ vector difference equations (11.5.21)–(11.5.23) hold only at the $K-1$ interior mesh points. At x_0 and x_K, $[\Phi_0]$ and $[\Phi_K]$ are required to vanish; however the vectors $[\Phi_0^{1-}]$ and $[\Phi_K^{1-}]$ are not. Thus two extra relationships involving $[\Phi_0^{1+}]$ and $[\Phi_K^{1-}]$ must be found. This can be done by requiring the variational expression (11.5.7) to vanish for arbitrary variations $[\delta\Phi_0^{1+*}]$ and $[\delta\Phi_K^{1-*}]$. Then there will be $3(K-1)+2$ homogeneous vector equations, exactly the number required. These may be solved directly for the fundamental eigenvalue λ and the corresponding eigenvectors $[\Phi_k]$, $[\Phi_k^{1+}]$, and

$[\Phi_k^{1-}]$, or (11.5.22) and (11.5.23) may first be used to eliminate the G-element column vectors $[\Phi_k^{1+}]$ and $[\Phi_k^{1-}]$ from (11.5.21) so that the usual three-point form relating each $[\Phi_k]$ to its two nearest neighbors results.

The accuracy of finite-difference approximations is generally specified in terms of the rate of convergence of some parameter (such as the eigenvalue or the flux at a point) to its asymptotic value as the $h_k \to 0$. If all the h_k are made proportional to some basic mesh spacing h ($h_k \equiv \alpha_k h$), it is usually possible to prove that the errors in these parameters approach zero as some power of h. In a one-group problem, for example, the error in the eigenvalue λ of the linear finite-element equations (11.5.13) approaches zero as $C_l h^2$, where C_l is some constant that must be determined empirically. This error for the corresponding cubic-element equations (11.5.21)–(11.5.23) approaches zero as $C_c h^6$.

It is tempting to conclude that the cubic approximation is always to be preferred. However such a conclusion is not justified for every particular case. For example error bounds depend on the details of the theoretical model. Thus, for a two-group problem (essentially because the group equations are no longer self-adjoint), the linear and cubic finite-element approximations have eigenvalue errors proportional only to h^1 and h^3. In two and three dimensions (because of the singular corner points), theory predicts an even slower decrease in error with the size of h. In addition the rate at which errors in the fluxes at a point decrease with h usually differs from that of errors in λ, and we must be concerned with both errors.

Two other considerations are even more important. The first is that the magnitudes of the constant coefficients C_l and C_c must be determined empirically and are quite problem-dependent. Yet this magnitude is most important. In the one-group finite-element case, for example, if $C_l = 10^{-3}$ cm^{-2} and $C_c = 10^{-10}$ cm^{-6}, the error in λ with the cubic approximation is 10^{-4} for $h = 10$ cm, whereas that in the linear is 10^{-1}. Clearly the cubic scheme is to be preferred. On the other hand, if $C_l = 10^{-6}$ cm^{-2} and $C_c = 10^{-8}$ cm^{-6}, the respective errors are 10^{-4} and 10^{-2}, and, even though the cubic approximation will improve in accuracy much faster than the linear one as h is made smaller, the linear scheme is already adequate for $h = 10$ cm and there is no reason to go to the expense of decreasing h further. One must thus determine the magnitude of the error at some particular value of h, as well as its rate of change with h, before favoring one method over the other.

Finally it must be remembered that the finite-element schemes will only be valid if the correct flux shapes can be accurately approximated by the element functions. If the material between mesh points is homogeneous or nearly so, this condition is generally met. However, if it is not and if equivalent homogenized constants cannot be found, there is little reason to incur the extra cost of using a cubic approximation since its capacity to decrease the number of mesh points cannot be taken advantage of and since the gain in accuracy it achieves with small mesh intervals is already beyond the basic accuracy of group diffusion theory.

Scrutinizing the cubic finite-element scheme with regard for these considerations indicates, fortunately, that, for many cases of practical concern, it is cheaper than the conventional finite-difference approach. For example light-water-moderated power reactors are usually composed of square fuel clusters having dimensions 18–20 cm on a side and loaded with fuel of different enrichments (the "checkerboard" design). Conventional design practice is to determine one- or two-group homogenized diffusion-theory parameters for these clusters and to predict criticality and cluster-averaged power distributions by solving the diffusion equations for the overall reactor employing these homogenized constants. Both one- and two-dimensional two-group studies indicate that the cubic finite-element scheme (compared with fine-mesh finite-difference solutions) predicts eigenvalues and cluster-averaged power distributions adequately well when employing 20 cm mesh intervals in both the core and the light-water reflector. To achieve comparable accuracy with the conventional finite-difference method one must use 4 or 5 cm mesh spacings in the core region and 1 or 2 cm spacings in the reflector. Of course the cubic finite-element equations have more unknowns per mesh point, and there is less literature available dealing with numerical methods for solving them. Nevertheless they appear to provide a net gain in efficiency.

Two-Dimensional Finite-Element Procedures

As the last paragraph implies, the finite-element method employing linear or cubic Hermite interpolating polynomials can be extended to two-dimensional XY geometry. To do so we define piecewise-polynomial functions that are products of Hermite interpolating polynomials in the X and Y directions; we might, for example, use products of the basis functions (11.5.14)–(11.5.16) such as

$$u_k^0(x)u_j^0(y) \equiv \left\{3\left(\frac{x - x_{k-1}}{h_{k-1}^{(x)}}\right) - 2\left(\frac{x - x_{k-1}}{h_{k-1}^{(x)}}\right)^3\right\}\left\{\left(\frac{y_{j+1} - y}{h_j^{(y)}}\right)^2 - \left(\frac{y_{j+1} - y}{h_j^{(y)}}\right)\right\}h_j^{(y)}.$$

We thus form functions of the type $u_k^\alpha(x)u_j^\beta(y)[\Phi_{kj}^{\alpha\beta}]$ and use these to construct trial and weight functions for the variational expression (11.5.7).

In carrying out this procedure, we must bear in mind (see Section 9.4) that, when a mesh rectangle of one composition has neighbors on two adjacent sides of a different composition, the mesh point common to all three mesh rectangles is a "singular point." That is, the derivatives $\partial[\Phi]/\partial x$ and $\partial[\Phi]/\partial y$ are infinite (although $[\Phi]$ itself remains finite). Attempting to relate the flux derivatives across such a singular point as in (11.5.19) (and thereby to reduce the number of unknowns in the finite-element equations for the $[\Phi_{kj}^{\alpha\beta}]$)

can then lead to absurd results. The simplest solution to this difficulty is to permit the flux derivatives to be discontinuous across such a point and let the magnitude of the discontinuity be found automatically when the variational expression (11.5.7) is applied. Alternate modes of attack that reduce the number of unknown elements to be determined are possible, but these must be applied with great care.

Rather than going through the details of extending the finite-element method to XY geometry, we shall sketch a *triangular* finite-element approximation for solving the two-dimensional group diffusion equations. That is, the region over which the diffusion equations are to be solved will be partitioned into triangles, and the element functions will be piecewise-polynomial functions defined over these triangles.

The chief advantage of a triangular mesh is that mesh points can be clustered in regions where the flux is varying rapidly and spread out in regions where it is varying slowly. Figure 11.4 illustrates this situation for the simple case of a bare cylindrical reactor controlled by a single, central, cross-shaped rod.

If it is assumed that reflecting boundary conditions for directions perpendicular to the 45° line can be imposed for both methods, the triangular layout has 37 mesh points and the rectangular layout has 66. Yet the triangular mesh yields more flux detail near and within the control rod and is clearly much better able to match the cylindrical outer boundary of the reactor. Since, with a triangular layout, fewer mesh points can provide geometric detail, small heterogeneous regions of a reactor can be represented explicitly, and the need for determining homogeneous cross sections that correctly represent the combined behavior of several heterogeneously distributed, isotopically different materials is greatly diminished.

We shall call the triangles into which the reactor layout is partitioned *basic mesh triangles*. To implement the finite-element method for a triangular mesh we shall represent the flux within each basic mesh triangle as a polynomial of degree m in x and y.

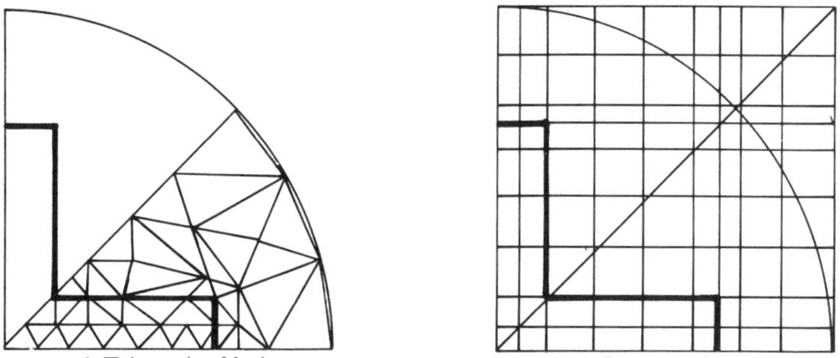

a. A Triangular Mesh b. A Rectangular Mesh

Figure 11.4 Mesh layouts for a cylindrical reactor controlled by a cross-shaped control rod.

In doing so, we shall find it necessary—except for the linear case—to subdivide the basic triangles into "subtriangles" of equal size. We then form tent functions having unit heights at the vertices common to adjacent subtriangles and construct a trial function $[U(x, y)]$ out of such tents. This trial function and a corresponding adjoint trial function can be used in the variational equation (11.5.7) to determine equations for the group flux vectors at each mesh point.

The first step in the procedure is to represent the group fluxes within any given basic mesh triangle as polynomials in x and y.

If we let $\Phi_g(x, y)$ be the polynomial representation of the flux for group g and, for simplicity, consider a quadratic expansion, we have

$$\Phi_g(x, y) = a_1 + a_2x + a_3y + a_4x^2 + a_5xy + a_6y^2, \tag{11.5.24}$$

where the a_i are coefficients to be determined.

As in the one-dimensional finite-element examples, it is convenient to transform (11.5.24) so that the coefficients of the "monomials" (x, y, xy, etc.) can be interpreted physically. Previously we did this by using Hermite interpolating polynomials, in which case the coefficients were interpreted as the fluxes and their derivatives at mesh points. In the present case it is more convenient to use Lagrange interpolating polynomials. These are constructed so that the coefficients have the values of the fluxes at different designated points throughout the triangle. Thus, in addition to the three corner points at the vertices of the basic mesh triangles, it is necessary to specify additional mesh points so that the coefficients in the transformed polynomial expansion will have the meaning of the flux at these points.

Figure 11.5 shows how a basic mesh triangle is subdivided into subtriangles for polynomials of degree 1, 2, and 3. A regular triangular submesh is simply imposed on the basic mesh triangle. (For $m = 1$, the linear case, there are only three coefficeints; hence the three vertices alone will suffice to specify the Lagrange interpolating polynomial, and no subdivision is necessary.)

We now consider the problem of transforming (11.5.24) so that the unknown coeffi-

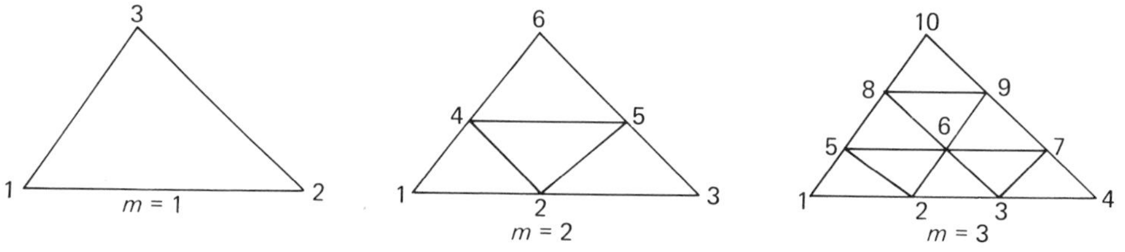

Figure 11.5 Interpolating points for polynomials of degree $m = 1, 2, 3$.

cients become, for example, the fluxes at the six points shown in Figure 11.5 for $m = 2$. Matrix manipulation greatly facilitates this operation.

We first note that (11.5.24) for $\Phi_g(x, y)$ may be rewritten

$$\begin{aligned} &\Phi_g(x, y) = [P(x, y)]^{\mathrm{T}}[a], \\ &[a] \equiv \mathrm{Col}\{a_1, a_2, \ldots, a_6\}, \\ &[P(x, y)]^{\mathrm{T}} \equiv \mathrm{Row}\{1, x, y, x^2, xy, y^2\}. \end{aligned} \tag{11.5.25}$$

It follows that, if (x_i, y_i) represents any one of the six mesh points in Figure 11.5 for $m = 2$,

$$\Phi_g(x_i, y_i) = [P(x_i, y_i)]^{\mathrm{T}}[a] \qquad (i = 1, 2, \ldots, 6). \tag{11.5.26}$$

Thus we get the matrix equation

$$\begin{bmatrix} \Phi_g(x_1, y_1) \\ \Phi_g(x_2, y_2) \\ \vdots \\ \Phi_g(x_6, y_6) \end{bmatrix} = \begin{bmatrix} 1 & x_1 & y_1 & x_1^2 & x_1y_1 & y_1^2 \\ 1 & x_2 & y_2 & x_2^2 & x_2y_2 & y_2^2 \\ \vdots & \vdots & \vdots & \vdots & \vdots & \vdots \\ 1 & x_6 & y_6 & x_6^2 & x_6y_6 & y_6^2 \end{bmatrix} \begin{bmatrix} a_1 \\ a_2 \\ \vdots \\ a_6 \end{bmatrix} \tag{11.5.27}$$

or, in condensed notation defined by comparison with (11.5.27),

$$[\Phi_g] = [\mathscr{P}][a]. \tag{11.5.28}$$

As a result, since the square matrix $[\mathscr{P}]$ is nonsingular and contains fixed numbers, we have from (11.5.25)

$$\Phi_g(x, y) = [P(x, y)]^{\mathrm{T}}[\mathscr{P}]^{-1}[\Phi_g]. \tag{11.5.29}$$

Thus the flux for group g throughout the basic mesh triangle is expressed as a quadratic polynomial in x and y having coefficients written in terms of the specific values of the flux at the six mesh points indicated in Figure 11.5 for $m = 2$.

This result can be readily extended to polynomial expansions of any order. Moreover, since the column vector $[P(x, y)]$ and the matrix $[\mathscr{P}]$ are group-independent, they may be used for all groups.

We now employ the interpolating polynomials to construct element functions for basic mesh triangles such as the three shown in Figure 11.5. These element functions, designated by $u_i(x, y)$, are to be polynomials of degree m over the basic mesh triangles. There are to be as many of them as there are mesh points formed by the vertices of subtriangles (3, 6, and 10 for cases shown in Figure 11.5). If we let $u_i(x, y)$ designate the element function to be associated with mesh point i, we require that $u_i(x_j, y_j) = \delta_{ij}$. Thus, for $m = 2$, $u_5(x, y)$ is to be a quadratic function of x and y having unit value at (x_5, y_5) and vanishing at (x_1, y_1), (x_2, y_2), etc. Except at these mesh points, $u_5(x, y)$ will in general be nonzero. Note, however, that, since it is a quadratic and vanishes at points 1, 2, and 3 and points 1, 4, and 6, which lie on straight lines, it will be zero at *all* points on these lines.

Algebraic expressions for the $u_i(x, y)$ can be obtained immediately from (11.5.29). We simply let the column vector $[\Phi_g]$ of flux values at the mesh points be unity at mesh point i and zero at all other mesh points. If $[E_i]$ is used to designate such a vector, we have

$$\begin{aligned} u_i(x, y) &= [P(x, y)]^{\mathrm{T}}[\mathscr{P}]^{-1}[E_i] \\ &= [E_i]^{\mathrm{T}}\{[\mathscr{P}]^{-1}\}^{\mathrm{T}}[P(x, y)] \\ &= \sum_j \{[\mathscr{P}]^{-1}\}_{ji} P_j(x, y), \end{aligned} \tag{11.5.30}$$

where $\{[\mathscr{P}]^{-1}\}_{ji}$ is the ijth element of the transpose of $[\mathscr{P}]^{-1}$.

The linear case provides the simplest example of this procedure. When $m = 1$, we have

$$[P(x, y)] = \begin{bmatrix} 1 \\ x \\ y \end{bmatrix}, \qquad [\mathscr{P}] = \begin{bmatrix} 1 & x_1 & y_1 \\ 1 & x_2 & y_2 \\ 1 & x_3 & y_3 \end{bmatrix}. \tag{11.5.31}$$

Thus, from (11.5.30),

$$u_i(x, y) = \frac{1}{\mathscr{D}}[E_i]^{\mathrm{T}} \begin{bmatrix} (x_2y_3 - x_3y_2) + (y_2 - y_3)x + (x_3 - x_2)y \\ (x_3y_1 - x_1y_3) + (y_3 - y_1)x + (x_1 - x_3)y \\ (x_1y_2 - x_2y_1) + (y_1 - y_2)x + (x_2 - x_1)y \end{bmatrix}, \tag{11.5.32}$$

where

$$\mathscr{D} \equiv x_2y_3 + x_1y_2 + x_3y_1 - x_2y_1 - x_1y_3 - x_3y_2.$$

It is easy to verify that, if $i = 2$ so that $[E_i] = \mathrm{Col}\{0, 1, 0\}$, (11.5.32) gives $u_2(x_2, y_2) = 1$ and $u_2(x_i, y_i) = 0$ $(i = 1, 3)$.

We must now construct tent functions $T_i(x, y)$ such that: (1) at mesh point i (whether at the vertex of a basic mesh triangle or at that of a subtriangle), $T_i(x_i, y_i) = 1$; (2) $T_i(x, y) = 0$ on some polygon containing point i; and (3) $T_i(x, y) = 0$ outside that polygon. We shall first do this construction for the linear case, for which there are no subtriangles.

Figure 11.6 depicts part of a basic mesh layout to be considered. With the points and basic mesh triangles labeled as shown, enough of the lattice is depicted so that all the nonzero portion of the tent functions $T_4(x, y)$ and $T_5(x, y)$ is visible. ($T_5(x, y)$ has unit value at (x_5, y_5) and is zero on and outside the polygon (1–2–4–7–6–1).) To represent

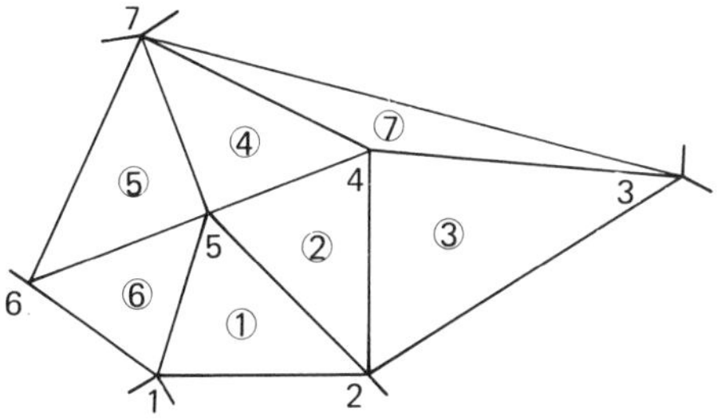

Figure 11.6 A mesh layout for linear elements.

the $T_i(x, y)$ mathematically we add a superscript (j) to the element functions $u_i(x, y)$ to indicate the basic mesh triangle over which it is defined. Thus $u_i^{(j)}$ is an element function of mesh point i, triangle (j), and $u_5^{(4)}$, $u_7^{(4)}$, and $u_4^{(4)}$ are the element functions associated with triangle (4), while $u_5^{(5)}$ $u_7^{(5)}$, and $u_6^{(5)}$ are those associated with triangle (5). We then have

$$T_5(x, y) = \begin{cases} u_5^{(4)}(x, y) \text{ for } (x, y) \text{ in triangle } (4), \\ u_5^{(5)}(x, y) \text{ for } (x, y) \text{ in triangle } (5), \\ u_5^{(6)}(x, y) \text{ for } (x, y) \text{ in triangle } (6), \\ u_5^{(1)}(x, y) \text{ for } (x, y) \text{ in triangle } (1), \\ u_5^{(2)}(x, y) \text{ for } (x, y) \text{ in triangle } (2), \\ 0 \qquad\qquad \text{ for all other } (x, y). \end{cases} \tag{11.5.33}$$

The tent function $T_4(x, y)$ is composed in an analogous fashion from $u_4^{(2)}$, $u_4^{(3)}$, $u_4^{(7)}$, and $u_4^{(4)}$. Tent functions for all other mesh points are constructed in a similar way. Notice that the tents have a variable number of sides depending on the mesh layout.

Since the $u_i^{(j)}$ are linear functions of x and y, the tent functions are continuous functions of x and y across the seams connecting their sides (the lines (1–5), (6–5), (7–5), etc., for $T_5(x, y)$), and, since they are zero on and outside the polygon surrounding their central pole (the polygon (1–2–4–7–6–1) for $T_5(x, y)$), they are continuous over the entire reactor. (Note, however, that derivatives of the flux are discontinuous across all seams and across the base.) The $T_i(x, y)$ are thus mathematically acceptable expansion functions for use in the variational expression (11.5.7). Before constructing a trial function from them, however, we shall examine what complications arise when polynomials of order two and higher are used for the $u_i(x, y)$.

Figure 11.7 shows part of a mesh layout for the case of quadratic element functions. The basic mesh triangles are drawn with solid lines, their vertices and circled identification numbers being the same as those in Figure 11.6. Each basic mesh triangle is partitioned by the dashed lines into four congruent subtriangles, and the extra mesh points

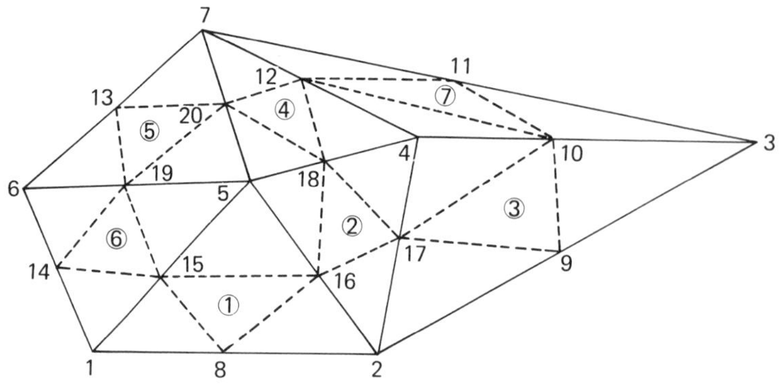

Figure 11.7 Basic mesh triangles and subtriangles for a quadratic finite-element approximation.

thereby created are also labeled. The unknowns of the problem now include the fluxes at *all twenty* resultant mesh points.

As before, the element functions $u_i^{(j)}$ are defined for all mesh points i belonging to basic mesh triangle (j). Thus $u_6^{(5)}(x, y)$ is a quadratic function of x and y defined over *all* of basic mesh triangle (5) and having unit value at mesh point 6 and zero value at mesh points 19, 5, 20, 7, and 13. Note that since $u_6^{(5)}(x, y)$ is quadratic and vanishes at three points (5, 20, and 7) lying on a straight line, it vanishes at all points on that line. Note also that $u_6^{(6)}(x, y)$, belonging to basic mesh triangle (6), is unity at mesh point 6 and vanishes at 19 and 5; thus, since $u_6^{(6)}$ and $u_6^{(5)}$ are both quadratics and have the same values at three points along a straight line, they are identical at *all* points along that line. As a result the combination of $u_6^{(6)}$ and $u_6^{(5)}$ is a continuous quadratic function of x and y vanishing on the line (1–15–5–20–7).

These observations should make it clear how to construct the tent function $T_i(x, y)$ for the quadratic case. We simply construct a unit-height tent pole at every mesh point i and make the perimeter of the tent the polygon composed of all basic mesh triangles common to point i. Thus (11.5.33) again specifies $T_5(x, y)$; it is unity at point 5 and vanishes at all (x, y) on and outside the polygon (1–2–4–7–6–1). (One might think that the polygon (15–16–18–20–19) could serve as a perimeter for $T_5(x, y)$. However, although T_5 vanishes *at* the points 15, 16, 13, etc., defining this polygon, being a quadratic function, it need not vanish at intermediate points on that polygon.)

The tent function associated with mesh point 18 involves only two basic mesh triangles. It is defined by

$$T_{18}(x, y) \equiv \begin{cases} u_{18}^{(4)}(x, y) \text{ for } (x, y) \text{ in triangle 4,} \\ u_{18}^{(2)}(x, y) \text{ for } (x, y) \text{ in triangle 2,} \\ 0 \qquad\qquad\quad \text{for all other } (x, y). \end{cases} \tag{11.5.34}$$

(Again, one might think the polygon (5–16–17–4–12–20) could serve as the perimeter of T_{18}; but, although $u_{18}^{(2)}$ and $u_{18}^{(4)}$ vanish at points 16, 17, 12, and 20, they do not vanish on the lines (16–17) or (20–12).)

The extension of this method of generating the $T_i(x, y)$ to element functions composed of higher-order polynomials is straightforward. Thus we are left only with the problem of constructing trial and weight functions from the tent functions. This is extremely simple; we merely write

$$
\begin{aligned}
[U(x, y)] &= \sum_{i=1}^{I} T_i(x, y)[\Phi_i], \\
[U^*(x, y)] &= \sum_{i=1}^{I} T_i(x, y)[\Phi_i^*],
\end{aligned}
\tag{11.5.35}
$$

where the $[\Phi_i]$ are G-element column vectors of group fluxes at the mesh points i and the $[\Phi_i^*]$ are corresponding adjoint fluxes.

The substitution of these trial and weight functions into the variational expression (11.5.7) and the subsequent derivation of equations for the $[\Phi_i]$ and $[\Phi_i^*]$ from that expression is conceptually straightforward but mechanically extremely complicated. The complexity involves finding integrals of the type $\int dx \int dy\, T_i(x, y)T_j(x, y)$ and $\int dx \int dy\, \nabla T_i(x, y)\cdot\nabla T_j(x, y)$. It is reduced considerably by taking the group parameters $[A]$, $[\nu\Sigma_f]$, and $[D]$ to be constant over basic mesh triangles. We shall not outline any of the details except to say that, for a given mesh layout, integrals involving any pair of triangles need be found only once. Sums of such integrals weighted by local diffusion-theory parameters (which may change from case to case) can be constructed quickly. The result is a set of equations for the $[\Phi_i]$ or $[\Phi_i^*]$ that are readily solved by standard methods. Once the $[\Phi_i]$ are found, the group fluxes at all points throughout the reactor are given by (11.5.35).

Numerical tests indicate that this finite-element procedure for determining group fluxes yields results equal in accuracy to a fine-mesh finite-difference computation, and does so at less cost.

Conclusions

It should be reemphasized that the discussion of the finite-element method in the foregoing section has been quite narrow. The principal aim has been to show the basic point of view that motivates the approach and to describe schemes for generating element functions and deriving the algebraic equations to which the group diffusion equations reduce when the finite-element method is applied. The fact that the method has a firm mathematical foundation which permits the establishment of error bounds for the solution has perhaps not been emphasized sufficiently. It also bears repeating that the technique is applicable to problems concerned with spectrum effects, time-dependent reactor behavior, and transport phenomena, as well as to problems in stress analysis and heat transfer.

Finally it is interesting to note that, even limiting discussion to attacks on the static group diffusion equations, we exceeded the initial goal of finding an approach that would be cheaper than the finite-difference method when applied to cores composed of *large* homogenized regions. The triangular scheme using Lagrange interpolating polynomials as element functions is able to deal with geometries involving explicitly treated *small* heterogeneous regions and, in doing so, uses fewer mesh points than would be required by the conventional finite-difference methods.

11.6 Techniques Not Restrained by the Diffusion-Theory Model

Finite-element methods and synthesis methods are conceptually quite similar. Both are based on the representation of the group fluxes by a general trial function constructed from a sum of known expansion functions with unknown (possibly space-dependent) coefficients, the coefficients being determined by variational or weighted-residual methods.

In synthesis schemes the expansion functions are strongly problem-dependent; a given set of them can be used only for a rather narrow class of reactor configurations, and they must, in general, be determined by special auxiliary computations. However their specificity produces powerful advantages, namely that relatively few expansion functions are required to analyze a given situation and extremely detailed flux shapes reflecting the effects of specifically represented small material regions can be found in a strikingly short time. Once the expansion functions are determined, group fluxes at literally millions of mesh points can be found in a few minutes of computer time.

On the other hand the rigidity of the synthesis expansion functions causes concern. Because of ignorance concerning the true behavior of the solution, functions needed to account for that behavior may be missing. This possibility is especially serious since no systematic way of estimating the error of a synthesized solution is available. In particular there is no guarantee that increasing the number of the expansion functions will improve the accuracy of the result.

The expansion functions used for the finite-element method lead to almost opposite strengths and weaknesses. They are low-order polynomials that are nonzero over relatively small regions of the reactor. The same functions are used for all problems, and, because of their simple mathematical form, it is possible to prove that decreasing the mesh spacing increases the accuracy of the result. Thus one is assured that any unforeseen behavior of the solution will appear in a finite-element result. On the other hand the equations that determine the expansion coefficients can be far more cumbersome than those resulting from synthesis schemes, and, since the basic mesh layout is usually determined by the geometry of the problem, long running times can result if a great many significantly different material regions must be represented explicitly. (A triangular mesh layout helps in this regard only if large homogeneous regions occupy a significant portion of the reactor volume.)

Thus the two classes of expansion-function method are at their best for different situations. If great detail is required for a narrow class of problems for which the synthesis method has already been tested, that scheme is strongly favored. Analyses associated with the final stages of a reactor design are of this nature. If the reactor can be partitioned into large homogeneous or homogenized zones and if little is known beforehand of the nature of the flux shape, the finite-element procedure is likely to be superior. Survey studies, fuel-management analyses, and problems for which confidence in the accuracy of the solution is overriding fall into this category. Taken together, the two schemes provide a calculational capability that a growing amount of evidence indicates is acceptably accurate and cheaper—sometimes orders of magnitude cheaper—than the conventional finite-difference procedure.

Synthesis and finite-element schemes have one common feature: they are both approximate procedures for solving the group diffusion equations. Thus, even though both schemes may yield very accurate and detailed results, we cannot have complete confidence that these results will correspond to physical reality unless we are certain that the input group diffusion-theory parameters are valid. Part of this validity will, of course, depend on how accurately the concentrations and basic nuclear data of the isotopes comprising the reactor are known. Reactor physics can do little about these problems except to recognize that they impose one practical limitation on the accuracy required of a theoretical analysis.

There is, however, another consideration pertaining to the validity of the group diffusion-theory parameters that is very much the concern of reactor physics. We saw in Chapter 10 that, in practical cases, it is almost always necessary to use effective homogenized diffusion-theory parameters for reactor computations. Material regions are usually so small that diffusion theory is not valid within them, and, even if it were, the number of mesh points required to specify the heterogeneous geometry of the reactor in detail would be excessively large. Accordingly we were led to define equivalent parameters for both small, isotopically homogeneous regions and for composite cells or subassemblies. These hopefully reproduce reaction rates correctly when used in the group diffusion equations.

Unfortunately we were unable to provide a firm theoretical foundation for these procedures. In fact we examined some evidence suggesting that the conventional flux-weighting method for finding equivalent constants for fuel clusters containing control rods may be inaccurate. Perhaps more serious was the fact that the investigation cast doubt on whether, for extended regions in two and three dimensions, exact equivalent constants even exist. The mathematical form of the group diffusion equations may be sufficiently restricted that homogenized parameters, constant over a heterogeneous subassembly, are not capable of reproducing correct subassembly-averaged reaction rates. Thus there is reason to examine methods for analzing reactors that avoid, either partially or completely, the use of equivalent diffusion-theory parameters.

One way of avoiding equivalent diffusion-theory parameters is to analyze reactor problems using some higher-order approximation to the transport equation. We discussed such methods in Chapter 8 but saw that they were very costly if applied to two- and three-dimensional situations, particularly if the reactor has a complex geometrical structure. We shall avoid this direct attack in the present section by assuming essentially that diffusion theory is valid on certain internal surfaces throughout the reactor. By connecting neutron currents on such surfaces to the average scalar fluxes in the interior regions or to the currents on neighboring surfaces, we shall be able to determine criticality and (at some cost) detailed flux distributions throughout the reactor.

Up to a point the blackness-theory methods discussed in Section 10.3 tend to treat regions where diffusion theory is invalid by internal boundary conditions. Equation (10.3.35) connects the flux-current vector at one side of an absorbing slab to that at another. We went on to find equivalent diffusion-theory parameters which, when inserted in the diffusion equation within such a region, make the same connection. But there was really no need to take this extra step. We could have used a matrix to make the connection directly.

A method called *source-sink theory* also involves this procedure of dealing with certain internal regions through the use of boundary conditions. Source-sink theory was originally invented to analyze large natural-uranium-fueled reactors consisting of widely separated cylindrical fuel rods uniformly dispersed in a graphite or D_2O moderator. Diffusion theory was assumed to be valid in the moderator, and (in the most elementary form of the theory) the fuel rods were treated as line absorbers (sinks) of thermal neutrons and line sources of fast neutrons. The strengths to be assigned to these sinks and sources were determined by auxiliary computations. We shall not describe the method other than to note its avoidance of the use of equivalent diffusion-theory parameters.

Blackness theory and source-sink theory both deal with only small regions of the core through the use of internal boundary conditions; the diffusion-theory model is used for the remainder of the reactor. In the sections which follow we shall sketch two classes of approximations, neither of which explicitly uses diffusion theory *anywhere* in the reactor.

For both classes the reactor is partitioned into large zones which we shall call *nodes*. The geometrical layout of the core usually suggests a natural way for this subdivision to be accomplished. The subdivision is also subject to two general rules (the basis of which will become apparent shortly): diffusion theory is to be valid on the surfaces between nodes; and the nodes are to be as large as possible, under the restriction that the scalar fluxes should be fairly flat over the nodal surfaces. Sections, fifteen or twenty centimeters in length, of a cluster of fuel subassemblies symmetrically grouped about a control rod or group of control fingers (see Figure 1.11) constitute a typical node for a light-water-moderated BWR or PWR.

The two classes of approximation schemes we shall discuss are generally referred to as *nodal methods* and *response-matrix methods*. The essential idea of the former is to relate neutron currents across an interface between two nodes to the average flux levels in those two nodes; that of the latter is to relate the fluxes and currents on the nodal surfaces directly to each other.

Nodal Methods

The formal basis of the nodal approach is the exact result (10.2.6) relating the column vector $[\mathbf{J}(\mathbf{r})]$ of net currents for the several energy groups to the column vector $[\Phi(\mathbf{r})]$ of group fluxes:

$$-\nabla\cdot[\mathbf{J}(\mathbf{r})] - [A(\mathbf{r})][\Phi(\mathbf{r})] + \frac{1}{\lambda}[M(\mathbf{r})][\Phi(\mathbf{r})] = 0. \tag{10.2.6}$$

We obtained this result by a formal integration of the transport equation over all directions and a subsequent formal reduction to the energy-group form. The resultant variables $[\mathbf{J}(\mathbf{r})]$ and $[\Phi(\mathbf{r})]$ are still functions of position; however they contain less information than does the directional flux density $\Psi(\mathbf{r}, \mathbf{\Omega}, E)$. In a nodal approximation we reduce this information content even further by integrating (10.2.6) over large nodes of volume V_{n} into which the core has been partitioned. The variable of interest in a nodal scheme is the *average* group flux vector $[\overline{\Phi}_{\mathrm{n}}]$ within a node. With this quantity formally defined by

$$[\overline{\Phi}_{\mathrm{n}}] = \mathrm{Col}\{\overline{\Phi}_{1\mathrm{n}}, \overline{\Phi}_{2\mathrm{n}}, \ldots, \overline{\Phi}_{G\mathrm{n}}\} \equiv \frac{1}{V_{\mathrm{n}}}\int_{V_{\mathrm{n}}} [\Phi(\mathbf{r})]\, dV, \tag{11.6.1}$$

integration of (10.2.6) over V_{n} and application of the divergence theorem yields

$$-\frac{1}{V_{\mathrm{n}}}\int_{S_{\mathrm{n}}} [\mathbf{J}(\mathbf{r})]\cdot\mathbf{n}_{\mathrm{n}}\, dS - [\overline{A}_{\mathrm{n}}][\overline{\Phi}_{\mathrm{n}}] + \frac{1}{\lambda}[\overline{M}_{\mathrm{n}}][\overline{\Phi}_{\mathrm{n}}] = 0, \tag{11.6.2}$$

where $\mathbf{n}_{\mathrm{n}}$ is the outward normal to the surface S_{n} of V_{n} and elements $\overline{A}_{\mathrm{n},gg'}$ and $\overline{M}_{\mathrm{n},gg'}$ of the matrices $[\bar{A}_{\mathrm{n}}]$ and $[\bar{M}_{\mathrm{n}}]$ are defined by

$$\begin{aligned}\bar{A}_{\mathrm{n},gg'}\bar{\Phi}_{g'\mathrm{n}} &\equiv \frac{1}{V_{\mathrm{n}}}\int_{V_{\mathrm{n}}} A_{gg'}(\mathbf{r})\Phi_{g'}(\mathbf{r})\, dV,\\ \bar{M}_{\mathrm{n},gg'}\bar{\Phi}_{g'\mathrm{n}} &\equiv \frac{1}{V_{\mathrm{n}}}\int_{V_{\mathrm{n}}} M_{gg'}(\mathbf{r})\Phi_{g'}(\mathbf{r})dV.\end{aligned} \tag{11.6.3}$$

The final step in deriving formally exact nodal equations is to define *coupling constants* that will permit us to express the net currents across the surfaces of V_{n} in terms of $[\Phi_{\mathrm{n}}]$ itself and the corresponding average flux vectors in neighboring nodes. To simplify notation we shall illustrate this procedure for the case of two-dimensional rectangular geometry only. Thus we consider the "mesh square" $h_i^{(x)}h_j^{(y)}$ illustrated in Figure 11.8.

Values of $[\bar{\Phi}]$, $[\bar{A}]$, and $[\bar{M}]$ within the square will be labeled by superscripts ij. The volume V_{n} becomes the area $h_i^{(x)}h_j^{(y)}$, and dV becomes $dx\,dy$.

The simplest scheme for relating the net current across a nodal interface to the average fluxes in the nodes which that interface separates is to define a constant of proportionality such that the net current across the interface equals that constant multiplied by the difference in the two nodal fluxes. Such an expression would reduce to Fick's Law in the limit of zero nodal mesh sizes. However the coupling constant so defined would depend equally on the properties of *both* adjacent nodes. Since the success of the nodal scheme will depend on being able to estimate the coupling constants in some approximate manner, we would much prefer to introduce, if possible, quantities that depend only on the properties of a *single* node. To achieve this end we shall divide the net current across a nodal interface into two *partial currents*, one consisting only of neutrons moving from the $(-)$ to the $(+)$ side of the surface and the other, only of neutrons moving from the $(+)$ to $(-)$ side:

Number of group-g neutrons per second crossing, from left to right, a strip of unit width and length dy at $x = x_{i+1}$

$$= J_g^{(x)+}(x_{i+1}, y)dy. \tag{11.6.4a}$$

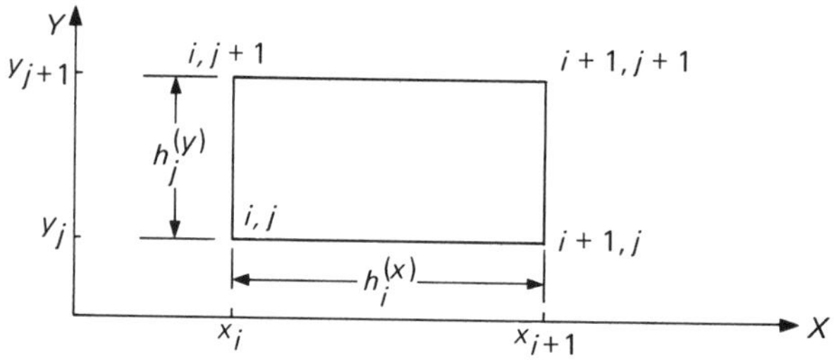

Figure 11.8 A mesh layout for a two-dimensional nodal scheme.

Number of group-g neutrons per second crossing, from right to left, a strip of unit width and length dy at $x = x_{i+1}$

$$= J_g^{(x)-}(x_{i+1}, y)dy. \tag{11.6.4b}$$

Analogous definitions apply to $J_g^{(y)\pm}(x, y_{j+1})$.

Clearly the *net* current in the $+X$ direction across dy at x_{i+1} for group-g neutrons is

$$J_g^{(x)}(x_{i+1}, y)dy = [J_g^{(x)+}(x_{i+1}, y) - J_g^{(x)-}(x_{i+1}, y)]dy. \tag{11.6.5}$$

Then, if we define coupling coefficients $\alpha_g^{(ij)(i'j')}$ by

$$\frac{1}{h_j^{(y)}}\int_{h_j} J_g^{(x)+}(x_{i+1}, y)dy = \alpha_g^{(ij)(i+1,j)}\bar{\Phi}_g^{ij} \tag{11.6.6}$$

and

$$\frac{1}{h_j^{(y)}}\int_{h_j} J_g^{(x)-}(x_{i+1}, y)\,dy = \alpha_g^{(i+1,j)(ij)}\bar{\Phi}_g^{i+1,j}, \tag{11.6.7}$$

with similar expressions for the Y direction, we obtain (by substituting into (11.6.2)) the formally exact equations

$$\begin{aligned}&\frac{1}{h_i^{(x)}}\{[\alpha^{(i+1,j)(ij)}][\bar{\Phi}^{i+1,j}] - [\alpha^{(ij)(i+1,j)}][\bar{\Phi}^{ij}]\} - \frac{1}{h_i^{(x)}}\{[\alpha^{(ij)(i-1,j)}][\bar{\Phi}^{ij}] - [\alpha^{(i-1,j)(ij)}][\bar{\Phi}^{i-1,j}]\}\\&+ \frac{1}{h_j^{(y)}}\{[\alpha^{(i,j+1)(ij)}][\bar{\Phi}^{i,j+1}] - [\alpha^{(ij)(i,j+1)}][\bar{\Phi}^{ij}]\} - \frac{1}{h_j^{(y)}}\{[\alpha^{(ij)(i,j-1)}][\bar{\Phi}^{ij}] - [\alpha^{(i,j-1)(ij)}][\bar{\Phi}^{i,j-1}]\}\\&- \left[\bar{A}^{ij} - \frac{1}{\lambda}\bar{M}^{ij}\right][\bar{\Phi}^{ij}] = 0,\end{aligned} \tag{11.6.8}$$

where $[\alpha^{(ij)(i'j')}]$ is a $G \times G$ diagonal matrix with elements $\alpha_g^{(ij)(i'j')}$.

Equation (11.6.8) is a formally exact relationship between the average group flux vector in an interior node (ij) and the corresponding flux vectors in the four nearest neighboring nodes. The result is still valid for the nodes that border the outer surface of the reactor provided the fluxes at nodes that would be outside the reactor according to the notation of (11.6.8) are set to zero. (The partial currents *into* the reactor will then correctly vanish.)

Although formally exact, (11.6.8) will be useful only if some acceptably accurate and efficient way of obtaining the matrices $[\alpha^{(ij)(i'j')}]$, $[\bar{A}^{ij}]$, and $[\bar{M}^{ij}]$ is available. The cross-sectional matrices $[\bar{A}^{ij}]$ and $[\bar{M}^{ij}]$ present no particular problem. If the material making up the node is homogeneous, they can be formed at once from the (spatially constant) group parameters. If the node is composed of heterogeneous regions, a discrete-ordinates or (if legitimate) a diffusion-theory solution based on zero-net-leakage boundary conditions can be found for the node alone, and the definitions (11.6.3) can be applied using the group scalar fluxes constituting that solution.

Determining the coupling coefficients $[\alpha^{(ij)(i'j')}]$ is much more difficult. If the material composition of the nodes in nuclearly homogeneous, diffusion theory is usually valid and it is not difficult to show that

$$\begin{aligned}\alpha_g^{(i+1,j)(ij)}\bar{\Phi}_g^{i+1,j} - \alpha_g^{(ij)(i+1,j)}\bar{\Phi}_g^{ij} &= \frac{1}{h_j^{(y)}}\int D_{gij}\frac{\partial\Phi_g^-(x_{i+1},y)}{\partial x}dy \\ &= \frac{1}{h_j^{(y)}}\int D_{g,i+1,j}\frac{\partial\Phi_g^+(x_{i+1},y)}{\partial x}dy.\end{aligned} \tag{11.6.9}$$

If the derivatives in (11.6.9) are then approximated by

$$\begin{aligned}\frac{1}{h_j^{(y)}}\int\frac{\partial\Phi_g^-(x_{i+1},y)}{\partial x}dy &\approx \frac{(1/h_j^{(y)})\int\Phi_g(x_{i+1},y)dy - \bar{\Phi}_g^{ij}}{h_i^{(x)}/2}, \\ \frac{1}{h_j^{(y)}}\int\frac{\partial\Phi_g^+(x_{i+1},y)}{\partial x}dy &\approx \frac{\bar{\Phi}_g^{i+1,j} - (1/h_j^{(y)})\int\Phi_g(x_{i+1},y)\,dy}{h_{i+1}^{(x)}/2},\end{aligned} \tag{11.6.10}$$

the nodal equation (11.6.8) reduces to the standard finite-difference equation for the $[\bar{\Phi}^{ij}]$.

However, for the usual mesh spacings encountered when the reactor is partitioned into nodes, (11.6.10) is a poor approximation. Thus expressions having a higher-order accuracy than (11.6.10) must be used. Several such expressions have been worked out. However we shall not discuss them other than to remark that they usually lead to nonlinear, five-point difference equations for the $[\bar{\Phi}^{ij}]$. The potential for numerical difficulties arises with nonlinear equations although none have been reported. With the nodes homogeneous and diffusion theory thus valid, the finite-element schemes are also readily applicable to the problem. These yield linear difference equations. However many more unknowns are coupled in a single equation, and it is not clear that such schemes offer a computing-cost advantage over nonlinear higher-order difference equations.

The real difficulty of finding the $[\alpha^{(ij)(i'j')}]$ appears when the nodes are not nuclearly homogeneous. Values of the coupling constants can, of course, be determined from the definitions (11.6.6) and (11.6.7) using fluxes obtained from full-core discrete-ordinates or (if acceptably accurate) full-core diffusion-theory solutions. But obtaining such solutions may be very expensive. Moreover experience indicates that the $[\alpha^{(ij)(i'j')}]$ are quite sensitive to changes in the material properties of the nodes and thus may have to be redetermined every time a significantly different case is to be analyzed.

Thus an essential feature of nodal schemes has been a group of procedures for computing the coupling constants that do not depend on full-core calculations. These procedures are almost always applied to a one-group nodal model and are based on expressing $[\alpha^{(ij)(i'j')}]$ in terms of neutron escape and capture probabilities. To see the motivation for trying to determine the coupling constants in this manner we use (11.6.6) for the one-group case (subscript g omitted) to define

$$W^{(ij)(i+1,j)} \equiv \frac{\int_{h_j^{(y)}} J^{(x)+}(x_{i+1}, y)\, dy}{h_i^{(x)} h_j^{(y)} \overline{\nu\Sigma_f^{ij}\Phi^{ij}}} = \frac{\alpha^{(ij)(i+1,j)}}{h_i \nu\Sigma_f^{ij}}, \tag{11.6.11}$$

with similar expressions for all the $W^{(ij)(i'j')}$.

Clearly $W^{(ij)(i+1,j)}$ is the ratio of the number of neutrons per second crossing face $h_j^{(y)}$ at x_{i+1} (from node (ij) to node $(i + 1, j)$) to the number of neutrons produced per second by fission within node (ij). If all the neutrons leaking into node $(i + 1, j)$ are absorbed there, $W^{(ij)(i+1,j)}$ is just the escape probability across the face $h_j^{(y)}$ at x_{i+1} for a neutron born in node (ij). If one then assumes some spatial distribution for the neutrons born in node (ij)—recall that we are *not* assuming node (ij) to have a homogeneous composition— this escape probability can be computed readily by Monte Carlo methods. Tables can be constructed of such escape probabilities as a function of the range of material characteristics typical of node (ij) and of all nodes making up the reactor (many of which will be identical). The solution of (11.6.8) for a large range of cases then becomes a very simple matter.

Unfortunately the assumption that all neutrons entering a node (ij) are absorbed there is sufficiently invalid that this simple procedure for computing the $W^{(ij)(i'j')}$ is usually unjustifiable. Moreover extending the scheme leads immediately to severe problems. Partial currents of the type $J^{(x)+}(x_{i+1}, y)$ now include neutrons produced by fissions in other nodes along with those, originally born in node (ij), that have returned to (ij) and are escaping a second time (with a different probability since their distribution and spectrum have changed). Thus the $W^{(ij)(i'j')}$ depend on the properties of other nodes in addition to those of (ij).

Acceptably accurate theoretical evaluations of the $W^{(ij)(i'j')}$ still appear to be possible. It is reasonable to assume that the nodes $(i'j')$ that contribute most are the nearest neighbors to (ij). Thus Monte Carlo, discrete-ordinates, or (if sufficiently accurate) group diffusion-theory calculations involving only nearest neighbors and zero-net-current boundary conditions with respect to the rest of the reactor can be used to determine the $W^{(ij)(i'j')}$. To do such calculations one must guess at the shape of the fission source over the entire cluster of nodes, and this implies some prior knowledge of the relative levels of the $[\overline{\Phi}^{i'j'}]$ adjacent to $[\overline{\Phi}^{ij}]$. But such knowledge *can* be attained by experience or, at worst, by iteration of (11.6.8). Thus it appears that, even when fissions from neighboring nodes contribute, the $W^{(ij)(i'j')}$ can be determined quite accurately. The calculations, however, would not be cheap.

In practice such elaborate computations are avoided, and, instead, attempts are made to account in a simple fashion for the effect of neighbors on values of $W^{(ij)(i'j')}$. Again one assumes that the secondary sources contributing to $J^{(x)+}$ are restricted to nearest neighboring nodes. In addition the fission rates in a node and its nearest neighbors are assumed to be flat and equal; escape and capture probabilities based on one-speed cross sections (and hence independent of the past history of the neutron) are used, and the nodes are tacitly assumed to be homogeneous.

It is difficult to establish confidence in such a procedure on theoretical grounds. At best the algebraic form of the resultant expressions for $W^{(ij)(i'j')}$ may provide a means of fitting to accurate calculations or to experiment. The most severe logical criticism of such simple prescriptions is that they neglect heterogeneity effects within the nodes. If the nodes are truly homogeneous (or if they can be homogenized), diffusion theory should be a good approximation to transport theory, and higher-order difference methods for solving the group diffusion equations ought to provide a more accurate answer than those based on crude, one-speed transport theory. Moreover, since the one-group nodal equation (11.6.8) is identical in form with one-group diffusion theory, running times for the two types of approximation should be comparable.

We conclude that the nodal scheme can be made quite accurate, but at a cost which, while not as severe as that incurred for a full-core solution of the heterogeneous problem, is nonetheless forbidding. We further conclude that simple nodal schemes based on nuclear homogenization of the material constituents of the nodes are conceptually inferior to one-group diffusion theory.

Response-Matrix Techniques

The high computing cost associated with the determination of theoretically precise values for the coupling matrices $[\alpha^{(ij)(i'j')}]$ is due in large measure to the fact that these matrices do not depend exclusively on the geometrical constituents and material properties of a single node. Including the effects of other nodes, determining how many neighboring nodes to account for, and estimating the fission-source shape within these nodes make the evaluation of the $[\alpha^{(ij)(i'j')}]$ more costly and more subject to uncertainty. There is thus considerable motivation for finding a method for coupling nodes together that leads to coupling constants dependant only on the properties of an individual node.

In Chapter 10 we encountered a way of finding such constants for slab geometry. It is easy to show from the definition (10.3.41) for the group blackness coefficients α_g and β_g that the flux-current vector at mesh location x_i (which we shall now symbolize by

$\text{Col}\{\Phi_{gi}, J_{gi}\}$) is related to that at x_{i+1} by

$$\begin{bmatrix}\Phi_{gi}\\ J_{gi}\end{bmatrix} = \frac{1}{\beta_{gi} - \alpha_{gi}}\begin{bmatrix}\alpha_{gi} + \beta_{gi} & 2\\ 2\alpha_{gi}\beta_{gi} & \alpha_{gi} + \beta_{gi}\end{bmatrix}\begin{bmatrix}\Phi_{g,i+1}\\ J_{g,i+1}\end{bmatrix}, \tag{11.6.12}$$

where α_{gi} and β_{gi} are the blackness-theory coefficients determined by transport theory for the region between x_{i+1} and x_i. Similarly, if diffusion theory is valid in a sequence of mesh intervals $h_i, h_{i+1}, \ldots, h_I$, (10.3.46) shows that

$$\begin{bmatrix}[\Phi_i]\\ [J_i]\end{bmatrix} = \exp[N_i]h_i \exp[N_{i+1}]h_{i+1} \cdots \exp[N_I]h_I \begin{bmatrix}[\Phi_{I+1}]\\ [J_{I+1}]\end{bmatrix}, \tag{11.6.13}$$

where $[\Phi_i]$ and $[J_i]$ are the G-element column vectors of group fluxes and net currents at x_i, and $[N_i]$, expressed in terms of the group net absorption, fission-production, and diffusion-constant matrices $[A]$, $[M]$, and $[D]$, is

$$[N_i] = \begin{bmatrix}0 & [D_i]^{-1}\\ [A_i - M_i] & 0\end{bmatrix}. \tag{11.3.45}$$

In Chapter 10 our object was to use the transmission matrices connecting the flux-current vectors to infer spatially constant equivalent diffusion-theory parameters for the composite region from x_i to x_I. This turned out to be satisfactory for symmetric arrangements of materials in that the equivalent homogenized constants were independent of the external boundary conditions. For unsymmetric arrangements, however, the equivalent constants *did* depend on the input flux-current vectors (i.e., on external conditions). Moreover, even in the symmetric case the homogenized constants preserved only the net production rate of neutrons in any given group and not the individual reaction rates. Finally we saw no way of extending the method to the "real" problem—that of finding equivalent homogenized parameters for two- and three-dimensional configurations.

Most of these difficulties arose when we attempted to infer equivalent diffusion-theory parameters from the transition matrices. Yet, from the point of view of determining criticality and flux shape, there is no reason to take this step. As we shall see, a series of algebraic equations connecting the flux-current vectors at mesh interfaces is sufficient to determine criticality; and, at some cost, once the flux-current vector is known over an internal surface bounding a node, the detailed flux within the node can be determined from the results of computations involving that node alone.

The response-matrix method is a procedure that exploits this point of view. The essential feature of the scheme is to connect the partial currents $J_g^{\pm}$ crossing nodal surfaces by means of transmission and reflection matrices. We shall first develop the method for the one-dimensional case and then sketch how it can be extended to two and three dimensions.

The Response-Matrix Equation in One Dimension

In one-dimensional slab geometry the notation for partial currents (11.6.4a) can be simplified so that J_{gi}^+ is the total number of group-g neutrons per second crossing a unit surface perpendicular to the X direction from left to right at the point x_i; J_{gi}^- is the analogous partial current from right to left. We shall speak of J_{gi}^+ and J_{gi}^- as partial currents in the $+X$ and $-X$ directions. The corresponding G-element column vectors of group partial currents will be represented by $[J_i^+]$ and $[J_i^-]$. We then define four transmission and reflection matrices, $[T_i^\pm]$ and $[R_i^\pm]$ such that:

The contribution to the partial-current vector in the $+X$ direction at x_{i+1} due to the partial current $[J_i^+]$ at x_i

$$= [T_i^+][J_i^+] \equiv [J_{i+1}^+]_T. \tag{11.6.14a}$$

The contribution to the partial-current vector in the $-X$ direction at x_i due to the partial current $[J_i^+]$ at x_i

$$= [R_i^+][J_i^+] \equiv [J_i^-]_R. \tag{11.6.14b}$$

Similarly

$$\begin{aligned}&[T_i^-][J_i^-] \equiv [J_{i-1}^-]_T,\\ &[R_i^-][J_i^-] \equiv [J_i^+]_R.\end{aligned} \tag{11.6.15}$$

It follows, for example, that the element $T_{i,gg'}^+$ of $[T_i^+]$ is the contribution to the group-g partial current in the $+X$ direction at x_{i+1} due to a unit-input, group-g partial current in the $+X$ direction at x_i. Note that this contribution includes neutrons due to any fissions in the region x_{i+1}–x_i caused by the original input sample. Note also that we must know the angular distribution of the incoming directional flux before these transmission and reflection matrices can be determined. We shall, however, defer the problem of finding $[R]$ and $[T]$ for a moment.

If there is no external source in the interval x_{i+1}–x_i, all neutrons belonging to $[J_i^+]$ can ultimately be traced back to ancestors crossing either the surface at x_i or the surface at x_{i+1}, and we thus have the fundamental relationships

$$\begin{aligned}&[J_{i+1}^+] = [J_{i+1}^+]_T + [J_{i+1}^+]_R = [T_i^+][J_i^+] + [R_{i+1}^-][J_{i+1}^-],\\ &[J_i^-] = [J_i^-]_T + [J_i^-]_R = [T_{i+1}^-][J_{i+1}^-] + [R_i^+][J_i^+].\end{aligned} \tag{11.6.16}$$

Assuming on physical grounds that the necessary inverses exist, we may solve (11.6 .16) for the $2G$-element column vector $\text{Col}\{[J_i^+], [J_i^-]\}$:

$$\begin{bmatrix}[J_i^+]\\ [J_i^-]\end{bmatrix} = \begin{bmatrix}[T_i^+]^{-1} & -[T_i^+]^{-1}[R_{i+1}^-]\\ [R_i^+][T_i^+]^{-1} & -[R_i^+][T_i^+]^{-1}[R_{i+1}^-] + [T_{i+1}^-]\end{bmatrix}\begin{bmatrix}[J_{i+1}^+]\\ [J_{i+1}^-]\end{bmatrix} \equiv [\mathscr{R}_i]\begin{bmatrix}[J_{i+1}^+]\\ [J_{i+1}^-]\end{bmatrix}. \tag{11.6.17}$$

Knowing the partial currents at x_{i+1}, we can immediately find them at x_i. We call the $2G \times 2G$ matrix $[\mathscr{R}_i]$ defined by (11.6.17) the response matrix for node (i). It follows that the partial-current vector at x_{i-1} can be written as $[\mathscr{R}_{i-1}]\text{Col}\{[J_i^+], [J_i^-]\}$ and then as $[\mathscr{R}_{i-1}][\mathscr{R}_i]\text{Col}\{[J_{i+1}^+], [J_{i+1}^-]\}$. Continuing this process leads to the result

$$\begin{bmatrix}[J_1^+]\\ [J_1^-]\end{bmatrix} = [\mathscr{R}]\begin{bmatrix}[J_I^+]\\ [J_I^-]\end{bmatrix} \equiv \begin{bmatrix}[r_{11}] & [r_{12}]\\ [r_{21}] & [r_{22}]\end{bmatrix}\begin{bmatrix}[J_I^+]\\ [J_I^-]\end{bmatrix}, \tag{11.6.18}$$

where subscripts 1 and I refer to the external boundary points of the reactor and the $2G \times 2G$ matrix $[\mathscr{R}]$ is the product of the response matrices of all the intervening regions.

The external boundary conditions at x_1 and x_I are that there are no neutrons entering the reactor from the outside. Thus $[J_1^+] = [J_I^-] = 0$, and (11.6.18) yields

$$0 = [r_{11}][J_I^+], \tag{11.6.19}$$

which will have a nontrivial solution only if

$$\text{Det}\{r_{11}\} = 0. \tag{11.6.20}$$

The vanishing of the $G \times G$ determinant (11.6.20) is thus the mathematical statement of the criticality condition in terms of response matrices.

Thus it is a straightforward problem to analyze for criticality once the response matrices for the nodes making up the reactor have been found. We now turn to that problem.

Determination of Transmission and Reflection Matrices

The first thing to note about the determination of transmission and reflection matrices is that, if we isolate the region x_{i+1}–x_i in a vacuum and introduce a source of neutrons at x_i, *all* the transmitted current $[J_{i+1}^+]_\text{T}$ of (11.6.14a) and *all* the reflected current $[J_i^-]_\text{R}$ will be due to the partial current $[J_i^+]$ associated with the source. Thus, for this case, we can drop subscripts T and R in (11.6.14). Similarly, if we isolate the region x_i–x_{i-1} in a vacuum and introduce a source at x_i that leads to a partial current $[J_i^-]$, the partial currents $[J_{i-1}^-]_\text{T}$ and $[J_i^+]_\text{R}$ will be due entirely to this source, and we may drop subscripts T and R in (11.6.15). Consequently, if we introduce a sequence of neutron sources at x_i leading to partial currents of unit magnitude,

$$
\begin{aligned}
[I_i^+]_1 &\equiv \mathrm{Col}\{1, 0, 0, \ldots, 0\},\\
[I_i^+]_2 &\equiv \mathrm{Col}\{0, 1, 0, \ldots, 0\},\\
&\vdots\\
[I_i^+]_G &\equiv \mathrm{Col}\{0, 0, \ldots, 1\},
\end{aligned}
\tag{11.6.21}
$$

and if the column vectors $[J_{i+1}^+]_g$ and $[J_i^-]_g$ are defined as the outgoing partial currents at x_{i+1} and x_i for the unit input vector $[I_i^+]_g$, we have immediately from (11.6.14) that $[J_{i+1}^+]_g$ is the gth column of $[T_i^+]$ and $[J_i^-]_g$ is the gth column of $[R_i^+]$. Thus G separate calculations determine both $[T_i^+]$ and $[R_i^+]$ (see (11.6.14) and (11.6.15)).

A similar procedure involving unit source vectors $[I_{i+1}^-]_g$ ($g = 1, 2, \ldots, G$) and node (i) again isolated in a vacuum permits us to find $[T_{i+1}^-]$ and $[R_{i+1}^-]$. (If the materials making up node (i) are unsymmetrically located with respect to its center plane, $[T_{i+1}^-]$ will not equal $[T_i^+]$ and $[R_{i+1}^-]$ will not equal $[R_i^+]$.)

The crucial step in this process is thus the determination of outgoing partial currents at x_i and x_{i+1} for a known input partial current at one of the faces of node (i). It is at this point that we must make an essential approximation.

It follows from the physical definition (11.6.4) of the partial currents that, for each energy group g,

$$
\begin{aligned}
J_{gi}^+ &= \frac{1}{2}\int_0^1 \mu\Psi_g(x_i, \mu)\, d\mu,\\
J_{gi}^- &= \frac{1}{2}\int_0^{-1} \mu\Psi_g(x_i, \mu)\, d\mu,
\end{aligned}
\tag{11.6.22}
$$

where $\Psi_g(x_i, \mu)$ is the directional flux density for group g at point x_i. Thus to get J_{gi}^- and $J_{g,i+1}^+$ we must know $\Psi_g(x_i, \mu)$ and $\Psi_g(x_{i+1}, \mu)$. These can readily be found by solving multigroup P_n or discrete-ordinates equations for the unit-input partial currents. But to set up such a problem it is insufficient to know only that the partial input current is unity for group g. We must also know the *angular* distribution of the neutrons comprising that current. Unfortunately, unless a multigroup transport solution for the whole reactor is available, the angular distribution of the neutrons traveling in the $+X$ direction at x_i or in the $-X$ direction at x_{i+1} will be unknown. Moreover it cannot even be estimated from the solutions of the response-matrix equations (11.6.17) since these provide only the integrals (11.6.22). Accordingly we must either guess at the angular distributions of neutrons on the surfaces of the nodes or somehow generalize the response-matrix equations so that they relate unknowns specifying more than do the $[J^\pm]$ about the angular distribution of the neutrons on the nodal surfaces.

We can deal with both of these possibilities by expanding the angular flux on the surfaces of the nodes and assuming that only a few terms in the expansion are required to represent that distribution accurately. (If many terms are required, there is no escape from having to use a transport-theory method for the whole problem.)

For a great many cases of interest, even though the angular distribution may be strongly peaked at some locations in the interior of the node, it is nearly isotropic on the nodal surfaces. This circumstance suggests that we assume a low-order expansion in μ for the $\Psi_g(x, \mu)$ on the nodal surfaces. The P_n approximation (8.4.12) immediately comes to mind. For the case at hand, however, an alternate angular representation, called the *double-P_n approximation* (DP_n) turns out to be conceptually attractive for one-dimensional problems, and we shall digress to outline this approximation.

The Double-P_n Representation of Nodal Surface Fluxes

In the DP_n approximation two different Legendre expansions for the angular behavior of $\Psi_g(x, \mu)$ are made: one for neutrons having a component of motion in the $+X$ direction and one for neutrons having a component in the $-X$ direction. Specifically, the group fluxes are expanded as

$$\Psi_g(x, \mu) = \begin{cases} \sum_{n=0}^{\infty} 2(2n + 1)\Psi_{gn}^{+}(x)P_n(2\mu - 1) \text{ for } 0 < \mu \leq 1, \\ \sum_{n=0}^{\infty} 2(2n + 1)\Psi_{gn}^{-}(x)P_n(2\mu + 1) \text{ for } -1 \leq \mu < 0. \end{cases} \tag{11.6.23}$$

The orthogonality relationship (8.4.9) for Legendre polynomials gives at once

$$\begin{aligned} &\int_0^1 (2n + 1)P_l(2\mu - 1)P_n(2\mu - 1)\, d\mu = \delta_{ln}, \\ &\int_{-1}^{0} (2n + 1)P_l(2\mu + 1)P_n(2\mu + 1)\, d\mu = \delta_{ln}, \end{aligned} \tag{11.6.24}$$

as a result of which (11.6.23) yields

$$\begin{aligned} \Psi_{gn}^{+}(x) &= \int_0^1 \frac{d\mu}{2} P_n(2\mu - 1)\Psi_g(x, \mu), \\ \Psi_{gn}^{-}(x) &= \int_{-1}^{0} \frac{d\mu}{2} P_n(2\mu + 1)\Psi_g(x, \mu). \end{aligned} \tag{11.6.25}$$

The scalar flux, net current, and partial currents are given by substituting (11.6.23) into the expressions that define them:

$$\Phi_g(x) \equiv \int_{-1}^{+1} \frac{d\mu}{2} \Psi_g(x, \mu) = \Psi_{g0}^{+}(x) + \Psi_{g0}^{-}(x), \tag{11.6.26}$$

$$J_g(x) \equiv \int_{-1}^{+1} \mu \frac{d\mu}{2} \Psi_g(x, \mu) = \frac{1}{2}\{\Psi_{g0}^{+}(x) - \Psi_{g0}^{-}(x)\} + \frac{1}{2}\{\Psi_{g1}^{+}(x) + \Psi_{g1}^{-}(x)\}, \tag{11.6.27}$$

$$J_g^{+}(x) \equiv \int_{0}^{+1} \mu \frac{d\mu}{2} \Psi_g(x, \mu) = \frac{1}{2}\{\Psi_{g0}^{+}(x) + \Psi_{g1}^{+}(x)\}, \tag{11.6.28}$$

$$J_g^{-}(x) \equiv \int_{0}^{-1} \mu \frac{d\mu}{2} \Psi_g(x, \mu) = \frac{1}{2}\{\Psi_{g0}^{-}(x) - \Psi_{g1}^{-}(x)\}. \tag{11.6.29}$$

If the expansion is terminated at the Nth term, the DP_N expression for $\Psi_g(x, \mu)$ will have twice as many unknowns as a conventional P_N expression (8.4.12). Thus we expect roughly that a DP_0 approximation will be equivalent to a P_1 approximation; a DP_1 will be equivalent to a P_3; etc. Actually the DP_n representation is more accurate near an interface between two media, where the outward-directed and inward-directed angular distributions may be quite different. Conversely the regular P_n approximation is better asymptotically far from boundaries.

For the purposes of making assumptions about the nature of the angular distribution of the $\Psi_g(x_i, \mu)$, a DP_n expansion is particularly convenient since the solutions we seek for the isolated node in order to determine the transmission and reflection matrices require only that we supply the angular distributions of the unit group sources (11.6.21) for *inward* directions. It is milder to assume a behavior for *just* the inward directions than for *both* the inward and outward directions.

The simplest approximation of the double-P variety that we can make about the angular distribution of neutrons crossing nodal surfaces is that it can be represented by a DP_0 expansion, so that on the nodal surfaces x_i

$$\Psi_g(x_i, \mu) = \begin{cases} 2\Psi_{g0}^{+}(x_i) \text{ for } 0 < \mu \leq 1, \\ 2\Psi_{g0}^{-}(x_i) \text{ for } -1 \leq \mu < 0. \end{cases} \tag{11.6.30}$$

Thus we assume that $\Psi_g(x_i, \mu)$ is isotropic in both the $+X$ and the $-X$ directions but permit the magnitudes of the two isotropic distributions to differ from each other. Clearly the approximation (11.6.30) can better match the correct distribution than can a *completely* isotropic distribution. Indeed the assumption (11.6.30) is expected to be acceptable for a great many practical cases, since, with an altered definition of the group diffusion constants, the DP_0 approximation yields the group diffusion equations. (For $\Sigma_{s1}(x, E' \to E) = \Sigma_{s1}(x, E')\delta(E' - E)$, D_g for a DP_0 approximation is $1/4(\Sigma_{tg} - \frac{3}{4}\Sigma_{s1g})$ rather than $1/3(\Sigma_{tg} - \Sigma_{s1g})$.) Thus the approximation (11.6.30) should be adequate if diffusion theory can reasonably be applied on the nodal surfaces.

Since all components $\Psi_{gn}^{\pm}$ for $n > 0$ are neglected in a DP_0 approximation, (11.6.26)–

(11.6.29) reduce to

$$\begin{aligned}
\Phi_g(x) &= \Psi^+_{g0}(x) + \Psi^-_{g0}(x), \\
J_g(x) &= \frac{1}{2}\{\Psi^+_{g0}(x) - \Psi^-_{g0}(x)\}, \\
J^+_g(x) &= \frac{1}{2}\Psi^+_{g0}(x), \\
J^-_g(x) &= \frac{1}{2}\Psi^-_{g0}(x).
\end{aligned} \tag{11.6.31}$$

Thus, for a DP_0 approximation,

$$\begin{aligned}
\Phi_g(x) &= 2\{J^+_g(x) + J^-_g(x)\}, \\
J_g(x) &= J^+_g(x) - J^-_g(x).
\end{aligned} \tag{11.6.32}$$

It is not difficult to show that a regular P_1 expansion of $\Psi_g(x, \mu)$ leads to these same relationships between Φ_g, J_g, and the two partial currents.

Summary of Procedure for Finding Response Matrices

The simplest overall method for finding a response matrix $[\mathscr{R}_i]$ defined by (11.6.17) is as follows:

1. Assume that the $\Psi_g(x, \mu)$ can be represented by the DP_0 form (11.6.30) at the nodal surface location x_i.
2. For G input sources at x_i having the form (11.6.21) and isotropic angular distributions, determine the corresponding G values of $[J^+_{i+1}]$ and $[J^-_i]$ for node (i) in a vacuum. (Generally the solution of a high-order approximation to the transport equation will accomplish this purpose by yielding values of the $\Psi_g(x, \mu)$ to be inserted into (11.6.22); however a Monte Carlo calculation may be used to find the outgoing partial currents directly.)
3. Each such $[J^-_i]_g$ so determined is the gth column of the reflection matrix $[R^+_i]$, and each $[J^+_{i+1}]_g$ is the gth column of the transmission matrix $[T^+_i]$. The matrices $[R^-_{i+1}]$ and $[T^-_{i+1}]$ are found by an analogous procedure. However, for symmetrically constructed nodes, $[T^-_{i+1}] = [T^+_i]$ and $[R^-_{i+1}] = [R^+_i]$.
4. Compute the response matrix $[\mathscr{R}_i]$ from its definition (11.6.17).

Note that assumption (1) is the only significant one in this whole process. Also note that only the properties of node (i) are needed to determine $[\mathscr{R}_i]$.

The particular response matrix we have found by the process just summarized connects the partial-current vectors $\text{Col}\{[J^+], [J^-]\}$ at the two surfaces of a node. It is easy to transform the result in order to find a response matrix connecting the corresponding vectors of the flux and total current. Thus, writing (11.6.32) in matrix form, we have

$$\begin{bmatrix} [\Phi_i] \\ [J_i] \end{bmatrix} = \begin{bmatrix} 2I & 2I \\ I & -I \end{bmatrix} \begin{bmatrix} [J_i^+] \\ [J_i^-] \end{bmatrix} \tag{11.6.33}$$

or

$$\begin{bmatrix} [J_i^+] \\ [J_i^-] \end{bmatrix} = \begin{bmatrix} \frac{1}{4}I & \frac{1}{2}I \\ \frac{1}{4}I & -\frac{1}{2}I \end{bmatrix} \begin{bmatrix} [\Phi_i] \\ [J_i] \end{bmatrix}, \tag{11.6.34}$$

where I is the $G \times G$ diagonal unit matrix.

Equation (11.6.17) leads immediately to

$$\begin{bmatrix} [\Phi_i] \\ [J_i] \end{bmatrix} = \begin{bmatrix} 2I & 2I \\ I & -I \end{bmatrix} [\mathscr{R}_i] \begin{bmatrix} \frac{1}{4}I & \frac{1}{2}I \\ \frac{1}{4}I & -\frac{1}{2}I \end{bmatrix} \begin{bmatrix} [\Phi_{i+1}] \\ [J_{i+1}] \end{bmatrix}. \tag{11.6.35}$$

Comparison with (11.6.12) or (11.6.13) permits us to establish the relationship between blackness coefficients, equivalent diffusion-theory parameters, and the transmission and reflection matrices.

It is possible, by doubling the dimensions of the response matrices, to remove the restriction that the $\Psi_g(x, \mu)$ on the nodal surfaces be well represented by a DP_0 expansion. We can instead deal with the four DP_1 components $\Psi_{g0}^{\pm}$ and $\Psi_{g1}^{\pm}$ as unknowns on each surface (rather than the two partial currents $J_g^{\pm}$). This generalization is closely equivalent to assuming that $\Psi_g(z, \mu)$ can be represented by a P_3 approximation on those surfaces. It should be a very good assumption for almost all cases of practical interest. The cost of the more accurate formulation is that the unknown column vectors become twice as large, e.g., $\text{Col}\{[\Psi_{0i}^+],[\Psi_{0i}^-],[\Psi_{1i}^+],[\Psi_{1i}^-]\}$, and twice the number of calculations must be performed to determine the elements of the transmission and reflection matrices.

Once the $[J_i^{\pm}]$ (or $[\Psi_{0i}^{\pm}]$ and $[\Psi_{1i}^{\pm}]$) are determined by solution of the response-matrix equations, the detailed solution within each node may be reconstructed from the (now known) values of the $[J_i^{\pm}]$ and the solutions of the unit-source problems that were run originally to determine the reflection and transmission matrices. If less detailed information is all that is required, appropriate averages over these unit-input fluxes can first be made. Data processing of this kind requires considerable programming effort, but it is not costly in machine running time.

Extension to Two Dimensions

A very practical virtue of the response-matrix technique—one not shared by many other methods—is that it can be extended readily to two and three dimensions. The mechanical part of the extension is, conceptually, almost trivial. One merely defines additional transmission matrices so that the partial current out of each face of the node is connected to all the input partial currents. It is, however, necessary to make a physical assumption about the spatial distribution of input currents over a given face of the node, and this ultimately limits the accuracy of the method.

To illustrate the additional complexity of the analysis associated with the extension beyond slab geometry we shall consider the two-dimensional rectangular case for the node depicted in Figure 11.8. With $[J^{\pm}_{i,j+1/2}]$ representing a G-element vector of partial currents averaged over y_j–y_{j+1} and crossing the surface y_j–y_{j+1} in the $+X$ or $-X$ direction at x_i and $[J^{\pm}_{i+1/2,j}]$ representing the analogous currents averaged over x_i–x_{i+1} and crossing the surface x_i–x_{i+1} in the $+Y$ or $-Y$ direction at y_j, we define reflection and transmission matrices $[R^{\pm}_{i,j+1/2}]$, $[R^{\pm}_{i+1/2,j}]$, $[T^{\pm}_{i,j+1/2}]$, and $[T^{\pm}_{i+1/2,j}]$ such that, for example, if node (i) is in a vacuum, $[R^{+}_{i,j+1/2}]$ is the space-averaged vector of group partial currents crossing the face y_j–y_{j+1} at x_i and traveling in the $-X$ direction, these partial currents being due to an input vector $[J^{+}_{i,j+1/2}]$ of partial currents crossing the surface y_j–y_{j+1} at location x_i and traveling in the $+X$ direction. In addition we introduce transmission matrices $[S_{(i,j+1/2)(i\pm 1/2,j)}]$ and $[S_{(i+1/2,j)(i,j\pm 1/2)}]$ such that, for example, if node (i) is in a vacuum, $[S_{(i,j+1/2)(i+1/2,j)}][J^{+}_{i,j+1/2}]$ is the average partial-current vector in the $-Y$ direction crossing the nodal face x_i–x_{i+1} at y_j due to an input average partial-current vector $[J^{+}_{i,j+1/2}]$ in the $+X$ direction crossing y_j–y_{j+1} at x_i. In terms of these reflection and transmission matrices, the fundamental current-balance conditions for node (i) are

$$
\begin{aligned}
[J^{-}_{i,j+1/2}] &= [R^{+}_{i,j+1/2}][J^{+}_{i,j+1/2}] + [S_{(i+1/2,j+1)(i,j+1/2)}][J^{-}_{i+1/2,j+1}] \\
&\quad + [T^{-}_{i+1,j+1/2}][J^{-}_{i+1,j+1/2}] + [S_{(i+1/2,j)(i,j+1/2)}][J^{+}_{i+1/2,j}], \\
[J^{-}_{i+1/2,j}] &= [R^{+}_{i+1/2,j}][J^{+}_{i+1/2,j}] + [S_{(i,j+1/2)(i+1/2,j)}][J^{+}_{i,j+1/2}] \\
&\quad + [T^{-}_{i+1/2,j+1}][J^{-}_{i+1/2,j+1}] + [S_{(i+1,j+1/2)(i+1/2,j)}][J^{-}_{i+1,j+1/2}], \\
[J^{+}_{i+1,j+1/2}] &= [R^{-}_{i+1,j+1/2}][J^{-}_{i+1,j+1/2}] + [S_{(i+1/2,j+1)(i+1,j+1/2)}][J^{-}_{i+1/2,j+1}] \\
&\quad + [T^{+}_{i,j+1/2}][J^{+}_{i,j+1/2}] + [S_{(i+1/2,j)(i+1,j+1/2)}][J^{+}_{i+1/2,j}], \\
[J^{+}_{i+1/2,j+1}] &= [R^{-}_{i+1/2,j+1}][J^{-}_{i+1/2,j+1}] + [S_{(i,j+1/2)(i+1/2,j+1)}][J^{+}_{i,j+1/2}] \\
&\quad + [T^{+}_{i+1/2,j}][J^{+}_{i+1/2,j}] + [S_{(i+1,j+1/2)(i+1/2,j+1)}][J^{-}_{i+1,j+1/2}].
\end{aligned}
\tag{11.6.36}
$$

We can think of every internal node as giving rise to four unknown, outgoing partial currents and as generating four coupling equations of the type (11.6.36) for these unknowns. If a node has a face that is part of the external surface of the reactor, the incoming partial current across that face has a zero value, and we need not write an equation for the outgoing partial current, since, while it is nonzero, it contributes nothing to any other expression for a partial current. Thus a node with one face part of the external surface introduces only three unknowns that need to be found and supplies three equations for these unknowns. An analogous reduction in the number of coupled unknowns and coupling equations occurs for nodes having more than one face that is part of the external surface of the reactor.

We conclude, then, that there are exactly as many equations of the type (11.6.36) as there are unknown partial currents. A variety of standard methods are available for solving these coupled homogeneous, linear equations.

To determine the $G \times G$ transmission and reflection matrices one must perform two-dimensional, G-group transport calculations for unit-source vectors of the type (11.6.21), treated as input to node (i) in a vacuum. A separate series of such problems must be run for each face of node (i) unless the node has 90° rotational symmetry about its center axis, in which case all $[R]$'s are equal, all $[T]$'s are equal, and all $[S]$'s are equal. As in the slab case, it is simplest to assume in performing these transport calculations that the directional fluxes $\Psi_g(\mathbf{r}, \mathbf{\Omega})$ can be approximated by a DP_0 expansion on each nodal surface, the directional fluxes always being assumed isotropic in the individual hemispheres pointing into and out of the node. Also as in the one-dimensional case, detailed fluxes may be reconstructed by multiplying the partial currents on the nodal surfaces into the unit-source solutions used to obtain the transmission and reflection matrices.

At the cost of doubling the number of unknowns, the assumption that the angular shape of flux on nodal surfaces is well represented by a DP_0 expansion can again be replaced by the much less restrictive assumption that a DP_1 expansion is adequate. Since the planes across which these double-P expansions are discontinuous always coincide with nodal surfaces, there is some conceptual ambiguity about the angular shape of the flux at corner points. But this does not appear to be a matter of practical concern.

We defined the partial current vectors $[J^{\pm}_{i'j'}]$ to be spatial averages along the nodal surfaces. Unfortunately this is not a sufficiently precise statement to permit the reflection and transmission matrices to be determined unambiguously. Clearly they will depend on the specific spatial shape of the input partial current along the nodal face. The simplest assumption to make about that current is that it is spatially flat. If the nodes are part of a very large reactor, acceptably accurate criticality and detailed flux distributions may result. If not, one may assume some shape based on experience or on some approximate full-core calculation (possibly using the conventional homogenized, flux-weighted parameters). However a more theoretically sound (but expensive) procedure would be to increase the number of unknown partial currents to be determined. For example currents that are spatially flat along each *half* of a nodal surface could be found (i.e., $[J^{\pm}_{i,j+1/2}]$ could be replaced by independent vectors $[J^{\pm}_{i,j+1/4}]$ and $[J^{\pm}_{i,j+3/4}]$). Twice the number of nodal transport-theory computations would have to be performed to determine the associated reflection and transmission matrices. A conceptually more satisfying alternative may be to assume that the partial currents $[J^{\pm}_i(y)]$ and $[J^{\pm}_i(x)]$ along nodal faces y_j–y_{j+1} and x_i–x_{i+1} are composed of two overlapping linear functions (see Figure 11.2), so that, for example, $[J^{+}_i]$ along surface y_j–y_{j+1} is a sum of two terms, $[J^{+}_{i,(j)^+}]((y_{j+1} - y)/h^{(y)}_j)$ for $y_j < y \leq y_{j+1}$ and $[J^{+}_{i,(j+1)^-}]((y - y_j)/h^{(y)}_j)$ for $y_j \leq y < y_{j+1}$. The "half-tent" functions $(y_{j+1} - y)/h_j^{(y)}$ and $(y - y_j)/h_j^{(y)}$ are then used as the source shapes in determining the necessary reflection and transmission matrices. Note that we permit $[J_{i,(j)+}^+]$, which specifies the magnitude of $(y_{j+1} - y)/h_j^{(y)}$, and $[J_{i,(j)-}^+]$, which specifies the magnitude of $(y - y_{j-1})/h_{j-1}^{(y)}$, to be independent vectors. This permits us to avoid dealing with currents *at* nodal corner points (i, j) where they may be singular.

Extension of the response-matrix method to other geometries—for example hexagonal, or three-dimensional rectangular geometries—is straightforward. The resultant equations for the partial currents and the transport-theory calculations necessary to determine the response matrices become more complex. Solving them, however, remains simpler than determining a transport-theory solution for the overall reactor.

In this connection it should be noted that there may be many instances when a diffusion-theory solution for the individual, heterogeneous nodes may be acceptably accurate for the determination of the nodal response functions. In such cases the response-matrix scheme provides an economical way of "stitching together" the detailed nodal flux shapes into an overall detailed flux solution for the entire reactor.

The response-matrix technique thus emerges as a potentially very powerful scheme for analyzing a large number of commonly encountered reactor configurations. It is particularly attractive for reactors composed of only a few different kinds of nodes on the surfaces of which the directional flux is a smooth function of position and direction. However the extent to which this potential can be realized in practical applications has not yet been established.

References

Kang, C. M., and Hansen, K. F., 1973. "Finite Element Methods for Reactor Analysis," *Nuclear Science and Engineering* **51**: 456–495.

Kaper, H. G., Leaf, G. K., and Linderman, A. J., 1972. *Applications of Finite Element Methods in Reactor Mathematics*, ANL-7925, Argonne National Laboratory.

Kaplan, S., 1962. "Some New Methods of Flux Synthesis," *Nuclear Science and Engineering* **13**: 22–35.

Mathematical Models and Computational Techniques for Analysis of Nuclear Systems, 1973. Proceedings of a conference held in Ann Arbor, Michigan, 9–11 April 1973, National Technical Information Service, U.S. Department of Commerce, Springfield, Virginia, CONF-730414-P1.

Numerical Reactor Calculations, 1972. Proceedings of a seminar held in Vienna, Austria, 17–21 January 1972, Vienna IAEA, STI/Pub/307, Sessions V–VIII.

Pfeiffer, W., and Shapiro, J. L., 1969. "Reflection and Transmission Functions in Reactor Physics," *Nuclear Science and Engineering* **38**: 253–264.

Shimuzu, A., and Aoki, K., 1972. *Application of Invariant Embedding to Reactor Physics* (New York: Academic Press).

Stacey, W. M., Jr., 1974. *Variational Methods in Nuclear Reactor Physics* (New York: Academic Press).

Problems

1. Suppose that a reactor is constructed of a single homogeneous composition and that (9.5.2) has been solved for the (unnormalized) P_1 components $F_0(E)$ and $F_1(E)$ of the asymptotic spectrum associated with that composition. What choice of expansion functions $u_1(E)$, $u_2(E)$, $v_1(E)$, $v_2(E)$ and of weight functions $u_1^*(E)$, $u_2^*(E)$, $v_1^*(E)$, $v_2^*(E)$ will reduce (11.2.12) to the non-overlapping two-group equations ((9.4.18) with $G = 2$) for that reactor?

Notice that the conventional form (with the $[D]$ matrix diagonal) does not appear in a natural fashion as a special case of (11.2.12).

2. Consider a slab reactor extending from $z = -Z$ to $z = +Z$. Suppose that for a one-group model the flux is represented by the trial function

$$\Phi(z) \approx T^{(1)} \cos\frac{\pi z}{2Z} + T^{(2)} \sin\frac{\pi z}{Z}$$

and that the values of the one-group parameters are:

$$\nu\Sigma_f = 0.01 \text{ cm}^{-1}, \qquad Z = 50\pi \text{ cm}, \qquad D = 2.5 \text{ cm},$$
$$\Sigma = \begin{cases} 0.00975 \text{ cm}^{-1} & \text{for } -Z \le z \le 0, \\ 0.00970 \text{ cm}^{-1} & \text{for } 0 < z \le Z. \end{cases}$$

Using $\cos(\pi z/2Z)$ and $\sin(\pi z/Z)$ as weight functions, find by the weighted-residual method the two values of λ implied by this choice of weight and trial functions. Find the corresponding ratios $T^{(1)}/T^{(2)}$ and sketch the two solutions $\Phi(z)$ that result.

3. Consider the following situation: We are trying to solve approximately the two-group diffusion equations

$$\nabla \cdot D_1 \nabla \Phi_1 - \Sigma_1 \Phi_1 + \frac{1}{\lambda_0}(\nu\Sigma_{f1}\Phi_1 + \nu\Sigma_{f2}\Phi_2) = 0,$$
$$\nabla \cdot D_2 \nabla \Phi_2 - \Sigma_2 \Phi_2 + \Sigma_{21}\Phi_1 = 0,$$

for the fundamental solution Col $\{\Phi_1, \Phi_2\}$ corresponding to the eigenvalue λ_0, and we are using as trial functions

$$\Phi_1(\mathbf{r}) = \Psi_1^1(\mathbf{r})T_1^1 + \Psi_2^1(\mathbf{r})T_2^1,$$
$$\Phi_2(\mathbf{r}) = \Psi_1^2(\mathbf{r})T_1^2 + \Psi_2^2(\mathbf{r})T_2^2,$$

where the expansion functions $\Psi_k^g(\mathbf{r})$ $(g, k = 1, 2)$ are given and the numbers $T_k^g(g, k = 1, 2)$ are to be found. We use the weighted-residual method to find the T_k^g with an arbitrary set of linearly independent weight functions $W_j^g(\mathbf{r})$ $(g, j = 1, 2)$.

Now suppose we accidently select for Col $\{\Psi_1^1(\mathbf{r}), \Psi_2^1(\mathbf{r})\}$ the column vector Col $\{\alpha_1\Phi_1(\mathbf{r}), \alpha_2\Phi_2(\mathbf{r})\}$, where Col $\{\Phi_1, \Phi_2\}$ happens to be the solution we are seeking corresponding to eigenvalue λ_0 and α_1 and α_2 are numbers, and suppose further that the second expansion function, Col $\{\Psi_2^1, \Psi_2^2\}$, is linearly independent of Col $\{\Psi_1^1, \Psi_1^2\}$. Using the theorem that a set of m homogeneous linear algebraic equations in n unknowns has a solution other than the trivial one $(0, 0, \ldots, 0)$ if and only if the matrix of the coefficients is of rank less than n, show that one solution to the equation (11.2.18) for the coefficients T_k^g exists for $\lambda = \lambda_0$ and is

$$\frac{T_1^1}{T_1^2} = \frac{\alpha_2}{\alpha_1}, \qquad T_2^1 = T_2^2 = 0.$$

4. Show that the matrix equation (11.2.30) is also valid for the collapsed-group synthesis approximation (the generalization of (11.2.20)) if we replace the $G \times G$ diagonal matrix $[W_j]$ in the definitions (11.2.28) of the elements $\mathscr{L}_{jk}$ and $\mathscr{M}_{jk}$ by $[W_j]^{\mathrm{T}}$, the transpose of $[W_j]$, a G-element column vector, and interpret the elements T_k of $[T]$ as scalars rather than as G-element column vectors.

5. For the slab-geometry diffusion-theory model, derive difference equations analogous to (11.3.27) for an interior point, using as the space of trial and weight functions

$$\left.\begin{array}{l}[U(x)] = [\Phi_n] \\ [U^*(x)] = [\Phi_n^*]\end{array}\right\} \text{ for } x_n < x < x_{n+1} \qquad (n = 1, 2, \ldots, N-2),$$

$$\left.\begin{array}{l}[V(x)] = [J_n] \\ [V^*(x)] = [J_n^*]\end{array}\right\} \text{ for } x_n - \frac{h_{n-1}}{2} < x < x_n + \frac{h_n}{2} \qquad (n = 1, 2, \ldots, N-1).$$

6. For the slab-geometry diffusion-theory model, derive difference equations analogous to (11.3.27), using as the space of trial and weight functions

$$[u_n(x)] = \left\{\begin{array}{ll} \dfrac{x_{n+1} - x}{h_n}[\Phi_n] & \text{for } x_n \le x \le x_{n+1} \\ \dfrac{x - x_{n-1}}{h_{n-1}}[\Phi_n] & \text{for } x_{n-1} \le x \le x_n \\ 0 & \text{for all other } x \end{array}\right\} (n = 1, 2, \ldots, N-1),$$

$$[U(x)] = \sum_{n=1}^{N-1} [u_n(x)],$$

$$[u_n^*(x)] = \left\{\begin{array}{ll} \dfrac{x_{n+1} - x}{h_n}[\Phi_n^*] & \text{for } x_n \le x \le x_{n+1} \\ \dfrac{x - x_{n-1}}{h_{n-1}}[\Phi_n^*] & \text{for } x_{n-1} \le x \le x_n \\ 0 & \text{for all other } x \end{array}\right\} (n = 1, 2, \ldots, N-1),$$

$$[U^*(x)] = \sum_{n=1}^{N-1} [u_n^*(x)],$$

$$\left.\begin{array}{l}[V(x)] = [J_n] \\ [V^*(x)] = [J_n^*]\end{array}\right\} \text{ for } x_n < x < x_{n+1} \qquad (n = 0, 1, 2, \ldots, N-1).$$

7. Simplify the functional (11.3.16) under the assumption that $[U]$ and $[\mathbf{V}] \cdot \mathbf{N}$ are continuous across internal interfaces and that $[\mathbf{V}]$ is not an independent function but is, rather, related to $[U]$ by Fick's law.

8. A simple method for estimating the flux shape throughout a reactor in which the tips of all control rods are at a height $z = Z_1$ is to use one set of expansions of the type (11.4.2) and (11.4.4) with $K = 1$ for the lower portion of the reactor ($z < Z_1$) and another set with $K = 1$ for $z > Z_1$. Thus (11.4.2) is replaced by

$$[U] = \left\{\begin{array}{l}[\Psi_l(x, y)]\,[T_l(z)] \text{ for } z < Z_1, \\ [\Psi_u(x, y)]\,[T_u(z)] \text{ for } z > Z_1,\end{array}\right.$$

and similarly for $[\mathbf{V}]$. ($[\Psi_l(x, y)]$ and $[\Psi_u(x, y)]$ are known, but differ from each other.)

Suppose that this is done and, further, that analogous expansions for $[U^*]$ and $[\mathbf{V}^*]$, also discontinuous at $z = Z_1$, are assumed. Show that requiring the first-order variation of (11.3.15) to vanish leads to physically unacceptable conditions relating $[T_l(z)]$ to $[T_u(z)]$ at $z = Z_1$. Show further that, if the discontinuity in $[U^*]$ and $[\mathbf{V}^*]$ is placed at some other elevation $Z_2 \neq Z_1$, acceptable (i.e., not obviously absurd) relations between $[T_l(z)]$ and $[T_u(z)]$ at $z = Z_1$ result.

9. Verify (11.4.18).

10. Derive (11.5.21)–(11.5.23).

11. Derive equations analogous to (11.5.21)–(11.5.23) using (11.5.20) as the trial function.

12. Although the correspondence is somewhat disguised by the use of the weak form of the diffusion equation, the defining equations (such as (11.5.13)) of the finite-element method can also be obtained by the weighted-residual method. To do so $[U(x)]$ of (11.5.10) is substituted into the weak form of the group diffusion equations (11.5.7), and the result is required to be valid when each of the individual functions (11.5.11) is used as weight function.

Using expansions similar to (11.5.10) applied to the *time* domain as trial functions for $T(t)$ and $C(t)$ and the $u_k^*(t)$ of (11.5.11) as weight functions, find time-differenced equations for the point-kinetics equations (7.3.13) assuming one group of delayed neutrons and no external source.

13. Derive (11.6.12) from (10.3.41). (Note that the notation of (10.3.41) differs from that of (11.6.12).)

14. Derive the criticality condition analogous to (11.6.20) under the conditions that the *net* currents at x_1 and x_{I} vanish.

15. The double-P_0 approximation for the directional flux density for slab geometry is

$$\Psi(z, \mu, E) \approx \begin{cases} 2\Psi_0^+(z, E) \text{ for } 0 < \mu \leq 1, \\ 2\Psi_0^-(z, E) \text{ for } -1 \leq \mu < 0. \end{cases}$$

By inserting the approximation into (8.4.11) with $\Sigma_s(z, E' \to E, \mu_0)$ approximated by $\Sigma_{s0}(z, E' \to E) + 3\,\mu\mu'\Sigma_{s1}(z, E')\,\delta(E' - E)$, and using the weighted-residual method, show that

$$J(z, E) = -\frac{1}{4[\Sigma_t(z, E) - \frac{3}{4}\Sigma_{s1}(z, E)]}\frac{\partial}{\partial z}\Phi(z, E),$$

where J and Φ are related to Ψ_0^+ and Ψ_0^- by relations analogous to (11.6.31).

16. For a one-group model applied to a homogeneous slab of material find the relationship between the matrix elements of the matrix $[\mathscr{R}_i]$ defind by (11.6.17) and:
a. blackness coefficients α and β for the slab;
b. its diffusion parameters D and $\nu\Sigma_f - \Sigma$.

Index